GENETIC DIAGNOSIS OF ENDOCRINE DISORDERS

SECOND EDITION

GENETIC DIAGNOSIS OF ENDOCRINE DISORDERS

SECOND EDITION

Edited by

ROY E. WEISS
Department of Medicine, University of Miami Miller School of Medicine, Miami, FL, USA

SAMUEL REFETOFF
Departments of Medicine, Pediatrics and Committee on Genetics,
University of Chicago, Chicago, IL, USA

AMSTERDAM • BOSTON • HEIDELBERG • LONDON
NEW YORK • OXFORD • PARIS • SAN DIEGO
SAN FRANCISCO • SINGAPORE • SYDNEY • TOKYO
Academic Press is an Imprint of Elsevier

Academic Press is an imprint of Elsevier
125, London Wall, EC2Y 5AS, UK
525 B Street, Suite 1800, San Diego, CA 92101-4495, USA
225 Wyman Street, Waltham, MA 02451, USA
The Boulevard, Langford Lane, Kidlington, Oxford OX5 1GB, UK

Medical Disclaimer
Medicine is an ever-changing field. Standard safety precautions must be followed, but as new
research and clinical experience broaden our knowledge, changes in treatment and drug therapy may
become necessary or appropriate. Readers are advised to check the most current product information
provided by the manufacturer of each drug to be administered to verify the recommended dose,
the method and duration of administrations, and contraindications. It is the responsibility of the
treating physician, relying on experience and knowledge of the patient, to determine dosages and
the best treatment for each individual patient. Neither the publisher nor the authors assume any
liability for any injury and/or damage to persons or property arising from this publication.

British Library Cataloguing-in-Publication Data
A catalogue record for this book is available from the British Library

Library of Congress Cataloging-in-Publication Data
A catalog record for this book is available from the Library of Congress

ISBN: 978-0-12-800892-8

For information on all Academic Press publications
visit our website at http://store.elsevier.com/

Typeset by Thomson Digital

Printed and bound in the United States of America

Working together
to grow libraries in
developing countries

www.elsevier.com • www.bookaid.org

Contents

I

INTRODUCTION

1. Mechanisms of Mutation

BERNARD S. STRAUSS

II

PANCREAS

2. A Clinical Guide to Monogenic Diabetes

DAVID CARMODY, JULIE STØY, SIRI ATMA W. GREELEY,
GRAEME I. BELL, LOUIS H. PHILIPSON

3. Hypoglycemia

DORIT KOREN, ANDREW PALLADINO

III

PITUITARY

4. Functioning Pituitary Adenomas

ALBERT BECKERS, LILIYA ROSTOMYAN, ADRIAN F. DALY

5. Diabetes Insipidus

JANE HVARREGAARD CHRISTENSEN, SØREN RITTIG

6. States of Pituitary Hypofunction

CHRISTOPHER J. ROMERO, ANDREA L. JONES, SALLY RADOVICK

IV

THYROID

7. Congenital Defects of Thyroid Hormone Synthesis

HELMUT GRASBERGER, SAMUEL REFETOFF

8. Developmental Abnormalities of the Thyroid

JOACHIM POHLENZ, GUY VAN VLIET, JOHNNY DELADOËY

VII
REPRODUCTIVE

VIII
ADIPOCYTE

IX
MULTISYSTEM DISORDERS

X

GROWTH

26. Genetic Diagnosis of Growth Failure

RON G. ROSENFELD, VIVIAN HWA

XI

MISCELLANEOUS

27. Cost-Effectiveness of Genetic Testing for Monogenic Diabetes

ROCHELLE N. NAYLOR, SIRI ATMA W. GREELEY,
ELBERT S. HUANG

28. Genetic Counseling: The Role of Genetic Counselors on Healthcare Providers and Endocrinology Teams

SARAH M. NIELSEN, SHELLY CUMMINGS

29. Setting Up a Laboratory

LOREN J. JOSEPH

30. Introduction to Applications of Genomic Sequencing

STEPHAN ZUCHNER

List of Contributors

Milad Abusag   University of Pittsburgh Medical Center – Horizon (UPMC-Horizon), Pittsburgh, PA, USA

Valerie Arboleda   Department of Pathology and Human Genetics, David Geffen School of Medicine, University of California, Los Angeles, CA, USA

Andrew Arnold   Center for Molecular Medicine and Division of Endocrinology and Metabolism, University of Connecticut School of Medicine, Farmington, CT, USA

Guillaume Assié   Département Hospitalo-Universitaire, Faculté de Médecine Paris Descartes, Université Paris, Paris, France; Service des Maladies Endocriniennes et Métaboliques, Centre de Recherche des Maladies Rares de la Surrénale, Hôpital Cochin, India

Albert Beckers   Department of Endocrinology, Centre Hospitalier Universitaire de Liège, University of Liège, Domaine Universitaire du Sart-Tilman, Liège, Belgium

Graeme I. Bell   Section of Endocrinology, Diabetes and Metabolism, The University of Chicago, Chicago, IL, USA

Xavier Bertagna   Département Hospitalo-Universitaire, Faculté de Médecine Paris Descartes, Université Paris, Paris, France; Service des Maladies Endocriniennes et Métaboliques, Centre de Recherche des Maladies Rares de la Surrénale, Hôpital Cochin, India

Jérôme Bertherat   Département Hospitalo-Universitaire, Faculté de Médecine Paris Descartes, Université Paris, Paris, France; Service des Maladies Endocriniennes et Métaboliques, Centre de Recherche des Maladies Rares de la Surrénale, Hôpital Cochin, India

Gemma V. Brierley   Metabolic Research Laboratories, University of Cambridge Institute of Metabolic Science, Addenbrooke's Hospital, Cambridge, UK

Silvia Cantara   Department of Medical, Surgical and Neurological Sciences, University of Siena, Siena, Italy

David Carmody   Section of Endocrinology, Diabetes and Metabolism, The University of Chicago, Chicago, IL, USA

Jane Hvarregaard Christensen   Department of Biomedicine, Aarhus University, Aarhus, Denmark

Karine Clément   Institute of Cardiometabolism and Nutrition (ICAN), Nutriomique, University Pierre et Marie Curie-Paris, Pitie-Salpêtrière Hospital, Paris; Nutrition Department, Pitié-Salpêtrière Hospital, Assistance Publique Hôpitaux de Paris, Paris, France

Shelly Cummings   Myriad Genetic Laboratories, Inc., Salt Lake City, UT, USA

Adrian F. Daly   Department of Endocrinology, Centre Hospitalier Universitaire de Liège, University of Liège, Domaine Universitaire du Sart-Tilman, Liège, Belgium

Johnny Deladoëy   Endocrinology Service and Research Center, Sainte-Justine Hospital and Department of Pediatrics, Universite de Montreal; Department of Biochemistry, Universite de Montreal, Montreal, Quebec, Canada

Koen M. Dreijerink   Department of Clinical Endocrinology, University Medical Center, Utrecht, the Netherlands

Béatrice Dubern   Institute of Cardiometabolism and Nutrition (ICAN), Nutriomique, University Pierre et Marie Curie-Paris, Pitie-Salpêtrière Hospital, Paris; Nutrition and Gastroenterology Department, Armand-Trousseau Hospital, Assistance Publique Hôpitaux de Paris, Paris, France

Alexandra M. Dumitrescu   Department of Medicine, The University of Chicago, Chicago, IL, USA

David A. Ehrmann   Section of Endocrinology, Diabetes, and Metabolism, The University of Chicago, Maryland, Chicago, IL, USA

Douglas B. Evans   Medical College of Wisconsin, Milwaukee, WI, USA

Murray J. Favus   Section of Endocrinology, Diabetes and Metabolism, Department of Medicine, University of Chicago, Chicago, IL, USA

Abhimanyu Garg   Division of Nutrition and Metabolic Diseases, Department of Internal Medicine, Distinguished Chair in Human Nutrition Research, Center for Human Nutrition, UT Southwestern Medical Center, Dallas, TX, USA

Jennifer L. Geurts   Medical College of Wisconsin, Milwaukee, WI, USA

Helmut Grasberger   Department of Medicine, University of Michigan, Ann Arbor, MI, USA

Siri Atma W. Greeley   Section of Endocrinology, Diabetes and Metabolism, The University of Chicago; Departments of Pediatrics and Medicine, Section of Adult and Pediatric Endocrinology, Diabetes and Metabolism, The University of Chicago, Chicago, IL, USA

Lionel Groussin   Département Hospitalo-Universitaire, Faculté de Médecine Paris Descartes, Université Paris, Paris, France; Service des Maladies Endocriniennes et Métaboliques, Centre de Recherche des Maladies Rares de la Surrénale, Hôpital Cochin, India

Jo W. Höppener   Department of Molecular Cancer Research, Laboratory of Translational Immunology, University Medical Center Utrecht, Utrecht, the Netherlands

Leslie Hoffman Columbus Endocrinology, Columbus, OH, USA

Michael F. Holick Department of Medicine, Section of Endocrinology, Nutrition, and Diabetes, Vitamin D, Skin and Bone Research Laboratory, Boston University Medical Center, Boston, MA, USA

Elbert S. Huang Department of Medicine, Section of General Internal Medicine, The University of Chicago, Chicago, IL, USA

Hélène Huvenne Department of Pediatrics, Saint-Vincent de Paul Hospital, GHICL, Lille; Institute of Cardiometabolism and Nutrition (ICAN), Nutriomique, University Pierre et Marie Curie-Paris, Pitie-Salpêtrière Hospital, Paris, France

Vivian Hwa Cincinnati Children's Hospital, Cincinnati, OH, USA

Andrea L. Jones Division of Pediatric Endocrinology, Department of Pediatrics, The Johns Hopkins University School of Medicine, Baltimore, MD, USA

Loren J. Joseph Molecular Diagnostics Laboratory, The University of Chicago Medical Center, The University of Chicago, Chicago, IL, USA

George J. Kahaly Department of Medicine I, Johannes Gutenberg University Medical Center, Mainz, Germany

Dorit Koren Section of Adult and Pediatric Endocrinology, Diabetes and Metabolism, University of Chicago, Chicago, IL, USA

Kelly Lauter Department of Medicine, Endocrine Division, Massachusetts General Hospital, Boston, MA, USA

Rossella Libé Département Hospitalo-Universitaire, Faculté de Médecine Paris Descartes, Université Paris, Paris, France; Service des Maladies Endocriniennes et Métaboliques, Centre de Recherche des Maladies Rares de la Surrénale, Hôpital Cochin, India

Thera P. Links Department of Clinical Endocrinology, University Medical Center, Groningen, Groningen, the Netherlands

Cornelis J. Lips Department of Clinical Endocrinology, University Medical Center, Utrecht, the Netherlands

Michael J. McPhaul Division of Endocrinology, University of Texas Southwestern Medical Center, Dallas, TX

Rochelle N. Naylor Departments of Pediatrics and Medicine, Section of Adult and Pediatric Endocrinology, Diabetes and Metabolism, The University of Chicago, Chicago, IL, USA

Maria I. New Mount Sinai School of Medicine, Department of Pediatrics, New York, NY, USA

Sarah M. Nielsen Center for Clinical Cancer Genetics and Global Health, Department of Medicine, University of Chicago, Chicago, IL, USA

Saroj Nimkarn Bumrungrad International Hospital, Department of Pediatrics, Bangkok, Thailand

Stephen O'Rahilly Metabolic Research Laboratories, University of Cambridge Institute of Metabolic Science, Addenbrooke's Hospital, Cambridge, UK

Furio Pacini Department of Medical, Surgical and Neurological Sciences, University of Siena, Siena, Italy

Andrew Palladino Perelman School of Medicine at the University of Pennsylvania, Philadelphia, PA, USA

Louis H. Philipson Section of Endocrinology, Diabetes and Metabolism, The University of Chicago, Chicago, IL, USA

Joachim Pohlenz Pediatric Endocrinology, Department of Pediatrics, Johannes Gutenberg University, Mainz, Germany

Sally Radovick Division of Pediatric Endocrinology, Department of Pediatrics, The Johns Hopkins University School of Medicine, Baltimore, MD, USA

Margarita Raygada Section on Endocrinology & Genetics, Program on Developmental Endocrinology & Genetics, (PDEGEN), Eunice Kenney Shriver National Institute of Child Health and Human Development (NICHD), National Institutes of Health (NIH), Bethesda, MD, USA

Samuel Refetoff Departments of Medicine, Pediatrics and Committee on Genetics, University of Chicago; Department of Medicine, The University of Chicago, Chicago, IL, USA

Thereasa A. Rich Myriad Genetic Laboratories, Denver, Co, USA

Inne Borel Rinkes Department of Surgery, University Medical Center, Utrecht, the Netherlands

Søren Rittig Department of Pediatrics, Aarhus University Hospital, Aarhus, Denmark

Christopher J. Romero Division of Pediatric Endocrinology, Department of Pediatrics, The Johns Hopkins University School of Medicine, Baltimore, MD, USA

Ron G. Rosenfeld Oregon Health & Science University, Portland, OR, USA

Liliya Rostomyan Department of Endocrinology, Centre Hospitalier Universitaire de Liège, University of Liège, Domaine Universitaire du Sart-Tilman, Liège, Belgium

David B. Savage Metabolic Research Laboratories, University of Cambridge Institute of Metabolic Science, Addenbrooke's Hospital, Cambridge, UK

Robert K. Semple Metabolic Research Laboratories, University of Cambridge Institute of Metabolic Science, Addenbrooke's Hospital, Cambridge, UK

Julie Støy Department of Clinical Medicine – Medical Research Laboratory, Aarhus University, Aarhus, Denmark

Constantine A. Stratakis   Section on Endocrinology & Genetics, Program on Developmental Endocrinology & Genetics, (PDEGEN), Eunice Kenney Shriver National Institute of Child Health and Human Development (NICHD), National Institutes of Health (NIH), Bethesda, MD, USA

Bernard S. Strauss   Department of Molecular Genetics and Cell Biology, The University of Chicago, Chicago, IL, USA

Marc Timmers   Department of Molecular Cancer Research, University Medical Center, Utrecht, the Netherlands

Patrick Tounian   Institute of Cardiometabolism and Nutrition (ICAN), Nutriomique, University Pierre et Marie Curie-Paris, Pitie-Salpêtrière Hospital, Paris; Nutrition and Gastroenterology Department, Armand-Trousseau Hospital, Assistance Publique Hôpitaux de Paris, Paris, France

Gerlof D. Valk   Department of Clinical Endocrinology, University Medical Center, Utrecht, the Netherlands

Anouk N.A. van der Horst-Schrivers   Department of Clinical Endocrinology, University Medical Center, Groningen, the Netherlands

Rob B. van der Luijt   Department of Medical Genetics, University Medical Center, Utrecht, the Netherlands

Bernadette P.M. van Nesselrooij   Department of Medical Genetics, University Medical Center, Utrecht, the Netherlands

Guy Van Vliet   Endocrinology Service and Research Center, Sainte-Justine Hospital and Department of Pediatrics, Universite de Montreal, Montreal, Quebec, Canada

Eric Vilain   Department of Human Genetics, Pediatrics, and Medical Genetics, David Geffen School of Medicine, University of California, Los Angeles, CA, USA

Menno Vriens   Department of Surgery, University Medical Center, Utrecht, the Netherlands

Tracy S. Wang   Medical College of Wisconsin, Milwaukee, WI, USA

Roy E. Weiss   Department of Medicine, University of Miami Miller School of Medicine, Miami, FL, USA

Mabel Yau   Mount Sinai School of Medicine, Department of Pediatrics, New York, NY, USA

Stephan Zuchner   Department of Human Genetics and Neurology, Hussman Institute for Human Genomics, University of Miami Miller School of Medicine, Miami, FL, USA

Preface to the First Edition

Imagine the skepticism of a physician of the 1950s or 1960s if told that genetic testing would be used to diagnose specific complex endocrine disorders. Until relatively recently major abnormalities of the endocrine system were diagnosed, treated, and monitored with clinical assessment only. It was the clinical acumen of the astute physician that enabled correct diagnosis and determined which gland was responsible for producing too much or too little of a given hormone. The clinically pertinent markers of endocrine diseases were physiological measurements of basal metabolic rates, body weight, and urine output, which we now term the "physiologic" era of endocrinology.

Despite the discovery of insulin by Banting, Best, Macleod, and Collip in 1921 and its use for humans in 1923, it was only with Rosalyn Yalow and Salomon Berson's seminal report on the immunoassay of endogenous plasma insulin in 1960 that the "assay" period of endocrinology was introduced. This momentous methodological breakthrough enabled the endocrinologist to assay hormones previously impossible to measure at physiologically or pathologically relevant levels. The competitive protein-binding assay facilitated measurement of nanomolar or picomolar concentrations of hormones in plasma and tissues. Adaptation for other compounds further extended the field until 20 years ago, when "molecular" or "genetic" endocrinology evolved with the discovery of genes for insulin and growth hormone. Despite Paul Wermer's 1954 publication of the first clearly inherited endocrine disease "familial adenomatosis" (*American Journal of Medicine*, 1954, pp. 363–371), it is only with access to the genetic tools of the new millennium that the clinician can now precisely identify the genetic defects causing a disease and apply rational therapy. In *Genetic Diagnosis of Endocrine Disorders* we present to the clinician a straightforward, clinically relevant review of important genetic tests currently in use for the diagnosis of endocrine disorders, and practical information as to where these tests are performed.

Some endocrine disorders follow familial patterns of inheritance, while others may represent sporadic mutations. In both cases, identification of the mutation associated with the particular disease ideally allows the physician to test other family members, who may be asymptomatic. For example, in the case of a mutation with potential for adverse outcome such as medullary thyroid cancer. Physician and patient can now consider prophylactic thyroidectomy, for harboring such a gene prior to actual presentation of clinical disease; an approach possible only with the endocrine "genetic" revolution. Correlation of phenotype and genotype is frequently concordant, though in some instances the same genotype may cause different subtle or obvious phenotypes. An example would be patients with resistance to thyroid hormone who, despite identical mutations in the thyroid hormone receptor gene, may present with different phenotypes. The contrary example would be where the same phenotype may be due to different genotypes, as is occasionally the case in adrenal hyperplasias. Even in "truly" monogenic diseases, the genetic background of affected individuals may substantially modulate the phenotype. Thus, while genetic diagnosis is a critical part of the armamentarium of the modern-day Banting and Best, correlation of the genetic abnormality with its physiological manifestations remains crucial. Knowledge of which genetic tests to order must be supported by a full understanding of the genetic information they provide. The health care team responsible for diagnosis and follow-up, should ensure inclusion of the patient's family/primary care physician and a genetic counselor.

Genetic Diagnosis of Endocrine Disorders was initially conceived for purely selfish reasons. For our own use we needed a comprehensive clinical practice handbook for the genetic diagnoses of endocrine diseases. We therefore invited world experts to summarize the full range of currently available genetic endocrine diagnoses for our text. Initially we had to justify to ourselves taking time from our research to edit yet another endocrine book. However, the real advantage of editing this compilation of excellent reviews by renowned experts is the knowledge we have acquired in doing so. We hope that the reader will as well.

Our many thanks to those who have so graciously contributed to *Genetic Diagnosis of Endocrine Disorders* as well as to Fay, Heather, and our children who have been unswerving in their support of our careers. We would also like to acknowledge the support of the National Institute of Health (grants DK15070, DK07011, DK20595, and RRO4999), the Abrams and Esformes Endowments, and the Sherman family.

Roy E. Weiss, MD, PhD, FACP, FACE
Samuel Refetoff, MD, PhD

Preface to the Second Edition

The first edition of Genetic Diagnosis of Endocrine Disorders was published five years ago. Since then, the revolution of Precision Medicine has taken center stage in medical diagnosis, and those treating endocrine diseases need the appropriate ammunition to approach their patient's problems. There is rarely an article in the medical literature involving endocrine diseases, which does not mention a new gene or mutation. An encyclopedic compilation of the totality of genes involved in endocrine diagnosis does not lend itself to a printed text, which by nature is static. Therefore, the purpose of this book is to present emerging concepts to practicing pediatric and adult endocrinologists, students in the field, and genetic counselors, and review the most common genetic causes for endocrine disorders. Given the increasing affordability and availability of whole exome sequencing, the genetic cause of many more diseases will be identified, and there will be additional genes to know. New genetic conditions will emerge. In addition the role of epigenetics miRNAs, and enhancers in the cause of genetic endocrine disorders is quickly being recognized.

The purpose of having a genetic diagnosis should enable the patient and physician to understand the basis for the disease and thereby apply rationale and targeted therapy. In addition, based on the genetic information, decisions can be made regarding risks in asymptomatic relatives, allowing for preemptive.

Another reason for the second edition was based on the favorable reviews of the first edition and being urged by those reviewers to write a new edition.

We thank our returning and new authors for their outstanding contributions to this second edition.

Roy E. Weiss, Miami
Samuel Refetoff, Chicago

PART I

INTRODUCTION

1

Mechanisms of Mutation

Bernard S. Strauss

Department of Molecular Genetics and Cell Biology, The University of Chicago, Chicago, IL, USA

INTRODUCTION

Darwin realized the need for variation to provide a basis for natural selection, but he had no way of understanding the mechanisms by which such variation arose. Vague ideas of the origin of variation in the late nineteenth century gave way to the term mutation, coined by deVries to describe the discontinuous variation associated with Mendelian traits.[1] Genes were first recognized and defined by mutations with an extreme phenotype (see Table 1.1). Further progress led to conceptualizing the gene as a more complex structure with multiple "sites" for mutation available within the same gene. The nature of such sites was not clear, nor was the relationship between the physiological effects of gene mutations and the structural change involved. A typical pre-Watson–Crick examination question, "what is a gene?" could be answered in terms of function, of mutation, or of recombination and the question for geneticists was the relationship between these definitions.

Modern understanding of the mechanism of mutation is based on the Watson–Crick DNA structure. Recognizing the importance of nucleotide sequence followed by the deciphering of the genetic code led to a change from a biological and formalistic or mathematical view of mutation to a more biochemical approach. This chapter presents the problem of mutation as mainly one of biochemistry.

The immediate response of investigators to the Watson–Crick structure was to focus attention on the base changes that resulted in mutation and on the chemical changes that might alter base pairing.[2] The specificity of particular mutagenic agents was initially ascribed to chemical changes in either the incoming or template nucleotide, resulting in altered pairing properties, mainly involving hydrogen bonding. Benzer and Freese[3] and Brenner et al.[4] defined mutation in terms of substitutions, additions, and deletions of nucleotides. DNA in eukaryotes is organized into discrete chromosomes.

Changes in the structure (rearrangements and translocations) and numerical distribution of these chromosomes that leave a viable organism are also mutation, but it is only recently, with the availability of extensive DNA sequence information, that these changes can even be partially accounted for biochemically.

Advances in our understanding of the complex biochemistry of DNA replication and its interaction with the various DNA repair and recombination pathways has led to a more mechanistic approach to understanding mutation (Fig. 1.1). The discovery in the late 1990s of a series of DNA polymerases with altered fidelity and ability to replicate past damaged sites in DNA[5] advanced a view of mutation as an event involving both initial changes in the DNA and the interaction of these changes with the protein complement of the cell. Most recently, the advent of rapid and relatively inexpensive DNA sequencing technology has permitted a direct measurement of normal human mutation rates and the recognition that *de novo* mutation plays a role in human disease. The identification of thousands of mutational changes in individual tumors, only a small minority of which are involved as "drivers" in the etiology of the tumors, has permitted recognition of a set of mutational "signatures" that implicate particular repair processes in the generation of the mutations. The observation of numerous closely linked mutations in tumor cells suggests the operation of unique mutagenic mechanisms whose operation in normal cells remains an open question.

THE TYPES OF MUTATION

Mutations are defined in this chapter as changes in the parental sequence of the DNA (Table 1.1). This definition is not without problems, since it is sometimes difficult to distinguish such changes from the normal process of recombination. Mutations include single

Genetic Diagnosis of Endocrine Disorders. http://dx.doi.org/10.1016/B978-0-12-800892-8.00001-4

TABLE 1.1   Special Abbreviations and Definitions

Term/Abbreviations	Definition
Abasic (apurinic/apyrimidinic) site	Site in DNA missing a base attached to the 1′-position of the sugar
APOBEC/AID	Apolipoprotein B mRNA editing enzyme (APOBEC) and activation-induced deaminase (AID). A family of cytidine deaminases
Aneuploidy	Eucaryotic cells with the normal diploid ($2n$) number of chromosomes are "euploid." "Haploid" cells are n. "Polyploid" cells are $3n$, $4n$, etc. "Aneuploid" cells are $2n \pm$ a number other than n
Base excision repair (BER)	A repair mechanism in which single nucleotide bases are removed and replaced by a patch of one or, at most, a few nucleotides
Chromothripsis	Multiple localized chromosome rearrangements occurring in a single event and in one or a few chromosomes
Copy number variation (CNV)	Altered number of copies of a gene or extended DNA sequence present in the genome
Double-strand break repair (DSBR)	Joining together of two DNA fragments to make a single molecule
Epigenetic	Heritable changes in gene expression that cannot be tied to DNA sequence variation and involving the active perpetuation of local chromatin states
Fidelity	A measure of the relative ability of DNA polymerases to insert the "correct" complementary base
Frameshift mutation	The insertion or deletion of a number of nucleotides not divisible by 3, properly speaking in a coding region of a gene. Largely replaced by the term "indel."
Genome	(1) The complete set of genetic material present in an organism. (2) The complete sequence of the DNA in an organism
Genotype	The genetic constitution of an organism
Holliday junction	A mobile junction formed in recombination between four strands of DNA. It is "resolved" by specific enzymes to regenerate two double-stranded molecules
Homologous recombination (HR)	A DSBR process involving the use of an allelic DNA sequence as a source of information
Indel	Insertion or deletion of a small number of nucleotides in the DNA structure
Insertional mutagenesis	Mutation by insertion of one or more nucleotides. Often used to denote inactivation of the genes by insertion of large transposable elements
Inversion	A rearrangement of the chromosome so that the order of the nucleotide pairs is reversed: if the normal order is ABCDEF, the order AEDCBF would constitute an inversion
Kataegis	Multiple, localized mutations, mostly C→T
L1 element	A common retrotransposon found in the human genome
Microhomology-mediated end joining (MMEJ)	An end joining DSBR mechanism utilizing the homology of a relatively few bases to orient the broken strands
Mismatch repair (MMR)	An excision repair process mainly devoted to correcting errors in replication
Missense mutation	A change in a gene, which results in a change in the meaning of a codon, e.g., the change from GAA (glutamic acid) to GUA (valine)
Mobile element insertion (MEI)	See transposon below. Mutational event in which a mobile element is inserted at a new position in the genome
Mutator	A mutation, often of a repair gene, that has the effect of increasing the spontaneous mutation rate
Nucleotide excision repair (NER)	The paradigm of an excision repair pathway. NER recognizes a wide range of damage and proceeds by cutting out and replacing an extensive series of nucleotides
Nonallelic homologous recombination (NAHR)	HR in which the complement is a homologous sequence other than the normal allele and which can lead to chromosome aberrations
Nonsense mutation	A mutation that results in one of the termination codons UAA, UAG, or UGA
Phenotype	The observable traits of an organism
Point mutation	A mutation involving one or a few nucleotides as distinguished from insertions, deletions, and duplications involving hundreds, thousands, or more nucleotides

TABLE 1.1 Special Abbreviations and Definitions *(cont.)*

Term/Abbreviations	Definition
Proofreading	In DNA synthesis, the process where an exonuclease checks a newly-inserted nucleotide for goodness of fit. Sometimes referred to as editing
Pseudogene	A copy of a gene made inactive by the accumulation of mutations and often devoid of introns
Retrotransposon	A transposable element that can shift its position in DNA via an RNA intermediate
Reactive oxygen species (ROS)	Chemically reactive radicals containing oxygen formed in metabolism and produced in clusters by ionizing radiation
Somatic hypermutation (SHM)	Process producing multiple mutations in mature B cells, mostly but not exclusively in the immunoglobulin gene during antibody maturation
Single nucleotide variation (SNV)	A point mutation involving a single nucleotide pair
Syn/anti base configuration	In the *anti* configuration, the bulky part of the base of a nucleoside or nucleotide rotates away from the sugar. In the *syn* configuration the bulky part rotates over the sugar
Synonymous/silent mutation	A nucleotide change that does not change the meaning of a codon, e.g., the change from GGU (glycine) to GGA (glycine). Not all synonymous mutations are silent, that is, without phenotypic effect
Translocation	Attachment of a segment of one chromosome to a different (nonhomologous) chromosome
Transposition	The movement of a transposable element from one position in the genome to another
Transition	The mutational change from a purine to another purine or a pyrimidine to another pyrimidine. G↔A and C↔T are the possible transitions
Transposon, mobile element (ME)	A DNA sequence able to move from one position to another within the genome. Movements are generally rare and are catalyzed by special enzymes coded for by the transposon
Transversion	The mutational change from a purine to a pyrimidine or a pyrimidine to a purine. A↔T, G↔T, C↔G, C↔A are possible transversions
Translesion synthesis (TLS)	Synthesis of DNA by specialized polymerases utilizing a damaged template
Transcription coupled nucleotide excision repair (TC-NER)	Specialized NER mechanism targeted to genes in the process of transcription
Ubiquitin	A conserved small (76 amino acids in humans) protein, which when covalently added to proteins in single or multiple copies serves as a signal for processes such as degradation and/or changes in conformation

nucleotide variation (SNV), insertion or deletion of small numbers of nucleotides (indels, frameshifts), rearrangements of the DNA sequence, change in the number of copies of larger stretches of DNA (copy number variation, CNV), and changes in the structure (inversions and translocations) or number of chromosomes (aneuploidy). The insertion or movement of transposable elements may affect phenotype and be obviously mutagenic. Nucleotide changes may occur outside the exome, the protein coding region of the genome, and these may or may not have an observable effect on phenotype. This view of mutation as a sequence change anywhere in the genome[6] is a product of the sequencing revolution, since the recognition of mutation in the pre-sequencing era required some observable change in the phenotype. The definition of mutation as any change in DNA sequence results in classifying sequence changes

that have no obvious phenotypic effect as mutations. There are about 20,000–25,000 human genes, and the exome comprises somewhere about 2% of the total number of nucleotides.[7] Much of the remainder of the DNA is transcribed into RNA,[8] and some of this plays an important regulatory role in gene function, but as yet there is no automatic way to predict whether or what a change in DNA sequence will mean for physiology.

The possible single base changes were first cataloged by Ernst Freese and Seymour Benzer.[3,9] Freese coined the term "transition" to denote the change from one purine to another, or of one pyrimidine to another. The four possible transitions are cytosine (C) to thymine (T) and its reverse, and adenine (A) to guanine (G) and its reverse. Freese defined "transversions" as changes from a purine to a pyrimidine or the reverse. Change from an A or a G to a C or a T, and the reverse C or T to A or G was defined

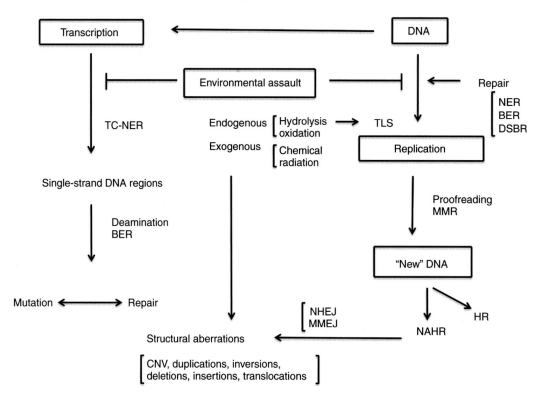

FIGURE 1.1   The interactions between repair and mutation. Abbreviations: TC-NER, transcription-coupled repair; BER, base excision repair; MMR, mismatch repair; HR, homologous recombination; NAHR, nonallelic homologous recombination; TLS, translesion synthesis; NER, nucleotide excision repair; DSBR double-strand break repair; NHEJ, nonhomologous end joining; MMEJ, microhomology-mediated end joining; CNV, copy number variations. See Table 1.1 for definitions.

as a transversion. These definitions remain even though some of the putative transversions described were actually insertions or deletions of a few nucleotides.[4] These were later called frameshifts because of the discovery that the genetic code was read in groups of three nucleotides to specify a particular amino acid. The addition (or deletion) of any number of nucleotides not divisible by three results in a change in the reading "frame," thereby changing the amino acid composition of all amino acids downstream of the coding change. The details of the genetic code, as elucidated in the 1960s, also indicated that such frameshifts might not only result in major changes in the amino acid composition of a protein but might also produce unexpected termination codons as a result of the shift. Point mutations that resulted in protein terminations were at first termed "nonsense" mutations, as opposed to missense mutations, that resulted in the substitution of one amino acid for another. The nonsense mutations did not make "sense," that is, did not specify any amino acid. There are three such codons, now called termination (ter) codons: UAA, UAG, and UGA. Since messenger RNA is the molecule that is actually read by the protein-synthesizing machinery, the code is an RNA code with U(racil) substituted for T(hymine). One of the stop codons, UGA, is read as tryptophan in mitochondria, and the mitochondrial code includes a few

other variations; AGG and AGA are mitochondrial stop codons instead of coding for arginine and AUA codes for methionine instead of isoleucine. There are 64 possible codons but only 20 (or 21 including selenocysteine) natural amino acids incorporated into protein, and the codes are degenerate, or redundant, in that several codons can specify the same amino acid. Selenocysteine is synthesized from a special sertRNA and is coded for by an in-frame UGA stop codon. The mechanism by which particular UGA sites are selected for selenocysteine incorporation requires particular transcription factors and a *cis*-acting insertion sequence (SECIS) on the mRNA.[10]

Point mutations within genes that do not change the meaning (amino acid coded) of the codon are termed synonymous, or "silent," as opposed to nonsynonymous changes. "Silent" mutations may actually affect physiology when they are part of splice sites and because of the different availability of various tRNAs.[11] "Synonymous" is probably the better term. Although there was some initial confusion about the necessity of punctuation between the triplet codons, it was realized that if reading of the code began at a fixed site, and if the reading "frame" was designed to read three nucleotides at a time, the correct sequence of amino acids would be automatically produced. This terminology was developed before it was realized that there were large amounts of noncoding DNA. "Frameshift" has no meaning for mutations within such noncoding regions. A more recent term for small *in*sertions or *de*letions, regardless of their physiological effect is "indel," although the frameshift terminology continues to be used when appropriate.

Change in the structure of other cellular constituents (e.g., membranes, prions) may also be heritable and alter physiology. Methylation of the DNA base cytosine and modifications of the histones constitute a set of markers that can alter cell physiology and direct patterns of differentiation.[12] Some of these modifications are propagated through mitosis and constitute a mechanism for somatic inheritance and for tissue differentiation. Certain of these markers can survive meiosis,[12] but the situation is complicated in large part because of the massive removal of tissue specific markers in the period between the formation of the gametes and early differentiation followed by their replacement. Cytosine methylation in DNA is carried out by specific maintenance and *de novo* methylation enzymes.[13] The signal for maintenance is the presence of a methyl group on the parental strand of a 5'CpG sequence. Methylated promoters are usually associated with gene inactivation. Environmental influences can affect DNA methylation and gene function, and such environmental effects can persist through several generations.[12] Perhaps the most well-known study is the demonstration of low birth weight in offspring of Dutch mothers pregnant during the 1945 famine.[14] The general biological significance of such epigenetic marks (see Table 1.1) is clearly a matter of current interest,[15] since it reintroduces a Lamarckian cast to molecular biology. Epigenetic changes can mimic mutational ones since they affect phenotype. In addition, the rate of somatic mutation can vary as much as 10-fold according to tissue, and the determining factor is apparently the distribution of (epigenetic) markers.[16]

THE MECHANISMS OF MUTATION

Base Pairing and the Action of Mutagenic Agents

It was first assumed that the fidelity of normal replication stemmed from the stability of the A:T and G:C base pairs resulting from hydrogen bonding. The early workers on the molecular nature of mutations accounted for the specificity of a variety of base analogs and other mutagenic agents by drawing acceptable alternative base pairings, resulting from the incorporation of these compounds into DNA or by their reaction with DNA nucleotides.[2] Alkylating agents, such as methylnitrosourea and the chemotherapeutically active mustard gas derivatives, were shown to react with individual nucleotides to produce multiple changes. Production of O^6-methylguanine by agents such as methylnitrosourea or methylnitronitrosoguanidine was shown to promote mistaken base pairing, making understandable the highly mutagenic characteristics of such compounds. A major development was the discovery that metabolic systems in the host activated ingested compounds, making it possible for them to react with DNA. Carcinogenic polycyclic hydrocarbons, including those present in tobacco smoke and aflatoxins, are converted to epoxide derivatives with the participation of the cytochrome p450 system. These epoxides react directly with DNA, producing mutagenic adducts.[2]

We live in an environment that is not friendly to DNA. We are essentially 55.6 M water. Given the law of mass action, hydrolytic reactions are inevitable. It has been estimated that we lose about 18,000 bases per cell in every 24-h period as a result of spontaneous hydrolysis of the glycosidic bond.[17] The abasic sites so created are targets for base excision repair (see Modifiers of Mutation Not Associated with Replication p 11) but those that survive are mutagenic.

The most reactive mutagen in our environment is undoubtedly oxygen. Breathing, however unavoidable, is inherently dangerous! The electron transport chain, by which adenosine triphosphate (ATP) is generated, results in the generation of reactive oxygen species (ROS) that produce the hydroxyl radical OH. When formed in proximity to DNA, this species produces a variety of oxidation products, of which a guanine with a saturated imidazole ring (8-oxoguanine) and altered hydrogen bonding properties are the most important.

We are dependent on the sun both as a primary source of energy and for our feeling of well-being. But sunlight is a major source of DNA damage and results in the production of pyrimidine dimers and altered pyrimidines, all of which may be mutagenic and carcinogenic.

The Role of Enzymes in Mutation

It is not surprising that organisms have developed a set of enzymes to protect themselves from such damage and from mutation. Most mutations, at least in the exome, may have a deleterious effect, but it is essential that they occur at a sufficient rate in germ cells to ensure the variation on which natural selection is based. The problem organisms have had to solve is to adjust the mutation rate to some presumably optimal rate. A group of enzymes accomplishes this purpose.

The free energy differences between correct and incorrect base pairs are very small, at most 0.4 kcal/mol. This means that in a water solution, in which there is much competitive hydrogen bonding, a correct base pair is only about twice as likely to form as is a mismatch. The major contribution to specificity is provided by the replicative polymerases: the structural nature of

TABLE 1.2 Eucaryotic DNA Polymerases

Pol	Family	Exo(?)	Error rate	Function	References
Alpha (α)	B	No	10^{-4}–10^{-5}	Replication	*Biochemistry* 1991;30:11751–59
Beta (β)	X	No	5×10^{-4}	Gap filling BER	*Biochemistry* 1991;30:11751–59
Gamma (γ)	A	Yes	1×10^{-5}	Mitochondrial	*J Biol Chem* 2001;276:38555–62
Delta (δ)	B	Yes	10^{-5}–10^{-6}	Replication	*Nat Rev Genet* 2008;9:594–604
Epsilon (ε)	B	Yes	4.4×10^{-5}	Replication	*Nucleic Acids Res* 2011;39:1763–73
Eta (η)	Y	No	0.26–6×10^{-2}	TLS (UV)	*Nature* 2000;404:1011–1013
Iota (ι)	Y	No	Template and metal ion dependent, SHM		*J Biol Chem* 2007;282:24689–96
Kappa (κ)	Y	No	5.8×10^{-3}	TLS	*J Mol Biol* 2001;312:335–346
Lambda (λ)	X	No	9×10^{-4}	DSBR	*J Biol Chem* 2002;277:13184–191
Mu (μ)	X	No	10^{-3}–10^{-5}	NHEJ	*Biochemistry* 2004;43:13827–38
Theta (τ)	A	No	2.4×10^{-3}	SHM, TLS	*Nucleic Acids Res* 2008;36:3847–56
Zeta (ζ)	B	No	1.3×10^{-3} (yeast)	TLS	*Nucleic Acids Res* 2006;34:4731–42
Rev1	Y	No	dCMP transferase	TLS accessory to pol zeta	*Nucleic Acids Res* 2010;38:5036–48
Nu (ν)	A	No	2.4×10^{-3}	T opposite template G, TLS, SHM	*DNA Repair* 2007;6:213–223

the pockets into which incoming nucleotides fit and the kinetic interactions between elongation of the chain and reversal of the reaction accounting for this specificity. Humans (and other organisms) have developed a variety of polymerases with vastly different specificities (Table 1.2). The free energy differences between correct and mismatched bases for a reaction catalyzed by a replicative polymerase (*Drosophila melanogaster* polymerase alpha) indicated a difference of 4.9 kcal/mol, equivalent to a discrimination factor of about 1 in 3,000.[18] The *in vitro* measured error frequency of the different polymerases (Table 1.2) varies from a low of about 1 in 100,000 for the different B family replicative polymerases to more than 3.5 per 100 for human polymerase eta.[19] Polymerase iota (pol iota) confronted with a template T will actually incorporate a G three times more frequently than the "correct" A![20]

DNA synthesis can be considered as a series of steps in which the growing chain is elongated. As a first step, the base to be inserted must fit into the appropriate pocket of the enzyme. The various DNA polymerases can be classified into "families" based on sequence homology. The Y family polymerases characteristically have pockets that are less restrictive[21] than those of the replicative polymerases of the B family. The nucleotides may have either a *syn* or *anti* configuration, depending on the orientation of the purine/pyrimidine relative to the sugar; the most common orientation is *anti*. The pocket of pol iota is so constructed that it forces template dA into a *syn* configuration where it is restricted

to pairing with T. Similar pol iota interactions keep template dT in the *anti* configuration, where structural considerations make it more likely to pair with incoming dG than with the "correct" dA.[22,23]

After formation of a phosphodiester bond with an incoming base, the newly elongated chain is "proofread" to determine if it meets built-in pairing specifications. Terminal bases that do not fit are removed by exonucleolytic action. The requisite 3′→5′ nuclease activity is either built into the structure of B class polymerases or exists as a separate but closely associated protein(s). Y-family polymerase members are devoid of exonuclease activity. Proofreading results in about a 100-fold increase in the fidelity of replication. Mutations in either the exonuclease domains or the separate exonucleases may have mutator properties, producing additional mutations in every round of replication. Organisms can fine tune their proofreading, and mutants (of bacterial viruses) have been isolated in which the rate of spontaneous mutation is lowered because of an increase in the efficiency of proofreading. Such an increase comes at a cost in energy, since the ATPs required to provide the pyrophosphates required for polymerization are wasted. Even replication events with normal bases involve proofreading. Measurements made many years ago[24] indicate about 6–13% of the polymerization events result in an excised (proofread) base. The replication process can be depicted as a competition between proofreading and further elongation (Fig. 1.2), since once the chain has been elongated five or six nucleotides beyond a

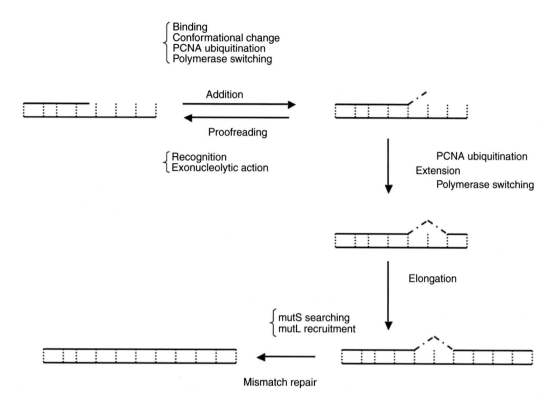

FIGURE 1.2   Schematic outline of the competition between extension and proofreading in DNA synthesis.

mismatch, it appears immune to proofreading. The elongation step is distinct from the initial addition opposite any particular base. Some of the enzymes of the Y series are relatively efficient in the addition of a nucleotide opposite a nonpairing template, but are unable to elongate the resulting product. It has been suggested that polymerase zeta, a B-family polymerase, has as a function the elongation of mismatched bases inserted by iota and other error-prone polymerases.[25–27]

Many agents that damage DNA block DNA synthesis but are nonetheless mutagenic. Since production of viable offspring requires replication of the DNA, cells must be able to overcome this inhibition. Several mechanisms are available to cells, not all of which are mutagenic. However, one of the mechanisms used is direct translesion synthesis (TLS), in which the damaged base serves as a template for replication. Mistakes in replication are likely to occur during such TLS. As just described, the Y-family polymerases are likely to be involved because of the relaxation in the specificity of their combining site. The events in translesion synthesis can be described as follows: the replicative complex recognizes a mismatch or altered base. TLS requires displacement of the replicative complex followed by recruitment of nonprocessive (generally Y family) polymerase(s) and then, after the bypass, reassociation of the normal replication complex. A key player in this process is proliferating cell nuclear antigen

(PCNA). This protein forms a trimeric clamp encircling replicating DNA and is essential for processive DNA synthesis. PCNA binds numerous proteins, and ubiquitination of the PCNA clamp results in dissociation of the replicative complex from the DNA and allows access of a Y-family polymerase to the growing point.[21,28] A deoxynucleotide is added, and in at least some cases before the replicative complex and its associated proofreading activity can access the mismatch, polymerase zeta replaces the Y-family polymerase and elongates for a few nucleotides.[25,27] Polymerase zeta then falls off and is replaced by the normal replication complex. Ubiquitination of PCNA plays a critical regulatory role in the process[28] as do other gene products.[21]

Superimposed on these events must be the availability of the different deoxynucleotides used for synthesis. Alterations in the pool size of the different DNA constituents can affect the selection of bases, and altering relative pool sizes can be mutagenic.[29,30] The result of any particular elongation attempt is determined by the various competitions for access to the nucleotide at the growing point.

Mismatch Repair

Neither hydrogen bonding, the innate specificity of the replicative polymerase, nor the efficiency of proofreading alone or together can account for the low *in vivo* mutation rate. Newly synthesized DNA is subject to yet another inspection by the set of proteins constituting the mismatch repair (MMR) system. These proteins detect mismatches in the DNA: both base pair mismatches and mismatches due to small additions or deletions. In bacteria, the detection is carried out by a single protein acting as a homodimer, the mutS protein, which when bound to the mismatch, recruits a second protein dimer, mutL. In the enteric bacteria, this ATP-dependent complex activates the endonuclease activity of a third protein, mutH, which makes a single stranded break in the error-containing strand. The nicked strand is unwound by a helicase encoded by the UvrD gene, and the displaced strand is degraded by an exonuclease. The resulting single-stranded gap is then filled by the replicative polymerase. The key to the successful operation of this scheme is making sure that the newly synthesized strand, including the "error," is the one removed. In *Escherichia coli*, this trick is accomplished by a special methylation mechanism. Adenines at GATC sites are methylated on both strands, but the methylation of the newly inserted adenine is accomplished only after replication. Immediately after replication, the newly synthesized strand is nonmethylated. It is this hemimethylated DNA, which is the substrate for mismatch repair, and it is the nonmethylated, that is, newly synthesized, strand which is removed.[17]

The MMR system of enteric bacteria has served as a paradigm, but eukaryotic cells have a more complex, although clearly similar, mismatch repair mechanism.[31,32] Instead of a single mutS protein, eukaryotes have five, three of which (MSH2 [MutS homolog], MSH3, and MSH6) form dimers with slightly different specificities. There are four MutL homologs (MLH1, MLH2, PMS1 [postmeiotic segregation protein], and PMS2) that also function as heterodimers. The MSH2:MSH6 heterodimer recognizes base–base mismatches and small insertions or deletions; the MSH2:MSH3 complex specializes in recognition of larger insertions and deletions. As the names indicate, certain of these proteins also play an important role in meiosis. There is no MutH analog. The adenine methylation recognition mechanism appears confined to enteric bacteria. *In vitro* reconstructions of the eukaryotic mismatch repair system use a free 3'OH end (i.e., a nick in the DNA) to identify the newly synthesized strand, and it appears likely that *in vivo*, it is the growing point of the DNA (or an unligated Okazaki fragment) that provides the MMR signal. Eukaryotic MMR is more closely tied to replication as compared to the enteric bacteria. The MSH proteins have been shown to bind to PCNA, which locates them at the site of the DNA growing fork.[33]

Organisms deficient in their ability to make one of the mismatch repair proteins have increased mutation rates. The medical interest in MMR dates from the discovery that an inherited colon carcinoma syndrome (Lynch syndrome/hereditary nonpolyposis colorectal cancer) can be traced to a deficiency in the MMR proteins.[34,35] The most frequent culprits are the hMLH1 (human mutL homolog) and hMSH2 genes, followed by hPMS2 and hMSH6. Analyses of tumor tissue show that the promoters of these MMR genes are frequent targets of epigenetic inactivation by methylation.[36] The absence of a functional MMR system is often signaled by an increase in microsatellite instability. The "microsatellites" are regions of mono- or dinucleotide repeats (e.g., CACACACACA) that are polymorphic, that is, in which the actual number of repeat units at a particular location differs among individuals. The number of repeat units at each locus is inherited and is the basis of much DNA "fingerprinting." "Instability" is observed as a detectable increase in the number of such repeats, easily demonstrated by gel electrophoresis. Individuals deficient in MMR may have thousands of microsatellite instabilities throughout the genome, but a panel of five selected loci is generally used for testing. Instability at two loci serves as a positive signal.[6]

Bound MMR proteins may also serve as a signal for apoptosis. Organisms deficient in the O^6-methylguanine methyltransferase protein are exquisitely sensitive to killing by methylating agents. The cells become much less sensitive when made MMR defective, possibly because of a loss of a signal from the MMR proteins combined at the O^6-methylguanine:T mismatch. The mechanism is important because MMR deficient cells with mutagenic lesions that should be signals for apoptosis survive, reproduce, and propagate mutations.[37]

The MMR repair system not only serves as a suppressor of mutation but also may be a component of systems producing mutation. Error-prone replication of single-stranded regions exposed by MMR is associated with the generation of mutations in the process of somatic hypermutation of immunoglobulin genes[31,38] (see, Somatic Hypermutation section) and in some not yet precisely defined way in the pathogenic expansion of triplet nucleotide repeats.[31]

About 30 mostly neurodegenerative diseases including Fragile X syndrome, Huntington's chorea, myotonic dystrophy, and Friedreich's ataxia are due to multiple expansions of simple repeats in the genome.[39] For example, a CAG sequence, which occurs from six to about 35 times in normal individuals, expands to up to 100 repeats in individuals affected with Huntington's chorea. In Friedreich's ataxia, a normal GAA sequence occurring from seven to 22 times in an intron may expand to 200–1700 units. There is a role for the mismatch repair system based on the requirement for mutS and mutL proteins in a mouse model of Huntington's disease,[40,41] but exactly what that role is and what sets off the changes are unknown.

Modifiers of Mutation Not Associated with Replication

Most endogenous and exogenous damage suffered by DNA is removed before the passage of the replication complex. This removal is accomplished by the action of one or more of the DNA repair pathways: base excision repair (BER), nucleotide excision repair (NER), and its cousin transcription-coupled nucleotide repair (TC-NER). The general tactic of all excision repair pathways is to use the information conserved in the complementary strand as a guide, an argument for the maintenance of the genome as a double-stranded entity. Cells also need to deal with breaks in DNA, some of which occur as the result of normal physiological processes. Double-strand breaks in the DNA would be lethal if not repaired by either recombination repair or the inherently mutagenic nonhomologous end joining (NHEJ). The details of these repair processes are the subject of an extensive text (see Ref. [17]). Over 175 structural genes have been identified with DNA repair functions.[42] The excision repair pathways have a general similarity with each other and with the general mismatch repair system, albeit utilizing different components. In general, the damage must be recognized and removed. The removal leaves a single-stranded region of variable length, and this single stranded region serves as a template for DNA synthesis, sometimes using specialized polymerases and leaving single-stranded breaks that must be ligated. Recombination involving double-strand breaks is a consequence of the need for precise segregation of chromosomes in meiosis, and double-strand breaks are necessary intermediates in class switching in the immune response. Repair of double-strand breaks occurs by either NHEJ or recombination repair[43,44] and, as will be seen, these complex processes can lead to mutations and chromosome aberrations.

Somatic Hypermutation

The frequency of new single nucleotide mutations in the human germline is about 1.2×10^{-8}.[45] During the development of a mature antibody response, the frequency of mutation in the variable (V) region of immunoglobulin may reach 10^{-3} per nucleotide as a result of a process called somatic hypermutation (SHM).[38] Mature B cells, migrating to the dark zone germinal centers of lymphoid organs, become centroblasts. These are the cells in which SHM is observed, mainly in the immunoglobulin genes but also at a lower frequency in a number of genes implicated in oncogenesis.[46] SHM requires transcription, is limited to a particular region of the affected gene, and is critically dependent on the activity of a cytidine deaminase, activation-induced cytidine deaminase (AID), which is expressed in high amounts in the mature B cells. AID requires single-stranded DNA as a substrate; hence, presumably, the requirement for transcription. The requirement for transcription is complex since SHM appears restricted to portions of the transcribed gene closer to the promoter.[47]

AID is one of a family of cytidine deaminases. The deaminated site is the target for repair. Base excision repair, error-prone DNA polymerases, and mismatch repair proteins are all involved.[38] Although a major protection against mutation, the MMR system plays an error-prone role in SHM and is involved in the production of the double-strand breaks that are intermediates in chromosome translocation.[48] It is not yet clear how the switch between the mutagenic and antimutagenic roles of the MMR proteins is managed.

THE ROLE OF TECHNOLOGY

Advances in DNA technology have made it possible to observe events that were previously inaccessible. New, rapid, and relatively inexpensive sequencing technologies promise to reduce the cost of sequencing individual human genomes to $1,000 or less. However, such rapid sequencing technologies are more prone to error than those based on the Sanger technique and require both numerous repetitions and sophisticated statistical techniques for their interpretation.[49] Determination of the genetic alterations present in tumors has provided not only information on possible "driver" mutations involved in the carcinogenic process, but also has illuminated the processes (possibly) operative in all cells by revealing multiple "passenger" mutations. The recognition of widespread CNVs in individual genomes is one example. Another phenomenon, possibly (but not necessarily[50]) related to SHM because of its dependence on cytosine deaminases, is a phenomenon termed kataegis. It is observed as a region of closely linked mutations at cytidines located in 5'TCW motifs (where W is an A or T).[51,52]

As a result of massive sequencing studies, classifying mutations by base change in the context of 3' and 5' bases, transcriptional strand, and dinucleotide mutations, a set of 20 mutational "signatures" was identified.[53] Some of these signatures have been correlated with known mechanisms; e.g., lung cancer mutations with etiologically-known carcinogens and deamination of cytosine. Others represent unknown mechanisms. It is a question whether these phenomena are observed only in cancer cells or whether the examination of the expanded single cell clones that occur in cancer provides insight into normal but very rare processes occurring in normal cells.

The new data also permit a more nuanced answer to the question of whether cancer genomes are inherently more mutable. The thought that tumors are hypermutable is an old suggestion put in modern terms in a series of papers by L. Loeb.[54] A parallel series of papers, summarized by W. Bodmer,[55] argued that the spontaneous human mutation rate is sufficient to account for the mutations involved. Mismatch repair deficiency and its associated increased mutation rate are certainly associated with a subclass of colon carcinomas. Determination of the frequency of mutations in a variety of tumors suggests the conclusion that both views are correct in part! The frequency of mutations observed in tumors can vary by over five logs from about 0.001 to 400 per megabase.[53] The difference is partly due to the number of divisions before the tumorigenic event;[56] juvenile tumors have fewer mutations than those in adults. Cancer incidence in about two-thirds of tumors can be correlated to the number of stem cell divisions in the tumor progenitor, suggesting a major role for spontaneous mutation in cancer etiology. Many of the mutations observed in adults involve repair genes and do have a mutator effect. The increased load of mutations induced would certainly be expected to direct the course of tumor progression.

DOUBLE-STRAND BREAK REPAIR-RELATED MECHANISMS

In spite of the attention given to single nucleotide changes, the quantitatively most important sources of genome variation are structural variations (see, Somatic Hypermutation section). These include CNVs of more than about 100 bp and the deletions, duplications, insertions, inversions, and translocations traditionally defined by cytological analysis. Analysis of cancer genomes has led to the discovery of complex structural changes involving multiple events.[57] DNA replication errors such as slippage or template switching may account for a minority of such events, but the great majority is due to the formation and (aberrant) repair of double-strand breaks. Study of the mechanism of such repair is now an active field of research with numerous reviews covering aspects of the subject (e.g., Refs [44,58,59]).

Double-strand breaks, catalyzed in part by the protein Spo11, are necessary for homologous recombination and chromosome segregation in meiosis and also for the VDJ joining and the formation of the different classes of antibodies. They are also produced by environmental stress, such as ionizing radiation and clustered oxidative free radicals as examples. Two major mechanisms are involved in their repair: NHEJ (Fig. 1.3) and homologous recombination (HR) (Fig. 1.4).

The first, and possibly more common, repair process in humans, NHEJ, is essentially the bringing together

and sealing of two broken DNA (chromosome) ends. It commonly occurs during the G0 and G1 stages of the cell cycle and involves the Ku70/80 protein, a barrel-shaped dimer that fits over the ends of the broken DNA and helps bring the ends together, as well as polymerase, ligase, kinase/phosphatase, and nuclease proteins.[58] Sequencing of the break point junctions reveals that there are often short regions of homology, leading to the classification of a subclass of events known as microhomology-mediated end joining (MMEJ). About 40% of the CNV display microhomology and in about 13%, there were a small number of inserted (nontemplated) bases.[60]

HR requires homologies of about 300 bp or more and involves Rad51 and numerous other proteins to sequentially connect and then break the connection between homologous chromosomes. Given the presence of pseudogenes and other repeated sequences in the human genome, aberrant recombination can occasionally occur between runs of homologous DNA sequence that are not in the same gene. This process of nonallelic homologous recombination (NAHR) is an obvious source of chromosomal change (Fig. 1.5).

A mutational phenomenon described in some cancer cells has been called "chromothripsis" and refers to a massive rearrangement of small chromosome segments from a few chromosomes occurring as a single event in cancer cells. One, or at most, a few chromosomes are involved and each may bear hundreds of rearrangements.[61] The lack of sequence homology in the joined

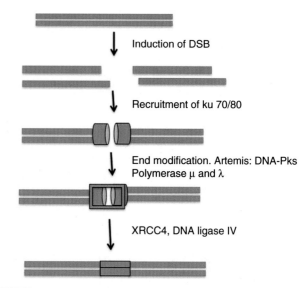

FIGURE 1.3 Cartoon illustration of the steps in nonhomologous end joining.[58]

Double strand break repair by homologous recombination (greatly simplified)

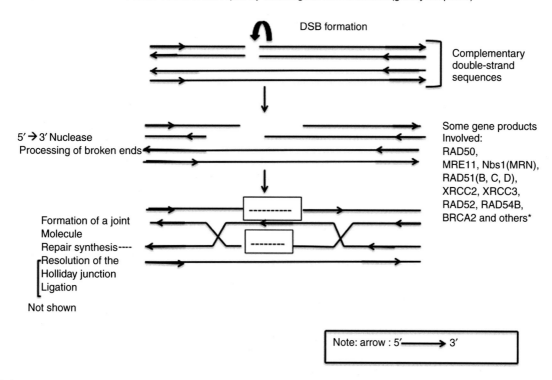

FIGURE 1.4   **A greatly simplified view of the initial changes in double-strand break repair (DSBR) by homologous recombination (HR) or nonallelic homologous recombination (NAHR).** In order to provide the missing information, either process requires the presence of a stretch of undamaged, homologous DNA of at least several hundreds of nucleotides either from an allele (HR), pseudoallele, or other repeat region in the genome. The RAD51 gene is critical for the invasion of the intact double-strand by the nuclease exposed single-stranded regions. Repair synthesis and ligation creates two complete strands, and then the "X" structure (a Holliday junction) needs to be cut (resolved) to free the separate DNA molecules.

fragments suggests NHEJ as a major mechanism, but the complex nature of the joints as determined by sequencing has suggested that mechanisms related to the restoration of DNA synthesis after replication fork collapse may be involved. When encountered by the replicating complex, a single-strand nick becomes a double-strand break. It is not clear how the original chromosome shattering comes about. One suggestion[61] is that the chromosomes are shattered in micronuclei during mitosis. Chromothripsis is not restricted to cancer cells but has been detected in the cells of some individuals with congenital disorders.[62,63] The result in individuals with such chromosomes is a relatively balanced state without extensive gains or losses of DNA, in spite of the large number of repair events that must be required. The events leading to chromothripsis appear mainly to occur in the generation of sperm.[63] The ability to carry out massive sequencing coupled with modern bioinformatics permits recognition of the extremely complex nature of what had previously appeared as simple chromosome structural changes.

MOBILE INSERTION ELEMENTS

Movement of transposable elements results in a significant proportion of germline structural variation; one estimate was that L1-mediated transposition alone accounted for 19% of the changes (NAHR accounts for 22%, and microhomology-mediated (3–20 bp) processes account for 28%).[64] Insertional mutagenesis, due to the introduction of a transposed element into a gene, can result in disruption of the gene product. The most prevalent transposable elements are the SINEs (short interspersed nuclear elements) and the LINEs (long interspersed elements). Full-sized LINE elements are about 6.1 kb in size and contain coding sequences for gene products essential for their transposition within the genome. Humans may have 200,000–500,000 of these elements in their genome, of which the most frequent is the L1 (LINE-1). Only a few such elements retain their capacity to catalyze movement. The others have suffered a variety of sequence changes in the course of history, resulting in their inactivation. The SINE elements, of which

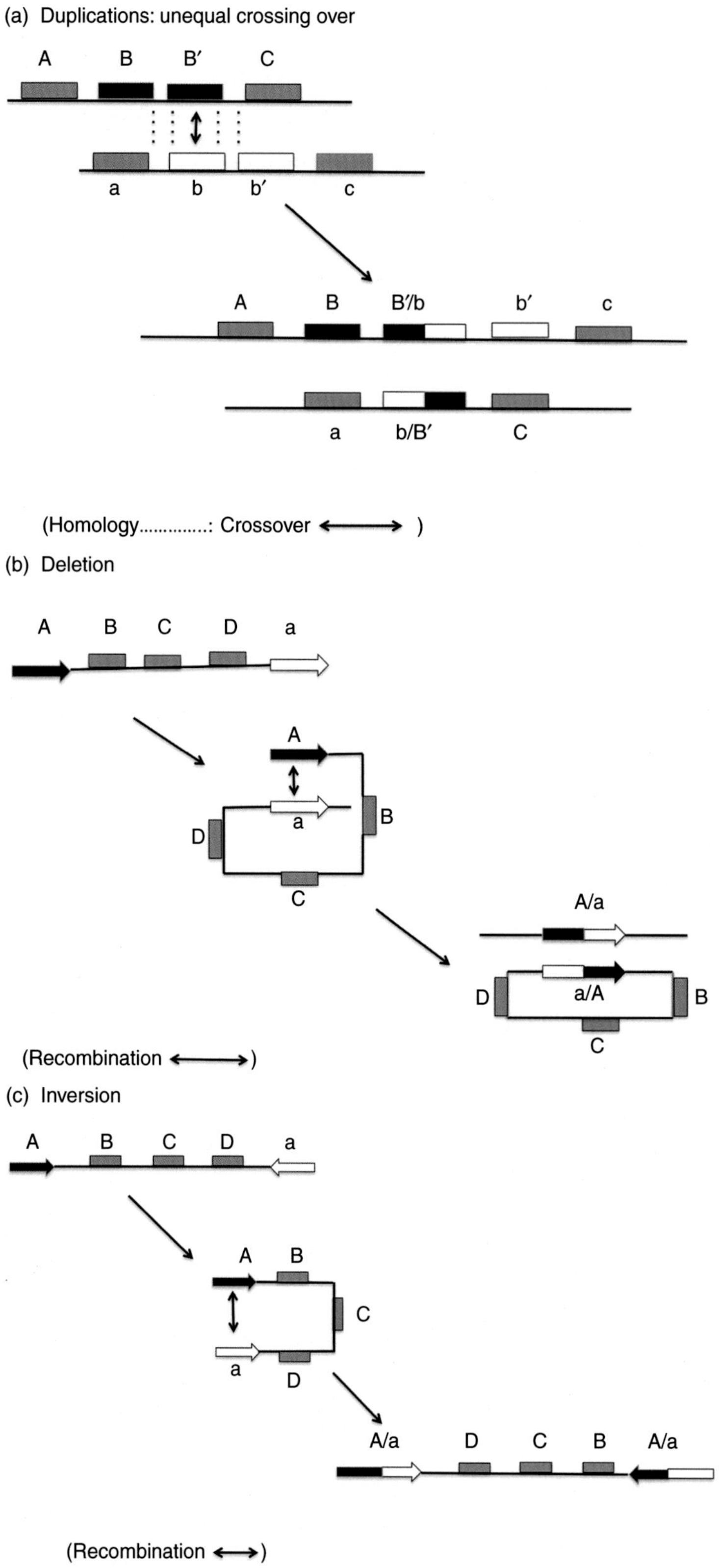

FIGURE 1.5   **Cartoon showing how repeated elements in the genome can lead to chromosome aberrations.** (a) Duplication due to unequal crossing over. (b) Deletion due to two transposable elements (repeated) inserted in the same chromosome in the same orientation. The loop including a copy of the transposon and the genes intervening is presumed to be lost in replication. Recombination between nonhomologous chromosomes containing such repeated sequences can lead to translocations. (c) Inversion due to two transposable elements inserted into the same chromosome in opposite orientation. Recombination will result in inversion of the chromosome order.

the most prevalent are the Alu sequences of about 300 bases, may be present in approximately 1,000,000 copies throughout the human genome. These elements cannot catalyze their own movement but can apparently transpose by utilizing some of the enzymatic machinery produced by LINE elements. Transposons are recognizable by the target site duplications that are found at either end of the insertion sites. Such duplications of about two to ten bases occur as a result of the insertion mechanism, which at one point involves making a staggered nick in the double helix, somewhat similar to the staggered cuts made by restriction enzymes, although there seems to be no specific sequence recognition for the insertion sites.

A 2008 compilation listed 33 pathological conditions associated with new insertions of Alu sequences and 11 with insertion of L1 elements. Eight of the L1 elements are reported on the X chromosome and nine of the Alu sequence are X-linked.[65] This unexplained excess on the X does not seem to be accounted for by the bias derived from the necessary dominance of all mutations in males. It has been estimated that there is a new Alu transposition about once in every 20 births, and between 1/100 and 1/200 carry a *de novo* Line 1 insertion. However, most of the structural aberrations occur as a result of recombination events between repeated sequences.

THE HUMAN MUTATION RATE

Does *de novo* mutation, that is, mutation occurring during the formation of germ cells and appearing in the soma of offspring but not parents, have clinical significance? As with the analysis of structural rearrangements, new DNA sequencing technology makes it possible to answer the question quantitatively, although the results for single nucleotide variation are not very different from those historically obtained by more indirect methods.[66,67] Estimates of mutation rates can be obtained from the analysis of trios, both parents and a child, and also by comparison between species based on the estimated divergence times. Based on a recent compilation,[45] the overall mutation rate for SNV is an average of about 1.2×10^{-8} per nucleotide per generation. Campbell and Eichler (*loc. cit.*[45]) have tabulated the best current values rates for CNV (1.2×10^{-2} per generation for large CNVs), small insertions or deletions (0.2×10^{-9} per site per generation for insertions and $0.53-0.58 \times 10^{-9}$ per site per generation for deletions) and transposon movement (MEI, mobile element insertion, 2.5×10^{-2} per genome per generation). Very roughly, every individual will differ from her/his parents by about 100 nucleotides due to new SNVs and indels, one out of 40 individuals will have a transposon movement, and one out of 80 a CNV of 100 kb or more. In addition, there is a low frequency of new (viable) aneuploidies.

The overall mutation rates do not reflect the actual heterogeneity of mutation probabilities across the genome with some sites being more mutable than others,[68] microsatellites, for example, being good substrates for mutation. CpG sites are particularly susceptible to C→T transitions (and G→A transitions on the other strand) because of the likelihood of deamination of methylated Cs at these sites to yield a T. "Islands" of CpG sequences are not so affected, perhaps because of a deficiency in methylation. Comparison of the average number of mutations with the average number of nucleotides affected (Fig. 1.6) indicates that although SNV occurs at the highest frequency with about 70–100 new mutations affecting one

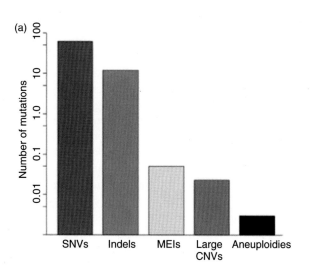
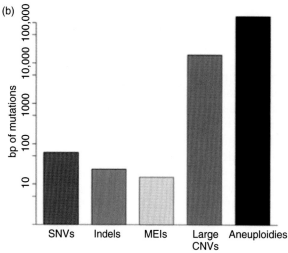

FIGURE 1.6 **Comparison of the frequency and magnitude of different forms of genetic variation.** (a) Average number of mutations of each type of variant per birth. (b) Average number of mutated bases contributed by each type of variant per birth. Vertical axis is log₁₀ scaled in both (a) and (b). SNV, single nucleotide variation; CNV, copy number variations; MEI, mobile element insertion. *From Fig. 1.1 in Ref. [45].*

base pair each diploid generation, the rarer large CNV and aneuploidies affect the greater number of base pairs, on the order of 20,000 per generation.

Sequencing has also confirmed an earlier conclusion by Haldane that the mutation frequency is higher in sperm from older fathers.[68] It has long been recognized that the rate of aneuploidy for chromosome 21 increases with maternal age. The increase in mutation with paternal age may have a similar foundation in the biology of gamete formation. Whereas the female gametes are fixed in number, the spermatagonia continue to divide throughout life, accumulating new mutations at each division. Actually the situation is more complex, since this explanation predicts only a modest paternal effect. For a few conditions, termed parental age effect disorders,[69] the effect is much greater. These conditions include the developmental Apert, Crouzon, Pfeiffer, and Muenke syndromes, and also mutations in the RET proto-oncogene involved in multiple endocrine neoplasia types 2A and 2B. The conditions are caused by dominant point mutations in growth-regulating pathways with the causative mutations occurring in the fathers. The mutations give the spermatagonia in which they occur a positive growth advantage, so that sperm are more likely to be formed from these cells.

The roughest calculation indicates that offspring are likely to differ from their parents by about 100 single nucleotide changes and small insertions or deletions. One out of 40 births will have had a variation in copy number of over 100 kb, and one out of 20 will have had a transposition event in which a mobile element is inserted at a different location.[45,70] Even genetically identical twins have been shown to vary in copy number, indicating the importance of mutational events occurring after fertilization.

THE PHENOTYPIC EFFECT OF MUTATIONS

Modern technology confirms the obvious: we are all different! Does the discovery of a particular (new) mutation have any clinical significance? This question turns out to be harder to answer than might be expected. Humans may carry numerous loss-of-function mutations in their genomes without noticeable phenotypic effect. MacArthur et al. estimate about 100 loss-of-function mutations with about 20 completely inactivated genes in a typical human genome.[71] An inactivating mutation in the relevant gene in an individual with clinically recognizable disease can be considered causative,[72] but the reverse is not necessarily so. In the absence of clinically recognizable signs it is not (yet) clear what the exact phenotypic result will be of any particular DNA change.[73]

Why should this be so? It turns out that in spite of great progress in understanding the interaction of different biochemical pathways, the phenotypic effect of a particular change in a particular individual is difficult, if not impossible, to predict.[74] Many functions are redundant. The experience over the past years with so-called "knockout" mutants of mice with no noticeable phenotypic effect is anecdotal evidence of that redundancy. The effect of other genes, the genetic background, and of the environment, broadly construed, provide a situation in which two individuals may have the same mutation while only one displays a phenotype. Even conditions clearly related to a specific mutation in a specific gene (e.g., sickle cell anemia) show clinical heterogeneity indicating the involvement of other factors, either environmental, genetic, or both.[75] Presumably, our ability to predict the importance of particular changes will increase. DECIPHER (https://decipher.sanger.ac.uk/) is one of several databases that attempts to correlate genetic variation and phenotype in patients with rare disorders.[76] The January 2014 issue of *Nucleic Acids Research* (as well as other January data issues of this journal) includes information about DECIPHER and related databases.

CONCLUSION AND SUMMARY

Each human genome is unique in some respect. The normal processes of replication are subject to error at a rate sufficient to generate 50–100 new point mutations in each generation, and environmental damage to the DNA can lead to additional changes. These changes are subject to surveillance by a variety of repair mechanisms, with the end result determined by the result of the interaction between the mutation and repair processes. Double-strand breaks occur in the DNA as a result of both normal physiological processes and environmental damage. Such breaks are often essential to cells, but they are also the initiating events for changes in gene copy number and for chromosome structural changes, such as inversions and translocations. The human genome is host to a variety of mobile elements; the vestiges of past viral attack and the movement of such elements are also agents for change in the genome. It is not yet possible to predict the phenotypic effects of all mutational, change since the genome is often redundant and equipped with a variety of control mechanisms that modulate the effects of many changes. It is apparent that drastic changes will be quickly eliminated by selection, but the results of sequencing show the tolerance of the genome.

This chapter version was written in 2014 and slightly modified in 2015, only six years after the first edition. These intervening years have seen an explosive expansion in our knowledge (if not understanding) of the variability of the human genome and of the mutational mechanisms that produce this variability. In large part, this development is due to the development of methods for sequencing DNA and bioinformatics techniques for analyzing the results. A major need is better quality control of the data being produced and published. It will be interesting to see how the material reported in this chapter is treated in some future edition.

References

1. Carlson EA. *Mutation: The history of an idea from Darwin to genomics.* Cold Spring Harbor: Cold Spring Harbor Press, 2011.
2. Singer B, Grunberger D. *Molecular biology of mutagens and carcinogens.* New York: Plenum Press, 1983.
3. Benzer S, Freese E. Induction of specific mutations with 5-bromouracil. *Proc Natl Acad Sci USA* 1958;**44**(2):112–9.
4. Brenner S, Barnett L, Crick FHC, et al. The theory of mutagenesis. *J Mol Biol* 1961;**3**:121–4.
5. Lange SS, Takata K, Wood RD. DNA polymerases and cancer. *Nat Rev Cancer* 2011;**11**(2):96–110.
6. Abdel-Rahman WM, Mecklin JP, Peltomaki P. The genetics of HNPCC: application to diagnosis and screening. *Crit Rev Oncol Hematol* 2006;**58**(3):208–20.
7. International Human Genome Sequencing C. Finishing the euchromatic sequence of the human genome. *Nature* 2004;**431**(7011):931–45.
8. Hangauer MJ, Vaughn IW, McManus MT. Pervasive transcription of the human genome produces thousands of previously unidentified long intergenic noncoding RNAs. *PLoS Gen* 2013;**9**(6):e1003569.
9. Freese E. The difference between spontaneous and base-analogue induced mutations of phage T4. *Proc Natl Acad Sci USA* 1959;**45**(4):622–33.
10. Labunskyy VM, Hatfield DL, Gladyshev VN. Selenoproteins: molecular pathways and physiological roles. *Physiol Rev* 2014;**94**(3):739–77.
11. Hunt RC, Simhadri VL, Iandoli M, et al. Exposing synonymous mutations. *Trends Genet* 2014;**30**(7):308–21.
12. Richards EJ. Inherited epigenetic variation – revisiting soft inheritance. *Nat Rev Genet* 2006;**7**(5):395–401.
13. Smith ZD, Meissner A. DNA methylation: roles in mammalian development. *Nat Rev Genet* 2013;**14**(3):204–20.
14. Heijmans BT, Tobi EW, Stein AD, et al. Persistent epigenetic differences associated with prenatal exposure to famine in humans. *Proc Natl Acad Sci USA* 2008;**105**(44):17046–9.
15. Soubry A. Epigenetic inheritance and evolution: a paternal perspective on dietary influences. *Prog Biophys Mol Biol* 2015;**118**(1–2):79–85.
16. Makova KD, Hardison RC. The effects of chromatin organization on variation in mutation rates in the genome. *Nat Rev Genet* 2015;**16**(4):213–23.
17. Friedberg EC, Walker GC, Siede W, et al. *DNA repair and mutagenesis.* 2nd ed. Washington, DC: ASM Press; 2006.
18. Petruska J, Goodman MF, Boosalis MS, et al. Comparison between DNA melting thermodynamics and DNA polymerase fidelity. *Proc Natl Acad Sci USA* 1988;**85**(17):6252–6.
19. McCulloch SD, Kunkel TA. The fidelity of DNA synthesis by eukaryotic replicative and translesion synthesis polymerases. *Cell Res* 2008;**18**(1):148–61.
20. Tissier A, McDonald JP, Frank EG, et al. Pol iota, a remarkably error-prone human DNA polymerase. *Gen Dev* 2000;**14**(13):1642–50.
21. Sale JE, Lehmann AR, Woodgate R. Y-family DNA polymerases and their role in tolerance of cellular DNA damage. *Nat Rev Mol Cell Biol* 2012;**13**(3):141–52.
22. Kirouac KN, Ling H. Structural basis of error-prone replication and stalling at a thymine base by human DNA polymerase iota. *EMBO J* 2009;**28**(11):1644–54.
23. Nair DT, Johnson RE, Prakash S, et al. Replication by human DNA polymerase-iota occurs by Hoogsteen base-pairing. *Nature* 2004;**430**(6997):377–80.
24. Fersht AR, Knill-Jones JW, Tsui WC. Kinetic basis of spontaneous mutation. Misinsertion frequencies, proofreading specificities and cost of proofreading by DNA polymerases of *Escherichia coli. J Mol Biol* 1982;**156**(1):37–51.
25. Johnson RE, Washington MT, Haracska L, et al. Eukaryotic polymerases iota and zeta act sequentially to bypass DNA lesions. *Nature* 2000;**406**(6799):1015–9.
26. Prakash S, Johnson RE, Prakash L. Eukaryotic translesion synthesis DNA polymerases: specificity of structure and function. *Ann Rev Biochem* 2005;**74**:317–53.
27. Wood RD, Lange SS. Breakthrough for a DNA break-preventer. *Proc Natl Acad Sci USA* 2014;**111**(8):2864–5.
28. Lehmann AR, Niimi A, Ogi T, et al. Translesion synthesis: Y-family polymerases and the polymerase switch. *DNA Repair* 2007;**6**(7):891–9.
29. Kumar D, Abdulovic AL, Viberg J, et al. Mechanisms of mutagenesis *in vivo* due to imbalanced dNTP pools. *Nucleic Acids Res* 2011;**39**(4):1360–71.
30. Waisertreiger IS, Liston VG, Menezes MR, et al. Modulation of mutagenesis in eukaryotes by DNA replication fork dynamics and quality of nucleotide pools. *Environ Mol Mutagen* 2012;**53**(9):699–724.
31. Pena-Diaz J, Jiricny J. Mammalian mismatch repair: error-free or error-prone? *Trends Biochem Sci* 2012;**37**(5):206–14.
32. Jiricny J. The multifaceted mismatch-repair system. *Nat Rev Mol Cell Biol* 2006;**7**(5):335–46.
33. Moldovan GL, Pfander B, Jentsch S. PCNA, the maestro of the replication fork. *Cell* 2007;**129**(4):665–79.
34. Leach FS, Nicolaides NC, Papadopoulos N, et al. Mutations of a mutS homolog in hereditary nonpolyposis colorectal cancer. *Cell* 1993;**75**(6):1215–25.
35. Fishel R, Lescoe MK, Rao MR, et al. The human mutator gene homolog MSH2 and its association with hereditary nonpolyposis colon cancer. *Cell* 1993;**75**(5):1027–38.
36. Jacinto FV, Esteller M. Mutator pathways unleashed by epigenetic silencing in human cancer. *Mutagenesis* 2007;**22**(4):247–53.
37. Kaina B, Christmann M, Naumann S, et al. MGMT: key node in the battle against genotoxicity, carcinogenicity and apoptosis induced by alkylating agents. *DNA Repair* 2007;**6**(8):1079–99.
38. Peled JU, Kuang FL, Iglesias-Ussel MD, et al. The biochemistry of somatic hypermutation. *Ann Rev Immunol* 2008;**26**:481–511.
39. Mirkin SM. Expandable DNA repeats and human disease. *Nature* 2007;**447**(7147):932–40.
40. McMurray CT. Hijacking of the mismatch repair system to cause CAG expansion and cell death in neurodegenerative disease. *DNA Repair* 2008;**7**(7):1121–34.
41. McMurray CT. Mechanisms of trinucleotide repeat instability during human development. *Nat Rev Genet* 2010;**11**(11):786–99.
42. Wood R. Human DNA Repair Genes <http://scienceparkmdandersonorg/labs/wood/DNA_Repair_Geneshtml; http://scienceparkmdandersonorg/labs/wood/DNA_Repair_Geneshtml;> 2013.
43. Goodarzi AA, Jeggo PA. The repair and signaling responses to DNA double-strand breaks. *Adv Genet* 2013;**82**:1–45.
44. Davis AJ, Chen DJ. DNA double strand break repair via nonhomologous end-joining. *Trans Cancer Res* 2013;**2**(3):130–43.
45. Campbell CD, Eichler EE. Properties and rates of germline mutations in humans. *Trends Genet* 2013;**29**(10):575–84.
46. Gazumyan A, Bothmer A, Klein IA, et al. Activation-induced cytidine deaminase in antibody diversification and chromosome translocation. *Adv Cancer Res* 2012;**113**:167–90.

47. Storb U. Why does somatic hypermutation by AID require transcription of its target genes? *Adv Immunol* 2014;**122**:253–77.

48. Chahwan R, Edelmann W, Scharff MD, et al. AIDing antibody diversity by error-prone mismatch repair. *Semin Immunol* 2012;**24**(4):293–300.

49. Allhoff M, Schonhuth A, Martin M, et al. Discovering motifs that induce sequencing errors. *BMC Bioinform* 2013;**14**(Suppl. 5):S1.

50. Bacolla A, Cooper DN, Vasquez KM. Mechanisms of base substitution mutagenesis in cancer genomes. *Genes* 2014;**5**(1):108–46.

51. Roberts SA, Lawrence MS, Klimczak LJ, et al. An APOBEC cytidine deaminase mutagenesis pattern is widespread in human cancers. *Nat Genet* 2013;**45**(9):970–6.

52. Taylor BJ, Nik-Zainal S, Wu YL, et al. DNA deaminases induce break-associated mutation showers with implication of APOBEC3B and 3A in breast cancer kataegis. *Elife* 2013;**2**:e00534.

53. Alexandrov LB, Nik-Zainal S, Wedge DC, et al. Signatures of mutational processes in human cancer. *Nature* 2013;**500**(7463):415–21.

54. Loeb LA, Bielas JH, Beckman RA. Cancers exhibit a mutator phenotype: clinical implications. *Cancer Res* 2008;**68**(10):3551–7 discussion 7.

55. Bodmer W. Genetic instability is not a requirement for tumor development. *Cancer Res* 2008;**68**(10):3558–60.

56. Tomasetti C, Vogelstein B. Cancer etiology. Variation in cancer risk among tissues can be explained by the number of stem cell divisions. *Science* 2015;**347**(6217):78–81.

57. Quinlan AR, Hall IM. Characterizing complex structural variation in germline and somatic genomes. *Trends Genet* 2012;**28**(1):43–53.

58. Lieber MR. The mechanism of double-strand DNA break repair by the nonhomologous DNA end-joining pathway. *Ann Rev Biochem* 2010;**79**:181–211.

59. Ottaviani D, Lecain M, Sheer D. The role of microhomology in genomic structural variation. *Trends Genet* 2014;**30**(3):85–94.

60. Conrad DF, Bird C, Blackburne B, et al. Mutation spectrum revealed by breakpoint sequencing of human germline CNVs. *Nat Genet* 2010;**42**(5):385–91.

61. Forment JV, Kaidi A, Jackson SP. Chromothripsis and cancer: causes and consequences of chromosome shattering. *Nat Rev Cancer* 2012;**12**(10):663–70.

62. Chiang C, Jacobsen JC, Ernst C, et al. Complex reorganization and predominant non-homologous repair following chromosomal breakage in karyotypically balanced germline rearrangements and transgenic integration. *Nat Genet* 2012;**44**(4):390–7 S1.

63. Kloosterman WP, Guryev V, van Roosmalen M, et al. Chromothripsis as a mechanism driving complex *de novo* structural rearrangements in the germline. *Hum Mol Genet* 2011;**20**(10):1916–24.

64. Kidd JM, Graves T, Newman TL, et al. A human genome structural variation sequencing resource reveals insights into mutational mechanisms. *Cell* 2010;**143**(5):837–47.

65. Belancio VP, Hedges DJ, Deininger P. Mammalian non LTR-retrotransposons: for better or worse, in sickness and in health. *Genome Res* 2008;**18**:343–58.

66. Kondrashov AS. Direct estimates of human per nucleotide mutation rates at 20 loci causing Mendelian diseases. *Hum Mutat* 2003;**21**(1):12–27.

67. Haldane JBS. The rate of spontaneous mutation of a human gene. *J Genet* 1935;**31**:317–26.

68. Segurel L, Wyman MJ, Przeworski M. Determinants of mutation rate variation in the human germline. *Ann Rev Genom Hum Genet* 2014;**15**:47–70.

69. Goriely A, Wilkie AO. Paternal age effect mutations and selfish spermatogonial selection: causes and consequences for human disease. *Am J Hum Genet* 2012;**90**(2):175–200.

70. Itsara A, Wu H, Smith JD, et al. *De novo* rates and selection of large copy number variation. *Genome Res* 2010;**20**(11):1469–81.

71. MacArthur DG, Balasubramanian S, Frankish A, et al. A systematic survey of loss-of-function variants in human protein-coding genes. *Science* 2012;**335**(6070):823–8.

72. Gilissen C, Hoischen A, Brunner HG, Veltman JA. Unlocking Mendelian disease using exome sequencing. *Genome Biol* 2011;**12**(9):228.

73. Wright CF, Middleton A, Burton H, et al. Policy challenges of clinical genome sequencing. *BMJ* 2013;**347**:f6845.

74. Lehner B. Genotype to phenotype: lessons from model organisms for human genetics. *Nat Rev Genet* 2013;**14**(3):168–78.

75. Beutler E. Discrepancies between genotype and phenotype in hematology: an important frontier. *Blood* 2001;**98**(9):2597–602.

76. Bragin E, Chatzimichali EA, Wright CF, et al. DECIPHER: database for the interpretation of phenotype-linked plausibly pathogenic sequence and copy-number variation. *Nucleic Acids Res* 2014;**42**(Database issue):D993–D1000.

PART II

PANCREAS

2

A Clinical Guide to Monogenic Diabetes

David Carmody*, Julie Støy**, Siri Atma W. Greeley*,
Graeme I. Bell*, Louis H. Philipson*

*Section of Endocrinology, Diabetes and Metabolism, The University of Chicago,
Chicago, IL, USA
**Department of Clinical Medicine – Medical Research Laboratory,
Aarhus University, Aarhus, Denmark

INTRODUCTION

Diabetes mellitus is a genetically and phenotypically heterogeneous set of metabolic diseases. The most common forms, type 1 and type 2, have susceptibility determined by variation at a number of different genes, none of which alone is sufficient to cause diabetes.[1,2] However, it is estimated that ~1% of this population have a form of monogenic diabetes caused by a single gene mutation.[3] In the United States, where the prevalence of diabetes mellitus is soaring, with 9.3% of the population affected and a total health cost of $245 billion, monogenic causes could account for as many as 3–500,000 cases of diabetes.[4,5] The clinical challenge is to identify those who should have genetic testing and how those genes should be tested.

Clinicians are confronted by an expanding and confusing list of diverse monogenic causes of diabetes.[6] This chapter is a guide to clinical diagnosis of monogenic causes of diabetes mellitus affecting children and adults. We aim to briefly review specific situations from the point of view of the clinician, focusing on: clinical presentation, genetic testing, therapeutic interventions, mutation-specific complications, and family counseling.

The term MODY (maturity onset diabetes of the young) has been used to describe a number of monogenic forms. MODY is a confusing term, since it now includes at least 11 separate genes that differ in important ways as to presentation, treatment, and associated conditions. For the purposes of the following discussion, we will employ the correct monogenic names of these forms of early onset diabetes.

Several recent reviews have discussed the genetics of neonatal diabetes and other forms of monogenic diabetes, and they can be consulted for more detailed information.[3,5,7–9] In addition, several excellent websites provide overviews and new information, for example, www.diabetesgenes.org, www.monogenicdiabetes.org and http://diabetes.niddk.nih.gov/dm/pubs/mody/.

CLINICAL PRESENTATION

Monogenic diabetes in children and adults can present clinically in a variety of ways. The clinical phenotype can be markedly different, despite many of these genes being globally referred to as monogenic, neonatal diabetes, or MODY genes. These genes and clinical presentations, due to their deleterious mutations, are summarized in Table 2.1. We advocate a simple clinical approach to the newly diagnosed patient, and the presumed etiology of each patient's diabetes should be challenged at each clinical visit. This is particularly important as the number of individuals affected by diabetes increases worldwide and with fewer patients with diabetes attending subspecialty care.[10] Table 2.2 outlines the genetic tests available for monogenic forms of diabetes.

The term MODY has been used to describe a monogenic form of nonketotic, noninsulin dependent diabetes with onset before 25 years of age with two to three consecutively affected generations. MODY is a confusing term, as it now includes at least 11 separate genes that differ in important ways as to presentation, treatment, and associated conditions. Additionally, use of traditional criteria accounts for less than 50% of individuals with a genetic diagnosis of MODY that fit this classic description.[9,11] We suggest a slightly broader criteria based on children and young adults diagnosed with diabetes mellitus with atypical features with presumed

Genetic Diagnosis of Endocrine Disorders. http://dx.doi.org/10.1016/B978-0-12-800892-8.00002-6

TABLE 2.1 Highly Penetrant Genes Causes of Monogenic Diabetes

Gene	Inheritance	Clinical features
GENETIC DISORDER OF GLUCOSE METABOLISM		
GCK	AD	MODY2: stable, nonprogressive elevated fasting blood glucose; typically does not require treatment; microvascular complications are rare; small rise in 2-h glucose level on OGTT (<54 mg/dL; <3 mmol/L)
GENETIC DISORDERS OF GENE EXPRESSION		
HNF1A	AD	MODY3: progressive insulin secretory defect with presentation in adolescence or early adulthood; lowered renal threshold for glucosuria; large rise in 2-h glucose level on OGTT (>90 mg/dL; >5 mmol/L); sensitive to sulfonylureas
HNF4A	AD	MODY1: progressive insulin secretory defect with presentation in adolescence or early adulthood; may have large birth weight and transient neonatal hypoglycemia; sensitive to sulfonylureas
HNF1B	AD	MODY5: developmental renal disease (typically cystic); genitourinary abnormalities; atrophy of the pancreas; hyperuricemia; gout
PDX1	AD AR	MODY4: rare, diabetes appears to be mild NDM: SGA; diarrhea; malnutrition; parents have PDX1-MODY (MODY4)
NEUROD1	AD AR	MODY6: rare NDM: cerebellar hypoplasia; deafness
KLF11	AD	MODY7: rare
PAX4	AD	MODY9: rare
BLK	AD	MODY11: rare
PCBD1	AR	MODY-like: rare; progressive insulin secretory defect
GATA6	AD	NDM: pancreatic hypoplasia; gastrointestinal and cardiac malformations
GLIS3	AR	NDM: congenital hypothyroidism; glaucoma; kidney cysts; hepatic fibrosis
PAX6	AR	NDM: brain malformations; microcephaly; microphthalmia; panhypopituitarism
NEUROG3	AR	NDM: SGA; severe intractable congenital diarrhea
PTF1A	AR	NDM: pancreatic hypoplasia; cerebellar hypoplasia
RFX6	AR	NDM: SGA, intestinal atresias; gall bladder hypoplasia/aplasia; diarrhea
NKX2.2	AR	NDM: rare
MNX1	AR	NDM: rare; pancreatic hypoplasia
GENETIC DISORDER OF ION CHANNELS		
ABCC8	AD	NDM: IUGR; responsive to sulfonylureas
KCNJ11	AD	NDM: IUGR; possible developmental delay and seizures; responsive to sulfonylureas
GENETIC DISORDER OF INSULIN SYNTHESIS		
INS	AD	NDM: IUGR MODY10: rare
GENETIC DISORDERS OF ER STRESS/CELL DEATH		
EIF2AK3	AR	NDM: Wolcott–Rallison syndrome; epiphyseal dysplasia; exocrine pancreatic insufficiency
WFS1	AR	Variable age of DM diagnosis (3 weeks–14 years): rare; optic atrophy; deafness; diabetes insipidus
IER3IP1	AR	NDM: rare; microcephaly with simplified gyral pattern; infantile epileptic encephalopathy
TRMT10A	AR	MODY: rare; short stature; microcephaly
EPIGENETIC DISORDERS OF THE BETA CELL		
6q24 (PLAGL1, HYMA1)	AD for paternal duplications	Transient NDM: IUGR; macroglossia; umbilical hernia. Mechanisms include UPD6, paternal duplication or maternal methylation defect
ZFP57	AR	Transient NDM: rare; IUGR; macroglossia

TABLE 2.1 Highly Penetrant Genes Causes of Monogenic Diabetes (*cont.*)

Gene	Inheritance	Clinical features
GENETIC DISORDER OF GLUCOSE TRANSPORT		
SLC2A2	AR	NDM: rare; hepatomegaly; proximal tubular nephropathy
PRIMARY GENETIC DISORDER OF AUTO-IMMUNITY		
FOXP3	X-linked	NDM: IPEX syndrome (immunodysregulation polyendocrinopathy enteropathy X-linked); autoimmune diabetes; autoimmune thyroid disease; exfoliative dermatitis
STAT3	AR	NDM: autoimmune diabetes; autoimmune enteropathy; short stature; eczema
GENETIC DISORDER OF INSULIN SECRETION		
SLC19A2	AR	NDM: rare; thiamine-responsive megaloblastic anemia; deafness
DISEASE OF THE EXOCRINE PANCREAS		
CFTR	AR	Variable age of DM diagnosis: pulmonary disease; exocrine pancreatic dysfunction; decreased insulin production; insulin resistance
CEL	AD	MODY8: rare; atrophy of the pancreas; exocrine pancreatic dysfunction
MITOCHONDRIAL DISORDERS		
MT-TL1, MT-TK, MT-TE	Maternally inherited	MIDD: maternally inherited diabetes and deafness
IRON DEPOSITION WITHIN BETA CELLS		
HFE	AR	Variable age of DM diagnosis: hepatic disease; skin pigmentation; arthropathy; hypogonadism in males; cardiomegaly

AD, autosomal dominant; AR, autosomal recessive; DM, diabetes mellitus; IUGR, intrauterine growth restriction; NDM, neonatal diabetes mellitus; MODY, maturity onset diabetes of the young; SGA, small for gestational age. Rare diseases associated with variable diabetes penetrance were not included.

TABLE 2.2 Genetic Tests for Monogenic Forms of Diabetes

Tests	Advantages	Disadvantages
Sanger sequencing	Cheap Limited interpretation challenges	Limited to the target gene(s) Large heterozygous deletions may be missed
Dedicated monogenic diabetes NGS gene panels	Excellent coverage of known monogenic DM genes Allow for expansion as new genes are identified	Higher cost Larger interpretation challenges Some panels may not identify deletions
Whole exome sequencing and whole genome sequencing	Potential for novel gene discovery Increasing coverage of exons (and introns with WGS)	More complex informed consent Higher costs Increased complexity of analysis/ interpretation Increased risk of incidental findings Increased reporting obligations Incomplete coverage of some known monogenic diabetes genes Difficulty in assessing areas of high G and C content and regions with repetitive elements
Deletion/duplication analysis (e.g., multiplex ligation-dependent probe amplification)	Cheap Limited interpretation challenges	Limited to the target gene(s)
Methylation-dedicated PCR for the 6q24-TND	Cheap	Limited to a few phenotypes Parental samples needed to identify the exact mechanism

type 1 or type 2 diabetes, which may be a more effective strategy for identifying patients for screening.[12]

The term neonatal diabetes mellitus (NDM) has been used to describe hyperglycemia requiring insulin therapy within the first 6 months of life, and it invariably has monogenetic etiology.[13] A growing list of over 20 gene causes have been identified[6] and, albeit at a much lower frequency, mutations in some of these genes may result in hyperglycemia that presents after 6 months of age.[14,15]

In the following sections, we briefly describe the more common clinical presentations of monogenic diabetes, the genes associated with each, and an approach to special aspects or to their treatment.

Familial Fasting Hyperglycemia: GCK-Related Hyperglycemia (also Known as GCK-MODY or MODY2)

A history of stable asymptomatic mild fasting hyperglycemia is most consistent with a heterozygous inactivating mutation in the glucokinase gene (GCK).[16] The population prevalence is estimated at 1 in 1000 individuals.[17] Glucokinase catalyzes the conversion of glucose to glucose-6-phosphate, a key step in glucose metabolism and the regulation of insulin secretion. While mutations within this gene are not typically associated with significant complications, misdiagnosis can result in inappropriate dietary restriction and pharmacotherapeutic intervention.[18] Other features often also include hyperglycemia in multiple first-degree family members, autosomal dominant inheritance, hyperglycemia in nonobese individuals, and hemaglobin A1c (HbA1c) in the range of 5.5–7%. While usually stable, hyperglycemia may sometimes progress with time, and elevated glucose levels (5.5–8 mmol/L, 100–144 mg/dL) are usually initially noted during routine screening of blood or urine or during oral glucose tolerance testing for gestational diabetes. Family members are often treated with widely varying treatments before diagnosis, from intensive insulin therapy to diet alone, yet the HbA1c values are usually not significantly altered by any treatment.[19] Drug therapy can usually be withdrawn with the possible benefits of intervention during pregnancy if the fetus does not inherit the maternal GCK mutation.[20,21] In individuals in which diabetes progresses, there is usually another cause, such as coexisting obesity or another form of insulin resistance associated with type 2 diabetes. Other forms of diabetes, including transcription factor forms of monogenic diabetes, may also present initially with mild fasting hyperglycemia. Careful attention should be paid to the family history, and a search for other causes is indicated if DNA sequencing does not reveal any significant mutations that alter the function of GCK.

Diet and exercise is an approach to management but will not make a significant impact on glycemic control. Glucose monitoring should be limited to occasional capillary blood glucose and HbA1c measurements. Obesity and other conditions leading to increased insulin resistance may require treatment if there is an increase in HbA1c. Therapeutic targets should be individualized, taking the patient's baseline HbA1c into account, along with age and other underlying factors.

Transcription Factor Diabetes

The more common transcription factor forms of monogenic diabetes include HNF4A-, HNF1A-, and HNF1B-related diabetes (also known as HNF4A, HNF1A, and HNF1B-MODY or MODY1/3/5). Other rare transcription factor forms of monogenic diabetes are listed in Table 2.1. The key features are: young adult onset, absence of autoimmunity associated with type 1 diabetes, usually nonobese, dominant inheritance if familial, and usually detectable C-peptide. Progression does occur, with all of the complications associated with type 1 and type 2 diabetes, if poor control of blood sugar has been present. Since these genes have a wide expression in many tissues, extrapancreatic features can be found. These include developmental abnormalities of genitourinary organs associated with HNF1B mutations and hepatic adenomas with HNF1A mutations.

The most frequent causes of transcription factor diabetes in most series are mutations in HNF1A, followed by HNF4A. These forms of diabetes are often misdiagnosed and incorrectly treated with insulin.[22] The financial cost of genetic testing, limited access to testing, and a lack of knowledge about monogenic diabetes likely contribute to the misdiagnosis.[9]

Kidney involvement, such as congenital abnormalities or cysts, should lead to analysis of HNF1B. In addition to missense and nonsense mutations, small insertions, deletions, and exon duplications, as well as complete or partial gene deletions and large genomic rearrangements, have been reported for HNF1B, missed by conventional sequencing methods.[23] If these genes lack diabetes-associated mutations, consideration can then be given to the more rare causes of nonobese, antibody-negative young adult onset diabetes. Some of the rare forms of monogenic diabetes are listed in Table 2.1.

In the absence of a significant family history and relatively recent onset, latent autoimmune diabetes in adults should also be considered and anti-GAD65 and anti-IA2 antibody levels obtained. Large pedigrees, novel mutations, and specific questions should be referred to an academic center with specific expertise. It is also helpful to obtain c-peptide measurements with simultaneous glucose levels. Typically these patients have low c-peptide levels, whereas individuals with high c-peptide levels may be demonstrating a failure to compensate for insulin resistance associated with such conditions as Cushing's syndrome, acromegaly, or HCV infection.[24,25]

HNF1A and *HNF4A* mutations show particular sensitivity to low dose oral sulfonylurea therapy.[26] Some *HNF1A* and *4A* patients report that such therapy was tried and discontinued due to hypoglycemia, but careful low-dose therapy with these oral agents has significant advantages over insulin therapy.[27] Older patients may no longer respond to sulfonylureas, necessitating the initiation of insulin. Since these patients are not usually insulin resistant, insulin sensitizers are unlikely to be of benefit and may even be contraindicated due to off-target effects. The efficacy of incretin agonists or dipeptidylpeptidase IV inhibitors has not yet been fully evaluated in this situation, but they have been used successfully to supplement sulfonylurea therapy in some patients[28] and in unpublished observations. Insulin is likely to be the best therapy for mutations *HNF1B* and other genes in this category, but few of the other mutations have been carefully studied for response.

Syndromic Forms of Monogenic Diabetes

In these conditions, diabetes is part of a broader clinical syndrome rather than being the sole presenting manifestation of the gene defect. The reader is referred to that literature for more detailed information about the complex and varied manifestations due to mutations within each of these genes.[6,29] We will briefly outline some of the more recognizable forms that present outside of infancy.

Mitochondrial mutations, principally the tRNA mutations that are known to cause MELAS syndrome (mitochondrial encephalopathy with stroke-like episodes), and maternally inherited diabetes and deafness (MIDD), are distinguished by evidence of maternal inheritance. The varied clinical manifestations due to heteroplasmy, defined as the unequal distribution of mitochondria to organs during early development, can result in widely differing phenotypes even within a single pedigree. Sequencing the mitochondrial genome should pay particular attention to the mtDNA A3243G mutation.[30,31]

Wolfram syndrome, previously termed DIDMOAD because of the occurrence of diabetes insipidus, diabetes mellitus, optic atrophy, and deafness, is now associated with mutations within *WFS1*.[32,33] This gene encodes a ubiquitously expressed transmembrane protein, which is located primarily in the endoplasmic reticulum (ER). High levels in the brain, retina, and pancreatic islet account for the neurological and endocrine manifestations. *WFS1* mutations cause a variety of progressive features, with CNS involvement being most pronounced, in which progression of brainstem and optic atrophy is inexorable. Insulin remains the mainstay of diabetes treatment in these conditions. A second form associated with mutations in *WFS2*, a different protein that may also be

associated with ER function, has been reported in a very small number of families.[34]

Three rare monogenic causes of insulin resistance are also associated with diabetes. These are mutations in the insulin receptor gene, forms of lipoatrophic diabetes, and mutations in the peroxisome proliferator-activated receptor γ (PPARγ). Mutations in the insulin receptor gene (*INSR*) result in congenital insulin resistance syndromes. Typically heterozygous mutations result in type A insulin resistance, while the homozygous form can cause Donohue and Rabson–Mendenhall syndromes.[35,36] Individuals with Donohue and Rabson–Mendenhall syndromes rarely survive beyond infancy, type A insulin resistance is typically present in adolescence marked insulin resistance. Some evidence suggests that treatment with recombinant leptin or IGF-1 may be helpful. Mutations in the lamin A/C gene *LMNA* are associated with the autosomal dominant lipoatrophy (familial partial lipodystrophy of the Dunnigan type (FPLD2)), or recessive forms of congenital generalized lipoatrophy (Seip–Berardinelli syndrome), and other syndromes as well, important to recognize for possible associated conditions and the possibility of treatment with thiazoladinediones.[37–40] Familial type 2 diabetes of early onset with insulin resistance may be caused by mutations in the gene encoding PPARγ. These rare mutations can cause severe insulin resistance, partial lipodystrophy, type 2 diabetes, and hypertension and may be treatable with leptin.[41–43]

Neonatal Diabetes Mellitus

Persistent hyperglycemia occurring at a very early age has long been termed neonatal diabetes mellitus (NDM). Genetic causes predominate as the etiology of diabetes diagnosed before the age of 6 months, but a small fraction of those diagnosed at later ages may also be found to have similar single gene mutations.[44] The onset of diabetes often goes undetected, and a delayed diagnosis is then made in a catastrophic setting with marked hyperglycemia with or without ketoacidosis and a high morbidity. Transient forms where insulin-requiring diabetes spontaneously resolves between 6 months and 18 months of life make up approximately half of cases.[45] Autoimmune type 1 diabetes is unusual in infancy, so while NDM is a rare condition, the yield from targeted genetic testing is high.[46] Clinicians are faced with an expanding list of over 20 identified genetic causes of NDM, but many of these are syndromic, with clinical clues that can direct genetic testing.[6] NDM was historically treated with insulin, but studies over the last decade have in many cases clarified the underlying molecular mechanism and pointed to disease-specific therapy.

Certain features, such as birth weight below the 10th percentile (especially in the presence of maternal diabetes), also suggest a genetic cause. Features may also include, depending on the gene and the precise mutation, developmental delay, learning disorders, speech disorders, muscle weakness (such as with climbing stairs), and seizures.[47,48] Some children have been diagnosed with attention deficit hyperactivity disorder as well.[49] Occasionally multiple family members may be found to also have either early onset, relapsing, or nonobese young adult appearance of diabetes, but most cases are sporadic.

Transient Neonatal Diabetes

The term transient neonatal diabetes (TND) is misleading as many patients develop permanent hyperglycemia later in life, usually in the teen years. The most common cause of TND is due to abnormalities of chromosome 6q24. 6q24-TND is caused by genetic or epigenetic changes affecting the expression of genes at the 6q24 locus. Overexpression of *PLAGL1* and *HYMAI* can result from a number of mechanisms including uniparental disomy of chromosome 6 (UPD6), submicroscopic duplication of the paternal 6q24 allele, or isolated loss of maternal methylation at the differentially-methylated region at 6q24. UPD6 does not carry an increased risk of recurrence in future offspring, as it is a sporadic event in embryonic development. There is a 50% chance of offspring inheriting a submicroscopic duplication of the paternal 6q24 allele, while loss of maternal methylation is a poorly understood phenomenon with an unknown etiology and an unknown inheritance risk. In addition to recurrence of hyperglycemia in later years, those with 6q24 may sometimes develop significant hypoglycemia during the remission period.[50]

In the neonatal period patients are typically treated with insulin therapy. When 6q24-TND recurs later in life, the best treatment is uncertain, but many patients are managed with insulin.[51] There is a paucity of data on endogenous insulin production and the use of insulin secretagogues or other oral agents in the treatment of this condition.[52–54]

Rare autosomal recessive mutations in *ZFP57* can give rise to multiple loci with loss of methylation.[55,56] Dedicated sequencing techniques must be employed to identify abnormalities of chromosome 6q24 to identify 6q24-TND, as exome sequencing will not identify these changes.

Transient neonatal diabetes may also be due to mutations in the *ABCC8* and *KCNJ11*, but *INS* and *HNF1B* can also rarely give rise to transient neonatal diabetes.[57,58]

Permanent Neonatal Diabetes

Activating heterozygous mutations in *KCNJ11* and *ABCC8* genes, encoding the two subunits of the ATP-sensitive potassium (K_{ATP}) channel, are the most common causes of diabetes in the first months of life.[59–61] The K_{ATP}-sensitive K^+ channel is well described as a key molecular switch in the beta cell, closing in response to generation of ATP following glucose metabolism. Identifying subjects with heterozygous activating mutations of this gene

is of significant clinical value, as oral SU treatment can routinely replace insulin therapy with improvement of glycemic control.[44,59,62–65] Mutations can disrupt the K_{ATP} channel to varying degrees and result in a broad range of phenotypes. Clinical severity ranges from isolated transient neonatal diabetes to the most severe cases affected by neurodevelopmental disability, seizures, and insensitivity to SU treatment.[59,62,65–68] Most but not all of the mutations in *ABCC8* and *KCNJ11* can be treated by high dose sulfonylurea therapy (in off-label use in patients under 18 years of age in all countries). Incremental increases in doses of sulfonylureas, up to relatively high doses depending on the specific mutation (0.5–2.5 mg/kg/day), are administered in divided doses with gradual insulin withdrawal to stimulate beta cell function.

Some patients with K_{ATP}-related NDM also exhibit a spectrum of neurodevelopmental and behavioral problems. These vary from mild learning disorders to more severe mutations resulting in cognitive dysfunction or seizures.[69] This syndrome of developmental delay, epilepsy, and neonatal diabetes is referred to as DEND or when less severe, intermediate DEND (iDEND).

These impairments are likely to be due to mutated K_{ATP} channels themselves that are widely expressed in the brain.[70] There is a correlation between the degree of the inability of the channels to close, resistance to block with sulfonylureas *in vitro*, and the degree of neurological disorder.[71] Mutations in these genes that result in decreased channel function cause the opposite condition, hyperinsulinemia with hypoglycemia, which is usually homozygous or compound heterozygous and recessive.[7,72,73]

After K_{ATP} channel mutations due to alterations in KCNJ11, the next most common known genetic cause of permanent neonatal diabetes is mutations in the insulin gene (*INS*) itself. These mutations are also rare causes of young adult onset type 2 diabetes and ketosis-prone T1b diabetes, usually negative for the presence of diabetes-associated islet autoantibodies.[14] *INS* mutations are presumed to result in a mutated translational product giving rise to ER stress and beta cell death.[74]

Mutations within an expanding list of other genes may also lead to insulin insufficiency and NDM through failure of normal development of the pancreas, or development of early onset autoimmunity and mutated translational products causing ER stress (Table 2.1).

GENETIC TESTING

Exome Testing Options

In addition to the increasing number of monogenic diabetes genes, there are a variety of ways the affected genes can be altered. Dedicated methylation polymerase chain reaction (PCR) or deletion analysis may be needed in addition to other modalities. The most common forms of MODY and NDM are typically tested using Sanger sequencing, which is based on selective incorporation of chain-terminating dideoxynucleotides by DNA polymerase during DNA replication. Genetic testing options have expanded greatly with the advent of next-generation sequencing (NGS) technologies. NGS platforms facilitate high-throughput sequencing by producing thousands or millions of sequences in parallel.

Traditional Sanger sequencing is the least costly option for patients and clinicians on a per gene basis. However, Sanger sequential sequencing of multiple genes can be time-consuming and costly while limiting the diagnosis to a few selected genes. GCK-MODY has a distinctive clinical phenotype and may be the best candidate for targeted Sanger sequencing of GCK as the initial test. While we recommend targeted sequencing when the clinical phenotype is clear, clinicians should consider a broader approach if the phenotype is not. Genetic testing for individuals with diabetes is likely to be sought more frequently given the modeled cost effectiveness of testing selected populations, the falling costs, and faster reporting times associated with next-generation techniques. We would encourage clinical testing laboratories to ensure testing options are made clear, given the complex nature of genetic testing.

As highlighted previously, the selection of the appropriate gene(s) to test can be challenging when based on clinical grounds alone. A number of dedicated monogenic diabetes panels have been described.[75–77] Pathogenic variants within multiple genes can present with a similar phenotype, while other pathognomonic features may not yet have manifested at the time of initial presentation with hyperglycemia. Some of the available genetic tests for monogenic diabetes are outlined in Table 2.2. Ordering clinicians should be aware that there are a number of important types of mutations that may be missed or not assessed by exome sequencing alone (e.g., assessment of methylation of the 6q24 locus, intronic mutations, and large deletions).

Additional Testing Considerations

6q24-TND Testing

Genetic testing for overexpression of the imprinted genes PLAGL1 and HYMAI is based on ratiometric measurement of methylated and unmethylated DNA within the differentially 6q24 region using methylation specific PCR.[55] Parental samples are needed to identify the underlying genetic mechanism (hypomethylation of the PLAGL1/HYMAI promoter region, uniparental disomy of chromosome 6, and paternal duplication of the 6q24 allele).

Deletion Analysis

Sanger sequencing can readily identify small indels (insertions or deletions). While partial or whole gene deletions make up a minority of monogenic diabetes cases, these may be missed using Sanger sequencing alone.[78] Medium-sized deletion longer than the PCR amplicons also are not detected because they cannot be amplified and may not always be identified through multiplex ligation-dependent probe amplification, the most commonly used test to screen for deletions.[79] Dedicated studies to identify deletions are particularly important when assessing families with an HNF1B-MODY (MODY5) phenotype as deletions are more common in this gene.[80] As NGS technologies improve they may be used to detect large deletions or duplications if deep; even coverage of the target genes can be maintained.[77]

Intronic Mutations

Mutations within noncoding regions will often be missed if sequencing is directed at exons alone. About 98% of the human genome is noncoding DNA and targeting coding regions dramatically reduces the cost and analysis time while still identifying the majority of potential deleterious mutations. Disease-causing intronic mutations within genes associated with diabetes have been identified, and clinicians must balance the merit of testing noncoding regions when assessing patients with suspected monogenic diabetes.[81,82] There are significant analytical challenges associated with the large amount of data derived from whole genome sequencing (WGS). Selectively including introns and promoter regions of established monogenic diabetes genes in NGS gene panels may be a reasonable compromise.

Interpreting Results

We strongly encourage clinicians to seek expert help when ordering tests and interpreting results. A negative report can be falsely reassuring and previously unreported variants may be of uncertain clinical significance. When confronted with a negative testing report clinicians should reevaluate the likelihood of an underlying genetic etiology and examine if further testing is warranted. As highlighted in the earlier section, the limitations of the test performed should be considered if no mutation is identified. For those patients diagnosed with diabetes less than 6 months, a genetic etiology is highly likely and further testing is prudent. The rapid advancement in the fields of high-throughput capture and NGS has made whole exome sequencing and WGS more affordable, and when parental samples are available, they should be considered for these select patients without a known cause. In those with a MODY phenotype or children diagnosed after 6 months of age, the yield from testing genes not included in most of the target panels is limited.

Many of the mutations identified in monogenic diabetes genes are novel.[83,84] Thus, clinicians are regularly faced with previously unreported variants of uncertain clinical significance, and they should try to stratify the likely pathogenicity based on a few key questions, as follows:

- Does the phenotype match previous reports of mutations within that gene? It is important to determine a pretest probability, even if conceptual, for a mutation within specific genes associated with diabetes.
- Has this variant been described previously? A comprehensive search of the established databases, published literature, and monogenic diabetes registries is invaluable to determine if this variant is truly novel and to determine relative frequency in large populations.
- Is the change predicted to be damaging to the protein? An assessment of how well a nucleotide is conserved and *in silico* analysis of protein function can help interpretation. This can be accomplished with several commercial software packages. Similarly, splicing prediction software can help as some mutations create new intron splice donor or acceptor sites or eliminate critical existing sites.
- How was the variant inherited? To assess this, other family members will need to be tested. Family members may share the genotype but have differing phenotypes. Ensuring relatedness is critical to this approach. Occasionally cryptic relatedness of the parents is uncovered, which must be considered carefully when reporting such information.

Genetic Counseling

Genetic counselors are an invaluable resource for any family being considered for genetic testing. Unfortunately, the majority of families with monogenic forms of diabetes may not have access to a dedicated counselor and this onus then falls on the other clinical staff to supervise pre- and posttesting counseling. We encourage clinicians and families to engage with national and international teams with experience in these rare conditions. These teams will usually offer to help inform families/clinicians of the medical, psychological, and familial implications of each genetic disease in addition to guiding the appropriate approach to testing and interpretation of results. Often the ordering physician will be primarily supervising the patient's diabetes care, but it is important for families to realize that many of the monogenic forms of diabetes have extrapancreatic manifestations.

CONCLUSIONS

Most of the gene mutations discussed here cause dysfunction of the pancreatic beta cells or lead to their apoptotic death, while for some genetic alterations, the pathophysiology is not yet well understood, and in other rare cases, the main effect is severe insulin resistance. We encourage clinicians to challenge the diagnostic labels of patients and families with atypical diabetes. A careful evaluation age of diagnosis, family history, patterns of inheritance, antibody testing, and history of ketosis can help to determine if a genetic etiology exists. By starting with the clinical presentation this guide will assist the clinician in determining the priority with which diabetes-related genes should be evaluated by DNA sequencing. Since recognition of monogenic causes may have a major impact on treatment, their identification by genetic testing is critical for appropriate management. Appropriate testing is also necessary to guide genetic counseling even when treatment is not affected. It is transformational for the patient, the family, and the treating physician to identify these individuals and correctly identify these less common forms of diabetes.

Given the rarity of many of the genetic forms of diabetes outlined, it is important to follow patients longitudinally. A number of international registries for monogenic diabetes offer guidance to patients and clinicians while following affected families longitudinally. We encourage clinicians with questions to contact these centers (United States – www.monogenicdiabetes.org, United Kingdom – www.diabetesgenes.org).

References

1. Giardiello FM, Brensinger JD, Petersen GM, Luce MC, Hylind LM, Bacon JA, et al. The use and interpretation of commercial APC gene testing for familial adenomatous polyposis. *N Engl J Med* 1997;**336**(12): 823–7.
2. Hattersley AT. Unlocking the secrets of the pancreatic beta cell: man and mouse provide the key. *J Clin Invest* 2004;**114**(3):314–6.
3. Fajans SS, Bell GI. MODY: history, genetics, pathophysiology, and clinical decision making. *Diabetes Care* 2011;**34**(8):1878–84.
4. Centers for Disease Control and Prevention. *National Diabetes Statistics Report: Estimates of Diabetes and Its Burden in the United States.* Atlanta, Georgia: US Department of Health and Human Services; 2014.
5. Murphy R, Ellard S, Hattersley AT. Clinical implications of a molecular genetic classification of monogenic beta-cell diabetes. *Nat Clin Pract Endocrinol Metab* 2008;**4**(4):200–13.
6. Greeley SAW, Naylor RN, Philipson LH, Bell GI. Neonatal diabetes: an expanding list of genes allows for improved diagnosis and treatment. *Curr Diab Rep* 2011;**11**(6):519–32.
7. Aguilar-Bryan L, Bryan J. Neonatal diabetes mellitus. *Endocr Rev* 2008;**29**(3):265–91.
8. Ellard S, Bellanné-Chantelot C, Hattersley AT. European Molecular Genetics Quality Network (EMQN) MODY Group. Best practice guidelines for the molecular genetic diagnosis of maturity-onset diabetes of the young. *Diabetologia* 2008;**51**(4):546–53.
9. Shields BM, Hicks S, Shepherd MH, Colclough K, Hattersley AT, Ellard S. Maturity-onset diabetes of the young (MODY): how many cases are we missing? *Diabetologia* 2010;**53**(12):2504–8.
10. Vigersky RA, Fish L, Hogan P, Stewart A, Kutler S, Ladenson PW, et al. The Clinical Endocrinology Workforce: current status and future projections of supply and demand. *J Clin Endocrinol Metab* 2014; jc.2014–2257.
11. Tattersall RB, Fajans SS. A difference between the inheritance of classical juvenile-onset and maturity-onset type diabetes of young people. *Diabetes* 1975;**24**(1):44–53.

12. Thanabalasingham G, Pal A, Selwood MP, Dudley C, Fisher K, Bingley PJ, et al. Systematic assessment of etiology in adults with a clinical diagnosis of young-onset type 2 diabetes is a successful strategy for identifying maturity-onset diabetes of the young. *Diabetes Care* 2012;**35**(6):1206–12.

13. Carmody D, Bell CD, Hwang JL, Dickens JT, Sima DI, Felipe DL, et al. Sulfonylurea treatment before genetic testing in neonatal diabetes: pros and cons. *J Clin Endocrinol Metab* 2014; jc20142494.

14. Støy J, Edghill EL, Flanagan SE, Ye H, Paz VP, Pluzhnikov A, et al. Insulin gene mutations as a cause of permanent neonatal diabetes. *Proc Natl Acad Sci USA* 2007;**104**(38):15040–4.

15. Rubio-Cabezas O, Flanagan SE, Damhuis A, Hattersley AT, Ellard S. KATP channel mutations in infants with permanent diabetes diagnosed after 6 months of life. *Pediatr Diabetes* 2012;**13**(4):322–5.

16. Byrne MM, Sturis J, Clément K, Vionnet N, Pueyo ME, Stoffel M, et al. Insulin secretory abnormalities in subjects with hyperglycemia due to glucokinase mutations. *J Clin Invest* 1994;**93**(3):1120–30.

17. Chakera AJ, Spyer G, Vincent N, Ellard S, Hattersley AT, Dunne FP. The 0.1% of the population with glucokinase monogenic diabetes can be recognized by clinical characteristics in pregnancy: the Atlantic Diabetes in Pregnancy Cohort. *Diabetes Care* 2014;**37**:1230–6.

18. Steele AM, Shields BM, Wensley KJ, Colclough K, Ellard S, Hattersley AT. Prevalence of vascular complications among patients with glucokinase mutations and prolonged, mild hyperglycemia. *JAMA* 2014;**311**(3):279–86.

19. Stride A, Shields B, Gill-Carey O, Chakera AJ, Colclough K, Ellard S, et al. Cross-sectional and longitudinal studies suggest pharmacological treatment used in patients with glucokinase mutations does not alter glycaemia. *Diabetologia* 2013;**57**:54–6.

20. Chakera AJ, Carleton VL, Ellard S, Wong J, Yue DK, Pinner J, et al. Antenatal diagnosis of fetal genotype determines if maternal hyperglycemia due to a glucokinase mutation requires treatment. *Diabetes Care* 2012;**35**(9):1832–4.

21. Spyer G, Hattersley AT, Sykes JE, Sturley RH, MacLeod KM. Influence of maternal and fetal glucokinase mutations in gestational diabetes. *Am J Obstet Gynecol* 2001;**185**(1):240–1.

22. Pihoker C, Gilliam LK, Ellard S, Dabelea D, Davis C, Dolan LM, et al. Prevalence, characteristics and clinical diagnosis of maturity onset diabetes of the young due to mutations in HNF1A, HNF4A, and glucokinase: results from the SEARCH for Diabetes in Youth. *J Clin Endocrinol Metab* 2013;**98**(10):4055–62.

23. Bellanné-Chantelot C, Clauin S, Chauveau D, Collin P, Daumont M, Douillard C, et al. Large genomic rearrangements in the hepatocyte nuclear factor-1beta (TCF2) gene are the most frequent cause of maturity-onset diabetes of the young type 5. *Diabetes* 2005;**54**(11):3126–32.

24. Lecube A, Hernández C, Genescà J, Simó R. Glucose abnormalities in patients with hepatitis C virus infection: epidemiology and pathogenesis. *Diabetes Care* 2006;**29**:1140–9.

25. Pivonello R, De Leo M, Vitale P, Cozzolino A, Simeoli C, et al. Pathophysiology of diabetes mellitus in Cushing's syndrome. *Neuroendocrinology* 2010;**92**(Suppl. 1):77–81.

26. Pearson ER, Starkey BJ, Powell RJ, Gribble FM, Clark PM, Hattersley AT. Genetic cause of hyperglycaemia and response to treatment in diabetes. *Lancet* 2003;**362**:1275–81.

27. Pearson ER, Pruhova S, Tack CJ, Johansen A, Castleden HA, et al. Molecular genetics and phenotypic characteristics of MODY caused by hepatocyte nuclear factor 4alpha mutations in a large European collection. *Diabetologia* 2005;**48**:878–85.

28. Katra B, Klupa T, Skupien J, Szopa M, Nowak N, et al. Dipeptidyl peptidase-IV inhibitors are efficient adjunct therapy in HNF1A maturity-onset diabetes of the young patients – report of two cases. *Diabetes Technol Ther* 2010;**12**:313–6.

29. Barrett TG. Mitochondrial diabetes, DIDMOAD and other inherited diabetes syndromes. *Best Pract Res Clin Endocrinol Metab* 2001;**15**:325–43.

30. Ballinger SW, Shoffner JM, Hedaya EV, Trounce I, Polak MA, Koontz DA, et al. Maternally transmitted diabetes and deafness associated with a 10.4 kb mitochondrial DNA deletion. *Nat Genet* 1992;**1**(1):11–5.

31. van den Ouweland JM, Lemkes HH, Ruitenbeek W, Sandkuijl LA, de Vijlder MF, Struyvenberg PA, et al. Mutation in mitochondrial tRNA (Leu) (UUR) gene in a large pedigree with maternally transmitted type II diabetes mellitus and deafness. *Nat Genet* 1992;**1**(5):368–71.

32. Inoue H, Tanizawa Y, Wasson J, Behn P, Kalidas K, Bernal-Mizrachi E, et al. A gene encoding a transmembrane protein is mutated in patients with diabetes mellitus and optic atrophy (Wolfram syndrome). *Nat Genet* 1998;**20**(2):143–8.

33. Scolding NJ, Kellar-Wood HF, Shaw C, Shneerson JM, Antoun N. Wolfram syndrome: hereditary diabetes mellitus with brainstem and optic atrophy. *Ann Neurol* 1996;**39**(3):352–60.

34. Amr S, Heisey C, Zhang M, Xia XJ, Shows KH, et al. A homozygous mutation in a novel zinc-finger protein, ERIS, is responsible for Wolfram syndrome 2. *Am J Hum Genet* 2007;**81**:673–83.

35. Taylor SI, Cama A, Accili D, Barbetti F, Quon MJ, la Luz Sierra de M, et al. Mutations in the insulin receptor gene. *Endocr Rev* 1992;**13**(3):566–95.

36. Musso C, Cochran E, Moran SA, Skarulis MC, Oral EA, Taylor S, et al. Clinical course of genetic diseases of the insulin receptor (type A and Rabson-Mendenhall syndromes): a 30-year prospective. *Medicine (Baltimore)* 2004;**83**(4):209–22.

37. Caron M, Auclair M, Donadille B, Béréziat V, Guerci B, Laville M, et al. Human lipodystrophies linked to mutations in A-type lamins and to HIV protease inhibitor therapy are both associated with prelamin A accumulation, oxidative stress and premature cellular senescence. *Cell Death Differ* 2007;**14**(10):1759–67.

38. Decaudain A, Vantyghem M-C, Guerci B, Hécart A-C, Auclair M, Reznik Y, et al. New metabolic phenotypes in laminopathies: LMNA mutations in patients with severe metabolic syndrome. *J Clin Endocrinol Metab* 2007;**92**(12):4835–44.

39. Moreau F, Boullu-Sanchis S, Vigouroux C, Lucescu C, Lascols O, Sapin R, et al. Efficacy of pioglitazone in familial partial lipodystrophy of the Dunnigan type: a case report. *Diabetes Metab* 2007;**33**(5):385–9.

40. Vantyghem MC, Faivre-Defrance F, Marcelli-Tourvieille S, Fermon C, Evrard A, Bourdelle-Hego MF, et al. Familial partial lipodystrophy due to the LMNA R482W mutation with multinodular goitre, extrapyramidal syndrome and primary hyperaldosteronism. *Clin Endocrinol (Oxf)* 2007;**67**(2):247–9.

41. Barroso I, Gurnell M, Crowley VE, Agostini M, Schwabe JW, Soos MA, et al. Dominant negative mutations in human PPARgamma associated with severe insulin resistance, diabetes mellitus and hypertension. *Nature* 1999;**402**(6764):880–3.

42. Meirhaeghe A, Amouyel P. Impact of genetic variation of PPARgamma in humans. *Mol Genet Metab* 2004;**83**(1–2):93–102.

43. Guettier J-M, Park JY, Cochran EK, Poitou C, Basdevant A, Meier M, et al. Leptin therapy for partial lipodystrophy linked to a PPAR-gamma mutation. *Clin Endocrinol (Oxf)* 2008;**68**(4):547–54.

44. Greeley SAW, Tucker SE, Naylor RN, Bell GI, Philipson LH. Neonatal diabetes mellitus: a model for personalized medicine. *Trends Endocrinol Metab* 2010;**21**(8):464–72.

45. Flanagan SE, Edghill EL, Gloyn AL, Ellard S, Hattersley AT. Mutations in KCNJ11, which encodes Kir6.2, are a common cause of diabetes diagnosed in the first 6 months of life, with the phenotype determined by genotype. *Diabetologia* 2006;**49**(6):1190–7.

46. Edghill EL, Dix RJ, Flanagan SE, Bingley PJ, Hattersley AT, Ellard S, et al. HLA genotyping supports a nonautoimmune etiology in patients diagnosed with diabetes under the age of 6 months. *Diabetes* 2006;**55**(6):1895–8.

47. Gloyn AL, Diatloff-Zito C, Edghill EL, Bellanné-Chantelot C, Nivot S, et al. KCNJ11 activating mutations are associated with developmental delay, epilepsy and neonatal diabetes syndrome and other neurological features. *Eur J Hum Genet* 2006;**14**:824–30.

48. Støy J, Greeley SA, Paz VP, Ye H, Pastore AN, et al. Diagnosis and treatment of neonatal diabetes: a United States experience. *Pediatr Diabetes* 2008;**9**:450–9.

49. Vaxillaire M, Dechaume A, Busiah K, Cave H, Pereira S, et al. New ABCC8 mutations in relapsing neonatal diabetes and clinical features. *Diabetes* 2007;**56**:1737–41.

50. Flanagan SE, Mackay DJG, Greeley SAW, McDonald TJ, Mericq V, Hassing J, et al. Hypoglycaemia following diabetes remission in patients with 6q24 methylation defects: expanding the clinical phenotype. *Diabetologia* 2012;**56**:218–21.

51. Temple IK, Shield JPH. 6q24 transient neonatal diabetes. *Rev Endocr Metab Disord* 2010;**11**(3):199–204.

52. Schimmel U. Long-standing sulfonylurea therapy after pubertal relapse of neonatal diabetes in a case of uniparental paternal isodisomy of chromosome 6. *Diabetes Care* 2009;**32**(1):e9.

53. Søvik O, Aagenaes O, Eide SÅ, Mackay D, Temple IK, Molven A, et al. Familial occurrence of neonatal diabetes with duplications in chromosome 6q24: treatment with sulfonylurea and 40-yr follow-up. *Pediatr Diabetes* 2012;**13**(2):155–62.

54. Yorifuji T, Hashimoto Y, Kawakita R, Hosokawa Y, Fujimaru R, Hatake K, et al. Relapsing 6q24-related transient neonatal diabetes mellitus successfully treated with a dipeptidyl peptidase-4 inhibitor: a case report. *Pediatr Diabetes* 2014;**15**:606–10.

55. Mackay DJG, Temple IK, Shield JPH, Robinson DO. Bisulphite sequencing of the transient neonatal diabetes mellitus DMR facilitates a novel diagnostic test but reveals no methylation anomalies in patients of unknown aetiology. *Hum Genet* 2005;**116**(4):255–61.

56. Mackay DJG, Callaway JLA, Marks SM, White HE, Acerini CL, Boonen SE, et al. Hypomethylation of multiple imprinted loci in individuals with transient neonatal diabetes is associated with mutations in ZFP57. *Nat Genet* 2008;**40**(8):949–51.

57. Garin I, Edghill EL, Akerman I, Rubio-Cabezas O, Rica I, Locke JM, et al. Recessive mutations in the INS gene result in neonatal diabetes through reduced insulin biosynthesis. *Proc Natl Acad Sci USA* 2010;**107**:3105–10.

58. Yorifuji T, Kurokawa K, Mamada M, Imai T, Kawai M, et al. Neonatal diabetes mellitus and neonatal polycystic, dysplastic kidneys: phenotypically discordant recurrence of a mutation in the hepatocyte nuclear factor-1beta gene due to germline mosaicism. *J Clin Endocrinol Metab* 2004;**89**:2905–8.

59. Zung A, Glaser B, Nimri R, Zadik Z. Glibenclamide treatment in permanent neonatal diabetes mellitus due to an activating mutation in Kir6.2. *J Clin Endocrinol Metab* 2004;**89**(11):5504–7.

60. Pearson ER, Flechtner I, Njølstad PR, Malecki MT, Flanagan SE, Larkin B, et al. Switching from insulin to oral sulfonylureas in patients with diabetes due to Kir6.2 mutations. *N Engl J Med* 2006;**355**(5):467–77.

61. Rafiq M, Flanagan SE, Patch A-M, Shields BM, Ellard S, Hattersley AT, et al. Effective treatment with oral sulfonylureas in patients with diabetes due to sulfonylurea receptor 1 (SUR1) mutations. *Diabetes Care* 2008;**31**(2):204–9.

62. Proks P, Antcliff JF, Lippiat J, Gloyn AL, Hattersley AT, Ashcroft FM. Molecular basis of Kir6.2 mutations associated with neonatal diabetes or neonatal diabetes plus neurological features. *Proc Natl Acad Sci USA* 2004;**101**(50):17539–44.

63. Hattersley AT. Molecular genetics goes to the diabetes clinic. *Clin Med* 2005;**5**(5):476–81.

64. Bonnefond A, Durand E, Sand O, De Graeve F, Gallina S, Busiah K, et al. Molecular diagnosis of neonatal diabetes mellitus using next-generation sequencing of the whole exome. *PLoS ONE* 2010;**5**(10):e13630.

65. Edghill EL, Flanagan SE, Ellard S. Permanent neonatal diabetes due to activating mutations in ABCC8 and KCNJ11. *Rev Endocr Metab Disord* 2010;**11**(3):193–8.

66. Ashcroft FM. ATP-sensitive potassium channelopathies: focus on insulin secretion. *J Clin Invest* 2005;**115**(8):2047–58.

67. Flanagan SE, Clauin S, Bellanné-Chantelot C, de Lonlay P, Harries LW, Gloyn AL, et al. Update of mutations in the genes encoding the pancreatic beta-cell K (ATP) channel subunits Kir6.2 (KCNJ11) and sulfonylurea receptor 1 (ABCC8) in diabetes mellitus and hyperinsulinism. *Hum Mutat* 2009;**30**(2):170–80.

68. McTaggart JS, Clark RH, Ashcroft FM. The role of the KATP channel in glucose homeostasis in health and disease: more than meets the islet. *J Physiol (Lond)* 2010;**588**(Pt. 17):3201–9.

69. Busiah K, Drunat S, Vaivre-Douret L, Bonnefond A, Simon A, Flechtner I, et al. Neuropsychological dysfunction and neurodevelopmental defects associated with genetic changes in infants with neonatal diabetes mellitus: a prospective cohort study. *Lancet Diabetes Endocrinol* 2013;**1**(3):199–207.

70. Ashcroft FM. Adenosine 5′-triphosphate-sensitive potassium channels. *Annu Rev Neurosci* 1988;**11**:97–118.

71. Ashcroft FM. New uses for old drugs: neonatal diabetes and sulphonylureas. *Cell Metab* 2010;**11**:179–81.

72. Babenko AP, Polak M, Cavé H, Busiah K, Czernichow P, Scharfmann R, et al. Activating mutations in the ABCC8 gene in neonatal diabetes mellitus. *N Engl J Med* 2006;**355**(5):456–66.

73. Bryan J, Muñoz A, Zhang X, Düfer M, Drews G, Krippeit-Drews P, et al. ABCC8 and ABCC9: ABC transporters that regulate K+ channels. *Pflugers Arch* 2007;**453**(5):703–18.

74. Støy J, Steiner DF, Park S-Y, Ye H, Philipson LH, Bell GI. Clinical and molecular genetics of neonatal diabetes due to mutations in the insulin gene. *Rev Endocr Metab Disord* 2010;**11**(3):205–15.

75. Alkorta-Aranburu G, Carmody D, Cheng YW, Nelakuditi V, Ma L, Dickens JT, et al. Phenotypic heterogeneity in monogenic diabetes: the clinical and diagnostic utility of a gene panel-based next-generation sequencing approach. *Mol Genet Metab* 2014;**113**:315–20.

76. Bonnefond A, Philippe J, Durand E, Muller J, Saeed S, Arslan M, et al. Highly sensitive diagnosis of 43 monogenic forms of diabetes or obesity through one-step PCR-based enrichment in combination with next-generation sequencing. *Diabetes Care* 2014;**37**(2):460–7.

77. Ellard S, Lango Allen H, De Franco E, Flanagan SE, Hysenaj G, Colclough K, et al. Improved genetic testing for monogenic diabetes using targeted next-generation sequencing. *Diabetologia* 2013;**56**(9):1958–63.

78. Ellard S, Thomas K, Edghill EL, Owens M, Ambye L, Cropper J, et al. Partial and whole gene deletion mutations of the GCK and HNF1A genes in maturity-onset diabetes of the young. *Diabetologia* 2007;**50**(11):2313–7.

79. Herman S, Varga D, Deissler HL, Kreienberg R, Deissler H. Medium-sized deletion in the BRCA1 gene: limitations of Sanger sequencing and MLPA analyses. *Genet Mol Biol* 2012;**35**(1):53–6.

80. Edghill EL, Oram RA, Owens M, Stals KL, Harries LW, Hattersley AT, et al. Hepatocyte nuclear factor-1beta gene deletions – a common cause of renal disease. *Nephrol Dial Transplant* 2008;**23**(2):627–35.

81. Garin I, Pérez de Nanclares G, Gastaldo E, Harries LW, Rubio-Cabezas O, Castaño L. Permanent neonatal diabetes caused by creation of an ectopic splice site within the INS gene. *PLoS ONE* 2012;**7**(1):e29205.

82. Flanagan SE, Xie W, Caswell R, Damhuis A, Vianey-Saban C, Akcay T, et al. Next-generation sequencing reveals deep intronic cryptic ABCC8 and HADH splicing founder mutations causing hyperinsulinism by pseudoexon activation. *Am J Hum Genet* 2013;**92**(1):131–6.

83. Osbak KK, Colclough K, Saint-Martin C, Beer NL, Bellanné-Chantelot C, Ellard S, et al. Update on mutations in glucokinase (GCK), which cause maturity-onset diabetes of the young, permanent neonatal diabetes, and hyperinsulinemic hypoglycemia. *Hum Mutat* 2009;**30**(11):1512–26.

84. Colclough K, Bellanné-Chantelot C, Saint-Martin C, Flanagan SE, Ellard S. Mutations in the genes encoding the transcription factors hepatocyte nuclear factor 1 alpha and 4 alpha in maturity-onset diabetes of the young and hyperinsulinemic hypoglycemia. *Hum Mutat* 2013;**34**(5):669–85.

3

Hypoglycemia

Dorit Koren*, Andrew Palladino**

*Section of Adult and Pediatric Endocrinology, Diabetes and Metabolism,
University of Chicago, Chicago, IL, USA
**Perelman School of Medicine at the University of Pennsylvania, Philadelphia, PA, USA

INTRODUCTION

Overview of Glucose Homeostasis

Glucose is the primary metabolic fuel – that is, the main substrate for energy metabolism. Thus, maintaining its levels in an optimal range is crucial for health and survival, and is regulated by a complex interplay of metabolic processes controlling glucose utilization and glucose production. The key metabolic pathways involved in the glucose utilization in the fed state are glycolysis (the breakdown of glucose to generate ATP and energy) and the Krebs cycle, also known as the citric acid cycle. The key anabolic pathways ensuring that excess glucose is stored and can later be utilized for energy are glycogen synthesis, protein synthesis, and lipogenesis; these pathways are stimulated by the actions of insulin.[1] In addition, insulin causes the translocation of the GLUT4 glucose transporters from the cytosol to the cell membrane of myocytes and adipocytes to increase glucose uptake.[2]

Endogenous glucose supply in the fasting state depends on the recognition of dropping glucose levels and consequent suppression of insulin and triggering of the hormonal counterregulatory response, stimulating the following key endogenous glucose production pathways:

- Glycogenolysis: Breakdown of hepatic glycogen. This suffices to maintain euglycemia from several hours in a child to up to 12 h in an adult).
- Gluconeogenesis: The generation of glucose from noncarbohydrate precursors – primarily lactate, pyruvate, amino acids (principally alanine and glutamine), and glycerol (from fat). Once the glycogen supply is exhausted, gluconeogenesis becomes the predominant glucose production pathway until its supplies are exhausted.
- Lipolysis: The generation of free fatty acids and glycerol from triglycerides (triacylglycerol).

- Fatty acid oxidation and ketogenesis: The generation of acetoacetate and beta-hydroxybutyrate through β-oxidation from free fatty acids.

These processes, discussed further, normally help to assure that sufficient glucose is available to all body tissues to fulfill their metabolic needs; a defect in one or more of these can lead to hypoglycemia. The endogenous glucose production pathways are suppressed by insulin and stimulated by the counterregulatory hormones, namely, glucagon, epinephrine, cortisol, and growth hormone (GH). Equilibrium is maintained by the central nervous system (CNS), which integrates systemic signaling and helps upregulate or suppress insulin and the counterregulatory responses, as well as coordinating the behavioral, neuroendocrine, and autonomic responses to hypoglycemia.[3,4] A schema of the counterregulatory hormone response to hypoglycemia is illustrated in Fig. 3.1.

Background and Incidence/Prevalence

The definition of hypoglycemia is controversial, especially in neonates and infants,[5-7] where operational thresholds for abnormally low glucoses range from 18 mg/dL to 70 mg/dL (1–4 mmol/L). An ideal operational definition for neonatal hypoglycemia would be defined as the blood glucose concentration below which value significant morbidity (especially neurologic sequelae) would be seen and therefore at which intervention should be undertaken. However, the plasma glucose level and duration of hypoglycemia associated with adverse neurodevelopmental sequelae has not been conclusively established.[5] It should be also borne in mind that hypoglycemia on the day of birth or during the first few days of life may not necessarily have a pathological etiology. The transition between

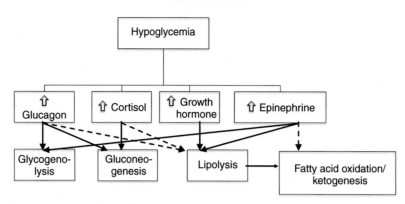

FIGURE 3.1   Counterregulatory response to hypoglycemia.

intrauterine and extrauterine life involves an inherent breaking off of the previously continuous supply of maternal glucose, and the consequent need for the neonate to adapt to fasting. This transition is marked by a number of other changes, including a 3–10-fold[8] surge in catecholamine secretion[9] at the time of delivery; a 3–5-fold increase in glucagon concentration, with a concomitant elevation in GH and cortisol;[8] and, under ordinary circumstances, a drop in insulin secretion.[8] All of these endocrine changes facilitate activation of pathways of fasting adaptation – namely, glycogenolysis, gluconeogenesis, lipolysis, fatty acid oxidation, and ketogenesis. It is noteworthy that these fasting mechanisms are not mature at the time of birth; therefore, a transient drop in blood glucose levels one to two h after birth is common,[10] even as low as 30 mg/dL,[7] and does not necessarily lead to adverse sequelae.[11] It is the infants with persistent hypoglycemia who are of concern and who must be evaluated for potential genetic etiologies.

Because different definitions of hypoglycemia are in use and there is no consensus definition, true incidence and prevalence of neonatal hypoglycemia is difficult to determine. Prevalence also depends upon the population in question – in at-risk groups, such as premature infants, small for gestational age infants, infants of diabetic mothers, or exclusively breastfed infants, prevalence of hypoglycemia in the first three days of life may be as high as 51%[12] without necessarily indicating a genetic etiology. Prevalence of asymptomatic hypoglycemia (defined as blood glucose <45 mg/dL) in normal neonates in the first days of life may be as high as 5–15%.[5] In a more recent case–control study of hypoglycemia incidence in full-term neonates weighing more than 2500 g that were products of nondiabetic, singleton pregnancies, using a cutoff for hypoglycemia of below 50 mg/dL, hypoglycemia prevalence in the first 24 h was 2.4%. In this study, hypoglycemia was found to be more common in infants whose mothers had puerperal fever, public versus private insurance, and/or who were earlier in gestation. It should be noted that serial glucose levels were not measured in this study, and that the true prevalence

of hypoglycemia could have been underestimated. Although the prevalence estimates vary, by all measures, neonatal hypoglycemia is relatively common and, if persistent, can lead to devastating sequelae, including seizures, visual disturbances,[13] motor and speech delay, cerebral palsy,[14] and white matter injury.[15] Thus, it is as incumbent upon care providers of infants to recognize this clinical entity and perform the appropriate diagnostic evaluation as it is upon care providers of older children and adults. Hypoglycemia has many different potential etiologies, ranging from inborn errors of metabolism to deficiencies in counterregulatory hormones, insulin excess, and many others. In addition to inborn errors, there are also a large variety of environmental exposures and other potential noninherent causes of hypoglycemia to consider in the differential diagnosis (e.g., exogenous insulin or insulin-sensitizing agents, salicylate poisoning, and others); while this chapter focuses only on inborn genetic and epigenetic entities causing hypoglycemia, the existence of a wider differential diagnosis should not be forgotten. Finally, it should be borne in mind that although many hypoglycemic disorders of genetic etiology manifest in the neonatal period, not all do so; some only manifest with more prolonged fasting in later infancy and the toddler years, and some mild disorders may not manifest until adulthood.

Clinical Presentation

Clinical presentation of hypoglycemia varies by age. In general, symptoms of hypoglycemia may be divided into two categories, adrenergic and neuroglycopenic. More about the different symptoms can be found in the following list, along with a separate discussion of presentation in infants:[1]

1. *Adrenergic symptoms*: Sweating, pupillary dilatation, tachycardia, shakiness or tremors, anxiety, nervousness, hunger, nausea/emesis.
2. *Neuroglycopenic symptoms*: Dizziness, confusion, lethargy, headache, visual disturbances, difficulty

concentrating, speech difficulty, somnolence/prolonged sleep, hypothermia, seizures or twitches, personality changes, temper outbursts, bizarre or manic behavior, depression or psychosis, and at the extreme, loss of consciousness and coma, with risk of permanent neurological damage or death.

3. *Presentation in neonates*: This requires a higher index of suspicion, because symptoms in infants may not be specific to hypoglycemia. In neonates and infants, symptoms and signs may include poor feeding or feed refusal, cyanosis or apneas, somnolence or lethargy, seizures or myoclonic jerks, emesis, sweating, low temperatures, and hypotonia.

Beyond these general symptoms, the clinical presentation of hypoglycemia varies by the underlying disorder, and will thus be discussed separately under each disorder in the subsequent sections.

Diagnostic Evaluation

The initial part of hypoglycemia evaluation lies in obtaining a history of the circumstances – fasting versus postprandial hypoglycemia, duration of fast required that elicits the hypoglycemia, associated symptoms, clinical circumstances (whether associated with intercurrent illness or other catabolic stress or not), and other associated clinical features. Nongenetic causes of hypoglycemia (e.g., salicylate or alcohol intoxication, exogenous insulation, or sulfonylurea exposure) should also be considered. Physical examination should be performed to examine for potential clinical symptoms that would suggest an etiology, such as hepatomegaly, which is common in glycogen storage diseases (GSDs), or characteristic dysmorphisms seen in certain syndromes associated with hypoglycemia such as Beckwith–Wiedemann syndrome. An acylcarnitine profile should be obtained to screen for disorders of fatty acid oxidation; individuals with such disorders should not undergo a diagnostic fast, or else a metabolic crisis may be triggered. Total and free carnitine levels should also be sent, as individuals with carnitine deficiency will have hypoglycemia, and these levels are not affected by fasting.

The most critical component of the hypoglycemia evaluation is the obtaining of a critical sample of blood and urine at the time of hypoglycemia (serum glucose <50); the hypoglycemia may be spontaneous, or may need to be elicited through a diagnostic fast of age-appropriate length. The diagnostic sample should include the following labs: serum glucose, insulin, C-peptide, CO_2, beta-hydroxybutyrate, free fatty acids, lactate, ammonia, GH, cortisol, serum amino acids, and urine organic acids. Finally, after a critical sample has been obtained, a glucagon stimulation test should be performed as further evaluation for insulin-mediated hypoglycemia: 1 mg glucagon is administered intravenously, and glucose levels are checked every 10 min for 40 min. A positive response to the glucagon stimulation test is a rise in glucose of greater than 30 mg/dL in 20 min or 40 mg/dL in 40 min.[1,16]

GENETIC PATHOPHYSIOLOGY

Hyperinsulinism

Hyperinsulinism (HI), also known as hyperinsulinemic hypoglycemia, is a disorder characterized by inappropriate insulin secretion by pancreatic β-cells, resulting in mild to severe hypoglycemia. This is then divided into two categories. The first is nongenetic perinatal stress-induced HI, a poorly understood disorder in which infants exposed to prenatal or perinatal stress such as preeclampsia, hypertension, perinatal asphyxia, prematurity, or intrauterine growth retardation manifest transient hyperinsulinemic hypoglycaemia,[17] which self-resolves (median time to resolution is 6 months[18]). The second is congenital HI, genetically and clinically heterogeneous, which results from various mutations along the insulin secretory pathway – primarily involving genes that encode the ATP-sensitive potassium (K_{ATP}) channel involved in insulin secretion as well as enzymes involved insulin secretion or, less commonly, gain of function mutations, which give rise to aberrant functioning proteins, which can impact insulin secretion. Fig. 3.2 illustrates normal glucose stimulated insulin secretion and the various mutations that can lead to HI, as well as the sites action for potentially therapeutic agents diazoxide and octreotide (a somatostatin) and, conversely, of sulfonylurea medications.

The prevalence of congenital HI is estimated to be between 1/30,000[19] and 1/50,000,[20] although the prevalence is higher in some populations, from ~1/11,000 in Ashkenazi Jews[21] to as high as 1/2500 in Saudi Arabia. The most common mutations are in the *ABCC8* and *KCNJ11* genes, which respectively encode the SUR1 and Kir6.2 subunits of the β-cell plasma membrane K_{ATP} channel. In a recent study of 417 children with HI, 91% of diazoxide-unresponsive patients had identifiable mutations (89% in either *ABCC8* or *KCNJ11*, and 2% in *GCK*, which encodes glucokinase, the enzyme that controls the rate-limiting step in glycolysis within the β-cell). Of patients with diazoxide-responsive HI, 47% had identifiable mutations (42% in *GLUD1*, which encodes glutamate dehydrogenase, 41% were dominant K_{ATP} channel mutations, and 16% were rare gene mutations).[22] Genetic defects leading to HI are outlined in Table 3.1 and discussed further in the subsequent section.

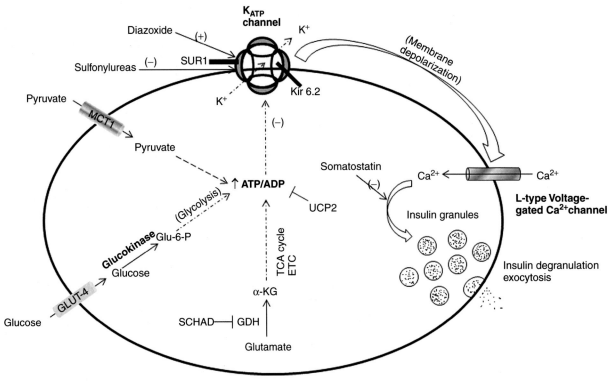

FIGURE 3.2 Beta-cell insulin secretion.

Hyperinsulinism Resulting from Mutations in ABCC8 or KCNJ11 (Hyperinsulinemic Hypoglycemia, Familial Types 1 and 2) – K_{ATP}-HI

ABCC8 and *KCNJ11* encode SUR1 and Kir6.2, respectively, which together form the subunits of the heterooctameric K_{ATP} channel in the β-cell plasma membrane. This channel is comprised of four inner Kir6.2 subunits, which form the channel's ion pore, and four regulatory SUR1 subunits, which surround the Kir6.2 core.[26] These subunits are assembled in the endoplasmic reticulum.[27] Normal insulin secretion results from a fuel stimulated increase in the ATP/ADP ratio within the β-cell leading to closure of the K_{ATP} channel with resultant depolarization of the β-cell membrane and opening of a voltage-gated calcium channel with subsequent Ca^{2+} influx. This influx leads to insulin degranulation and exocytosis. Inactivating mutations of either the K_{ATP} or SUR1 moieties of the K_{ATP} channel result in reduced or absent K_{ATP} channel activity, leading to permanent depolarization, spontaneous and independent opening of the voltage-gated Ca^{2+} channels,[28] and consequent constitutive insulin secretion irrespective of serum glucose.

Over 300 mutations in *ABCC8* and over 30 mutations in *KCNJ11* have been described in association with HI. Mutations may lead to an inability of the altered channel to exit the endoplasmic reticulum and be trafficked to the cell membrane,[29–31] a disruption of the nucleotide-binding domain of SUR1[32] (i.e., lack of response to increased Mg-ADP concentrations[33]), an alteration of K_{ATP} channel density, or a complete abolishment of channel activity.[33]

Recessively inherited K_{ATP}-HI is both the most common and most severe form of HI. Clinically, children present with severe hyperinsulinemic hypoglycemia, often, though not always, are in the neonatal period. The initial medical management for a child diagnosed with HI is with diazoxide, a medication that acts on the K_{ATP} channel to prevent closure of the channel. Diazoxide unresponsive HI suggests a K_{ATP} channel defect and a potential need for surgery. Genetic testing of the child and parents is recommended to further delineate the type of HI. K_{ATP}-HI is divided into two histopathological morphologies: diffuse disease and focal disease. Diffuse K_{ATP}-HI is, in most cases, due to autosomal recessive mutations in *ABCC8*[32] or, less commonly, *KCNJ11*.[34,35]

Children with a homozygous or compound heterozygous recessive mutation will have diffuse disease, whereas children with a paternally inherited heterozygous recessive mutation most likely have focal disease. Focal HI occurs as a result of inheriting a paternally derived mutation in *ABCC8* or *KCNJ11* and experiencing a somatic loss of the maternal 11p15 allele,[36] resulting in paternal uniparental isodisomy. The maternal loss of heterozygosity entails loss of tumor suppressor genes, including CDKN1c [p57(kip2)] and H19;[37] there is also upregulation of paternal IGF2.[36] The combination of

TABLE 3.1 Genetic Forms of Hyperinsulinism

HI type	Hyperinsulinemic hypoglycemia, familial #	Gene/s	Affected protein/ enzyme	Chromosomal location	Inheritance	Clinical features	Treatment
A. More common forms of hyperinsulinemic hypoglycemia							
Hyperinsulinism relating to mutations in the KATP channel (K_{ATP}-HI)	HHF-1 (ABCC8) HHF2 (KCNJ11)	ABCC8 KCNJ11	SUR1 (*ABCC8*) or Kir6.2 (*KCNJ11*)	11p15.1 11p15.1	Diffuse: autosomal recessive Focal: sporadic. "2-hits": inheritance of paternal mutation + loss of maternal heterozygosity	• Most common form of genetic hyperinsulinism • No response to diazoxide therapy • Typically severe hypoglycemia	1. Pancreatectomy: a. Subtotal – for diffuse disease b. Partial – for focal disease 2. Other: Continuous gastrostomy-tube feedings and octreotide
Dominant K_{ATP}-HI		ABCC8 KCNJ11	SUR1 or Kir6.2	11p15.1 11p15.1	Autosomal dominant	Milder hypoglycemia than recessive K_{ATP}-HI Responsive to diazoxide	Diazoxide-responsive
Hyperinsulinism-hyperammonemia (HI-HA, or GDH-HI)	HHF-6	*GLUD1*	GDH	10q23.3	• 80% *de novo* • 20% autosomal dominant	• Second most common form of genetic hyperinsulinism • Causes both fasting and postprandial (protein-sensitive) hypoglycemia • Hyperammonemia is asymptomatic	Diazoxide-responsive
Glucokinase-hyperinsulinism (GCK-HI)	HHF-3	*GCK*	Glucokinase (hexokinase 4)	7p15.3–7p15.1	Autosomal dominant; may also be *de novo*	• Phenotype varies – ranges from mild, easy-to-treat hypoglycemia to severe hypoglycemia	Diazoxide-responsive in most cases; may require subtotal pancreatectomy in severe cases
SCHAD-HI	HHF-4	*HADH*	Hydroxyacyl dehydrogenase	4q25	Autosomal recessive	Rare. Hypoglycemia ranges from mild to severe; abnormal acylcarnitine profile	Diazoxide-responsive
MCT1 (exercise-induced HI, or EIHI)	HHF-7	*SLC16A1*	Proton-linked monocarboxylate transporter, member 1 (MCT-1)	1p13.2	Autosomal dominant	Exercise-induced hyperinsulinemic hypoglycemia	Carbohydrate intake during exercise; limit anaerobic exercise
HNF4A-HI		*HNF4A*	Hepatocyte nuclear factor 4α	20q13.12	Autosomal dominant	• Macrosomia, neonatal hypoglycemia, varying degrees of severity and persistence • At risk for developing monogenic diabetes (MODY1) in adulthood	Diazoxide-responsive

(Continued)

TABLE 3.1 Genetic Forms of Hyperinsulinism (*cont.*)

HI type	Hyperinsulinemic hypoglycemia, familial #	Gene/s	Affected protein/ enzyme	Chromosomal location	Inheritance	Clinical features	Treatment
B. Exceedingly rare forms of hyperinsulinemic hypoglycemia, or hypoglycemia with insulin-mimicking mechanisms							
INSR-HI	HHF-5	INSR	Insulin receptor	19p13.2	Autosomal dominant	Postprandial hypoglycemia, fasting hyperinsulinemia	Long-active somatostatin analog
HNF1A-HI		HNF1A	Hepatocyte nuclear factor 1α	12q24.2	Autosomal dominant	Rare	Diazoxide-responsive, partially
UCP2-HI		UCP2	Uncoupling protein 2	11q13.4	Autosomal dominant	Rare. Tends to self-resolve within a few years	Diazoxide-responsive
		HK1	Hexokinase 1	10q22.1	Autosomal dominant (in some cases, somatic mosaicism)	Rare	Diazoxide-responsive. Somatic mutations with focal lesions can be cured by partial pancreatectomy.
		KCNQ1	Kv7.1	11p15.5	Autosomal dominant	One case series.[23] Prolonged QT interval; postprandial hyperinsulinemic hypoglycemia; hypokalemia after oral glucose load	Unknown if diazoxide-responsive
		AKT2	AKT2 – part of the insulin receptor signal transduction cascade	19q13.2	Autosomal dominant	Two case reports[24,25] • Mild-severe fasting hypoglycemia *Autonomous of insulin*	Diazoxide unresponsive; require regular carbohydrate feeds to maintain euglycemia

HI, hyperinsulinism; ATP, adenylyl triphosphate; SUR1, sulfonylurea receptor 1; Kir6.2, inward rectifier potassium channel; GCK, glucokinase; GDH, glutamate dehydrogenase; SCHAD, short-chain-3-hydroxyacyl-CoA dehydrogenase; SLC16A1, solute carrier family 16 (monocarboxylate transporter), member 1; MCT-1, monocarboxylase transporter-1; HNF, hepatocyte nuclear factor; 1A, 1-alpha; 4A, 4-alpha; UCP2, uncoupling protein 2.

loss of tumor suppressor genes and upregulation of IGF2 leads to adenomatous expansion of the affected region. Thus, somatic regions within the pancreas in which a paternal mutation has been inherited and maternal loss of heterozygosity has occurred become foci of HI (focal adenomatosis). While focal and diffuse HI resemble each other clinically much of the time, a recent article reviewed 223 children with diffuse or focal HI and reported several significant clinical differences between the two (all values in parentheses are means and all differences reached statistical significance).[38] Compared to children with focal disease, children with diffuse disease were born at an earlier age (38 weeks vs. 39 weeks) and were born with larger birth weights (3,963 vs. 3,717 g). Additionally, children with diffuse HI had higher insulin levels at the time of diagnosis (31.8 µU/mL vs. 12 µU/mL) and had higher maximal GIR requirements (19.2 mg/kg/min vs. 16.1 mg/kg/min). Children with focal HI presented at a later age compared to those with diffuse disease (0.3 months vs. 0 months) and were more likely to present with a seizure (50% vs. 25%). Distinguishing focal from diffuse disease is of utmost importance as focal disease can be cured by surgical resection of the lesion, whereas diffuse disease requires a near-total pancreatectomy. In addition to genetic testing, an ^{18}F-DOPA PET scan can help in localization of a focal lesion. In the review of 223 patients, those with diffuse disease had a median percent pancreatectomy of 98%, compared to 27% in those with focal disease. Of the children that had surgery for diffuse disease, 41% had persistent hypoglycemia requiring treatment with enteral dextrose and/or octreotide. Comparatively, 94% of the children that had surgery for focal disease required no further management.

Autosomal dominant mutations in *KCNJ11* and *ABCC8* have been reported.[26] They are characterized by normal subunit trafficking but impaired K_{ATP} channel activity, which overall can result in a milder hypoglycemia compared to the recessive mutations earlier, and many are responsive to therapy with diazoxide, unlike the recessively inherited mutations.

Hyperinsulinism Resulting from Mutations in GLUD-1 (Hyperinsulinemic Hypoglycemia, Familial 6) – GDH-HI

GLUD1 encodes glutamate dehydrogenase (GDH), a mitochondrial matrix enzyme[39] that catalyzes the oxidative deamination of glutamate to alpha-ketoglutarate and ammonia with the assistance of either NAD^+ or $NADP^+$ as a cofactor.[40] GDH is an important regulator of amino acid and ammonia metabolism; it is expressed at high levels in the brain, hepatocytes, pancreatic β-cells, and the kidneys, but not in muscle.[41] GDH enzyme activity is allosterically inhibited by ATP and GTP and allosterically activated by ADP, GDP, and leucine.[42]

GDH-HI results from gain-of-function mutations in *GLUD1*.[17] The insulin secretion may be triggered by glucose or by amino acids, and is potentiated by leucine.[43] Eighty percent of GDH-HI cases are due to *de novo* mutations in *GLUD1*; the remaining 20% are inherited in an autosomal dominant pattern. Over 20 disease-causing mutations have been identified. Most heterozygous missense amino acid substitutions in *GLUD1* affect the GTP-binding site, decreasing the sensitivity of GDH to inhibition by GTP to varying degrees, increased sensitivity to leucine stimulation, and leading to excess insulin secretion; the degree of sensitivity to GTP inhibition is inversely related to severity of hypoglycemia phenotype.[39] Other mutations affect the antenna region of GDH, which communicates with adjacent enzyme subunits. Increased GDH activity results in increased glutamate oxidation to alpha-ketoglutarate, which enters the Krebs cycle and leads to increased ATP levels. The increased ATP/ADP ratio leads to K_{ATP} channel closing and insulin secretion as discussed earlier. The mutations result in increased enzyme activity in pancreatic β-cells, hepatocytes, in the kidneys, and in the brain. These children also manifest asymptomatic persistent elevations in serum ammonia levels due to increased hepatic GDH activity[17] (and with some contribution from the kidney[44]); thus, this syndrome is referred to as hyperinsulinism-hyperammonemia syndrome, or HI-HA.

Clinically, children present recurring episodes of both fasting and postprandial hypoglycemia, especially after consuming protein-rich meals.[45] This hypoglycemia is typically milder than that seen in K_{ATP}-HI and is thus often not diagnosed at birth. It is responsive to diazoxide. Although ammonia levels are usually (though not always) quite elevated, typically two to five times above the usual normal limits, affected patients typically do not manifest hyperammonemia symptoms.[17] That said, it should be noted that generalized seizures, especially absence seizures, may be seen in patients with GDH-HI even without concurrent hypoglycemia;[46] this may be due to hypoglycemic brain damage, chronically elevated ammonia levels, or CNS hyperexcitability due to GDH mutations.[47] The exact etiology of the seizures, however, is unclear. Additionally, children with HI-HA have a higher incidence of developmental delays and behavior problems.

Hyperinsulinism Resulting from Mutations in GCK (Hyperinsulinemic Hypoglycemia, Familial 3) – GK-HI

GCK encodes glucokinase (GK), an isoform of hexokinase expressed in pancreatic β-cells, hepatocytes, glucose-sensing hypothalamic neurons (predominantly in the ventromedial nucleus and arcuate nucleus), and some enteroendocrine cells.[48] GK acts as a glucose sensor[48] and catalyzes the initial and rate-limiting step in glycolysis, the conversion of D-glucose to D-glucose-6-phosphate at physiologic glucose concentrations using MgATP as a second substrate.[49] GK exists in three different conformations, which control catalytic function, generating a sigmoidal response to plasma glucose concentrations; the transitions between these conformations are controlled by glucose concentrations and by allosteric interactions with modulators of GK activity[50] such as biotin, cyclic AMP, and insulin,[48] as well as glucokinase regulatory protein (GKRP). GK acts as the β-cell's glucose sensor because it has a relatively low affinity for its substrate, glucose; thus, it determines the glucose threshold for insulin release.[51] The typical set-point for glucose-induced GK expression in β-cells is ~5 mmol/L, or 72 mg/dL.[48]

Autosomal dominant activating mutations in *GCK* are relatively rare causes of congenital HI;[51] as of this writing, only 12 activating mutations have been identified.[52] That said, one recent multicenter study found that *GCK* mutations were responsible for up to 1.2% of cases of congenital HI and nearly 7% of non-K_{ATP}-channel cases of HI.[53] These activating missense mutations typically occur in or near the allosteric binding sites for either GK activators or the competitive inhibitor, GKRP; the net result of activating mutations is to increase the affinity of GK for its substrate glucose, effectively lowering the threshold for insulin release,[50] sometimes to as low as 1 mmol/L or 9 mg/dL.[54] Of interest, mutations that conversely decrease GK's binding affinity for glucose and increase the threshold for insulin secretion are responsible for a form of monogenic diabetes, MODY2; it is noteworthy that inactivating mutations are far more numerous than activating *GCK* mutations.

Clinically, the phenotype of GK-HI ranges from severe neonatal HI[51,55] to mild fasting hypoglycemia, which may not be recognized until later in childhood[51] or even into adulthood.[49] Many patients will respond to diazoxide,[53] but the degree of responsiveness can be quite variable. Patients may demonstrate an initially good response, but over time become less responsive. Those that do not respond to diazoxide or lose responsiveness over time may require a partial or near-total pancreatectomy in order to treat the HI.[55]

Hyperinsulinism Resulting from Mutations in HADH (Hyperinsulinemic Hypoglycemia, Familial 4) – SCHAD-HI

HADH encodes medium/short-chain-3-hydroxyacyl-CoA dehydrogenase (SCHAD), a mitochondrial matrix enzyme that catalyzes the third and penultimate step in β-oxidation of fatty acids into acetyl-CoA,[56] the conversion of 3-hydroxyacyl-CoA to 3-ketoacyl-CoA using NAD[+] as a cofactor.[57] There are different isoforms of the various enzymes that catalyze the β-oxidation of medium- and short-chain fatty acids. In addition to its

actions in the β-oxidation pathway, SCHAD also acts as an allosteric inhibitor of GDH, resulting in decreased affinity of GDH for its substrate, exerting an overall inhibitory effect on GDH in pancreatic islets resulting in excessive insulin secretion.[56]

HADH loss-of-function mutations are a very rare cause of HI,[57,58] although they are more common in children born to consanguineous parents[59] and in children of Irish or Turkish populations, wherein there is a founder mutation.[60] These mutations have an autosomal recessive inheritance mode. Several mutations have now been reported. The mechanism by which SCHAD mutations cause hyperinsulinemic hypoglycemia (as opposed to noninsulin mediated hypoketonemic hypoglycemia) entails loss of its inhibitory effect on GDH, leading to increased GDH affinity for leucine, alanine, and glutamate. Mechanisms by which the mutations cause SCHAD loss of function include causing deletions or alternate splice variants within the gene,[59] mutations affecting protein folding,[57] or deep intronic mutations introducing a frameshift in the gene.[60]

Clinically, children present with hyperinsulinemic hypoglycemia ranging from mild to severe in degree, accompanied by elevated levels of plasma 3-hydroxy-butyrylcarnitine in both the fasting and fed states[57] and elevated urine 3-hydroxyglutaric acid.[61] Unlike most other fatty acid oxidation disorders, children with SCHAD-HI typically do not have myopathies of skeletal and/or cardiac muscle and also typically do not have hepatotoxicity.[61] SCHAD-HI is typically responsive to diazoxide.

Hyperinsulinism Resulting from Mutations in SLC16A1 (Hyperinsulinemic Hypoglycemia, Familial 7) – Exercise-Induced-HI

Proton-linked monocarboxylate transporter, member 1 (MCT-1) is an enzyme that catalyzes the movement of several monocarboxylates (e.g., lactate, pyruvate, valine, leucine-derived branched-chain oxo-acids, and ketone bodies) across the plasma membrane. It is widely expressed across multiple tissues,[62] but its expression in pancreatic α- and β-cell membranes is typically low to absent.[63] In addition, expression of lactate dehydrogenase-A, which is responsible for the conversion of pyruvate to lactate as the final step of anaerobic metabolism, is relatively low in β-cells; this helps channel pyruvate preferentially toward mitochondrial oxidative metabolism rather than toward lactate, and keeps glucose rather than pyruvate or lactate as the preferential trigger for insulin secretion.[64] As a result of the low β-cell MCT-1 expression and LDH-A activity, MCT-1 activity does not ordinarily contribute to β-cell insulin secretion – in other words, changes in extracellular concentrations of lactate or pyruvate (e.g., during exercise or during a catabolic state) do not trigger insulin secretion.[65]

In exercise-induced hyperinsulinism (EIHI, or MCT-HI), a gain-of-function mutation in the regulatory regions of *SLC16A1* in the binding sites of several transcription factors allows for unusually high levels of expression of MCT-1 in β-cell membranes, enabling β-cell pyruvate uptake and pyruvate-stimulated insulin release in the presence of elevated lactate or pyruvate levels, such as in the setting of intense exercise or anaerobic exercise, leading to triggering of the usual K_{ATP}-mediated insulin secretory pathway and thus to ensuing hypoglycemia.[65] Three different autosomal dominant mutations have been identified – two in Finnish families and one in a German family.[65]

Clinically, patients present with hyperinsulinemic hypoglycemia of varying degrees in the setting of exercise; response to diazoxide is partial. Treatment involves limiting anaerobic exercise and frequent intake of carbohydrate-containing snacks during aerobic exercise.

Hyperinsulinism Resulting from Mutations in INSR (Hyperinsulinemic Hypoglycemia, Familial 5) – Insulin Receptor Activating Mutations

INSR encodes the insulin tyrosine kinase receptor, which is a heterotetrameric protein composed of two extracellular α subunits to which insulin binds and two transmembrane kinase β subunits. The binding of insulin to the α subunits stimulates the kinase, which triggers autophosphorylation of the β subunits on tyrosine residues, which in turn leads to the phosphorylation of a number of intracellular substrates, including the insulin receptor substrates. These phosphorylated substrates in turn activate two signaling cascades: the phosphotydilinositol-3-kinase-AKT/protein kinase B pathway, which mediates the majority of insulin's metabolic actions, and the Ras-mitogen-activated protein (MAP) kinase pathway.[2]

In 2004, an activating, autosomal dominant mutation in *INSR* was reported in three generations of a Danish family.[23] This was a missense mutation in the tyrosine kinase domain of the insulin receptor gene. This is the only reported mutation in *INSR* linked to hypoglycemia – all other known pathogenic mutations cause variable degrees of insulin resistance and a diabetic phenotype.

Affected children and adults present with fasting hyperinsulinemia, elevated insulin: C-peptide ratios, and moderate to severe postprandial hypoglycemia coexisting in some cases with impaired glucose tolerance (first a significant rise in blood glucose levels postglucose challenge, then a lower-than-normal nadir). All affected patients manifested insulin resistance as measured by hyperinsulinemic-euglycemic clamp. Thus, the mutation caused a divergent phenotype of insulin resistance and insulin-mediated hypoglycemia. Patients responded to treatment with a long-acting formulation of octreotide,[23] a somatostatin analog that acts downstream of the voltage-gated calcium channel in the β-cell to inhibit insulin secretion.[66]

Hyperinsulinism Resulting from Mutations in HNF1A (Hepatocyte Nuclear Factor-1α)

HNF1A encodes the protein HNF-1α, a transcription factor required for the expression of several liver-specific genes. More specifically, the homodimer protein binds to a promoter sequence that is required for hepatocyte-specific transcription of hepatic-specific genes, including fibrinogen, alpha-1-antitrypsin and others.[67] HNF-1α's dimerization affects gene transcription.

HNF1A mutations are typically associated with a form of autosomal dominant monogenic diabetes (MODY3); however, in 2011, one case was described where an adult with MODY3 was found to have had fetal macrosomia and childhood hypoglycaemia, which self-resolved prior to developing diabetes mellitus at the age of 19.[68] This mutation was heterozygous (implying dominant inheritance) and affected the *HNF1A* gene's DNA-binding domain. Two other mutations have since been reported, one a heterozygous nonsense mutation affecting *HNF1A*'s dimerization domain and another a heretozygous missense mutation.[69] The mechanism by which loss of function mutations in *HNF1A* lead to hypoglycemia alone or the dual phenotype of hypoglycemia in infancy/childhood and monogenic diabetes in adulthood has not been well elucidated.

Clinically, the reported hypoglycemia phenotypes range from isolated incidents in childhood[68] to full-blown neonatal HI.[69] Variable degrees of responsiveness to diazoxide have been reported.

Hyperinsulinism Resulting from Mutations in HNF4A (Hepatocyte Nuclear Factor 4-Alpha)

HNF4A encodes the protein HNF4-α, a nuclear transcription factor that binds DNA (to HNF4-α response elements and with the involvement of coactivator peptides) as a homodimer. HNF4-α is the most abundant hepatic DNA-binding protein, influencing the expression of approximately 40% of actively transcribed hepatic genes, including *HNF1A* and other genes involved in gluconeogenesis and hepatic liver metabolism. HNF4-α is also expressed in the pancreas, and controls the expression of approximately 11% of pancreatic islet genes.[70]

Most pathogenic mutations in *HNF4A* cause monogenic diabetes (MODY1). Heterozygous loss-of-function mutations in *HNF4A* have also been linked with hyperinsulinemic hypoglycemia. Most of these mutations affect the DNA-binding domain; some affect the ligand-binding domain.[70] In one recent series, *HNF4A* mutations were found to be the third most common mutations to cause diazoxide-responsive HI – mutations were seen in 11/220 cases, with each case being caused by a different mutation.[71] Inheritance is generally autosomal dominant, though *de novo* mutations were seen in

4/11 of the cases above. MODY1 and the hyperinsulinemic hypoglycemia may not be entirely unrelated – in one report, ~14% of patients with MODY1 had a dual phenotype, having experienced transient hypoglycemia in infancy, in addition to fetal macrosomia (which was also reported in a larger subset of patients with *HNF4A* mutations but without hypoglycemia[72]).

Clinically, affected children often manifest fetal macrosomia.[72] The infantile hyperinsulinemic hypoglycemia has a range of severity, from mild and transient hypoglycemia requiring brief intravenous glucose support to more severe, persistent hypoglycemia requiring pharmacotherapy for several years. Cases of HNF4A-HI are typically responsive to diazoxide.[69,71] One reported case also developed hypophosphatemic rickets, renal Fanconi syndrome, and hepatic glycogenosis.[69] *HNF4A* mutations may have a dual phenotypic manifestation – patients experiencing infantile hypoglycemia are at risk of developing MODY1, a form of monogenic diabetes, as described earlier.

Hyperinsulinism Resulting from Mutations in UCP2 (Uncoupling Protein 2)

UCP2 encodes a mitochondrial uncoupling protein that facilitates anion transfer from the inner to the outer mitochondrial membrane in exchange for protons. It is part of a larger family of uncoupling proteins that separate ("uncouple") oxidative phosphorylation from ATP synthesis; the resulting energy is dissipated as heat. UCP2 is widely expressed in many tissues, including pancreatic islets.[73] UCP2 acts as a negative regulator of insulin secretion in pancreatic β-cells by decreasing ATP content and thus inhibiting glucose-stimulated insulin secretion. Two heterozygous, autosomal dominant loss-of-function *UCP2* mutations have been reported to cause congenital HI.[73] Both cases were diazoxide responsive.

Other Rare Forms of Hypoglycemia Due to Hyperinsulinemia or Related Causes

1. Hexokinase 1 (*HK1*): Hexokinase 1 is another member of the mitochondrial hexokinase family, like glucokinase; it catalyzes the first step in glycolysis, namely, the conversion of glucose into glucose-6-phosphate. Several cases of fasting and postprandial HI due to autosomal dominant *HK1* mutations have been reported in a single family.[74] All cases presented with moderate to severe hypoglycemia in infancy; 40% of those presented with a seizure. All responded well to diazoxide treatment. In another report, a form of HI involved islet cell mosaicism, with some localized hyperfunctioning islets with low insulin storage along with high proinsulin production, and a larger number of atrophied islets; the hyperfunctioning was found to be due to either elevated expression of low-Km *HK1* (which is normally expressed at minimal levels in β-cells) substituting for GCK in β-cell glycolysis, or

to an activating mutation in *GCK*.[54] These cases of HI responded to diazoxide and/or were cured by focal pancreatectomy.[75]

2. KCNQ1: KCNQ1 encodes a voltage-gated potassium channel with six transmembrane regions. It is located on a region of chromosome 11 along with a cluster of other genes involved in Beckwith–Wiedemann syndrome. KCNQ1 is primarily found in the plasma cell membrane of both cardiac myocytes, where activity of this channel is required for the repolarization phase of the cardiac action potential; loss-of-function mutations cause long QT syndrome, a disease of prolonged cardiac repolarization predisposing to reentry arrhythmias. However, KCNQ1 is also expressed in other tissues, including pancreatic β-cells, where it regulates potassium current and thus, potentially, regulates insulin secretion. Single-nucleotide polymorphisms associated with increased KCNQ1 activity have been linked to susceptibility to type 2 diabetes mellitus due to reduced insulin secretion in several Asian populations.[76] The converse has very recently been reported – fourteen individuals from six families with long QT syndrome who were discovered to have postprandial and symptomatic hyperinsulinemic hypoglycemia along with hypokalemia after oral glucose challenge.[77] It is unknown how people with hypoglycemia would respond to medical therapy; however, the mechanism of the hypoglycemia suggests that they may be diazoxide unresponsive.

3. *AKT2*: *AKT2* encodes one of three serine/threonine protein kinases that regulate a number of processes – cell proliferation, metabolism, survival, and angiogenesis. *AKT2* is also involved in insulin signal processing downstream of the insulin receptor. Three known cases of hypoglycemia due to gain-of-function *AKT2* mutations (heretozygous, meaning autosomal dominant, in two-thirds of cases, and mosaic in the remaining case) resulting in hypoglycemia have been reported.[24] All three presented with severe hypoglycemia beginning at around 6 months of age. As the hypoglycemia in their cases was not caused by insulin but rather by a mediator of insulin's effects, it was not amenable to treatment with diazoxide or octreotide; all required gastrostomy tube placement for frequent feeds to prevent hypoglycemia.

Summary

As the most common cause of hypoglycemia in infants and children, it is imperative that the diagnosis of HI be made swiftly in order to prevent further hypoglycemia and possible brain damage. Once the diagnosis has been made, the child should be started on dextrose-containing intravenous fluids while initiating therapy with diazoxide. Genetic testing will further aid in the management of the patient with HI, as the results can suggest possible focal disease, which can be cured with surgery, versus results that confirm diffuse disease. It is critical to understand the different genetic forms of HI in order to manage the patient with HI correctly. Lastly, for the surgical HI patient, it is crucial that they be cared for in a center that specializes in the management of HI and that has a team of endocrinologists, radiologists, pathologists, and surgeons that are all trained in the management of this disease.

Insulinomas

Insulinomas are insulin-producing islet-cell tumors. They are rare, with an estimated incidence of 1–4/1,000,000. While the majority of cases are sporadic, 10% are associated with a multiple endocrine neoplasia (MEN) type 1 syndrome. MEN1 results from inactivating mutations, deletions, or duplications in the *MEN1* gene on chromosome 11q13.1, which encodes the tumor suppressor protein menin.[78] Affected individuals present with episodic hypoketotic hyperinsulinemic hypoglycemia.[16] MEN1 should be in the differential diagnosis for any insulinoma presenting in a child or young adult. Detectable insulin levels or a rise in blood sugar in response to glucagon at the time of hypoglycemia, along with other findings discussed in the previous section, establish excess insulin as the etiology for the hypoglycemia; a diagnostic fast of up to 72 h may be necessary to elicit the hypoglycemia. The vast majority of insulinomas are intrapancreatic; diagnostic imaging studies such as helical or multislice computerized tomography and magnetic resonance imaging (MRI) with gadolinium are good at identifying larger lesions, though endoscopic ultrasonography may be required to identify smaller lesions. Sequence analysis of 11q13.1 can identify mutations, deletions, or duplications in the *MEN1* gene. Treatment of the insulinoma involves surgical excision in the 90% of cases in which the disease is localized to the pancreas; the hypoglycemia may also respond to diazoxide. Screening for other MEN1-associated neoplasias (pituitary tumors, parathyroid tumors, and other gastrointestinal neuroendocrine tumors) is a must in affected individuals.

Disorders of Glycogenolysis and Gluconeogenesis – "Glycogen Storage Diseases"

Disorders of glycogenolysis and gluconeogenesis, also called glycogen storage diseases (GSDs), are a group of disorders characterized by an inability to break down hepatic glycogen to form D-glucose or disorders in the hepatic process of gluconeogenesis – that is, converting noncarbohydrate precursors to D-glucose. As the liver does not have capacity to release free D-glucose, it no longer functions as a major player in glucose homeostasis. Depending upon the degree of impairment, the affected child is predisposed to hypoglycemia after varying durations of fasting. These disorders may also have

a myriad of other features. In order to understand the GSDs, glycogenolysis and gluconeogenesis are reviewed briefly in the subsequent section.

Glycogen is a multibranched glucose polymer comprised of linear chains of glucose residues linked by $\alpha(1\rightarrow4)$ glycosidic bonds. Branch points occur approximately every 10 residues, wherein a glucose residue on the stem chain is linked by $\alpha(1\rightarrow6)$ glycosidic bonds to glucose on a new branch, which begins a new linear chain.[1] Glycogen synthesis involves several steps, catalyzed by different enzymes. The liver is permeable to glucose; glucose is then converted into glucose-6-phosphate (G6P), preventing its exit, and then glucose-1-phosphate (G1P), which is the starting point for glycogen synthesis. In the initial step of glycogen synthesis, catalyzed by uridine-diphospate (UDP)-glucose pyrophosphorylase, G1P reacts with uridine-triphosphate (UTP) to form UDP-glucose. UDP-glucose monomers are then linked to each other via $\alpha(1\rightarrow4)$ glycosidic bonds in a reaction catalyzed by glycogen synthase. The $\alpha(1\rightarrow6)$ glycosidic bonds are created by the actions of glycogen branching enzyme. Glycogenolysis also entails the actions of several enzymes. Glycogen phosphorylase cleaves $\alpha(1\rightarrow4)$ glycosidic bonds, yielding G1P and a linear glycogen chain shortened by one glucose residue; this does not work closer than four glucose residues to any branching point. Glycogen debranching enzyme transfers three of the four remaining glucose residues to the end of another glycogen branch, leaving only the branch-point glucose. Finally, the $\alpha(1\rightarrow6)$ glycosidic bonds are cleaved by $\alpha(1\rightarrow6)$ glucosidase, yielding a glucose molecule. G1P is converted by phosphoglucomutase to G6P, which in turn is hydrolyzed by G6P to yield free glucose and a phosphate residue. Finally (technically a step in gluconeogenesis rather than glycogenolysis), G6P is hydrolyzed to glucose via glucose-6-phosphatase (G6Pase). This step is crucial because phosphorylated glucose is charged and cannot easily cross the cell membrane to become available for use as energy in other tissues, while free glucose can easily diffuse across the cell membrane. The GSDs, or glycogenoses, are several inherited diseases due to abnormalities of the above-mentioned enzymes regulating the synthesis or degradation of glycogen.[79] A deficiency in each of these enzymes is responsible for a different GSD, with variable impact upon fasting tolerance and various sequelae. The overall GSD incidence is estimated to be 1/20,000–1/43,000 live births.[80] Glycogen is primarily stored in liver and muscle. Hypoglycemia is the primary manifestation of the hepatic glycogenoses; muscle glycogenoses are predominantly characterized by muscle weakness and muscle cramps with variable cardiac manifestations. Only the glycogenoses that cause hypoglycemia are discussed in this chapter; a detailed review of the muscle glycogenoses is beyond the scope of this review.

Gluconeogenesis is the metabolic pathway consisting of 11 enzyme-catalyzed reactions by which D-glucose is generated from noncarbohydrate substrates. Gluconeogenesis may begin in the mitochondria or cytoplasm, depending on the substrate. The primary amino acid substrates are alanine and glutamine, though others may contribute via the citric acid cycle. The other main gluconeogenic substrates are glycerol and lactate. In humans, the primary site of gluconeogenesis is the liver. The initial substrate for most, though not all, gluconeogenic reactions is pyruvate, which is generated from the aforementioned precursors (except glycerol) either by direct conversion of certain amino acids (alanine, cysteine, others), via the Cori cycle (lactate), α-ketoglutarate, and the citric acid cycle (glutamate and other amino acids), or via acetyl-CoA and the citric acid cycle (branched-chain amino acids). Many of the reactions are catalyzed by the same enzymes that catalyze the reversible steps in glycolysis; there are, however, four enzymatic steps that are unique to gluconeogenesis. The first is the mitochondrial carboxylation of pyruvate to oxaloacetate, catalyzed by pyruvate carboxylase (PC) using 1 ATP. Oxaloacetate is then reduced to malate using NADH, transported out of the mitochondrion into the cytosol, and oxidized back to oxaloacetate via NAD^+. Oxaloacetate is then decarboxylated and phosphorylated to phosphoenolpyruvate (PEP) via the enzyme PEP-carboxykinase (PEP-CK), which is the second unique enzymatic step in gluconeogenesis is the decarboxylation and phosphate; one GTP molecule is hydrolyzed to GDP. Several steps then follow, which are the reverse of the glycolytic steps catalyzed by the same enzymes as in glycolysis; in one of these steps, glycerol (a byproduct of lipolysis) enters the gluconeogenic pathway. The third unique step is the conversion of fructose 1,6-bisphosphate to fructose-6-phosphate (F6P) by fructose 1,6-bisphosphatase (FBPase), releasing one phosphate molecule. This is the rate-limiting step in gluconeogenesis. One more reversible step follows, the conversion of F6P to G6P, then G6P is hydrolyzed to glucose via G6Pase, the final step of gluconeogenesis; as described earlier, this is where glucose released from glycogen enters the gluconeogenic pathway. Deficiencies in these enzymes are responsible for a different gluconeogenic disorder (generally considered in tandem with the GSDs), with variable impact upon fasting tolerance and various sequelae.

Disorders of Glycogenolysis

Table 3.2 provides a brief overview of GSDs; they are described in further detail in the subsequent section.

Glycogen Storage Disease Type "0": Glycogen Synthase Deficiency

Glycogen synthase, as discussed earlier, catalyzes the rate-limiting step in glycogen synthesis in the liver and in skeletal muscle, namely, the transfer of glucose monomers from UDP-glucose to the terminal branch of the growing glycogen chain via the formation of α(1→4) glycosidic bonds.[79] Several glycogen synthase isoforms exist – one specific to skeletal muscle (encoded by *GYS1*), and one specific to the liver (encoded by *GYS2*). Mutations in *GYS2* cause glycogen synthase deficiency leading to GSD 0. Of note, the appellation of "glycogen storage disease" for GSD 0 is actually a misnomer; this is a disorder of glycogen formation rather than of aberrant glycogen breakdown.

In GSD 0, the lack of glycogen synthase activity leads to a decrease in hepatic glycogen content. Since excess dietary carbohydrates cannot be converted into glycogen, shunting into the glycolytic pathway and consequent lactate formation occur, resulting in postprandial lactic acid elevation, hyperglycemia, and sometimes postprandial glycosuria. However, once all dietary carbohydrate is consumed, gluconeogenesis and ketosis are triggered. Initially, gluconeogenesis blunts the impact of the hypoglycemia, but eventually, the ketosis and free fatty acid elevation due to lipolysis, fatty acid oxidation, and ketogenesis inhibit alanine release from skeletal muscle. Thus, the end result is postprandial ketotic hypoglycemia accompanied by elevated triglycerides and low alanine.[82] To date, 15 different missense mutations in *GYS2* leading to glycogen synthase deficiency have been reported to date; all are autosomal recessive in inheritance.[82]

This disease may first manifest as infants transition to longer periods of overnight fasting in between feeds or may remain asymptomatic until a period of illness (e.g., gastroenteritis) leads to a prolonged fast, which is why some cases remain undiagnosed well outside of infancy and the toddler years. Fasting tolerance typically improves as children get older – adolescents may be able to fast as long as 18 h – but hyperketonemia is still frequently seen in the fasting state if the individual is untreated.[82] Hypoglycemia can also occur even into adulthood in circumstances of prolonged fasting, pregnancy, gastrointestinal, or other illness leading to poor enteral carbohydrate intake, and upon strenuous exercise. Other manifestations in untreated individuals include short stature and osteopenia (frequently seen), cognitive delay in 22% of cases, and rarely seizures. These sequelae are preventable if hypoglycemia and hyperketonemia are prevented. Hepatomegaly is absent, as are other complications seen in other GSDs.

The diagnosis of GSD 0 is suggested by the presence of postprandial hyperglycemia, elevated lactates, and postprandial hyperketotic hypoglycemia on oral glucose tolerance testing. Definitive diagnosis of this condition can be done noninvasively via mutation analysis of the *GYS2* gene. Liver biopsies are unnecessary.

Treatment of GSD 0 entails prevention of hypoglycemia by frequent daytime snacking (every two to four h)

TABLE 3.2 Disorders of Glycogenolysis

GSD type	Eponym/ other name	Gene/s	Affected protein/ enzyme	Chromosomal location and inheritance[81]*	Affected tissue	Diagnosis	Clinical features	Treatment
GSD Ia**	Von Gierke's disease	*G6PC*	Glucose-6-phosphatase, catalytic subunit	17q21.31 AR	• Liver • Kidney • Intestine		If undiagnosed/untreated • Recurrent asymptomatic hypoglycemia • Hepatomegaly • Chronic lactic acidosis • Hyperuricemia • Hyperlipidemia • Hypertriglyceridemia • Xanthomas • "Doll-like" face • Thin extremities • Short stature Long-term complications • Gout • Hepatic adenomas • Hepatic adenocarinomas	Frequent carb-containing meals during the day • Overnight gastrostomy feeds when young • Cornstarch (1.75 g/kg) every 3–5 h during the day, and every 4–6 h overnight • Avoidance of galactose and fructose containing foods • Allopurinol and occasionally triglyceride-lowering agents (e.g., gemfibrozil) as adjunct therapy
GSD Ib**	Von Gierke's disease	*SLC37A4/ G6PT1*	Glucose-6-phosphate transporter	11q23.3 AR	• Liver only		• All of the above • Impaired neutrophil, monocyte function • Chronic neutropenia • Recurrent bacterial infections • Oral and intestinal mucosal ulcers	• As above • Consider filgastrim (recombinant granulocyte colony-stimulating factor), but monitor dosage due to risk of leukemia
GSD IIIa	Cori disease; glycogen debranching enzyme deficiency	*AGL*	Glycogen debranching enzyme	1p21.2 AR	IIIa (80–85%) • Liver • Heart • Skeletal muscle		This enzyme has two independent catalytic activities, which occur at different sites on the protein: a 4-α-glucotransferase activity and an amylo-1,6-glucosidase activity If undiagnosed/untreated • Ketotic fasting hypoglycemia • Hepatomegaly and elevated transaminases in childhood • Hyperlipidemia	Similar to GSD I • Avoid prolonged fasting • Frequent carb-containing meals during the day • Overnight gastrostomy feeds when young • Overnight cornstarch (1.75 g/kg every 6 h) • IIIa only: periodic EKGs/ Echos
GSD IIIb					IIIb (mainly sephardic Jews) • Liver only		• *Normal* lactate, urate levels • Poor linear growth (if poor control) • Type IIIa only: progressive myopathy (3rd and 4th decades of life) – large proximal muscles (hips, shoulders), sometimes distal muscles (e.g., hand) • Cardiomyopathy (due to limited dextrin accumulation in heart) • Hepatic adenomas (~25%)	

(Continued)

II. PANCREAS

TABLE 3.2 Disorders of Glycogenolysis (*cont.*)

GSD type	Eponym/other name	Gene/s	Affected protein/enzyme	Chromosomal location and inheritance[81]*	Affected tissue	Diagnosis	Clinical features	Treatment
GSD VI	Hers disease	*PYGL*	Glycogen phosphorylase	14q21.1 AR	Liver	Sequencing of *PYGL*	Very rare except in Mennonites (affects 0.1% of Mennonite population) If undiagnosed/untreated • Hepatomegaly • Ketotic hypoglycemia after overnight or other prolonged fast • Poor linear growth • ±Mild elevation of triglycerides, total cholesterol • ±Mildly high transaminases • Normal lactate, urate	• Avoid prolonged fasting • Bedtime snack • Rarely: 2 g/kg cooked cornstarch at bedtime to prevent ketotic hypoglycemia overnight
GSD IX 25% of all GSD cases	Fanconi–Bickel disease	*PHK*	Phosphorylase kinase – various isoforms	Varies	Varies	Sequence analysis of genes encoding affected subunits	Causes failure of hepatic glycogen phosphorylase activation Milder than GSD I or III If undiagnosed/untreated in early childhood: • Protruberant abdomen • Hepatomegaly • Poor linear growth • Hypoglycemia (with ketosis) rare except with prolonged fasting or exercise • ±Mild elevation of triglycerides, total cholesterol	
GSD IXa1, or GSD9A1 (formerly GSD 8)	Several subtypes	*PHKA2*	Phosphorylase kinase α2 subunit – hepatic isoform	Xp22.13 X-linked	• Liver • RBCs • WBCs			
GSD IXb, or GSD9B	Most common form of GSD IX – 75% of all cases	*PHKB*	Phosphorylase kinase β subunit – hepatic and muscle isoform	16q12.1 AR	• Liver • RBCs • WBCs • Muscle		• ±Mildly high transaminases • Normal lactate, urate • Hepatomegaly usually decreases with puberty • Often asymptomatic in adulthood	

	Gene	Protein	Locus	Tissue	Testing	Clinical features	Treatment
GSD IXc, or GSD9C	PHKG2	Phosphorylase kinase γ subunit – hepatic and testis isoform	16p11.2 AR	• Liver only		Rare • Proximal renal tubular acidosis (RTA) • CNS symptoms (seizures, cognitive delay, speech delay, peripheral sensory neuropathy) • Cirrhosis	
GSD 0 (or GSD 0a)	GYS2	Glycogen synthase 2 (hepatic)	12p12.1 AR	• Liver	Sequence analysis of GYS2	• Ketotic hypoglycemia upon fasting • Short stature • Osteopenia • Occasionally seizures • Cognitive delay in 22% • Postprandial hyperglycemia, glycosuria, elevated lactate • ±Mildly high transaminases	• Frequent daytime feeds • Uncooked cornstarch at bedtime for younger children • Avoid simple carbohydrate • Low glycemic index diet, protein supplementation

AR, autosomal recessive; AD, autosomal dominant; GSD, glycogen storage disease; RBCs, red blood cells (erythrocytes); WBCs, white blood cells (leukocytes); RTA, renal tubular acidosis; CNS, central nervous system.
* See also http://www.genecards.org.
** A disorder of both glycogenolysis and gluconeogenesis.

and, in young children, administering uncooked cornstarch (1–1.5 g/kg) at bedtime. Uncooked cornstarch should also be administered every 6 h during times of intercurrent illness and prior to strenuous or prolonged exercise. These interventions also have the effect of minimizing the ketosis and accumulation of free fatty acids to prevent hyperlipidemia and the suppression of skeletal muscle alanine release. Finally, affected individuals should consume a low-glycemic index diet with complex carbohydrates (as simple carbohydrates are more prone to trigger lactate elevation) and with protein supplementation to provide a greater quantity of gluconeogenic precursors in between meals.[82]

Glycogen Storage Disease Type 1: Von Gierke's Disease (Glucose-6-Phosphatase Deficiency)

G6Pase, as discussed earlier, catalyzes the final step of both glycogenolysis and gluconeogenesis in hepatocytes, intestinal mucosa, and renal epithelial cells – namely, the dephosphorylation of D-G6P to D-glucose and orthophosphate.[81] G6Pase is a 35-kD endoplasmic reticulum (ER) integral membrane protein with nine transmembrane domains; the active domain faces into the ER. G6Pase is comprised of a catalytic subunit, which performs the dephosphorylation, and a transporter subunit encoded by SLC37A4 (chromosome 11q23.3), which handles the transport of G6P and of inorganic phosphate across the ER membrane.[83] There are three G6Pase catalytic subunit-encoding genes in humans – G6PC, G6PC2, and G6PC3.[84] Mutations in G6PC, a 12.5-kb single-copy gene comprised of five exons located on chromosome 17q21.31, are responsible for GSD 1a, while mutations in SLC37A4 are responsible for GSD 1b.[85,86] GSD type 1 accounts for ~25% of all GSDs diagnosed in the United States and Europe; overall prevalence is between 1/100,000 and 1/400,000 live births, with increased prevalence in some populations such as Ashkenazi Jews (prevalence 1/20,000).[80] GSD 1a is far more common than 1b; 80% of GSD 1 patients have mutations in G6PC.[79]

In both GSD 1a and 1b, hepatic production of glucose from both glycogenolysis and gluconeogenesis is impaired, because G6P cannot be converted to glucose. Since G6P cannot freely cross out of the liver and into the systemic circulation, the liver effectively cannot provide glucose to other tissues to use for energy. This results in severe hypoglycemia – as low as 40 mg/dL – within two to three h of feeding, as soon as the last bit of ingested carbohydrates has been metabolized. The hypoglycemia is accompanied by increased production of lactate, triglycerides, and uric acid.[79] Onset of hypoglycemia typically occurs in infancy once the interval between feeding is lengthened, the infant starts feeding through the night, or an intercurrent illness interrupts the feeds. As glycogen synthesis is not impaired, glycogen accumulates in the liver, resulting in massive hepatomegaly and a protruberant abdomen as well as motor delay. Poor linear growth is frequently seen.

Cognitive delay is infrequent, but may occur if recurrent hypoglycemia leads to cerebral damage.

GSD 1a

GSD 1a is far more common than 1b; 80% of GSD 1 patients have mutations in G6PC, with resulting absence of catalytic activity of G6Pase. GSD 1a is inherited in an autosomal recessive fashion, with either homozygous or compound heterozygous mutations; more than 80 different mutations have been described.[87] The result of these mutations is an inability to generate glucose from carbohydrate or noncarbohydrate precursors, resulting in severe hypoglycemia and increased production of lactic acid, triglyceride, and uric acid,[79] as above.

The hypoglycemia in GSD 1a is recurrent, seen multiple times daily in untreated patients, and is often asymptomatic due to the formation and utilization of other energy sources (e.g., lactic acid); symptomatic hypoglycemia is less common. In addition, because fructose and galactose must be converted to glucose via G6Pase prior to being available for use for energy systemically, their intake (and consequently intake of sucrose and lactose, which contain fructose and galactose, respectively) should be avoided or at least minimized in these patients – any ingestion does not aid systemic energy balance and only contributes to hepatomegaly and the laboratory abnormalities noted earlier. Thus, the diet for patients with GSD 1a and 1b is quite restrictive.

Metabolic manifestations of GSD 1a are as follows:

- Excess lactic acid production: This is the result of buildup of excess G6P in hepatocytes, with consequent buildup of glycolytic metabolites, exceeding the oxidative capacity of the citric acid cycle. The excess pyruvate gets shunted into production of lactic acid, a fuel the brain can utilize but which can contribute to metabolic acidosis.
- Hypertriglyceridemia: Some of the pyruvate is oxidized to acetyl-CoA in preparation for entry into the tricarboxylic acid (TCA) cycle; however, the levels of acetyl-CoA still exceed the capacity of the TCA cycle, and excess acetyl-CoA serves as substrate for cytosolic production of free fatty acids and eventually triglycerides, resulting in hypertriglyceridemia. Triglyceride elevations into the 400–800 mg/dL range and higher may be seen in poor metabolic control.
- Hyperuricemia: Excess G6P is produced, and some is shunted down the pentose phosphate pathway; this leads to increased production of ribose-5-phosphate and the subsequent *de novo* synthesis of the purine nucleotides. These excess purine nucleotides are subsequently degraded; this catabolic process produces uric acid. The uric acid also builds up due to decreased renal uric acid clearance in the presence of competition from lactate.

- Hyperalaninemia: Some excess pyruvate is also transaminated, forming alanine and resulting in hyperalaninemia.

Clinical manifestations of GSD 1a in untreated individuals include failure to thrive and consequent short stature (frequently seen), massive hepatomegaly and a protruberant abdomen (sometimes described as a Cushingoid appearance), "doll-like" facies in infants, motor developmental delay, metabolic acidosis (due to lactic acidosis) with consequent hyperventilation and apparent respiratory distress (infrequently seen), and hypoglycemic seizures (less commonly seen, as the brain can also utilize lactate as fuel). Social and cognitive development are only affected if the patient suffers cerebral damage from recurrent hypoglycemic seizures (uncommon).[79] If metabolic control is not established and maintained, numerous complications may ensue,[79] including:

- Metabolic acidosis: Impairment of gluconeogenesis results in chronic elevations of lactic acid (4–10 mM) even when the child is well. During intercurrent illness, prolonged fasting or during an episode of metabolic decompensation (which can mimic gastroenteritis), lactic acid levels can rise abruptly, sometimes exceeding 15 mM, and can produce severe metabolic acidosis. (The anion gap may be further increased by the production of uric acid, free fatty acids, and ketones). Symptoms include hyperpnea, emesis, and dehydration (all accompanied by hypoglycemia).
- Pancreatitis: May result from chronic hypertriglyceridemia.[88] (Of note, pseudohyponatremia may also be seen, and serum may be lipemic.)
- Cardiovascular: Moderate elevations of phospholipids, total, and LDL cholesterol can be seen. The presence of these atherogenic lipoprotein profiles has been reported to be associated with the presence of cardiovascular disease risk measures, included carotid intima-media thickness, a measure of atherosclerosis, and increased central pressure wave augmentation, which predicts increased risk of cardiovascular events.[89]
- Gout: This occurs due to chronic uric acid elevation; it is sometimes accompanied by nephrolithiasis.[79] The nephrolithiasis also may have other etiologies – decreased citrate secretion in GSD 1a is seen with age; consequently, hypercalciuria develops, which can predispose to nephrocalcinosis and nephrolithiasis.[90]
- Renal: Renal dysfunction may occur due to a combination of risk factors. Nephrolithiasis can develop as a complication of chronic uric acid elevation. In addition, proximal tubular dysfunction may be seen when metabolic control is poor, and includes glycosuria, phosphaturia, hypokalemia, and generalized aminoaciduria. These are reversible with improved metabolic control; however,

chronic glomerular hyperfiltration is also seen in GSD 1a. In adulthood, long-term clinically silent hyperfiltration may lead to the development of chronic renal disease; this may include proteinuria, hypertension, and varying degrees of renal insufficiency. Renal biopsy in these cases may show focal glomerulosclerosis.[91]
- Short stature and delayed puberty were frequently seen in patients with uncontrolled GSD 1 (a and b) in the past. Normal growth and pubertal development are typically seen in children who adhere to the specific GSD diet and consequently good metabolic control. However, GH secretory abnormalities and consequent lower GH levels may be seen even if linear growth is normal.[92]
- Ovarian: Polycystic ovaries are frequently seen along with oligomenorrhea, thought to relate to insulin resistance; fertility may be preserved.[93]
- Musculoskeletal: Decreased bone density is seen in over 50% of GSD 1a patients, and may relate to poor metabolic control as well as presence of other disease complications;[94] decreased grip strength may also be seen.[95]
- Platelet function: Abnormalities of platelet function are a common feature of GSD 1a, and include abnormal aggregation and low adhesiveness, decreased prothrombin consumption, and consequent prolonged bleeding time.[96] These abnormalities correlate with degree of metabolic abnormalities and are reversible if good metabolic control is reached.[79]
- Anemia: Anemia is frequently seen in GSD 1a. In preadolescents, iron-deficiency anemia is the predominant form, and may occur because uncooked cornstarch reduced the bioavailability of ingested iron. Over 50% of adults with GSD 1a may have anemia, of which anemia of chronic disease is the most common type. Severe anemia in GSD 1a adults may be associated with hepatocellular adenomas, most likely relating to the adenomas' expression of hepcidin (which inhibits iron transport across gut mucosa); the anemia frequently improves following resection of the hepatic lesions.[97]
- Liver: Hepatocellular adenomas, which predominantly develop during and after puberty, afflict up to 75% of patients with GSD 1a. Poor metabolic control, especially hypertriglyceridemia, may increase risk of developing the adenomas.[98] Adenomas may be associated with chronic iron-resistant anemia. The adenomas can undergo malignant degeneration into hepatocellular carcinomas; hemorrhage may also occur.[79]

The diagnosis of GSD 1a is suggested by the presence of severe recurrent hypoglycemia beginning three to four h postprandially, accompanied by the metabolic abnormalities listed earlier (elevated lactates, hypertri-glyceridemia, and hyperuricemia), along with a rise in lactate in response to postprandial glucagon administration. On physical examination, massive hepatomegaly may be seen.

Diagnosis was formerly made via liver biopsy; GSD 1 (a and b) hepatic histopathologic findings include hepatocyte distension by glycogen and fat, uniform glycogen distribution, and nuclear hyperglycogenation. Large lipid vacuoles are frequently seen.[99] However, definitive diagnosis of this condition can now be made noninvasively via mutation analysis of the *G6PC* gene. *Liver biopsies are unnecessary.*

The mainstay of therapy for GSD 1a is twofold – prevention of hypoglycemia and strict adherence to a specialized diet:

- Hypoglycemia prevention: In infants (<6–12 months of age), this can be accomplished with frequent meals or snacks (every two to four h) during the daytime and by continuous infusions of dextrose via gastrostomy tubes overnight. In older infants and children, similar results can be obtained by administering uncooked cornstarch (1–1.5 g/kg, dissolved in formula, water or sweetened beverages such as Kool-Aid) every three to five h during the day and every four to six h overnight.
- Other dietary interventions: Carbohydrates typically provide 60–65% of calories. In order to minimize hypertriglyceridemia, dietary fat is restricted to no more than 20% of total energy intake, and cholesterol intake is restricted to <300 mg/day. Foods that contain fructose and/or galactose (including sucrose and lactose containing foods) must be restricted, in order to minimize intake substrates that require action of G6Pase in order to be available for systemic use.
- When hypoglycemia and hyperlacticacidemia are prevented, hepatic size decreases, linear growth improves, and serum uric acid and lipids (especially triglycerides) normalize. Allopurinol may be used as adjunct therapy if severe hyperuricemia persists despite good metabolic control. Therapy with triglyceride-lowering agents (e.g., gemfibrozil) is seldom required.[79] Serum alpha-fetoprotein and hepatic ultrasound are performed annually to screen for hepatic adenomas.

GSD 1b

GSD 1b is far less common than 1a, comprising only 20% of cases of GSD 1. GSD 1b results from a mutation in *SLC37A4*, the gene-encoding G6P transporter 1 (*G6PT1*), which regulates the rate-limiting step of G6P transport through the ER membrane and also plays a role in calcium sequestration in the ER lumen.[100] GSD 1b is also inherited in an autosomal recessive fashion, with either homozygous or compound heterozygous mutations; more than 35 different mutations have been described.[81] The result of these mutations is inability to generate glucose from carbohydrate or noncarbohydrate precursors, resulting in severe hypoglycemia and increased production of lactic acid, triglyceride, and uric acid,[79] as in GSD 1a.

The hypoglycemia in GSD 1b is very like that seen in GSD 1a – recurrent, seen multiple times daily in untreated patients, and often asymptomatic. The other clinical features of GSD 1b and the long-term complications are very similar to those of GSD 1a. However, GSD 1b carries with it a few additional comorbidities. Most GSD 1b patients suffer from neutropenia, either cyclic or constant, accompanied by neutrophil and monocyte dysfunction, which results in an impaired respiratory burst, defective bactericidal activity, and increased susceptibility to infection,[101] especially skin and pulmonary infections. The neutropenia is thought to result from enhanced ER stress as a consequence of loss of *G6PT* activity; this in turns leads to increased oxidative stress and apoptosis.[102] In addition, GSD 1b patients may also display symptoms of inflammatory bowel disease, including oral aphthous ulcers, perianal abscess or fistulas, and colitis, which may be indistinguishable from Crohn's disease.[103] Finally, an increased risk of autoimmune disorders – including thyroiditis, GH deficiency, and myasthenia gravis – in GSD 1b has occasionally been reported.[104]

The diagnosis of GSD 1b, as in 1a, is suggested by the presence of severe recurrent hypoglycemia beginning 3–4 h postprandially, accompanied by the metabolic abnormalities listed earlier (elevated lactates, significant hypertriglyceridemia, and hyperuricemia) and a rise in lactate in response to postprandial glucagon administration. On physical examination, massive hepatomegaly may be seen, along with short stature and failure to thrive. The diagnosis of GSD 1b rather than 1a is suggested by the presence of neutropenia. Histopathological findings are similar to those seen in GSD 1a; however, definitive diagnosis of this condition can be done noninvasively via mutation analysis of the *SLC37A4* gene, rendering liver biopsies unnecessary.

The mainstay of treatment of GSD 1b is very similar to that of GSD 1a: it entails adherence to a fructose- and galactose-free diet, frequent feeds (every 2–4 h) during the day with overnight gastrostromy tube feeds or dextrose in infants, and uncooked cornstarch 1–1.5 g/kg every 3–5 h during the day or every 4–6 h at night. Liver transplantation may be indicated for those with hepatic malignancy. Granulocyte colony stimulating factor (G-CSF) has been widely used since the 1990s to treat the chronic neutropenia associated with GSD 1b;[105] however, caution must be utilized, as there have been case reports of acute myelogenous leukemia developing as a consequence of G-CSF therapy.[106,107]

GSD III (Cori Disease) – Glycogen Debrancher Enzyme Deficiency

As discussed earlier, glycogenolysis entails the actions of several different enzymes. The major one is glycogen phosphorylase, which cleaves $\alpha(1\to4)$ glycosidic bonds between adjacent glucose moieties in glycogen. However, 1 in 10 glucose residues in glycogen are branch points, existing as $\alpha(1\to6)$ linkages. As glycogen phosphorylase stops cleaving when it reaches a point four residues from a branching point, the actions of glycogen phosphorylase alone do not suffice to mobilize glycogen stores. To mobilize the rest, the actions of another enzyme are required – glycogen debranching enzyme, encoded by the AGL gene and located on chromosome 1p21.2. The AGL gene is 85 kbp long and is composed of 35 exons. Six major mRNA isoforms have been identified. All isoforms share the same translation initiation site in exon 3, but differ in the 5′-untranslated regions due to differential transcription relating to specific cryptic splice sites located upstream from the translation initiation site and which enable exon removal or retention.[108] Isoform 1, the major isoform, contains exon 1 and lacks exon 2; it is the most widely expressed, and is found in liver, kidney, and lymphoblastoid cells, as well as in myocytes; it encodes a 1532-amino acid residue protein.[99] Isoforms 2–4 lack exon 1 and are exclusively expressed in cardiac and skeletal myocytes.[79] All major isoforms contain exons 4–35.

Debranching enzyme is a 165-kD protein that acts as a 1,4-alpha-D-glucan:1,4-alpha-D-glucan 4-alpha-D-glycosyltransferase and amylo-1,6-glucosidase in glycogen. The enzyme has two different, independent catalytic activities that take place at different sites and can function independently of each other – 4-alpha-glucotransferase activity and amylo-1,6-glucosidase activity. GSD III results from an autosomal recessive mutation in AGL, due either to homozygous or to compound heterozygous mutations or deletions. Missense, splice site and nonsense mutations, small frameshift deletions and insertions, and large gene deletions and duplications have all been described in the AGL gene. The majority of GSD III patients have deficient enzyme activity in both liver and muscle (GSD IIIa), while ~15% have insufficient enzyme activity in the liver only (GSD IIIb).[109] GSD IIIa mutations possess considerable allelic heterogeneity; GSD IIIb, on the other hand, is primarily associated with two mutations in exon 3 of AGL – c.18_19delGA (p.Gln6HisfsX20), formerly described as c.17_18delAG, and c.16C > T (p.Gln6X).[109] Most patients with GSD III lack enzymatic activity at both catalytic sites (both the 4-α-glucotransferase and the amylo-1,6-glucosidase sites); deficiencies only of the glucosidase activity (GSD IIIc) or of the glucotransferase activity (GSD IIId)[110,111] have been reported, but are rare. The incidence of GSD III in the US population is 1/100,000, although it is more common in certain subpopulations, such as individuals of North African Jewish descent, where the prevalence is estimated to be 1/5400.[112]

Hypoglycemia in GSD III occurs due to an inability to fully utilize glycogen. Glycogen debrancher enzyme (GDE) deficiency leads to accumulation of abnormal glycogen possessing short outer chains in the liver, phosphorylase limit dextrin, in affected tissues. It is typically seen after more prolonged fasts than in GSD I, is accompanied by ketosis, and is usually milder than in GSD I because gluconeogenesis is not impacted, and hepatic glycogen can be metabolized to a certain extent via glycogen phosphorylase. However, some patients with GSD III have hypoglycemia equal in severity to patients with GSD I. Other biochemical abnormalities include hyperlipidemia and transaminase elevation of over 500 U/L are often seen (the latter primarily in childhood). Uric acid and lactate concentrations are typically normal, in contrast to GSD 1a and 1b.[79,113] Profound ketosis can develop after overnight fast.

The clinical phenotype of GSD III is somewhat variable. However, the main underlying clinical features in childhood are hepatomegaly, fasting hypoglycemia as above, and hyperlipidemia; if treatment is not initiated, growth failure and pubertal delay can result. Renal disease is not typically seen. In GSD IIIa, where skeletal and/or cardiac muscle is involved, a vacuolar myopathy may develop, and serum creatine kinase (CK) elevations of 5–45x above the upper limit of normal may be seen.[99] Long-term complications of GSD III can include:[99]

- Hepatic cirrhosis: The hepatomegaly associated with GSD III usually resolves postpuberty, which may be attributable to a combination of lower glucose requirements but also to possible development of livery cirrhosis, which can progress to end-stage liver disease.
- Hepatic adenomas, whose prevalence in GSD III is estimated at between 4% and 25%, and rarely hepatocellular carcinomas.
- Skeletal myopathy: Motor delays may be the initial signs, followed by slowly developing weakness and wasting of both proximal and distal muscles. Nerve conduction studies may be abnormal. Muscles involved in respiration are typically spared. Skeletal myopathy may occur in isolation, or in combination with cardiomyopathy.
- Osteoporosis: Seen primarily in GSD III patients with skeletal myopathy.
- In women, polycystic ovaries may develop at higher rates; however, fertility is not typically impacted.[113]
- Cardiomyopathy: Ventricular hypertrophy is commonly seen; overt cardiac dysfunction is rare, but congestive heart failure may occasionally develop.

- Sudden death: Rarely seen; thought to be secondary to cardiomyopathy-related arrhythmias.

The diagnosis of GSD III is suggested by fasting hypoglycemia accompanied by ketosis in the setting of hepatomegaly and hyperlipidemia, especially if CK elevation and/or transaminitis is seen. The absence of uric acid and lactate elevation help to differentiate from GSD I (though uric acid elevations may be seen following exertion) muscle. Glucagon administration after an overnight fast does not trigger a rise in blood glucose levels, but if glucagon is administered 2 h following a carbohydrate-rich meal, blood glucose levels do increase normally, again in contrast to GSD I. In most GSD III individuals, liver biopsy (or muscle biopsy in those patients with myopathy) shows cytosolic vacuolar accumulation of nonmembrane-bound glycogen. The excess glycogen that accumulates is structurally abnormal (short outer branches); however, lipid vacuoles are less common than in GSD I. Hepatic fibrosis ranging in severity from periportal fibrosis to micronodular cirrhosis may be seen. Diagnosis of GSD III may be based on demonstration of deficient GDE activity in biopsied hepatocytes and/or myocytes. *However, biopsy is no longer necessary to establish a diagnosis; molecular genetic testing for pathogenic AGL mutations on both alleles is sufficient to establish the diagnosis noninvasively.*[99]

Treatment of GSD III entails avoidance of prolonged fasting, typically involving uncooked cornstarch (1.75 g/kg), though at longer intervals than in GSD I – every 6 h. This avoids hypoglycemia, improves linear growth, and reduces the transaminitis. For patients too young for cornstarch or who have significant growth failure and/or myopathy, continuous tube feeds overnight of a mixture of glucose and protein (or glucose oligosaccharides and amino acids) is beneficial. There is no need to restrict fructose or galactose in GSD III as in GSD I, as gluconeogenesis is intact. Serum alphafetoprotein and hepatic ultrasound are performed annually to biannually to screen for hepatic adenomas; liver transplants are performed, but rarely. Patients with skeletal myopathies should have intermittent electrocardiograms and echocardiograms. The prognosis for patients with myopathy is less favorable than for those with isolated hepatic involvement. Physical therapy and aerobic training may be beneficial for patients with myopathies.

Glycogen Storage Disease VI (Hers Disease, Glycogen Phosphorylase Deficiency) and GSD IX (Phosphorylase Kinase Deficiency)

In the liver, glycogen phosphorylase cleaves $\alpha(1{\rightarrow}4)$ glycosidic bonds, yielding G1P and a linear glycogen chain shortened by one glucose residue. Glycogen phosphorylase has two configurations – the inactive configuration, phosphorylase B, and the active configuration, phosphorylase A. The conformational shift to the more active configuration is triggered by phosphorylation of two serine residues by phosphorylase kinase (PhK). Glycogenoses caused by reduced glycogen phosphorylase activity (GSD VI and IX) are a heterogeneous group, clinically indistinguishable but of different etiologies; some result from loss of function of glycogen phosphorylase itself (GSD VI), while others are due to hepatic phosphorylase kinase deficiency (GSD IX) resulting from mutations in one of the genes encoding PhK subunits.

Glycogen phosphorylase has three different isoforms in humans encoded by three different genes; the isoforms are respectively expressed primarily in the liver, brain, and muscle. The hepatic isoform is encoded by the *PYGL* gene, located on chromosome 14q22.1; the classic form of GSD VI results from loss-of-function mutations in or deletions of *PYGL* (homozygous or compound heterozygous mutations, deletions or duplications of one or more exons or whole-gene deletions or duplications; at least 12 different pathogenic mutations are known). Glycogen phosphorylase mutations are autosomally recessively inherited, and the deficiency is relatively rare except in the Mennonite population, where they are relatively common due to a founder mutation, affecting 0.1% of the population.[114]

PhK is a serine-threonine phosphorylase comprised of a homotetramer of tetramers, for a total of 16 subunits; each of the four tetramers is composed of four different subunits, each encoded by a different gene: α (encoded by *PHKA1* and *PHKA2*), β (encoded by *PHKB*), γ (encoded by different *PHKG* genes for different tissues – the hepatic/testis-specific isoform is encoded by *PHKG2*), and δ (encoded by *CALM1*).[79,115] The enzyme's catalytic activity occurs at the γ subunit; the α and β subunits have regulatory functions; the δ subunit, a calmodulin, mediates the calcium dependence of the enzyme. PhK is encoded by several different genes:

- *PHKA1*, located on chromosome Xq13.2, encodes a muscle isoform of subunit α; mutations therein result in X-linked muscle PhK deficiency, a rare X-linked disorder.
- *PHKA2*, located on chromosome Xp22.13, also encodes a hepatic isoform of subunit α; *PHKA2* mutations cause hepatic PhK deficiency, or GSD9A, also known as X-linked liver glycogenosis (XLG). XLG is the most common variant of GSD IX, responsible for ~75% of cases of GSD IX. These mutations are subdivided into several categories, including GSD IXa1, where there is no PhK activity in hepatocytes or erythrocytes, and GSD IXa2, where the PhK activity is diminished or absent in hepatocytes but preserved in erythrocytes. At least 28 different *PHKA2* mutations have been identified.

Both GSD9A1 and GSD9A2 are caused by mutations in *PHKA2*; however, GSD9A2 is caused by either small in-frame deletions or insertions or by missense mutations, explaining the residual enzymatic expression in erythrocytes.[116,117]

- *PHKB*, located on chromosome 16q12.1, encodes subunit β in liver and muscle; mutations, which are rare, cause autosomal recessive GSD9B. Fourteen different pathogenic variants have been described, including missense and nonsense mutations, splice site variations, and small insertions or deletions.

- *PHKG2*, located on chromosome 16p11.2, encodes the catalytic δ subunit of PhK in the liver and testis. Mutations cause autosomal recessive hepatic PhK deficiency, GSD9C. Nine different pathogenic mutations have been described.

Phosphorylase kinase mutations are considerably more common than glycogen phosphorylase mutations, and are collectively responsible for ~25% of all GSDs,[79] with an estimated prevalence of 1/100,000.[118] GSD9A, 9B, and 9C are clinically indistinguishable from each other, as well as from GSD XI.

GSD VI and XI have milder phenotypes than GSD I or III and a more favorable prognosis. Symptomatic hypoglycemia is uncommon, and primarily occurs after prolonged fasting or intensive exercise; it is accompanied by ketosis. Relatively mild hyperlipidemia (triglycerides and other lipids) and mild transaminase elevation may be seen; lactic acid and uric acid are typically normal, and thus metabolic acidosis is uncommon. Affected individuals usually present in infancy or early childhood due to growth retardation and hepatomegaly. If muscle as well as hepatic glycogen phosphorylase is impacted (*PHKA1*, *PHKB*), hypotonia and motor delay may be present. As patients age, the laboratory abnormalities and hepatomegaly typically improve, as does linear growth – often without treatment. Adults are often asymptomatic. Long-term complications have been reported – liver cirrhosis, CNS complications such as impaired cognition, speech impairment, seizures, peripheral sensory neuropathy, and/or renal impairments such as tubular dysfunction and proximal renal tubular acidosis. However, these long-term complications are all rare.[79]

Diagnosis of GSD VI can be made by assaying phosphorylase activity in erythrocytes. *PHKB* activity can also be assessed in erythrocytes or leukocytes, but normal PhK activity in blood does not definitively rule out GSD9, given the possibility of XLG due to *PHKA2*. Hepatic biopsies for enzyme activity are possible, but again unnecessary; molecular genetic diagnosis of mutations or deletions in *PYGL*, *PHKA1*, *PHKA2*, *PHKB*, or *PHKG2* is the test of choice.

GSD VI and IX are easier to treat than GSD I or III. No special diet is required, and there is no need to avoid fructose- or galactose-containing foods. Prolonged fasting should be avoided, and plenty of carbohydrates should be ingested if strenuous exercise is undertaken. Bedtime snacks often suffice to prevent nocturnal hypoglycemia and ketosis; if not, 2 g/kg of uncooked cornstarch can also be administered at bedtime. Overnight feeds are not required.[119]

Disorders of Gluconeogenesis

Gluconeogenic disorders are summarized in Table 3.3.

Fructose-1,6-Bisphosphatase (FBPase) Deficiency

FBPase is a key gluconeogenic enzyme, catalyzing the hydrolysis of fructose-1,6-bisphosphate to fructose-6-phosphate and inorganic phosphate. Two different paralogs exist in humans; FBP1 is found in the liver, and FBP2 is found in muscle. FBP1 is encoded by the gene *FBP1*, a 31-kbp 7-exon gene located on chromosome 9q22.2.[121] FBPase deficiency, due to autosomal recessive mutations or insertions or deletions in the gene, is uncommon; the prevalence is ~1/20,000 worldwide.[122] FBPase deficiency does not allow for glucose formation from gluconeogenic precursors, including fructose, protein, and fat; thus, patients with FBPase deficiency are dependent upon dietary glucose and galactose and upon glycogenolysis to maintain euglycemia. Hypoglycemia is thus likely to occur in situations of limited hepatic glycogen reserves, such as in newborns, or after glycogen reserves are exhausted, such as after prolonged fasting (~12 h in adults). This hypoglycemia is accompanied by elevations of gluconeogenic substrates such as pyruvate, alanine, glycerol, glycerol-3-phosphate, as well as of lactic acid; these occur because gluconeogenesis is blocked at the level of fructose-1,6-bisphosphate. The conversion of 1,3-bisphosphoglycerate into glyceraldehyde-3-phosphate is secondarily impaired, resulting in accumulation of NADH and relatively decreased pyruvate formation (though it does still accumulate above normal levels), and a relatively elevated lactate/pyruvate ratio. The hypoglycemia drives counterregulatory response, stimulating lipolysis, fatty acid oxidation, and ketogenesis. Thus, the hypoglycemia is often accompanied by ketosis. However, pyruvate can accumulate and result in a buildup of oxaloacetate, with consequent diversion of acetyl-CoA into citrate synthesis, which can cause impaired ketogenesis and cause an absence of ketosis and accumulation of fatty acids in liver and plasma, which can in turn lead to hyperlipidemia and hypertriglyceridemia.[120] Also, affected individuals can develop severe hypoglycemia with metabolic acidosis upon ingestion of fructose, which can lead to hepatic and renal impairment.

Treatment of FBPase deficiency involves management of acute life-threatening episodes (metabolic crises) and

TABLE 3.3 Disorders of Gluconeogenesis[4,120]

Disorder	Gene/s	Affected protein/enzyme	Chromosomal location and inheritance[81]	Diagnosis	Clinical features	Treatment
Fructose-1,6-bis-phosphatase (FBPase) deficiency	FBP1	Fructose-1,6-bisphosphatase	9q22.2–9q22.3 AR	• FBP1 sequence analysis • Liver biopsy with direct enzymatic assay of hepatic FDPase activity from liver specimens	(May manifest in infancy or in times of illness/prolonged fast) • Episodic severe hypoglycemia + lactic/metabolic acidosis – lactate elevation to 15–25 mmol/L • Elevated lactate/pyruvate ratio (up to 30) If undiagnosed • Myopathy • Moderate hepatomegaly If diagnosed • Normal growth and development • Fasting tolerance improves with aging	• During metabolic crisis episodes: IV glucose and sodium bicarbonate • Avoid fasting • Uncooked cornstarch or overnight feeds as necessary • Young children: restrict fructose, sucrose, sorbitol intake • Restrict protein to 10% of energy intake • Restrict fat to 20–25% of energy intake
Cytosolic phosphoenolpyruvate carboxykinase (PEPCK) deficiency	PCK1	Phosphoenolpyruvate carboxykinase	20q13.31 AR	Sequence analysis of PCK1 or PCK2	1. Episodic hypoglycemia (less severe than with FBPase deficiency) a. Lactic acidosis b. Elevated alanine c. Ketosis d. Transaminase elevation e. Hyperlipidemia/hypertriglyceridemia 2. Progressive multisystemic damage: a. Myopathy b. Hypotonia – eventual spasticity c. RTA d. Microcephaly e. Developmental delay f. Seizures g. Heptic dysfunction h. Cardiomyopathy	Supportive treatment • IV glucose and sodium bicarbonate during metabolic acidosis episodes • Avoid fasting • Uncooked cornstarch or overnight feeds as necessary
Mitochondrial PEPCK deficiency	PCK2		14q11.2 AR	Biochemical analysis of fibroblast cells (PCK2) or hepatocytes (PCK1 and PCK2)		• Frequent feeds (slowly-absorbed carbohydrates) • Restrict protein and fat intake • No need to restrict fructose Prognosis: tolerance to fasting improves; myopathy, neurological, and hepatic damage persist
Pyruvate carboxylase deficiency Type A North American Type B French Type C benign 1/250,000 births	PC	Pyruvate carboxylase	11q13.2	• Enzyme functional assay (Leukocytes, fibroblasts or biopsied hepatocytes) • Sequence analysis of PC	• Hypoglycemia • Ketosis • Elevated lactate, pyruvate, alanine • Metabolic (lactic) acidosis • Abnormal eye movements • Seizures • Developmental delay • Hypotonia • Severe encephalopathy • Death in early infancy (type B) or early childhood (type A)	• High-protein, high-carbohydrate diet • Citrate supplement (counteract acidosis) • Biotin supplement (type C) • Aspartic acid supplementation

AR, autosomal recessive.

long-term dietary management. Acute episodes should be treated with intravenous dextrose infusions at high rates (10–12 mg/kg/min for newborns) along with sodium bicarbonate to treat the hypoglycemia and acidosis, respectively. Maintenance therapy involves avoidance of prolonged fasting, particularly during periods of increased metabolic demand (e.g., fever) via frequent feeds, and, if necessary, overnight tube feeds and/or uncooked cornstarch. In younger children, restriction of fructose, sucrose, and sorbitol is highly recommended, as well as restriction of dietary nutrients that must undergo gluconeogenesis to be utilized for energy; protein should be restricted to no more than 10% and fat to no more than 20–25% of energy intake. Long-term, postdiagnosis, growth and development are usually normal, and fasting tolerance improves with age and increased ability to store glycogen in the liver; adults are often asymptomatic.[120]

Phosphoenolpyruvate Carboxykinase (PEPCK) Deficiency

PEPCK is another key rate-limiting gluconeogenic enzyme, catalyzing the conversion of oxaloacetate to phosphoenolpyruvate and water. There are two PEPCK paralogs, cytosolic (encoded by PCK1, a 20 kbp 9-exon gene located on chromosome 20q13.31) and mitochondrial (encoded by the PCK2 gene, also comprised of a maximum of 10 exons but with 16 splice variants, located on chromosome 14q11.2). Cytosolic PEPCK activity is stimulated by the actions of two counterregulatory hormones, cortisol and glucagon. There are two forms of PEPCK deficiency – cytosolic and mitochondrial. Both are autosomal recessive in inheritance, and both are exceedingly rare – <1/1,000,000. PEPCK deficiency impairs gluconeogenesis from protein (alanine, most other gluconeogenic amino acids), lactate, and pyruvate, but gluconeogenesis can proceed using glycerol (fat), fructose, glycine, and serine; thus, the hypoglycemia associated with PEPCK deficiency is less profound than that seen in some other disorders of gluconeogenesis.

Patients present with episodes of hypoglycemia accompanied by elevated levels of lactate and consequent lactic acidosis. Alanine levels may also be elevated. Transaminases may be elevated. Ketosis may be absent at times of hypoglycemia, because pyruvate accumulation leads to increased synthesis of citrate and consequently of malonyl-CoA, which inhibits the entry of long-chain fatty acids into the mitochondria for ketogenesis; this subsequently leads to free fatty acid accumulation in plasma and to hepatic, muscle, and renal steatosis. Urinalysis may show elevations of lactate and alanine, along with generalized aminoaciduria.[120] PCK2 (mitochondrial) deficiency has a more severe presentation than cytosolic PEPCK deficiency, possibly due to defects in the mitochondrial respiratory chain. The disease usually progresses to multiorgan system damage or failure, presenting in infancy with failure to thrive and hypotonia and progressing to myopathy, developmental (cognitive and motor) delay, spasticity, microcephaly, hepatomegaly and hepatocellular dysfunction, RTA, and cardiomyopathy. Diagnosis is based upon sequence analysis of PCK1 or PCK2, or upon enzyme activity assay in affected tissues (hepatocytes for PCK1 or PCK2, though activity of the latter can also be assayed in cultured fibroblasts or in lymphocytes). However, deficiency may also be a secondary finding, so results of testing should be interpreted within the clinical context.

Treatment of metabolic crises is similar to the treatment of crises in patients with FBPase deficiency – intravenous infusion of dextrose at high rates, combined with bicarbonate to combat the acidosis. Similarly, maintenance therapy involves avoidance of prolonged fasting, frequent meals or snacks, uncooked cornstarch or overnight tube feeds if necessary, and restriction of dietary fat and protein. Unlike FBPase patients, there is no need to restrict fructose intake. Long-term prognosis is poor – hypoglycemia may become rarer in adulthood as fasting tolerance improves, but the myopathy, hepatic, and neurological damage persist.

Pyruvate Carboxylase Deficiency

Pyruvate carboxylase (PC) is a biotin- and ATP-dependent mitochondrial enzyme that catalyzes the anaplerotic carboxylation of pyruvate to oxaloacetate, a substrate for gluconeogenesis. This is the point at which not only pyruvate, but also lactate and alanine, enter the gluconeogenic pathway. PC requires magnesium or manganese and acetyl-CoA to perform its function. PC is expressed in a tissue-specific fashion, with high levels of expression in liver, kidney, adipose tissue, pancreatic islets, and mammary glands. PC is encoded by the PC gene, located on chromosome 11q13.2, whose coding region spans ~16 kb and contains 19 exons.

PC deficiency is a rare disorder, with an incidence of 1/250,000 worldwide, although higher in some Algonkian native tribes in eastern Canada due to a postulated founder effect.[123] The disorder is autosomal recessive in inheritance, and is attributable to mutations (missense or nonsense), splice site substitution, deletions or insertions in the PC gene.[124] Eighteen known mutations have been described. Gluconeogenesis from lactate and from protein sources is impacted, but gluconeogenesis from other substrates – fructose, glycerol (fat), glycine, and serine – and glycogenolysis are unimpacted. Pyruvate is shunted toward alternate metabolic pathways, leading to increased lactate, alanine, and acetyl-CoA; the latter is shunted toward ketogenesis. The citric acid cycle cannot proceed without oxaloacetate, so energy is primarily extracted from glucose via glycolysis. The hypoglycemia that results is accompanied by lactic acidosis, ketosis, and elevation of pyruvate and alanine. Tissues that are more heavily dependent on glucose, such as the brain, are more vulnerable to damage.

Clinically, PC deficiency falls into three different phenotypic subgroups. "Group A" are patients from North America; they typically present in late infancy with hypoglycemia, lactic acidosis and motor delay, severe developmental delay, seizures, and necrotizing encephalopathy. Prognosis is poor – affected individuals may live into adulthood, but typically suffer global neurocognitive dysfunction. Type A PC deficiency is caused by two missense mutations in either the biotin carboxylase or the N-terminal carboxyltransferase domain of PC.[123] "Group B" patients hail from France and the United Kingdom (also described in Canada, Egypt, and Saudi Arabia). These individuals present in infancy, and also have elevations of ammonia (occurs due to decreased aspartate production from oxaloacetate and thus decreased ammonia disposal), citrulline and lysine in addition to lactic acidosis and hypoglycemia. Clinical features of this group include seizures, abnormal eye movements, hypotonia, and coma; prognosis is grim, with death typically occurring within the first three to six months of life due to severe energy deficits in the brain. Type B PC phenotype results from truncating mutations, typically in the C-terminal carboxyltransferase or in the biotin carboxyl carrier protein domains. "Group C" patients present with a more benign phenotype. These individuals present with mild episodes of metabolic acidosis with increased lactate, lysine, and proline concentrations, and with normal plasma citrulline concentrations; they typically manifest only minimal CNS impact.

Diagnosis of PC deficiency is typically made through either enzyme functional assay performed on leukocytes, fibroblasts, or biopsied hepatocytes or by PC sequence analysis. Prenatal diagnosis is possible through assay of PC activity in cultured amniotic fluid cells obtained via amniocentesis.[125] Therapy for PC deficiency aims to provide alternative metabolic fuels, limit the dependence on gluconeogenesis, and stimulate the pyruvate dehydrogenase complex. Providing cofactor supplementation (biotin, thiamin, lipoic acid, dichloroacetate[126,127]) optimizes any residual PC enzyme activity, provides other pathways to metabolize pyruvate, and thus limits the accumulation of pyruvate and lactate. Citrate supplementation is used to provide substrate further down the Krebs cycle and to reduce acidosis. Aspartate supplementation helps allow the urea cycle to function and thus to reduce ammonia levels.[128] Dietary intervention – limiting fasting to limit dependence on gluconeogenesis, high-carbohydrate, and high-protein intake. The ketogenic diet should be avoided. Lactate levels should be monitored regularly. However, treatment does not substantially alter the prognosis or ameliorate the progressive CNS damage.

Other Disorders of Carbohydrate Metabolism

Hereditary Fructose Intolerance

Fructose can enter cells via the insulin-independent GLUT5 fructose transporter. In hepatocytes, fructose is phosphorylated to fructose-1-phosphate (F1P) by the enzyme fructokinase. F1P is then hydrolyzed by the enzyme aldolase B to dihydroxyacetone phosphate (DHAP) and glyceraldehyde; the latter is phosphorylated to glyceraldehyde-3-phosphate. Glyceraldehyde-3-phosphate and DHAP are then converted into fructose-1,6-bisphosphate (F1,6BP) by the enzyme aldolase A, and F1,6BP then enters the gluconeogenic pathway. Aldolase B, one of the key enzymes in this process that is preferentially expressed in the liver, is a tetramer formed of four identical 40-kD subunits.[4] In humans, aldolase B is encoded in humans by the *ALDOB* gene, a 14.5-kbp 9-exon gene, variously reported to be located on chromosome 9q23.3 or 9q31.1.[81] Hereditary fructose intolerance is an autosomal recessive disorder caused by loss of function of aldolase B, resulting in accumulation of F1P.[129] At least 54 mutations and two large intragenic deletions have been reported in the *ALDOB* gene.[130] Worldwide incidence is difficult to estimate, but has been estimated at 1/20,000–1/22,000 live births in reports from Switzerland and the United Kingdom, respectively.[131,132]

Hereditary fructose intolerance (HFI) typically manifests in infancy when the infant is offered fructose-containing foods – whether formulas sweetened with fructose or actual fruits or vegetables. If diagnosis is delayed, aversion to fruits and to sweet foods often develops. Ingestion of fructose or of fructose-containing substances such as sucrose, or sorbitol, which is metabolized into fructose by sorbitol dehydrogenase leads to accumulation of F1P in hepatocytes and subsequent sequestration of inorganic phosphate, increased degradation of adenine nucleotides,[133] and inhibition of phosphorylase A.[134] The result is hypoglycemia (accompanied by hypophosphatemia), which is typically refractory to glucagon (or glycerol), but which does improve with galactose administration.[135] Purine nucleotide degradation leads to hyperuricemia and to release of Mg-ATP complexes, transient hypermagnesemia, and up to 60% loss of cellular ATP content, along with fructosemia and fructosuria.[135] Biochemically, then, HFI results in hypoglycemia accompanied by hypophosphatemia, hyperuricemia, and sometimes hypermagenesemia; nonspecific transaminitis may also be seen. Symptoms include vomiting and abdominal pain; if severe liver damage has occurred, jaundice and hemorrhage can occur. Delay in diagnosis or dietary nonadherence (intentional or otherwise) can result in poor linear growth, hepatomegaly and chronic liver disease, as well as episodic hypoglycemia as above. If untreated, HFI can result in hepatic and renal failure and even death.[136]

Diagnosis of HFI can be made either by *ALDOB* sequence analysis or by functional assay of aldolase B from biopsied hepatocytes (though the latter is unnecessary as the former can be done less invasively). *Fructose challenge should be avoided*; the test results are not always diagnostic, and hypoglycemia and its potentially severe or even life-threatening sequelae can result. Treatment is to the limiting or exclusion (to less than 40 mg/kg/day) of all foods containing fructose or its precursors, such as sucrose or sorbitol, from the diet, and the replacement in the diet by sugars that are tolerated, such as glucose, galactose, or maltose. Although adherence to the diet can be quite difficulty (fructose is found in or added to many foods), those who successfully adhere can lead normal, healthy lives.

Galactosemia

Galactose is metabolized by means of the Leloir pathway, which allows it to undergo either gluconeogenesis or glycolysis. The initial step in this pathway is the conversion of β-D-galactose to α-D-galactose by mutarotase. α-D-galactose is then phosphorylated by galactokinase (encoded by the gene *GALK1*, located on chromosome 17q25.1) into galactose-1-phosphate (Gal1P). Gal1P uridyltransferase (encoded by *GALT*, on chromosome 9p13.3) then transfers a uridylmonophosphate (UMP) group from uridyldiphosphate (UDP)-glucose to Gal1P, forming UDP-galactose. Subsequently, UDP galactose-4'-epimerase (encoded by *GALE*, on chromosome 1p36.11) converts UDP-galactose to UDP-glucose.[137]

Galactosemia is an autosomal recessive disorder of galactose metabolism resulting from defects in any of the enzymes involved in the galactose metabolism pathway. Classic galactosemia, also known as galactosemia type 1, due to a homozygous or compound heterozygous mutation/deletion in the *GALT* gene, which results in a decrease in enzyme function of over 95%, is very common. Incidence of classical galactosemia is between 1/30,000 and 1/60,000 in western countries,[138,139] with higher incidence in Irish Travelers (possibly due to increased consanguinity). It is rarer in Asian populations. Over 120 mutations in GALT have been identified as causing galactosemia; an additional 173 pathogenic mutations have been identified.[81] Of these, one study found that two mutations – Q188R and K285N – accounted for more than 70% of classic galactosemia associated with *GALT* dysfunction in the Caucasian population and that one mutation, S135L, was responsible for 62% of galactosemia in the Black population; the latter was associated with better clinical outcomes.[140]

Mutations that reduce rather than eliminate *GALT* function lead to a milder clinical variant of galactosemia. Galactosemia type II results from mutation in *GALK1*, and galactosemia type III results from mutation in *GALE*. Both are considerably rarer than galactosemia

type I in western populations, though the incidence of *GALE* mutations in Japan has been reported to be as high as 1/23,000.[141] Inactivating mutations in any of the above lead to abnormal accumulation of galactose, which then enters the polyol pathway of carbohydrate metabolism; aldose reductase reduces galactose to galactitol, a toxic metabolite, which accumulates and is excreted in the urine. Galactose may also be oxidized to galactonate by galactose dehydrogenase; galactonate is less harmful than galactitol. Clinical presentation varies somewhat depending on the degree of residual enzyme activity. Variations are most notable in populations of African descent, where affected individuals may have 10% of *GALT* activity in the liver but no activity in erythrocytes.[142] Another, fairly benign variant is recognized – the Duarte variant, characterized by mutations c.940 A>G and a GTCA deletion in the promoter region of *GALT*, which leave 5–20% of residual enzyme function. Neonates with the Duarte variant may or may not be caught by the newborn screen; most can tolerate a normal diet, and may or may not have elevated serum levels of galactitol or galactonate.[143] Classical galactosemia presents as severe hypoglycemia and vomiting in the neonatal period following ingestion of milk accompanied by elevated serum levels of galactose, galactitol, and galactonate; if treatment is not instituted, failure to thrive, hepatocellular damage and bleeding, and increased susceptibility to *Escherichia coli* sepsis are also seen. Long-term sequelae of untreated galactosemia include hepatic dysfunction and jaundice, hepatosplenomegaly, renal tubular dysfunction, hypotonia, seizures, cataracts, and neurocognitive sequelae.

Because of the need to intervene promptly in the neonatal period to prevent some of the more serious adverse sequelae such as hepatic failure, galactosemia is now included as part of the disorders tested for in the US newborn screening program in all 50 states. The infant's blood is transferred to filter paper and the dried specimen undergoes a fluorometric assay for total serum galactose (measures both galactose and Gal1P), whose levels are typically elevated (though there may be false negatives if the infant has been transfused or fed soy formula). If elevated galactose levels are detected, a Beutler assay of the specimen is subsequently performed to measure GALT enzyme activity. The Duarte variant of *GALT* deficiency may be missed by the newborn screen, as galactose serum levels are not as elevated. If the newborn screen detects low levels of *GALT* activity suggesting either affected or carrier status, molecular testing for common *GALT* mutations is then performed; if those are negative, the testing algorithm calls for sequencing analysis of less common *GALT* mutations and for mutations in *GALE* and *GALK1*. Treatment of galactosemia entails avoidance of galactose-containing foods; specifically, infants are switched to a soy-based or protein hydrolysate formula (both cow's milk formulas and human breast milk contain galactose).[144] If dietary treatment is instituted within the first 10 days of life, the signs quickly resolve and the life-threatening sequelae are prevented; however, affected individuals remain at higher risk of developmental delay and speech problems – verbal dyspraxia and dysarthria – and the majority of females with classical galactosemia develop premature ovarian failure (hypergonadotrophic hypogonadism).[137] Patients should be referred for early intervention given the high likelihood of speech difficulties and screened for ovarian failure. Calcium supplementation is also recommended.

Defects in Amino Acid Metabolism

The branched-chain amino acids (BCAA) – valine, leucine, and isoleucine – are amino acids that have aliphatic side chains with a branch and are essential amino acids. Although most essential amino acids are oxidized predominantly in the liver, BCAA oxidation in mammals primarily occurs in skeletal myocytes, where they are catabolized to branched-chain keto acids (BCKAs) in a reaction catalyzed by the enzyme branched-chain aminotransferase (BCAT). In humans, the branched-chain alpha-keto acid dehydrogenase (BCKDH) complex catalyzes the second step in branched-chain amino acid metabolism, oxidative decarboxylation of BCKAs; this reaction occurs in the liver. BCKHD is a mitochondrial enzyme that has four subunits – E1α, E1β, E2, and E3 – and requires thiamine pyrophosphate as a coenzyme.[145] E3 (lipoamide dehydrogenase) is also found in the pyruvate dehydrogenase complex and in the alpha-ketoglutarate dehydrogenase complex. The first two metabolic steps are common to all three BCAAs. In the following steps of BCAA catabolism, enzymes specific to each BCKA byproduct (isovaleryl-CoA, α-methylbutyryl-CoA, and isobutyryl-CoA) catalyze the reaction converting the BCKA byproduct to intermediates and finally to succinyl-CoA and acetyl-CoA, which can then enter the Krebs cycle or serve as substrates for lipogenesis. The BCKDH complex is regulated by two enzymes: BCKDH kinase inactivates the complex by phosphorylating its E1α subunit, while BCKDH phosphatase activates the enzyme by phosphorylating the said E1α subunit. Maple syrup urine disease (MSUD), methylmalonic acidemia (MMA), and propionic acidemia are three inborn errors of protein metabolism that are caused by an inability to catabolize BCAA, leading to accumulation of organic acids.

Maple Syrup Urine Disease (MSUD)/ Branched-Chain Amino Aciduria

MSUD is an autosomal recessive disorder resulting from homozygous or compound heterozygous mutations in one of three genes encoding the aforementioned

subunits of the BCKDH complex, also known as BCKDC. MSUD type IA results from mutations in *BCKDHA* (chromosome 19q13.2), which encodes the E1α subunit; MSUD type IB results from mutations in *BCKDHB* (chromosome 6q14.1), which encodes the E1β subunit; and MSUD type II *DBT* on chromosome 1p21.1, which encodes the E2 subunit. Mutations in the gene encoding the E3 subunit, *DLD* (chromosome 7q31.1), causes a broader disorder called dihydrolipoamide dehydrogenase deficiency, or DLDD, sometimes referred to as MSUD type III.[146] All of these various mutations result in accumulation of both the branched-chain amino acids as well as alloisoleucine and organic acids. The disorder derives its name from the distinctive sweet odor of affected individuals' urine. One analysis found that 33% of MSUD was due to *BCKDHA mutations*, 38% due to *BCKDHB* mutations, and 19% due to *DBT* mutations (10% of cases were not found to have mutations in any of the above).[147] The incidence of MSUD worldwide is ~1/185,000 live births,[148] with a higher frequency noted in certain populations – 1/290,000 live births in New Englanders of mostly Caucasian descent,[149] and in old-order Mennonites in Pennsylvania, as high as 1/176 live births,[150] likely due to a founder effect. *BCKHB* mutations are more common in Ashkenazi Jews, with a carrier frequency of 1/113; most of these individuals carried the same mutation, R183P.[151]

Clinically, MSUD is divided into five phenotypes. Of note, there is considerable genotype-phenotype heterogeneity, with multiple mutations (over 60 have been identified) in the above-mentioned genes found in the various different phenotypes; the phenotypic manifestations depend more upon the degree of residual enzyme function than upon which subunit is impaired, with specific exceptions discussed in the subsequent section. The classic form is the most severe,[152] with less than 2% residual activity of the BCKDH complex. Untreated infants appear normal at birth, but elevated BCAA are detectable within 24 h of life, and by the third day of life, poor feeding, ketosis, and irritability develop. By days of life four to seven, a severe encephalopathy develops, manifesting as lethargy, alternating hypotonia, and hypotonia, seizures, intermittent apneas, opisthotonus, and stereotyped movements such as "fencing" and "bicycling" develop. Coma and central respiratory failure soon follow by 7–10 days of age.[146,148] In intermediate and intermittent types of MSUD, the residual BCKDC activity is higher, so the phenotype is milder. Intermediate MSUD typically presents in the toddler years with a somewhat milder phenotype ophthalmoplegia to severe developmental delay. Intermittent-type MSUD has late onset and is episodic, with transient episodes of neurological symptoms (e.g., ataxia, lethargy) accompanied by elevated urinary branched-chain keto acids. Cognitive delay may be seen in intermittent MSUD, but may be partially reversible with dietary therapy.[153] Thiamine-responsive MSUD is due to decreased BCKDC enzyme (usually E1α subunit) affinity for thiamine pyrophosphate and is ameliorable by thiamine supplementation.[154–156] Finally, E3-deficiency MSUD with lactic acidosis, also known as MSUD type III or DLDD, features combined keto acid dehydrogenase deficiencies and clinically manifests with various neurological symptoms (developmental delay, irregular respirations, and hypotonia), failure to thrive, and a combined metabolic acidosis (lactic and pyruvic acidemia) along with organic aciduria;[146] this particular form of MSUD may respond to lipoic acid administration.

In states that include MSUD among the conditions tested for in the newborn screening programs, tandem mass spectrometry is used to measure whole blood leucine/isoleucine concentrations relative to other amino acids (e.g., alanine and phenylalanine). Otherwise, the presence of clinical features, BCAA and BCKA elevations in urine and BCAA elevation in plasma along with presence of alloisoleucine, suggest the disorder. Sequence analyses of the various genes can confirm the diagnosis. Treatment of MSUD entails adherence to a special diet-restricting intake of BCAA, especially leucine, BCAA-free medical foods, and limited supplementation of isoleucine and valine. Frequent biochemical and clinical monitoring is recommended. Thiamine-responsive MSUD is treated with thiamine supplementation. Metabolic decompensation episodes are treated by correcting the underlying predisposing stressor and delivering calories, free amino acids, isoleucine, valine, and insulin, in order to restore an anabolic state.

Propionic Acidemia

Propionic acid is an intermediate metabolite of the BCAA valine and isoleucine, and of threonine and methionine. These four amino acids are carboxylated into D-methylmalonyl-CoA (which eventually enters the Krebs cycle as succinyl-CoA – see the "Methylmalonic Acidemia" section) by propionyl-CoA carboxylase (PCC), a biotin-dependent carboxylase located in the inner mitochondrial space. PCC is a heterododecamer made up of six α and six β subunits; the biotin-binding site is located on α subunits. The α subunits are encoded by the gene *PCCA*, located on chromosome 13q32.3. The β subunits are encoded by *PCCB*, located on chromosome 3q22.3.

Propionic acidemia (PA) results from loss or absence of function of PCC; over 45 mutations have been identified in *PCCA* and over 55 mutations identified in *PCCB*. The worldwide incidence of propionic acidemia is estimated to be between 1/50,000 and 1/100,000 live births, with higher incidence in certain ethnic groups such as the Inuit in Greenland, where the incidence is as high as 1/1000 live births.[157] As a result of the deficient PCC function, metabolism is profoundly deranged, likely due to a combination of decreased production of Krebs cycle intermediates and inhibition of oxidative phosphorylation due to inhibition by excess propionyl-CoA, resulting in decreased energy production,[158] and the toxic impact of hyperammonemia and of accumulated free organic acids.

Phenotypic severity varies; affected individuals may present with vomiting, dehydration, and encephalopathy and are predisposed to basal ganglia infarcts, but in some individuals, these complications may be absent.[159] Severely affected individuals have neonatal-onset PA: they present in the first six weeks of life, have poor feeding and lethargy accompanied by anion-gap metabolic acidosis, ketosis and ketonuria, and hyperammonemia (in most); biochemical abnormalities seen include hypoglycemia, hyperglycinemia, elevated propionylcarnitine, elevated 3-OH-propionate and methylcitric acid, neutropenia, and thrombocytopenia. Clinically, they progress to seizures, progressive encephalopathy, coma, and death if untreated. Late-onset PA presents after six weeks of age and has a variety of phenotypic manifestations depending on degree of residual enzyme activity, occurrence of metabolic stressors, and intake of BCAA and other propiogenic precursors. Acute intermittent late-onset PA presents as encephalopathy, seizures, or even coma precipitated by catabolic stressors such as illness or surgery. Chronic progressive PA manifests as failure to thrive, hypotonia, developmental regression, movement disorders, and/or dystonia. Finally, some patients with PA manifest only isolated cardiomyopathy. Biochemical abnormalities seen may include elevated 3-OH-propionate and methylcitric acid, hyperammonemia, metabolic acidosis, and hyperglycinemia. MRI may reveal basal ganglia lesions.[160,161]

Some US states include propionic acidemia in their expanded newborn screens; in affected infants, testing reveals elevations of propionylcarnitine (C3). Otherwise, testing urine of symptomatic individuals will reveal elevations of 3-OH-propionate above the normal levels and presence of methylcitrate, propionylglycine, and tiglylglycine, none of which are normally present in urine. Serum amino acids typically show elevated glycine levels. Confirmation of the diagnosis is obtained by demonstrating the presence of biallelic mutations in either the *PCCA* or *PCCB* genes, or by demonstration of deficient PCC activity in the liver. Treatment entails strict adherence to a diet low in propiogenic substrates and increased caloric intake (or nasogastric/gastrostomy tube feeds, or intravenous dextrose at high rates) during periods of illness to prevent catabolism. Other medications may include N-carbamoylglutamate, L-carnitine, and/or metronidazole to reduce propionic acid production by intestinal bacteria. Acute PA decompensation is a medical emergency of life-threatening metabolic acidosis and hypoglycemia; it should be treated by arresting the catabolic state by means of intravenous infusion of dextrose at high rates and/or high-calorie diets plus significant fluid intake, minimization of protein intake, and intravenous carnitine supplementation, as well as treatment of the acidosis. Orthotopic liver transplantation may be indicated in individuals with very poor linear growth, frequent decompensations, or difficult to control hyperammonemia.

Methylmalonic Acidemia (MMA)

As discussed earlier, propionic acid is converted by PCC to D-methylmalonyl-CoA. D-Methylmalonyl-CoA is then converted by methylmanolyl-CoA epimerase (encoded by the gene *MCEE* on chromosome 2p13.3) into L-methylmalonyl-CoA, which is in turn converted in a reversible reaction to succinyl-CoA in a reaction catalyzed by the mitochondrial 5-adenosylcobalamin (B12)-dependent enzyme, methylmalonyl-CoA mutase (MCM), encoded by the *MUT* gene on chromosome 6p12.3. Succinyl-CoA then enters the Krebs cycle and can be used as a substrate for oxidative phosphorylation or for gluconeogenesis.

Failure to convert methylmalonyl-CoA into succcinyl-CoA in the mitochondrial matrix results in methylmalonic acidemia, an autosomal recessive group of inborn errors of protein metabolism in which methylmalonic acid concentrations in blood and urine are elevated without a concomitant elevation in homocysteine. MMA can result from a reduction or complete absence of function of *MUT*; indeed, over 200 mutations, deletions, and insertions in *MUT* have been identified in individuals with MMA. However, as MCM is a cobalamin-dependent enzyme, defects in the synthesis of 5'-deoxyadenosylcobalamin from cobalamin (involving three enzymes, *cblA*, *cblB*, and *cblD*, respectively encoded by the genes *MMAA* on chromosome 4q31.21, *MMAB* on chromosome 12q24.11, and *MMADHC* on chromosome 2q23.2). The worldwide incidence of MMA is uncertain, though estimates suggest a prevalence of 1/25,000–1/48,000; of all cases, ~50% of affected individuals have mutations in *MUT* (the incidence of those mutations is estimated to be between 1/80,000 and 1/100,000 live births;[162] it is more common in Japan, where the prevalence is estimated to be as high as 1/50,000.[163] Clinically, the phenotype of the disorder is similar regardless of the genetic defect involved.

Onset of MMA ranges from the neonatal period through adulthood, and the phenotypic severity varies considerably. All affected individuals have episodic metabolic decompensation, usually associated with intercurrent infections or other catabolic states (e.g., surgery). These episodes entail anion gap metabolic acidosis (ketoacidosis and lactic acidosis), hypoglycemia, hyperammonemia, and hyperglycinemia.

Neonatal presentation of the disease involves lethargy, hypoglycemia accompanied by ketoacidosis, hyperglycinemia, hyperammonemia, vomiting, hypotonia,

hypothermia, respiratory distress, neutropenia, and thrombocytopenia; death can result. The infantile/non-B12-responsive subtype is the most common MMA phenotype; affected infants are normal at birth but subsequently develop lethargy, dehydration, hypotonia, hepatomegaly, and encephalopathy. An intermediate B12-responsive MMA phenotype typically presents later in infancy or early childhood; affected individuals typically manifest hypotonia, failure to thrive, developmental delay, and may exhibit protein aversion. Finally, the atypical and adult-onset forms of MMA are associated with only a mild methylmalonic acidemia and aciduria; these individuals may or may not develop other symptoms. Long-term complications of MMA include neurological damage, manifesting as variable degrees of developmental delay, dystonia choreoathetosis, and paresis; growth failure; tubulointerstitial nephritis, and progressive renal failure; "metabolic stroke" with basal ganglia involvement; optic nerve atrophy; and functional immunodeficiency.

Some states include methylmalonic acidemia testing among their expanded panel of newborn screening tests; if so, elevated levels of methylmalonyl-CoA are seen, and follow-up testing is required. Analysis of plasma and urine organic acids and serum amino acids helps make the diagnosis of MMA; methylmalonic acidemia and aciduria, hyperammonemia, hyperglycinemia, elevations of 3-OH-propionate, 2-methylcitrate and tiglylglycine in urine, and elevated plasma concentrations of propionyl-carnitine (C3) accompanied by variable elevations of C4-dicarboxylic acid or methylmalonic/succinyl-carnitine (C4DC) in the plasma. Neutropenia, thrombocytopenia, and anemia may also be present. Establishing the specific enzyme subtype of MMA involves 14C propionate incorporation assays, B12 responsiveness studies, cobalamin distribution assays, and complementation analysis; alternatively, demonstrating the presence of biallelic mutations in *MUT, MMAA, MMAB, MCEE, and MMAD-HC* suffices to establish the diagnosis. Acute metabolic decompensation episodes are treated as discussed for propionic acidemia above. Long-term treatment of MMA involves avoidance of fasting, implementing a protein-restricted diet (0.5–1.5 g/kg/day), and supplementation with L-carnitine and with cobalamin. If cobalamin is not helpful, intake of threonine, methionine, valine, and isoleucine should be restricted.

Tyrosinemia

There are several different enzymes involved in the tyrosine catabolic pathway – tyrosine transaminase, *p*-hydroxyphenylpyruvate dioxygenase, homogentisate dioxygenase, maleylacetoacetate *cis-trans*-isomerase, and 4-fumarylacetotacetate hydrolase. Defects in 4-fumarylacetotacetase, tyrosine transaminase, and *p*-hydroxyphenylpyruvate each result in tyrosinemia types I, II, and III, respectively; defects in homogentisate dioxygenase results in alcaptonuria. However, of all of these aforementioned inborn errors of amino acid metabolism, only tyrosinemia I presents with hypoglycemia; the others are not discussed further here.

Hereditary tyrosinemia I results from an autosomal recessive mutation, deletion, or insertion leading to loss of function of the enzyme 4-fumarylacetotacetate hydrolase (FAH, encoded by the *FAH* gene on chromosome 15q25.1) and thus failure to complete the last step of tyrosine degradation, namely, cleavage of 4-fumarylacetoacetate into fumarate and acetoacetate, leading to an accumulation of 4-fumarylacetoacetate.[164] This disorder is very rare, with a worldwide incidence of ~1/100,000, but is more common in Quebec, where the incidence is ~1/16,000. The 4-fumarylacetoacetate accumulates in hepatocytes and proximal renal tubular cells, leading to oxidative damage, dysfunctional gene expression, and cell death. Other tyrosine byproducts also accumulate systemically, including tyrosine itself, which is hepatotoxic, nephrotoxic, dermatotoxic, and neurotoxic, leading to neurodevelopmental sequelae. This disorder typically manifests in infancy in more severe cases through to adolescence in less severe cases. In the absence of fumarate, tyrosine cannot enter either the Krebs cycle or the gluconeogenic pathway, leading to hypoglycemia. The clinical presentation involves failure to thrive, hepatomegaly, and progressive hepatorenal dysfunction. Other biochemical markers include elevated alphafetoprotein, conjugated hyperbilirubinemia, and coagulation pathway abnormalities. Liver failure and cirrhosis can result. A renal Fanconi syndrome is also seen, with renal tubular acidosis, hypophosphatemia, and aminoaciduria. More chronic tyrosinemia may manifest as hypophosphatemic rickets. Cardiomyopathy, dermopathy, and neuropathy have also been reported, as has hepatocellular carcinoma. Elevations of succinylacetone in serum and urine are typically seen. Diagnosis is made by sequence analysis of *FAH*. The primary treatment for tyrosinemia type I is nitisinone, which inhibits the second step of tyrosine catabolism (the conversion of 4-OH-phenylpyruvate to homogentisic acid by 4-hydroxyphenylpyruvate dioxygenase), preventing the accumulation of fumarylacetoacetate. Dietary restriction of tyrosine and phenylalanine is also implemented. Liver transplantation is curative in those in whom nitisinone fails.

Fatty Acid Oxidation Disorders

During fasting, lipolysis and fatty acid oxidation with ketogenesis become important sources of energy, both by providing glycerol as a substrate for gluconeogenesis and by an alternative fuel, which can be utilized by the brain for energy (along with lactic acid). Mitochondrial fatty acid oxidation provides up to 80% of the fuel for cardiac and hepatic function, and is an important source of fuel in times of increased metabolic demands (e.g., illness, after surgery, or during intense exercise) in addition to fasting.[165] Initially, once hepatic glycogen stores are depleted, gluconeogenesis is the predominant source of energy, but after the main muscle mass is depleted, fatty acid oxidation becomes the primary source of fuel.[166] During lipolysis, a cytosolic process, triacylglycerols (triglycerides) are converted into free fatty acids and glycerol. Glycerol enters the gluconeogenic pathway. The nonesterified free fatty acids are converted into acyl-CoA esters by acyl-CoA synthetases, which are ATP dependent, and must then enter the mitochondria to undergo β-oxidation, a process in which acyl-CoAs are gradually converted by the actions of several enzymes into acetyl-CoA, which can then either enter the Krebs cycle or enter the ketogenic pathway to generate acetoacetate and beta-hydroxybutyrate. Medium- and short-chain fatty acids can enter the mitochondrial membrane unassisted; long-chain fatty acids (acyl-CoAs with hydrocarbon tails longer than 12 carbons) must enter via the carnitine shuttle as follows:

- Carnitine palmitoyltransferase-I (CPT-I), already present in the outer mitochondrial membrane, catalyzes the formation of long-chain acylcarnitine from carnitine and an acyl-CoA. This is the rate-limiting step of the carnitine shuttle process.
- Acylcarnitine enters the inner mitochondrial membrane via the carnitine:acylcarnitine translocase (CACT).
- Inside the mitochondrial matrix, carnitine palmitoyltransferase-II (CPT-II) catalyzes the reverse reaction compared to CPT-I, reforming the acyl-CoA, freeing the carnitine to be shuttled back across the inner mitochondrial membrane by CACT, and combining with another acyl-CoA to start the cycle over again.

Once inside the mitochondrial matrix, the acyl-CoAs undergo a β-oxidation cycle to yield an acyl-carbon chain that is two carbons shorter and a molecule of acetyl-CoA via a four-stage process involving dehydrogenation, hydration, oxidation, and cleavage by a thiolase; each of the enzymes involved in the process only acts upon fatty acids possessing chain length (very long-chain fatty acids, long-chain fatty acids, medium-chain fatty acids, and short-chain fatty acids – e.g., medium-chain acyl-CoA dehydrogenase only dehydrogenates acyl-CoAs possessing 6–10 carbons). The acetyl-CoA then undergoes ketogenesis to yield acetoacetate and beta-hydroxybutyrate. The processes of β-oxidation and ketogenesis are summarized in Fig. 3.3a,b.[1,165]

Defects in β-oxidation and ketogenesis are all inherited in an autosomal recessive manner. The disorders are generally quiescent in the fed state, but fasting may

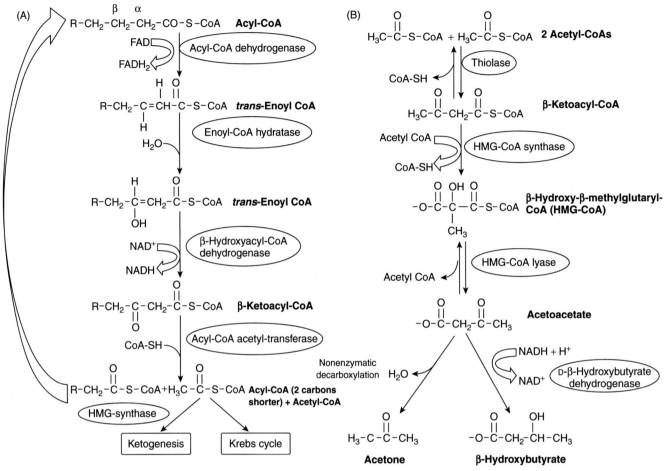

FIGURE 3.3   (A) Fatty acid oxidation. (B) Ketogenesis.

provoke a life-threatening exacerbation in which hypoglycemia is associated with deficient ketone body formation. The clinical presentation varies with the specific disorder, but can include muscular, cardiac, and hepatic manifestations:

- Hepatic: Hypoglycemia accompanied by coma. Medium-chain hydroxyacyl-CoA dehydrogenase deficiency is the classical disorder that has this manifestation. These episodes of hypoglycemia are not accompanied by ketosis, individuals are usually not acidemic, and biochemical abnormalities include elevated urea, ammonia, and uric acid, as well as transaminases. Hepatic steatosis may be seen. If the catabolic state is not treated quickly, death can result.
- Cardiac (usually seen in the long-chain fatty acid oxidation defects): Acute or chronic dilated or hypertrophic cardiomyopathy, which can progress to heart failure.
- Muscular: Rhabdomyolysis.

Specific defects in β-oxidation and ketogenesis, which cause hypoglycemia, are discussed in the subsequent section.

A note regarding diagnosis: fasting individuals with disorders of fatty acid oxidation can precipitate a metabolic decompensation, which can have long-term adverse sequelae or even be fatal. All individuals who are scheduled to undergo a diagnostic fast should always have an acylcarnitine profile drawn prior to screening for fatty acid oxidation disorders; this can be drawn in a fed state. If the acylcarnitine profile is abnormal, under no circumstances should a diagnostic fast be undertaken.

Long-Chain Fatty Acid Oxidation Defects

This group of disorders contains several different subgroups of disorders in different enzymes. These defects can be divided into four categories:[167]

- Carnitine transporter defect (organic carnitine transporter 2, or OCTN2).
- Carnitine shuttle defects – mutations in the genes encoding CPT-I, CACT, or CPT-II, which together enable entry of very-long-chain and long-chain fatty acids into the mitochondrial matrix.
- Defects in one of the long-chain fatty acid-specific isoforms of the enzymes involved in β-oxidation of

long-chain fatty acids: very long-chain hydroxyacyl dehydrogenase (VLCAD) deficiency, long-chain 3-hydroxyacyl-CoA dehydrogenase (LCHAD) deficiency, long-chain 3-ketoacyl-CoA thiolase (LKAT) deficiency, acyl-CoA dehydrogenase 9 (ACAD9) deficiency, and mitochondrial trifunctional protein (TFP) deficiency.
- Multiple acyl-CoA dehydrogenase deficiency.

Organic Carnitine Transporter 2 (OCTN2) Deficiency

Organic carnitine transporter 2 (OCTN2) is an enzyme encoded by the 10-exon, 26-kb *SLC22A5* gene[168] located on chromosome 5q31.1. OCTN2 is an integral plasma membrane protein that is a sodium-dependent high-affinity carnitine transporter involved in cellular uptake of carnitine. Deficiency is due to homozygous or compound heterozygous *SLC22A5* mutations, deletions, or duplications; 49 have been reported to cause primary systemic carnitine deficiency (CDSP).[81] This condition is typically rare, with incidence ranging from 1/40,000 in Japan, 1/50,000 in the United States to 1/120,000 in Australia; however, the incidence is remarkably high in the Faroe Islands, namely 1/300 live births.[169] There are several different phenotypic presentations:[168]

- Infantile metabolic (hepatic) presentation: Affected individuals typically present between a few months and 2 years of age with episodes of metabolic decompensation triggered by fasting or illness (e.g., gastroenteritis or upper respiratory infection) characterized clinically by poor feeding, irritability, lethargy, and biochemically by hypoketotic hypoglycemia, transaminase elevation, and hyperammonemia. Hepatomegaly is frequently seen. Treatment during these episodes is with intravenous dextrose at high glucose infusion rates to disrupt the catabolic state; if treatment is not initiated promptly, coma and even death may result.
- Childhood myopathic presentation: These individuals typically present between 2 and 4 years of age with hypotonia, skeletal muscle weakness, and dilated cardiomyopathy. Serum CK may be elevated. This is believed to be most likely due to energy deficiency, can be fatal due to cardiac failure if not diagnosed, and is reversible with treatment. Individuals with the infantile form may also develop the myopathic features later in life.
- Adulthood presentation: These are people who are either asymptomatic or who complain of easy fatigability. Presentation during pregnancy may also occur. Affected individuals are sometimes diagnosed after their children present with more severe forms.

Carnitine deficiency is first detected by low plasma carnitine levels, either on filter paper with the newborn screen or in blood samples. Acylcarnitine profile is often unrevealing of anomalies because all acylcarnitine levels may be low. Urine organic acid analysis may reveal dicarboxylic aciduria, which is seen in many FA oxidation disorders during metabolic decompensation and is thus nonspecific. Secondary causes of carnitine deficiency (e.g., prematurity, malnutrition, renal Fanconi syndrome, certain antiepileptic medications) must be excluded. Diagnosis is confirmed by sequence analysis revealing biallelic pathogenic variants of *SLC22A5* or by demonstrating decreased carnitine transport in fibroblasts. Treatment of episodes of metabolic decompensation is with infusions of dextrose at high concentrations as above. Chronic treatment entails supplementation with oral levocarnitine (L-carnitine) at doses of 100–400 mg/kg/day to maintain normal plasma carnitine concentrations and prevent sequelae, especially the cardiomyopathy (can be successful if supplementation is started early), avoidance of fasting for longer than age-appropriate duration, and consultation with appropriate specialists. Echocardiograms should be performed annually during childhood and periodically during adulthood. Carnitine levels should be monitored every 4 months during infancy and early childhood, biannually in older children, and annually in adulthood, with special attention paid in pregnancy. Prognosis is good if treatment is initiated early. Siblings and parents of affected probands should have screening carnitine levels measured.[170]

Carnitine Shuttle Defects

This is a group of disorders characterized by deficiencies in the hepatic form of carnitine palmitoyl-CoA transferase-I (CPT1A), CPT-2, and carnitine acylcarnitine translocase (CACT) deficiency. These genes are respectively encoded by the *CPT1A* gene on chromosome 11q13.3, *CPT2* gene on chromosome 1q32.3, and the *SLC25A20* gene on chromosome 3p21.31.[81] The function of the carnitine shuttle was described earlier. Deficiencies are due to homozygous or compound heterozygous mutations, deletions, or insertions in the three genes above, resulting in autosomal recessive disorders. CPT1A deficiency is very rare,[171] as are CPT-I deficiency (although over 300 cases worldwide have been described)[172] and CACT deficiency.[173] Clinically, CPT1A deficiency has three phenotypes – hepatic encephalopathy, which typically presents in childhood following illness or fasting with episodic hypoketotic hypoglycemia accompanied by elevated transaminases and ammonia, *elevated* serum carnitine (specifically free carnitine and the ratio of free carnitine to very long-chain fatty acids C16/18), and possible sudden-onset hepatic failure. Individuals may function normally between episodes; however, if not treated promptly, long-term damage may result. The second phenotype, adult-onset myopathy, was reported in one adult of Inuit descent, and the third, acute steatohepatitis of pregnancy, is seen in women carrying fetuses homozygous for *CPT1A* mutations.[171] CPT-II deficiency also has three phenotypic presentations: lethal neonatal form, severe infantile hepatocardiomuscular form, and myopathic form.[172] The most severe is the lethal neonatal form, characterized by hypoketotic hypoglycemia and hepatic failure, hepatic calcifications, cystic dysplastic kidney, respiratory distress, and cardiomyopathy sometimes accompanied by arrhythmias. It is characterized biochemically by low plasma total and free carnitine levels and increased concentrations of very long chain fatty acids and long-chain acylcarnitines. Death typically occurs in days to weeks. The hepatocardiomuscular form typically presents within the first year of life and is characterized by episodes of hypoketotic hypoglycemia possibly accompanied by seizures, abdominal pain and headaches, hepatic failure, peripheral myopathy, and cardiomyopathy. Fatal arrhythmias may result. The myopathic presentation has variable onset (ranging from the first few years to the sixth decade of life), and is characterized by recurrent myalgias precipitated by prolonged exercise (especially after fasting), stress/illness or cold exposure and accompanied by weakness, myoglobinuria, and CK elevations; affected individuals are typically asymptomatic clinically and biochemically between episodes. Very long-chain fatty acids, specifically C16 and 18:1, are typically elevated in CPT-II deficiency. CPT-II deficiency is the most common lipid metabolism disorder to affect skeletal muscle as well as the most common etiology of hereditary myoglobinuria. CACT deficiency typically presents in infancy with hypoketotic hypoglycemia after fasting, hepatic dysfunction, skeletal myopathies, cardiomyopathy and possible arrhythmias, accompanied biochemically by elevated ammonia and transaminases, elevated CK, dicarboxylic aciduria, low free carnitine and abnormal acylcarnitine profile marked by elevated long-chain acylcarnitines[173] – very similar to the biochemical presentation of CPT-II deficiency. Typically, progressive deterioration and death in infancy is seen; however, milder presentations have occasionally been reported.

The presence of a carnitine shuttle defect should be suspected when one of the above clinical presentations is seen; total and free carnitine levels (elevated in CPT-I deficiency, low in CPT-II and CACT deficiencies), acylcarnitine profiles, and urine organic acids should be measured. Definitive diagnosis is made based upon either sequence analysis showing biallelic pathogenic variants in *CPT1A*, *CPT2*, or *SLC25A20*, or by demonstrating low activity of the enzyme in question in cultured fibroblasts or lymphocytes. Treatment of metabolic decompensation episodes entails intravenous dextrose at high infusion rates to halt lipolysis and fatty acid oxidation. Chronically, treatment entails avoidance of fasting (infants and children may require tube feeds and/or uncooked cornstarch at night), sometimes avoidance of prolonged exercise, adhering to a low-fat high-carbohydrate diet, L-carnitine and medium-chain triglycerides (MCT)-oil supplementation, avoidance of certain hepatotoxic or potentially triggering agents (e.g., valproic acid, salicylates, ibuprofen, and others), and treatment of any complications (e.g., hydration for rhabdomyolysis). Carrier screening should be undertaken for siblings of affected probands.

Deficiencies of Mitochondrial Long-Chain Fatty Acid β-Oxidation Enzymes

The long-chain fatty acid oxidation disorders that cause disorders featuring hypoglycemia are very long-chain acyl-CoA dehydrogenase (VLCAD) deficiency; mitochondrial trifunctional protein (MTP) deficiencies, of which long-chain 3-hydroxyacyl-CoA dehydrogenase (LCHAD) deficiency is the most common; long-chain 3-ketoacyl-CoA thiolase (LKAT) deficiency; acyl-CoA dehydrogenase 9 (ACAD9) deficiency; and mitochondrial trifunctional protein (TFP) deficiency. These are briefly discussed further in the subsequent section.

Very Long-Chain Acyl-CoA Dehydrogenase (VLCAD) Deficiency VLCAD is an enzyme encoded by the *ACADVL* gene on chromosome 17p13.1.[174] Deficiency is due to homozygous or compound heterozygous *ACADVL* mutations. More severe presentations are due to null mutations; less severe mutations with residual enzyme activity are associated with a milder phenotype; affected individuals may even be asymptomatic. Incidence in the United States is suspected to be 1/30,000 live births.[174] The clinical presentation of VLCAD deficiency clinical presentation is somewhat variable:

- Severe early-onset cardiac and multiorgan failure VLCAD deficiency: Manifests within the first few months of life with hypertrophic or dilated cardiomyopathy, arrhythmias, pericardial effusions, and hypotonia and hepatomegaly along with intermittent episodes of hypoglycemia. The arrhythmias may be life-threatening – atrioventricular block, ventricular tachycardia, and fibrillation have all been reported. Cardiomyopathies may be reversible if treated early.
- Hepatic or hypoketotic hypoglycemic VLCAD deficiency: Typically presents similarly to MCAD deficiency, with episodes of hypoketotic hypoglycemia; cardiomyopathy is typically absent in these individuals.
- Late-onset episodic myopathic VLCAD deficiency: Typically presents with intermittent muscular pain/cramps, exercise intolerance, and at the extreme, rhabdomyolysis. Hypoglycemia is typically not seen. This may be the most common phenotype.

Elevations of 12- and 14-hydrocarbon fatty acids (C14:1, C14:2, C14, and C12:1) on acylcarnitine profile on tandem mass spectrometry of plasma or on dried blood spot on filter paper, especially if collected during periods of fasting or metabolic stress, should raise suspicion for VLCAD deficiency. Diagnosis rests on molecular genetic testing for *ACADVL* mutations/deletions, though functional analysis of VLCAD activity in lymphocytes or cultured fibroblasts can also be performed. *Fasting individuals with suspected VLCAD deficiency should not be done.* Treatment entails placing affected infants on a low-fat formula supplemented with MCT for additional calories. Fasting, dehydration, and a high-fat diet should be avoided. Acute rhabdomyolysis should be treated with fluid and urine alkalinization. Acute metabolic decompensation episodes should be treated with intravenous dextrose at high rates to switch the individual from a catabolic to an anabolic state. Arrhythmias, should they occur, should be treated appropriately; cardiac dysfunction should be treated with intensive supportive care and dietary modification, and is often reversible. First-degree relatives of affected probands should be screened.

Mitochondrial Trifunctional Protein (MTP) Deficiency

MTP catalyzes the last three steps of mitochondrial β-oxidation; LCHAD, long-chain enoyl-CoA hydratase, and long-chain thiolase are all part of the MTP. The MTP is a membrane-bound complex composed of four α- and four β subunits. The α subunit, encoded by the *HADHA* gene on chromosome 2p23.3, catalyzes the LCHAD and enoyl-CoA hydratase activities. The β subunit, encoded by the *HADHB* gene that is also found on chromosome 2p23.3, catalyzes the thiolase activity. Compound heterozygous or homozygous mutations, deletions, or insertions in either *HADHA* or *HADHB* result in the very rare autosomal recessive complete mitochondrial trifunctional protein deficiency, in which all three enzymatic processes do not function; more common is LCHAD deficiency, due to a less-encompassing *HADHA* missense mutation. Both deficiencies are quite rare – the reported incidence of LCHAD deficiency in the United States ranges from 1/75,000[175] to 1/250,000,[176] with a higher incidence in Finland, and complete MTP deficiency is even rarer, such that the true incidence is unknown. Clinically, LCHAD deficiency manifests early, typically in the first year of life, with episodes of hypoketotic hypoglycemia accompanied by variable other manifestations, including poor feeding, vomiting, lethargy, hypotonia, hepatic failure, skeletal myopathy and possibly rhabdomyolysis, early-onset cardiomyopathy, sensory polyneuropathy, and sudden death in infancy.[177] MTP deficiency has three clinical phenotypes: severe lethal neonatal onset presenting as sudden-infant death syndrome; infantile hepatic onset presenting with hypoketotic hypoglycemia accompanied by transaminase elevation, lactic acidosis and CK elevation in individuals who may develop cardiomyopathy and arrhythmias; and an adolescent-onset skeletal myopathy.[178] Some individuals with MTP deficiency manifest with a slowly progressive myopathy with recurrent rhabdomyolysis and sensorimotor axonal neuropathy as well; they may survive into adulthood. If diagnosis is made prior to onset of hypoglycemic episodes (which are, as typical for other fatty acid oxidation disorders, triggered by fasting or illness), such as by newborn screening, then neurological damage may be avoided, though skeletal myopathies and retinopathy may progress. Of note, mothers of babies with LCHAD or MTP deficiency may be more at risk for acute fatty liver of infancy (AFLP) or for HELLP syndrome.

The diagnosis of LCHAD/MTP deficiency may be suspected by elevation of long-chain fatty acyl-CoAs (C14, C16, and C18), especially 16-carbon acyl-CoA (C16), along with elevated urinary dicarboxylic acids. Definitive diagnosis is made through molecular genetic studies of *HADHA* and *HADHB* or via demonstration of decreased enzyme activity in cultured fibroblasts. Treatment is similar to the treatment of VLCAD deficiency as discussed earlier, although this may not prevent the progression of myopathies.

Multiple Acyl-CoA Dehydrogenase (MAD) Deficiency

MAD deficiency is caused by a mutation in one of three genes involved in electron transfer in the mitochondrial respiratory chain: *ETFA* (chromosome 15q24.2-q24.3), *ETFB* (chromosome 19q13.41), and *ETFDH* (chromosome 4q32.1), respectively responsible for glutaric acidemia IIA, IIB, and IIC. *ETFA* and *ETFB* encode the α and β subunits of the electron transfer flavoprotein (ETF), which *ETFDH* encodes the electron transfer flavoprotein dehydrogenase,[179] also known as ETF ubiquinone oxidoreductase.[167] These autosomal recessive disorders are also characterized by disruptions of amino acid and choline metabolism as well as of fatty acid metabolism. Precise incidence is unknown.

Clinical phenotypes for glutaric acidemia IIA–IIC are similar; while there are three main clinical phenotypes of MAD deficiency, distinguished not by which gene is impacted but by the degree with which the enzyme function is disrupted. There are three general clinical phenotypes of MAD deficiency: a neonatal-onset form with congenital anomalies (type I), a neonatal-onset form without congenital anomalies (type II), and a late-onset form (type III). The two neonatal-onset forms (collectively known as "MADD-severe") are characterized by hepatomegaly, hypotonia, and severe hypoketotic hypoglycemia accompanied by metabolic acidosis and excretion of large amounts of fatty acid- and amino acid-derived metabolites, often presenting within the first 24–48 h of life. Affected infants usually have dysplastic multicystic kidneys. Infants with type I may also have a variety of dysmorphisms, from rocker-bottom feet to hypertelorism, low-set ears, and dysplastic midface. MAD deficiency type I is typically fatal within the first week of life; infants with MAD deficiency type II typically survive only slightly longer, typically dying within the first few weeks of life. Late-onset type III MAD deficiency is highly phenotyically variable, manifests any time between infancy and adulthood, and is characterized by hepatomegaly and intermittent episodes of vomiting, lethargy, nonketotic hypoglycemia, and metabolic acidosis typically triggered by fasting or illness (catabolic stress) accompanied by an organic aciduria. A lipid storage myopathy typically occurs, manifesting as pain and muscle weakness. The prognosis for type III MAD deficiency is more favorable.

The diagnosis should be suspected in an individual with a clinical presentation resembling one of the three phenotypes above along with elevated acylcarnitines of all lengths C4:C18, elevated urine dicarboxylic acids, lactic acid, ethylmalonic acid, butyric acid, isobutyric acid, 2-methyl-butyric acid, and isovaleric acid. Definitive diagnosis is made by molecular genetic diagnosis of biallelic mutations in any of the three genes; prenatal diagnosis is also possible. As for all fatty acid oxidation disorders, supportive care with the hypoketotic hypoglycemic episodes is paramount along with avoidance of fasting, restriction of fat and protein intake, and high carbohydrate intake. The late-onset type III MAD deficiency may also respond to supplementation with riboflavin 400 mg/day, and possibly to supplementation with coQ10 as well.

Medium-Chain Fatty Acid Oxidation Defects

Theoretically, medium-chain fatty acid oxidation defects could encompass any of the medium chain-specific β-oxidation enzymes. However, in practice, the only one seen with any frequency and thus the only one discussed here is medium chain Acyl-CoA dehydrogenase deficiency.

Medium-Chain Acyl-CoA Dehydrogenase (MCAD) Deficiency

MCAD is an enzyme encoded by the gene *ACADM* on chromosome 1p31.1. MCAD deficiency is the most common fatty oxidation defect, and is due to either a homozygous or a compound heterozygous mutation I *ACADM*. The majority of affected individuals are either homozygous for a Lys304Glu mutation or have one Lys-304Glu mutation and another rare mutation.[180] Incidence varies widely worldwide; it is more common in the western populations, with incidence (drawn from population newborn screening reports) ranging from ~1/5,000 live births in northern Germany[181] to 1/15,700 in the United States,[182] and less common in east Asian populations, with prevalence as low as 1/700,000 in Taiwan.[183]

Clinically, MCAD deficiency causes deficiency insufficiency of β-oxidation, leading to increased rate of utilization of glucose as a fuel to meet energy demands. Sudden unexplained death may be the initial manifestation of MCAD deficiency if it is not caught by newborn screening programs; otherwise, presentation is with hypoketotic hypoglycemia presenting with vomiting, seizures, and possibly coma during periods of fasting or catabolic stress as described earlier, typically occurring by six years of age. Mortality rate from these hypoglycemic episodes in previously undiagnosed individuals can be as high as 16–25% (sudden death after a minor illness), and 20–25% of survivors sustained significant adverse neurocognitive and other sequelae.[180] Thus, many countries and all US states (and Puerto Rico) include this disorder in the newborn screening program. First-degree relatives of affected probands should be screened.

Elevated ammonia levels and elevated plasma medium chain fatty acids (C8:0, C10:0, and C10:1) should raise suspicion for MCAD deficiency; the newborn screen detects these elevations. Plasma acylcarnitine profile is characterized by elevations of C6–C10 (6–10 hydrocarbon chain fatty acids). Diagnosis of MCAD deficiency entails sequencing analysis of the *ACADM* gene and/or measurement of MCAD enzyme activity in fibroblasts or other tissues. *Under no circumstances should an individual with suspected MCAD deficiency undergo a diagnostic fast, lest a metabolic decompensation be precipitated.* Treatment entails avoidance of fasting – frequent feedings in infants – and a low-fat diet in toddlers (<30% of energy from fat). Toddlers may require two g/kg uncooked cornstarch at bedtime to ensure a sufficiency supply of glucose overnight. Infant formulas high in medium-chain fatty acids should be avoided.

Short-Chain Fatty Acid Oxidation Defects

Again, defects in any one of the enzymes involved in β-oxidation of short-chain fatty acids could result in a hypoglycemic disorder. However, clinically, the main one that has been described is short-chain 3-hydroxyacyl-CoA dehydrogenase (SCHAD) deficiency, previously discussed in the "Hyperinsulinism" section.

Glycerol Metabolism Disorder

As discussed previously, lipolysis of triacylglycerols yields free fatty acids and glycerol. Glycerol enters the bloodstream and subsequently is taken up by either liver or kidneys. In these organs (primarily in the liver), glycerol is converted into glycerol-3-phosphate in a reaction catalyzed by glycerol kinase. Hepatic glycerol-3-phosphate is then oxidized into dihydroxyacetone phosphate (DHAP) in a reversible redox reaction catalyzed by glycerol-3-phosphate dehydrogenase; DHAP can then enter either the glycolytic or gluconeogenic pathways.

The main clinical disorder of glycerol metabolism that has been described is glycerol kinase deficiency. Glycerol kinase is encoded by the *GK* gene on chromosome Xp21.2; thus, in distinction to all of the fatty acid oxidation defects, it is an X-linked rather than autosomal disorder. Symptoms are due to the inability for glycerol to enter the gluconeogenic pathway; GK deficiency is a very rare disorder and has the following phenotypic presentations:[177,184]

1. Complex GK deficiency: This infantile-onset form involves deletion not only of the *GK* gene, but also of the gene causing Duchenne muscular dystrophy (DMD) and/or the gene causing hypoplasia adrenal congenita (AHC). Thus, it has a very severe presentation, either involving progressive muscular dystrophy or glucocorticoid and mineralocorticoid insufficiency. This form is characterized by severe developmental delay.
2. Isolated GK deficiency: This form involves isolated GK deficiency, without concomitant DMD or AHC. It is further subdivided into two phenotypes:
   a. Juvenile-onset GK deficiency: Affected children may present with severe hypoglycemic episodes accompanied by profound metabolic acidosis, vomiting, lethargy, and hypotonia, episodes triggered by the same catabolic stressors as described earlier – fasting, illness, or prolonged exercise. As fasting tolerance increases with age due to increased hepatic gluconeogenic capacity from nonglycerol precursors, symptoms often disappear.
   b. Adult-onset GK deficiency: Individuals with this form of GK deficiency are asymptomatic; this condition is often detected incidentally.

All forms of GK deficiency are characterized by pseudohypertriglyceridemia, since the elevated glycerol is detected by assays as triglycerides. Levels may be 40 times above the normal range.

Once the pseudohypertriglyceridemia is detected, definitive diagnosis is based upon detection of the GK mutation by sequence analysis. In individuals with the infantile and juvenile-onset forms, treatment is based upon avoidance of fasting combined with a low-fat, high-carbohydrate diet, and treatment of any metabolic exacerbations as discussed earlier with fatty acid oxidation disorders. Those with DMD or AHC are treated as per standard for those disorders. Individuals with adult-onset GK deficiency do not require treatment.

Disorders of Ketogenesis (Ketone Body Synthesis) and Ketolysis (Ketone Utilization)

The ketone bodies, acetoacetate (AcAc), 3-β-hydroxybutyrate (BOHB), and acetone, are the end-products of the pathways of mitochondrial fatty acid β-oxidation and ketogenesis; they can also be derived as metabolites of ketogenic amino acids (e.g., leucine). They are produced mainly in the liver. They provide a reservoir of glucose-sparing energy supply, especially during periods of prolonged starvation, when they can meet two-thirds of the brain's energy needs.[185] The ketogenic pathway is shown in Fig. 3.3b. The initial step of ketogenesis is the generation of acetoacetate from two acetyl-CoA molecules in a reaction catalyzed by mitochondrial acetoacetyl-CoA thiolase (MAT), also known as acetyl-CoA acetyltransferase 1, encoded by *ACAT1*, a 27-kb 12-exon gene on chromosome 11q22.3.[186] The second and rate-limiting step of ketogenesis is the generation of 3-hydroxy-3-methylglutaryl-coenzyme A (HMG-CoA) from acetoacetyl-CoA and acetyl-CoA. This reaction is catalyzed by HMG-CoA synthase, encoded by the *HMGCS2* gene (chromosome 1p12). Subsequent steps are the cleavage of HMG-CoA into acetyl-CoA and AcAc, catalyzed by HMG-CoA lyase (encoded by the *HMGCL* gene on chromosome 1p36.11), and the isomerization of AcAc into BOHB, catalyzed by mitochondrial 3-β-hydroxybutyrate dehydrogenase (*BDH1* gene, chromosome 3q29), an enzyme that requires phosphatidylcholine as an allosteric activator. Disorders of ketogenesis are caused by autosomal recessive inherited defects *HMGCS2* or *HMGCL*; *ACAT1* deficiency presents as a disorder of isoleucine metabolism and of ketone utilization and is discussed further in the subsequent section, and *BDH1* defects have not been reported. Defects in ketolysis involve defects in enzymes involved in ketone utilization. These disorders may present in the neonatal period in severe cases, or later in childhood in milder cases. Disorders of ketogenesis typically present with episodic hypoglycemia in times of catabolic stress, with or without hyperammonemia, often with vomiting and encephalopathy. Hepatomegaly may be seen. Disorders of ketolysis typically present as severe ketoacidosis and dehydration during times of catabolic stress.[185,187]

Disorders of Ketogenesis

Disorders of ketone body metabolism present either in the first few days of life or later in childhood, during an infection or some other metabolic stress. In defects of ketogenesis, decompensation leads to encephalopathy, with vomiting and a reduced level of consciousness, often accompanied by hepatomegaly. The biochemical features – hypoketotic hypoglycemia, with or without hyperammonemia – resemble those seen in fatty acid oxidation disorders. In defects of ketolysis, the clinical picture is dominated by severe ketoacidosis. This is often accompanied by decreased consciousness and dehydration. Treatment of disorders of *ketogenesis* involves avoidance of prolonged fasting, careful watchfulness with early hospitalization if necessary during other catabolic stressors (e.g., illness with fever), and infusions of intravenous dextrose at high rates during episodes of metabolic decompensation. Specifics of presentation of the various enzyme deficiencies are discussed in the subsequent section.

HMG-CoA Synthase Deficiency[185,188]

This is a very rare condition reported in ~100 individuals worldwide. Of note, the ability to conduct ketogenesis using leucine as a precursor is not impacted, since leucine enters the ketogenic pathway below this step. Affected individuals present, typically in infancy or childhood, with episodes of hypoketotic hypoglycemia, vomiting, encephalopathy, and hepatomegaly, resembling disorders of fatty acid oxidation; these episodes are usually precipitated by a catabolic stressor such as an intercurrent infection or prolonged fasting. During these metabolic crises (but not in between), elevations of urine dicarboxylic acid may be seen, along with elevations of serum free fatty acids, without ketosis. Acylcarnitine profiles are usually nonspecific, though elevations of acetylcarnitine (acylcarnitine C2) may be seen and should prompt suspicion of this disorder if present. Ammonia, CK, transaminase, and lactate levels are typically normal. If untreated, coma may result. Diagnosis is made via sequence analysis of *HMGCS2*. Early diagnosis is crucial – avoidance of fasting can help mitigate possible sequelae.

HMG-CoA Lyase Deficiency[185,189]

This is a very rare condition reported in ~100 individuals worldwide, mostly from Spain, Portugal, and Saudi Arabia. Degradation of leucine is also affected, unlike the case in HMG-CoA synthase deficiency. Affected individuals again present in infancy – sometimes even within the first week of life – or less commonly in early childhood. Clinical presentation is with episodic hypoketotic hypoglycemia, vomiting, lethargy, and hepatomegaly triggered by catabolic stress. During these episodes, metabolic acidosis is seen, along with dicarboxylic aciduria, hyperammonemia (sometimes), and because leucine degradation is impacted, elevated urine organic acid metabolites (3-hydroxy-3-methylglutarate, 3-hydroxyisovalerate, and 3-methylglutarate). Acylcarnitine profiles show elevation of acylcarnitines C5OH and C6DC. Surprisingly, although AcAc formation is impacted in this deficiency, AcAc levels in the urine may be normal, believed to be due to bacterial degradation of HMG-CoA. Diagnosis is confirmed by sequence analysis of *HMGCL*.

If untreated, episodes can result in seizures, apneas, coma, and death; repeated episodes have been reported to result in dilated cardiomyopathy, retinitis pigmentosa, hearing loss, pancreatitis, and developmental delay.

White matter abnormalities have been found in individuals with repeated episodes.[185] Treatment of episodes of metabolic decompensation is as described for HMG-CoA synthase deficiency above. Long-term treatment involves avoidance of prolonged fasting, a carbohydrate-rich diet with leucine restriction, and L-carnitine supplementation to avoid a secondary carnitine deficiency. If episodes are treated promptly and fasting is avoided, prognosis can be good.

Disorders of Ketolysis[185,190]

There are two enzymes involved in ketolysis: succinyl-CoA:3-ketoacid-CoA transferase, also known as acetoacetate succinyl-CoA transferase (SCOT, encoded by *OXCT1* on chromosome 5p13.1), which catalyzes the formation of acetoacetyl-CoA from acetoacetate, and β-ketothiolase, also known as mitochondrial acetoacetyl-CoA thiolase (MAT) and acetyl-CoA acetyltransferase 1, encoded by *ACAT1*, which catalyzes the cleavage of acetoacetyl-CoA into two acetyl-CoA molecules. Deficiencies of either enzyme result in a defect of ketone utilization; both disorders are autosomal recessive.

SCOT deficiency is a disorder in which extrahepatic tissues cannot utilize the AcAc and BOHB produced by the liver for energy. This is a very rare disorder – only ~20 cases have been reported. Affected individuals present with intermittent episodes of severe ketoacidosis with fasting; hyperglycemia is not seen (differentiating these patients from those with diabetic ketoacidosis), and hypoglycemia has occasionally been reported; thus, this disorder is on the differential diagnosis for ketotic hypoglycemia. Only AcAc and BOHB are elevated – acylcarnitine profile is normal. Postprandial ketosis is occasionally seen. Diagnosis entails sequence analysis of *OXCT1*. Treatment entails avoidance of prolonged fasting and of excessive dietary fat intake; if disease is recognized early, prognosis is usually good.

β-Ketothiolase (MAT) deficiency is a disorder of ketone utilization known as alpha-methylacetoacetic aciduria. Catabolism of isoleucine via ketolytic cleavage of 2-methylacetoacetyl-CoA is also impaired. This is also a very rare disorder – only ~60 known cases have been reported worldwide. Affected individuals typically present in infancy with episodes of intermittent severe ketoacidosis with the typical symptoms of emesis, dehydration, hypotonia, lethargy, dyspnea, seizures, and possible coma or even death if untreated. Both hyper- and hypoglycemia have been reported in this disorder; hyperammonemia may also be present. Suspicion is more easily raised in this disorder than in the case of SCOT deficiency, however, because levels of isoleucine metabolites are also raised. Urine organic acid analysis shows elevation of 2-methyl-3-hydroxybutyrate as the most common abnormality in MAT deficiency, often in combination with tiglylglycine elevation. 2-Ethylhydracrylic acid and 2-methylacetoacetate levels may also be elevated. Acylcarnitine profile may show increased concentrations of C5:1 and C5:OH (this may be seen on newborn screens). Because of allelic heterogeneity of *ACAT1*, sequence analysis may not yield a definitive diagnosis; enzyme activity testing in cultured fibroblasts may also be useful in confirming the diagnosis. Again, treatment entails avoidance of excessive dietary fat intake, of prolonged fasting, and of prompt treatment of ketoacidosis episodes to prevent metabolic decompensation and sequelae thereof. If diagnosis is made early and episodes are treated promptly, prognosis is good.

Hypoglycemia Due to Counterregulatory Hormone Deficiency

The counterregulatory hormones – glucagon, epinephrine, cortisol, and growth hormone – act to stimulate key endogenous glucose production pathways, as discussed in the Introduction and outlined in Fig. 3.1. Thus, deficiency of either the counterregulatory hormones or the factors that control these hormones' production and secretion (e.g., ACTH deficiency) may also predispose to hypoglycemia. These deficiencies may be single or may be seen in combination, as in hypopituitarism. Genetic defects leading to counterregulatory hormone deficiency are discussed individually in the subsequent sections.

Congenital Pituitary Hormone Deficiencies

The anterior pituitary gland contains five different populations of cells, which produce six different hormones: somatotrophs, which produce GH; gonadotrophs, which produce luteinizing hormone (LH), and follicle stimulating hormone (FSH), which acts on the gonads to produce sex steroids; corticotrophs, which produce adrenocorticotropic hormone (ACTH), which acts on the adrenal glands to stimulate glucocorticoids and sex steroid production; thyrotrophs, which produce thyroid stimulating hormone (TSH), which act on the thyroid gland to stimulate thyroid hormone production; and lactotrophs, which produce prolactin. The anterior pituitary hormones that are involved in counterregulation and glucose homeostasis are GH, which itself is one of the counteregulatory hormones, and ACTH, which as aforementioned stimulates cortisol production. Hypoglycemia in the neonatal period is a common presenting feature of deficiencies of GH or ACTH in isolation or of multiple pituitary hormones. A large number of pituitary-specific transcription factors are necessary to establish the developing anterior pituitary gland's initial structure and subsequently to allow the differentiation of the five specialized, hormone-secreting cells: somatotropes (GH), gonadotropes (FSH and LH), lactotropes (prolactin), corticotropes (ACTH), and thyrotropes (TSH). Many of these genes are homeobox

domain-containing transcription factors. A defect in one of the genes involved the anterior pituitary development cascade can lead to various combinations of pituitary hormone deficiencies, either in isolation or may occur as part of a wider syndrome, such as septooptic dysplasia. Genes involved later in the pituitary cascade manifest as multiple or single pituitary hormone deficiencies but not in deficiencies of all pituitary hormones. Details of the genetics of inheritable pituitary hormone deficiencies are reviewed in Chapter 8.

Disorders of the Adrenal Gland

The adrenal cortex generates three steroid hormones from cholesterol: aldosterone, the mineralocorticoid hormone produced in the zona glomerulosa; cortisol, the glucocorticoid hormone produced in the zona fasciculate; and androstenedione, the androgen produced in the zona reticularis. A number of enzymes are involved in this process, as well as a large number of intermediates; a detailed discussion of adrenal steroidogenesis is beyond the scope of this chapter. Cortisol, as discussed earlier, is one of the counterregulatory hormones that stimulates gluconeogenesis and lipolysis. A number of genetic disorders can disrupt adrenal steroidogenesis in general and glucocorticoid production in particular, leading to primary adrenal insufficiency (as opposed to the secondary adrenal insufficiency discussed in the prior section, which resulted from ACTH deficiency rather than cortisol deficiency). Some of the disorders manifest in the neonatal period in infancy, while others may manifest later in childhood or even in adulthood; some result in failure of the adrenal gland to develop properly and/or in a predisposition to adrenal cortical destruction; other disorders are the result of lack of functioning of one of the critical enzymes, leading to failure to make the end hormones and accumulation of precursors. Only the disorders that result in sufficient glucocorticoid deficiency, which result in hypoglycemia, are discussed here. Treatment of primary adrenal insufficiency and/or isolated glucocorticoid deficiency entails replacement of cortisol with synthetic equivalents (most commonly hydrocortisone, less commonly prednisone or other synthetic glucocorticoids) with physiologic replacement dosing at baseline and increased "stress dosing" in times of illness; the exception is congenital adrenal hyperplasia, in which somewhat supraphysiologic doses of glucocorticoids are typically given to suppress ACTH secretion and consequent adrenal androgen production. If mineralocorticoid deficiency coexists, replacement with synthetic mineralocorticoids (typically fludrocortisone) is also undertaken, and salt supplementation may be required. Specifics of genetic disorders of the adrenal glands are discussed in Chapter 16.

Other Counterregulatory Hormone Deficiency Disorders

Dopamine beta-hydroxylase (DβH) deficiency is a very rare autosomal recessive disorder resulting from a mutation in the *DBH* gene (chromosome 9q34.2), encoding the dopamine beta-hydroxylase protein, a key enzyme in the norepinephrine synthesis pathway. The prevalence is unknown, as only 20 cases have been reported. As epinephrine, a key counterregulatory hormone, is produced from norepinephrine, this disorder results in primary autonomic nervous system failure, elevated dopamine, and absence of norepinephrine and epinephrine and thus of epinephrine's counterregulatory effects.[191] Clinically, affected individuals often manifest in the perinatal period with hypoglycemia, hypotension, hypothermia, and hypotonia. Later on, symptoms include exercise intolerance (due to autonomic maladaptation – exercise results in hypotension and syncope), orthostatic hypotension, ptosis, nasal congestion, and sexual disorders. Blood urea nitrogen elevation may be seen. Treatment entails administration of 100–500 mg/day of L-threo-dihydroxyphenylserine, a norepinephrine precursor. With treatment, normal norepinephrine levels can be restored and can alleviate the hypotension. Surgical correction of ptosis may be required. Screening for renal dysfunction regularly is important. Little is known about the long-term prognosis of this disorder.

Miscellaneous Disorders Resulting in Hypoglycemia

A number of other syndromes exist, which can result in hypoglycemia by a variety of mechanisms. These are briefly discussed in the subsequent section.

Beckwith–Wiedemann Syndrome (BWS)

Beckwith–Wiedemann syndrome is a pediatric overgrowth syndrome caused by either mutation or deletion of imprinted genes within the chromosome 11p15.5 region. Specific genes in this region that have been found to cause BWS include *CDKN1* (chromosome 11p15.4), *KCNQ1OT1* (chromosome 11p15.5), *H19* (chromosome 11p15.5), and *LIT1* (chromosome 11p15.5). In addition, mutations in the *NSD1* gene (5q35.2-35.3), which is mutated in the Sotos syndrome, have been reported in two cases of patients who had clinically been diagnosed with BWS. The prevalence of BWS in North America has been reported to be as high as 1/10,000 live births.[192] BWS results from a number of different possible genetic pathways, including contiguous gene deletions, variable expression, and imprinting defects due to a defective or absent maternal allele, whether from epigenetic hypermethylation or from uniparental paternal isodisomy. Thus, the mode of inheritance is variable.[193] The classic clinical phenotype involves macroglossia, hemihypertrophy, and macrosomia, with increased risk of abdominal well defects (e.g., omphalocele) and visceromegaly (hepatomegaly, splenomegaly, and/or enlargement of other abdominal organs). Renal anomalies may also be seen, such as primary malformations, renal medullary dysplasia, nephrolithiasis and nephrocalcinosis. Other dysmorphisms may be seen, including helical posterior ear pits. Fetal adrenocortical cytomegaly may be seen. Affected individuals are at increased risk of embryonal malignancies, most commonly Wilms' tumor and hepatoblastoma. Individuals with BWS are at increased risk of mental retardation. The hypoglycemia associated with BWS is seen in 30–50% of affected infants. It is typically a hyperinsulinemic hypoglycemia, though the mechanism is unclear, and is usually responsive to diazoxide, with some rare exceptions where the reported cause was an abnormality in pancreatic β-cell K_{ATP} channels.[194] BWS diagnosis can be usually made through sequence analysis and methylation studies of the affected region on chromosome 11q15.4-5. Treatment is complex and involves multiple specialists; screening for complications, especially renal failure and embryonic tumors, is critical.

Russell–Silver Syndrome

Russell–Silver syndrome (RSS) is a pediatric "undergrowth" syndrome known to be caused by either epigenetic hypomethylation of the telomeric imprinting control region (ICR1) on chromosome 11p15, which impacts the H19 and IGF-2 genes (20–60% of cases) or by maternal uniparental isodisomy of chromosome 7, likely due to disruption of *GRB10* (growth factor receptor-bound protein 10, 7p11.2), which GRB10 exerts a growth-suppressive effect on through interacting with either the IGF-I receptor or the GH receptor.[195] Rarely, it is inherited in an autosomal dominant or recessive manner. Prevalence of RSS is estimated to be 1/100,000.[78] RSS is a clinically heterogeneous syndrome; common clinical features include intrauterine growth retardation, poor linear growth and short stature with sparing of head circumference, triangular facies with a broad forehead, fifth finger clinodactyly, and limb length discrepancy. Children with RSS are at increased risk of developing hypoglycemia, partially due to poor appetites and low fat stores, and partially attributable to GH deficiency, which is common in RSS.[78] Since RSS is genetically heterogeneous and an underlying etiology is not always identified, diagnosis is typically made based upon clinical features, though deletion/duplication analysis and/or methylation studies of 11p15 and deletion/duplication cytogenetic and uniparental isodisomy studies of 7p11.2-p12 may confirm the clinical diagnosis. The hypoglycemia is treated by avoidance of fasting, frequent meals with high content of complex carbohydrates, and in cases of GH deficiency, treatment with recombinant human GH.

Sotos Syndrome (Cerebral Gigantism)

Sotos syndrome is a pediatric overgrowth syndrome. Sotos syndrome 1 results from a mutation or deletion in the NSD1 gene (chromosome 5q35.2-35.3; 80–90% of cases). Sotos syndrome 2 results from a mutation in the NFIX gene (chromosome 9p13). Incidence is estimated at 1/14,000 live births.[196] Most cases are sporadic; when an inherited pattern is observed, it is typically an autosomal dominant pattern. Clinically, children present with rapid linear growth in the first few years of life, physical features reminiscent of acromegaly (prominent jaw, frontal bossing, large hands and feet, high-arched palate, coarse facial features, and others), and cognitive delay, motor delay, speech impairment, and hypotonia. Some affected individuals have significant macrocephaly. Adult height is not always excessive. There is an increased risk of congenital heart defects (primarily closure defects – ASD, VSD, PDA).[197] Transient neonatal hypoglycemia is sometimes seen in Sotos syndrome (<15% of affected individuals),[4] and this was recently reported to be a hyperinsulinemic hypoglycemia, usually transient and requiring only supplemental intravenous glucose and/or tube feeds; one affected individual required medical therapy, and his hypoglycemia proved diazoxide responsive.[198] Diagnosis is based upon a combination of possession of characteristic clinical features and molecular diagnostic testing of the NSD1 and NFIX genes. Treatment involves a team of multiple specialists to manage the many organ system manifestations (cardiac, renal, and others).

Congenital Disorders of Glycosylation (CDG)

Congenital disorders of N-linked glycosylation is a heterogeneous group of disorders of N-oligosaccharides (sugars lined together in a specific pattern, which are attached to lipids or proteins). These occur due to deficiencies in any of 42 different enzymes in the N-linked synthetic pathway. There are three subtypes of CDG that can cause hypoglycemia: the most common, CDG IA (due to mutation in phosphomannomutase2, or PMM2, chromosome 16p13.2), may have a prevalence as high as 1/20,000 individuals; the other two, CDG IB (due to mutations in phosphomannose isomerase or PMI gene, chromosome 15q24.1) and CDG IC (ALG6 gene, chromosome 1p31.3), have only been reported in 20 and 30 individuals worldwide, respectively.[192] The disorders have a heterogeneous clinical presentation, but most commonly present in infancy with severe developmental delay, hypotonia, protein-losing enteropathy, and hypoglycemia, and with a 20% rate of death in the first year due to liver failure, cardiomyopathy, or severe infections. That said, clinical presentation is highly variable, ranging from death in infancy to a mild phenotype in adults. Initial screening tests usually involve transferring isoform analysis to determine the number and/or composition of N-linked oligosaccharide residues linked to transferring; confirmation is obtained through sequence analysis of the aforementioned genes. The hypoglycemia is of the hyperinsulinemic type, and typically responds to diazoxide. Otherwise, treatment is complex, involving nutrition supplementation to maximize caloric intake (tube feeds are sometimes required). Hypothyroidism may be seen, and responds to treatment with levothyroxine. Other manifestations, such as gastroesophageal reflux, emesis, developmental delays, scoliosis, other orthopedic findings, and ocular findings, are treated as per clinical routine.

Laron Syndrome

Laron syndrome (LS) is an autosomal recessive disorder of GH resistance caused by a mutation, deletion, or insertion in the gene encoding the GH receptor (GHR, chromosome 5p13-p12). An LS-like phenotype is caused by a postreceptor defect (in the GH signaling cascade) due to a mutation in the STAT5B gene (chromosome 17q21.2), the acid-labile subunit (ALS), the IGF-1 gene, and the IGF-1 receptor. This is a very rare disorder except in certain groups in Ecuador and in Sephardic Jews of Moroccan descent. GH levels are typically normal or elevated; levels of insulin-like growth factor-1 and its binding proteins are typically very low. Affected individuals have characteristic faces (prominent forehead with frontal bossing, marcocephaly, frontotemporal hairline recession, hypoplastic nasal bridge, shallow orbits, small chin, and others) and dysmorphisms similar to those with severe GH deficiency; linear growth is very poor. Fasting hypoglycemia is often seen in infants due to the resistance to GH's counterregulatory activities; more frequent feedings may be required. Diagnosis involves a combination of clinical testing (elevated GH levels in a child with poor linear growth, failure of rise in IGF-1 on GH provocative testing) and molecular genetic testing of the genes mentioned earlier. Treatment of classic LS is with a synthetic IGF-1 analog, Increlex.[4,199]

Congenital Severe Insulin Resistance Syndromes

Biallelic missense or nonsense mutations in the insulin receptor (INSR gene, chromosome 19p13.2) result in very rare autosomal recessive disorders, Donohue syndrome (formerly called leprechaunism), and Rabson–Mendenhall syndrome. Both syndromes are extremely rare and are characterized by poor linear growth, severe acanthosis nigricans, fasting hypoglycemia (due to the very high insulin levels acting through the IGF-1 receptor) and postprandial severe hyperglycemia (in addition to extraordinarily elevated insulin levels), presenting in infancy. Donohue syndrome is the more severe of the two, and is characterized by lipoatrophy and underdevelopment of skeletal muscles as well as characteristic dysmorphisms in addition to the features mentioned earlier. Rabson–Mendenhall syndrome is a somewhat less severe syndrome also characterized by hypertrichosis, characteristic coarse features, predisposition to dental caries, nephrocalcinosis, pineal hypoplasia, and tissue overgrowth, as well as clitoromegaly or an enlarged penis in affected females and males, respectively. Beta-cell failure is inevitable and progression to diabetes mellitus is expected. Diagnosis relies upon molecular genetic sequencing of the *INSR* gene. Some patients respond to metformin or thiazolidinediones in the early course of the disease, with extraordinarily high insulin doses administered later as β-cell failure progresses; response to synthetic IGF-1 has also been reported to be good. However, prognosis remains poor, with death occurring within a few years with Donohue syndrome and often in the second decade of life in Rabson–Mendenhall syndrome.[4]

Primary Generalized Glucocorticoid Resistance

Primary generalized glucocorticoid resistance is due to mutations in the human glucocorticoid receptor gene (*HGR*, chromosome 5q31.3).[200] This is a very rare condition that is sometimes sporadic, sometimes inherited in an autosomal recessive manner and occasionally in an autosomal dominant manner (due to a dominant negative effect on the receptor). Target-organ resistance to glucocorticoids is typically partial; complete deficiency is seldom seen. Thus, the patients present with elevated levels of corticotropin releasing hormone, ACTH, and cortisol. As production of adrenal androgens and mineralocorticoids is also increased, affected individuals tend to present with symptoms of mineralocorticoid excess (hypertension, hyperkalemia, metabolic alkalosis) as well as androgen excess (virilization, precocious puberty). Symptoms of glucocorticoid deficiency are sometimes seen – chronic fatigue is one such sign. Hypoglycemia has been reported in one case. Diagnosis is through sequencing of the *hGR* gene. Treatment entails administering synthetic glucocorticoids mimics without mineralocorticoid activity, such as dexamethasone.

Timothy Syndrome

Timothy syndrome is an autosomal dominant disorder caused by mutations on exons 8 or 8A of the alpha subunit of the CaV1.2 (an L-type calcium channel encoded by the *CACNA1C* gene, chromosome 12p13.33). These mutations cause delayed channel closure and consequent increased excitability. Prominent signs and symptoms include syndactyly, hyperextensible joints, prolonged QT interval, and other arrhythmias such as atrial fibrillation, autism, facial dysmorphisms (e.g., flattened nose), and poor enamel coating and thus increased risk of caries. Episodic hypoglycemia may be seen. Diagnosis is via sequence analysis of *CACNA1C*. Death

typically occurs in the first few years of life due to ventricular tachyarrhythmias.[4]

AKT2 Mutations

AKT2 is a serine/threonine kinase that is an effector molecule in the insulin signaling pathway linked to insulin's metabolic effects. It is encoded by the *AKT2* gene (chromosome 19q13.1).[24] In this extraordinarily rare disorder, of which only several reports of sporadic cases are known, gain-of-function mutations result in severe symptomatic hypoglycemia upon fasts of longer than 3 h in duration, which clinically and biochemically is identical to hyperinsulinemic hypoglycemia, except that insulin levels are undetectable. Affected children may show hemihypertrophy.[4] Treatment entails continuous tube feeds to prevent fasting hypoglycemia.

Citrin Deficiency

Citrin is a hepatic mitochondrial carrier of aspartate and glutamate that transports aspartate from the mitochondria to the cytoplasm during the urea cycle, allowing citrulline to be converted to arginosuccinate. Citrin deficiency is due to autosomal recessive mutations in *SLC25A13* (chromosome 7q21.3). Citrin deficiency has three recognized phenotypes: neonatal intrahepatic cholestasis (NICCD), failure to thrive and dyslipidemia due to citrin deficiency (FTTDCD) of older children, and citrulilinemia type II, or CTLN2, which is not discussed further here. NICCD and FTTDCD are associated with hypoglycemia. NICCD is characterized by growth delay, hepatomegaly, fatty liver, hepatic fibrosis, and hepatic dysfunction. Hypoglycemia is seen due to disruption in gluconeogenesis. Most liver problems are resolved by 1 year of age, although some infants die of cirrhosis or infection and others must undergo hepatic transplantation. At 1 year of age, children often develop carbohydrate aversion and preference for high-protein and/or high-fat food. They may have growth retardation, fatigue, pancreatitis, fatty liver, and hepatoma in addition to hypoglycemia. Diagnosis is made through a combination of sequence analysis of *SLC25A12* and through detection of elevated ammonia, arginine, and citrulline levels. NICCD is treated with supplementation of fat-soluble vitamins and MCTs. FTTDCD is treated similar to NICCD, with possible pyruvate supplementation. Hypoglycemia is treated by limiting fasting duration. Liver transplantation may be required.

SUMMARY

Hypoglycemia can be caused by many different entities with a wide variety of genetic underpinnings. It may be seen in isolation, or as part of an underlying broader syndrome, and can have a very wide range of etiologies.

The diagnostic evaluation of hypoglycemia always starts with history and physical examination. Obtaining a critical sample at the time of hypoglycemia is crucial to making the correct diagnosis. However, if the critical sample is not obtained at the time of a spontaneous hypoglycemic episode, a diagnostic fast is then required. There is one important exception: individuals with fatty acid oxidation disorders should never be fasted, so as not to trigger a potentially fatal metabolic crisis. An acylcarnitine profile should *always* be obtained with results received prior to undertaking a diagnostic fast to exclude this possibility. Once the category of hypoglycemia is established, molecular genetic testing (sequence analysis, methylation studies, and others) can confirm the suspected underlying genetic diagnosis. Treatment of the hypoglycemia and prevention of sequelae depends on the underlying diagnosis; it is crucial to identify the proper method of treatment and prevent serious hypoglycemic sequelae, from seizures to potential neurological damage or even death.

References

1. De Leon DD, Stanley CA, Sperling MA. Hypoglycemia in neonates and infants. In: Sperling MA, editor. *Pediatric Endocrinology*. Philadelphia, PA: Saunders, an imprint of Elsevier Inc; 2008. p. 165–97.
2. Kahn CR, White MF. The insulin receptor and the molecular mechanism of insulin action. *J Clin Invest* 1988;**82**:1151–6.
3. Watts AG, Donovan CM. Sweet talk in the brain: glucosensing, neural networks, and hypoglycemic counterregulation. *Front Neuroendocrinol* 2010;**31**:32–43.
4. Chandran S, Yap F, Hussain K. Genetic disorders leading to hypoglycaemia. *J Genet Syndr Gene Ther* 2013;**4**:192.
5. Cornblath M, Hawdon JM, Williams AF, et al. Controversies regarding definition of neonatal hypoglycemia: suggested operational thresholds. *Pediatrics* 2000;**105**:1141–5.
6. Koh TH, Eyre JA, Aynsley-Green A. Neonatal hypoglycaemia – the controversy regarding definition. *Arch Dis Child* 1988;**63**:1386–8.
7. Adamkin DH. Postnatal glucose homeostasis in late-preterm and term infants. *Pediatrics* 2011;**127**:575–9.
8. Sperling MA, Ganguli S, Leslie N, Landt KAT Fetal-perinatal catecholamine secretion: role in perinatal glucose homeostasis. *Am J Physiol* 1984;**247**:E69–74.
9. Lucas A, Bloom SR, Aynsley-Green A. Metabolic and endocrine events at the time of the first feed of human milk in preterm and term infants. *Arch Dis Child* 1978;**53**:731–6.
10. Alkalay AL, Sarnat HB, Flores-Sarnat L, Elashoff JD, Farber SJ, Simmons CF. Population meta-analysis of low plasma glucose thresholds in full-term normal newborns. *Am J Perinatol* 2006;**23**:115–9.
11. Brand PL, Molenaar NL, Kaaijk C, Wierenga WS. Neurodevelopmental outcome of hypoglycaemia in healthy, large for gestational age, term newborns. *Arch Dis Child* 2005;**90**:78–81.
12. Harris DL, Weston PJ, Harding JE. Incidence of neonatal hypoglycemia in babies identified as at risk. *J Pediatr* 2012;**161**:787–91.
13. Per H, Kumandas S, Coskun A, Gumus H, Oztop D. Neurologic sequelae of neonatal hypoglycemia in Kayseri, Turkey. *J Child Neurol* 2008;**23**:1406–12.
14. Lucas A, Morley R, Cole TJ. Adverse neurodevelopmental outcome of moderate neonatal hypoglycaemia. *BMJ* 1988;**297**:1304–8.
15. Burns CM, Rutherford MA, Boardman JP, Cowan FM. Patterns of cerebral injury and neurodevelopmental outcomes after symptomatic neonatal hypoglycemia. *Pediatrics* 2008;**122**:65–74.
16. Langdon DR, Stanley CA, Sperling MA. Hypoglycemia in the infant and child. In: Sperling MA, editor. *Pediatric Endocrinology*. Philadelphia, PA: Saunders, an imprint of Elsevier Inc; 2008. p. 422–43.
17. Palladino AA, Bennett MJ, Stanley CA. Hyperinsulinism in infancy and childhood: when an insulin level is not always enough. *Clin Chem* 2008;**54**:256–63.
18. Bauer A, Demetz F, Bruegger D, et al. Effect of high altitude and exercise on microvascular parameters in acclimatized subjects. *Clin Sci (Lond)* 2006;**110**:207–15.
19. Dekelbab BH, Sperling MA. Recent advances in hyperinsulinemic hypoglycemia of infancy. *Acta Paediatr* 2006;**95**:1157–64.
20. Arnoux JB, Verkarre V, Saint-Martin C, et al. Congenital hyperinsulinism: current trends in diagnosis and therapy. *Orphanet J Rare Dis* 2011;**6**:63.
21. Glaser B, Blech I, Krakinovsky Y, et al. ABCC8 mutation allele frequency in the Ashkenazi Jewish population and risk of focal hyperinsulinemic hypoglycemia. *Genet Med* 2011;**13**:891–4.
22. Snider KE, Becker S, Boyajian L, et al. Genotype and phenotype correlations in 417 children with congenital hyperinsulinism. *J Clin Endocrinol Metab* 2013;**98**:E355–63.
23. Hojlund K, Hansen T, Lajer M, et al. A novel syndrome of autosomal-dominant hyperinsulinemic hypoglycemia linked to a mutation in the human insulin receptor gene. *Diabetes* 2004;**53**:1592–8.
24. Hussain K, Challis B, Rocha N, et al. An activating mutation of AKT2 and human hypoglycemia. *Science* 2011;**334**:474.
25. Arya VB, Flanagan SE, Schober E, Rami-Merhar B, Ellard S, Hussain K. Activating AKT2 mutation: hypoinsulinemic hypoketotic hypoglycemia. *J Clin Endocrinol Metab* 2014;**99**:391–4.
26. Pinney SE, MacMullen C, Becker S, et al. Clinical characteristics and biochemical mechanisms of congenital hyperinsulinism associated with dominant KATP channel mutations. *J Clin Invest* 2008;**118**:2877–86.
27. Zerangue N, Schwappach B, Jan YN, Jan LY. A new ER trafficking signal regulates the subunit stoichiometry of plasma membrane K(ATP) channels. *Neuron* 1999;**22**:537–48.
28. Dunne MJ, Cosgrove KE, Shepherd RM, Aynsley-Green A, Lindley KJ. Hyperinsulinism in infancy: from basic science to clinical disease. *Physiol Rev* 2004;**84**:239–75.
29. Cartier EA, Conti LR, Vandenberg CA, Shyng SL. Defective trafficking and function of KATP channels caused by a sulfonylurea receptor 1 mutation associated with persistent hyperinsulinemic hypoglycemia of infancy. *Proc Natl Acad Sci USA* 2001;**98**:2882–7.
30. Taschenberger G, Mougey A, Shen S, Lester LB, LaFranchi S, Shyng SL. Identification of a familial hyperinsulinism-causing mutation in the sulfonylurea receptor 1 that prevents normal trafficking and function of KATP channels. *J Biol Chem* 2002;**277**:17139–46.
31. Yan FF, Lin YW, MacMullen C, Ganguly A, Stanley CA, Shyng SL. Congenital hyperinsulinism associated ABCC8 mutations that cause defective trafficking of ATP-sensitive K+ channels: identification and rescue. *Diabetes* 2007;**56**:2339–48.
32. Thomas PM, Cote GJ, Wohllk N, et al. Mutations in the sulfonylurea receptor gene in familial persistent hyperinsulinemic hypoglycemia of infancy. *Science* 1995;**268**:426–9.
33. Glaser B, Thornton P, Otonkoski T, Junien C. Genetics of neonatal hyperinsulinism. *Arch Dis Child Fetal Neonatal Ed* 2000;**82**:F79–86.
34. Nestorowicz A, Inagaki N, Gonoi T, et al. A nonsense mutation in the inward rectifier potassium channel gene, Kir6.2, is associated with familial hyperinsulinism. *Diabetes* 1997;**46**:1743–8.
35. Thomas P, Ye Y, Lightner E. Mutation of the pancreatic islet inward rectifier Kir6.2 also leads to familial persistent hyperinsulinemic hypoglycemia of infancy. *Hum Mol Genet* 1996;**5**:1809–12.

36. Otonkoski T, Nanto-Salonen K, Seppanen M, et al. Noninvasive diagnosis of focal hyperinsulinism of infancy with [18F]-DOPA positron emission tomography. *Diabetes* 2006;**55**:13–8.

37. Kassem SA, Ariel I, Thornton PS, et al. p57(KIP2) expression in normal islet cells and in hyperinsulinism of infancy. *Diabetes* 2001;**50**:2763–9.

38. Lord K, Dzata E, Snider KE, Gallagher PR, De Leon DD. Clinical presentation and management of children with diffuse and focal hyperinsulinism: a review of 223 cases. *J Clin Endocrinol Metab* 2013;**98**:E1786–9.

39. Stanley CA, Lieu YK, Hsu BY, et al. Hyperinsulinism and hyperammonemia in infants with regulatory mutations of the glutamate dehydrogenase gene. *N Engl J Med* 1998;**338**:1352–7.

40. Smith TJ, Peterson PE, Schmidt T, Fang J, Stanley CA. Structures of bovine glutamate dehydrogenase complexes elucidate the mechanism of purine regulation. *J Mol Biol* 2001;**307**:707–20.

41. Fang J, Hsu BY, MacMullen CM, Poncz M, Smith TJ, Stanley CA. Expression, purification and characterization of human glutamate dehydrogenase (GDH) allosteric regulatory mutations. *Biochem J* 2002;**363**:81–7.

42. Zhang T, Li C. Mechanisms of amino acid-stimulated insulin secretion in congenital hyperinsulinism. *Acta Biochim Biophys Sin (Shanghai)* 2013;**45**:36–43.

43. Kelly A, Li C, Gao Z, Stanley CA, Matschinsky FM. Glutaminolysis and insulin secretion: from bedside to bench and back. *Diabetes* 2002;**51**(Suppl. 3):S421–6.

44. Treberg JR, Clow KA, Greene KA, Brosnan ME, Brosnan JT. Systemic activation of glutamate dehydrogenase increases renal ammoniagenesis: implications for the hyperinsulinism/hyperammonemia syndrome. *Am J Physiol Endocrinol Metab* 2010;**298**:E1219–25.

45. Hsu BY, Kelly A, Thornton PS, Greenberg CR, Dilling LA, Stanley CA. Protein-sensitive and fasting hypoglycemia in children with the hyperinsulinism/hyperammonemia syndrome. *J Pediatr* 2001;**138**:383–9.

46. Kapoor RR, Flanagan SE, Fulton P, et al. Hyperinsulinism-hyperammonaemia syndrome: novel mutations in the GLUD1 gene and genotype-phenotype correlations. *Eur J Endocrinol* 2009;**161**:731–5.

47. Raizen DM, Brooks-Kayal A, Steinkrauss L, Tennekoon GI, Stanley CA, Kelly A. Central nervous system hyperexcitability associated with glutamate dehydrogenase gain of function mutations. *J Pediatr* 2005;**146**:388–94.

48. Matschinsky FM. Regulation of pancreatic beta-cell glucokinase: from basics to therapeutics. *Diabetes* 2002;**51**(Suppl. 3):S394–404.

49. Glaser B, Kesavan P, Heyman M, et al. Familial hyperinsulinism caused by an activating glucokinase mutation. *N Engl J Med* 1998;**338**:226–30.

50. Kamata K, Mitsuya M, Nishimura T, Eiki J, Nagata Y. Structural basis for allosteric regulation of the monomeric allosteric enzyme human glucokinase. *Structure* 2004;**12**:429–38.

51. Sayed S, Langdon DR, Odili S, et al. Extremes of clinical and enzymatic phenotypes in children with hyperinsulinism caused by glucokinase activating mutations. *Diabetes* 2009;**58**:1419–27.

52. Marquard J, Palladino AA, Stanley CA, Mayatepek E, Meissner T. Rare forms of congenital hyperinsulinism. *Semin Pediatr Surg* 2011;**20**:38–44.

53. Christesen HB, Tribble ND, Molven A, et al. Activating glucokinase (GCK) mutations as a cause of medically responsive congenital hyperinsulinism: prevalence in children and characterisation of a novel GCK mutation. *Eur J Endocrinol* 2008;**159**: 27–34.

54. Henquin JC, Sempoux C, Marchandise J, et al. Congenital hyperinsulinism caused by hexokinase I expression or glucokinase-activating mutation in a subset of beta-cells. *Diabetes* 2013;**62**: 1689–96.

55. Cuesta-Munoz AL, Huopio H, Otonkoski T, et al. Severe persistent hyperinsulinemic hypoglycemia due to a *de novo* glucokinase mutation. *Diabetes* 2004;**53**:2164–8.

56. Li C, Chen P, Palladino A, et al. Mechanism of hyperinsulinism in short-chain 3-hydroxyacyl-CoA dehydrogenase deficiency involves activation of glutamate dehydrogenase. *J Biol Chem* 2010;**285**:31806–18.

57. Clayton PT, Eaton S, Aynsley-Green A, et al. Hyperinsulinism in short-chain L-3-hydroxyacyl-CoA dehydrogenase deficiency reveals the importance of beta-oxidation in insulin secretion. *J Clin Invest* 2001;**108**:457–65.

58. Hussain K, Clayton PT, Krywawych S, et al. Hyperinsulinism of infancy associated with a novel splice site mutation in the SCHAD gene. *J Pediatr* 2005;**146**:706–8.

59. Flanagan SE, Patch AM, Locke JM, et al. Genome-wide homozygosity analysis reveals HADH mutations as a common cause of diazoxide-responsive hyperinsulinemic-hypoglycemia in consanguineous pedigrees. *J Clin Endocrinol Metab* 2011;**96**:E498–502.

60. Flanagan SE, Xie W, Caswell R, et al. Next-generation sequencing reveals deep intronic cryptic ABCC8 and HADH splicing founder mutations causing hyperinsulinism by pseudoexon activation. *Am J Hum Genet* 2013;**92**:131–6.

61. Molven A, Matre GE, Duran M, et al. Familial hyperinsulinemic hypoglycemia caused by a defect in the SCHAD enzyme of mitochondrial fatty acid oxidation. *Diabetes* 2004;**53**:221–7.

62. Halestrap AP, Price NT. The proton-linked monocarboxylate transporter (MCT) family: structure, function and regulation. *Biochem J* 1999;**343**(Pt. 2):281–99.

63. Zhao C, Wilson MC, Schuit F, Halestrap AP, Rutter GA. Expression and distribution of lactate/monocarboxylate transporter isoforms in pancreatic islets and the exocrine pancreas. *Diabetes* 2001;**50**:361–6.

64. Alcazar O, Tiedge M, Lenzen S. Importance of lactate dehydrogenase for the regulation of glycolytic flux and insulin secretion in insulin-producing cells. *Biochem J* 2000;**352**(Pt. 2):373–80.

65. Otonkoski T, Kaminen N, Ustinov J, et al. Physical exercise-induced hyperinsulinemic hypoglycemia is an autosomal-dominant trait characterized by abnormal pyruvate-induced insulin release. *Diabetes* 2003;**52**:199–204.

66. Thornton PS, Alter CA, Katz LE, Baker L, Stanley CA. Short- and long-term use of octreotide in the treatment of congenital hyperinsulinism. *J Pediatr* 1993;**123**:637–43.

67. Courtois G, Morgan JG, Campbell LA, Fourel G, Crabtree GR. Interaction of a liver-specific nuclear factor with the fibrinogen and alpha 1-antitrypsin promoters. *Science* 1987;**238**:688–92.

68. Dusatkova P, Pruhova S, Sumnik Z, et al. HNF1A mutation presenting with fetal macrosomia and hypoglycemia in childhood prior to onset of overt diabetes. *J Pediatr Endocrinol Metab* 2011;**24**:187–9.

69. Stanescu DE, Hughes N, Kaplan B, Stanley CA, De Leon DD. Novel presentations of congenital hyperinsulinism due to mutations in the MODY genes: HNF1A and HNF4A. *J Clin Endocrinol Metab* 2012;**97**:E2026–30.

70. Chandra V, Huang P, Potluri N, Wu D, Kim Y, Rastinejad F. Multidomain integration in the structure of the HNF-4alpha nuclear receptor complex. *Nature* 2013;**495**:394–8.

71. Flanagan SE, Kapoor RR, Mali G, et al. Diazoxide-responsive hyperinsulinemic hypoglycemia caused by HNF4A gene mutations. *Eur J Endocrinol* 2010;**162**:987–92.

72. Pearson ER, Boj SF, Steele AM, et al. Macrosomia and hyperinsulinaemic hypoglycaemia in patients with heterozygous mutations in the HNF4A gene. *PLoS Med* 2007;**4**:e118.

73. Gonzalez-Barroso MM, Giurgea I, Bouillaud F, et al. Mutations in UCP2 in congenital hyperinsulinism reveal a role for regulation of insulin secretion. *PLoS One* 2008;**3**: e3850.

74. Pinney SE, Ganapathy K, Bradfield J, et al. Dominant form of congenital hyperinsulinism maps to HK1 region on 10q. *Horm Res Paediatr* 2013;**80**:18–27.

75. Sempoux C, Capito C, Bellanne-Chantelot C, et al. Morphological mosaicism of the pancreatic islets: a novel anatomopathological form of persistent hyperinsulinemic hypoglycemia of infancy. *J Clin Endocrinol Metab* 2011;**96**:3785–93.

76. Tan JT, Nurbaya S, Gardner D, Ye S, Tai ES, Ng DP. Genetic variation in KCNQ1 associates with fasting glucose and beta-cell function: a study of 3,734 subjects comprising three ethnicities living in Singapore. *Diabetes* 2009;**58**:1445–9.

77. Torekov SS, Iepsen E, Christiansen M, et al. KCNQ1 long QT syndrome patients have hyperinsulinemia and symptomatic hypoglycemia. *Diabetes* 2014;**63**:1315–25.

78. *Russell-Silver Syndrome. University of Washington, Seattle.* http://www.ncbi.nlm.nih.gov/books/NBK1324/; 2002 (accessed 08.07.2014).

79. Wolfsdorf JI, Weinstein DA. Glycogen storage diseases. *Rev Endocr Metab Disord* 2003;**4**:95–102.

80. Ozen H. Glycogen storage diseases: new perspectives. *World J Gastroenterol* 2007;**13**:2541–53.

81. Safran M, Chalifa-Caspi V, Shmueli O, et al. Human gene-centric databases at the Weizmann Institute of Science: GeneCards, UDB, CroW 21 and HORDE. *Nucleic Acids Res* 2003;**31**:142–6.

82. Weinstein DA, Correia CE, Saunders AC, Wolfsdorf JI. Hepatic glycogen synthase deficiency: an infrequently recognized cause of ketotic hypoglycemia. *Mol Genet Metab* 2006;**87**:284–8.

83. Lei KJ, Shelly LL, Pan CJ, Sidbury JB, Chou JY. Mutations in the glucose-6-phosphatase gene that cause glycogen storage disease type 1a. *Science* 1993;**262**:580–3.

84. Hutton JC, O'Brien RM. Glucose-6-phosphatase catalytic subunit gene family. *J Biol Chem* 2009;**284**:29241–5.

85. Yang Chou J, Mansfield BC. Molecular genetics of type 1 glycogen storage diseases. *Trends Endocrinol Metab* 1999;**10**(3):104–13.

86. Veiga-da-Cunha M, Gerin I, Chen YT, et al. A gene on chromosome 11q23 coding for a putative glucose-6-phosphate translocase is mutated in glycogen-storage disease types Ib and Ic. *Am J Hum Genet* 1998;**63**:976–83.

87. Gu LL, Li XH, Han Y, Zhang DH, Gong QM, Zhang XX. A novel homozygous no-stop mutation in G6PC gene from a Chinese patient with glycogen storage disease type Ia. *Gene* 2014;**536**:362–5.

88. Kikuchi M, Hasegawa K, Handa I, Watabe M, Narisawa K, Tada K. Chronic pancreatitis in a child with glycogen storage disease type 1. *Eur J Pediatr* 1991;**150**:852–3.

89. Bernier AV, Correia CE, Haller MJ, Theriaque DW, Shuster JJ, Weinstein DA. Vascular dysfunction in glycogen storage disease type I. *J Pediatr* 2009;**154**:588–91.

90. Weinstein DA, Somers MJ, Wolfsdorf JI. Decreased urinary citrate excretion in type 1a glycogen storage disease. *J Pediatr* 2001;**138**:378–82.

91. Reitsma-Bierens WC. Renal complications in glycogen storage disease type I. *Eur J Pediatr* 1993;**152**(Suppl. 1):S60–2.

92. Melis D, Pivonello R, Parenti G, et al. The growth hormone-insulin-like growth factor axis in glycogen storage disease type 1: evidence of different growth patterns and insulin-like growth factor levels in patients with glycogen storage disease type 1a and 1b. *J Pediatr* 2010;**156**:663–70 e1.

93. Sechi A, Deroma L, Lapolla A, et al. Fertility and pregnancy in women affected by glycogen storage disease type I, results of a multicenter Italian study. *J Inherit Metab Dis* 2013;**36**:83–9.

94. Minarich LA, Kirpich A, Fiske LM, Weinstein DA. Bone mineral density in glycogen storage disease type Ia and Ib. *Genet Med* 2012;**14**:737–41.

95. Schwahn B, Rauch F, Wendel U, Schonau E. Low bone mass in glycogen storage disease type 1 is associated with reduced muscle force and poor metabolic control. *J Pediatr* 2002;**141**:350–6.

96. Czapek EE, Deykin D, Salzman EW. Platelet dysfunction in glycogen storage disease type I. *Blood* 1973;**41**:235–47.

97. Wang DQ, Carreras CT, Fiske LM, et al. Characterization and pathogenesis of anemia in glycogen storage disease type Ia and Ib. *Genet Med* 2012;**14**:795–9.

98. Wang DQ, Fiske LM, Carreras CT, Weinstein DA. Natural history of hepatocellular adenoma formation in glycogen storage disease type I. *J Pediatr* 2011;**159**:442–6.

99. Kishnani PS, Austin SL, Arn P, et al. Glycogen storage disease type III diagnosis and management guidelines. *Genet Med* 2010;**12**:446–63.

100. Hiraiwa H, Pan CJ, Lin B, Moses SW, Chou JY. Inactivation of the glucose 6-phosphate transporter causes glycogen storage disease type 1b. *J Biol Chem* 1999;**274**:5532–6.

101. Jun HS, Weinstein DA, Lee YM, Mansfield BC, Chou JY. Molecular mechanisms of neutrophil dysfunction in glycogen storage disease type Ib. *Blood* 2014;**123**:2843–53.

102. Chou JY, Jun HS, Mansfield BC. Neutropenia in type Ib glycogen storage disease. *Curr Opin Hematol* 2010;**17**:36–42.

103. Dieckgraefe B, Korzenik J, Husain A, Dieruf L. Association of glycogen storage disease 1b and Crohn disease: results of a North American survey. *Eur J Pediatr* 2002;**161**:S88–92.

104. Melis D, Della Casa R, Balivo F, et al. Involvement of endocrine system in a patient affected by glycogen storage disease 1b: speculation on the role of autoimmunity. *Ital J Pediatr* 2014;**40**:30.

105. Bolyard AA, Marrero TM, Kelley ML, et al. Neutropenia in glycogen storage disease 1b (GSD1b). *Blood* 2013;**122**:2265.

106. Pinsk M, Burzynski J, Yhap M, Fraser RB, Cummings B, Ste-Marie M. Acute myelogenous leukemia and glycogen storage disease 1b. *J Pediatr Hematol Oncol* 2002;**24**:756–8.

107. Schroeder T, Hildebrandt B, Mayatepek E, Germing U, Haas R. A patient with glycogen storage disease type Ib presenting with acute myeloid leukemia (AML) bearing monosomy 7 and translocation t(3;8)(q26;q24) after 14 years of treatment with granulocyte colony-stimulating factor (G-CSF): a case report. *J Med Case Rep* 2008;**2**:319.

108. Bao Y, Yang BZ, Dawson Jr TL, Chen YT. Isolation and nucleotide sequence of human liver glycogen debranching enzyme mRNA: identification of multiple tissue-specific isoforms. *Gene* 1997;**197**:389–98.

109. Shen J, Bao Y, Liu HM, Lee P, Leonard JV, Chen YT. Mutations in exon 3 of the glycogen debranching enzyme gene are associated with glycogen storage disease type III that is differentially expressed in liver and muscle. *J Clin Invest* 1996;**98**:352–7.

110. Van Hoof F, Hers HG. The subgroups of type 3 glycogenosis. *Eur J Biochem* 1967;**2**:265–70.

111. Ding JH, de Barsy T, Brown BI, Coleman RA, Chen YT. Immunoblot analyses of glycogen debranching enzyme in different subtypes of glycogen storage disease type III. *J Pediatr* 1990;**116**:95–100.

112. Parvari R, Moses S, Shen J, Hershkovitz E, Lerner A, Chen YT. A single-base deletion in the 3'-coding region of glycogen-debranching enzyme is prevalent in glycogen storage disease type IIIA in a population of North African Jewish patients. *Eur J Hum Genet* 1997;**5**:266–70.

113. Lee PJ, Patel A, Hindmarsh PC, Mowat AP, Leonard JV. The prevalence of polycystic ovaries in the hepatic glycogen storage diseases: its association with hyperinsulinism. *Clin Endocrinol (Oxf)* 1995;**42**:601–6.

114. Chang S, Rosenberg MJ, Morton H, Francomano CA, Biesecker LG. Identification of a mutation in liver glycogen phosphorylase in glycogen storage disease type VI. *Hum Mol Genet* 1998;**7**:865–70.

115. *Glycogen Storage Disease IXa1; GSD9A1; #306000.* http://www.omim.org/entry/306000?search=GSD%20IX&highlight=ix%20gsd#contributors-shutter; 1986 (accessed 28.06.2014).

116. Hendrickx J, Coucke P, Hors-Cayla MC, et al. Localization of a new type of X-linked liver glycogenosis to the chromosomal re-

gion Xp22 containing the liver alpha-subunit of phosphorylase kinase (PHKA2). *Genomics* 1994;**21**:620–5.

117. Beauchamp NJ, Dalton A, Ramaswami U, et al. Glycogen storage disease type IX: high variability in clinical phenotype. *Mol Genet Metab* 2007;**92**:88–99.

118. Maichele AJ, Burwinkel B, Maire I, Sovik O, Kilimann MW. Mutations in the testis/liver isoform of the phosphorylase kinase gamma subunit (PHKG2) cause autosomal liver glycogenosis in the gsd rat and in humans. *Nat Genet* 1996;**14**:337–40.

119. Nakai A, Shigematsu Y, Takano T, Kikawa Y, Sudo M. Uncooked cornstarch treatment for hepatic phosphorylase kinase deficiency. *Eur J Pediatr* 1994;**153**:581–3.

120. van den Berghe G. Disorders of gluconeogenesis. *J Inherit Metab Dis* 1996;**19**:470–7.

121. el-Maghrabi MR, Lange AJ, Jiang W, et al. Human fructose-1,6-bisphosphatase gene (FBP1): exon-intron organization, localization to chromosome bands 9q22.2-q22.3, and mutation screening in subjects with fructose-1,6-bisphosphatase deficiency. *Genomics* 1995;**27**:520–5.

122. Baker L, Winegrad AI. Fasting hypoglycaemia and metabolic acidosis associated with deficiency of hepatic fructose-1,6-diphosphatase activity. *Lancet* 1970;**2**:13–6.

123. Carbone MA, MacKay N, Ling M, et al. Amerindian pyruvate carboxylase deficiency is associated with two distinct missense mutations. *Am J Hum Genet* 1998;**62**:1312–9.

124. Wang D, Yang H, De Braganca KC, et al. The molecular basis of pyruvate carboxylase deficiency: mosaicism correlates with prolonged survival. *Mol Genet Metab* 2008;**95**:31–8.

125. Tsuchiyama A, Oyanagi K, Hirano S, et al. A case of pyruvate carboxylase deficiency with later prenatal diagnosis of an unaffected sibling. *J Inherit Metab Dis* 1983;**6**:85–8.

126. Maesaka H, Komiya K, Misugi K, Tada K. Hyperalaninemia hyperpyruvicemia and lactic acidosis due to pyruvate carboxylase deficiency of the liver; treatment with thiamine and lipoic acid. *Eur J Pediatr* 1976;**122**:159–68.

127. Baal MG, Gabreels FJ, Renier WO, et al. A patient with pyruvate carboxylase deficiency in the liver: treatment with aspartic acid and thiamine. *Dev Med Child Neurol* 1981;**23**:521–30.

128. Ahmad A, Kahler SG, Kishnani PS, et al. Treatment of pyruvate carboxylase deficiency with high doses of citrate and aspartate. *Am J Med Genet* 1999;**87**:331–8.

129. Ali M, Rellos P, Cox TM. Hereditary fructose intolerance. *J Med Genet* 1998;**35**:353–65.

130. Esposito G, Imperato MR, Ieno L, et al. Hereditary fructose intolerance: functional study of two novel ALDOB natural variants and characterization of a partial gene deletion. *Hum Mutat* 2010;**31**:1294–303.

131. Gitzelmann R, Baerlocher K. Vorteile und Nachteile der Frucose in der Nahrung. *Padiat Fortbildk Praxis* 1973;**37**:40–55.

132. James CL, Rellos P, Ali M, Heeley AF, Cox TM. Neonatal screening for hereditary fructose intolerance: frequency of the most common mutant aldolase B allele (A149P) in the British population. *J Med Genet* 1996;**33**:837–41.

133. Van den Berghe H, Bruntman M, Vannestes R, Hers HG. The mechanism of adenine triphosphate depletion in the liver after a load of fructose. A kinetic study of liver adenylate deaminase. *Biochem J* 1977;**134**:637–45.

134. Kaufmann U, Froesch ER. Inhibition of phosphorylase-a by fructose-1-phosphate, alpha-glycerophosphate and fructose-1,6-diphosphate: explanation for fructose-induced hypoglycaemia in hereditary fructose intolerance and fructose-1,6-diphosphatase deficiency. *Eur J Clin Invest* 1973;**3**:407–13.

135. Cox TM. Aldolase B and fructose intolerance. *Faseb J* 1994;**8**:62–71.

136. Cox TM. Iatrogenic deaths in hereditary fructose intolerance. *Arch Dis Child* 1993;**69**:413–5.

137. Bosch AM. Classical galactosaemia revisited. *J Inherit Metab Dis* 2006;**29**:516–25.

138. Murphy M, McHugh B, Tighe O, et al. Genetic basis of transferase-deficient galactosaemia in Ireland and the population history of the Irish Travellers. *Eur J Hum Genet* 1999;**7**:549–54.

139. *National Newborn Screening and Genetics Resource Center; 2002.* Newborn Screening and Genetic Testing Symposium.

140. Elsas II LJ, Lai K. The molecular biology of galactosemia. *Genet Med* 1998;**1**:40–8.

141. Misumi H, Wada H, Kawakami M, et al. Detection of UDP-galactose-4-epimerase deficiency in a galactosemia screening program. *Clin Chim Acta* 1981;**116**:101–5.

142. Segal S. Galactosemia unsolved. *Eur J Pediatr* 1995;**154**:S97–S102.

143. Berry GT. Is prenatal myo-inositol deficiency a mechanism of CNS injury in galactosemia? *J Inherit Metab Dis* 2011;**34**:345–55.

144. Jumbo-Lucioni PP, Garber K, Kiel J, et al. Diversity of approaches to classic galactosemia around the world: a comparison of diagnosis, intervention, and outcomes. *J Inherit Metab Dis* 2012;**35**:1037–49.

145. Shimomura Y, Honda T, Shiraki M, et al. Branched-chain amino acid catabolism in exercise and liver disease. *J Nutr* 2006;**136**:250S–3S.

146. Chuang DT, Chuang JL, Wynn RM. Lessons from genetic disorders of branched-chain amino acid metabolism. *J Nutr* 2006;**136**:243S–9S.

147. Nellis MM, Danner DJ. Gene preference in maple syrup urine disease. *Am J Hum Genet* 2001;**68**:232–7.

148. Chuang DTSV. Maple syrup urine disease (branched-chain ketoaciduria). In: Scriver CRBA, Sly WS, Valle D, editors. *The Metabolic and Molecular Bases of Inherited Disease*. 8th ed. New York, NY: McGraw-Hill; 2001. p. 1971–2006.

149. Levy HL. Genetic screening: notes added in proof. *Adv Hum Genet* 1973;**4**:389–94.

150. DiGeorge AM, Rezvani I, Garibaldi LR, Schwartz M. Prospective study of maple-syrup-urine disease for the first four days of life. *N Engl J Med* 1982;**307**:1492–5.

151. Edelmann L, Wasserstein MP, Kornreich R, Sansaricq C, Snyderman SE, Diaz GA. Maple syrup urine disease: identification and carrier-frequency determination of a novel founder mutation in the Ashkenazi Jewish population. *Am J Hum Genet* 2001;**69**:863–8.

152. Menkes JH, Hurst PL, Craig JM. A new syndrome: progressive familial infantile cerebral dysfunction associated with an unusual urinary substance. *Pediatrics* 1954;**14**:462–7.

153. Van der Horst JLWS. A variant form of branched-chain keto aciduria: case report. *Acta Paediatr Scand* 1971;**60**:594–9.

154. Scriver CR, Mackenzie S, Clow CL, Delvin E. Thiamine-responsive maple-syrup-urine disease. *Lancet* 1971;**1**:310–2.

155. Chuang DT, Ku LS, Cox RP. Biochemical basis of thiamin-responsive maple syrup urine disease. *Trans Assoc Am Physicians* 1982;**95**:196–204.

156. Zhang B, Wappner RS, Brandt IK, Harris RA, Crabb DW. Sequence of the E1 alpha subunit of branched-chain alpha-ketoacid dehydrogenase in two patients with thiamine-responsive maple syrup urine disease. *Am J Hum Genet* 1990;**46**:843–6.

157. Ravn K, Chloupkova M, Christensen E, et al. High incidence of propionic acidemia in greenland is due to a prevalent mutation, 1540insCCC, in the gene for the beta-subunit of propionyl CoA carboxylase. *Am J Hum Genet* 2000;**67**:203–6.

158. de Keyzer Y, Valayannopoulos V, Benoist JF, et al. Multiple OX-PHOS deficiency in the liver, kidney, heart, and skeletal muscle of patients with methylmalonic aciduria and propionic aciduria. *Pediatr Res* 2009;**66**:91–5.

159. Pena L, Franks J, Chapman KA, et al. Natural history of propionic acidemia. *Mol Genet Metab* 2012;**105**:5–9.

160. *Propionic Acidemia. University of Washington.* http://www.ncbi.nlm.nih.gov/books/NBK929; 2012 (accessed 30.06.2014).

161. Surtees RA, Matthews EE, Leonard JV. Neurologic outcome of propionic acidemia. *Pediatr Neurol* 1992;**8**:333–7.

162. *Methylmalonic Acidemia. University of Washington, Seattle.* http://www.ncbi.nlm.nih.gov/books/NBK1231/; 2005 (accessed 30.06.2014).

163. Shigematsu Y, Hirano S, Hata I, et al. Newborn mass screening and selective screening using electrospray tandem mass spectrometry in Japan. *J Chromatogr B Analyt Technol Biomed Life Sci* 2002;**776**:39–48.

164. *MIM#276700; Tyrosinemia, Type I; TYRSN1. Johns Hopkins University.* http://omim.org/entry/276700; 1986 (accessed 06.07.2014).

165. Rinaldo P, Matern D, Bennett MJ. Fatty acid oxidation disorders. *Annu Rev Physiol* 2002;**64**:477–502.

166. Cahill Jr GF. Starvation in man. *N Engl J Med* 1970;**282**:668–75.

167. Spiekerkoetter U. Mitochondrial fatty acid oxidation disorders: clinical presentation of long-chain fatty acid oxidation defects before and after newborn screening. *J Inherit Metab Dis* 2010;**33**:527–32.

168. Wu X, Prasad PD, Leibach FH, Ganapathy V. cDNA sequence, transport function, and genomic organization of human OCTN2, a new member of the organic cation transporter family. *Biochem Biophys Res Commun* 1998;**246**:589–95.

169. Rasmussen J, Nielsen OW, Janzen N, et al. Carnitine levels in 26,462 individuals from the nationwide screening program for primary carnitine deficiency in the Faroe Islands. *J Inherit Metab Dis* 2014;**37**:215–22.

170. *Systemic Primary Carnitine Deficiency. University of Washington, Seattle.* http://www.ncbi.nlm.nih.gov/books/NBK84551/; 2012 (accessed 04.07.2014).

171. *Carnitine Palmitoyltransferase 1A Deficiency. University of Washington, Seattle* 1993–2014. http://www.ncbi.nlm.nih.gov/books/NBK1527/; 2005 (accessed 04.07.2014).

172. *Carnitine Palmitoyltransferase II Deficiency. University of Washington, Seattle.* http://www.ncbi.nlm.nih.gov/books/NBK1253/; 2004 (accessed 04.07.2014).

173. *MIM #212138: Carnitine-Acylcarnitine Translocase Deficiency; CACTD Johns Hopkins University.* http://www.omim.org/entry/212138#contributors-shutter; 1992 (accessed 04.07.2014).

174. *Very Long-Chain Acyl-Coenzyme A Dehydrogenase Deficiency. University of Washington, Seattle.* http://www.ncbi.nlm.nih.gov/books/NBK6816/; 2009 (accessed 02.07.2014).

175. Lindner M, Hoffmann GF, Matern D. Newborn screening for disorders of fatty-acid oxidation: experience and recommendations from an expert meeting. *J Inherit Metab Dis* 2010;**33**:521–6.

176. Schulze A, Lindner M, Kohlmuller D, Olgemoller K, Mayatepek E, Hoffmann GF. Expanded newborn screening for inborn errors of metabolism by electrospray ionization-tandem mass spectrometry: results, outcome, and implications. *Pediatrics* 2003;**111**:1399–406.

177. *MIM# 609016; Long-Chain 3-Hydroxyacyl-CoA Dehydrogenase Deficiency. Johns Hopkins University.* http://www.omim.org/entry/609016; 2004 (accessed 04.07.2014).

178. *MIM#609015; Trifunctional Protein Deficiency. Johns Hopkins University.* http://www.omim.org/entry/609015; 2004 (accessed 04.07.2014).

179. *MIM# 231680; Multiple Acyl-CoA Dehydrogenase Deficiency; MADD. Johns Hopkins University.* http://www.omim.org/entry/231680; 1986 (accessed 04.07.2014).

180. Wilcken B. Fatty acid oxidation disorders: outcome and long-term prognosis. *J Inherit Metab Dis* 2010;**33**:501–6.

181. Sander S, Janzen N, Janetzky B, et al. Neonatal screening for medium chain acyl-CoA deficiency: high incidence in Lower Saxony (northern Germany). *Eur J Pediatr* 2001;**160**:318–9.

182. Chace DH, Kalas TA, Naylor EW. The application of tandem mass spectrometry to neonatal screening for inherited disorders of intermediary metabolism. *Annu Rev Genom Hum Genet* 2002;**3**:17–45.

183. Niu DM, Chien YH, Chiang CC, et al. Nationwide survey of extended newborn screening by tandem mass spectrometry in Taiwan. *J Inherit Metab Dis* 2010;**33**:S295–305.

184. Fodor E, Hellerud C, Hulting J, et al. Glycerol kinase deficiency in adult hypoglycemic acidemia. *N Engl J Med* 2011;**364**:1781–2.

185. Sass JO. Inborn errors of ketogenesis and ketone body utilization. *J Inherit Metab Dis* 2012;**35**:23–8.

186. *MIM #307030; Glycerol Kinase Deficiency. Johns Hopkins University.* http://omim.org/entry/307030; 1986 (accessed 04.07.2014).

187. Morris AAM. Disorders of ketogenesis and ketolysis. In: Saudubray J-M, van den Berghe G, Walter J, editors. *Inborn metabolic diseases.* Berlin Heidelberg: Springer; 2012. p. 217–22.

188. *MIM# 605911; 3-Hydroxy-3-Methylglutaryl-CoA Synthase 2 Deficiency. Johns Hopkins university.* http://omim.org/entry/605911; 2001 (accessed 05.07.2014).

189. *MIM #246450; 3-Hydroxy-3-Methylglutaryl-CoA Lyase Deficiency; HMGCLD. Johns Hopkins University.* http://www.omim.org/entry/246450; 1986 (accessed 05.07.2014).

190. Chandran Suresh YF, Hussain, Khalid. Genetic disorders leading to hypoglycaemia. *J Genet Syndr Gene Ther* 2013;**4**:192.

191. Senard JM, Rouet P. Dopamine beta-hydroxylase deficiency. *Orphanet J Rare Dis* 2006;**1**:7.

192. Mussa A, Russo S, De Crescenzo A, et al. Prevalence of Beckwith-Wiedemann syndrome in North West of Italy. *Am J Med Genet A* 2013;**161**:2481–6.

193. MIM #130650; Beckwith-Wiedemann Syndrome; BWS. Johns Hopkins University, 1986 June 4. (Accessed July 6, 2014, at http://www.omim.org/entry/130650.)

194. Hussain K, Cosgrove KE, Shepherd RM, et al. Hyperinsulinemic hypoglycemia in Beckwith-Wiedemann syndrome due to defects in the function of pancreatic beta-cell adenosine triphosphate-sensitive potassium channels. *J Clin Endocrinol Metab* 2005;**90**:4376–82.

195. *MIM #180860; Silver-Russell Syndrome; SRS. Johns Hopkins University.* http://www.omim.org/entry/180860; 1994 (accessed 07.07.2014).

196. *Sotos Syndrome. University of Washington, Seattle.* http://www.ncbi.nlm.nih.gov/books/NBK1479/; 2004 (accessed 06.07.2012).

197. *MIM #117550; Soto syndrome 1; SOTOS1. Johns Hopkins University.* www.omim.org/entry/117550; 1896 (accessed 06.07.2014)

198. Matsuo T, Ihara K, Kinjo T, et al. Hyperinsulinemic hypoglycemia of infancy in Sotos syndrome. *Am J Med Genet* 2013;**61A**:34–7.

199. *MIM #262500; Laron syndrome. Johns Hopkins University.* http://omim.org/entry/262500; 1986 (accessed 06.07.2014).

200. Chrousos G. Q&A: Primary generalized glucocorticoid resistance. *BioMed Central Medicine* 2011;9.

PART III

PITUITARY

CHAPTER

4

Functioning Pituitary Adenomas

Albert Beckers, Liliya Rostomyan, Adrian F. Daly

Department of Endocrinology, Centre Hospitalier Universitaire de Liège, University of Liège,
Domaine Universitaire du Sart-Tilman, Liège, Belgium

INTRODUCTION

Many molecular genetic abnormalities have been recognized in the setting of anterior pituitary adenomas. However, pituitary adenomas with heritable genetic causes are rare and have been described most often in the setting of an endocrine tumor syndrome, such as multiple endocrine neoplasia type 1 (MEN1) (see Chapter 23) and Carney complex (CNC). MEN1 is an autosomal dominant condition that is associated with the occurrence of parathyroid, enteropancreatic, and anterior pituitary tumors.[1] Endocrine-inactive tumors, such as lipomas and angiofibromas, are also frequently seen in MEN1 patients. The *MEN1* gene on chromosome 11q13[2] encodes the nuclear protein, menin.[3] *MEN1* appears to act as a tumor suppressor gene, and recent data suggest that menin can potentially interact with thousands of genes, 3′ sites and in chromatin.[4,5] Up to 2010, over 500 different individual mutations in the *MEN1* gene have been described,[6] most of which predict a truncated menin protein. However, in 20–30% of cases suggestive of MEN1 clinically, the *MEN1* sequence is normal. About 40% of patients with MEN1 have pituitary adenomas, and 17% of cases present with a pituitary tumor.[7,8] Patients with pituitary tumors may present earlier than patients presenting with other MEN1 tumors. In families with MEN1, pituitary tumors are more frequent than in sporadic MEN1. From a clinical perspective, MEN1-associated pituitary adenomas are more aggressive and less responsive to treatment than sporadic non-MEN1 tumors.

CNC is rare, with less than 1000 cases reported to date.[9] CNC, usually familial, consists of patients with a complex of skin pigmentation, cardiac myxomas, endocrine hypersecretion, and schwannomas.[10] The primary genetic cause of CNC is mutation of the *protein kinase A regulatory subunit 1 A gene* (*PRKAR1A*) on chromosome 17q22-24.[11] Acromegaly occurs in the setting of CNC, although infrequently. Most patients have evidence of abnormal growth

hormone (GH), insulin-like growth factor-1 (IGF-1), or prolactin levels.

A MEN1-like syndrome (MEN4) has been reported recently in rats and in humans, and relates to mutations in the *CDKN1B* gene that encodes $p27^{kip1}$.[12,13] To date less than 10 patients with MEN4 that have pituitary adenomas have been described in the literature and it is a rare cause of pituitary adenomas.[14]

Isolated pituitary adenomas also occur in an inherited setting in the absence of MEN1 and CNC. Familial isolated pituitary adenomas (FIPA; >/=2 pituitary adenoma patients occurring in related members of the same kindred in the absence of MEN1, MEN4, or CNC) have been described in more than 200 since 1999. Within FIPA, patients may have pituitary adenomas of any functional or nonfunctional type, and different tumor types may occur within different members of the same family. Previous studies on familial cases of acromegaly pointed to a region of chromosome 11q13.1-q13.3 as being involved.[15–17] Subsequently, mutations in the *aryl hydrocarbon receptor interacting protein* gene (*AIP*) were noted to occur in association with familial acromegaly/prolactinoma kindreds.[18] *AIP* mutations account for about 15% of FIPA families and 50% of familial acromegaly kindreds, indicating that other genetic factors remain to be described. *AIP* mutations are also associated with larger/more aggressive pituitary tumors that occur at an earlier age than non-*AIP* mutated cases; a relatively high frequency of *AIP* mutations occur in sporadic (nonfamilial) cohorts with acromegaly or other pituitary adenomas with a young age at onset and an aggressive clinical history.[19]

Recently a novel X-linked genomic disorder consisting of early childhood-onset gigantism with excessive secretion of GH due to pituitary hyperplasia or adenoma was described. This gigantism cohort was characterized by heritable duplications in chromosome Xq26.3 containing four duplicated genes (CD40LG, ARHGEF6, RBMX, and

Genetic Diagnosis of Endocrine Disorders. http://dx.doi.org/10.1016/B978-0-12-800892-8.00004-X

GPR101). GPR101 is the gene most likely responsible for GH oversecretion and gigantism in these patients.[20,21]

GENETIC PATHOPHYSIOLOGY OF PITUITARY ADENOMAS

Most pituitary adenomas arise as a clonal expansion from a single mutated anterior pituitary cell, which can be accompanied by a wide variety of genetic and molecular alterations in adenomatous pituitary tissue. At the tissue level, however, the picture can be somewhat complicated, as a single pituitary can contain multiple tumors or hyperplastic areas, each with its own clonal origin and specific pattern of growth, apoptosis, and pathological features. The development of a pituitary adenoma is dependent on a variety of tumor suppressor genes and oncogenes (Table 4.1). The most important oncogene involved in sporadic pituitary tumorigenesis is *gsp*, which encodes the Gsα subunit, a stimulatory guanine binding protein that regulates hypothalamic GH-releasing hormone effects in somatotropes. Biallelic expression of mutated *gsp* can lead to endogenous activation of adenylate cyclase and elevated levels of cyclic adenosine monophosphate (cAMP). Mutations in *gsp* have been most closely associated with somatotropinomas, and they are found to occur in up to 40% of these tumors. The oncogene *ras* has also been implicated in pituitary tumorigenesis, although in a very small number of cases. Mutations in *ras* appear to be associated with high levels of tumor aggression and have been noted to occur among rare pituitary carcinomas.[22,23] Pituitary tumor transforming gene (PTTG) is a gene that is usually poorly expressed in normal pituitary, but is upregulated in most pituitary tumor types.[24,25]

Mutations in tumor suppressor genes have been identified in the setting of pituitary adenomas and tumorigenesis. The best known among these is the gene *MEN1* that is responsible for MEN1, which is discussed in detail in the subsequent section, as it may lead to inherited disease and is relevant for screening. Heterozygotic retinoblastoma gene (*Rb*) mutation status is associated with the development of pituitary adenomas in mice.[26] However, in humans the role of *Rb* is less certain. Somatic *Rb* loss may occur in occasional pituitary adenomas, while *Rb* promoter hypermethylation has also been reported.[27,28]

Cell cycle regulators have also been implicated in pituitary tumorigenesis and development. Cyclin D1, which regulates the transition from G_1 to S-phase is overexpressed in nearly 40% of somatotropinomas (and 70% of nonfunctional pituitary adenomas). *CCND1* genotypes are related to tumor grades seen in pituitary adenomas.[29,30] In pituitary adenomas, the

TABLE 4.1 Germline and Somatic Genetic Abnormalities Associated with Pituitary Adenomas

Genes	Defects
Cyclin D1	Overexpression in nonsecreting adenomas and somatotropinomas
Gsp	Somatic activating mutations in up to 40% of somatotropinomas
	Mosaicism in McCune–Albright syndrome (somatotropinoma, somatomammotropinoma, and Cushing's syndrome in association with precocious puberty, hyperthyroidism, and dermal and bony lesions)
PRKAR1A	Truncation mutations in Carney's complex leading to somatolactotrope hyperplasia and adenomas
Pdt-FGFR4	Alternative transcription initiation in pituitary adenomas
PTTG	Increased expression in more aggressive pituitary tumors
BMP-4	Diminished expression in prolactinoma
GADD45G	Promoter methylation in nonsecreting adenomas, prolactinomas, and somatotropinomas
MEG3a	Promoter methylation in nonsecreting adenomas and gonadotropinomas
MEN1	Inactivating mutations in all pituitary adenoma types
PKC	Point mutations in invasive pituitary adenomas
p16	Promoter methylation in pituitary adenomas
CDKN1B (p27Kip1)	Germline heterozygous nonsense mutation in MEN4, a novel, rare MEN1-like syndrome
Retinoblastoma	Promoter methylation in pituitary adenomas
ZAC	Promoter methylation in nonfunctioning adenomas
AIP	Germline mutations and loss of heterozygosity in 15% of FIPA cases. Seen in familial/sporadic somatotropinomas, somatolactotrope adenomas, prolactinomas, nonsecreting adenomas, and Cushing's disease (sporadic only)
GPR101	Duplication in early onset gigantism patients with somatotropinoma or pituitary hyperplasia; a pathogenic cause in certain AIP negative FIPA kindred with acrogigantism. Mutation p.E308D in some sporadic acromegaly patients, mostly in tumor DNA.
USP8	Somatic recurrent mutations seen in Cushing disease

cyclin-dependent kinase inhibitor p16 is heavily downregulated due to gene promoter hypermethylation.[31,32] The cyclin dependent kinase inhibitor p27[kip1] appears to play an important role in pituitary tumorigenesis, as evidenced by data from a knockout mouse that show the development of specific patterns of pituitary adenomas and other abnormalities.[33] The relevance of mutations in the CDKN1B gene that encodes p27[kip1] are discussed in the subsequent section as they may be rarely associated with inherited endocrine tumors, including pituitary adenomas.

The protein ZAC is normally expressed at high levels in healthy pituitary tissue. In pituitary adenomas (predominantly nonsecreting tumors), ZAC expression is strongly reduced. The somatostatin analog, octreotide, functions in somatotropinomas in part via ZAC as it increases the expression of the gene Zac1.[34,35] Fibroblast growth factor receptors (FGFR) play a role in the growth and development of many tissues. A truncated pituitary tumor-derived form of FGFR4 has been identified in humans and was reported be associated with invasive pituitary tumorigenesis in a transgenic mouse model.[36]

MEG3 appears to play a role as a growth suppressor in pituitary tissue; a pituitary-derived variant is absent from both functional and nonsecreting pituitary adenomas, potentially due to promoter hypermethylation.[37,38] Expression of the growth arrest and DNA damage-inducible gene (GADD45G) is decreased in somatotropinomas, prolactinomas, and nonsecreting adenomas.[39] Bone morphogenetic protein-4, which may indirectly stimulate c-myc expression, is overexpressed in prolactinomas as compared with other tissues.[40]

While a wide variety of genetic mutations and molecular abnormalities have been implicated in tumorigenesis in functioning pituitary adenomas, only a handful are of clinical relevance for genetic screening and testing (Table 4.2). These clinically relevant genes for human screening are associated with a variety of presentations of pituitary adenomas in combination with other tumors, or occasionally as isolated pituitary adenomas. The status of these conditions from a screening perspective is also heterogeneous, as some are relatively recent discoveries over the last few years, while others have been recognized for more than a decade and have consensus guidelines.

GENETIC SCREENING IN FUNCTIONING PITUITARY ADENOMAS

Currently, genetic screening for specific mutations in patients with functioning pituitary adenomas is limited to a minority of cases in which suggestive pathological features are present. Pituitary adenomas that occur in a familial setting account for no more than 4–5% of all pituitary adenomas. Scheithauer et al. estimated that 2.7% of pituitary adenomas were due to multiple endocrine neoplasia type I (MEN1).[41] Our data suggest that a further 2% of pituitary tumor cases have family links. Widespread genetic screening is, therefore, not warranted in the vast majority of patients with sporadic (i.e., nonfamilial) pituitary adenomas that do not have associated endocrine/nonendocrine tumors in the same patient or in their family. This is particularly relevant from a healthcare resource utilization perspective, as clinically active pituitary adenomas are not rare, and occur with a prevalence of approximately one in every 1000 of the population in developed countries.[42]

MEN1

The history of pituitary tumors in the setting of multiple endocrine neoplasia dates back to as early as 1903, with the description by Erdheim of a patient with adenomas in the parathyroid and pituitary.[43] Later in the

TABLE 4.2 Familial Pituitary Adenoma Syndromes

Condition	Gene	Molecular pathology	Pituitary tumor
MEN1	MEN1 (Ch11q13)	Decreased menin expression/function	All pituitary tumor types (prolactinomas, nonsecreting adenomas, and GH-secreting adenomas most frequent)
MEN4	CDKN1B (Chr 12p13)	Decreased p27 levels in tumor	Associated with only acromegaly and Cushing's disease in two patients to date
CNC	PPKR1A (Ch17q22-24) ? (Ch2p16)	Decreased protein kinase A regulatory subunit Ia expression/function	GH and GH/prolactin secreting adenomas
FIPA	AIP (Ch11q13.32) in 15% of cases (50% of familial acromegaly) X-LAG syndrome (chromosome Xq26.3 microduplications; GPR101)	Decreased mRNA and protein in some mutated tissues. ?Altered regulation of AhR or phosphodiesterase function Increased expression of GPR101 and growth hormone secretion	All pituitary adenoma subtypes involved; AIP mutation-associated cases include somatotropinomas, prolactinomas, mixed GH/prolactin tumors, and nonsecreting adenomas Early onset acrogigantism, somatotropinomas/somatotroph hyperplasia with prolactin cosecretion

AIP, aryl hydrocarbon receptor interacting protein; CNC, Carney complex; FIPA, familial isolated pituitary adenoma; MEN1, multiple endocrine neoplasia type 1; PRKAR1A, protein kinase A type I regulatory subunit Iα.

1950s, Wermer described a family with four sisters affected with pituitary adenomas (one had acromegaly), hypercalcemia, and adenomatosis of the pancreas and gut.[44] Further investigation revealed that the father had evidence of multiple pancreatic and gut tumors on autopsy. After analyzing this and many isolated cases, Wermer correctly posited an autosomal dominant mode of inheritance for this condition, which would later be termed MEN1. For further information on MEN1, see Chapter 23.

MEN1-RELATED PITUITARY TUMORS

About 40% of patients with MEN1 have pituitary adenomas,[45–47] and this was the presenting tumor in about 17% of cases in one large series.[7] Such patients that presented with a pituitary adenoma did so 7 years before patients presenting with enteropancreatic lesions. Among familial MEN1 cases, pituitary disease was significantly more frequent than in nonfamilial MEN1 cases (59% vs. 34%, respectively). Females with MEN1 have a somewhat increased chance of having a pituitary adenoma. The characteristics and features of MEN1 and non-MEN1-related pituitary tumors are outlined in Table 4.3.

Prolactinomas predominate among both MEN1-associated and non-MEN1 pituitary adenomas, and the proportions of prolactinomas, GH-secreting, ACTH-secreting, nonsecreting, and cosecreting adenomas are similar between the MEN1 and non-MEN1 patients. MEN1-related prolactinomas are predominantly macroadenomas (84%) and higher rates of invasion are seen than in non-MEN1 prolactinomas. The response of MEN1-related prolactinomas to dopamine agonists is poor, with only 44% of patients being normalized.

Pituitary tumors in MEN1 appear to be larger and more aggressive than in patients without MEN1,[8] with macroadenomas being present in 85% of the former, compared with only 42% of the sporadic cases. MEN1-associated pituitary tumors are significantly more likely to cause signs due to tumor size and have a significantly lower rate of hormonal normalization than non-MEN1 pituitary tumors.

Mutations of the *MEN1* gene are not an important factor in the tumorigenesis of non-MEN1 sporadic pituitary adenomas.[48–55] Theodoropoulou et al. found that menin was detectable in 67 of 68 sporadic non-MEN1 pituitary tumors.[56] There is no recognized relationship between the site or type of genetic mutation in the *MEN1* gene and the expressed MEN1 disease phenotype,[16] although disease clustering and variations in severity have been recognized.[57] Such clusters include the "prolactinoma variant" seen in kindreds from the Burin peninsula in Canada (MEN1$_{BURIN}$).[58,59]

Recommendations for Testing in MEN1

As MEN1 has been recognized clinically and genetically for some time, consensus guidelines for its investigation and management have been developed.[60] These guidelines were updated in 2012 and advances in the availability of and access to DNA testing may allow for some practical simplification. Assessment of which patients to test for *MEN1* germline mutations depends on their meeting the criteria for the disease (practically, two of the three constituent major affected tissues: parathyroid, enteropancreatic, or pituitary tumors). Assessment of family history is also useful to detect previously unrecognized contributory information; however, it is often practically difficult to exclude a familial case of MEN1 in small kindreds with few living relatives. For index cases in which MEN1 is suspected as a cause for their pituitary tumor, germline DNA analysis of the *MEN1* gene would be recommended. In cases that are negative for mutations on first screening, the potential for further testing for large gene deletions in an investigational laboratory could be considered. In the case where a *MEN1* mutation is found, then full family screening for clinical features and biochemical abnormalities (particularly hypercalcemia) is a good first step. However, as MEN1 is an autosomal dominant disease, it may be more cost effective to construct a genealogy and undertake genetic screening

TABLE 4.3   Pituitary Tumor Characteristics in MEN1 and Non-MEN1 Patients

Characteristics	Pituitary adenoma		
	MEN1	Non-MEN1	P
Age (year)	38.0 ± 15.3	36.2 ± 14.6	NS
Mean follow-up (year)	11.1 ± 8.7	10.0 ± 6.3	NS
Adenoma type			
Prolactinoma	85	68	NS
GH-secreting	12	15	NS
ACTH-secreting	6	7	NS
Cosecreting	13	2	NS
Nonsecreting	20	18	NS
Tumor size			
Micro (*n*, %)	19 (14%)	64 (58%)	0.001
Macro (*n*, %)	116 (85%)	46 (42%)	
Clinical signs due to tumor size (*n*, %)	39 (29%)	15 (14%)	0.01
Normalization of pituitary hypersecretion (*n*, %)	49 (42%)	83 (90%)	0.001

Micro, microadenomas; macro, macroadenoma.

for the particular mutation discovered. This has the advantage of definitively identifying carriers quickly, and focusing resources and time on assessing their potential tumor expression. Also, mutation negative individuals and their descendents can be excluded from unnecessary further work-up. Carriers should be followed closely with regular biochemical, endocrine, and appropriate radiological screening for nascent tumors.

In familial MEN1, screening of potential carriers of a known mutation should not be age limited, as MEN1 can occasionally present in the pediatric setting or occult disease can be present for some time in some individuals. Again, DNA sequencing for the specific familial mutation, while potentially more expensive, is more definitive than biochemical or other clinically based methods and allows noncarriers to be excluded from further follow-up. Completion of MEN1 genealogies may benefit from sourcing and testing of relevant stored surgical pathological material in deceased members.

CARNEY COMPLEX (CNC)

J. Aidan Carney described a complex of myxomas, spotty pigmentation, and endocrine overactivity that included pituitary adenomas causing acromegaly in four of a total of 40 cases in his original series.[61] This condition, termed Carney complex (CNC), is rare and has been described in about 500 people in the largest database.[9] CNC is familial in 70% of cases, occurs in all racial groups, and has a slight female preponderance.[62] Two gene loci have been identified, one on chromosome 17q22-24;[63] the other is on chromosome 2p16.[64] The former is associated with the gene encoding the protein kinase A regulatory subunit Iα (PRKAR1A); mutations in PRKAR1A have been identified in up to 60% of CNC patients.[65] Most PRKAR1A mutations lead to mRNA instability, decreased or absent protein expression and PRKAR1A haploinsufficiency in CNC tumors.[66] Loss of heterozygosity at 17q22-24 and allelic loss have been shown in CNC tumors, while the loss of PRKAR1A function enhances intracellular response to cAMP in CNC tumors.[67] In knockout mouse models, the $Prkar1a^{-/-}$ state is lethal in embryonic life.[68,69] In heterozygous $Prkar1a^{+/-}$ mice, no typical CNC features are encountered. However, a transgenic mouse with an antisense PRKAR1A exon 2 construct develops multiple endocrine abnormalities similar to CNC.

CNC-Related Pituitary Tumors

The main endocrine abnormalities seen in CNC are primary pigmented nodular adrenocortical disease (PP-NAD), thyroid tumors and nodules, testicular tumors (large cell calcifying Sertoli cell tumor (LCCSCT), Leydig cell tumors) and acromegaly due to a pituitary adenoma.[70] Acromegaly itself is seen in a minority of patients with CNC (10% of cases), but about 75% of patients exhibit asymptomatic elevations in GH, IGF-1 or prolactin levels, or abnormal responses to dynamic pituitary testing. A histologic analysis of pituitary tumors in CNC patients with acromegaly reported that all tumors were prolactin and GH positive, while a minority also stained for thyroid-stimulating hormone, luteinizing hormone, or alpha-subunit.[71] A distinguishing feature of CNC-related acromegaly is multifocal hyperplasia of somatomammotropic cells that included nonadenomatous pituitary tissue within the tumors of CNC patients. The zones of hyperplasia were not well demarcated and exhibited increased cellularity and altered reticulin staining that merged with normal pituitary tissue. No consistent genetic abnormalities were seen on comparative genome hybridization. Electron microscopy showed that tumors from acromegalic patients with CNC demonstrate heterogeneous intracellular structure.[72] Acromegaly in CNC develops insidiously and may begin in apparently normal somatomammotrope tissue that undergoes multifocal hyperplasia to form GH/prolactin-secreting adenomas. As in MEN1, sporadic pituitary tumors do not exhibit mutations of the PRKAR1A gene.[73]

Recommendations for Testing in CNC

The diagnosis of CNC is clinical in the first instance, and patients displaying two or more of the following are generally considered to have CNC: PPNAD, cardiac myxoma, cutaneous myxoma, lentigines, blue nevi, LCC-SCT, thyroid nodules/tumor, ovarian cysts, acromegaly, melanotic Schwann cell tumor, and osteochondromyxoma. In patients that meet these criteria for CNC, germline DNA testing for mutations in the PRKAR1A gene should be undertaken. In familial cases, patients with a diagnosed close relative with CNC and even one of the above manifestations can be considered as having CNC. Based on current experience, it is expected that one-half to two-thirds of patients with clinical CNC will have a germline PRKAR1A mutation in the heterozygous state. Up to 80% of those with familial CNC will have a PRKAR1A mutation. On diagnosing a new incident case of CNC, kindred clinical assessment should be undertaken, but as with the case in MEN1 described earlier, it may be more cost effective to undertake DNA screening. A detailed genealogical tree may allow for targeted testing for a known PRKAR1A mutation beginning with first-degree relatives. This permits resource allocation to those identified as carriers and allows those with a wild-type genotype to avoid costly and repetitive endocrine and cardiac interventions. Genetic and clinical studies should begin as early as possible and in infancy in at-risk individuals as the disease can manifest at a young age. Accurate and early diagnosis is important mainly due to the threat of sudden death caused by cardiac myxoma (potentially multiple, and occurring in any cardiac chamber), rather than by endocrine manifestations. Carriers of a *PRKAR1A* mutation should be screened clinically, hormonally, and/or with imaging studies at least yearly for all manifestations of CNC. With respect to the pituitary, patients should have GH, IGF-I, and prolactin secretion assessed on a yearly basis as a minimum. A baseline pituitary magnetic resonance image (MRI) should be undertaken, but the timing of repeat imaging should be based on clinical progression, or the appearance of new hormonal alterations. Diagnosis and management of acromegaly in the setting of CNC is the same as for sporadic non-CNC-related disease, with transsphenoidal neurosurgery and medical therapy (somatostatin analogs, dopamine agonists) representing the mainstream forms of treatment.

MULTIPLE ENDOCRINE NEOPLASIA 4 (MEN4)

A MEN-like syndrome (MENX) that occurred spontaneously in the rat was reported between 2002 and 2004.[74,75] The rat phenotype consisted of multiple neuroendocrine cancers that included pheochromocytoma, medullary thyroid cell neoplasia, parathyroid adenomas, paragangliomas, pancreatic hyperplasia, and pituitary adenomas. These were preceded by the development of early cataracts within a few weeks of life. MENX was initially mapped to a chromosome 4 locus and was later revealed to occur due to a mutation in the *cyclin dependent kinase n1b (cdkn1b)* gene.[76] In humans the corresponding *CDKN1B* gene (which codes for p27[kip1]) is on chromosome 12 and Pellegata et al. identified a nonsense mutation in the *CDKN1B* gene in a German family exhibiting acromegaly, primary hyperparathyroidism, renal angiomyolipoma, and testicular cancer among various members. A Dutch patient with a pituitary adenoma (Cushing's disease), a cervical carcinoid tumor, hyperparathyroidism, and no *MEN1* mutation was found to have a *CDKN1B* mutation.[77] A study of a population with parathyroid and pituitary tumors and no *MEN1* mutation noted no abnormalities in *CDKN1B*.[78] Although it appears to be a very rare syndrome, given the multiple endocrine neoplastic features of *CDKN1B* mutations in the human, it has been proposed to call this condition MEN4.

Recommendations for Testing in MEN4

MEN4 is a very rare condition, and testing remains in the investigational setting. Testing of *MEN1* mutation negative families and cohorts with clinical disease highly suggestive of *MEN1* has shown that *CDKN1B* mutations occur only in exceptional cases. Currently there are no specific clinical features that would indicate a greater likelihood of the presence of MEN4.

FAMILIAL ISOLATED PITUITARY ADENOMAS (FIPA)

Pituitary tumors of all types can occur in multiple members of a single kindred in the absence of MEN1/ CNC, a condition termed *familial isolated pituitary adenomas* (FIPA). To date we have identified >200 FIPA kindreds in our collaborative series, and FIPA families have also been reported by separate research groups.[79–83] Mutations in the *AIP* gene in familial acromegaly kindreds has explained the pathophysiology of a proportion of cases. FIPA is not limited to the phenotype of acromegaly, and represents a clinical framework for further genetic study.

In FIPA, pituitary tumors of the same type can present in all affected family members (homogeneous presentation), or affected members can have different types of tumors (heterogeneous presentation). In an international study performed from 2000 to 2005, we identified a total of 64 FIPA families.[84] To date, FIPA kindreds with up to four affected members (i.e., subjects with pituitary tumors) have been described. The cohort is comprised equally of families with homogeneous and heterogeneous tumor types in affected members. The frequencies of the various different tumor types in FIPA are: prolactinoma (41%), somatotropinoma (30%), nonsecreting tumor (13%), somatolactotropinoma (7%), gonadotropinoma (4%), Cushing's disease (4%), and thyrotropinoma (1%). First-degree relationships between affected members within families occurs in approximately 75% of FIPA families. FIPA patients present with pituitary tumors 4 years earlier than their sporadic counterparts. In families with multiple affected generations, the children/grandchildren presented significantly earlier (20 years) than their parents/grandparents. Macroadenomas are seen in 63% of cases in FIPA kindreds. In terms of specific tumor types, prolactinomas in FIPA are mainly microadenomas occurring in women, while males invariably have macroadenomas, which largely reflect the characteristics of sporadic prolactinomas.[85] Prolactinomas in heterogeneous FIPA have higher rates of extension and invasion as compared with sporadic cases. In somatotropinoma patients from FIPA families, half occur as homogeneous acromegaly (familial acromegaly) families, and 50% in combination with other tumor types (heterogeneous families). Nonsecreting pituitary tumors occur in heterogeneous FIPA families, are diagnosed 8 years earlier, and have a higher rate of extension/invasion than sporadic tumors. Gonadotropinomas and Cushing's disease can occur rarely in a homogeneous FIPA setting.

In 2006, Vierimaa et al. reported the results of a comprehensive genetic study that identified mutations in the AIP gene as being associated with the familial presentation of somatotropinomas and prolactinomas.[18] Loss of heterozygosity at the AIP locus in tumor samples indicated that these tumors had lost the function of the normal allele in a "second hit," according to the Knudson model. Other families tested negative for AIP mutations. In the FIPA cohort we studied 73 FIPA families from nine countries, and 15% of the cohort had germline mutations in AIP.[86] Ten separate mutations were found, one of which (R304X) was found in a FIPA family that is apparently unrelated to a family from the same country (Italy) with the same mutation reported by Vierimaa et al. Patients with AIP mutations were significantly younger at diagnosis (12 years) than FIPA patients without AIP mutations. Tumors were larger in the AIP mutation positive groups versus the remainder of the cohort. Only 50% of those with homogeneous acromegaly had AIP mutations. Importantly, kindreds with strong familiarity for pituitary tumors (three or four affected) can be negative for mutations in AIP (and CDKN1B), which in-

dicates strongly that other genes may be involved in the causation of FIPA.[87]

Further analysis of the disease characteristics of FIPA patients with AIP mutations indicates that tumor and hormonal data are heterogeneous. Over 60% of AIP mutation positive patients with somatotropinomas had increased GH/IGF-I only, and the remaining 38% also had elevated prolactin. Somatotropinoma patients with AIP mutations can be immunohistochemically positive for GH alone (59%), GH and prolactin (33%), or GH and FSH (8%).

Since these initial studies, many AIP mutations have been described in the FIPA setting (Fig. 4.1). FIPA families with AIP mutations have also been reported by other groups.[88,89] A Q14X mutation, although found with high frequency in Finland, was not found in populations of sporadic adenomas from across the world, indicating it is a founder mutation.[90] Sporadic pituitary tumor patients infrequently have AIP mutations, although they are not entirely absent.[91] Overall, sporadic pituitary tumor patients with AIP mutations seem to present at a young age and mainly with somatotropinomas, although other pituitary tumor types do occur.[92] In particular, 10% of young patients (age at diagnosis less than 30 years) and 20% of children with sporadic macroadenomas have AIP

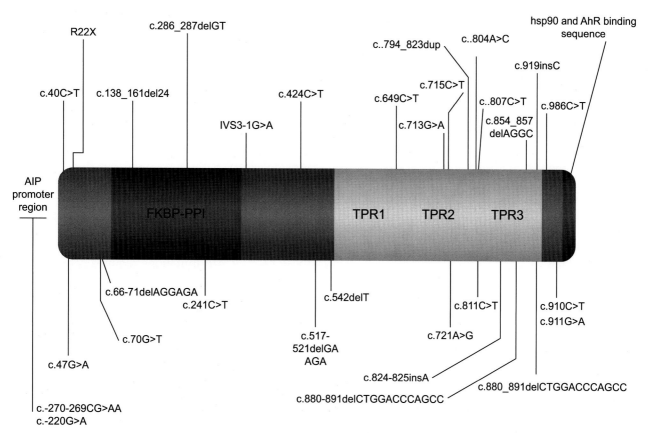

FIGURE 4.1 Mutations in the *AIP gene* reported in FIPA and sporadic pituitary tumor patients. FKBP-PPI, FK506 binding protein-type peptidyl-prolyl *cis–trans*-isomerase; TPR, tetratricopeptide repeat domain; hsp90, heat-shock protein 90; AhR, aryl hydrocarbon receptor.

mutations.[93] Cazabat et al. reported that in a total of 154 sporadic patients with acromegaly, five patients (3.2%) demonstrated *AIP* mutations. Studies in other tumors have revealed no firm evidence of germline *AIP* mutations as a potentially causative or contributory factor.[94]

Patients with *AIP* mut-somatotropinomas show male predominance and have larger and more aggressive tumors that are presented more than 20 years before than in *AIP* negative cases. This explains significantly higher frequency of gigantism among acromegaly patients with germline mutation in *AIP* gene than in those without mutations in *AIP* (32% vs. 6.5%).[19] Consecutively, we performed a study for assessing the role of genetic abnormalities in somatotropinoma patients with gigantism abnormal growth for age or final height >2SD above country local standards. *AIP* mutations account for about one third of cases with pituitary gigantism (29%), while other known genetic pituitary diseases, such as McCune-Albright syndrome (5%), Carney complex (1%), and MEN1 (1%) are rarer.[91] Thus, *AIP* mutations are more frequent among specific subgroups of patients (such as children and young adults or pituitary gigantism cases or FIPA families) than in sporadic cases.[95–99]

The manner by which *AIP* mutations cause pituitary adenomas in FIPA and apparently sporadic cases is largely unknown. Many *AIP* mutations described to date would involve truncations of the AIP protein, with the loss of a tetratricopeptide repeat domain and the carboxy terminal that are important for interactions with the other proteins such as heat shock protein 90 (hsp90) and the aryl hydrocarbon receptor.[100–104] Other missense mutations (e.g., R271W) involve highly conserved amino acids, which may alter AIP function in other ways. Whether various mutated versions of AIP are actually expressed or undergo mRNA degradation is unknown at this time. A variety of cellular effects are potentially related to AIP activity, of which modulation of phosphodiesterase PDE4A5 and phosphodiesterase PDE2A activities are of interest.[105,106] Leontiou et al. found that overexpression of wild-type AIP in HEK 293, human embryonic lung fibroblast (TIG 3), and the rat somatomammotroph (GH3) cell lines led to marked reductions in measures of cell proliferation.[81] When a variety of mutated forms of AIP were expressed in the cell lines (including a number of mutations described in the setting of FIPA), suppression of cell proliferation was negated. Also protein–protein interactions between AIP and PDE4A5 were disrupted by mutations in *AIP*. Immunohistochemical data revealed that in normal pituitary, AIP colocalized only with GH and prolactin secreting cells and was found in association with secretory granules. In sporadic tumors, AIP protein is expressed in all tumor types; however, it was only expressed in cytoplasm in prolactinomas, nonfunctioning, and Cushing disease tumors; AIP appeared to colocalize with secretory granules in somatotropinomas. Study of

Aip knockout models is at an early stage and no specific information on pituitary status has been reported.[107,108] Given the important role of AhR in mediating the biological effects on dioxin, the potential for pituitary tumorigenesis having its roots in environmental toxins has been suggested. Data from one of the most exhaustively studied industrial accidents involving dioxin exposure would argue otherwise. Pesatori et al. studied the incidence of pituitary tumors in the Seveso population exposed to 2,3,7,8-tetrachlorodibenzo-*para*-dioxin after an accident in 1976.[109] They found no significant increase in the incidence of pituitary tumors in this region, although given the often indolent nature of pituitary adenoma formation, this may require longer follow-up. Furthermore, a more intensive study of patients with very high clinical exposure to dioxins (e.g., those with chloracne and high dioxin titers) may be useful to finally discount the link between environmental dioxin exposure and increased rates of pituitary tumorigenesis.

Recommendations for AIP Mutation Testing in FIPA and Sporadic Pituitary Adenomas

Patients with FIPA are negative for *MEN1* and *PRKAR1A* mutations; *CDKN1B* mutations causing MEN4 are exceedingly rare. The study of FIPA and its molecular and genetic characteristics has progressed rapidly over the past decade. From a clinical perspective, it should be emphasized that clinically relevant pituitary adenomas are more common than previously thought (1:1000) and occur in a familial setting in about 5% of cases overall. Therefore, careful questioning regarding family history of pituitary disease should be part of the work-up of all patients with pituitary adenomas. The discovery linking mutations in *AIP* to pituitary tumors occurring in a family setting has provided particularly important impetus, but molecular pathophysiology remains very unclear. Families bearing *AIP* mutations have more aggressive pituitary tumors in affected members, and are often seen at a much younger age than is usual in the sporadic setting. Genetic screening for *AIP* mutations in patients with sporadic pituitary adenomas and in relatives of those bearing *AIP* mutations requires careful consideration. There remains divergence about the penetrance of pituitary adenomas among kindreds with *AIP* mutations, making the true risk of disease in mutation carriers unclear. We suggest that the penetrance of pituitary disease in *AIP* mutation-bearing FIPA kindreds may be relatively high, at least 33% in the largest kindreds.[110] For newly recognized FIPA families, we would recommend initial sequencing of the full length of the *AIP* gene in at least one affected individual. This remains an investigational study that is not offered commercially at this time, although the relatively short length of the *AIP* gene makes sequencing relatively uncomplicated. Sequencing of the *AIP* gene from somatic DNA in patients that have undergone pituitary neurosurgery is somewhat more complex, but has led to the identification of *AIP* mutations using the multiplex ligation-dependent probe amplification analysis (MLPA) methodology.[100] MLPA of germline DNA is also useful for the initial identification of large genetic deletions of the *AIP* gene or more extensive deletions, including the *AIP* gene.[111] In the case of relatives of patients with *AIP* mutation-related pituitary adenomas, we suggest constructing a careful genealogical tree and undertaking targeted germline *AIP* screening to identify carriers. As tumors in individuals with *AIP* mutations are more aggressive and occur at an earlier age, there is potential value in identifying carriers for the purpose of performing MRI and hormonal testing; the ultimate aim would be to diagnose tumors at as small a size as possible in order to permit potentially curative pituitary neurosurgical resection. In the absence of a tumor on MRI, follow-up of mutation carriers can be performed on a regular basis (yearly), relying predominantly on clinical symptoms and basal hormonal tests (IGF-I and prolactin).

Widespread *AIP* screening in unselected patients with sporadic pituitary adenomas is not warranted at this time. Young patients with aggressive pituitary tumors are also more likely to carry *AIP* mutations, and testing for *AIP* mutations among apparently sporadic populations should at this time be limited to such young cases.

Immunohistochemistry for *AIP* itself has been suggested as a useful diagnostic tool in pituitary tumor banks and in routine pathological practice. However, while the absence of AIP protein is likely with early truncating mutations, it is less clear how strongly other mutated forms of AIP are expressed at the protein level in pituitary tumors. In a recent study, Jaffrain-Rea et al. reported that cytoplasmic AIP was detected by immunohistochemistry in 84% of pituitary adenomas, with the highest AIP expression being observed in somatotropinomas and nonsecreting pituitary adenomas.[112] In somatotropinomas, significantly lower AIP immunostaining was seen in invasive as compared with noninvasive adenomas. Crucially, AIP expression was abolished in a minority of pituitary adenomas from patients with germline *AIP* mutations (including FIPA families). In general, immunohistochemistry is a poorly sensitive tool for screening for *AIP* mutations, and the field would benefit from the design of new antibodies with a range of epitopes that are appropriate for detecting a range of mutated AIP proteins.

X-Linked Acrogigantism (X-LAG) Syndrome

Recently, we described a new pediatric syndrome, X-linked acrogigantism syndrome (X-LAG) caused by a chromosome Xq26.3 microduplication and characterized by the early onset of gigantism due to excessive secretion of growth hormone by adenomas and hyperplasia of somatotropes. To date we observed Xq26.3 microduplication in samples from >20 patients with gigantism. Four were obtained from members of two FIPA families, and others were sporadic cases. The characterization of Xq26.3 region on X chromosome by high density array comparative genomic hybridization (aCGH) and breakpoint sequencing showed the precise limits of the duplication. All sporadic cases had an original duplication, while familial cases had inherited identical duplications.[20] This syndrome affects mainly females (71%); pituitary tumors are large and almost all are mixed tumors that cosecrete growth hormone and prolactin. In all patients, the disease appeared in infancy. Among patients with gigantism that do not bear the Xq26.3 microduplication, none has grown excessively before the age of 5 years. Somatic overgrowth was accompanied by acral enlargement and facial acromegalic changes. In about 25% increased appetite was reported, and several cases had signs of insulin resistance (e.g., acanthosis nigricans). Baseline GH and IGF1 levels were extremely elevated with high levels of hyperprolactinemia being usual. Control of the pituitary disease and linear growth by different treatment options (including traditional somatostatin analogs and neurosurgery) was notably poor and postoperative hypopituitarism was frequent.[21] All patients with X-LAG shared common area of overlap in the duplicated region on chromosome X, involving *CD40L*, *ARHGEF6*, *RBMX*, and *GPR101* genes. The expression of these four genes has been studied by quantitative PCR in pituitary tumors removed during surgery in two patients with X-LAG syndrome. Only one gene (GPR101), which encodes a receptor with seven transmembrane domains and coupled to a G-protein was strongly overexpressed. And it is the likely cause of the phenotype in X-LAG syndrome. In addition, a mutation (p.E308D) of *GPR101*, possibly activating, has been identified in several patients with sporadic pituitary adenomas (eleven mutated cases in 248 patients studied), mostly in tumor DNA.[20,93]

Recommendations for Testing in X-LAG

To date Xq26.3 microduplications were detected only in patients with a very specific phenotype of infantile gigantism.[21] According to the results of the largest international study on pituitary gigantism X-LAG syndrome explained most of cases with early-onset disease in this cohort.[92] Therefore, we recommend that children with markedly increased height and weight (>2 SD) and growth hormone hypersecretion, and adult patients with a history early childhood-onset gigantism with excessive secretion of GH by pituitary hyperplasia or adenoma should be tested. As it was noted above, some X-LAG patients had inherited pituitary disease. Thus, Xq26.3 microduplication represent a newly described cause in certain FIPA kindreds. Therefore, we would recommend performing array comparative genomic hybridization in *AIP* –negative FIPA with acrogigantism.[92–95]

References

1. Agarwal SK, Lee Burns A, Sukhodolets KE, et al. Molecular pathology of the MEN1 gene. *Ann NY Acad Sci* 2004;**1014**:189–98.

2. Larsson C, Skogseid B, Oberg K, et al. Multiple endocrine neoplasia type 1 gene maps to chromosome 11 and is lost in insulinoma. *Nature* 1988;**332**(6159):85–7.

3. Chandrasekharappa SC, Guru SC, Manickam P, et al. Positional cloning of the gene for multiple endocrine neoplasia-type 1. *Science* 1997;**276**(5311):404–7.

4. Scacheri PC, Davis S, Odom DT, et al. Genome-wide analysis of menin binding provides insights into MEN1 tumorigenesis. *PLoS Genet* 2006;**2**(4):e51.

5. Agarwal SK, Impey S, McWeeney S, et al. Distribution of menin-occupied regions in chromatin specifies a broad role of menin in transcriptional regulation. *Neoplasia* 2007;**9**(2):101–7.

6. Lemos MC, Thakker RV. Multiple endocrine neoplasia type 1 (MEN1): analysis of 1336 mutations reported in the first decade following identification of the gene. *Hum Mutat* 2008;**29**(1):22–32.

7. Vergès B, Boureille F, Goudet P, et al. Pituitary disease in MEN type 1 (MEN1): data from the France-Belgium MEN1 multicenter study. *J Clin Endocrinol Metab* 2002;**87**(2):457–65.

8. Beckers A, Betea D, Valdes Socin H, et al. The treatment of sporadic versus MEN1-related pituitary adenomas. *J Int Med* 2003;**253**: 599–605.

9. Boikos SA, Stratakis CA. Carney complex: the first 20 years. *Curr Opin Oncol* 2007;**19**:24–9.

10. Stratakis CA, Kirschner LS, Carney JA. Clinical and molecular features of the Carney complex: diagnostic criteria and recommendations for patient evaluation. *J Clin Endocrinol Metab* 2001;**86**(9): 4041–6.

11. Casey M, Mah C, Merliss AD, et al. Identification of a novel genetic locus for familial cardiac myxomas and Carney complex. *Circulation* 1998;**98**:2560–6.

12. Fritz A, Walch A, Piotrowska K, et al. Recessive transmission of a multiple endocrine neoplasia syndrome in the rat. *Cancer Res* 2002;**62**(11):3048–51.

13. Pellegata NS, Quintanilla-Martinez L, Siggelkow H, et al. Germline mutations in p27Kip1 cause a multiple endocrine neoplasia syndrome in rats and humans. *Proc Natl Acad Sci USA* 2006;**103**(42): 15558–63.

14. Georgitsi M, Raitila A, Karhu A, et al. Germline CDKN1B/p27Kip1 mutation in multiple endocrine neoplasia. *J Clin Endocrinol Metab* 2007;**92**(8):3321–5.

15. Teh BT, Kytölä S, Farnebo F, et al. Mutation analysis of the MEN1 gene in multiple endocrine neoplasia type 1, familial acromegaly and familial isolated hyperparathyroidism. *J Clin Endocrinol Metab* 1998;**83**:2621–6.

16. Poncin J, Abs R, Velkeniers B, et al. Mutation analysis of the MEN1 gene in Belgian patients with multiple endocrine neoplasia type 1 and related diseases. *Hum Mutat* 1999;**13**:54–60.

17. Luccio-Camelo DC, Une KN, Ferreira RE, et al. A meiotic recombination in a new isolated familial somatotropinoma kindred. *Eur J Endocrinol* 2004;**150**(5):643–8.

18. Vierimaa O, Georgitsi M, Lehtonen R, et al. Pituitary adenoma predisposition caused by germline mutations in the AIP gene. *Science* 2006;**312**(5777):1228–30.

19. Daly AF, Tichomirowa MA, Petrossians P, Heliovaara E, Jaffrain-Rea ML, Barlier A, et al. Clinical characteristics and therapeutic responses in patients with germ-line AIP mutations and pituitary adenomas: an international collaborative study. *J Clin Endocrinol Metab* 2010;**95**(11):E373–83.

20. Trivellin G, Daly AF, Faucz FR, Yuan B, Rostomyan L, Larco DO, Schernthaner-Reiter MH, Szarek E, Leal LF, Caberg JH, Castermans E, Villa C, Dimopoulos A, Chittiboina P, Xekouki P, Shah N, Metzger D, Lysy PA, Ferrante E, Strebkova N, Mazerkina N, Zatelli MC, Lodish M, Horvath A, de Alexandre RB, Manning AD, Levy I, Keil MF, Sierra Mde L, Palmeira L, Coppieters W, Georges M, Naves LA, Jamar M, Bours V, Wu TJ, Choong CS, Bertherat J, Chanson P, Kamenický P, Farrell WE, Barlier A, Quezado M, Bjelobaba I, Stojilkovic SS, Wess J, Costanzi S, Liu P, Lupski JR, Beckers A, Stratakis CA. Gigantism and acromegaly due to Xq26 microduplications and GPR101 mutation. *N Engl J Med* 2014;**371**(25): 2363–74.

21. Beckers A, Lodish MB, Trivellin G, Rostomyan L, Lee M, Faucz FR, Yuan B, Choong CS, Caberg JH, Verrua E, Naves LA, Cheetham TD, Young J, Lysy PA, Petrossians P, Cotterill A, Shah NS, Metzger D, Castermans E, Ambrosio MR, Villa C, Strebkova N, Mazerkina N, Gaillard S, Barra GB, Casulari LA, Neggers SJ, Salvatori R, Jaffrain-Rea ML, Zacharin M, Santamaria BL, Zacharieva S, Lim EM, Mantovani G, Zatelli MC, Collins MT, Bonneville JF, Quezado M, Chittiboina P, Oldfield EH, Bours V, Liu P, W de Herder W, Pellegata N, Lupski JR, Daly AF, Stratakis CA. X-linked acrogigantism syndrome: clinical profile and therapeutic responses. *Endocr Relat Cancer* 2015;**22**(3):353–67.

22. Karga HJ, Alexander JM, Hedley-Whyte ET, et al. Ras mutations in human pituitary tumors. *J Clin Endocrinol Metab* 1992;**74**: 914–9.

23. Pei L, Melmed S, Scheithauer B, et al. H-ras mutations in human pituitary carcinoma metastases. *J Clin Endocrinol Metab* 1994;**78**: 842–6.

24. Zhang X, Horwitz GA, Heaney AP, et al. Pituitary tumor transforming gene (PTTG) expression in pituitary adenomas. *J Clin Endocrinol Metab* 1999;**84**:761–7.

25. Minematsu T, Suzuki M, Sanno N, et al. PTTG overexpression is correlated with angiogenesis in human pituitary adenomas. *Endocr Pathol* 2006;**17**:143–53.

26. Hu N, Gutsmann A, Herbert DC, et al. Heterozygous Rb-1 delta 20/+mice are predisposed to tumors of the pituitary gland with a nearly complete penetrance. *Oncogene* 1994;**9**:1021–102.

27. Zhu J, Leon SP, Beggs AH, et al. Human pituitary adenomas show no loss of heterozygosity at the retinoblastoma gene locus. *J Clin Endocrinol Metab* 1994;**78**:922–7.

28. Simpson DJ, Hibberts NA, McNicol AM, et al. Loss of pRb expression in pituitary adenomas is associated with methylation of the RB1 CpG island. *Cancer Res* 2000;**60**:1211–6.

29. Hibberts NA, Simpson DJ, Bicknell JE, et al. Analysis of cyclin D1 (CCND1) allelic imbalance and overexpression in sporadic human pituitary tumors. *Clin Cancer Res* 1999;**5**:2133–9.

30. Simpson DJ, Fryer AA, Grossman AB, et al. Cyclin D1 (CCND1) genotype is associated with tumour grade in sporadic pituitary adenomas. *Carcinogenesis* 2001;**22**:1801–7.

31. Woloschak M, Yu A, Xiao J, et al. Frequent loss of the P16INK4a gene product in human pituitary tumors. *Cancer Res* 1996;**56**: 2493–6.

32. Farrell WE, Simpson DJ, Bicknell JE, et al. Chromosome 9p deletions in invasive and noninvasive nonfunctional pituitary adenomas: the deleted region involves markers outside of the MTS1 and MTS2 genes. *Cancer Res* 1997;**57**:2703–9.

33. Fero ML, Rivkin M, Tasch M, et al. A syndrome of multiorgan hyperplasia with features of gigantism, tumorigenesis, and female sterility in p27(Kip1)-deficient mice. *Cell* 1996;**85**: 733–44.

34. Pagotto U, Arzberger T, Theodoropoulou M, et al. The expression of the antiproliferative gene ZAC is lost or highly reduced in non-functioning pituitary adenomas. *Cancer Res* 2000;**60**:6794–9.

35. Theodoropoulou M, Zhang J, Laupheimer S, et al. Octreotide, a somatostatin analogue, mediates its antiproliferative action in pituitary tumor cells by altering phosphatidylinositol 3-kinase signaling and inducing Zac1 expression. *Cancer Res* 2000;**66**: 1576–82.

36. Ezzat S, Asa SL. Mechanisms of disease: the pathogenesis of pituitary tumors. *Nat Clin Pract Endocrinol Metab* 2006;**2**:220–30.

37. Zhang X, Zhou Y, Mehta KR, et al. A pituitary-derived MEG3 isoform functions as a growth suppressor in tumor cells. *J Clin Endocrinol Metab* 2003;**88**:5119–26.

38. Zhao J, Dahle D, Zhou Y, et al. Hypermethylation of the promoter region is associated with the loss of MEG3 gene expression in human pituitary tumors. *J Clin Endocrinol Metab* 2005;**90**:2179–86.

39. Zhang X, Sun H, Danila DC, et al. Loss of expression of GADD45 gamma, a growth inhibitory gene, in human pituitary adenomas: implications for tumorigenesis. *J Clin Endocrinol Metab* 2002;**87**:1262–7.

40. Paez-Pereda M, Giacomini D, Refojo D, et al. Involvement of bone morphogenetic protein 4 (BMP-4) in pituitary prolactinoma pathogenesis through a Smad/estrogen receptor crosstalk. *Proc Natl Acad Sci USA* 2003;**100**:1034–9.

41. Scheithauer BW, Laws Jr ER, Kovacs K, et al. Pituitary adenomas of the multiple endocrine neoplasia type I syndrome. *Semin Diagn Pathol* 1987;**4**(3):205–11.

42. Daly A, Rixhon M, Adam C, et al. High prevalence of pituitary adenomas: a cross sectional study in the Province of Liege, Belgium. *J Clin Endocrinol Metab* 2006;**91**:4769–75.

43. Erdheim J. Zur normalen und pathologischen Histologic der Glandula thyreoidea, parathyreoidea und Hypophysis. *Beitr z path Anat u z allg Path* 1903;**33**:158–233.

44. Wermer P. Genetic aspects of adenomatosis of endocrine glands. *Am J Med* 1954;**16**(3):363–71.

45. Skogseid B, Eriksson B, Lundqvist G, et al. Multiple endocrine neoplasia type 1: a 10-year prospective screening study in four kindreds. *J Clin Endocrinol Metab* 1991;**73**:281–7.

46. Burgess JR, Shepherd JJ, Parameswaran V, et al. Spectrum of pituitary disease in multiple endocrine neoplasia type 1 (MEN1): clinical, biochemical, and radiological features of pituitary disease in a large MEN1 kindred. *J Clin Endocrinol Metab* 1996;**81**: 2642–6.

47. Marx S, Spiegel AM, Skarulis MC, et al. Multiple endocrine neoplasia type 1: clinical and genetic topics. *Ann Intern Med* 1998;**129**: 484–94.

48. Zhuang Z, Ezzat SZ, Vortmeyer AO, et al. Mutations of the MEN1 tumor suppressor gene in pituitary tumors. *Cancer Res* 1997;**57**(24): 5446–51.

49. Poncin J, Stevenaert A, Beckers A. Somatic MEN1 gene mutation does not contribute significantly to sporadic pituitary tumorigenesis. *Eur J Endocrinol* 1999;**140**(6):573–6.

50. Wenbin C, Asai A, Teramoto A, et al. Mutations of the MEN1 tumor suppressor gene in sporadic pituitary tumors. *Cancer Lett* 1999;**142**(1):43–7.

51. Schmidt MC, Henke RT, Stangl AP, et al. Analysis of the MEN1 gene in sporadic pituitary adenomas. *J Pathol* 1999;**188**(2):168–73.

52. Fukino K, Kitamura Y, Sanno N, et al. Analysis of the MEN1 gene in sporadic pituitary adenomas from Japanese patients. *Cancer Lett* 1999;**144**(1):85–92.

53. Asa SL, Somers K, Ezzat S. The MEN-1 gene is rarely down-regulated in pituitary adenomas. *J Clin Endocrinol Metab* 1998;**83**(9): 3210–2.

54. Tanaka C, Kimura T, Yang P, et al. Analysis of loss of heterozygosity on chromosome 11 and infrequent inactivation of the MEN1 gene in sporadic pituitary adenomas. *J Clin Endocrinol Metab* 1998;**83**(8): 2631–4.

55. Tanaka C, Yoshimoto K, Yamada S, et al. Absence of germ-line mutations of the multiple endocrine neoplasia type 1 (MEN1) gene in familial pituitary adenoma in contrast to MEN1 in Japanese. *J Clin Endocrinol Metab* 1998;**83**(3):960–5.

56. Theodoropoulou M, Cavallari I, Barzon L, et al. Differential expression of menin in sporadic pituitary adenomas. *Endocr Relat Cancer* 2004;**11**(2):333–44.

57. Beckers A, Abs R, Reyniers E, et al. Variable regions of chromosome 11 loss in different pathological tissues of a patient with the multiple endocrine neoplasia type I syndrome. *J Clin Endocrinol Metab* 1994;**79**(5):1498–502.

58. Farid NR, Buehler S, Russell NA, et al. Prolactinomas in familial multiple endocrine neoplasia syndrome type I. Relationship to HLA and carcinoid tumors. *Am J Med* 1980;**69**(6):874–80.

59. Olufemi SE, Green JS, Manickam P, et al. Common ancestral mutation in the MEN1 gene is likely responsible for the prolactinoma variant of MEN1 (MEN1Burin) in four kindreds from Newfoundland. *Hum Mutat* 1998;**11**(4):264–9.

60. Thakker R, et al. Clinical practice guidelines for multiple endocrine neoplasia Type 1 (MEN1). *J Clin Endocrinol Metab* 2012;**97**: 2990–3011.

61. Carney JA, Hruska LS, Beauchamp GD, et al. Dominant inheritance of the complex of myxomas, spotty pigmentation and endocrine overactivity. *Mayo Clinic Proc* 1985;**61**:165–72.

62. Stratakis CA, Kirschner LS, Carney JA. Clinical and molecular features of the Carney complex: diagnostic criteria and recommendations for patient evaluation. *J Clin Endocrinol Metab* 2001;**86**(9): 4041–6.

63. Casey M, Mah C, Merliss AD, et al. Identification of a novel genetic locus for familial cardiac myxomas and Carney complex. *Circulation* 1998;**98**:2560–6.

64. Stratakis CA, Carney JA, Lin JP, et al. Carney complex, a familial multiple endocrine neoplasia and lentiginosis syndrome. Analysis of 11 kindreds and linkage to the short arm of chromosome 2. *J Clin Invest* 1996;**97**:699–705.

65. Veugelers M, Wilkes D, Burton K, et al. Comparative PRKAR1A genotype–phenotype analyses in humans with Carney complex and prkar1a haploinsufficient mice. *Proc Natl Acad Sci USA* 2004;**101**(39):14222–7.

66. Kirschner LS, Sandrini F, Monbo J, et al. Genetic heterogeneity and spectrum of mutations of the PPKAR1A gene in patients with the Carney complex. *Hum Mol Genet* 2000;**9**:3037–46.

67. Bossis I, Stratakis CA. Minireview: PRKAR1: normal and abnormal functions. *Endocrinology* 2004;**145**:5452–8.

68. Griffin KJ, Kirschner LS, Matyakhina L, et al. Down-regulation of regulatory subunit type 1A of protein kinase A leads to endocrine and other tumors. *Cancer Res* 2004;**64**(24):8811–5.

69. Griffin KJ, Kirschner LS, Matyakhina L, et al. A transgenic mouse bearing an antisense construct of regulatory subunit type 1A of protein kinase A develops endocrine and other tumours. *J Med Genet* 2004;**41**(12):923–31.

70. Stergiopoulos SG, Stratakis CA. Human tumors associated with Carney complex and germline PPKAR1A mutations: a protein kinase A disease. *FEBS Lett* 2003;**546**:59–64.

71. Pack SD, Kirschner LS, Pak E, et al. Genetic and histological studies of somatomammotropic tumors in patients with the "Complex of spotty skin pigmentation, myxomas, endocrine overactivity and schwannomas" (Carney complex). *J Clin Endocrinol Metab* 2000;**85**: 3860–5.

72. Kurtkaya-Yapicier O, Scheithauer BW, Carney JA, et al. Pituitary adenoma in Carney complex: an immunohistochemical, ultrastructural, and immunoelectron microscopic study. *Ultrastruct Pathol* 2002;**26**(6):345–53.

73. Kaltsas GA, Kola B, Borboli N, et al. Sequence analysis of the PRKAR1A gene in sporadic somatotroph and other pituitary tumours. *Clin Endocrinol (Oxf)* 2002;**57**(4):443–8.

74. Fritz A, Walch A, Piotrowska K, et al. Recessive transmission of a multiple endocrine neoplasia syndrome in the rat. *Cancer Res* 2002;**62**(11):3048–51.

75. Piotrowska K, Pellegata NS, Rosemann M, et al. Mapping of a novel MEN-like syndrome locus to rat chromosome 4. *Mamm Genome* 2004;**15**(2):135–41.

76. Pellegata NS, Quintanilla-Martinez L, Siggelkow H, et al. Germline mutations in p27Kip1 cause a multiple endocrine neoplasia syndrome in rats and humans. *Proc Natl Acad Sci USA* 2006;**103**(42):15558–63.

77. Georgitsi M, Raitila A, Karhu A, et al. Germline CDKN1B/p27Kip1 mutation in multiple endocrine neoplasia. *J Clin Endocrinol Metab* 2007;**92**(8):3321–5.

78. Ozawa A, Agarwal SK, Mateo CM, et al. The parathyroid/pituitary variant of multiple endocrine neoplasia type 1 usually has causes other than p27Kip1 mutations. *J Clin Endocrinol Metab* 2007;**92**(5):1948–51.

79. Beckers A, Daly AF. The clinical, pathological, and genetic features of familial isolated pituitary adenomas. *Eur J Endocrinol* 2007;**157**(4):371–82.

80. Villa C, Magri F, Morbini P, et al. Silent familial isolated pituitary adenomas: histopathological and clinical case report. *Endocr Pathol* 2008;**19**(1):40–6.

81. Leontiou CA, Gueorguiev M, van der Spuy J, et al. The role of the aryl hydrocarbon receptor-interacting protein gene in familial and sporadic pituitary adenomas. *J Clin Endocrinol Metab* 2008;**93**(6):2390–401.

82. Beckers A, Aaltonen LA, Daly AF, Karhu A. Familial isolated pituitary adenomas (FIPA) and the pituitary adenoma predisposition due to mutations in the aryl hydrocarbon receptor interacting protein (AIP) gene. *Endocr Rev* 2013;**34**(2):239–77.

83. Daly AF, Beckers A. Familial isolated pituitary adenomas (FIPA) and mutations in the aryl hydrocarbon receptor interacting protein (AIP) gene. *Endocrinol Metab Clin North Am* 2015;**44**(1):19–25.

84. Daly AF, Jaffrain-Rea ML, Ciccarelli A, et al. Clinical characterization of familial isolated pituitary adenomas. *J Clin Endocrinol Metab* 2006;**91**(9):3316–23.

85. Ciccarelli A, Daly AF, Beckers A. The epidemiology of prolactinomas. *Pituitary* 2005;**8**(1):3–6.

86. Daly AF, Vanbellinghen JF, Khoo SK, et al. Aryl hydrocarbon receptor-interacting protein gene mutations in familial isolated pituitary adenomas: analysis in 73 families. *J Clin Endocrinol Metab* 2007;**92**(5):1891–6.

87. Tichomirowa MA, Lee M, Barlier A, Daly AF, Marinoni I, Jaffrain-Rea ML, Naves LA, Rodien P, Rohmer V, Faucz FR, Caron P, Estour B, Lecomte P, Borson-Chazot F, Penfornis A, Yaneva M, Guitelman M, Castermans E, Verhaege C, Wémeau JL, Tabarin A, Fajardo Montañana C, Delemer B, Kerlan V, Sadoul JL, Cortet Rudelli C, Archambeaud F, Zacharieva S, Theodoropoulou M, Brue T, Enjalbert A, Bours V, Pellegata NS, Beckers A. Cyclin-dependent kinase inhibitor 1B (CDKN1B) gene variants in AIP mutation-negative familial isolated pituitary adenoma kindreds. *Endocr Relat Cancer* 2012;**19**(3):233–41.

88. Iwata T, Yamada S, Mizusawa N, et al. The aryl hydrocarbon receptor-interacting protein gene is rarely mutated in sporadic GH-secreting adenomas. *Clin Endocrinol (Oxf)* 2007;**66**(4):499–502.

89. Toledo RA, Lourenco Jr DM, Liberman B, et al. Germline mutation in the aryl hydrocarbon receptor interacting protein gene in familial somatotropinoma. *J Clin Endocrinol Metab* 2007;**92**(5):1934–7.

90. Yu R, Bonert V, Saporta I, et al. AIP variants in sporadic pituitary adenomas. *J Clin Endocrinol Metab* 2006;**91**(12):5126–9.

91. Barlier A, Vanbellinghen JF, Daly AF, et al. Mutations in the aryl hydrocarbon receptor interacting protein gene are not highly prevalent among subjects with sporadic pituitary adenomas. *J Clin Endocrinol Metab* 2007;**92**(5):1952–5.

92. Georgitsi M, Raitila A, Karhu A, et al. Molecular diagnosis of pituitary adenoma predisposition caused by aryl hydrocarbon receptor-interacting protein gene mutations. *Proc Natl Acad Sci USA* 2007;**104**(10):4101–5105.

93. Tichomirowa MA, Barlier A, Daly AF, Jaffrain-Rea ML, Ronchi C, Yaneva M, et al. High prevalence of AIP gene mutations following focused screening in young patients with sporadic pituitary macroadenomas. *Eur J Endocrinol/Eur Fed Endocr Soc* 2011;**165**(4):509–15.

94. Georgitsi M, Karhu A, Winqvist R, et al. Mutation analysis of aryl hydrocarbon receptor interacting protein (AIP) gene in colorectal, breast, and prostate cancers. *Br J Cancer* 2007;**96**(2):352–6.

95. Cuny T, Pertuit M, Sahnoun-Fathallah M, Daly A, Occhi G, Odou MF, et al. Genetic analysis in young patients with sporadic pituitary macroadenomas: besides AIP don't forget MEN1 genetic analysis. *Eur J Endocrinol/Eur Fed Endocr Soc* 2013;**168**(4):533–41.

96. Stratakis CA, Tichomirowa MA, Boikos S, Azevedo MF, Lodish M, Martari M, et al. The role of germline AIP, MEN1, PRKAR1A, CDKN1B and CDKN2C mutations in causing pituitary adenomas in a large cohort of children, adolescents, and patients with genetic syndromes. *Clin Genet* 2010;**78**(5):457–63.

97. Jennings JE, Georgitsi M, Holdaway I, Daly AF, Tichomirowa M, Beckers A, et al. Aggressive pituitary adenomas occurring in young patients in a large Polynesian kindred with a germline R271W mutation in the AIP gene. *Eur J Endocrinol* 2009;**161**(5):799–804.

98. Rostomyan L, Daly AF, Petrossians P, Natchev E, Lila AR, Lecoq AL, Lecumberri B, Trivellin G, Salvatori R, Moraitis A, Holdaway I, Kranenburg-Van Klaveren D, Zatelli MC, Palacios N, Nozieres C, Zacharin M, Ebeling TM, Ojaniemi M, Rozhinskaya L, Verrua E, Jaffrain Rea ML, Filipponi S, Guskova D, Pronin V, Bertherat J, Belaya Z, Ilovaiskaya I, Sahnoun Fathallah M, Sievers C, Stalla GK, Castermans E, Caberg JH, Sorkina E, Auriemma R, Mittal S, Kareva M, Lysy P, Emy P, de Menis E, Choong C, Mantovani G, Bours V, de Herder WW, Brue T, Barlier A, Neggers S, Zacharieva S, Chanson P, Shah N, Stratakis CA, Naves LA, Beckers A. Clinical and genetic characterization of pituitary gigantism: an international study in 208 patients. *Endocr Relat Cancer* 2015 Jul 17; pii: ERC-15-0320. [Epub ahead of print] PubMed PMID: 26187128.

99. Daly AF, Trivellin G, Stratakis CA. Gigantism, acromegaly, and GPR101 mutations. *N Engl J Med* 2015;**372**(13):1265.

100. Petrulis JR, Perdew GH. The role of chaperone proteins in the aryl hydrocarbon receptor core complex. *Chem Biol Interact* 2002;**141**(1–2):25–40.

101. Bell DR, Poland A. Binding of aryl hydrocarbon receptor (AhR) to AhR-interacting protein. The role of hsp90. *J Biol Chem* 2000;**275**(46):36407–14.

102. Meyer BK, Petrulis JR, Perdew GH. Aryl hydrocarbon (Ah) receptor levels are selectively modulated by hsp90-associated immunophilin homolog XAP2. *Cell Stress Chaper* 2000;**5**(3):243–54.

103. Carver LA, LaPres JJ, Jain S, et al. Characterization of the Ah receptor-associated protein, ARA9. *J Biol Chem* 1998;**273**(50):33580–7.

104. Meyer BK, Perdew GH. Characterization of the AhR-hsp90-XAP2 core complex and the role of the immunophilin-related protein XAP2 in AhR stabilization. *Biochemistry* 1999;**38**(28):8907–17.

105. Bolger GB, Peden AH, Steele MR, et al. Attenuation of the activity of the cAMP-specific phosphodiesterase PDE4A5 by interaction with the immunophilin XAP2. *J Biol Chem* 2003;**278**(35):33351–63.

106. de Oliveira SK, Hoffmeister M, Gambaryan S, et al. Phosphodiesterase 2A forms a complex with the co-chaperone XAP2 and regulates nuclear translocation of the aryl hydrocarbon receptor. *J Biol Chem* 2007;**282**(18):13656–63.

107. Lin BC, Sullivan R, Lee Y, et al. Deletion of the aryl hydrocarbon receptor-associated protein 9 leads to cardiac malformation and embryonic lethality. *J Biol Chem* 2007;**282**(49):35924–32.

108. Lin BC, Nguyen LP, Walisser JA, et al. A hypomorphic allele of aryl-hydrocarbon receptor-associated-9 produces a phenocopy of the Ahr null. *Mol Pharmacol* 2008;**74**(5):1367–71.

109. Pesatori A, Baccarelli A, Consonni D, et al. Aryl hydrocarbon receptor interacting protein and pituitary adenomas: a population-based study on subjects exposed to dioxin after the Seveso, Italy, accident. *Eur J Endocrinol* 2008;**159**(6):699–703.

110. Naves LA, Daly AF, Vanbellinghen JF, et al. Variable pathological and clinical features of a large Brazilian family harboring a muta-tion in the aryl hydrocarbon receptor-interacting protein gene. *Eur J Endocrinol* 2007;**157**(4):383–91.

111. Georgitsi M, Heliövaara E, Paschke R, et al. Large genomic deletions of aryl hydrocarbon receptor interacting protein (AIP) gene in pituitary adenoma predisposition. *J Clin Endocrin Metab* 2009;**93**(10):4146–51.

112. Jaffrain Rea ML, Angelini M, Gargano D, et al. Expression of aryl hydrocarbon receptor (AHR) and aryl hydrocarbon receptor interacting protein (AIP) in pituitary adenomas: pathological and clinical implications. *Endocr Relat Cancer* 2009;**16**:1029–43.

5

Diabetes Insipidus

Jane Hvarregaard Christensen*, Søren Rittig**

*Department of Biomedicine, Aarhus University, Aarhus, Denmark
**Department of Pediatrics, Aarhus University Hospital, Aarhus, Denmark

INTRODUCTION

Disorders of water balance are common, and practicing physicians are often met with complaints of increased urination and thirst. Only a minority of such patients, however, suffers from diabetes insipidus and in even fewer the symptoms are caused by genetic defects in any one of the genes being essential in ensuring proper water homeostasis of the human body. Nevertheless, such patients should be identified and subjected to proper clinical and genetic testing in order to secure correct diagnosis and optimal treatment. The clinical differential diagnosis of diabetes insipidus can be challenging, and there are multiple examples of misdiagnosis, especially when the disease presents in a partial form. Although familial forms of diabetes insipidus were recognized more than 150 years ago, the genetic causes have only been revealed in recent decades. At present, genetic testing represents not only an important tool in the differential diagnosis of diabetes insipidus, but also a major advance in clinical care, as it has to some extent eliminated the problem of trying to identify at birth which offspring are likely to develop the disease.

The term "diabetes insipidus" is derived from the Greek "diabainein," meaning to pass through, and from Latin "insipidus," meaning tasteless. A distinction between diabetes associated with sweet tasting urine (diabetes mellitus) and diabetes associated with insipid (tasteless) urine (diabetes insipidus) seems to have been firmly established in 1831,[1,2] but as early as in 1790, a case of diabetes in which "the urine was perfectly insipid" was recorded in the literature and noted to be "very rare; and that the other (read: diabetes mellitus) is by much the more common."[2,3]

The prevalence of diabetes insipidus is yet not clearly established but has been estimated to be approximately 1:25,000 and with an annual incidence of approximately 0.01%.[4] Of these, less than 10% are inherited.[5]

Diabetes insipidus is characterized by the excretion of abnormally large volumes of dilute urine under conditions of *ad libitum* fluid intake.[6–10] Diabetes insipidus is defined clinically by the following two criteria:

1. 24-h urine volume exceeding 50 mL/kg body weight (in children 75–100 mL/kg body weight, due to higher water content in their food)
2. Urinary osmolality < 300 mosmol/kg

The polyuria is distinguishable from the osmotic diuresis of uncontrolled diabetes mellitus or other forms of solute diuresis by the absence of glucosuria and a normal rate of solute excretion (10–20 mosmol/kg per day).

This chapter aims to provide a brief overview of clinical and genetic aspects of different types of diabetes insipidus and the differential diagnosis and information about the indications and practicalities of genetic testing in diabetes insipidus available today.

TYPES OF DIABETES INSIPIDUS

Diabetes insipidus can be divided into four different types that are caused by any one of four fundamentally different defects (Fig. 5.1): 1. *pituitary, central, neurogenic,* or *neurohypophyseal* diabetes insipidus, the most common type, results from a deficiency in the production of the antidiuretic hormone arginine vasopressin (AVP); 2. *renal or nephrogenic* diabetes insipidus is caused by renal insensitivity to the antidiuretic effects of AVP, for example, due to impairment of the renal vasopressin V2 receptor or aquaporin-2 water channel; 3. *primary polydipsia* is due to suppression of AVP secretion as a result of excessive fluid intake. Depending on whether the excessive fluid intake is due to abnormal thirst or

Genetic Diagnosis of Endocrine Disorders. http://dx.doi.org/10.1016/B978-0-12-800892-8.00005-1

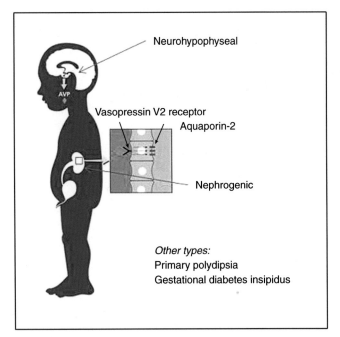

FIGURE 5.1   **Four different types of diabetes insipidus.** The neurohypophyseal form is caused by a deficiency in the production of the antidiuretic hormone, arginine vasopressin (AVP) and the nephrogenic form by insensitivity to the antidiuretic action of AVP. A third type is due to suppression of AVP secretion as a result of excessive fluid intake (primary polydipsia) and a fourth is caused by increased AVP metabolism in pregnant women (gestational diabetes insipidus). The pituitary/neurohypophyseal and renal/nephrogenic types of diabetes insipidus exist in both acquired and genetic (familial) forms.

due to a psychological disorder, primary polydipsia is subdivided into, respectively, *dipsogenic* diabetes insipidus[11,12] *psychogenic* diabetes insipidus,[13,14] and 4. *gestational* diabetes insipidus,[15–18]which is primarily due to increased metabolism of AVP by circulating vassopressinase produced by the placenta in the pregnant woman

but may also involve renal resistance and/or subclinical deficiency in AVP production.

Complete diabetes insipidus is defined by persistently low urine osmolality (<300 mosmol/kg) during a fluid deprivation test providing plasma osmolality rises above 295 mosmol/kg. *Partial* diabetes insipidus is defined by a subnormal increase in urine osmolality (300–600 mosmol/kg) during a fluid deprivation test with the same rise in plasma osmolality.[4]

FAMILIAL TYPES OF DIABETES INSIPIDUS

Neurohypophyseal and nephrogenic diabetes insipidus are in rare cases inherited and thus referred to as *familial neurohypophyseal diabetes insipidus* (FNDI) and *congenital nephrogenic diabetes insipidus* (CNDI or NDI). Based upon further clinical characteristics and inheritance patterns, FNDI can be subdivided into six different forms and NDI into three (Table 5.1). The underlying genetic defects have been identified in the majority of these different forms, and in FNDI they include mutations in either the *AVP* gene (Entrez Gene: 551) encoding the AVP prohormone (vasopressin-neurophysin 2-copeptin),[6] the *WFS1* gene (Entrez Gene: 7466) encoding wolframin[19] or in the *PCSK1* gene (Entrez Gene: 5122) encoding proprotein convertase 1/3 (neuroendocrine convertase 1).[20–22] In NDI, they include mutation in either the *AVPR2* gene (Entrez Gene: 554) encoding the vasopressin V2 receptor[8] or the *AQP2* gene (Entrez Gene: 359) encoding aquaporin-2.[8] In the following sections, we will summarize the main clinical characteristics and genetic aspects of the four most common forms of inherited diabetes insipidus: 1. *autosomal dominant FNDI*, 2. *autosomal recessive FNDI*,

TABLE 5.1   Overview of the Familial Types of Diabetes Insipidus

Disease	Inheritance pattern	Chromosomal location	Affected gene	OMIM	Number of kindreds
FNDI	Autosomal dominant	20p13	*AVP*	125700	>100
	Autosomal recessive, type a	20p13	*AVP*	125700	2
	Autosomal recessive, type b	20p13	*AVP*	NR	1
	Autosomal recessive, type c	4p16.1	*WFS1*	222300	>175*
	Autosomal recessive, type d	5q15	*PCSK1*	NR	8**
	X-linked recessive	Xq28	Unknown	NR	1
NDI	X-linked recessive	Xq28	*AVPR2*	304800	>300
	Autosomal dominant	12q13.12	*AQP2*	125800	NR
	Autosomal recessive	12q13.12	*AQP2*	125800	NR

NR, not reported; OMIM, Online Mendelian Inheritance in Man (http://www.ncbi.nlm.nih.gov/omim/)
* As reviewed by Rigoli et al.[19]. According to Yu et al.,[23] diabetes insipidus occurs in 52.8% of the patients.
** According to Martin et al.[21] and Yourshaw et al.,[22] diabetes insipidus occurs in at least 8/14 patients.

type c (in Wolfram syndrome), 3. *X-linked recessive NDI*, and 4. *autosomal dominant/recessive NDI* (Table 5.1).

Autosomal Dominant FNDI

The familial occurrence of severe polyuria and polydipsia (up to 28 L/24 h), segregating in an autosomal dominant pattern, and responding readily to exogenous dDAVP (autosomal dominant FNDI) (OMIM: 125700) shows several intriguing features that separate it from other familial forms of diabetes insipidus (Table 5.2). Usually, affected family members show no signs of disturbed water balance at birth and during early infancy but develop progressive symptoms of excessive drinking and polyuria at some point during infancy or early childhood.[7] Even at this stage, the deficiency of AVP, although marked, is often incomplete, and can be stimulated by hypertonic dehydration, leading to concentration of the urine. This may lead to delayed recognition of the condition and even misdiagnosis. In the few cases in which it has been studied by repetitive fluid-deprivation tests, AVP production is normal before the onset of FNDI but diminishes progressively during early childhood.[7,24–26] Once fully developed, the diabetes insipidus with severe thirst, polydipsia, and polyuria (8–20 L/day) is similar to the diabetes insipidus in other complete forms. In a few patients, yet, the AVP deficiency remains partial for decades, even though it is complete in other affected members of the same family with the same disease causing mutation, implying that other factors, genetic and/or environmental, are in play to modulate the penetrance of the mutations.[27] That could, for example, be differences between families (and individuals) in their ability/awareness to ensure early fluid replacement in affected children, which would be expected to delay the time of onset and slow down disease progression.[27] In some middle-aged male patients, diabetes insipidus symptoms decrease markedly without treatment and with preserved AVP deficiency and normal glomerular filtration.[28] The mechanism of these remissions is currently unexplained. Consistent with potential loss of AVP-producing magnocellular neurons,[9] the hyperintense MRI (magnetic resonance imaging) signal normally emitted by the posterior pituitary is absent or very small in patients with autosomal dominant FNDI, at least by the time AVP deficiency becomes clinically recognizable. Since the same lack of signal is characteristic in patients with NDI,[29] the usefulness of this investigation in the differential diagnosis is questionable.

The clinical characteristics of the rare autosomal recessive forms of FNDI (type a and type b) and autosomal recessive FNDI in congenital malabsorptive diarrhea (type d) (Table 5.1) differ in many aspects from those of autosomal dominant FNDI, for example, regarding age of onset, plasma levels of AVP during fluid deprivation, inter-/intrafamily variation, and cooccurrence with a complex picture of other symptoms.[21,22,30–32] For a thorough discussion of some of these rare conditions we refer to Refs [9,21].

Nonsyndromic FNDI has been reported in more than 100 kindreds worldwide, and in the majority of these, the disease is caused by mutations in the *AVP* gene (Table 5.1). With only a few well-documented exceptions, FNDI is transmitted by autosomal dominant inheritance and appears to be largely, if not completely, penetrant with age.[33] Based upon the initial report in 1945 by Forssman[34,35] it has been listed in databases (OMIM: 304900) that an X-linked recessive form of FNDI exists. However, a reinvestigation, including genetic and clinical examinations on descendants of

TABLE 5.2 Comparison of Clinical Features in Nonsyndromic Forms of Familial Diabetes Insipidus

Clinical features	FNDI Autosomal dominant	NDI X-linked recessive	NDI Autosomal dominant/recessive
Affected gene	*AVP*	*AVPR2*	*AQP2*
Male:female ratio	1:1	Males only*	1:1
Debut of symptoms	6 months to 6 years	From birth**	From birth†
Plasma AVP during thirst	Low/undetectable	High	High
Decrease of symptoms at middle age	Yes, in some cases	NR	NR
Antidiuretic response to dDAVP	>50% increase in U-osm	<50% increase in U-osm	<50% increase in U-osm
Extrarenal response to dDAVP (e.g., factor VIII)	Normal	Reduced	Normal
Posterior pituitary bright signal on MRI	Lacking	Lacking	NR

FNDI, familial neurohypophyseal diabetes insipidus; NDI, nephrogenic diabetes insipidus; AVP, arginine vasopressin; NR, not reported; dDAVP, desmopressin; U-osm, urine osmolality; MRI, magnetic resonance imaging.

Females may have mild symptoms of diabetes insipidus due to skewed X-chromosome inactivation.

**In some patients with partial (mild) diabetes insipidus debut has been reported later in life.*

† *Debut is usually reported from the second half of the first year or later.*

the original patients, has revealed that the original patients more likely had a partial NDI phenotype, and consistent with this the reinvestigated patients actually carried a mutation in the *AVPR2* gene (g.310C > T, p.Arg104Cys). On the other hand, in another report of X-linked recessive inheritance of FNDI (Table 5.1), the clinical phenotype in one family is clearly consistent with that of neurohypophyseal diabetes insipidus, and although the disease seems to be linked to the same chromosomal location as the *AVPR2* gene (Xq28), mutations have been found neither in this gene nor in the *AVP* gene.[7,36] For a more thorough discussion of this rare condition, we refer to Ref. [9].

Until now, FNDI has been associated with >70 different mutations in the *AVP* gene.[9] All but one of these mutations (g.1919 + 1delG) disturbs the coding region of the gene. The type and location of the mutations differ widely and affect amino acid residues ranging from the N-terminal residue of the signal peptide (the start codon), through the AVP moiety and into the C-terminal part of the neurophysin 2 domain of the AVP preprohormone. Of note, however, FNDI has not been attributed to any mutation affecting solely copeptin, the glycosylated peptide encoded by exon 3. A copeptin mutation (p.Ala159Thr) has been identified in one family in which autosomal dominant FNDI was segregating. It did not, however, segregate with the disease and all affected family members were heterozygous for another mutation (p.Gly96Asp) previously reported to cause FNDI. This strongly suggests that the p.Ala159Thr mutation had no pathological effect.[37]

There is no real prevalent mutation in the *AVP* gene in autosomal dominant FNDI but one mutation is nonetheless more frequent than all others, namely, the g.279G > A substitution (predicting an p.Ala19Thr amino acid substitution), which has been identified in 11 kindreds with no known relationship to each other.[38]

The genetic basis of autosomal dominant FNDI remains unknown in one reported kindred.[39] In this Chinese family, the disease showed linkage to a 7-cM interval on chromosome 20p13 containing the *AVP* gene; however, unexpectedly, no mutations could be found in the coding regions, the promoter, or the introns of the *AVP* gene. In addition, the coding regions of the nearby *UBCE7IP5* gene and the *AQP2* gene on chromosome 12 were analyzed, but no mutations were detected. Thus, the cause of FNDI in this specific family remains unexplained.

The diversity of the mutations identified in the *AVP* gene in autosomal dominant FNDI could indicate diverse pathogenic mechanisms. However, the fact that different disease alleles result in rather uniform clinical presentations calls for a single unifying explanation. One such explanation suggests that mutations in the *AVP* gene exert a dominant-negative effect by leading to the production of a mutant AVP prohormone that fails to fold and/or

dimerize properly in the endoplasmatic reticulum (ER) and, as a consequence, is retained by the ER protein quality control resulting in cytotoxic accumulation of protein in the magnocellular neurons that produce AVP (i.e., a misfolding-neurotoxicity hypothesis).[6]

Autosomal Recessive FNDI, Type C

Autosomal recessive inheritance of FNDI has been described and is by far most common in conjunction with Wolfram syndrome (OMIM: 222300) (Table 5.1). The syndrome is a rare, multifaceted, and progressive neurodegenerative disease also referred to as DIDMOAD, abbreviating its symptom spectrum, including diabetes insipidus, early-onset, insulin-dependent diabetes mellitus, progressive optic atrophy, and sensorineural deafness, combined with many other neurological abnormalities.[19] It has been estimated that approximately 75% of patients with Wolfram syndrome present with partial neurohypophyseal diabetes insipidus at an average age of 14 years (range: three months to 40 years of age). This is about the same time as they lose their hearing, but after onset of diabetes mellitus and loss of vision. The polyuria in Wolfram syndrome is due to a partial or severe deficiency in AVP production,[40] and seems to be associated with posterior pituitary degeneration and impaired processing of the AVP prohormone.[41] In general, the patients respond well to dDAVP administration. Accordingly, the diabetes insipidus observed in Wolfram syndrome seems to be clinically distinguishable from autosomal dominant FNDI only by its mode of inheritance, a tendency toward higher age of onset, and most obvious by its cooccurrence with a complex picture of other symptoms. Besides, in clear contrast to the rather mild course of autosomal dominant FNDI, patients with Wolfram syndrome usually die from central respiratory failure as a result of brainstem atrophy in the third or fourth decade of life.

Patients with the Wolfram syndrome are often offspring from consanguineous marriages between unaffected parents, and they have affected siblings, strongly supporting a recessive pattern of inheritance. Most commonly, patients are homozygous or compound heterozygous for mutations in the *WFS1* gene located on chromosome 4 (4p16.1). Infrequently, mutations in the *CISD2* gene can also result in Wolfram syndrome; however in these cases diabetes insipidus seems to be absent.[42]

The protein product Wolframin, encoded by *WFS1*, has been shown to play a crucial role in the negative regulation of a feedback loop of the ER stress–signaling network, induced by the accumulation of misfolded and unfolded proteins in the organelle. More specifically, it seems to prevent secretory cells, such as pancreatic β-cells, from death caused by dysregulation of this signaling pathway.[43] It could be that Wolframin plays a similar role in the AVP producing magnocellular neurons. These cells may experience ER stress as an effect of the large biosynthetic load placed on the ER by the exceptional high demands for AVP prohormone production under physiological conditions that stimulate the neurons (elevated osmotic pressure), exactly as pancreatic β-cells during postprandial stimulation of proinsulin biosynthesis. Loss of Wolframin function under such conditions would lead to unregulated ER stress signaling and probably neuronal cell death, exactly as has been shown in pancreatic β-cells.[44] This mechanism links directly to the ER stress, possibly elicited by the accumulation of misfolded mutant AVP prohormone in the magnocellular neurons of patients with autosomal dominant FNDI.

X-Linked Recessive NDI

NDI is characterized by the same symptoms of diabetes insipidus as seen in FNDI but shows renal insensitivity to the antidiuretic effect of endogenously produced AVP as well as to dDAVP treatment (Table 5.2). As mentioned earlier, it is inherited either in an X-linked recessive pattern (in approximately 90% of NDI cases) (OMIM: 304800) and caused by mutations in the *AVPR2* gene or in an autosomal dominant or recessive pattern (in approximately 10% of NDI cases) (OMIM: 125800) and caused by mutations in the *AQP2* gene (Table 5.2). Inherited NDI additionally exists in complex forms characterized by loss of water and ions, for example, Bartter syndrome (OMIM: 601678, 241200, 607364, and 602522). These rare variant forms of NDI are caused by mutation in genes (*KCNJ1*, *SLC12A1*, *CLCNKB*, *CLCNKB*, and *CLCNKA* in combination, or *BSND*) encoding other renal tubular transporters, resulting in a disturbance of the inner medullary concentration mechanism and thereby polyuria and polydipsia together with characteristic electrolyte disturbances.[45] The electrolyte disturbances, however, clearly distinguish these conditions from genuine diabetes insipidus.

The clinical characteristics of NDI attributable to mutations in the *AVPR2* or *AQP2* gene, regardless of the mode of inheritance or specific mutation involved, include hypernatremia, hyperthermia, mental retardation, and repeated episodes of dehydration if patients cannot obtain enough water. In contrast to autosomal dominant FNDI, the urinary concentration defect is present at birth, and newborns and infants are especially prone to dehydration (Table 5.2). Furthermore, the clinical picture of dehydration is often difficult to interpret at this age (usually manifested as failure to thrive), and therefore severe cases of prolonged dehydration are seen. Mental retardation was prevalent in 70–90% of the patients reported in the original studies of NDI and was thought to be part of the disease.[46] However, early recognition of NDI based on genetic screening of at-risk children and early treatment of the disease permit such children to

have normal physical and mental development[47] and suggest that the mental retardation reported in the original studies probably resulted from repeated episodes of severe dehydration rather than from pathologies directly caused by the genetic defect.

The X-linked recessive and most frequent form of familial NDI is caused by loss-of-function mutations in the *AVPR2* gene, encoding the renal vasopressin V2 receptor. Currently, at least 214 putative disease-causing mutations in the *AVPR2* gene have been identified in more than 300 families.[7,10,48] Usually, heterozygous females are asymptomatic but may in some cases have variable degrees of polyuria and polydipsia because of skewed X-chromosome inactivation.[47] Recent reports of patients having mutations in the *AVPR2* gene and initially suspected of having FNDI due to in some cases an impressive antidiuretic response to exogenous AVP[49] have emphasized the importance of genetic testing and have added to the complexity of clinical differential diagnosis in diabetes insipidus.

The molecular mechanism underlying the renal AVP insensitivity in NDI differs among disease alleles. Hence, *AVPR2* mutations have been divided into five different classes according to their pathogenic effect at the cellular levels,[50] exactly as the system used for classification of mutant low-density lipoprotein (LDL) receptors. Most mutations in the *AVPR2* gene (>50%) result in vasopressin V2 receptors that are trapped inside the cell due to impaired intracellular trafficking, and they are consequently unable to reach the plasma membrane and bind the ligand (AVP) (Type 2 mutations). Other mutations lead to an unstable mRNA, and thereby no protein product (Type 3 mutations) or mutant vasopressin V2 receptors that reach the cell surface, but either cannot bind AVP (the ligand) efficiently (Type 1 mutations) or properly trigger an increase in intracellular cyclic AMP (cAMP) production (Type 4 mutations).

Treatment of patients with NDI with dDAVP is usually not effective, although females having symptoms due to skewed X-chromosome inactivation and patients with partial NDI can be alleviated from their diabetes insipidus by using high doses, possibly because such patients retain some functional vasopressin V2 receptors. Low sodium diets to reduce the solute load to the kidneys and treatment with diuretic thiazide eventually in combination with cyclooxygenase inhibitor (indomethacin) or hydrochlorothiazide in combination with amiloride represents treatment regimens that to some extent can relieve symptoms of NDI.[8] Novel strategies suggested for the treatment of NDI are: vasopressin V2 receptor antagonists and agonists, pharmacological chaperones, nonpeptide agonists, cyclic guanosine monophosphate pathway activation, cAMP pathway activation, statins, prostaglandins, and molecular chaperones. For a detailed discussion of these approaches, we refer to Ref. [8].

Autosomal Dominant/Recessive NDI

As mentioned earlier, the clinical characteristics of NDI attributable to mutations in the *AVPR2* or *AQP2* gene, regardless of the mode of inheritance or specific mutation involved, are similar (Table 5.2). *Autosomal dominant* or *autosomal recessive* modes of inheritance occur in approximately 10% of the families with inherited NDI, with the autosomal recessive form being the most common among the two (90% vs. 10%). Typically, these families have mutations in the *AQP2* gene, encoding the aquaporin-2 water channel. Currently, at least 51 putative disease-causing mutations in the AQP2 gene have been identified, mostly in children of consanguineous parents.[8]

In vitro studies show that different mutations in the *AQP2* gene have different pathological effects at the cellular level. Misfolding of mutant aquaporin-2 proteins and subsequent degradation in the ER is likely the major mechanism underlying autosomal recessive NDI,[51] whereas mutations causing autosomal dominant NDI generally affect the carboxy-terminus of aquaporin-2, causing misrouting of both mutant and wild-type aquaporin-2 proteins.[8] Thus, these mutations probably act by a dominant negative mechanism by preventing the normal aquaporin-2 protein from reaching the apical surface of the collecting duct epithelial cells. Some of these mutations result in a partial NDI phenotype, which has also been observed in patients with mutations in the *AVPR2* gene. For a detailed discussion of partial NDI in patients with *AVPR2* or *AQP2* mutations, we refer to Ref. [7].

CLINICAL DIAGNOSIS

Differentiating between the types of diabetes insipidus is relatively easy if the patient has a severe deficiency in either the production or renal action of AVP. In either condition, dehydration induced by fluid deprivation fails to result in concentration of the urine. Because this result excludes primary polydipsia, measuring the urinary response to a subsequent injection of AVP or dDAVP will differentiate nephrogenic diabetes insipidus from the neurohypophyseal and gestational forms. Neurohypophyseal and gestational diabetes insipidus can usually be distinguished on clinical grounds. If fluid deprivation results in concentration of the urine, other tests are necessary to determine whether the patient has primary polydipsia or a less severe ("partial") deficiency in the secretion or action of AVP.[4] The most reliable way to make this distinction is to measure plasma AVP and to relate the results to the concurrent plasma and urine osmolality during a fluid deprivation and/or hypertonic saline infusion test. A satisfactory alternative is to closely monitor changes in fluid balance during a therapeutic trial of dDAVP.

In addition to the renal vasopressin V2 receptor, which mediates the antidiuretic effect of AVP and dDAVP, there is evidence for an extrarenal vasopressin V2 receptor, which mediates extrarenal effects such as an increase in factor VIII, von Willebrand factor, and tissue-type plasminogen activator.[52] Although rarely necessary, a dDAVP infusion test can be used to distinguish NDI caused by mutations in the *AVPR2* gene from other causes (Table 5.2).

GENETIC TESTING

The four genes associated with the most common forms of diabetes insipidus as described above are the *AVP* gene (autosomal dominant FNDI), the *WFS1* gene (autosomal recessive FNDI, type c in Wolfram syndrome), the *AVPR2* gene (X-linked recessive NDI), and the *AQP2* gene (autosomal dominant or recessive NDI). The *AVP* gene has three exons covering a 2.2 kb genomic region at chromosome 20p13, the *WFS1* gene spans approximately 33.4 kb of genomic DNA at chromosome 4p16.1 and consists of eight exons, the *AVPR2* gene has three exons and two small introns covering a 4.6 kb genomic region at chromosome Xq28, and the *AQP2* gene has four exons covering a 8.1 kb genomic region at chromosome 12q13.12. Thus, all genes involved in the common forms of diabetes insipidus are relatively small genes with few exons. This makes genetic testing by direct sequence analysis of the entire coding region feasible in most cases, and clinical molecular genetics tests based upon this approach are available for all fours genes (Table 5.3). The same goes for deletion/duplication analysis (Table 5.3) whereas sequence analysis of select exons and targeted variant analysis is only available as clinical tests in the case of the *WFS1* gene (Table 5.3).

TABLE 5.3   Molecular Genetics Tests Available for the Four Genes Most Commonly Being Affected in Diabetes Insipidus

Molecular genetics tests*	*AVP* gene**	*WFS1* gene	*AVPR2* gene**	*AQP2* gene**
Sequence analysis of select exons	NA	+	NA	NA
Sequence analysis of entire coding region	+	+	+	+
Deletion/duplication analysis	+	+	+	+
Targeted variant analysis	NA	+	NA	NA

NA, not available as clinical tests; +, available as clinical tests.
* According to GTR: Genetic Testing Registry (https://www.ncbi.nlm.nih.gov/gtr/).
** All molecular genetics test approaches are available at a research basis (see contact information in the last section of this chapter).

All molecular genetics test approaches in the relation to the *AVP*, *AVPR2*, and *AQP2* genes are, however, available at a research basis.

Which Gene to Test?

Genetic testing of patients with clinical characteristics in accordance with nonsyndromic FNDI usually involves sequence analysis of the entire coding region of the *AVP* gene, as the large majority of cases have mutations in this gene. If no mutations are found, it should be considered whether the clinical characteristics and the eventual inheritance of the disease in the family is in accordance with partial NDI[49] indicating sequence analysis of either the *AVPR2* or the *AQP2* genes. Since most NDI cases are caused by mutations in the *AVPR2* gene, molecular genetics testing of a symptomatic individual, male or female, usually starts with sequencing of the entire coding region of the *AVPR2* gene. If no mutations are found, sequencing of the coding region of the *AQP2* gene should be performed. In children with NDI (male or female) from consanguineous parents, sequencing of the *AQP2* gene should be performed first, followed by sequencing of the *AVPR2* gene if no mutations in the *AQP2* gene are identified. Genetic testing in syndromic forms of inherited diabetes insipidus involves analysis of the *WFS1* gene in patients with Wolfram syndrome, and in the rare cases of autosomal recessive FNDI in congenital malabsorptive diarrhea (type d), it should include analysis of the *PCSK1* gene. Locus heterogeneity (*KCNJ1*, *SLC12A1*, *CLCNKB*, *CLCNKB*, and *CLCNKA* in combination or *BSND*) in the complex form of NDI, Bartter syndrome, calls for careful clinical evaluation and alternative approaches in genetic testing (reviewed in Ref. [53]).

Diagnostic Genetic Testing

There are several reasons why all patients with familial occurrence of diabetes insipidus should be properly diagnosed using molecular genetics testing. As mentioned earlier, there are now multiple examples where genetic testing resolved the differential diagnosis and changed treatment options. Furthermore, confirming the presence of a mutation in the *AVP* gene in patients with neurohypophyseal diabetes insipidus relieves the physician from further exploration of the cause of diabetes insipidus, for example, searching for a hypothalamic lesion/ tumor, testing anterior pituitary function, and so forth. In NDI, early diagnosis in newborns is essential to ensure prompt treatment in order to reduce morbidity from hypernatremia, dehydration, and dilation of the urinary tract. Therefore, in all patients with familial occurrence of diabetes insipidus regardless of age and the results of clinical tests, genetic testing should be performed.

Genetic testing should also be considered in all patients with NDI without an identifiable clinical cause regardless of family history as *de novo* mutations are not uncommon. It has not been clear whether patients with idiopathic neurohypophyseal diabetes insipidus should be tested genetically for *de novo* mutations in the *AVP* gene. These patients constitute a relatively large proportion of neurohypophyseal diabetes insipidus cases (20–50%), and some occur during childhood.[54] In our experience approximately 5% of these have abnormalities in the *AVP* gene[33] and therefore we suggest that patients with neurohypophyseal diabetes insipidus occurring during childhood, without a family history, and without an identifiable cause (e.g., thickening of the pituitary stalk), should be tested genetically.

Presymptomatic Genetic Diagnosis in FNDI

Once the molecular diagnosis is established in families with FNDI, it is feasible to screen other family members for the specific mutation in question. This is particularly relevant in infants at risk of inheriting the mutation because presymptomatic diagnosis thereby is possible, relieving years of parental concern about the carrier status of their offspring, as well as repeated clinical diagnostic examinations. Usually, the parents request a genetic test already when the children are newborn.

Genetic Testing in Individuals at Risk of Being Carriers

In all recessive forms of diabetes insipidus (autosomal recessive FNDI, types a–d, X-linked recessive NDI, and autosomal recessive NDI) testing of relatives at risk of being carriers would be possible if the mutation has been identified in the proband. This is particularly useful in X-linked recessive NDI both to explain possible mild symptoms in female carriers and for genetic counseling. Carrier testing in autosomal dominant FNDI is relevant due to the age-dependent penetrance whereby presymptomatic genetic testing is both possible and useful for the reasons mentioned earlier.

Prenatal Genetic Diagnosis

Prenatal genetic diagnosis seems not to be indicated in most forms of FNDI, first of all because it rarely presents at birth whereby the risk of severe central nervous system complications are low compared with congenital NDI, and second because the disease, when fully developed, is associated with few symptoms that are compatible with an almost normal quality of life at least when treated appropriately. However, it could be indicated in autosomal recessive FNDI, types c and d, due to the severe cause of Wolfram syndrome and congenital malabsorptive diarrhea.

In conditions such as NDI that do not affect intellect and have available treatment options, requests for prenatal genetic testing are not common. Prenatal testing is available for pregnancies at increased risk if the *AVPR2* mutation has been identified in an affected family member. The usual procedure is to determine fetal sex by performing chromosome analysis on fetal cells obtained either by chorionic villus sampling or by amniocentesis. If the karyotype is 46,XY, DNA from fetal cells can be analyzed for known mutations.

Prenatal diagnosis is available for pregnancies at increased risk for autosomal recessive NDI by the methods described above. Both disease-causing alleles within the family must have been identified to allow efficient prenatal testing.

Differences in perspective may exist among medical professionals and in families regarding the use of prenatal genetic testing, particularly if the testing is being considered for the purpose of pregnancy termination rather than timely diagnosis. Although most centers would consider decisions about prenatal genetic testing to be the choice of the parents, careful discussion of these issues is appropriate.

Preimplantation genetic diagnosis may be available for families in which the disease-causing allele(s) has been identified.

Available Laboratories and Resources

Below listed is a few contact addresses to laboratories that have been actively involved in research-based molecular genetics testing in diabetes insipidus for many years and where different approaches for the analysis of the *AVP*, *AVPR2*, and *AQP2* genes can be performed. Additional contacts and more details can be found at GTR: Genetic Testing Registry (https://www.ncbi.nlm.nih.gov/gtr/).

Søren Rittig, MD, DMSc, Pediatric Research Laboratory, Department of Pediatrics, Aarhus University Hospital, Skejby Palle Juul-Jensens Boulevard 99, DK-8200 Aarhus N, Denmark; Email: soren.rittig@skejby.rm.dk

Daniel G. Bichet, MD, Centre de Recherche, Hopital du Sacre-Coeur de Montreal, 5400 Blvd. Gouin Ouest, Montreal, Quebec, Canada, H4J 1C5; Email: daniel.bichet@umontreal.ca

Peter M.T. Deen, PhD, Department of Physiology, Radboud University Medical Center (RUMC), Geert Grooteplein Zuid 30, 6525 GA Nijmegen, The Netherlands; Email: peter.deen@Radboudumc.nl

An important diabetes insipidus information resource is the NDI Foundation that has been formed to support education, research, treatment, and cure for diabetes insipidus. Their contact details are:

NDI Foundation, Main Street, PO Box 1390, Eastsound, WA 98245, USA; Phone: 1-888-376-6343, Fax: 1-888-376-6356, E-mail: info@ndif.org; www.ndif.org.

References

1. Eberle J. Treatise on the practice of medicine. 2nd ed. Grigg J., Philadelphia; 1831. p. 381–396.
2. Robertson GL. History. In: Robertson, GL, ed. Translational Endocrinology & Metabolism: Posterior Pituitary Update. Chevy Chase: The Endocrine Society; 2012. p. 15–26.
3. Cullen W. First lines of the practice of physic. Isaiah Thomas, Worcester, Massachusetts; 1790. vol. 3, p. 153–155.
4. Robertson GL. Diabetes insipidus. *Endocrinol Metab Clin North Am* 1995;**24**(3):549–5472.
5. Fujiwara TM, Bichet DG. Molecular biology of hereditary diabetes insipidus. *J Am Soc Nephrol* 2005;**16**(10):2836–46.
6. Christensen JH, Rittig S. Familial neurohypophyseal diabetes insipidus – an update. *Semin Nephrol* 2006;**26**(3):209–23.
7. Babey M, Kopp P, Robertson GL. Familial forms of diabetes insipidus: clinical and molecular characteristics. *Nat Rev Endocrinol* 2011;**7**(12):701–14.
8. Moeller HB, Rittig S, Fenton RA. Nephrogenic diabetes insipidus: essential insights into the molecular background and potential therapies for treatment. *Endocr Rev* 2013;**34**(2):278–301.
9. Christensen JH., Robertson GL. Deficiencies of vasopressin and thirst. In: Robertson GL, ed. Translational Endocrinology & Metabolism: Posterior Pituitary Update. Chevy Chase: The Endocrine Society; 2012. p. 57–94.
10. Bichet DG. Renal actions of vasopressin. In: Robertson GL, ed. Translational Endocrinology & Metabolism: Posterior Pituitary Update. Chevy Chase: The Endocrine Society; 2012. p. 95–122.
11. Stuart CA, Neelon FA, Lebovitz HE. Disordered control of thirst in hypothalamic-pituitary sarcoidosis. *N Engl J Med* 1980;**303**(19):1078–82.
12. Robertson GL. Dipsogenic diabetes insipidus: a newly recognized syndrome caused by a selective defect in the osmoregulation of thirst. *Trans Assoc Am Physicians* 1987;**100**:241–9.
13. Barlow ED, De Wardener HE. Compulsive water drinking. *Q J Med* 1959;**28**(110):235–58.
14. Sleepert FH, Jellinek EM. A comparative physiologic, psychologic and psychiatric study of polyuric and non-polyuric schizophrenic patients. *J Nerv Ment Dis* 1936;**83**:557–63.
15. Durr JA, Hoggard JG, Hunt JM, Schrier RW. Diabetes insipidus in pregnancy associated with abnormally high circulating vasopressinase activity. *N Engl J Med* 1987;**316**(17):1070–4.
16. Barron WM, Cohen LH, Ulland LA, Lassiter WE, Fulghum EM, Emmanouel D, et al. Transient vasopressin-resistant diabetes insipidus of pregnancy. *N Engl J Med* 1984;**310**(7):442–4.
17. Ford Jr SM, Lumpkin III HL. Transient vasopressin-resistant diabetes insipidus of pregnancy. *Obstet Gynecol* 1986;**68**(5):726–8.
18. Iwasaki Y, Oiso Y, Kondo K, Takagi S, Takatsuki K, Hasegawa H, et al. Aggravation of subclinical diabetes insipidus during pregnancy. *N Engl J Med* 1991;**324**(8):522–6.
19. Rigoli L, Lombardo F, Di BC. Wolfram syndrome and WFS1 gene. *Clin Genet* 2011;**79**(2):103–17.
20. Farooqi IS, Volders K, Stanhope R, Heuschkel R, White A, Lank E, et al. Hyperphagia and early-onset obesity due to a novel homozygous missense mutation in prohormone convertase 1/3. *J Clin Endocrinol Metab* 2007;**92**(9):3369–73.
21. Martin MG, Lindberg I, Solorzano-Vargas RS, Wang J, Avitzur Y, Bandsma R, et al. Congenital proprotein convertase 1/3 deficiency causes malabsorptive diarrhea and other endocrinopathies in a pediatric cohort. *Gastroenterology* 2013;**145**(1):138–48.

22. Yourshaw M, Solorzano-Vargas RS, Pickett LA, Lindberg I, Wang J, Cortina G, et al. Exome sequencing finds a novel PCSK1 mutation in a child with generalized malabsorptive diarrhea and diabetes insipidus. *J Pediatr Gastroenterol Nutr* 2013;**57**(6):759–67.

23. Yu G, Yu ML, Wang JF, Gao CR, Chen ZJ. WS1 gene mutation analysis of Wolfram syndrome in a Chinese patient and a systematic review of literatures. *Endocrine* 2010;**38**(2):147–52.

24. Elias PC, Elias LL, Torres N, Moreira AC, Antunes-Rodrigues J, Castro M. Progressive decline of vasopressin secretion in familial autosomal dominant neurohypophyseal diabetes insipidus presenting a novel mutation in the vasopressin-neurophysin II gene. *Clin Endocrinol (Oxf)* 2003;**59**(4):511–8.

25. Mahoney CP, Weinberger E, Bryant C, et al. Effects of aging on vasopressin production in a kindred with autosomal dominant neurohypophyseal diabetes insipidus due to the DeltaE47 neurophysin mutation. *J Clin Endocrinol Metab* 2002;**87**(2):870–6.

26. McLeod JF, Kovacs L, Gaskill MB, Rittig S, Bradley GS, Robertson GL. Familial neurohypophyseal diabetes insipidus associated with a signal peptide mutation [see comments]. *J Clin Endocrinol Metab* 1993;**77**(3):599A–1599A.

27. Jendle J, Christensen JH, Kvistgaard H, Gregersen N, Rittig S. Late-onset familial neurohypophyseal diabetes insipidus due to a novel mutation in the AVP gene. *Clin Endocrinol (Oxf)* 2012;**77**(4):586–92.

28. Robertson GL. The use of vasopressin assays in physiology and pathophysiology. *Semin Nephrol* 1994;**14**(4):368–83.

29. Sato N, Ishizaka H, Yagi H, Matsumoto M, Endo K. Posterior lobe of the pituitary in diabetes insipidus: dynamic MR imaging. *Radiology* 1993;**186**(2):357–60.

30. Willcutts MD, Felner E, White PC. Autosomal recessive familial neurohypophyseal diabetes insipidus with continued secretion of mutant weakly active vasopressin. *Hum Mol Genet* 1999;**8**(7):1303–7.

31. Abu LA, Levy-Khademi F, Abdulhadi-Atwan M, Bosin E, Korner M, White PC, et al. Autosomal recessive familial neurohypophyseal diabetes insipidus: onset in early infancy. *Eur J Endocrinol* 2010;**162**(2):221–6.

32. Christensen JH, Kvistgaard H, Knudsen J, Shaikh G, Tolmie J, Cooke S, et al. A novel deletion partly removing the AVP gene causes autosomal recessive inheritance of early-onset neurohypophyseal diabetes insipidus. *Clin Genet* 2013;**83**(1):44–52.

33. Christensen JH, Siggaard C, Corydon TJ, deSanctis L, Kovacs L, Robertson GL, et al. Six novel mutations in the arginine vasopressin gene in 15 kindreds with autosomal dominant familial neurohypophyseal diabetes insipidus give further insight into the pathogenesis. *Eur J Hum Genet* 2004;**12**(1):44–51.

34. Forssman H. On hereditary diabetes insipidus with special regard to a sex-linked form. *Acta Med Scand* 1945;**159**(Suppl):1–196.

35. Hansen LK, Rittig S, Robertson GL. Genetic basis of familial neurohypophyseal diabetes insipidus. *Trends Endocrinol Metab* 1997;**8**(9):363–72.

36. Habiby RL, Robertson GL, Kaplowitz PB, Morris M, Siggaard C, Rittig S. A novel X-linked form of familial neurohypophyseal diabetes insipidus. *J Invest Med* 1996;**44**:388A.

37. Hedrich CM, Zachurzok-Buczynska A, Gawlik A, Russ S, Hahn G, Koehler K, et al. Autosomal dominant neurohypophyseal diabetes insipidus in two families. Molecular analysis of the vasopressin-neurophysin II gene and functional studies of three missense mutations. *Horm Res* 2009;**71**(2):111–9.

38. Cizmarova M, Nagyova G, Janko V, Pribilincova Z, Virgova D, Ilencikova D, et al. Late onset of familial neurogenic diabetes insipidus in monozygotic twins. *Endocr Regul* 2013;**47**(4):211–6.

39. Ye L, Li X, Chen Y, Sun H, Wang W, Su T, et al. Autosomal dominant neurohypophyseal diabetes insipidus with linkage to chromosome 20p13 but without mutations in the AVP-NPII gene. *J Clin Endocrinol Metab* 2005;**90**:4388–93.

40. Thompson CJ, Charlton J, Walford S, Baird J, Hearnshaw J, McCulloch A, et al. Vasopressin secretion in the DIDMOAD (Wolfram) syndrome. *Q J Med* 1989;**71**(264):333–45.

41. Gabreels BA, Swaab DF, de KD, Dean A, Seidah NG, Van de Loo JW, et al. The vasopressin precursor is not processed in the hypothalamus of Wolfram syndrome patients with diabetes insipidus: evidence for the involvement of PC2 and 7B2. *J Clin Endocrinol Metab* 1998;**83**(11):4026–33.

42. Amr S, Heisey C, Zhang M, Xia XJ, Shows KH, Ajlouni K, et al. A homozygous mutation in a novel zinc-finger protein, ERIS, is responsible for Wolfram syndrome 2. *Am J Hum Genet* 2007;**81**(4):673–83.

43. Fonseca SG, Ishigaki S, Oslowski CM, Lu S, Lipson KL, Ghosh R, et al. Wolfram syndrome 1 gene negatively regulates ER stress signaling in rodent and human cells. *J Clin Invest* 2010;**120**(3):744–55.

44. Cnop M, Ladriere L, Hekerman P, Ortis F, Cardozo AK, Dogusan Z, et al. Selective inhibition of eukaryotic translation initiation factor 2 alpha dephosphorylation potentiates fatty acid-induced endoplasmic reticulum stress and causes pancreatic beta-cell dysfunction and apoptosis. *J Biol Chem* 2007;**282**(6):3989–97.

45. Kleta R, Bockenhauer D. Bartter syndromes and other salt-losing tubulopathies. *Nephron Physiol* 2006;**104**(2):73–80.

46. van Lieburg AF, Knoers NV, Monnens LA. Clinical presentation and follow-up of 30 patients with congenital nephrogenic diabetes insipidus. *J Am Soc Nephrol* 1999;**10**(9):1958–64.

47. Morello JP, Bichet DG. Nephrogenic diabetes insipidus. *Annu Rev Physiol* 2001;**63**:607–30.

48. Spanakis E, Milord E, Gragnoli C. AVPR2 variants and mutations in nephrogenic diabetes insipidus: review and missense mutation significance. *J Cell Physiol* 2008;**217**(3):605–17.

49. Faerch M, Christensen JH, Corydon TJ, Kamperis K, de ZF, Gregersen N, et al. Partial nephrogenic diabetes insipidus caused by a novel mutation in the AVPR2 gene. *Clin Endocrinol (Oxf)* 2008;**68**(3):395–403.

50. Robben JH, Knoers NV, Deen PM. Cell biological aspects of the vasopressin type-2 receptor and aquaporin 2 water channel in nephrogenic diabetes insipidus. *Am J Physiol Renal Physiol* 2006;**291**(2):F257–70.

51. Loonen AJ, Knoers NV, van Os CH, Deen PM. Aquaporin 2 mutations in nephrogenic diabetes insipidus. *Semin Nephrol* 2008;**28**(3):252–65.

52. Kaufmann JE, Oksche A, Wollheim CB, Gunther G, Rosenthal W, Vischer UM. Vasopressin-induced von Willebrand factor secretion from endothelial cells involves V2 receptors and cAMP. *J Clin Invest* 2000;**106**(1):107–16.

53. Jain G, Ong S, Warnock DG. Genetic disorders of potassium homeostasis. *Semin Nephrol* 2013;**33**(3):300–9.

54. Ghirardello S, Malattia C, Scagnelli P, Maghnie M. Current perspective on the pathogenesis of central diabetes insipidus. *J Pediatr Endocrinol Metab* 2005;**18**(7):631–45.

6

States of Pituitary Hypofunction

Christopher J. Romero, Andrea L. Jones, Sally Radovick

Division of Pediatric Endocrinology, Department of Pediatrics, The Johns Hopkins University School of Medicine, Baltimore, MD, USA

INTRODUCTION

Hypopituitarism is a condition that presents clinically with one or more deficiencies of hormones secreted from the adenohypophysis, neurohypophysis, or both. The annual incidence rate has been estimated to be 4.2 cases per 100,000 of the population, along with a prevalence of 45.5 per 100,000 in a 2001 population study from Spain.[1] The etiology of hypopituitarism can be attributed to several causes including head injury, neurosurgical sequelae, infiltrative disorders, and cranial radiotherapy; however, for many patients, in particular for those with congenital or idiopathic hypopituitarism, the etiology for hormone deficiency cannot be identified. There appear to be limited data on the true prevalence of idiopathic hypopituitarism. The advancements in understanding hypothalamic-pituitary development using animal models have helped to identify a cascade of several developmental factors necessary for proper pituitary function. Mutations in these factors found in both animal models and affected patients have been identified and linked to the development of hypopituitarism.

Pituitary hormone deficiency may present in a wide variety of ways; for example, it may manifest acutely in neonates as an adrenal crisis or insidiously in children with a poor growth velocity. The clinical presentation will depend on the deficient hormone(s) and may be relatively nonspecific; symptoms may include increased lethargy, cold intolerance, poor weight gain, decreased appetite, or abdominal pain. An evaluation of pituitary function in the newborn period typically takes place in the setting of persistent hypoglycemia or electrolyte imbalance. Although the phenotype of idiopathic hypopituitarism is varied, infants found to have midline defects and male infants with micropenis should be evaluated for possible pituitary hormone deficiency. In children and adolescents, poor growth or delayed puberty is often an indicator for care providers to evaluate the pituitary for possible deficiencies. It is prudent to recognize that the evaluation of some patients may initially reveal a single pituitary hormone deficiency; however, follow-up evaluations are required, as additional pituitary hormone deficiencies may develop over time.

GENETIC PATHOPHYSIOLOGY

The development of the pituitary occurs early during embryogenesis by the coordinated spatial and temporal expression of signaling molecules and transcription factors. Initially, the primordial Rathke's pouch develops by the thickening and invagination of the oral ectoderm that comes into contact with the ventral diencephalon. Proliferation of the cells through expression of various signaling factors drives oral ectoderm closure and forms a detached rudimentary gland. The expression of Sonic hedgehog (Shh) from the oral ectoderm and downstream effectors, such as those in the Gli family, are important in orchestrating the proliferation and formation of the gland.[2] In addition, the Wnt family of signaling molecules and bone morphogenetic protein 4 (BMP4), among others, also may have roles in guiding proliferation and specification of pituitary cell types.[3]

The mature anterior pituitary gland ultimately contains five cell types regulated by trophic hormones produced by the hypothalamus, as well as positive and negative feedback from peripherally secreted hormones. The cell types and hormones they secrete include: somatotrophs, growth hormone (GH); thyrotrophs, thyroid-stimulating hormone (TSH), lactotrophs, prolactin (PRL), gonadotrophs, lutenizing hormone (LH), and follicle-stimulating hormone (FSH), and corticotrophs, adrenocorticotrophic hormone (ACTH). The development of these pituitary cell types is regulated by the expression of transcription factors including Hesx homeobox 1, LIM homeobox protein 3 (Lhx3), paired-like

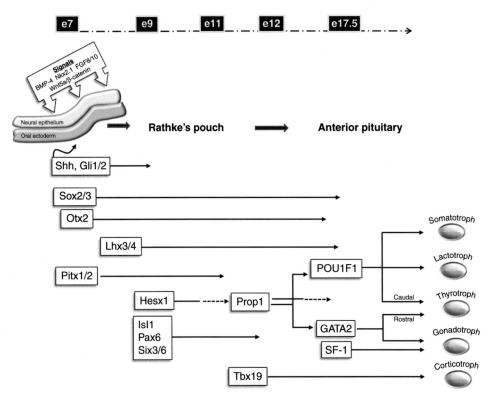

FIGURE 6.1   **An overview of major transcription and signaling factors involved in pituitary development.** Several signaling factors are expressed to regulate the development of the primordial pituitary associated with contact of the ventral diencephalon (neural epithelium) and oral ectoderm. The subsequent expression of transcription factors occurs in an orchestrated temporal and spatial sequence that leads to the development of Rathke's pouch and ultimately the mature anterior pituitary, which will contain the five mature pituitary cell types. The figure illustrates the major transcription factors involved in the cascade leading to the mature pituitary. The arrows represent the relative temporal appearance of these factors over the course of development as well as the attenuation of some of these factors (depicted as dashed lines), which have been shown to be important for appropriate development. The disruption of any of these factors has been suggested as a potential etiology for hypopituitarism.[5–7]

homeodomain 1 (Pitx1), and POU1F1, among others.[3,4] These factors play significant roles in the coordinated temporal and spatial development of specific pituitary cell types (Fig. 6.1). Since several of the pituitary cell types require common factors for their development, the early disruption in the cascade of events may ultimately affect several cell types. This may lead to loss of pituitary gene expression and secretion as well as abnormal structural development of the pituitary gland.[5]

Although a large part of our knowledge regarding pituitary organogenesis is based on nonhuman studies, careful clinical investigation of patients with hypopituitarism has yielded important insights into pituitary development. For example, murine studies of the Ames dwarf mouse, which bears a mutation in the PROP1 gene, revealed a hypocellular anterior pituitary generally lacking somatotrophs, lactotrophs, and thyrotrophs.[8,9] These findings led to studies in patients with a similar phenotype and the identification of human mutations in the PROP1 gene.[10,11] Two other naturally occurring inbred murine models, the Jackson and Snell dwarf mice, illustrated the importance of POU1F1 (formerly Pit1) in normal pituitary development.[12] Mutations in the

POU1F1 gene were subsequently identified in patients and revealed the importance of heterozygous point mutations encoding proteins acting as dominant negative inhibitors of pituitary gene expression.[13,14]

Over the last several decades, several research laboratories have shown that mutations in specific transcription factors can disrupt the balanced orchestration of pituitary development and ultimately the expression and function of the five pituitary cell types. This progress has resulted in the genetic characterization of hypopituitarism in many patients whose condition had been labeled as idiopathic or "congenital." Mutations in several of these factors leads to multiple pituitary hormone deficiencies, whereas mutations in others may only lead to loss of a single hormone.[15,16] Table 6.1 contains an outline of several genetic factors that have importance in anterior pituitary development. For each factor, the following are listed: a brief description of its function, the pituitary cell types affected by its disruption, a brief summary of the clinical features noted in patients with reported mutations, and the mode of inheritance. Several of these factors are associated with syndromes or are affected as a result of chromosomal abnormalities; thus, hypopituitarism may be one of the clues suggesting a genetic etiology. Finally, these factors are crucial to pituitary development, but may also be important for pituitary cell survival; therefore, the initial clinical phenotype in some patients may progress and additional pituitary hormone deficiencies can develop over time.

Numerous mutations associated with hypopituitarism have been reported in many of these factors. Table 6.1 is therefore not meant to be comprehensive, since ongoing genetic screening of patients with idiopathic hypopituitarism continues to identify and characterize new mutations in these factors. Researchers also continue to identify novel factors that contribute to pituitary development and therefore a complete up-to-date listing of transcription factor mutations is not possible. The references to specific studies that have characterized one or more mutations for the specific gene are noted. These references, many of which are reviews, provide

TABLE 6.1 II Genetic Pathophysiology: Developmental Factors Associated with Pituitary Development

Factor	Gene function	Affected cell types	Clinical phenotype	Mode of inheritance
One of more pituitary hormone deficiencies				
Hesx1[17–21]	• Paired-like homeobox gene • Early marker for pituitary primordium found in oral ectoderm of Rathke's pouch • Requires Lhx3 for maintenance and Propi for repression • Involved in development of forebrain, hypothalamus, optic nerve, and posterior pituitary	Somatotrophs, thyrotrophs, gonadotrophs, corticotrophs, posterior pituitary may also be affected	• Isolated GH deficiency or multiple hormone deficiency • Puberty may be unaffected or delayed • Associated with septooptic dysplasia • Possibly associated with Kallman syndrome • Abnormal MRI findings: anterior pituitary hypoplasia, ectopic posterior pituitary, midline forebrain abnormalities	AD, AR
Lhx3 (Lim3, P-LIM)[22–27]	• Member of LIM-homeodomain (LIM-HD) family of gene regulatory proteins • Required for survival/proliferation of Rathke's pouch cells and spinal cord motor neurons • Binds and activates α-GSU promoter • Acts with Pit-1 to activate TSH-β gene promoter • Three isoforms: Lhx3a, Lhx3b, M2-Lhx3	Somatrotrophs, lactotrophs, thyrotrophs, gonadotrophs, possibly corticotrophs	• Vertebral and skeletal malformations, including rigid cervical spine causing limited neck rotation • Hypoplastic anterior/intermediate pituitary lobe • Possible sensorineural hearing loss and/or intellectual disability	AR
Lhx4[28–33]	• A LIM protein with close resemblance to Lhx3 important for proliferation and differentiation of cell lineages • May have overlapping function with PROP1 and POU1F1 • Expressed during development of hindbrain, cerebral cortex, Rathke's pouch, and spinal cord	Somatrotrophs, lactotrophs, thyrotrophs, gonadotrophs, corticotrophs	• Combined pituitary hormone deficiencies with predominant GH deficiency • Severe anterior pituitary hypoplasia, ectopic neurohypophysis • Cerebellar abnormalities	AD
Six6 (Optx2)[34–36]	• Member of the SIX *sine oculis* family of homeobox genes • Expressed early in hypothalamus, later in Rathke's pouch, neural retina and optic chiasma	Somatotrophs, gonadotrophs	• Microphthalmia or anophthalmia • Pituitary hypoplasia • Associated with deletion at chromosome 14q22-23	Unclear
Pitx2 (RIEG1)[37–40]	• Bicoid-related homeobox gene expressed early in the nascent Rathke's pouch • Importance in maintaining expression of Hesx1 and Prop1	Somatrotrophs, lactotrophs, thyrotrophs, corticotrophs, reduced expression of gonadotrophs	• Associated with Rieger syndrome: • Anterior chamber eye anomalies • Craniofacial dysmorphism • Cardiovascular anomalies • Dental hypoplasia • Protuberant umbilicus • Mental retardation • Pituitary dysfunction	AD

(Continued)

TABLE 6.1   II Genetic Pathophysiology: Developmental Factors Associated with Pituitary Development (*cont.*)

Factor	Gene function	Affected cell types	Clinical phenotype	Mode of inheritance
Prop1 (prophet of Pit 1)[6,41–45]	• Paired-like homeodomain transcription factor required for Pit1 expression • Blocks expression of Hesx1 • Important for FSHβ expression, possibly even into adulthood	Somatrotrophs, lactotrophs, thyrotrophs, gonadotrophs, corticotrophs (delayed)	• Most common genetic cause of combined pituitary hormone deficiency (GH, TSH, PRL, and late onset ACTH reported) • Gonadotropin insufficiency or normal puberty with later onset of deficiency • Several mutations noted in nonconsanguineous families • Enlarged or hypoplastic anterior pituitary	AR
POU1F1 (Pit1)[10,46–50]	• A member of the POU transcription factor family • Important for activation of GH1, PRL, and TSHIβ genes	Somatotrophs, lactotrophs, thyrotrophs	• GH deficiency along with PRL and TSH dysregulation (TSH secretion may initially be normal) • Anterior pituitary hypoplasia	AD, AR
Otx2[51–58]	• Bicoid-type homeodomain gene required for forebrain, ear, nose, and eye development • Antagonizes *Fgf8* and *Shh* expression • May have importance in activation of Hesx1	Somatotrophs, thyrotrophs, corticotrophs, gonadotrophs	• Severe ocular malformations including anophthalmia • Developmental delay • Combined pituitary hormone deficiencies • Anterior pituitary hypoplasia with ectopic posterior pituitary	Unknown
SOX2[59–65]	• Member of SOXB1 subfamily as *Sox1* and *Sox3* expressed early in development	Somatotrophs, gonadotrophs, and in animal models thyrotrophs	• Hypogonadotrophic hypogonadism • Anterior pituitary hypoplasia • Bilateral anophthalmia/microphthalmia • Mid-brain defects including corpus callosum and hippocampus • Sensorineural hearing loss • Genital abnormalities • Esophageal atresia and learning difficulty • Associated with hypothalamic and pituitary tumors	*De novo*

SOX3[66–68]	• Member of SOX (SRY-related high mobility group (HMG) box)
	• Developmental factor expressed in developing infundibulum and hypothalamus
		Somatotrophs and/or possible deficiency in gonadotrophs and thyrotrophs	• Duplications of Xq26-27 in affected males (female carriers unaffected)
			• Variable mental retardation
			• Hypopituitarism with abnormal MRI:
			  • Anterior pituitary hypoplasia
			  • Infundibular hypoplasia
			  • Ectopic/undescended posterior pituitary
			  • Abnormal corpus callosum
			• Murine studies suggest SOX3 dosage critical for normal pituitary development
				X-linked

Isolated hormone deficiency

Factor	Gene function	Affected cell types	Clinical phenotype	Mode of inheritance
GLI2[69–71]	• Member of the Gli gene family; transcription factors that mediate Sonic hedgehog signaling	Somatotrophs (possible association with CPHD)	• Heterozygous loss-of-function mutations in patients with holoprosencephaly • Variable penetrance • Pituitary dysfunction accompanied by craniofacial abnormalities	
GHRHr[72–75]	• Encodes GHRH receptor	Somatotrophs	• Short stature • Anterior pituitary hypoplasia	AR

TABLE 6.1 II Genetic Pathophysiology: Developmental Factors Associated with Pituitary Development (*cont.*)

Factor	Gene function	Affected cell types	Clinical phenotype	Mode of inheritance
GH1[76–80]	• Encodes GH peptide • Several heterozygous mutations shown to affect GH secretion or function • Mutations reported leading to aberrant splicing	Somatotrophs	• Short stature • Abnormal fades • Associated with Kowarski syndrome (bioinactive GH)	AR, AD, or X-linked
KAL-1[81–84]	• Encodes protein anosmin-1 • Important in migration of both olfactory and GnRH neurons	Gonadotrophs	• Failed or arrested puberty • Anosmia • Renal abnormalities • Mutations implicated in hearing deficit • Associated with CPHD	X-linked
FGFR-1 (KAL-2)[85–87]	• Encodes fibroblast growth factor receptor • May interact with anosmin-1 (encoded by KAL1)	Gonadotrophs	• Failed or arrested puberty; typically less severe impairment of gonadotropin secretion vs. KAL-1 mutations • Associated with normosic hypogonadism • Associated with Pfeiffer syndrome (OMIM #101600) • Associated with Hartsfield syndrome (triad: holoprosencephaly, ectrodactyly, and cleft/lip palate; OMIM #615465) • Associated with CPHD	AD
FGF8[88]	• Critical role in formation of anterior/midline border • Regulates development of GnRH neurons • Animal studies demonstrate important role in midfacial integration and optic capsular development	Gonadotrophs	• Hypogonadotropic hypogonadism • Associated with CPHD	
PROKR2 PROK2[89–92]	• Encodes G protein-coupled prokineticin receptor-2 • Ligand is prokinectin-2	Gonadotrophs	• Hypogonadotropic hypogonadism and anosmia (Kallman syndrome) • Majority of patients have heterozygote mutations • Associated with CPHD	AR
GnRHRI[93,94]	• Encodes GnRH receptor in gonadotroph	Gonadotrophs	• Associated with broad range of reproductive phenotypes • May account for large proportion of familial and sporadic cases • No defects in olfaction	AR
TBX19 (TPIT)[95–98]	• Member of T-box family of transcription factors that contain a homologous DNA binding domain (the T-box) • Specific role in differentiation of POMC lineage • Several types of identified mutations leading to loss of function	Corticotrophs	• Neonatal isolated ACTH deficiency	AR
POMC[99–101]	• Encodes for peptide proopiomelano-cortin (POMC) • Functional role in skin pigmentation • Control of adrenal growth • Regulation of obesity and energy balance	Corticotrophs	• POMC deficiency syndrome: ○ Severe early-onset obesity ○ Adrenal insufficiency ○ Red hair	AR

further characterization of specific gene disruptions and the clinical phenotypes. The authors would also like to point out that since the previous edition, several studies have shown that for some genes once thought to be responsible for isolated hormone deficiencies, evidence suggests that their disruption may affect other pituitary cells and cause CPHD.

There continues to be a low percentage of patients with CPHD who are found to have genetic disruption in the genes that code for pituitary specific developmental factors; therefore, the improvement of screening techniques as well as advancing technology will aid the genetic characterization of these patients. In addition, some investigators have studied the effect of epigenetic changes within these genes in order to help better delineate possible mechanisms of pathology leading to pituitary maldevelopment and failure.[102,103]

DIAGNOSIS, GENETIC TESTING, AND INTERPRETATION

Currently, the identification of a genetic mutation in one of the developmental factors associated with pituitary hormone deficiency is rare. One study of 195 patients diagnosed with pituitary hormone deficiency that were tested for genetic mutations in five of the more commonly known factors revealed only 13.5% of patients had a mutation.[104] The prevalence, however, increased to 52.4% when the authors considered 21 familial cases. Most often, affected patients participate in studies conducted at academic institutions with institutional review board approval. Consequently, the results of these studies may not be available to study subjects. Despite the large number of pituitary developmental factors that could harbor mutations and that currently only a small number of patients are found to have mutations, genetic testing is generally low yield.

There have been efforts to stratify which factors to test using the patient's clinical phenotype and results from biochemical and radiological testing.[104] In addition, various syndromes have been noted to be associated with hypopituitarism and potentially guide investigators toward appropriate genetic screening. Figure 6.2 presents an algorithm to help identify the most commonly affected genes based on clinical presentation and laboratory evaluation in patients with idiopathic hypopituitarism.

Although the clinical and biochemical evaluation of hypopituitarism may help guide the clinician to a genetic evaluation of a specific developmental factor, a linear algorithm to guide an evaluation is not uniformly feasible given the spectrum of overlapping phenotypes associated with hypopituitarism. The paucity of well-defined phenotype-genotype studies in the literature contributes to the difficulty in assigning a specific phenotype to a

mutation in a specific factor. Finally, as seen in Fig. 6.2, patients presenting with isolated hormone deficiencies may develop additional hormone deficiencies over time and hence be appropriately considered for genetic testing for a larger or different subset of specific developmental factors. Currently, the clinical endocrinologist or geneticist as well as the researcher often do not have clear recognition of which developmental factor to evaluate based on the phenotypic or biochemical assessment. Such a dilemma emphasizes the need for not only additional traditional routine genetic screening of patients, but also the utilization of more advanced screening techniques such as whole genome sequencing (WGS) or whole exome sequencing (WES) to better understand the pathogenesis of congenital hypopituitarism.

Currently, very limited commercial testing is available to patients with hypopituitarism. We identified two testing sites, Athena Diagnostics (www.athenadiagnostics.com) and Gene Tests (www.geneclinics.org), that currently offer commercial testing for factors HESX1, PROP1, POU1F1, LHX3, LHX4, and growth hormone receptor. The strategy for how information from this testing is presented to the patient or what type of genetic counseling is offered is unclear.

The advent of next generation sequencing, which includes WGS and WES, offers much promise in helping to identify mutations that may affect pituitary development and function.[105,106] Given the complexity of pituitary development, these techniques may offer an opportunity to describe mutations in novel genes that may be implicated in the pathogenesis of hypopituitarism. There are several barriers, however, since these techniques result in a report including a large number of variants without known clinical consequence.[107] In particular, even though changes may be predicted to cause abnormal pituitary function, pathogenicity must still be shown through functional studies. Furthermore, the analysis of family members to demonstrate the heritability of predicted low frequency, deleterious changes helps to strengthen the importance of findings. Although caution must be taken with the interpretation of data, we believe this technology will continue to evolve as an effective screening tool.

TREATMENT

Regardless of the etiology of hypopituitarism, hormone replacement therapy continues as the mainstay of treatment. Currently there are various modes of delivery for replacement therapy including oral, injectable, or topical. For some deficiencies, only a single route of delivery is available to the patient, such as growth hormone deficiency, which requires injections. Patients with gonadotropin deficiency who require testosterone

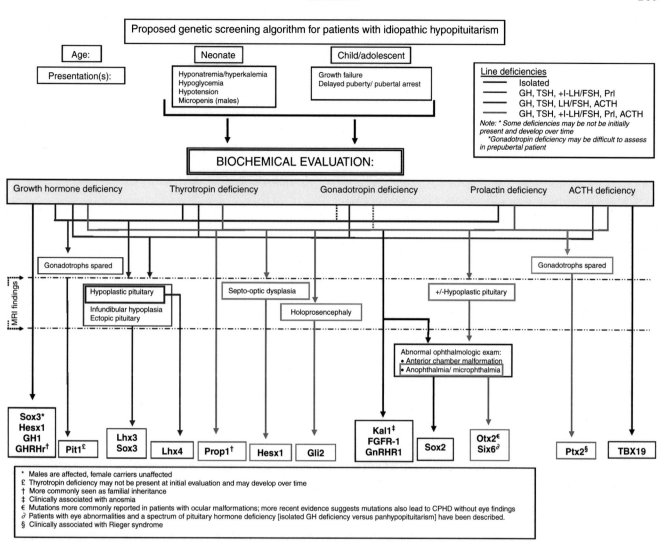

FIGURE 6.2 Genetic algorithm based on clinical presentation of hypopituitarism. Illustrated is a proposed clinically based algorithm to assist identifying the most likely developmental factor(s) to be affected based on phenotypic presentation of hypopituitarism in addition to clinical and/or radiological pathology. Isolated, as well as multiple, pituitary hormone deficiencies have been outlined in Fig. 6.2. In clinical practice, however, the patient evaluation may not necessarily coincide with one of the proposed deficiency scenarios outlined in the Fig. 6.2 as certain hormone deficiencies may have not yet manifested at the time of evaluation. A developing phenotype may raise suspicion to more than one developmental factor and it may be appropriate to investigate for abnormalities in several factors.

or estrogen replacement, however, have the option to choose replacement from all three types. Replacement therapy can involve prescribing the deficient hormone, such as recombinant growth hormone, or in the case of thyrotropin and ACTH deficiency, it is the target hormones, levothyroxine, and hydrocortisone, respectively, that are replaced. Replacement therapy for prolactin deficiency is not required.

Many patients may initially be diagnosed with one pituitary hormone deficiency, but over time may develop additional deficiencies. Therefore, appropriate follow-up and retesting of pituitary function is required. Growth hormone deficiency usually manifests at around 2–3 years of age with short stature or the deceleration of growth velocity. Growth hormone may also be used

at an earlier age or before the deceleration of growth in the subset of patients, usually newborns, who present with persistent hypoglycemia. In this scenario, growth hormone replacement therapy is required for metabolic homeostasis. Patients with gonadotropin deficiency present with the phenotype of delayed or absent pubertal development that does not manifest until late childhood, at which time testing and replacement with sex steroids is initiated.

Given the fact most genetic screening of patients with hypopituitarism occurs in the context of research protocols in academic institutions, the use of genetic counseling is limited. More importantly, the benefit of genetic counseling to predict either morbidity in the affected patient or the risk for future affected siblings remains unclear. Not surprisingly, the probability of familial transmission of a mutant pituitary developmental factor is rare and often not discovered unless a family is enrolled in a research study. Despite our understanding of pituitary development, the benefits are arguable in disclosing genetic findings to parents and patients, as this information will not necessarily change the mode of treatment or the sequelae of hypopituitarism. In contrast, since congenital hypopituitarism carries a risk for increased morbidity and mortality, recognizing that pituitary dysfunction may be of genetic etiology should heighten the awareness of physicians who care for families with a history of hypopituitarism. Ultimately, this may guide physicians to evaluate pituitary function more promptly and potentially initiate appropriate treatment, avoiding the complications that may result from pituitary hormone deficiency. Reports of vertical transmission of hypopituitarism from mother to child have reinforced the need for careful and appropriate monitoring of maternal thyroid function to ensure adequate delivery of thyroid hormone to the fetus.[108] Finally, another hope is that educating providers of patients with hypopituitarism will allow them to recognize the need to further study these patients at the genetic level. The advancements of genetic manipulation and the ability to target specific developmental factors will further elucidate the complexities of pituitary development and provide a mechanistic understanding of hypopituitarism in affected patients. In the long term, the anatomic accessibility of the pituitary may provide the option for gene therapy in patients with hypopituitarism. This goal, however, will require a highly sophisticated level of understanding of pituitary gene expression in normal children, as well as in those with defined mutations, in order for genetic interventional therapies to be successful.

References

1. Regal M, Páramo C, Sierra SM, Garcia-Mayor RV. Prevalence and incidence of hypopituitarism in an adult Caucasian population in northwestern Spain. *Clin Endocrinol (Oxf)* 2001;**55**:735–40.
2. Burgess R, Lunyak V, Rosenfeld M. Signaling and transcriptional control of pituitary development. *Curr Opin Genet Dev* 2002;**12**:534–9.
3. Scully KM, Rosenfeld MG. Pituitary development: regulatory codes in mammalian organogenesis. *Science* 2002;**295**:2231–5.
4. Kelberman D, Dattani MT. Hypothalamic and pituitary development: novel insights into the aetiology. *Eur J Endocrinol* 2007;**157**(Suppl. 1):S3–S14.
5. Zhu X, Wang J, Ju BG, Rosenfeld MG. Signaling and epigenetic regulation of pituitary development. *Curr Opin Cell Biol* 2007;**19**:605–11.
6. Kelberman D, Dattani MT. The role of transcription factors implicated in anterior pituitary development in the aetiology of congenital hypopituitarism. *Ann Med* 2006;**38**:560–77.
7. Dasen JS, Rosenfeld MG. Signaling and transcriptional mechanisms in pituitary development. *Annu Rev Neurosci* 2001;**24**:327–55.
8. Gage PJ, Lossie AC, Scarlett LM, Lloyd RV, Camper SA. Ames dwarf mice exhibit somatotrope commitment but lack growth hormone-releasing factor response. *Endocrinology* 1995;**136**:1161–7.
9. Gage PJ, Brinkmeier ML, Scarlett LM, Knapp LT, Camper SA, Mahon KA. The Ames dwarf gene, df, is required early in pituitary ontogeny for the extinction of Rpx transcription and initiation of lineage-specific cell proliferation. *Mol Endocrinol* 1996;**10**:1570–81.
10. Wu W, Cogan JD, Pfäffle RW, Dasen JS, Frisch H, O'Connell SM, Flynn SE, Brown MR, Mullis PE, Parks JS, Phillips 3rd JA, Rosenfeld MG. Mutations in PROP1 cause familial combined pituitary hormone deficiency. *Nat Genet* 1998;**18**:147–9.
11. Deladoëy J, Flück C, Büyükgebiz A, Kuhlmann BV, Eblé A, Hindmarsh PC, Wu W, Mullis PE. "Hot spot" in the PROP1 gene responsible for combined pituitary hormone deficiency. *J Clin Endocrinol Metab* 1999;**84**:1645–50.
12. Li S, Crenshaw 3rd EB, Rawson EJ, Simmons DM, Swanson LW, Rosenfeld MG. Dwarf locus mutants lacking three pituitary cell types result from mutations in the POU-domain gene pit-1. *Nature* 1990;**347**:528–33.
13. Tatsumi K, Miyai K, Notomi T, Kaibe K, Amino N, Mizuno Y, Kohno H. Cretinism with combined hormone deficiency caused by a mutation in the PIT1 gene. *Nat Genet* 1992;**1**:56–8.
14. Turton JP, Reynaud R, Mehta A, Torpiano J, Saveanu A, Woods KS, Tiulpakov A, Zdravkovic V, Hamilton J, Attard-Montalto S, Parascandalo R, Vella C, Clayton PE, Shalet S, Barton J, Brue T, Dattani MT. Novel mutations within the POU1F1 gene associated with variable combined pituitary hormone deficiency. *J Clin Endocrinol Metab* 2005;**90**:4762–70.
15. Kelberman D, Dattani MT. Hypopituitarism oddities: congenital causes. *Horm Res* 2007;**68**(Suppl. 5):138–44.
16. Toogood AA, Stewart PM. Hypopituitarism: clinical features, diagnosis, and management. *Endocrinol Metab Clin North Am* 2008;**37**:235–61.
17. Dattani MT, Martinez-Barbera JP, Thomas PQ, Brickman JM, Gupta R, Mårtensson IL, Toresson H, Fox M, Wales JK, Hindmarsh PC, Krauss S, Beddington RS, Robinson IC. Mutations in the homeobox gene HESX1/Hesx1 associated with septo-optic dysplasia in human and mouse. *Nat Genet* 1998;**19**:125–33.
18. Dattani MT, Robinson IC. HESX1 and septo-optic dysplasia. *Rev Endocr Metab Disord* 2002;**3**:289–300.
19. McNay DE, Turton JP, Kelberman D, Woods KS, Brauner R, Papadimitriou A, Keller E, Keller A, Haufs N, Krude H, Shalet SM, Dattani MT. HESX1 mutations are an uncommon cause of septooptic dysplasia and hypopituitarism. *J Clin Endocrinol Metab* 2007;**92**:691–7.
20. Alatzoglou KS, Dattani MT. Genetic forms of hypopituitarism and their manifestation in the neonatal period. *Early Hum Dev* 2009;**85**:705–12.
21. Newbern K, Natrajan N, Kim HG, Chorich LP, Halvorson LM, Cameron RS, Layman LC. Identification of HESX1 mutations in Kallman syndrome. *Fertil Steril* 2013;**99**:1831–7.
22. Netchine I, Sobrier ML, Krude H, Schnabel D, Maghnie M, Marcos E, Duriez B, Cacheux V, Moers A, Goossens M, Grüters A, Amselem S. Mutations in LHX3 result in a new syndrome revealed by combined pituitary hormone deficiency. *Nat Genet* 2000;**25**:182–6.
23. Savage JJ, Hunter CS, Clark-Sturm SL, Jacob TM, Pfaeffle RW, Rhodes SJ. Mutations in the LHX3 gene cause dysregulation of pituitary and neural target genes that reflect patient phenotypes. *Gene* 2007;**400**:44–51.
24. Pfaeffle RW, Savage JJ, Hunter CS, Palme C, Ahlmann M, Kumar P, Bellone J, Schoenau E, Korsch E, Brämswig JH, Stobbe HM, Blum WF, Rhodes SJ. Four novel mutations of the LHX3 gene cause combined pituitary hormone deficiencies with or without limited neck rotation. *J Clin Endocrinol Metab* 2007;**92**:1909–19.
25. Rajab A, Kelberman D, de Castro SC, Biebermann H, Shaikh H, Pearce K, Hall CM, Shaikh G, Gerrelli D, Grueters A, Krude H,

Dattani MT. Novel mutations in LHX3 are associated with hypopituitarism and sensorineural hearing loss. *Hum Mol Genet* 2008;**17**:2150–9.

26. Bonfig W, Krude H, Schmidt H. A novel mutation of LHX3 is associated with combined pituitary hormone deficiency including ACTH deficiency, sensorineural hearing loss, and short neck – a case report and review of the literature. *Eur J Pediatr* 2011;**170**:1017–21.

27. Pfäffle R, Klammt J. Pituitary transcription factors in the aetiology of combined pituitary hormone deficiency. *Best Pract Res Clin Endocrinol Metab* 2011;**25**:43–60.

28. Mullen RD, Colvin SC, Hunter CS, Savage JJ, Walvoord EC, Bhangoo AP, Ten S, Weigel J, Pfäffle RW, Rhodes SJ. Roles of the LHX3 and LHX4 LIM-homeodomain factors in pituitary development. *Mol Cell Endocrinol* 2007;**265-266**:190–5.

29. Raetzman LT, Ward R, Camper SA. Lhx4 and Prop1 are required for cell survival and expansion of the pituitary primordia. *Development* 2002;**129**:4229–39.

30. Machinis K, Pantel J, Netchine I, Léger J, Camand OJ, Sobrier ML, Dastot-Le Moal F, Duquesnoy P, Abitbol M, Czernichow P, Amselem S. Syndromic short stature in patients with a germline mutation in the LIM homeobox LHX4. *Am J Hum Genet* 2001;**69**:961–8.

31. Pfaeffle RW, Hunter CS, Savage JJ, Duran-Prado M, Mullen RD, Neeb ZP, Eiholzer U, Hesse V, Haddad NG, Stobbe HM, Blum WF, Weigel JF, Rhodes SJ. Three novel missense mutations within the LHX4 gene are associated with variable pituitary hormone deficiencies. *J Clin Endocrinol Metab* 2008;**93**:1062–71.

32. Prince KL, Walvoord EC, Rhodes SJ. The role of homeodomain transcription factors in heritable pituitary disease. *Nat Rev Endocrinol* 2011;**7**:727–37.

33. Takagi M, Ishii T, Inokuchi M, Amano N, Narumi S, Asakura Y, Muroya K, Hasegawa Y, Adachi M, Hasegawa T. Gradual loss of ACTH due to a novel mutation in LHX4: comprehensive mutation screening in Japanese patients with congenital hypopituitarism. *PLoS One* 2012;**7**:e46008.

34. Nolen LD, Amor D, Haywood A, St Heaps L, Willcock C, Mihelec M, Tam P, Billson F, Grigg J, Peters G, Jamieson RV. Deletion at 14q22-23 indicates a contiguous gene syndrome comprising anophthalmia, pituitary hypoplasia, and ear anomalies. *Am J Med Genet A* 2006;**140**:1711–8.

35. Aldahmesh MA, Khan AO, Hijazi H, Alkuraya FS. Homozygous truncation of SIX6 causes complex microphthalmia in humans. *Clin Genet* 2013;**84**:198–9.

36. Martínez-Frías ML, Ocejo-Vinyals JG, Arteaga R, Martinez-Fernández ML, Macdonald A, Pérez-Belmonte E, Bermejo-Sánchez E, Martínez S. Interstitial deletion 14q22.3-q23.2: genotype-phenotype correlation. *Am J Med Genet A* 2014;**164A**:639–47.

37. Perveen R, Lloyd IC, Clayton-Smith J, Churchill A, van Heyningen V, Hanson I, Taylor D, McKeown C, Super M, Kerr B, Winter R, Black GC. Phenotypic variability and asymmetry of Rieger syndrome associated with PITX2 mutations. *Invest Ophthalmol Vis Sci* 2000;**41**:2456–60.

38. Suh H, Gage PJ, Drouin J, Camper SA. Pitx2 is required at multiple stages of pituitary organogenesis: pituitary primordium formation and cell specification. *Development* 2002;**129**:329–37.

39. Tümer Z, Bach-Holm D. Axenfeld-Rieger syndrome and spectrum of PITX2 and FOXC1 mutations. *Eur J Hum Genet* 2009;**17**:1527–39.

40. Quentien MH, Vieira V, Menashe M, Dufier JL, Herman JP, Enjalbert A, Abitbol M, Brue T. Truncation of PITX2 differentially affects its activity on physiological targets. *J Mol Endocrinol* 2010;**46**:9–19.

41. Duquesnoy P, Roy A, Dastot F, Ghali I, Teinturier C, Netchine I, Cacheux V, Hafez M, Salah N, Chaussain JL, Goossens M, Bougnères P, Amselem S. Human Prop-1: cloning, mapping, genomic structure. Mutations in familial combined pituitary hormone deficiency. *FEBS Lett* 1998;**437**:216–20.

42. Vallette-Kasic S, Barlier A, Teinturier C, Diaz A, Manavela M, Berthezène F, Bouchard P, Chaussain JL, Brauner R, Pellegrini-Bouiller I, Jaquet P, Enjalbert A, Brue T. PROP1 gene screening in patients with multiple pituitary hormone deficiency reveals two sites of hypermutability and a high incidence of corticotroph deficiency. *J Clin Endocrinol Metab* 2001;**86**:4529–35.

43. De Moraes DC, Vaisman M, Congeição FL, Ortiga-Carvalho TM. Pituitary development: a complex, temporal regulated process dependent on specific transcriptional factors. *J Endocrinol* 2012;**215**:239–45.

44. Obermannova B, Pfaeffle R, Zygmunt-Gorska A, Starzyk J, Verkauskiene R, Smetanina N, Bezlepkina O, Peterkova V, Frisch H, Cinek O, Child CJ, Blum WF, Lebl J. Mutations and pituitary morphology in a series of 82 patients with PROP1 gene defects. *Horm Res Paediatr* 2011;**76**:348–54.

45. Kelberman D, Turton JPG, Woods KS, Mehta A, Al-Khawari M, Greening J, Swift PG, Otonkoski T, Rhodes SJ, Dattani MT. Molecular analysis of novel PROP1 mutations associated with combined pituitary hormone deficiency (CPHD). *Clin Endocrinol (Oxf)* 2008;**70**:96–103.

46. Cohen LE, Wondisford FE, Salvatoni A, Maghnie M, Brucker-Davis F, Weintraub BD, Radovick S. A "hot spot" in the Pit-1 gene responsible for combined pituitary hormone deficiency: clinical and molecular correlates. *J Clin Endocrinol Metab* 1995;**80**:679–84.

47. Tenenbaum-Rakover Y, Sobrier ML, Amselem S. A novel POU1F1 mutation (p.Thr168IlefsX7) associated with an early and severe form of combined pituitary hormone deficiency: functional analysis and follow-up from infancy to childhood. *Clin Endocrinol (Oxf)* 2011;**75**:214–9.

48. Turton JP, Strom M, Langham S, Dattani MT, Le Tissier P. Two novel mutations in the POU1F1 gene generate null alleles through different mechanisms leading to combined pituitary hormone deficiency. *Clin Endocrinol (Oxf)* 2012;**76**:387–93.

49. Lee NC, Tsai WY, Peng SF, Tung YC, Chien YH, Hwu WL. Congenital hypopituitarism due to POU1F1 gene mutation. *J Formos Med Assoc* 2011;**110**:58–61.

50. Kerr J, Wood W, Ridgway EC. Basic science and clinical research advances in the pituitary transcription factors: pit-1 and prop-1. *Curr Opin Endocrinol Diabetes Obes* 2008;**15**:359–63.

51. Chatelain G, Fossat N, Brun G, Lamonerie T. Molecular dissection reveals decreased activity and not dominant negative effect in human OTX2 mutants. *J Mol Med* 2006;**84**:604–15.

52. Ragge NK, Brown AG, Poloschek CM, Lorenz B, Henderson RA, Clarke MP, Russell-Eggitt I, Fielder A, Gerrelli D, Martinez-Barbera JP, Ruddle P, Hurst J, Collin JR, Salt A, Cooper ST, Thompson PJ, Sisodiya SM, Williamson KA, Fitzpatrick DR, van Heyningen V, Hanson IM. Heterozygous mutations of OTX2 cause severe ocular malformations. *Am J Hum Genet* 2005;**76**:1008–22.

53. Tajima T, Ishizu K, Nakamur A. Molecular and clinical findings in patins with LHX4 and OTX2 mutations. *Clin Pediatr Endocrinol* 2013;**22**:15–23.

54. Diaczok D, Romero C, Zunich J, Marshall I, Radovick S. A novel dominant negative mutation of OTX2 associated with combined pituitary hormone deficiency. *J Clin Endocrinol Metab* 2008;**93**:4351–9.

55. Dateki S, Fukami M, Sato N, Muroya K, Adachi M, Ogata T. OTX2 mutation in a patient with anophthalmia, short stature, and partial growth hormone deficiency: functional studies using the IRBP, HESX1, and POU1F1 promoters. *J Clin Endocrinol Metab* 2008;**93**:3697–702.

56. Schilter KF, Schneider A, Bardakjian T, Soucy JF, Tyler RC, Reis LM, Semina EV. OTX2 microphthalmia syndrome: four novel mutations and deliniation of a phenotype. *Clin Genet* 2011;**79**:158–68.

57. Gorbenko Del Blanco D, Romero CJ, Diaczok D, de Graaff LC, Radovick S, Hokken-Koelega AC. A novel OTX2 mutation in a patient with combined pituitary hormone deficiency, pituitary malformation, and an underdeveloped left optic nerve. *Eur J Endocrinol* 2012;**167**:441–52.

58. Dateki S, Kosaka K, Hasegawa K, Tanaka H, Azuma N, Yokoya S, Muroya K, Adachi M, Tajima T, Motomura K, Kinoshita E, Moriuchi H, Sato N, Fukami M, Ogata T. Heterozygous orthodenticle homeobox 2 mutations are associated with variable pituitary phenotype. *J Clin Endocrinol Metab* 2010;**95**:756–64.

59. Kelberman D, Rizzoti K, Avilion A, Bitner-Glindzicz M, Cianfarani S, Collins J, Chong WK, Kirk JM, Achermann JC, Ross R, Carmignac D, Lovell-Badge R, Robinson IC, Dattani MT. Mutations within Sox2/SOX2 are associated with abnormalities in the hypothalamo-pituitary-gonadal axis in mice and humans. *J Clin Invest* 2006;**116**:2442–55.

60. Kelberman D, de Castro SC, Huang S, Crolla JA, Palmer R, Gregory JW, Taylor D, Cavallo L, Faienza MF, Fischetto R, Achermann JC, Martinez-Barbera JP, Rizzoti K, Lovell-Badge R, Robinson IC, Gerrelli D, Dattani MT. SOX2 plays a critical role in the pituitary, forebrain, and eye during human embryonic development. *J Clin Endocrinol Metab* 2008;**93**:1865–73.

61. Macchiaroli A, Kelberman D, Auriemma RS, Drury S, Islam L, Giangiobbe S, Ironi G, Lench N, Sowden JC, Colao A, Pivonello R, Cavallo L, Gasperi M, Faienza MF. A novel heterozygous SOX2 mutation causing congenital bilateral anophthalmia, hypogonadotropic hypogonadism and growth hormone deficiency. *Gene* 2014;**2**:282–5.

62. Alatzoglou KS, Andoniadou CL, Kelberman D, Buchanan CR, Crolla J, Arriazu MC, Roubicek M, Moncet D, Martinez-Barbara JP, Dattani MT. SOX2 haploinsufficiency is associated with slow progressing hypothalamo-pituitary tumours. *Hum Mutat* 2011;**32**:1376–80.

63. McCabe MJ, Alatzoglou KS, Dattani MT. Septo-optic dysplasia and other midline defects: the role of transcription factors: HESX1 and beyond. *Best Pract Res Clin Endocrinol Metab* 2011;**25**:115–24.

64. Kelberman D, Dattani MT. Septo-optic dysplasia – novel insights into the aetiology. *Horm Res* 2008;**69**:257–65.

65. Jayakody SA, Andoniadou CL, Gaston-Massuet C, Signore M, Cariboni A, Bouloux PM, Le Tissier P, Pevny LH, Dattani MT, Martinez-Barbera JP. SOX2 regulates the hypothalamic-pituitary axis at multiple levels. *J Clin Invest* 2012;**122**:3635–46.

66. Solomon NM, Ross SA, Forrest SM, Thomas PQ, Morgan T, Belsky JL, Hol FA, Karnes PS, Hopwood NJ, Myers SE, Tan AS, Warne GL. Array comparative genomic hybridisation analysis of boys with X-linked hypopituitarism identifies a 3.9 Mb duplicated critical region at Xq27 containing SOX3. *J Med Genet* 2007;**44**:e75.

67. Alatzoglou KS, Kelberman D, Cowell CT, Palmer R, Arnhold IJ, Melo ME, Schnabel D, Grueters A, Dattani MT. Increased transactivation associated with SOX3 polyalanine tract deletion in a patient with hypopituitarism. *J Clin Endocrinol Metab* 2011;**96**:685–90.

68. Helle JR, Baroy T, Misceo D, Braaten O, Fannernel M, Frengen E. Hyperphagia, mild developmental delay, but apparently no structural brain anomalies in a boy without SOX3 expression. *Am J Med Genet A* 2013;**161A**:1137–42.

69. Roessler E, Du YZ, Mullor JL, Casas E, Allen WP, Gillessen-Kaesbach G, Roeder ER, Ming JE, Ruiz I, Altaba A, Muenke M. Loss-of-function mutations in the human GLI2 gene are associated with pituitary anomalies and holoprosencephaly-like features. *Proc Natl Acad Sci USA* 2003;**100**:13424–9.

70. Flemming GM, Klammt J, Ambler G, Bao Y, Blum WF, Cowell C, Donaghue K, Howard N, Kumar A, Sanchez J, Stobbe H, Pfäffle RW. Functional characterization of a heterozygous GLI2 missense mutation in patients with multiple pituitary hormone deficiency. *J Clin Endocrinol Metab* 2013;**98**:E567–75.

71. Franca MM, Jorge AA, Carvalho LR, Costalonga EF, Otto AP, Correa FA, Mendonca BB, Arnhold IJ. Relatively high frequency of non-synonymous GLI2 variants in patients with congenital hypopituitarism without holoprosencephaly. *Clin Endocrinol (Oxf)* 2013;**78**:551–7.

72. Maheshwari HG, Silverman BL, Dupuis J, Baumann G. Phenotype and genetic analysis of a syndrome caused by an inactivating mutation in the growth hormone-releasing hormone receptor: dwarfism of Sindh. *J Clin Endocrinol Metab* 1998;**83**:4065–74.

73. Soneda A, Adachi M, Muroya K, Asakura Y, Takagi M, Hasegawa T, Inoue H, Itakura M. Novel compound heterozygous mutations of the growth hormone-releasing hormone receptor gene in a case of isolated growth hormone deficiency. *Growth Horm IGF Res* 2013;**23**:89–97.

74. Marui S, Trarbach EB, Boguszewski MC, França MM, Jorge AA, Inoue H, Nishi MY, de Lacerda Filho L, Aguiar-Oliveira MH, Mendonca BB, Arnhold IJ. GH-releasing hormone receptor gene: a novel splice-disrupting mutation and study of founder effects. *Horm Res Paediatr* 2012;**78**:165–72.

75. Shohreh R, Sherafat-Kazemzadeh R, Jee YH, Blitz A, Salvatori R. A novel frame shift mutation in the GHRH receptor gene in familial isolated GH deficiency: early occurrence of anterior pituitary hypoplasia. *J Clin Endocrinol Metab* 2011;**96**:2982–6 Erratum in: *J Clin Endocrinol Metab* 2012, 97, 307.

76. Millar DS, Lewis MD, Horan M, Newsway V, Easter TE, Gregory JW, Fryklund L, Norin M, Crowne EC, Davies SJ, Edwards P, Kirk J, Waldron K, Smith PJ, Phillips 3rd JA, Scanlon MF, Krawczak M, Cooper DN, Procter AM. Novel mutations of the growth hormone 1 (GH1) gene disclosed by modulation of the clinical selection criteria for individuals with short stature. *Hum Mutat* 2003;**21**:424–40.

77. Besson A, Salemi S, Deladoey J, Vuissoz JM, Eble A, Bidlingmaier M, Burgi S, Honegger U, Fluck C, Mullis PE. Short stature caused by a biologically inactive mutant growth hormone (GH-C53S). *J Clin Endocrinol Metab* 2005;**90**:2493–9.

78. Babu D, Mellone S, Fusco I, Petri A, Walker GE, Bellone S, Prodam F, Momigliano-Richiardi P, Bona G, Giordano M. Novel mutations in the GH gene (GH1) uncover putative splicing regulatory elements. *Endocrinology* 2014;**155**:1786–92.

79. Juanes M, Marino R, Ciaccio M, Di Palma I, Ramirez P, Warman DM, De Dona V, Chaler E, Maceiras M, Rivarola MA, Belgorosky A. Presence of GH1 and absence of GHRHR gene mutations in a large cohort of Argentinian patients with severe short stature and isolated GH deficiency. *Clin Endocrinol (Oxf)* 2014;**80**:618–20.

80. Alatzoglou KS, Turton JP, Kelberman D, Clayton PE, Mehta A, Buchanan C, Aylwin S, Crowne EC, Christesen HT, Hertel NT, Trainer PJ, Savage MO, Raza J, Banerjee K, Sinha SK, Ten S, Mushtaq T, Brauner R, Cheetham TD, Hindmarsh PC, Mullis PE, Dattani MT. Expanding the spectrum of mutations in GH1 and GHRHR: genetic screening in a large cohort of patients with congenital isolated growth hormone deficiency. *J Clin Endocrinol Metab* 2009;**94**:3191–9.

81. Albuisson J, Pecheux C, Carel JC, Lacombe D, Leheup B, Lapuzina P, Bouchard P, Legius E, Matthijs G, Wasniewska M, Delpech M, Young J, Hardelin JP, Dode C. Kallmann syndrome: 14 novel mutations in KAL1 and FGFR1 (KAL2). *Hum Mutat* 2005;**25**:98–9.

82. Hardelin JP, Levilliers J, Blanchard S, Carel JC, Leutenegger M, Pinard-Bertelletto JP, Bouloux P, Petit C. Heterogeneity in the mutations responsible for X chromosome-linked Kallmann syndrome. *Hum Mol Genet* 1993;**2**:373–7.

83. Raivio T, Avbelj M, McCabe MJ, Romero CJ, Dwyer AA, Tommiska J, Sykiotis GP, Gregory LC, Diaczok B, et al. Genetic overlap in Kallmann syndrome, combined pituitary hormone deficiency, and septo-optic dysplasia. *J Clin Endocrinol Metab* 2012;**97**:E694–9.

84. Marlin S, Chantot-Bastaraud S, David A, Loundon N, Jonard L, Portnoï MF, Bonnet C, Louha M, Gherbi S, Garabedian EN,

Couderc R, Denoyelle F. Discovery of a large deletion of KAL1 in 2 deaf brothers. *Otol Neurotol* 2013;**34**:1590–4.

85. Salenave S, Chanson P, Bry H, Pugeat M, Cabrol S, Carel JC, Murat A, Lecomte P, Brailly S, Hardelin JP, Dode C, Young J. Kallmann's syndrome: a comparison of the reproductive phenotypes in men carrying KAL1 and FGFR1/KAL2 mutations. *J Clin Endocrinol Metab* 2008;**93**:758–63.

86. Vizeneux A, Hilfiger A, Bouligand J, Pouillot M, Brailly-Tabard S, Bashamboo A, McElreavey K, Brauner R. Congenital hypogonadotropic hypogonadism during childhood: presentation and genetic analyses in 46 boys. *PLoS One* 2013;**8**:e77827.

87. Dodé C, Levilliers J, Dupont JM, De Paepe A, Le Dû N, Soussi-Yanicostas N, Coimbra RS, Delmaghani S, Compain-Nouaille S, Baverel F, Pêcheux C, Le Tessier D, Cruaud C, Delpech M, Speleman F, Vermeulen S, Amalfitano A, Bachelot Y, Bouchard P, Cabrol S, Carel JC, Delemarre-van de Waal H, Goulet-Salmon B, Kottler ML, Richard O, Sanchez-Franco F, Saura R, Young J, Petit C, Hardelin JP. Loss-of-function mutations in FGFR1 cause autosomal dominant Kallmann syndrome. *Nat Genet* 2003;**33**:463–5.

88. Suzuki E, Yatsuga S, Igarashi M, Miyado M, Nakabayashi K, Hayashi K, Hata K, Umezawa A, Yamada G, Ogata T, Fukami M. *De novo* frameshift mutation in fibroblast growth factor 8 in a male patient with gonadotropin deficiency. *Horm Res Paediatr* 2014;**81**:139–44.

89. Dodé C, Rondard P. PROK2/PROKR2 signaling and Kallmann syndrome. *Front Endocrinol (Lausanne)* 2013;**4**:19.

90. McCabe MJ, Gaston-Massuet C, Gregory LC, Alatzoglou KS, Tziaferi V, Sbai O, Rondard P, Masumoto KH, Nagano M, Shigeyoshi Y, Pfeifer M, Hulse T, Buchanan CR, Pitteloud N, Martinez-Barbera JP, Dattani MT. Variations in PROKR2, but not PROK2, are associated with hypopituitarism and septo-optic dysplasia. *J Clin Endocrinol Metab* 2013;**98**:E547–57.

91. Tommiska J, Toppari J, Vaaralahti K, Känsäkoski J, Laitinen EM, Noisa P, Kinnala A, Niinikoski H, Raivio T. PROKR2 mutations in autosomal recessive Kallmann syndrome. *Fertil Steril* 2013;**99**:815–8.

92. Reynaud R, Jayakody SA, Monnier C, Saveanu A, Bouligand J, Guedj AM, Simonin G, Lecomte P, Barlier A, Rondard P, Martinez-Barbera JP, Guiochon-Mantel A, Brue T. PROKR2 variants in multiple hypopituitarism with pituitary stalk interruption. *J Clin Endocrinol Metab* 2012;**97**:E1068–73.

93. Beranova M, Oliveira LM, Bedecarrats GY, Schipani E, Vallejo M, Ammini AC, Quintos JB, Hall JE, Martin KA, Hayes FJ, Pitteloud N, Kaiser UB, Crowley Jr WF, Seminara SB. Prevalence, phenotypic spectrum, and modes of inheritance of gonadotropin-releasing hormone receptor mutations in idiopathic hypogonadotropic hypogonadism. *J Clin Endocrinol Metab* 2001;**86**:1580–8.

94. Beneduzzi D, Trarbach EB, Latronico AC, Mendonca BB, Silveira LF. Novel mutation in the gonadotropin-releasing hormone receptor (GNRHR) gene in a patient with normosmic isolated hypogonadotropic hypogonadism. *Arq Bras Endocrinol Metabol* 2012;**56**:540–4.

95. Lamolet B, Pulichino AM, Lamonerie T, Gauthier Y, Brue T, Enjalbert A, Drouin J. A pituitary cell-restricted T box factor, Tpit, activates POMC transcription in cooperation with Pitx homeoproteins. *Cell* 2001;**104**:849–59.

96. Vallette-Kasic S, Brue T, Pulichino AM, Gueydan M, Barlier A, David M, Nicolino M, Malpuech G, Dechelotte P, Deal C, Van

Vliet G, De Vroede M, Riepe FG, Partsch CJ, Sippell WG, Berberoglu M, Atasay B, de Zegher F, Beckers D, Kyllo J, Donohoue P, Fassnacht M, Hahner S, Allolio B, Noordam C, Dunkel L, Hero M, Pigeon B, Weill J, Yigit S, Brauner R, Heinrich JJ, Cummings E, Riddell C, Enjalbert A, Drouin J. Congenital isolated adrenocorticotropin deficiency: an underestimated cause of neonatal death, explained by TPIT gene mutations. *J Clin Endocrinol Metab* 2005;**90**:1323–31.

97. McEachern R, Drouin J, Metherell L, Huot C, Van Vliet G, Deal C. Severe cortisol deficiency associated with reversible growth hormone deficiency in two infants: what is the link? *J Clin Endocrinol Metab* 2011;**6**:2670–4.

98. Couture C, Saveanu A, Barlier A, Carel JC, Fassnacht M, Flück CE, Houang M, Maes M, Phan-Hug F, Enjalbert A, Drouin J, Brue T, Vallette S. Phenotypic homogeneity and genotypic variability in a large series of congenital isolated ACTH-deficiency patients with TPIT gene mutations. *J Clin Endocrinol Metab* 2012;**97**:E486–95.

99. Krude H, Biebermann H, Luck W, Horn R, Brabant G, Gruters A. Severe early-onset obesity, adrenal insufficiency and red hair pigmentation caused by POMC mutations in humans. *Nat Genet* 1998;**19**:155–7.

100. Krude H, Biebermann H, Schnabel D, Tansek MZ, Theunissen P, Mullis PE, Gruters A. Obesity due to proopiomelanocortin deficiency: three new cases and treatment trials with thyroid hormone and ACTH4-10. *J Clin Endocrinol Metab* 2003;**88**:4633–40.

101. Farooqi IS, Drop S, Clements A, Keogh JM, Biernacka J, Lowenbein S, Challis BG, O'Rahilly S. Heterozygosity for a POMC-null mutation and increased obesity risk in humans. *Diabetes* 2006;**55**:2549–53.

102. Malik RE, Rhodes SJ. The role of DNA methylation in regulation of the murine Lhx3 gene. *Gene* 2014;**534**:272–81.

103. Hunter CS, Malik RE, Witzmann FA, Rhodes SJ. LHX3 interacts with inhibitor of histone acetyltransferase complex subunits LANP and TAF-1β to modulate pituitary gene regulation. *PLoS One* 2013;**8**:e68898.

104. Reynaud R, Gueydan M, Saveanu A, Vallette-Kasic S, Enjalbert A, Brue T, Barlier A. Genetic screening of combined pituitary hormone deficiency: experience in 195 patients. *J Clin Endocrinol Metab* 2006;**91**:3329–36.

105. Wassner AJ, Cohen LE, Hechter E, Dauber A. Isolated central hypothyroidism in young siblings as a manifestation of PROP1 deficiency: clinical impact of whole exome sequencing. *Horm Res Paediatr* 2013;**79**:379–86.

106. Yang Y, Muzny DM, Reid JG, Bainbridge MN, Willis A, Ward PA, Braxton A, Beuten J, Xia F, Niu Z, Hardison M, Person R, Bekheirnia MR, Leduc MS, Kirby A, Pham P, Scull J, Wang M, Ding Y, Plon SE, Lupski JR, Beaudet AL, Gibbs RA, Eng CM. Clinical whole-exome sequencing for the diagnosis of Mendelian disorders. *N Engl J Med* 2013;**369**:1502–11.

107. Bromberg Y. Building a genome analysis pipeline to predict disease risk and prevent disease. *J Mol Biol* 2013;**425**:3993–4005.

108. Pine-Twaddell E, Romero CJ, Radovick S. Vertical transmission of hypopituitarism: critical importance of appropriate interpretation of thyroid function tests and levothyroxine therapy during pregnancy. *Thyroid* 2013;**23**:892–7.

PART IV

THYROID

7

Congenital Defects of Thyroid Hormone Synthesis

Helmut Grasberger*, Samuel Refetoff**

*Department of Medicine, University of Michigan, Ann Arbor, MI, USA
**Departments of Medicine, Pediatrics and Committee on Genetics,
University of Chicago, Chicago, IL, USA

INTRODUCTION

With a prevalence of one in 3,000–4,000 newborns, congenital hypothyroidism (CH) is the most common inborn endocrine disorder and one of the most common preventable causes of mental retardation. While most cases are sporadic and associated with abnormalities of thyroid gland development and migration (thyroid dysgenesis), approximately 15–20% are caused by inherited defects in one of the steps of thyroid hormone synthesis (thyroid dyshormonogenesis) (Fig. 7.1). When the synthesis defect results in reduced hormone secretion, the ensuing diminished negative feedback on the anterior pituitary thyrotrophs leads to an increase in thyrotropin (TSH) secretion stimulating the thyroid gland. Consequently, patients are born with an enlarged thyroid gland (goiter) or develop goiter postnatally, especially when diagnosis and treatment with levothyroxine (L-T_4) are delayed. With the exceptions indicated below, these defects are inherited in an autosomal recessive fashion and are amenable to detection by newborn screening for CH.

The etiological classification of CH is based on clinical and biochemical evaluation. Useful tests are measurement of serum TSH, thyroxine (T_4), triiodothyronine (T_3), and thyroglobulin [TG]; thyroid ultrasound and scintigraphy, using ^{99m}Tc or, preferably, ^{123}I^1 and, when indicated, the perchlorate (ClO_4^2) discharge test. Often infants with CH, confirmed by the TSH and T_4 values, are started on thyroid hormone replacement without detailed etiologic diagnosis. The latter is relegated to later years, usually two or three following a one-month withdrawal of L-T_4 replacement.

With the identification of the key steps involved in thyroid hormone synthesis, a molecular genetic diagnosis should be feasible for the vast majority of patients with dyshormonogenesis. A complete diagnostic work-up facilitates the selection of the most likely candidate gene(s) for etiologic confirmation. While the distinguishing features of each particular hormone synthesis defect are outlined in the following section, it should be noted that early genetic screening might be justified even without complete etiologic classification. Other than providing useful information for genetic counseling, there are additional benefits to a definite genetic diagnosis. One is the potential impact on treatment. For instance, patients with specific defects may be efficiently treated with iodide supplementation rather than L-T_4. Another benefit is in the identification of a subset of patients with transient CH due to partial hormonogenesis defects. Even if euthyroid at a particular point in time, increased demand for thyroid hormone synthesis may precipitate hypothyroidism. An early molecular diagnosis predicts the necessity for lifelong hormone replacement therapy. Finally, some genetic defects may not manifest at birth but produce hypothyroidism later in life. The definitive diagnosis of an index case will enable early identification of subsequent cases in the same family and help to avoid the negative consequences on mental development associated with delayed diagnosis and treatment of hypothyroidism.

PATHOPHYSIOLOGY AND GENETICS OF SPECIFIC DYSHORMONOGENESIS DEFECTS

Defect in Thyroidal Iodide Trapping – Gene: SLC5A5 (NIS)

The sodium-iodide symporter (NIS; official gene symbol: *SLC5A5*) is a 13-transmembrane domain glycoprotein that mediates the uptake of iodide through the basolateral

Genetic Diagnosis of Endocrine Disorders. http://dx.doi.org/10.1016/B978-0-12-800892-8.00007-5

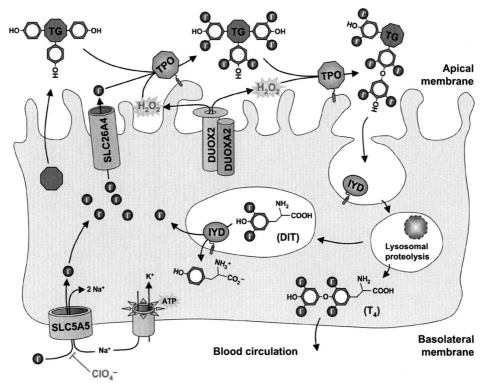

FIGURE 7.1 **Schematic drawing of a follicular thyroid cell illustrating the key players involved in thyroid hormonogenesis and iodide metabolism.** The distinct steps comprise (1) the active uptake of iodide from the blood circulation via the sodium–iodide symporter (SLC5A5); (2) the facilitated efflux of iodide into the colloid via an apical anion channel (SLC26A4); (3) iodination of tyrosine groups of thyroglobulin (TG) catalyzed by thyroid peroxidase (TPO); (4) subsequent coupling of iodinated tyrosines (mono- or diiodinated; MIT, DIT) within TG via ether-bond formation to iodothyronines; steps 3 and 4 require hydrogen peroxide as cosubstrate, which is provided by a hydrogen peroxide generating NADPH-oxidase constituted by dual oxidase 2 (DUOX2) and its maturation factor (DUOXA2); (5) following endocytosis, iodothyronines ($T_4 > T_3$) are liberated by lysosomal degradation of the TG matrix protein; (6) concomitantly released iodotyrosines are dehalogenated by iodotyrosine deiodinase (IYD) allowing "recycling" of iodide for further hormone synthesis. See plate section.

membrane of follicular thyroid cells (Fig. 7.1). Iodide is actively concentrated into these cells by the electrogenic symport of Na$^+$ (two Na$^+$ for one I^2) down the electrochemical gradient maintained by the activity of Na$^+$/K$^+$-ATPase. NIS is also expressed, but not regulated, by TSH in several other differentiated epithelia, notably salivary glands, lachrymal glands, gastric mucosa, choroid plexus, and lactating mammary glands. All these tissues can concentrate iodide, but do not retain it because they lack the ability to bind it to a protein matrix.

The first congenital iodide transport or trapping defect (ITD) was found in a child, born to consanguineous parents, with an inability to concentrate iodide in the thyroid, salivary glands, and gastric mucosa.[2] In 1997, a mutation in the gene-encoding NIS (*SLC5A5*)[3] was found to cause autosomal recessive ITD.[4] The incidence is probably very low, given the quite characteristic clinical findings in iodide trapping defects and the small number of reported mutations. It should be noted, however, that unless TG is measured, the lack of thyroidal iodide uptake could lead to the erroneous diagnosis of athyreosis. Since 1997, 15 families (eight Japanese, two Brazilian,

one Hispanic-Mexican, one Hutterite, one Spanish, one Italian, Argentinian, and one African from the Cameroon) with IDT have been genetically studied. Affected individuals were found to be homo- or compound heterozygous for 13 different *SLC5A5* inactivating mutations (V59E,[5] G93R,[6] R124H,[7] Q267E,[8] C272X,[9] G395R,[10] T354P,[4] fs515X, Y531X,[8] ΔM143-Q323,[11] ΔA439-P443,[12] G543E,[6] and g.-54C > T [13]). The age of onset of hypothyroidism is variable (neonatal, infancy, childhood) and seems to correlate with the residual *in vitro* activity of the mutant NIS.[7]

Scintigraphy reveals blunted or absent radioactive iodide uptake (RAIU) in a normally located thyroid gland. The RAIU is a direct measure for NIS activity *in vivo*. RAIU in ITD is 0–5% (normal 10–40%) and imaging may suggest an absent thyroid gland (apparent athyreosis), especially when goiter is not prominent. Since the loss of NIS function is generalized, it also involves reduced salivary glands and gastric parietal cell uptake of iodide. Hence, there is little or no early ^{123}I or ^{99m}TcO$_4^-$ uptake resulting in the absence of salivary glands or stomach images during scintillation scanning.[14] A simple and reliable test is the measurement of radioactivity in equal volumes of saliva and plasma obtained one h after the oral administration of 5 μCi of [125]I. A salivary-to-plasma ratio close to unity (normal 20) is pathognomonic of an NIS defect. The presence of TSH receptor blocking antibodies acquired transplacentally from a mother with autoimmune thyroid disease or inactivating mutations in the TSH receptor (causing resistance to TSH) will not affect the test or salivary gland and stomach imaging. Ultrasound examination further helps to distinguish defects in *SLC5A5* from other conditions with reduced iodide uptake as it shows, characteristically, a normally located enlarged thyroid gland, excluding athyreosis and thyroid gland hypoplasia due to TSH receptor defects.

Although mutations in *SLC5A5* appear to be a rare cause of dyshormonogenesis, genetic screening should be considered in all patients with permanent goitrous hypothyroidism in conjunction with low or absent RAIU. Without RAIU results, patients with goitrous hypothyroidism associated with high serum TG are better first evaluated for defects in *TPO* and *DUOX2*. Apart from establishing a definite diagnosis, finding *SLC5A5* mutations has further implications. Identification of an index case will allow subsequent prenatal diagnoses of other cases in the same family. The latter is critical, since patients with delayed onset of CH in ITD already had signs of developmental delay at time of diagnosis.[7] Concerning treatment options, iodide supplementation can improve thyroid function in patients with residual NIS activity and should be considered, either alone or as adjunctive therapy together with L-T_4 replacement.

Defect in Efflux of Iodide Across the Apical Thyroid Cell Membrane – Gene: SLC26A4 (PDS)

SLC26A4 is a member of the multifunctional SLC26 transporter family facilitating the passive efflux of iodide across the apical thyrocyte membrane into the follicular lumen (Fig. 7.1). In the inner ear, chloride/bicarbonate exchange by SLC26A4 is crucial for acid–base homeostasis of the endolymphatic fluid.

Biallelic inactivating mutations of *SLC26A4* are a cause of Pendred syndrome (PDS).[15] First described in 1896,[16] it is clinically defined by congenital bilateral sensorineural hearing loss (associated with vestibular dysfunction) combined with diffuse or multinodular goiter. The latter usually develops in late childhood or early adolescence and is the consequence of a partial iodide organification defect consistent with the function of SLC26A4 as an apical iodide efflux channel in thyrocytes. However, about half of patients with *SLC26A4* defects do not manifest thyroid abnormalities (designated nonsyndromic hearing loss with familial enlarged vestibular aqueduct or DFNB4). Nutritional iodide intake is an important modifier of the thyroid phenotype in PDS. High iodide intake

may even completely prevent thyroid enlargement. With sufficient iodide in their diet, about 90% of patients are clinically and biochemically euthyroid. In the remaining 10% with elevated TSH level, goiter is always present.[17]

Based on data for an English population, the incidence of *SLC26A4* mutations causing isolated hearing loss or complete PDS can be estimated at 1:60,000.[18] Note that in a substantial number of patients with clinically suspected PDS no *SLC26A4* mutations were found, presumably because of genetic heterogeneity and phenocopies. Both goiter (particularly in iodine deficient regions) and congenital hearing loss of other causes are quite common.

PDS is rarely detected by neonatal screening for congenital hypothyroidism.[19] Rather, patients present because of severe to profound congenital deafness. The finding of thyroid enlargement or a family history indicative of PDS would justify screening for *SLC26A4* mutations. In the absence of goiter, children with suspected autosomal recessive nonsyndromic hearing loss should first be evaluated for mutations in the *GJB2* (encoding gap junction connexin 26) gene, which account for up to 50% of all cases (*SLC26A4* mutations: up to 4%).[20,21] In the case of negative GJB2 screening, either computed tomography or magnetic resonance imaging of the temporal bones should be considered. About 80% of patients with bilateral dilatation of the vestibular aqueduct and/or Mondini dysplasia were subsequently shown to have *SLC26A4* mutations.[22] There seems to be little additional diagnostic value in performing a perchlorate discharge test, which has a relatively high false negative rate (5%) in demonstrating partial iodide organification defect (PIOD; 10–90% radioiodide discharge) in patients with *SLC26A4* mutations.[23] The perchlorate (ClO_4^2) discharge test is based on the following physiological and pathologic considerations. Iodide transported into the follicular lumen is immediately covalently bound to TG (organified) and therefore does not normally require the concentrating activity of NIS for its retention. The test involves the administration of radioiodide, the uptake of which is measured by counting over the neck using a Geiger counter. Two hours later, ClO_4^2 is given, which blocks further iodide uptake by competitive inhibition of NIS. While bound iodine is retained, any inorganic iodide remaining in the thyroid gland is discharged and detected over the ensuing hour by falling counts over the gland. This occurs only if there is an organification defect, as in PDS, or other causes affecting protein binding of iodide.

Before systematic mutation scanning, targeted screening for the most common, recurrent mutations can be considered. L236P, T416P, and IVS8+1G≥A account for 50% of known *SLC26A4* mutations in Caucasians of northern European descent,[22,24] whereas H723R represents 53% of reported mutant alleles among Japanese.[25] It has been suggested that, in the absence of a direct functional assessment *in vitro*, the pathogenicity of novel *SLC26A4* mutations can be confidently predicted in cases of addition or omission of proline or a charged amino acid.[26]

Defect in the Follicular Matrix Protein Providing Tyrosyl Groups for Iodide Organification – Gene: TG

Thyroglobulin (TG), a glycoprotein homodimer of 660 kDa, is the most abundantly expressed protein in the thyroid gland. It is secreted into the follicular lumen where it functions as matrix for hormone synthesis providing tyrosyl groups, the noniodide component of thyroid hormone (Fig. 7.1). Iodinated TG constitutes the storage pool for thyroid hormone and iodide.

Although the existence of congenital TG defects was known in 1959,[27] a demonstration of TG gene defect was first reported in 1991.[28] Since that report, at least 40 distinct inactivating TG gene mutations have been described (see Ref. [29] for a recent list of published mutations). Defects in TG as a cause of CH have been most extensively studied in Japanese subjects, for whom the estimated incidence is 1:67,000[30] equivalent to one-quarter to one-third of all cases with thyroid dyshormonogenesis.

Patients with biallelic TG gene defects typically manifest elevated serum TSH levels detectable on neonatal screening. As in other forms of dyshormonogenesis, free T_3 levels are usually disproportionately high compared to the low free T_4 serum concentrations, which have been explained by an increased intrathyroidal type 2 iodothyronine deiodinase activity converting T_4 to T_3.[31] Mild TG secretion defects can manifest as compensated hypothyroidism (isolated hyperthyrotropinemia). Goiters are often remarkably large, although early treatment of mild TG defects would prevent goitrogenesis. More revealing, serum TG is undetectable or very low in relation to the elevated serum TSH. Scintigraphy shows high uptake (due to induction of NIS expression by TSH stimulation) in a typically enlarged thyroid gland. Since the iodide organification process is not affected, there is, usually, no discharge after administration of ClO_4^2. In the absence of TG, iodide is covalently bound to other proteins, such as albumin. Collectively, screening for TG gene mutations may be justified when an absent or unexpectedly low serum TG level is found in a CH patient presenting with normal-sized or enlarged thyroid gland.

Defects in the Enzymes Required for Iodide Organification

Defect in the Key Enzyme Catalyzing the Iodination and Coupling of Tyrosyl Moieties – Gene: TPO

Thyroid peroxidase (TPO) is a thyroid-specific heme peroxidase anchored via a C-terminal transmembrane domain at the apical membrane surface of follicular thyroid cells (Fig. 7.1). Using hydrogen peroxide as oxidative equivalents, TPO catalyzes the iodination of tyrosyl residues in TG and the subsequent phenoxy ether bond formation between pairs of iodotyrosines to generate iodothyronines (T_4 and, to a lesser degree, T_3 and reverse T_3).

The first case of CH due to failure of iodide organification in the presence of hydrogen peroxide was reported in 1950 by Stanbury et al.[32] Four decades later, the human TPO gene was cloned,[33] followed shortly after by the description of a TPO mutation in a patient with CH.[34] Inactivating biallelic defects in the TPO gene appear to be the most frequent cause of inherited dyshormonogenesis with permanent CH[35,36] and the culprit in essentially all patients with permanent total iodide organification defects (TIOD; ≥90% ClO_4^- discharge).[37] Bakker et al. estimated the incidence of TIOD owing to biallelic TPO defects at 1:66,000 for a Dutch population.[37]

In about 20% of cases with TIOD, only monoallelic defects of TPO are found, presumably due to unidentified cryptic mutations in unexamined intronic or regulatory regions of the gene. Indeed, in a patient with TIOD and single allele mutation, TPO mRNA analysis of thyroid tissue obtained at surgery revealed monoallelic expression of only the mutant allele, indicating an unidentified defect on the other allele.[38] Although heterozygous TPO mutations do not directly result in abnormal thyroid function, such monoallelic defects may play a role as genetic susceptibility factors in transient hypothyroidism. In a Chinese population, heterozygosity for a common TPO founder mutation (2268insT) is 16 times more common in babies with transient neonatal hypothyroidism compared to normal babies.[39]

TPO is the indisputable candidate gene in patients with permanent TIOD.[40] In contrast, nonsyndromic PIOD is heterogeneous, including defects in the hydrogen peroxide generator driving TPO activity. When the ClO_4^- discharge test is not available, screening for TPO mutations is still reasonable. For instance, 10 of 53 unrelated patients from Portugal with permanent CH, orthotopic thyroid gland, and elevated serum TG levels were found to harbor TPO defects.[35] In an Eastern European population with an unusually high frequency of dyshormonogenesis (34% of CH), 18 of 39 apparently unrelated patients with permanent CH, normal or enlarged gland in situ, and normal or high serum TG were found to have TPO mutations. In 12 patients, only a single heterozygous mutation (1273_1276dupGGCC) was detected.[36]

Finding of TPO mutations in a neonate with CH indicates that the patient will require lifelong treatment with thyroid hormone and that future pregnancies should be carefully monitored for the presence of fetal goiter. The latter can be detected by ultrasonography and treated by a single intraamniotic injection of L-T_4 to prevent goiter-related dystocia and improve neurological development.[41]

Defect in the NADPH-oxidase Providing Hydrogen Peroxide for TPO – Gene: *DUOX2* (*THOX2*)

Dual oxidases (DUOX1 and DUOX2; formerly known as thyroid oxidases or THOX) are NADPH oxidases expressed at the apical membrane of follicular thyroid cells (Fig. 7.1).[42,43] They provide hydrogen peroxide, the essential electron acceptor for the TPO-catalyzed iodination and coupling reactions. DUOX2 is also expressed at high levels in other epithelia, particularly in the gastrointestinal tract and salivary glands, and is proposed to function in a host defense mechanism.

Since the 2002 description of *DUOX2* mutations in patients with CH,[44] 26 different mutations have been reported.[44–53] About half of these are nonsense, frameshift or splice site mutations predicting a dysfunctional enzyme lacking the C-terminal NADPH oxidase domain (G201fs, E327X, W414X, Y425X, R434X, L479fs, G488R, K530X, K628fs, Q686X, R701X, R842X, S965fs, Q1023X, Q1026X, g.IVS19-2A≥C). Of the missense mutations (Q36H, Y475C, A649E, H678R, E879K, R885Q, R110Q, D506N, R376W, G1518S), three have been studied *in vitro* and shown to cause either a complete or partial defect in trafficking of DUOX to the cell surface[54] or reduced expression of an inactive protein.[51]

Although most dyshormonogenesis defects are inherited in an autosomal recessive fashion, a single defective *DUOX2* allele suffices to cause CH. When these patients were reevaluated at three years of age after withdrawal of L-T$_4$, they often had normal thyroid function tests, indicating that the CH was transient.[44] In addition, adult heterozygotes in these and other families with *DUOX2* gene defects all had normal serum TSH concentrations. Since no evidence was found for dominant negative effects of the mutant DUOX2 proteins,[54] these patients appear to have *DUOX2* haploinsufficiency with manifestation limited to the neonatal period when thyroid hormone synthesis requirements are highest (from about 10 μg T$_4$/kg/day progressively decreasing to around 3 μg T$_4$/kg/day after the first year of life).

Several studies have linked biallelic *DUOX2* defects to permanent CH with PIOD. Based on the type of mutations (nonsense, frameshift) or the *in vitro* study of missense mutations found in a homozygous or compound heterozygous state in these patients, most are predicted to express no residual DUOX2 activity. In contrast to the complete inactivation of TPO, which consistently leads to TIOD, a complete loss of DUOX2 activity does not completely abrogate the ability to synthesize the thyroid hormone. Of all the patients with biallelic *DUOX2* defects, only one reportedly had TIOD. However, the results of

the ClO$_4^-$ discharge test in this patient are doubtful since L-T$_4$ treatment had not been discontinued.[44] Indeed, a recent case report from Japan describes several patients with complete loss of DUOX2 activity due to biallelic frameshift mutations, who all presented with only transient CH and normal thyroid function tests in childhood.[49] Limited iodide organification in these patients is likely maintained by the activity of DUOX1, which is also expressed in thyrocytes, albeit at a lower level compared to DUOX2.

With the increasing number of reported cases, phenotype–genotype correlations have become more complex than initially anticipated. The expressivity of *DUOX2* defects is likely influenced by genetic background (e.g., *DUOX1*) and may, at least in part, depend on the iodide intake. Since iodination by TPO requires both iodide and hydrogen peroxide, a diet containing excessive amounts of iodide, common in Japan, would lead to better utilization of hydrogen peroxide provided by DUOX1. Further credence to an important role of iodide intake in expressivity of *DUOX2* defects is provided by anecdotal reports showing that perinatal iodine overload completely normalized TSH levels in the early postnatal period,[45,55] indicating compensation of the defect. This is opposite to the situation in normal infants in whom such iodine overload induces a physiological organification blockade (Wolff–Chaikoff effect).

The incidence of *DUOX2* mutations in CH has not been determined. Certainly, *DUOX2* mutations are frequent in patients with PIOD. For instance, seven of 20 such unrelated patients from Italy were found to have pathogenetic *DUOX2* mutations.[50] Screening of *DUOX2* is therefore recommended in patients with nonsyndromic PIOD. In those with suspected nonsyndromic organification defect (normal sized to enlarged thyroid gland with high serum TG concentration) not confirmed by ClO$_4^-$ discharge test, *TPO* should be screened first, especially in cases where CH is profound. Concerning follow-up, patients with transient CH due to *DUOX2* haploinsufficiency are likely at risk for recurrent hypothyroidism at times of increased hormonogenesis requirements, such as pregnancy. With regard to treatment after the neonatal period, we would advocate assessment of iodide supplementation as an alternative to potentially life-long L-T$_4$ replacement therapy.

Defect in the DUOX2 Cofactor – Gene: *DUOXA2*

Two novel genes, called DUOX maturation factors (*DUOXA1* and *DUOXA2*), were recently identified in the *DUOX1*/*DUOX2* intergenic region.[56] These genes are oriented head-to-head to the *DUOX* genes and thus form bidirectional transcriptional units with their corresponding dual oxidase genes. This arrangement ensures coexpression of *DUOXA2* with *DUOX2* (and *DUOXA1* with *DUOX1*). The *DUOXA* genes encode integral membrane proteins essential for the endoplasmic reticulum-to Golgi transition, maturation, and translocation to the plasma membrane of functional DUOX enzymes [56] (Fig. 7.1).

In 2008, the first mutation in *DUOXA2* was described in a Chinese patient with PIOD and mild, permanent CH.[57] The patient was homozygous for an Y246X nonsense mutation that resulted in a complete loss of DUOXA2 function *in vitro*. A heterozygous Y246X carrier was also identified among 92 unrelated Han Chinese control individuals suggesting that this mutation could be relatively common in this population.[57] In fact, a recent report described another Chinese CH patient with gland *in situ* that was compound heterozygous for Y246X and another nonsense variant, Y138X.[58] In a patient with mild CH of European descent, Hulur et al. identified a DUOXA2 missense mutation (C189R) that abolished the functional expression of the protein *in vitro*.[59] This patient was again a compound heterozygote, harboring a large deletion of the DUOX2/DUOXA2/DUOXA1 region on the paternal allele. In all three reported cases, the loss of a single *DUOXA2* allele did not lead to abnormal thyroid function, in contrast to the haploinsufficiency caused by monoallelic *DUOX2* mutations. Apart from the intact DUOXA1/DUOX1 system, an additional mechanism for maintaining adequate hydrogen peroxide supply in patients with DUOXA2 deficiency is the partial activation of DUOX2 by DUOXA1, as demonstrated *in vitro*.[57] Since *DUOXA2* defects lead to secondary deficiency of functional DUOX2 enzyme, one can anticipate that expressivity will be similarly modulated by nutritional iodide as described for *DUOX2* defects.

Defect in Iodide Recycling with Secondary Iodide Deficiency – Gene: *IYD (DEHAL1)*

The lysosomal proteolysis of endocytosed iodinated TG liberates the iodothyronines ($T_4 \geq T_3$). However, most iodide contained in TG is released as uncoupled mono- and diiodotyrosines (MIT, DIT). MIT and DIT are subject to NADPH-dependent reductive deiodination by iodotyrosine deiodinase (IYD, or dehalogenase) leading to formation of free iodide and tyrosine, both of which can be reutilized in hormone synthesis (Fig. 7.1).

IYD contains an N-terminal membrane anchor, a less conserved intermediate domain, and a C-terminal domain resembling enzymes of the bacterial NADH oxidase/flavin reductase superfamily.[60] The protein is predominantly localized at the apical thyroid cell membrane and in subapical, endosomal compartments, with the catalytic domain facing outside the cell or into the endosomal lumen, respectively.[61] In addition to the thyrocyte, the enzyme is expressed in liver and kidney. The

expression in the latter tissue serves for the execution of a pathognomonic *in vivo* test.

A congenital defect in iodotyrosine deiodination was first described in 1953 in a consanguineous group of Scottish itinerant thinkers.[62] In 2008, molecular defects in *IYD* underlying impaired intrathyroidal dehalogenation were described in four unrelated consanguineous families.[63,64] Six affected individuals were homozygous for either a missense mutation (R101W, I116T, or A220T) or a combined missense/deletion mutation c.315delCAT (resulting in replacement of both F105 and I106 by leucine at position 105). All mutations map to the flavin-binding domain and virtually abolished the capacity of IYD to dehalogenate MIT and DIT *in vitro*. Notably, one heterozygous carrier of A220T developed nonautoimmune goitrous hypothyroidism at 15 years of age, pointing to a possible dominant behavior of the mutation in some individuals.[64]

Loss of IYD activity prevents the normal intrathyroidal iodide "recycling" and leads to excessive urinary secretion of MIT and DIT. Since the resulting iodide deficiency does not manifest at birth, patients with biallelic *IYD* mutations tested normal at neonatal screening for CH. They subsequently came to medical attention at 1.5–8 years of age because of sequelae of hypothyroidism. On scintigraphy, a very rapid and high initial uptake of 123I in the enlarged thyroid is observed, followed by a relatively rapid spontaneous decline of the accumulated iodine without the administration of ClO_4^2. A pathognomonic finding is the intact excretion in urine of intravenously administered MIT or DIT, without removal of the iodine. The detection of high urinary MIT and DIT by tandem mass spectroscopy may become a useful diagnostic test.

The incidence of *IYD* mutations is unknown. Although not a viable candidate gene for CH, a potential role for *IYD* variants in susceptibility to endemic goiter remains to be investigated. We suggest that screening of *IYD* may be considered in patients developing "idiopathic" diffuse or multinodular goiter between the neonatal period and adolescence, while early and late images during 123I scintigraphy are compatible with *IYD* defects. However, deficient nutritional iodide intake, as in areas of endemic goiter, dietary goitrogens, and autoimmune thyroid disease has to be excluded. An autosomal recessive inheritance pattern of the disorder, as well as consanguinity of the parents, would likely increase the yield of mutation screening.[65] Anecdotal evidence indicates that iodine supplementation (Lugol's solution) is an effective alternative to L-T_4 treatment.[66,67]

AVAILABILITY OF GENETIC TESTING

The following commercial laboratories currently provide genetic testing for selected genes involved in dyshormonogenesis:

TG, TPO, DUOX2, DUOXA2, IYD (analysis of entire coding region):

- Academic Medical Centre, University of Amsterdam
- DNA Diagnostic Laboratory
- Amsterdam, The Netherlands
- *SLC5A5* (analysis of entire coding region)
- Kyoto University School of Medicine
- Department of Clinical Genetics
- Kyoto, Japan

SLC26A4 (either mutations scanning or targeted mutation analysis, carrier testing, prenatal diagnosis).

Screening for *SLC26A4* defects is widely available in medical genetics laboratories. The reader may refer to GeneTests and Genetic Testing Registry (http://www.ncbi .nlm.nih.gov/gtr/tests) for an up-to-date list of available facilities.

In addition to commercial testing facilities, we and other individuals interested in research on these conditions could provide assistance in genetic studies. These include, among others, Gilbert Vassart (Brussels, Belgium), Joachim Pohlenz (Mainz, Germany), Massimo Tonacchera (Pisa, Italy), Akira Nishinuma (Tochigi, Japan), Hector Targovnik (Buenos Aires, Argentina), and Samuel Refetoff (Chicago, IL, USA).

CONCLUSIONS

This chapter provides a succinct outline of the pathophysiology and genetics of thyroid dyshormonogenesis. We are aware that tests most crucial for an etiologic diagnosis (e.g., scintigraphy) are frequently not performed, since the exact classification is widely assumed not to be essential for the management of CH. We have tried to take this situation into account by basing our recommendations for rational genetic screening on the incidence of the underlying defects and readily obtainable information (thyroid gland size and serum TG level).

For the clinician treating patients with thyroid dyshormonogenesis, a practical implication of a genetic diagnosis is the identification of those patients in whom the defect can likely be compensated by iodide supplementation avoiding life-long L-T$_4$ replacement therapy. Iodide supplementation should improve hormonogenesis in all defects that (1) only partially diminish vectorial iodide transport across thyrocytes (all *SLC26A4* defects and partial *SLC5A5* defects); (2) produce secondary iodide deficiency (all *IYD* defects); or (3) partially reduce the efficiency of iodide organification (all *DUOX2* and *DUOXA2* defects). While requiring L-T$_4$ replacement for CH in the neonatal period, we expect that many of these patients may eventually be managed with iodide alone. On the other hand, except for rare partial deficiency alleles, *TPO* or *TG* defects are not compensated by increased dietary iodide.

Acknowledgment

This work was supported in part by grants UL1TR 000430 and 4R37 DK15070 from the National Institutes of Health.

References

1. Clerc J, Monpeyssen H, Chevalier A, et al. Scintigraphic imaging of paediatric thyroid dysfunction. *Horm Res* 2008;**70**:1–13.
2. Stanbury JB, Chapman EM. Congenital hypothyroidism with goitre. Absence of an iodide-concentrating mechanism. *Lancet* 1960;**1**:1162–5.
3. Dai G, Levy O, Carrasco N. Cloning and characterization of the thyroid iodide transporter. *Nature* 1996;**379**:458–60.
4. Fujiwara H, Tatsumi K, Miki K, et al. Congenital hypothyroidism caused by a mutation in the Na$^+$/I$^-$ symporter. *Nat Genet* 1997;**16**:124–5.
5. Fujiwara H, Tatsumi K, Tanaka S, et al. A novel V59E missense mutation in the sodium iodide symporter gene in a family with iodide transport defect. *Thyroid* 2000;**10**:471–4.
6. Kosugi S, Inoue S, Matsuda A, et al. Novel, missense and loss-of-function mutations in the sodium/iodide symporter gene causing iodide transport defect in three Japanese patients. *J Clin Endocrinol Metab* 1998;**83**:3373–6.
7. Szinnai G, Kosugi S, Derrien C, et al. Extending the clinical heterogeneity of iodide transport defect (ITD): a novel mutation R124H of the sodium/iodide symporter gene and review of genotype–phenotype correlations in ITD. *J Clin Endocrinol Metab* 2006;**91**:1199–204.
8. Pohlenz J, Rosenthal IM, Weiss RE, et al. Congenital hypothyroidism due to mutations in the sodium/iodide symporter. Identification of a nonsense mutation producing a downstream cryptic 3′ splice site. *J Clin Invest* 1998;**101**:1028–35.
9. Pohlenz J, Medeiros-Neto G, Gross JL, et al. Hypothyroidism in a Brazilian kindred due to iodide trapping defect caused by a homozygous mutation in the sodium/iodide symporter gene. *Biochem Biophys Res Commun* 1997;**240**:488–91.
10. Kosugi S, Bhayana S, Dean HJ. A novel mutation in the sodium/iodide symporter gene in the largest family with iodide transport defect. *J Clin Endocrinol Metab* 1999;**84**:3248–53.
11. Kosugi S, Okamoto H, Tamada A, et al. A novel peculiar mutation in the sodium/iodide symporter gene in Spanish siblings with iodide transport defect. *J Clin Endocrinol Metab* 2002;**87**:3830–6.
12. Tonacchera M, Agretti P, de Marco G, et al. Congenital hypothyroidism due to a new deletion in the sodium/iodide symporter protein. *Clin Endocrinol (Oxf)* 2003;**59**:500–6.
13. Nicola JP, Nazar M, Serrano-Nascimento C, et al. Iodide transport defect: functional characterization of a novel mutation in the Na+/I- symporter 5′-untranslated region in a patient with congenital hypothyroidism. *J Clin Endocrinol Metab* 2011;**96**:1100–7.
14. Fukata S, Hishinuma A, Nakatake N, et al. Diagnosis of iodide transport defect: do we need to measure the saliva/serum radioactive iodide ratio to diagnose iodide transport defect? *Thyroid* 2010;**20**:1419–21.
15. Everett LA, Glaser B, Beck JC, et al. Pendred syndrome is caused by mutations in a putative sulphate transporter gene (PDS). *Nat Genet* 1997;**17**:411–22.
16. Pendred V. Deaf-mutism and goitre. *Lancet* 1896;**2**:532.
17. Reardon W, Coffey R, Phelps PD, et al. Pendred syndrome – 100 years of under-ascertainment? *QJM* 1997;**90**:443–7.

18. Hutchin T, Coy NN, Conlon H, et al. Assessment of the genetic causes of recessive childhood non-syndromic deafness in the UK – implications for genetic testing. *Clin Genet* 2005;**68**:506–12.

19. Banghova K, Al Taji E, Cinek O, et al. Pendred syndrome among patients with congenital hypothyroidism detected by neonatal screening: identification of two novel PDS/SLC26A4 mutations. *Eur J Pediatr* 2008;**167**:777–83.

20. Denoyelle F, Weil D, Maw MA, et al. Prelingual deafness: high prevalence of a 30delG mutation in the connexin 26 gene. *Hum Mol Genet* 1997;**6**:2173–7.

21. Albert S, Blons H, Jonard L, et al. SLC26A4 gene is frequently involved in nonsyndromic hearing impairment with enlarged vestibular aqueduct in Caucasian populations. *Eur J Hum Genet* 2006;**14**:773–9.

22. Campbell C, Cucci RA, Prasad S, et al. Pendred syndrome, DFNB4, and PDS/SLC26A4 identification of eight novel mutations and possible genotype–phenotype correlations. *Hum Mutat* 2001;**17**:403–11.

23. Ladsous M, Vlaeminck-Guillem V, Dumur V, et al. Analysis of the thyroid phenotype in 42 patients with Pendred syndrome and nonsyndromic enlargement of the vestibular aqueduct. *Thyroid* 2014;**24**:639–48.

24. Coyle B, Reardon W, Herbrick JA, et al. Molecular analysis of the PDS gene in Pendred syndrome. *Hum Mol Genet* 1998;**7**:1105–12.

25. Tsukamoto K, Suzuki H, Harada D, et al. Distribution and frequencies of PDS (SLC26A4) mutations in Pendred syndrome and nonsyndromic hearing loss associated with enlarged vestibular aqueduct: a unique spectrum of mutations in Japanese. *Eur J Hum Genet* 2003;**11**:916–22.

26. Pera A, Dossena S, Rodighiero S, et al. Functional assessment of allelic variants in the SLC26A4 gene involved in Pendred syndrome and nonsyndromic EVA. *Proc Natl Acad Sci USA* 2008;**105**:18608–13.

27. Degroot LJ, Stanbury JB. The syndrome of congenital goiter with butanol-insoluble serum iodine. *Am J Med* 1959;**27**:586–95.

28. Ieiri T, Cochaux P, Targovnik HM, et al. A 3′ splice site mutation in the thyroglobulin gene responsible for congenital goiter with hypothyroidism. *J Clin Invest* 1991;**88**:1901–5.

29. Targovnik HM, Citterio CE, Rivolta CM. Thyroglobulin gene mutations in congenital hypothyroidism. *Horm Res Paediatr* 2011;**75**:311–21.

30. Hishinuma A, Fukata S, Nishiyama S, et al. Haplotype analysis reveals founder effects of thyroglobulin gene mutations C1058R and C1977S in Japan. *J Clin Endocrinol Metab* 2006;**91**:3100–4.

31. Kanou Y, Hishinuma A, Tsunekawa K, et al. Thyroglobulin gene mutations producing defective intracellular transport of thyroglobulin are associated with increased thyroidal type 2 iodothyronine deiodinase activity. *J Clin Endocrinol Metab* 2007;**92**:1451–7.

32. Stanbury JB, Hedge AN. A study of a family of goitrous cretins. *J Clin Endocrinol Metab* 1950;**10**:1471–84.

33. Kimura S, Kotani T, McBride OW, et al. Human thyroid peroxidase: complete cDNA and protein sequence, chromosome mapping, and identification of two alternately spliced mRNAs. *Proc Natl Acad Sci USA* 1987;**84**:5555–9.

34. Abramowicz MJ, Targovnik HM, Varela V, et al. Identification of a mutation in the coding sequence of the human thyroid peroxidase gene causing congenital goiter. *J Clin Invest* 1992;**90**:1200–4.

35. Rodrigues C, Jorge P, Soares JP, et al. Mutation screening of the thyroid peroxidase gene in a cohort of 55 Portuguese patients with congenital hypothyroidism. *Eur J Endocrinol* 2005;**152**:193–8.

36. Avbelj M, Tahirovic H, Debeljak M, et al. High prevalence of thyroid peroxidase gene mutations in patients with thyroid dyshormonogenesis. *Eur J Endocrinol* 2007;**156**:511–9.

37. Bakker B, Bikker H, Vulsma T, et al. Two decades of screening for congenital hypothyroidism in The Netherlands: TPO gene mutations in total iodide organification defects (an update). *J Clin Endocrinol Metab* 2000;**85**:3708–12.

38. Fugazzola L, Cerutti N, Mannavola D, et al. Monoallelic expression of mutant thyroid peroxidase allele causing total iodide organification defect. *J Clin Endocrinol Metab* 2003;**88**:3264–71.

39. Niu DM, Lin CY, Hwang B, et al. Contribution of genetic factors to neonatal transient hypothyroidism. *Arch Dis Child Fetal Neonatal Ed* 2005;**90**:F69–72.

40. Ris-Stalpers C, Bikker H. Genetics and phenomics of hypothyroidism and goiter due to TPO mutations. *Mol Cell Endocrinol* 2010;**322**:38–43.

41. Agrawal P, Ogilvy-Stuart A, Lees C. Intrauterine diagnosis and management of congenital goitrous hypothyroidism. *Ultrasound Obstet Gynecol* 2002;**19**:501–5.

42. Dupuy C, Ohayon R, Valent A, et al. Purification of a novel flavoprotein involved in the thyroid NADPH oxidase. Cloning of the porcine and human cDNAs. *J Biol Chem* 1999;**274**:37265–9.

43. De Deken X, Wang D, Many MC, et al. Cloning of two human thyroid cDNAs encoding new members of the NADPH oxidase family. *J Biol Chem* 2000;**275**:23227–33.

44. Moreno JC, Bikker H, Kempers MJ, et al. Inactivating mutations in the gene for thyroid oxidase 2 (THOX2) and congenital hypothyroidism. *N Engl J Med* 2002;**347**:95–102.

45. Vigone MC, Fugazzola L, Zamproni I, et al. Persistent mild hypothyroidism associated with novel sequence variants of the DUOX2 gene in two siblings. *Hum Mutat* 2005;**26**:395.

46. Varela V, Rivolta CM, Esperante SA, et al. Three mutations (p.Q36H, p.G418fsX482, and g.IVS19-2A>C) in the dual oxidase 2 gene responsible for congenital goiter and iodide organification defect. *Clin Chem* 2006;**52**:182–91.

47. Pfarr N, Korsch E, Kaspers S, et al. Congenital hypothyroidism caused by new mutations in the thyroid oxidase 2 (THOX2) gene. *Clin Endocrinol (Oxf)* 2006;**65**:810–5.

48. Ohye H, Fukata S, Hishinuma A, et al. A novel homozygous missense mutation of the dual oxidase 2 (DUOX2) gene in an adult patient with large goiter. *Thyroid* 2008;**18**:561–6.

49. Maruo Y, Takahashi H, Soeda I, et al. Transient congenital hypothyroidism caused by biallelic mutations of the dual oxidase 2 gene in Japanese patients detected by a neonatal screening program. *J Clin Endocrinol Metab* 2008;**93**:4261–7.

50. Cortinovis F, Zamproni I, Persani L, et al. Prevalence of DUOX2 mutations among children affected by congenital hypothyroidism and dyshormonogenesis. *Hormone Res* 2008;**70**:40.

51. Hoste C, Rigutto S, Van Vliet G, et al. Compound heterozygosity for a novel hemizygous missense mutation and a partial deletion affecting the catalytic core of the H2O2-generating enzyme DUOX2 associated with transient congenital hypothyroidism. *Hum Mutat* 2010;**31**:1304–18.

52. Grasberger H. Defects of thyroidal hydrogen peroxide generation in congenital hypothyroidism. *Mol Cell Endocrinol* 2010;**322**:99–106.

53. Yoshizawa-Ogasawara A, Ogikubo S, Satoh M, et al. Congenital hypothyroidism caused by a novel mutation of the dual oxidase 2 (DUOX2) gene. *J Pediatr Endocr Met* 2013;**26**:45–52.

54. Grasberger H, De Deken X, Miot F, et al. Missense mutations of dual oxidase 2 (DUOX2) implicated in congenital hypothyroidism have impaired trafficking in cells reconstituted with DUOX2 maturation factor. *Mol Endocrinol* 2007;**21**:1408–21.

55. Kasahara T, Narumi S, Okasora K, et al. Delayed onset congenital hypothyroidism in a patient with DUOX2 mutations and maternal iodine excess. *Am J Med Genet Part A* 2013;**161**:214–7.

56. Grasberger H, Refetoff S. Identification of the maturation factor for dual oxidase. Evolution of a eukaryotic operon equivalent. *J Biol Chem* 2006;**281**:18269–72.

57. Zamproni I, Grasberger H, Cortinovis F, et al. Biallelic inactivation of the dual oxidase maturation factor 2 (DUOXA2) gene as a novel cause of congenital hypothyroidism. *J Clin Endocrinol Metab* 2008;**93**:605–10.

58. Yi R-H, Zhu W-B, Yang L-Y, et al. A novel dual oxidase maturation factor 2 gene mutation for congenital hypothyroidism. *Int J Mol Med* 2013;**31**:467–70.

59. Hulur I, Hermanns P, Nestoris C, et al. A single copy of the recently identified dual oxidase maturation factor (DUOXA) 1 gene produces only mild transient hypothyroidism in a patient with a novel biallelic DUOXA2 mutation and monoallelic DUOXA1 deletion. *J Clin Endocrinol Metab* 2011;**96**:841–5.

60. Friedman JE, Watson Jr JA, Lam DW, et al. Iodotyrosine deiodinase is the first mammalian member of the NADH oxidase/flavin reductase superfamily. *J Biol Chem* 2006;**281**:2812–9.

61. Gnidehou S, Caillou B, Talbot M, et al. Iodotyrosine dehalogenase 1 (DEHAL1) is a transmembrane protein involved in the recycling of iodide close to the thyroglobulin iodination site. *FASEB J* 2004;**18**:1574–6.

62. McGirr EM, Hutchison JH. Radioactive-iodine studies in non-endemic goitrous cretinism. *Lancet* 1953;**1**:1117–20.

63. Moreno JC, Klootwijk W, van Toor H, et al. Mutations in the iodotyrosine deiodinase gene and hypothyroidism. *N Engl J Med* 2008;**358**:1811–8.

64. Afink G, Kulik W, Overmars H, et al. Molecular characterization of iodotyrosine dehalogenase deficiency in patients with hypothyroidism. *J Clin Endocrinol Metab* 2008;**93**:4894–901.

65. Burniat A, Pirson I, Vilain C, et al. Iodotyrosine deiodinase defect identified via genome-wide approach. *J Clin Endocrinol Metab* 2012;**97**:1276–83.

66. De Maegd M, Van Nevel C. Hypothyroidism caused by congenital familial defect of iodotyrosine deiodation. *Acta Paediatr Belg* 1970;**24**:115–20.

67. Hirsch HJ, Shilo S, Spitz IM. Evolution of hypothyroidism in familial goitre due to deiodinase deficiency: report of a family and review of the literature. *Postgrad Med J* 1986;**62**:477–80.

8

Developmental Abnormalities of the Thyroid

Joachim Pohlenz*, Guy Van Vliet**, Johnny Deladoëy**,†

*Pediatric Endocrinology, Department of Pediatrics, Johannes Gutenberg University, Mainz, Germany
**Endocrinology Service and Research Center, Sainte-Justine Hospital and Department of Pediatrics,
Universite de Montreal, Montreal, Quebec, Canada
†Department of Biochemistry, Universite de Montreal, Montreal, Quebec, Canada

INTRODUCTION

Permanent primary congenital hypothyroidism (CH), as estimated from systematic biochemical screening of newborns, occurs in about one in 2500 births,[1] making it the most common congenital endocrine disorder. Approximately two-thirds of the cases are due to abnormalities in the development of the thyroid gland during organogenesis (which are collectively called thyroid dysgenesis), the remainder being due to functional disorders.[2] Classically, thyroid dyshormonogenesis leads to goiter, although exceptional cases with thyroid atrophy have been described, blurring the distinction between dyshormonogenesis and dysgenesis.[3,4] Although pseudodominant inheritance has been reported,[5] thyroperoxidase deficiency, the commonest form of dyshormonogenesis in most populations,[6] is typically inherited in an autosomal recessive fashion and is accordingly more common in populations with a high degree of consanguinity.[3]

In contrast to dyshormonogenesis, the mechanisms underlying CH due to thyroid dysgenesis remain largely unknown, and the proportion of patients in whom a genetic cause has been identified remains extremely small. Indeed, this condition was traditionally considered as sporadic until a systematic survey in France revealed that 2% of patients identified by neonatal screening had an affected relative.[7] This figure is 15-fold higher than what would be predicted by chance alone and clearly suggests a genetic mechanism *in those cases*. However, it should not be construed as evidence that there is a genetic component *in all patients* with CH from thyroid dysgenesis; that is, the 2% with a positive family history and the 98% without may be two discrete populations. It

is also noteworthy that a positive family history does not always imply a common genetic mechanism, as illustrated by a pedigree with two athyreotic individuals, only one of whom had mutations inactivating the TSH receptor.[8] The fact that CH caused by thyroid dysgenesis, especially ectopy, has a marked female predominance[9] is also incompatible with simple Mendelian inheritance. A greater prevalence among less genetically diverse populations, such as Caucasians, may suggest a genetic susceptibility.[10] However, the major argument against Mendelian mechanisms is the observation that at least 92% of monozygotic twin pairs are discordant for this condition.[11,12] This suggests that early postzygotic mutations or epigenetic modifications account for the vast majority of cases. A two-hit model integrating the discrepant observations of a higher-than-expected percentage of familial cases and the almost universal discordance of monozygotic twins has been proposed[13] but remains speculative at this stage. Therefore, the remainder of this chapter mostly reviews the single gene disorders that have been shown to cause CH due to thyroid dysgenesis, but the reader should be aware that these only account for a small proportion of cases. Mutations that activate the TSH receptor are also briefly reviewed, since one of their modes of presentation is congenital hyperthyroid goiter.

Before embarking on a description of these single-gene disorders, the reader should also be aware that thyroid dysgenesis is a heterogeneous condition. Because ultrasound examination generally fails to reveal its most common form (ectopic thyroid),[14] the gold standard for differentiating between the various forms of thyroid dysgenesis remains radionuclide scintigraphy with

Genetic Diagnosis of Endocrine Disorders. http://dx.doi.org/10.1016/B978-0-12-800892-8.00008-7

[99m]sodium pertechnetate or [123]iodine. In 50% of the cases, this will reveal only a small round or oval mass of ectopic thyroid tissue, usually at the base of the tongue, suggesting an arrest in thyroid migration during embryogenesis; in about 9% of newborns with thyroid ectopy, two clusters of cells are visible.[15] In the remainder, no uptake of isotope will be detectable. In this situation, an undetectable plasma thyroglobulin documents true athyreosis (a term to be preferred to agenesis, because there may have been a thyroid at some point during embryogenesis that later disappeared). We have suggested describing the situation in which there is no detectable uptake but a measurable thyroglobulin as "apparent athyreosis."[8] In only a small proportion of cases of overt CH (less than 5%) will imaging reveal a small thyroid gland in the normal location and with the normal bilobed shape ("orthotopic" hypoplasia) and in an even smaller proportion the absence of one thyroid lobe (generally the left) and sometimes of the isthmus. It is humbling that, after almost two decades of intensive research, the monogenic defects described later have been mostly documented in patients with the rarest variant of thyroid dysgenesis, that is, orthotopic thyroid hypoplasia, and that most cases of the commonest variant, ectopic thyroid, remain unexplained. Ectopic thyroid tissue seldom needs to be resected, which has hampered the search for somatic mutations. In only one case have mutations in the coding regions of the candidate genes been excluded.[16] On the other hand, a transcriptome analysis revealed the presence of calcitonin mRNA in lingual thyroids.[17] This was confirmed by RT-PCR and immunohistochemistry[18] and challenges the long-held dogma that parafollicular cells originate only from the lateral thyroid anlagen.

We give herein an overview of the mutations in genes that have an impact on thyroid development. For molecular analysis of a potentially affected patient, the clinician should contact individual research groups (see References) or consult https://www.nextgxdx.com/.

TSH RECEPTOR GENE MUTATIONS (LOSS OF FUNCTION)

The TSH receptor (TSHR) mediates all the effects of TSH on thyroid growth and function but is not required for thyroid migration. It is encoded by a gene that was sequenced in 1989[19] and mapped to chromosome 14q31 in 1990.[20] Mutations resulting in inactivation of the TSHR were first reported in 1995 by Sunthornthepvarakul et al. in three sisters with asymptomatic hyperthyrotropinemia, normal serum-free thyroxine (fT$_4$), and a thyroid gland of normal size and normal radioiodide uptake.[21] Since then, many different inactivating TSHR gene mutations have been reported (http://gris.ulb.ac.be). Detailed information on the characteristics of the TSHR gene and of all the other genes discussed in this chapter is given in Table 8.1. The phenotype of homozygotes and compound heterozygotes is very variable, ranging from asymptomatic hyperthyrotropinemia, as in the three sisters described originally, to severe CH with "apparent athyreosis," as defined above. Careful ultrasonography reveals a small thyroid of normal shape and in the normal position. The contrast between the small amount of thyroid tissue and the elevated plasma thyroglobulin (TG), which has been observed in the affected neonates, is thought to be due to "leakage" of TG from disorganized follicles,[8,22] as seen in the hyt/hyt mouse with a naturally occurring mutation that inactivates Tshr.[23] Sequencing of TSHR should therefore be considered in patients with the phenotypes described above, especially if there is parental consanguinity[24] or a family history suggestive of autosomal recessive transmission.

While TSH resistance from inactivating mutations in the TSHR gene is generally described as being transmitted in a recessive manner, it has now become apparent that heterozygotes can have a mild phenotype, with mild hyperthyrotropinemia and a normal

TABLE 8.1   Characteristics of the Genes that have been Studied in CH from Thyroid Dysgenesis

Gene (abbreviation)	Chromosomal location	Genomic size (in kb)	Exons	Transcript length (nt) starting at the "A" of the first ATG	Amino acids
Thyrotropin receptor (TSHR)	14q31	190	10	2295	764
Paired box gene 8 A (PAX8)	2q12-14	62	12	1353	450
Thyroid transcription factor 1 (TTF1, TITF1, NKX2.1 or T/EBP)	14q13	3.7	3	1206	401
Thyroid transcription factor 2 (TTF2, TITF2, FOXE1 or FKHL15)	9q22	3.4	1	1122	373
GLIS3	9p24.3-p23	328	10	2328	775
NKX2.5	5q34	3	2	975	324
Unknown	15q25.3-26.1	?	?	?	?

or small gland, which is transmitted in a dominant fashion. In a study from Italy, 12% of patients younger than 18 years of age with mild hyperthyrotropinemia, an orthotopic thyroid, and no evidence of autoimmunity had heterozygous *TSHR* gene mutations.[25] It has been argued that making this diagnosis is clinically relevant, because it allows stopping treatment and follow up.[26]

In another form of dominantly transmitted TSH resistance, which appears to be common,[9] the TSHR gene has been excluded[27] and a locus on chromosome 15 identified.[28]

PAX8 GENE MUTATIONS

PAX8, a transcriptional factor playing an important role in the initiation of thyrocyte differentiation and maintenance of the follicular cell, is encoded by a gene on chromosome 2q12-14.[29,30] Containing a highly conserved 128 amino acid-paired domain, PAX8 recognizes specific DNA response elements and regulates the expression of TG, thyroperoxidase (TPO), and the sodium iodide symporter (NIS) by binding to their promoter regions. Furthermore, PAX8 and the thyroid transcription factor 1 (TTF1) synergistically activate the promoter of human TG.[31] Whereas Pax8 deficient mice have hypoplastic thyroid glands with absent follicular cells, mice heterozygous for targeted disruption of the *Pax8* gene do not display an obvious thyroid phenotype.[32] The human *PAX8* gene consists of 11 exons encoding a 450 amino acid protein. In humans, heterozygous *PAX8* loss-of-function mutations can be associated with orthotopic, thyroid hypoplasia, or to no overt structural abnormality of the thyroid gland.[33–41] The functional status of affected individuals is very variable even within the same family, ranging from severe hypothyroidism to compensated hypothyroidism to euthyroidism.[37] This extreme variability supports the hypothesis that many factors modulate the phenotypic expression of *PAX8* gene mutations. This is supported by work in mice showing that the phenotype of Pax8 deficiency is strain specific.[42]

So far, 19 different mutations in the coding sequence of the *PAX8* gene have been reported.[33–41,43–48] The majority of these are located in the paired domain (L16P, F20S, P25R, R31C, R31H, Q40P, G41V, D46SfsX24, S48F, R52P, S54G, S54R, H55Q, C57Y, L62R, R108X, R133Q). With the exception of S48F, which has a normal binding affinity to DNA but an impaired capacity to recruit the general coactivator p300 and a dominant negative effect,[39] these mutants result in severe reduction of DNA binding activity of PAX8. Dominant inheritance of the phenotype together with a heterozygous inactivating mutation is the rule, although all mutations are not fully penetrant

and the mechanisms of disease expression in the presence of only one mutated allele are unknown. For one of the mutations (R133Q), which is located in the terminal portion of the DNA binding domain, the results regarding DNA binding ability are contradictory.[46,48] The mutations (del989-992AAAC and T225M) outside the paired domain have a normal DNA binding ability *in vitro*.[37,41] Whereas del989-992AAAC is truncated and transcriptionally inactive,[37] an impaired synergistic effect of p300 on T225M-mediated transactivation has been described.[41] Also sequence alterations in the 5′untranslated region of the *PAX8* gene may cause diminished transcriptional activity.[49] Hermanns et al. reported a girl with a monoallelic mutation in the *PAX8* promoter who also had a heterozygous mutation in the *NKX2-5* gene.[49] In addition, a mutation in the *PAX8* promoter at position-3, also causing impaired transcriptional activity *in vitro*, was found in a girl with Down's syndrome and CH.[50]

Except for three families with *de novo PAX8* gene mutations,[33,46] dominant inheritance with widely variable penetrance is the rule. The initial report of a patient with an ectopic gland and a *PAX8* mutation[33] was not based on scintigraphy, and this patient likely had orthotopic hypoplasia, as have the vast majority of the cases studied since. In a few patients normal thyroid volumes have been found and, in some, cystic thyroid rudiments have been reported on ultrasound.[33,36]

It is of particular interest that one patient with a heterozygous *PAX8* gene mutation was phenotypically normal at birth but his thyroid became hypoplastic postnatally probably because of deficient postnatal growth of the gland.[40] Therefore, even an unremarkable neonatal screening result for CH and a normally developed and located thyroid gland at birth does not rule out the possibility of a *PAX8* gene deficiency.

It should be noted that one individual with a heterozygous *PAX8* loss-of-function mutation had a positive perchlorate discharge test.[38] Another carrier of a *PAX8* gene mutation was suspected to have a defective sodium iodide symporter because he had a low thyroidal uptake of I^{123} and a low I^{123} saliva-to-plasma ratio.[44] This phenomenon, which might lead to the erroneous diagnosis of dyshormonogenesis, can be explained by the fact that *TPO* and *NIS* genes transcription is dependent on PAX8. Thus, impaired PAX8 function can cause reduced TPO or NIS expression and consequently to a partial organification defect.[38,44] The *PAX8* gene is also expressed in the kidney. Recently, a family with a *PAX8* gene mutation responsible for a severe form of dominantly inherited CH was described. The mutation seemed to be associated with abnormalities of the urogenital tract (incomplete horseshoe kidney, undescended testicles, hydrocele, and ureterocele).[47] However, renal abnormalities seem to be uncommon in individuals with *PAX8* gene mutations and have only been reported in two other affected cases.[38]

Given the extreme variability of the thyroid phenotype in patients with mutations in the *PAX8* gene, it is difficult to define in which patients these should be looked for. However, these mutations appear to be quite rare and we would suggest to screen only patients with a family history suggestive of dominant transmission or early-onset CH and an *in situ* gland of normal or reduced size.

TTF1/NKX2-1 GENE MUTATIONS

TTF1 (also known as NKX2-1, T/EBP (thyroid-specific-enhancer-binding protein) or TITF1) is a member of the homeobox domain type of transcription factors and is encoded by a gene on chromosome 14q13. Its role in the development of the thyroid gland and of other organs was clearly shown by the observation that *Ttf1[-/-]* mice had complete absence of follicular and parafollicular cells as well as agenesis of lung parenchyma, ventral forebrain, and anterior and posterior pituitary.[51] Reexamination of *Ttf1[+/-]* mice, initially reported to be normal,[51] revealed a mild thyroid and neurological phenotype,[52] demonstrating that "half a loaf is not enough."[53] Studies trying to correlate the severity of the thyroid and lung phenotypes with the effect of the mutation on organ-specific reporter genes *in vitro* have yielded conflicting results.[54,55]

Subsequently, the role of its human homolog, TTF1, in the pathophysiology of CH, was suggested by the observation of patients with CH who had chromosomal deletions encompassing the TTF1 locus,[56] including in a sib pair.[57] Next, point mutations in the TTF1 gene confirmed its implication in the phenotype,[52,58,59] which includes CH with a thyroid gland in place associated with respiratory distress syndrome and with neonatal hypotonia followed by choreoathetosis or ataxia. "Benign hereditary chorea" (BHC)[60] had been defined clinically and is now known to be due to dominantly inherited TTF1 mutations. The severity of the three components of the "brain-thyroid-lung"[61] syndrome is very variable.[62] Thyroid insufficiency is generally the least severe of the three, the neurological phenotype intermediate and the lung phenotype the most severe. Indeed, neonatal death from lung hypoplasia has been described.[55] Heterozygous *TTF1* gene mutations lead to a phenotype likely through haploinsufficiency.

In 2009, a systematic review of 46 cases with variants of *TTF1*[62] showed that *TTF1* mutations occur either *de novo* or via an autosomal dominant mode of inheritance. In addition, this review showed that only 50% of the cases had involvement of all three organs, 30% showed CH and a neurologic disorder, 13% presented with isolated BHC, and only 7% had no neurologic phenotype. Functionally, hyperthyrotropinemia was more common at diagnosis than overt hypothyroidism (61% vs. 39%). Thyroid morphology was normal in 55%, with hypoplasia or hemiagenesis in 35% and athyreosis in 10%.[62]

Sixteen percent died of severe lung disease. Recently, it has been shown that *TTF1* gene mutations might be associated with severe pulmonary disease as the only clinical manifestation in the absence of thyroid and neurologic abnormalities.[63] In CH, on the other hand, *TTF1* gene mutations have only been found in patients in association with another organ involvement (Table 8.2) and searching for *TTF1* gene mutations in patients with isolated CH from thyroid dysgenesis has yielded negative results.[64,65] Lastly, the "brain-thyroid-lung" syndrome has also been observed in one child bearing a 14q13.1-3 deletion adjacent to, but not involving *TTF1*,[66] and in another without identified molecular cause.[67]

TTF2 (FOXE 1 OR FKHL15) GENE MUTATIONS

Thyroid transcription factor 2 (FOXE1, TTF2, or FKHL15) is a member of the forkhead/winged helix domain protein family and is encoded by a gene on chromosome 9q22. It regulates transcription of TG and TPO. FOXE1 binds to specific regulatory DNA sequences in the promoter regions through its highly conserved forkhead domain.[30] The human gene is located in chromosome 9q22. It consists of a single exon, which encodes a 373 amino acid protein.[68] *Foxe1[-/-]* mice embryos have cleft palates and athyreosis or an ectopic sublingual gland, whereas heterozygous *Foxe1* knockout mice are phenotypically normal.[32]

So far, only five *FOXE1* gene mutations have been identified in humans.[68–72] In the first report of a human *FOXE1* gene mutation, two Welsh boys with athyreosis, cleft palates, spiky hair, bilateral choanal atresia, and hypoplastic bifid epiglottis (so-called Bamforth–Lazarus syndrome[73]) were identified to have a homozygous *FOXE1* gene mutation (A65V), which is located in the highly conserved forkhead DNA-binding domain. *In vitro* studies showed that this mutant FOXE1 protein displayed impaired DNA binding and no transcriptional function.[68] Interestingly, one of the parents had unilateral choanal atresia.[73] In the second family, two affected individuals of Tunisian origin with thyroid athyreosis, cleft palates, and spiky hair but neither choanal atresia nor bifid epiglottis were described. Both boys were also homozygous for a mutation in the forkhead domain of FOXE1 (S57N), but *in vitro* studies of the mutant protein showed only partial loss of DNA binding and retained some transcriptional activity, which might explain the normal epiglottis and choanae.[69] The third patient with a known *FOXE1* gene mutation (R102C) was of Turkish origin. She presented with CH and also had a cleft palate, bilateral choanal atresia, and spiky hair. Her mutation was located in the forkhead DNA-binding domain of FOXE1 and encodes a defective FOXE1, which has

TABLE 8.2 Monogenic Causes of Thyroid Dysgenesis

Thyroid phenotype	Other features	Gene	Transmission
From apparent athyreosis to normally appearing gland	None	*TSHR*	AR
Normally appearing gland or orthotopic hypoplasia	None	? on 15q25.3-26.1	AD
From apparent athyreosis to normal gland, usually mild ↑ TSH	• RDS • Hypotonia • Developmental delay • Ataxia/choreoathetosis	*TTF1*	*De novo* or AD
From apparent athyreosis to normally appearing gland	• Cysts within the thyroid • Unilateral kidney agenesis	*PAX8*	AD or *de novo*
True athyreosis	• Cleft palate • Choanal atresia • Kinky hair • Bifid epiglottis	*TTF2*	AR
Apparent athyreosis	• IUGR • Permanent neonatal diabetes • Congenital glaucoma • Hepatic fibrosis • Polycystic kidneys • Osteopenia • Bilateral deafness • Pancreatic exocrine deficit	*GLIS3*	AR

RDS, respiratory distress syndrome; IUGR, intrauterine growth retardation; AR, autosomal recessive; AD, autosomal dominant.

negligible DNA binding and transcriptional activity. Neck ultrasound and CT examination showed hyperechoic, soft, nonenhancing tissue at the site of the normal thyroid lobes. However, ^{123}I uptake was negligible, and plasma TG was very low. It is therefore likely that the structures seen on ultrasound and CT were the ultimobranchial bodies[74] and that, in spite of the authors' claim to the contrary,[70] the thyroid phenotype was true athyreosis, as in the two previously reported pedigrees.

Another mode of inheritance of *FOXE1* gene mutations was reported by Castanet et al. in a patient with CH due to severe thyroid hypoplasia, a cleft palate, and spiky hair who had a homozygous missense mutation (F137S) in the *FOXE1* gene. His mother was an unaffected heterozygous carrier of the mutation, but the father was homozygous for the wild-type sequence. Using microsatellite markers and multiplex ligation-dependent probe amplification, the authors demonstrated paternity and complete maternal uniparental disomy.[71] Therefore, *FOXE1* gene mutations should also be considered in nonconsanguineous families. Very recently, Carré et al. identified a new *FOXE1* gene mutation (R73S) in a patient with athyreosis, a cleft palate, and partial choanal atresia. Interestingly, this mutation led to *increased* thyroidal gene expression *in vitro* by enhancing the *TG*- and *TPO*-promoters.[72] In another study, Carré et al. suggested a different mechanism whereby FOXE1 may play a role in thyroid dysgenesis.[75] They found that the lengths of the polyalanine tract within the *FOXE1* gene differed between patients and controls. However, the length of the polyalanine tract was the same in the affected and the unaffected twin of four discordant monozygotic pairs. These seemingly contradictory observations underscore the complexity of the genetic and epigenetic mechanisms involved in thyroid dysgenesis.

Given the extreme rarity of mutations in the *FOXE1* gene, it seems reasonable to restrict the search for mutations in this gene to patients with at least three elements of the Bamforth–Lazarus syndrome. Indeed, sequencing of the *FOXE1* gene in patients with "only" CH and cleft lips/palates has yielded negative results.[9,76]

Nevertheless, as *Foxe1* is the only gene known to be associated with thyroid ectopy in animal models, a broader association between *FOXE1* gene polymorphisms and thyroid dysgenesis is suspected. As described earlier, the length of the polymorphic *FOXE1* gene alanine stretch is associated with thyroid dysgenesis.[75] In human leukocytes and thyroid tissues, the methylation status of a tissue-dependent differentially methylated region (DMR) in the *FOXE1* gene promoter shows an inverse correlation with FOXE1 expression.[77] Whether rare genetic variants in the *FOXE1* DMR are associated with CH due to thyroid ectopy is currently under investigation.

GLIS3 GENE MUTATIONS

In patients with permanent neonatal diabetes and CH, in association with intrauterine growth retardation, congenital glaucoma, hepatic fibrosis, and polycystic kidneys in some, mutations in *GLIS3*, the gene encoding the transcription factor GLI similar 3 (which maps to chromosome 9p24.3-p23), were initially identified in six individuals from three pedigrees. In the initial family, a consanguineous family from Saudi Arabia, the responsible gene was found by genome-wide linkage scanning of all family members available and a homozygous insertion (2067insC) was found in the affected, leading to a frameshift and a truncated protein (625fs703STOP). In the other two families (also consanguineous, one from Saudi Arabia and the other French Gypsy), distinct homozygous deletions in the *GLIS3* gene were found. The thyroid phenotype was apparent athyreosis.[78] Two cases were subsequently reported from the United Kingdom. One was the son of first cousins of Bangladeshi origin and the other the son of Welsh parents without known consanguinity. Both had homozygous partial deletions of the *GLIS3* gene. Additional features included osteopenia, bilateral sensorineural deafness, and pancreatic exocrine insufficiency.[79]

NKX2-5 GENE MUTATIONS

Aside from the rare syndromes described earlier, thyroid dysgenesis is typically isolated except for an increased incidence of mild congenital heart malformations, mostly septationl defects.[7,9,80] Dominantly transmitted mutations in the *NKX2-5* gene on chromosome 5q34 causing heart conduction defects have been known for 15 years.[81] NKX2-5 has therefore been proposed as a candidate gene for thyroid dysgenesis and sequence variants have been found in four of 241 patients with CH, some of whom had cardiac anomalies. Unfortunately, the imaging modality used to establish the etiology of CH in these four patients was not specified. Furthermore, these sequence variants were transmitted by one of the parents, who did not have CH and only one of whom had a heart defect.[82] The involvement of NKX2-5 in CH in humans therefore remains to be confirmed and sequencing of the *NKX2-5* gene in CH due to thyroid dysgenesis is not warranted.[83]

SYNDROMES ASSOCIATED WITH CH FROM THYROID DYSGENESIS

Of the many dysmorphic syndromes said to be associated with CH, the best studied and commonest syndromes are trisomy 21, DiGeorge syndrome, and Williams syndrome. The studies of van Trotsenburg et al. have clearly shown that, as a group, patients with trisomy 21 present a mild form of CH associated with orthotopic thyroid hypoplasia.[84] A single case of a girl with trisomy 21, overt CH, orthotopic thyroid hypoplasia, and

a mutation in the *PAX8* gene promoter has recently been described.[50] In DiGeorge syndrome (which results from a deletion of chromosome region 22q11), a case of severe CH has been reported but without scintigraphic diagnosis.[85] We are following a 10-year-old girl with DiGeorge syndrome and CH with normal thyroid morphology on technetium scanning (unpublished observation). In Williams syndrome (which results from a deletion of the elastin gene on chromosome 7q11.23), hypothyroidism is usually so mild that it is not detected by neonatal TSH screening. It is generally associated with orthotopic thyroid hypoplasia,[86] but single cases of Williams syndrome with sublingual thyroid ectopy on technetium scanning[87] or with hemiagenesis on ultrasound and scintigraphy[88] have been reported. In DiGeorge syndrome, a candidate gene is *TBX1* and the postulated mechanism is a disruption in the development of the arterial supply, which is essential for stabilization and growth of the thyroid lobes,[89,90] but in trisomy 21 and Williams syndrome, the link between the chromosomal lesion and thyroid dysgenesis is unknown.

TSHR GENE MUTATIONS (GAIN OF FUNCTION)

Activating or "gain-of-function" mutations in the TSHR gene can occur in the germline or can be somatic. Germline mutations result in congenital hyperthyroidism and goiter. This was first reported by Duprez et al. in two families with hyperthyroidism without signs of autoimmunity.[91] Since then, many more different activating TSHR germline mutations have been described (http://gris.ulb.ac.be) and characterized *in vitro*. These mutations are transmitted in a dominant fashion,[92] leading to persistent hyperthyroidism of variable severity, or occur *de novo*,[93] in which case hyperthyroidism is usually severe so that definite treatment is required.

Interestingly, activating mutations in the *TSHR* gene can also result in *hyperemesis gravidarum*. This was shown in a mother and her daughter who had a mutation in the *TSHR* gene in a highly conserved region, which encodes part of the extracellular N-terminal domain of the TSHR. Their *TSHR* gene mutation in codon 183 (K183R) resulted in a higher affinity to human chorionic gonadotrophin (hCG) of the mutant compared to the wild-type receptor. As a consequence, both women had thyrotoxicosis and *hyperemesis gravidarum* during pregnancy but were clinically euthyroid when not pregnant.[94]

In addition to germline mutations, somatic *TSHR* gene mutations account for about 80% of "hot" nodules causing nonautoimmune hyperthyroidism of adult or childhood onset.[95–97] In one case, fetal tachycardia suggested prenatal onset of hyperthyroidism.[98]

TREATMENT

Because of its potential impact on neurocognitive development, it is generally considered that CH should be treated as promptly as possible.[99] However, unless thyroid ectopy has been demonstrated on scintigraphy or thyroglobulin was undetectable in the absence of antibodies and in the face of a very high TSH before treatment (indicating true athyreosis), the need for lifelong treatment can be safely assessed by withdrawing thyroxine therapy for a month after age two to three years. If TSH rises but free T_4 and imaging are normal, a genetic diagnosis may help in deciding on the need for treatment (as in patients with a heterozygous mutation inactivating TSHR, see above Ref. [26]).

The treatment of patients with mutations that result in an increase in the constitutive activity of the TSHR will generally require thyroidectomy, although the time at which this is performed depends on the severity of the phenotype and one can "buy time" by administering antithyroid medications for a few years.[100] Even after "total" thyroidectomy, radioiodine ablation may be required.[93,101]

CONCLUSIONS

Although thousands of patients with CH due to thyroid dysgenesis have been investigated, only a few cases have been elucidated at the molecular level. In such cases, the genetic findings provide useful information for genetic counseling to patients and their families. When there are features other than CH, mutational screening of candidate genes is perhaps more likely to be successful (see Table 8.2). The existence of familial cases in which the genes discussed above have been excluded by linkage analysis suggests that other genes that are important for the development of the thyroid gland remain to be discovered.[102] However, the main reason for normal results of sequencing candidate genes in leukocyte DNA is that CH from thyroid dysgenesis is predominantly non-Mendelian. Possible non-Mendelian mechanisms are currently being investigated.[77]

References

1. Deladoey J, Ruel J, Giguere Y, Van Vliet G. Is the incidence of congenital hypothyroidism really increasing? A 20-year retrospective population-based study in Quebec. *J Clin Endocrinol Metab* 2011;**96**(8):2422–9.
2. Van Vliet G, Deladoey J. Hypothyroidism in infants and children. In: Braverman LE, Cooper DS, editors. *The Thyroid: A Fundamental and Clinical Text*. 10th ed. New York: Lippincott Williams & Wilkins; 2013. p. 787–802.
3. Cangul H, Aycan Z, Olivera-Nappa A, Saglam H, Schoenmakers NA, Boelaert K, Cetinkaya S, Tarim O, Bober E, Darendeliler F, Bas V, Demir K, Aydin BK, Kendall M, Cole T, Hogler W, Chatterjee VK, Barrett TG, Maher ER. Thyroid dyshormonogenesis is mainly caused by TPO mutations in consanguineous community. *Clin Endocrinol (Oxf)* 2013;**79**(2):275–81.
4. Kuhnen P, Turan S, Frohler S, Guran T, Abali S, Biebermann H, Bereket A, Gruters A, Chen W, Krude H. Identification of PENDRIN (SLC26A4) mutations in patients with congenital hypothyroidism and "apparent" thyroid dysgenesis. *J Clin Endocrinol Metab* 2014;**99**(1):E169–76.
5. Deladoey J, Pfarr N, Vuissoz JM, Parma J, Vassart G, Biesterfeld S, Pohlenz J, Van Vliet G. Pseudodominant inheritance of goitrous congenital hypothyroidism caused by TPO mutations: molecular and *in silico* studies. *J Clin Endocrinol Metab* 2008;**93**(2):627–33.
6. Belforte FS, Miras MB, Olcese MC, Sobrero G, Testa G, Munoz L, Gruneiro-Papendieck L, Chiesa A, Gonzalez-Sarmiento R, Targovnik HM, Rivolta CM. Congenital goitrous hypothyroidism: mutation analysis in the thyroid peroxidase gene. *Clin Endocrinol (Oxf)* 2012;**76**(4):568–76.
7. Castanet M, Polak M, Bonaiti-Pellie C, Lyonnet S, Czernichow P, Leger J. Nineteen years of national screening for congenital hypothyroidism: familial cases with thyroid dysgenesis suggest the involvement of genetic factors. *J Clin Endocrinol Metab* 2001;**86**(5):2009–14.
8. Gagne N, Parma J, Deal C, Vassart G, Van Vliet G. Apparent congenital athyreosis contrasting with normal plasma thyroglobulin levels and associated with inactivating mutations in the thyrotropin receptor gene: are athyreosis and ectopic thyroid distinct entities? *J Clin Endocrinol Metab* 1998;**83**(5):1771–5.
9. Devos H, Rodd C, Gagne N, Laframboise R, Van Vliet G. A search for the possible molecular mechanisms of thyroid dysgenesis: sex ratios and associated malformations. *J Clin Endocrinol Metab* 1999;**84**(7):2502–6.
10. Stoppa-Vaucher S, Van Vliet G, Deladoey J. Variation by ethnicity in the prevalence of congenital hypothyroidism due to thyroid dysgenesis. *Thyroid* 2011;**21**(1):13–8.
11. Perry R, Heinrichs C, Bourdoux P, Khoury K, Szots F, Dussault JH, Vassart G, Van Vliet G. Discordance of monozygotic twins for thyroid dysgenesis: implications for screening and for molecular pathophysiology. *J Clin Endocrinol Metab* 2002;**87**(9):4072–7.
12. Van Vliet G, Vassart G. Monozygotic twins are generally discordant for congenital hypothyroidism from thyroid dysgenesis. *Horm Res* 2009;**72**(5):320.
13. Deladoey J, Vassart G, Van Vliet G. Possible non-Mendelian mechanisms of thyroid dysgenesis. *Endocr Dev* 2007;**10**:29–42.
14. Jones JH, Attaie M, Maroo S, Neumann D, Perry R, Donaldson MD. Heterogeneous tissue in the thyroid fossa on ultrasound in infants with proven thyroid ectopia on isotope scan – a diagnostic trap. *Pediatr Radiol* 2010;**40**(5):725–31.
15. Wildi-Runge S, Stoppa-Vaucher S, Lambert R, Turpin S, Van Vliet G, Deladoey J. A High prevalence of dual thyroid ectopy in congenital hypothyroidism: evidence for insufficient signaling gradients during embryonic thyroid migration or for the polyclonal nature of the thyroid gland? *J Clin Endocrinol Metab* 2012;**97**(6):E978–81.
16. Stoppa-Vaucher S, Lapointe A, Turpin S, Rydlewski C, Vassart G, Deladoey J. Ectopic thyroid gland causing dysphonia: imaging and molecular studies. *J Clin Endocrinol Metab* 2010;**95**(10):4509–10.
17. Abu-Khudir R, Paquette J, Lefort A, Libert F, Chanoine JP, Vassart G, Deladoey J. Transcriptome, methylome and genomic variations analysis of ectopic thyroid glands. *PLoS One* 2010;**5**(10):e13420.
18. Vandernoot I, Sartelet H, Abu-Khudir R, Chanoine JP, Deladoey J. Evidence for calcitonin-producing cells in human lingual thyroids. *J Clin Endocrinol Metab* 2012;**97**(3):951–6.
19. Parmentier M, Libert F, Maenhaut C, Lefort A, Gerard C, Perret J, Van Sande J, Dumont JE, Vassart G. Molecular cloning of the thyrotropin receptor. *Science* 1989;**246**(4937):1620–2.

20. Libert F, Passage E, Lefort A, Vassart G, Mattei MG. Localization of human thyrotropin receptor gene to chromosome region 14q3 by *in situ* hybridization. *Cytogenet Cell Genet* 1990;**54**(1-2):82–3.

21. Sunthornthepvarakul T, Gottschalk ME, Hayashi Y, Refetoff S. Brief report: resistance to thyrotropin caused by mutations in the thyrotropin-receptor gene. *N Engl J Med* 1995;**332**(3):155–60.

22. Abramowicz MJ, Duprez L, Parma J, Vassart G, Heinrichs C. Familial congenital hypothyroidism due to inactivating mutation of the thyrotropin receptor causing profound hypoplasia of the thyroid gland. *J Clin Invest* 1997;**99**(12):3018–24.

23. Stein SA, Shanklin DR, Krulich L, Roth MG, Chubb CM, Adams PM. Evaluation and characterization of the hyt/hyt hypothyroid mouse. II. Abnormalities of TSH and the thyroid gland. *Neuroendocrinology* 1989;**49**(5):509–19.

24. Cangul H, Morgan NV, Forman JR, Saglam H, Aycan Z, Yakut T, Gulten T, Tarim O, Bober E, Cesur Y, Kirby GA, Pasha S, Karkucak M, Eren E, Cetinkaya S, Bas V, Demir K, Yuca SA, Meyer E, Kendall M, Hogler W, Barrett TG, Maher ER. Novel TSHR mutations in consanguineous families with congenital nongoitrous hypothyroidism. *Clin Endocrinol (Oxf)* 2010;**73**(5):671–7.

25. Calebiro D, Gelmini G, Cordella D, Bonomi M, Winkler F, Biebermann H, de MA, Marelli F, Libri DV, Antonica F, Vigone MC, Cappa M, Mian C, Sartorio A, Beck-Peccoz P, Radetti G, Weber G, Persani L. Frequent TSH receptor genetic alterations with variable signaling impairment in a large series of children with nonautoimmune isolated hyperthyrotropinemia. *J Clin Endocrinol Metab* 2012;**97**(1):E156–60.

26. Lucas-Herald A, Bradley T, Hermanns P, Jones J, Attaie M, Thompson E, Pohlenz J, Donaldson M. Novel heterozygous thyrotropin receptor mutation presenting with neonatal hyperthyrotropinaemia, mild thyroid hypoplasia and absent uptake on radioisotope scan. *J Pediatr Endocrinol Metab* 2013;**26**(5-6):583–6.

27. Grasberger H, Mimouni-Bloch A, Vantyghem MC, Van Vliet G, Abramowicz M, Metzger DL, Abdullatif H, Rydlewski C, Macchia PE, Scherberg NH, Van Sande J, Mimouni M, Weiss RE, Vassart G, Refetoff S. Autosomal dominant resistance to thyrotropin as a distinct entity in five multigenerational kindreds: clinical characterization and exclusion of candidate loci. *J Clin Endocrinol Metab* 2005;**90**(7):4025–34.

28. Grasberger H, Vaxillaire M, Pannain S, Beck JC, Mimouni-Bloch A, Vatin V, Vassart G, Froguel P, Refetoff S. Identification of a locus for nongoitrous congenital hypothyroidism on chromosome 15q25.3-26.1. *Hum Genet* 2005;**118**(3-4):348–55.

29. Plachov D, Chowdhury K, Walther C, Simon D, Guenet JL, Gruss P. Pax8, a murine paired box gene expressed in the developing excretory system and thyroid gland. *Development* 1990;**110**(2):643–51.

30. De Felice M, Di Lauro R. Thyroid development and its disorders: genetics and molecular mechanisms. *Endocr Rev* 2004;**25**(5):722–46.

31. Zannini M, Francis-Lang H, Plachov D, Di Lauro R. Pax-8, a paired domain-containing protein, binds to a sequence overlapping the recognition site of a homeodomain and activates transcription from two thyroid-specific promoters. *Mol Cell Biol* 1992;**12**(9):4230–41.

32. De Felice M, Ovitt C, Biffali E, Rodriguez-Mallon A, Arra C, Anastassiadis K, Macchia PE, Mattei MG, Mariano A, Scholer H, Macchia V, Di Lauro R. A mouse model for hereditary thyroid dysgenesis and cleft palate. *Nat Genet* 1998;**19**(4):395–8.

33. Macchia PE, Lapi P, Krude H, Pirro MT, Missero C, Chiovato L, Souabni A, Baserga M, Tassi V, Pinchera A, Fenzi G, Gruters A, Busslinger M, Di Lauro R. PAX8 mutations associated with congenital hypothyroidism caused by thyroid dysgenesis. *Nat Genet* 1998;**19**(1):83–6.

34. Congdon T, Nguyen LQ, Nogueira CR, Habiby RL, Medeiros-Neto G, Kopp P. A novel mutation (Q40P) in PAX8 associated with congenital hypothyroidism and thyroid hypoplasia: evidence for phenotypic variability in mother and child. *J Clin Endocrinol Metab* 2001;**86**(8):3962–7.

35. Komatsu M, Takahashi T, Takahashi I, Nakamura M, Takahashi I, Takada G. Thyroid dysgenesis caused by PAX8 mutation: the hypermutability with CpG dinucleotides at codon 31. *J Pediatr* 2001;**139**(4):597–9.

36. Vilain C, Rydlewski C, Duprez L, Heinrichs C, Abramowicz M, Malvaux P, Renneboog B, Parma J, Costagliola S, Vassart G. Autosomal dominant transmission of congenital thyroid hypoplasia due to loss-of-function mutation of PAX8. *J Clin Endocrinol Metab* 2001;**86**(1):234–8.

37. de Sanctis L, Corrias A, Romagnolo D, Di PT, Biava A, Borgarello G, Gianino P, Silvestro L, Zannini M, Dianzani I. Familial PAX8 small deletion (c.989_992delACCC) associated with extreme phenotype variability. *J Clin Endocrinol Metab* 2004;**89**(11):5669–74.

38. Meeus L, Gilbert B, Rydlewski C, Parma J, Roussie AL, Abramowicz M, Vilain C, Christophe D, Costagliola S, Vassart G. Characterization of a novel loss of function mutation of PAX8 in a familial case of congenital hypothyroidism with in-place, normal-sized thyroid. *J Clin Endocrinol Metab* 2004;**89**(9):4285–91.

39. Grasberger H, Ringkananont U, Lefrancois P, Abramowicz M, Vassart G, Refetoff S. Thyroid transcription factor 1 rescues PAX8/p300 synergism impaired by a natural PAX8 paired domain mutation with dominant negative activity. *Mol Endocrinol* 2005;**19**(7):1779–91.

40. Al Taji E, Biebermann H, Limanova Z, Hnikova O, Zikmund J, Dame C, Gruters A, Lebl J, Krude H. Screening for mutations in transcription factors in a Czech cohort of 170 patients with congenital and early-onset hypothyroidism: identification of a novel PAX8 mutation in dominantly inherited early-onset non-autoimmune hypothyroidism. *Eur J Endocrinol* 2007;**156**(5):521–9.

41. Tonacchera M, Banco ME, Montanelli L, Di CC, Agretti P, De MG, Ferrarini E, Ordookhani A, Perri A, Chiovato L, Santini F, Vitti P, Pinchera A. Genetic analysis of the PAX8 gene in children with congenital hypothyroidism and dysgenetic or eutopic thyroid glands: identification of a novel sequence variant. *Clin Endocrinol (Oxf)* 2007;**67**(1):34–40.

42. Amendola E, De LP, Macchia PE, Terracciano D, Rosica A, Chiappetta G, Kimura S, Mansouri A, Affuso A, Arra C, Macchia V, Di LR, De FM. A mouse model demonstrates a multigenic origin of congenital hypothyroidism. *Endocrinology* 2005;**146**(12):5038–47.

43. Di Palma T, Zampella E, Filippone MG, Macchia PE, Ris-Stalpers C, de VM, Zannini M. Characterization of a novel loss-of-function mutation of PAX8 associated with congenital hypothyroidism. *Clin Endocrinol (Oxf)* 2010;**73**(6):808–14.

44. Jo W, Ishizu K, Fujieda K, Tajima T. Congenital hypothyroidism caused by a PAX8 gene mutation manifested as sodium/iodide symporter gene defect. *J Thyroid Res* 2010;**2010**:619013.

45. Liu SG, Zhang SS, Zhang LQ, Li WJ, Zhang AQ, Lu KN, Wang MJ, Yan SL, Ma X. Screening of PAX8 mutations in Chinese patients with congenital hypothyroidism. *J Endocrinol Invest* 2012;**35**(10):889–92.

46. Narumi S, Araki S, Hori N, Muroya K, Yamamoto Y, Asakura Y, Adachi M, Hasegawa T. Functional characterization of four novel PAX8 mutations causing congenital hypothyroidism: new evidence for haploinsufficiency as a disease mechanism. *Eur J Endocrinol* 2012;**167**(5):625–32.

47. Carvalho A, Hermanns P, Rodrigues AL, Sousa I, Anselmo J, Bikker H, Cabral R, Pereira-Duarte C, Mota-Vieira L, Pohlenz J. A new PAX8 mutation causing congenital hypothyroidism in three generations of a family is associated with abnormalities in the urogenital tract. *Thyroid* 2013;**23**(9):1074–8.

48. Hermanns P, Grasberger H, Cohen R, Freiberg C, Dorr HG, Refetoff S, Pohlenz J. Two cases of thyroid dysgenesis caused by different novel PAX8 mutations in the DNA-binding region: *in vitro* studies reveal different pathogenic mechanisms. *Thyroid* 2013;**23**(7):791–6.

49. Hermanns P, Grasberger H, Refetoff S, Pohlenz J. Mutations in the NKX2.5 gene and the PAX8 promoter in a girl with thyroid dysgenesis. *J Clin Endocrinol Metab* 2011;**96**(6):E977–81.

50. Hermanns P, Shepherd S, Mansor M, Schulga J, Jones J, Donaldson M, Pohlenz J. A new mutation in the promoter region of the PAX8 gene causes true congenital hypothyroidism with thyroid hypoplasia in a girl with Down's syndrome. *Thyroid* 2014;**24**(6):939–44.

51. Kimura S, Hara Y, Pineau T, Fernandez-Salguero P, Fox CH, Ward JM, Gonzalez FJ. The T/ebp null mouse: thyroid-specific enhancer-binding protein is essential for the organogenesis of the thyroid, lung, ventral forebrain, and pituitary. *Genes Dev* 1996;**10**:60–9.

52. Pohlenz J, Dumitrescu A, Zundel D, Martine U, Schonberger W, Koo E, Weiss RE, Cohen RN, Kimura S, Refetoff S. Partial deficiency of thyroid transcription factor 1 produces predominantly neurological defects in humans and mice. *J Clin Invest* 2002;**109**(4):469–73.

53. Seidman JG, Seidman C. Transcription factor haploinsufficiency: when half a loaf is not enough. *J Clin Invest* 2002;**109**(4):451–5.

54. Moya CM, Perez de NG, Castano L, Potau N, Bilbao JR, Carrascosa A, Bargada M, Coya R, Martul P, Vicens-Calvet E, Santisteban P. Functional study of a novel single deletion in the TITF1/NKX2.1 homeobox gene that produces congenital hypothyroidism and benign chorea but not pulmonary distress. *J Clin Endocrinol Metab* 2006;**91**(5):1832–41.

55. Maquet E, Costagliola S, Parma J, Christophe-Hobertus C, Oligny LL, Fournet JC, Robitaille Y, Vuissoz JM, Payot A, Laberge S, Vassart G, Van Vliet G, Deladoey J. Lethal respiratory failure and mild primary hypothyroidism in a term girl with a *de novo* heterozygous mutation in the TITF1/NKX2.1 gene. *J Clin Endocrinol Metab* 2009;**94**(1):197–203.

56. Devriendt K, Vanhole C, Matthijs G, de Zegher F. Deletion of thyroid transcription factor-1 gene in an infant with neonatal thyroid dysfunction and respiratory failure. *N Engl J Med* 1998;**338**(18):1317–8.

57. Iwatani N, Mabe H, Devriendt K, Kodama M, Miike T. Deletion of NKX2.1 gene encoding thyroid transcription factor-1 in two siblings with hypothyroidism and respiratory failure. *J Pediatr* 2000;**137**(2):272–6.

58. Krude H, Schutz B, Biebermann H, von Moers A, Schnabel D, Neitzel H, Tonnies H, Weise D, Lafferty A, Schwarz S, DeFelice M, von Deimling A, van Landeghem F, DiLauro R, Gruters A. Choreoathetosis, hypothyroidism, and pulmonary alterations due to human NKX2-1 haploinsufficiency. *J Clin Invest* 2002;**109**(4):475–80.

59. Doyle DA, Gonzalez I, Thomas B, Scavina M. Autosomal dominant transmission of congenital hypothyroidism, neonatal respiratory distress, and ataxia caused by a mutation of NKX2-1. *J Pediatr* 2004;**145**(2):190–3.

60. Breedveld GJ, van Dongen JW, Danesino C, Guala A, Percy AK, Dure LS, Harper P, Lazarou LP, van der LH, Joosse M, Gruters A, MacDonald ME, de Vries BB, Arts WF, Oostra BA, Krude H, Heutink P. Mutations in TITF-1 are associated with benign hereditary chorea. *Hum Mol Genet* 2002;**11**(8):971–9.

61. Willemsen MA, Breedveld GJ, Wouda S, Otten BJ, Yntema JL, Lammens M, de Vries BB. Brain-thyroid-lung syndrome: a patient with a severe multi-system disorder due to a *de novo* mutation in the thyroid transcription factor 1 gene. *Eur J Pediatr* 2005;**164**(1):28–30.

62. Carré A, Szinnai G, Castanet M, Sura-Trueba S, Tron E, Broutin-L'Hermite I, Barat P, Goizet C, Lacombe D, Moutard ML, Raybaud C, Raynaud-Ravni C, Romana S, Ythier H, Leger J, Polak M. Five new TTF1/NKX2.1 mutations in brain-lung-thyroid syndrome: rescue by PAX8 synergism in one case. *Hum Mol Genet* 2009;**18**(12):2266–76.

63. Hamvas A, Deterding RR, Wert SE, White FV, Dishop MK, Alfano DN, Halbower AC, Planer B, Stephan MJ, Uchida DA, Williames LD, Rosenfeld JA, Lebel RR, Young LR, Cole FS, Nogee

LM. Heterogeneous pulmonary phenotypes associated with mutations in the thyroid transcription factor gene NKX2-1. *Chest* 2013;**144**(3):794–804.

64. Lapi P, Macchia PE, Chiovato L, Biffali E, Moschini L, Larizza D, Baserga M, Pinchera A, Fenzi G, Di Lauro R. Mutations in the gene encoding thyroid transcription factor-1 (TTF-1) are not a frequent cause of congenital hypothyroidism (CH) with thyroid dysgenesis. *Thyroid* 1997;**7**(3):383–7.

65. Perna MG, Civitareale D, De FV, Sacco M, Cisternino C, Tassi V. Absence of mutations in the gene encoding thyroid transcription factor-1 (TTF-1) in patients with thyroid dysgenesis. *Thyroid* 1997;**7**(3):377–81.

66. Barnett CP, Mencel JJ, Gecz J, Waters W, Kirwin SM, Vinette KM, Uppill M, Nicholl J. Choreoathetosis, congenital hypothyroidism and neonatal respiratory distress syndrome with intact NKX2-1. *Am J Med Genet A* 2012;**158A**(12):3168–73.

67. Shenoy A, Esquibies AE, Dunbar N, Dishop MK, Reyes-Mugica M, Langston C, Deladoey J, Abu-Khudir R, Carpenter T, Bazzy-Asaad A. A novel presentation of diffuse lung disease caused by congenital hypothyroidism. *J Pediatr* 2009;**155**(4):593–5.

68. Clifton-Bligh RJ, Wentworth JM, Heinz P, Crisp MS, John R, Lazarus JH, Ludgate M, Chatterjee VK. Mutation of the gene encoding human TTF-2 associated with thyroid agenesis, cleft palate and choanal atresia. *Nat Genet* 1998;**19**:399.

69. Castanet M, Park SM, Smith A, Bost M, Leger J, Lyonnet S, Pelet A, Czernichow P, Chatterjee K, Polak M. A novel loss-of-function mutation in TTF-2 is associated with congenital hypothyroidism, thyroid agenesis and cleft palate. *Hum Mol Genet* 2002;**11**(17):2051–9.

70. Baris I, Arisoy AE, Smith A, Agostini M, Mitchell CS, Park SM, Halefoglu AM, Zengin E, Chatterjee VK, Battaloglu E. A novel missense mutation in human TTF-2 (FKHL15) gene associated with congenital hypothyroidism but not athyreosis. *J Clin Endocrinol Metab* 2006;**91**(10):4183–7.

71. Castanet M, Mallya U, Agostini M, Schoenmakers E, Mitchell C, Demuth S, Raymond FL, Schwabe J, Gurnell M, Chatterjee VK. Maternal isodisomy for chromosome 9 causing homozygosity for a novel FOXE1 mutation in syndromic congenital hypothyroidism. *J Clin Endocrinol Metab* 2010;**95**(8):4031–6.

72. Carré A, Hamza RT, Kariyawasam D, Guillot L, Teissier R, Tron E, Castanet M, Dupuy C, El KM, Polak M. A novel FOXE1 mutation (R73S) in Bamforth-Lazarus syndrome causing increased thyroidal gene expression. *Thyroid* 2014;**24**(4):649–54.

73. Bamforth JS, Hughes IA, Lazarus JH, Weaver CM, Harper PS. Congenital hypothyroidism, spiky hair, and cleft palate. *J Med Genet* 1989;**26**(1):49–51.

74. Chanoine JP, Toppet V, Body JJ, Van Vliet G, Lagasse R, Bourdoux P, Spehl M, Delange F. Contribution of thyroid ultrasound and serum calcitonin to the diagnosis of congenital hypothyroidism. *J Endocrinol Invest* 1990;**13**(2):103–9.

75. Carré A, Castanet M, Sura-Trueba S, Szinnai G, Van Vliet G, Trochet D, Amiel J, Leger J, Czernichow P, Scotet V, Polak M. Polymorphic length of FOXE1 alanine stretch: evidence for genetic susceptibility to thyroid dysgenesis. *Hum Genet* 2007;**122**(5):467–76.

76. Tonacchera M, Banco M, Lapi P, Di CC, Perri A, Montanelli L, Moschini L, Gatti G, Gandini D, Massei A, Agretti P, De MG, Vitti P, Chiovato L, Pinchera A. Genetic analysis of TTF-2 gene in children with congenital hypothyroidism and cleft palate, congenital hypothyroidism, or isolated cleft palate. *Thyroid* 2004;**14**(8):584–8.

77. Abu-Khudir R, Magne F, Chanoine JP, Deal C, Van Vliet G, Deladoey J. Role for tissue-dependent methylation differences in the expression of FOXE1 in non-tumoral thyroid glands. *J Clin Endocrinol Metab* 2014;**99**(6):E1120–9.

78. Senee V, Chelala C, Duchatelet S, Feng D, Blanc H, Cossec JC, Charon C, Nicolino M, Boileau P, Cavener DR, Bougneres P, Taha D, Julier C. Mutations in GLIS3 are responsible for a rare syndrome

with neonatal diabetes mellitus and congenital hypothyroidism. *Nat Genet* 2006;**38**(6):682–7.

79. Dimitri P, Warner JT, Minton JA, Patch AM, Ellard S, Hattersley AT, Barr S, Hawkes D, Wales JK, Gregory JW. Novel GLIS3 mutations demonstrate an extended multisystem phenotype. *Eur J Endocrinol* 2011;**164**(3):437–43.

80. Olivieri A, Stazi MA, Mastroiacovo P, Fazzini C, Medda E, Spagnolo A, De Angelis S, Grandolfo ME, Taruscio D, Cordeddu V, Sorcini M. A population-based study on the frequency of additional congenital malformations in infants with congenital hypothyroidism: data from the Italian registry for congenital hypothyroidism (1991-1998). *J Clin Endocrinol Metab* 2002;**87**(2):557–62.

81. Schott JJ, Benson DW, Basson CT, Pease W, Silberbach GM, Moak JP, Maron BJ, Seidman CE, Seidman JG. Congenital heart disease caused by mutations in the transcription factor NKX2-5. *Science* 1998;**281**(5373):108–11.

82. Dentice M, Cordeddu V, Rosica A, Ferrara AM, Santarpia L, Salvatore D, Chiovato L, Perri A, Moschini L, Fazzini C, Olivieri A, Costa P, Stoppioni V, Baserga M, De FM, Sorcini M, Fenzi G, Di LR, Tartaglia M, Macchia PE. Missense mutation in the transcription factor NKX2-5: a novel molecular event in the pathogenesis of thyroid dysgenesis. *J Clin Endocrinol Metab* 2006;**91**(4):1428–33.

83. van Engelen K, Mommersteeg MT, Baars MJ, Lam J, Ilgun A, van Trotsenburg AS, Smets AM, Christoffels VM, Mulder BJ, Postma AV. The ambiguous role of NKX2-5 mutations in thyroid dysgenesis. *PLoS One* 2012;**7**(12):e52685.

84. van Trotsenburg AS, Kempers MJ, Endert E, Tijssen JG, de Vijlder JJ, Vulsma T. Trisomy 21 causes persistent congenital hypothyroidism presumably of thyroidal origin. *Thyroid* 2006;**16**(7):671–80.

85. Scuccimarri R, Rodd C. Thyroid abnormalities as a feature of DiGeorge syndrome: a patient report and review of the literature. *J Pediatr Endocrinol Metab* 1998;**11**(2):273–6.

86. Cambiaso P, Orazi C, Digilio MC, Loche S, Capolino R, Tozzi A, Faedda A, Cappa M. Thyroid morphology and subclinical hypothyroidism in children and adolescents with Williams syndrome. *J Pediatr* 2007;**150**(1):62–5.

87. Bini R, Pela I. New case of thyroid dysgenesis and clinical signs of hypothyroidism in Williams syndrome. *Am J Med Genet A* 2004;**127A**(2):183–5.

88. Cammareri V, Vignati G, Nocera G, Beck-Peccoz P, Persani L. Thyroid hemiagenesis and elevated thyrotropin levels in a child with Williams syndrome. *Am J Med Genet* 1999;**85**(5):491–4.

89. Fagman H, Andersson L, Nilsson M. The developing mouse thyroid: embryonic vessel contacts and parenchymal growth pattern during specification, budding, migration, and lobulation. *Dev Dyn* 2006;**235**(2):444–55.

90. Alt B, Elsalini OA, Schrumpf P, Haufs N, Lawson ND, Schwabe GC, Mundlos S, Gruters A, Krude H, Rohr KB. Arteries define the position of the thyroid gland during its developmental relocalisation. *Development* 2006;**133**(19):3797–804.

91. Duprez L, Parma J, Van Sande J, Allgeier A, Leclere J, Schvartz C, Delisle MJ, Decoulx M, Orgiazzi J, Dumont J. Germline mutations in the thyrotropin receptor gene cause non-autoimmune autosomal dominant hyperthyroidism. *Nat Genet* 1994;**7**(3):396–401.

92. Leclere J, Bene MC, Aubert V, Klein M, Pascal-Vigneron V, Weryha G, Faure G. Clinical consequences of activating germline mutations of TSH receptor, the concept of toxic hyperplasia. *Horm Res* 1997;**47**(4-6):158–62.

93. Kopp P, Van Sande J, Parma J, Duprez L, Gerber H, Joss E, Jameson JL, Dumont JE, Vassart G. Brief report: congenital hyperthyroidism caused by a mutation in the thyrotropin-receptor gene. *N Engl J Med* 1995;**332**(3):150–4.

94. Rodien P, Bremont C, Sanson ML, Parma J, van SJ, Costagliola S, Luton JP, Vassart G, Duprez L. Familial gestational hyperthyroidism caused by a mutant thyrotropin receptor hypersensitive to human chorionic gonadotropin. *N Engl J Med* 1998;**339**(25):1823–6.

95. Parma J, Duprez L, Van Sande J, Hermans J, Rocmans P, Van Vliet G, Costagliola S, Rodien P, Dumont JE, Vassart G. Diversity and prevalence of somatic mutations in the thyrotropin receptor and Gs alpha genes as a cause of toxic thyroid adenomas. *J Clin Endocrinol Metab* 1997;**82**(8):2695–701.

96. Schwab KO, Pfarr N, van der Werf-Grohmann N, Pohl M, Radecke J, Musholt T, Pohlenz J. Autonomous thyroid adenoma: only an adulthood disease? *J Pediatr* 2009;**154**(6):931–3.

97. Grob F, Deladoey J, Legault L, Spigelblatt L, Fournier A, Vassart G, Van Vliet G. Autonomous adenomas caused by somatic mutations of the thyroid-stimulating hormone receptor in children. *Horm Res Paediatr* 2014;**81**(2):73–9.

98. Kopp P, Muirhead S, Jourdain N, Gu WX, Jameson JL, Rodd C. Congenital hyperthyroidism caused by a solitary toxic adenoma harboring a novel somatic mutation (serine281→isoleucine) in the extracellular domain of the thyrotropin receptor. *J Clin Invest* 1997;**100**(6):1634–9.

99. Leger J, Olivieri A, Donaldson M, Torresani T, Krude H, Van Vliet G, Polak M, Butler G. European Society for Paediatric Endocrinology consensus guidelines on screening, diagnosis, and management of congenital hypothyroidism. *J Clin Endocrinol Metab* 2014;**81**(2):80–103.

100. Borgel K, Pohlenz J, Koch HG, Bramswig JH. Long-term carbimazole treatment of neonatal nonautoimmune hyperthyroidism due to a new activating TSH receptor gene mutation (Ala428Val). *Horm Res* 2005;**64**(4):203–8.

101. Singer K, Menon RK, Lesperance MM, McHugh JB, Gebarski SS, Avram AM. Residual thyroid tissue after thyroidectomy in a patient with TSH receptor-activating mutation presenting as a neck mass. *J Clin Endocrinol Metab* 2013;**98**(2):448–52.

102. Castanet M, Sura-Trueba S, Chauty A, Carre A, de Roux N, Heath S, Leger J, Lyonnet S, Czernichow P, Polak M. Linkage and mutational analysis of familial thyroid dysgenesis demonstrate genetic heterogeneity implicating novel genes. *Eur J Hum Genet* 2005;**13**(2):232–9.

9

Syndromes of Impaired Sensitivity to Thyroid Hormone

Roy E. Weiss*, *Alexandra M. Dumitrescu***, *Samuel Refetoff***,[†]

*Department of Medicine, University of Miami Miller School of Medicine, Miami, FL, USA
**Department of Medicine, The University of Chicago, Chicago, IL, USA
[†]Department of Pediatrics and the Committee on Genetics, The University of Chicago, Chicago, IL, USA

INTRODUCTION

Thyroid hormone (TH) synthesis and release is under feedback regulation from the hypothalamic input to the pituitary via thyroid stimulating hormone (TSH) and thyroid-releasing hormone (TRH). Clinicians generally diagnose thyroid disease based on the serum levels of the THs [(L-3,3',5-triiodothyronine (T_3) and L-3,3',5,5'-tetraiodothyronine thyroxine, (T_4)] in conjunction with the measurement of serum TSH. Patients with hypothyroidism have low serum THs and high TSH serum concentrations (see Chapters 7 and 8). This is usually a result of reduced production of TH due to gland destruction (such as in autoimmune thyroid disease) or impaired biosynthesis of TH (such as in organification defects). On the other hand, hyperthyroidism is usually diagnosed when the T_4 and T_3 are high and the TSH is low. The latter is commonly seen in autoimmune hyperthyroidism or autonomous TH secreting adenomas of the thyroid gland. When there is loss of the inverse relationship between TH levels and TSH concentrations or when T_4 and T_3 levels are markedly different (e.g., high T_3 and low T_4 in a TH transporter defect, or low T_3 and high T_4 as seen in TH metabolism defects, see further), the clinician must consider a broad differential diagnosis. Correct diagnosis can be made from clinical observations and confirmed by appropriate genetic testing, as discussed in this chapter. The correct diagnosis will lead to a rational therapy avoiding inappropriate thyroid gland ablation or hormone supplementation.

It is important to appreciate that the centrally regulated system described previously, is not affected by TH demands in cells not directly involved in the feedback control. Local requirement in TH is adjusted though additional mechanisms. One such system is the control of TH entry into the cell through active transmembrane transporters.[1] Another is the activation of the precursor T_4 by removal of the outer ring iodine (5'-deiodination) to form T_3 or, inactivate T_4 and T_3 by inner ring (5-deiodination) to form L-3,3',5'-triiodothyronine or reverse T_3 (rT_3) and L-3,3'-diiodothyronine (T_2), respectively (Fig. 9.1). Changing concentrations of deiodinases in various cell types, allows an additional local regulation of hormone supply.[2]

Finally, the presence and abundance of TH receptors (TRs), through which TH action is mediated, determines the type and degree of hormonal response. Action takes place in the cytosol as well as in the nucleus.[3] The latter, known as genomic effect, has been more extensively studied[4,5] (Fig. 9.1). TRs are transcription factors that are associated with DNA of genes whose expression they regulate.

The syndromes of impaired sensitivity to TH include a group of disorders with apparent discordance between serum TSH and TH levels and/or hormone action at the tissue and cell level. Resistance to thyroid hormone (RTH), a syndrome of reduced end-organ responsiveness to TH was identified in 1967.[6] With the recognition of *thyroid hormone receptor beta (THRB)* gene mutations[7,8] the term RTH became synonymous with defects of the TR.[9] Recent discoveries of genetic defects that reduce the effectiveness of TH through altered cell membrane transport[10,11] and metabolism[12] broadened the definition of resistance to TH to include all defects that alter the biological activity of an authentic hormone secreted in normal amounts.[13] It is suggested that use of the acronym RTH be limited to

Genetic Diagnosis of Endocrine Disorders. **http://dx.doi.org/10.1016/B978-0-12-800892-8.00009-9**

the syndrome produced by reduced intracellular action of the active TH, T_3. Impaired sensitivity to TH (ISTH) is used to describe altered effectiveness of TH in the broader sense.[13] While the clinician considers these defects when confronted with thyroid function tests that show a discordance of serum TH and TSH concentrations, each defect has its own constellation of test abnormalities and different clinical presentation.

OVERVIEW OF DESCRIBED AND PUTATIVE DEFECTS IN SYNDROMES OF IMPAIRED SENSITIVITY TO THYROID HORMONE

Thyroid Hormone Cell Transport Defect (THCTD)

Altered intracellular accumulation of hormone can be caused by mutations in cell membrane hormone transport proteins (Fig. 9.1a). These molecules belong to different families of solute carriers, organic anions, amino acids, and monocarboxylate transporters (MCT). A defective gene may fail to synthesize a protein, form a molecule that may not reach the cell membrane, or be defective in its ability to transport the hormone. In all these instances there is lack or insufficient hormone in cells dependent on the specific hormone transporter. A defect in one such transporter, MCT8, presents elevated serum concentrations of T_3 and low levels of T_4 and rT_3 accompanied by severe psychomotor deficit.[14]

Thyroid Hormone Metabolism Defect (THMD)

T_4, the major product secreted by the thyroid gland, is a prohormone requiring activation through its conversion to T_3 (Fig. 9.1b). Defects in any of the factors involved in this enzymatic reaction can cause a diminished production of T_3 and thus, reduced sensitivity to the hormone. Defects may include abnormalities in the synthesis or degradation of the various deiodinases. One such

Target cell

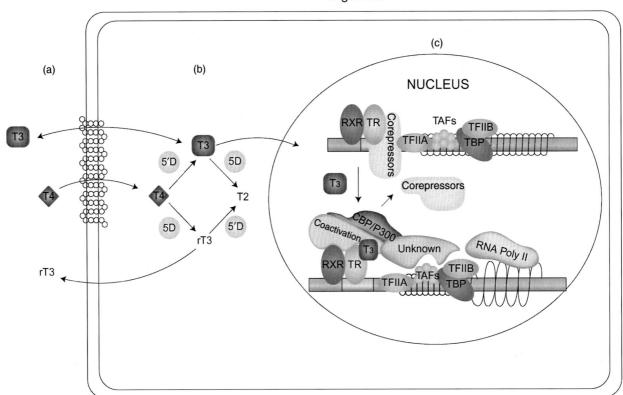

FIGURE 9.1   Regulation of TH transport, metabolism and genomic action. (a) Transport of TH into the cell via thyroid hormone transporter, MCT8. (b) Intracellular metabolism of TH, regulating TH bioactivity. (c) Genomic action of TH. For details see text. TRH, TSH-releasing hormone; TSH, thyroid-stimulating hormone (thyrotropin); T_4, 3,3',5,5'-tetraiodothyronine (thyroxine, T_4); T_3, 3,3',5-triiodothyronine (T_3); rT_3 3,3',5'-triiodothyronine (reverse T_3, rT_3); T_2, 3,3'-dioidothyronine; 5'D, deiodination by removal of an iodine from the 5' position of iodothyronine; 5D, 5-deiodination; T_2, 3,3'-dioidothyronine; TR, TH receptor; RXR, retinoid X receptor; CBP/P300, cAMP-binding protein/general transcription adaptor; TFIIA and TFIIB, transcription intermediary factor II, A and B; TBP, TATA-binding protein; TAF, TBP-associated factor.

defect has been recently identified reducing the synthesis of selenoproteins, a family of proteins to which deiodinases belong.[12] Patients present low serum T_3 and high T_4 and rT_3 concentrations.

Abnormal Hormone Transfer to the Nucleus

The main and best-studied TH effect requires the translocation of the hormone into the nucleus where it interacts with the TR (Fig. 9.1c). Putative defects in the transfer of hormone and/or its receptor to the nucleus are expected to reduce the hormonal action at the genomic level. No such defects have been yet described.

Thyroid Hormone Receptor Defects Resulting in RTH

Nuclear, also known as genomic action of TH is mediated through the TRs (Fig. 9.1c). Mutant TR proteins have reduced ability to bind cognate ligand or protein cofactors and possibly DNA. Mutations have been identified in each of the two THR genes, alpha and beta. Patients with *thyroid hormone receptor beta (THRB)* gene mutations have RTHβ that manifests persistent elevations of all three serum iodothyronines with nonsuppressed TSH. Clinical manifestation are usually minor and could suggest both hypo- and hyperthyroidism. Those with mutations in the *THRA* gene have subtle abnormalities in serum thyroid tests as TRα has a minor role in the feedback regulation of thyroid hormone synthesis and secretion. The cardinal features of sort stature, constipation and intellectual impairment reflect the organs that are TRα dependent.[15]

Abnormal Cofactors or Interfering Substances

Nuclear hormone receptors exert the hormone-mediated activity by forming a complex that involves accessory and modulatory proteins as well as the ligand (Fig. 9.1c). Putative defects in cofactors that normally stabilize the hormone-receptor complex or cofactors that repress or activate function could be responsible for reduced hormone sensitivity. Patients without *THRB* gene mutations but presenting a phenotype indistinguishable from that manifested in the presence of a mutation, are believed to belong to this category of abnormalities.

"Nongenomic" Abnormalities

Recent work has established that TH can also act at the level of the cytosol through nongenomic mechanisms.[3] It is anticipated that impairment of such activity would result in yet unrecognized disease states.

RESISTANCE TO THYROID HORMONE (RTH)

Background

With the recent discovery of TRα mutations in humans and their distinct phenotype, it has been proposed to classify the RTH syndromes based on their molecular etiology,[13] namely: (1) RTHβ for classic RTH due to mutations in the *THRB* gene; (2) NonTR-RTH for patients with an identical phenotype as RTHβ but lacking *THRB* gene mutations; (3) RTHα for those subjects with a distinct phenotype harboring *THRA* gene mutations.

Expression of TH effects requires the presence in the cell of sufficient amount of the active hormone, T_3. The best-studied effect involves the translocation of T_3 into the nucleus where it interacts with TRs to regulate (activate or repress) transcription of target genes.[4] There are two TRs, α and β, which are encoded by separate genes located on chromosomes 17 and 3, respectively. The receptors have structural and sequence similarities and exist as different isoforms with DNA-binding and T_3-binding domains (Fig. 9.2). Other regions of the molecules are involved in dimerization with another TR (homodimerization) or another type of nuclear receptor (heterodimerization), and in binding corepressor and coactivator protein cofactors (Fig. 9.1c).[16]

Dimers of unliganded (without T_3) receptors bind to TH-response elements(TRE), resulting in inhibition of the expression of genes that are positively regulated by T_3 through association with corepressor proteins. T_3-binding to the receptors produces a stearic change of the TR molecule that results in release of the corepressor protein, dissociation of the dimers, and formation of heterodimers of TR and retinoid X receptors that then bind coactivator proteins (Fig. 9.1c). These changes promote gene expression and ultimately increase the synthesis of specific proteins.

*RTH*β

RTHβ is an inherited syndrome characterized by reduced responsiveness of target tissues to TH. It is first suspected by the findings of high serum concentrations of free T_4 and usually also free T_3, accompanied by normal or slightly high serum TSH concentration.[9,29]

Incidence and Prevalence

RTH has been detected in 1 of 40,000 live births,[30] and it occurs with equal frequency in both sexes. With the exception of a single family, it is inherited as an autosomal dominant trait. In the past, RTH was subdivided into generalized, isolated pituitary and isolated peripheral

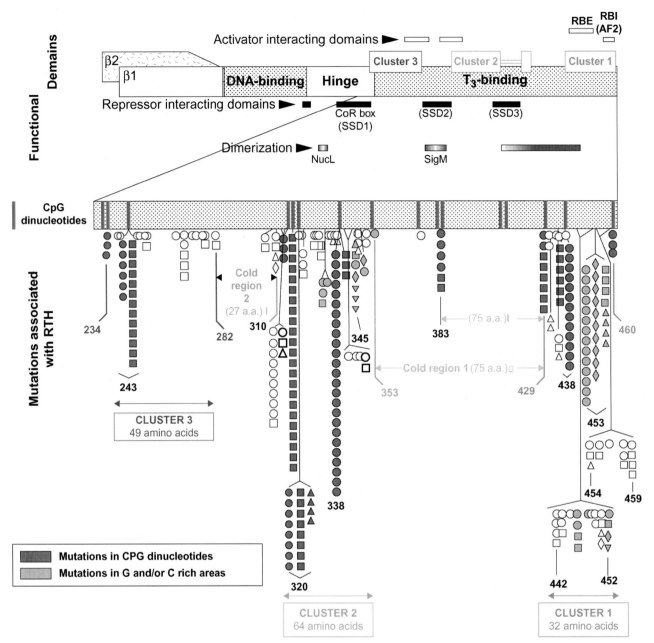

FIGURE 9.2   **Location of mutations in the TRβ molecule causing RTH.** (Top): Schematic representation of TRβ and its functional domains for interaction with TREs (DNA-binding), with hormone (T3-binding), with activating,[17] repressing[18–20] cofactors and with nuclear receptor partners (dimerization).[21–23] Note their relationship to the three clusters of natural mutations. (Bottom): The T₃-binding domain and distal end of the hinge region, which contain the three mutation clusters, are expanded and show the positions of CpG dinucleotide which are mutational "hot spots" in the *TRβ* gene. The location of 124 different mutations detected in 343 unrelated families (published and our unpublished data) are each indicated by a symbol. Identical mutations in members of unrelated families are indicated vertically by the same color and symbol. "Cold regions" are areas devoid of mutations associated with RTH. Amino acids are numbered consecutively starting at the amino terminus of the TRβ1 molecule according to the consensus statement of the First International Workshop on RTH.[24] TRβ2 has 15 additional residues at the aminoterminus. AF2, Hormone-dependent activation function (12th amphipatic helix)[25,26]; RBE, corepressor-binding enhancer; RBI, corepressor-binding inhibitor[26]; SSD, silencing subdomain[20]; NucL, nuclear localization[27]; SigM, signature motif.[28] *Reproduced from Chapter 16D in www.thyroidmanager.org, with permission.*

tissues resistance. This classification, based upon clinical findings alone, has no genetic basis as the former two can occur in individuals with the same mutation.[29] The latter represents the development of tolerance to the ingestion of excess TH.

Current published cases surpass 3000 and of the 460 families with RTH, 50% were investigated in the authors' institution. Of the latter, 34 families (15%) had no *THRB* gene mutations and 48 (21 %) of the mutations occurred *de novo*. The latter commonly occur in

CpG dinucleotide hot spots or in guanine or cytosine reach areas[31] (Fig. 9.2).

Clinical Features

The hallmark of RTHβ is the paucity of symptoms and signs of thyroid dysfunction despite the presence of high serum T_4 and T_3 concentrations. Affected individuals may have variable symptoms or signs of hypothyroidism or hyperthyroidism.

Growth retardation, delayed bone maturation, learning disabilities, mental retardation, sensorineural deafness, and nystagmus are compatible with hypothyroidism, while tachycardia, hyperactivity, and increased basal metabolic rate are suggestive of hyperthyroidism. Overt hypothyroidism is more common in those patients who, because of erroneous diagnosis of hyperthyroidism, received treatment to decrease thyroid secretion.[9,29,32] There is a high prevalence of attention deficit hyperactivity disorder and learning disabilities.[9,32,33] Hearing loss may be due in part to recurrent ear infections, which are more common in the patients with RTH,[32] but sensorineural deafness is typical of RTH due to TRβ gene deletion in humans.[34] In mice, absence of TRβ1 is sufficient to cause hearing loss.[35]

Among all clinical findings, goiter is by far the most common (65–95 %), followed by hyperactivity (33–68%), and tachycardia (33–75 %). It is these abnormalities that usually lead to evaluation of thyroid function. The subsequent finding of high serum T_4 and T_3 concentrations often results in the erroneous diagnosis of hyperthyroidism.

Laboratory Findings and Differential Diagnosis

In the untreated subject, a high serum free T_4 concentration and nonsuppressed TSH are *sine qua non* requirements for the diagnosis of RTHβ. Serum levels of T_3 and rT_3 are usually also high. Thyroglobulin concentration tends to be high, reflecting the level of TSH induced thyroid gland hyperactivity. The response of TSH to TSH-releasing hormone is normal or exaggerated, depending on the baseline TSH level. The suppressive effect of administered TH on TSH, cholesterol, and creatine kinase, and the stimulatory effect on sex-hormone binding globulin and ferritin are attenuated.

All causes of high serum T_4 and T_3 concentrations in association with normal to high serum TSH levels should be considered in the differential diagnosis, after the abnormalities are confirmed by repeat measurements several weeks later (Table 9.1). The next step is to measure serum free T_4 and T_3, preferably by equilibrium dialysis; normal values suggest a defect of TH transport in serum, not RTH.

The demonstration of similar abnormalities in serum T_4 and TSH concentrations in first-degree relatives obviates the need to exclude a TSH-producing pituitary adenoma (TSHoma). A high ratio of the alpha subunit to whole TSH is pathognomonic of TSHoma. The latter disorder is associated with a similar TH profile as RTH but most, if not all, patients are hyperthyroid.[36]

The ability to identify mutations in the *THRB* gene provides a means to confirm the diagnosis, to obtain prenatal diagnosis, and to prevent inappropriate antithyroid treatment of patients with high serum levels of free TH.

Genetic Pathophysiology

RTHβ is caused by mutations in the *THRB* gene, located on chromosome 3. Tissues in patients with RTHβ are resistant to action of T_3 to the extent this gene is expressed in the cells involved. In the majority of cases, mutations have been found in the carboxyl terminus of the TRβ covering the ligand-binding domain and adjacent hinge domain[25,37,38] (Fig. 9.2). They are contained within three clusters rich in CpG "hot spots", separated by areas devoid of mutations (cold regions). The latter are located between codons 282 and 310, and with the exception of 383, codons 353 and 429. No mutation has been reported upstream of codon 234. As cold regions are not devoid of "hot spots," the lack of mutations reflects the observation that mutations in the second cold region do not impair TR function and therefore, are not expected to produce a phenotype.[39]

THRB gene defects have been identified in 426 families comprising 132 distinct mutations. The authors have found mutations in 213 families and a partial listing is available from http://www.receptors.org/cgi-bin/nrmd/nrmd.py. Although mostly are missense mutations, nucleotide deletion, and insertions producing frameshifts have created in five instances nonsense proteins with two additional aminoacids. In four instances single nucleotide deletions have produced truncated receptors. In only one family complete *THRB* gene deletion resulted in recessively inherited RTH.[40] The following mutations have been identified in more than 10 unrelated families, often the consequence of *de novo* mutations: R243Q (15 families); A317T (29 families); R338W (30 families); R438H (17 families); P453T (17 families); and P453S (12 families).

The mutant TRβ molecules have either reduced affinity for T_3[25,37] or impaired interaction with one of the cofactors involved in the mediation of TH action.[37,41–44] As TR mutants are still able to bind to TREs on DNA and dimerize with normal TRs or the RXR partner, they interfere with the function of the normal TRs, explaining the dominant mode of inheritance. Therefore, it is not surprising that in the single family reported with a

TABLE 9.1   Conditions Associated with Reduced Sensitivity to Thyroid Hormone

Defect	Thyroid function tests						Prevalence
	T_4	T_3	rT_3	TSH	FT_4 direct	FT_4 dialysis	
Abnormal binding protein							
↑TBG	↑	↑	↑	N	N	N	1:100 males
↑TTR*	↑	N	↑	N	N	N	1:10,000
FDH	↑	↑ or N	↑	N	↑↑	N	1:600
Reduced sensitivity to TH							
RTHβ	↑	↑ or N	↑	sl↑ or N	↑	↑	1:40,000
TSHoma**	↑	↑	↑	sl↑ or N	↑	↑	Unknown
THCTD	↓	↑↑	↓↓	sl↑ or N	↑	↑	Unknown 108 families
THMD	↑↑	↓	↑↑	↑ or N	↑↑	↑↑	Unknown 9 families
RTHα	↓	↑ or N	↓↓	sl↑ or N	↓	↓	Unknown 5 families

* *Refers to TTR with increased affinity for T_4 and rT_3.*

** *To distinguish between RTH and TSHoma please see text. In brief: response to TRH stimulation, lower TSH value, lower alpha subunit, family history, and presence of mutation are consistent with RTH.*

sl, slight; ↑, increased; ↓, decreased; N, normal; TBG, T_4-binding globulin; TTR, transthyretin; FDH, familial dysalbuminemic hyperthyroxinemia; TSHoma, TSH-producing pituitary adenoma; THCTD, thyroid hormone cell transport defect; THMD, thyroid hormone metabolism defect.

deletion of all coding sequences of the *THRB* gene, only homozygotes manifest the phenotype of RTHβ.[34]

A family with two *de novo THRB* gene mutations occurring in the same nucleotide has been reported.[45] The proposita with apparent *de novo* missense mutation (GTG to GGG) in codon 458 of the *THRB* gene (V458G), transmitted this mutation to her affected son. The mutant allele underwent another *de novo* mutation transferred to her affected daughter as GAG (V458E). This apparent attempt of repair is more likely the result of the creation of a mutagenic prone three guanines sequence by the first mutation.

A poorly understood feature of RTH is the variable severity of hormonal resistance in different tissues in the same patient and among different patients having the same mutation. Studies of transgenic mice bearing *Thrb* gene mutations and gene deletions have provided some insight into these differences.[46] As an example, the tachycardia that occurs in patients with RTHβ is due to the unopposed activation of the TRα by the high serum T_4 and T_3 concentrations in the heart that expresses predominantly this TR isoform.[47] The variable manifestations of RTH among affected sibs with the same mutation is unexplained, but they likely result from genetic variability of cofactors involved in TH action.[38,48,49]

Treatment

There is no treatment that will correct the defect of TRβ function in subjects with RTHβ. Fortunately, in most patients, the hyposensitivity to TH seems to be adequately compensated by the increase in secretion of T_4 and generation of T_3. Thus, treatment is usually not needed. This is not the case in patients with limited thyroid reserve due to prior destructive therapy directed to the thyroid gland. These patients should be treated with sufficient amount of levothyroxine (L-T_4) to reduce their serum TSH concentrations to normal or near normal.

In some patients with RTHβ, several peripheral tissues may be relatively more resistant than the thyrotrophs. Thus, the compensation for the hormonal resistance in these tissues is incomplete and judicious administration of a dose of L-T_4 higher than that needed to restore TSH secretion to normal may be indicated. The dose must be individually determined by assessing tissue responses. In children, this should be done by regular assessment of growth, bone maturation, and mental development. L-T_4 should be given in incremental doses, and the basal metabolic rate, nitrogen balance, and serum sex hormone-binding globulin should be measured after treatment for 4–6 weeks before the dose is changed; bone age and growth should be followed on a longer term basis. Development of a catabolic state is an indication of overtreatment.

Management of TH levels during pregnancy in a mother with RTHβ or a normal mother carrying a fetus with RTHβ is not straightforward. A retrospective study of a large family with RTHβ demonstrated that the adverse effect of TH on the fetus was independent of that on the pregnant woman who, because of the resistance, is protected from the metabolic effect of the hormone.[50]

The prevalence of early pregnancy loss was increased fivefold in affected mothers, but not in couples

TABLE 9.2   THRA Gene Mutations in Humans

THRA gene nucleotide #	THRA protein	FT4 % lower limit of normal	FT3 % upper limit of normal	TrT3 % lower limit of normal	TSH (mU/L)	Known *THRB* gene mutations in corresponding codons	References
806 C > T	A263V	99*	90*	<63*	4.2*	A317V/T/S/D	62
1075 A > T	N359Y	114	100	121	0.3		61
1144 delG	A382fsM388X	100	140	91	5.8	A436T/fs, M442V/T	60
1176 C > A	C392X	107	148		2.8	C446X/G/R	63
1193 C > G	P398R	72	70		0.5	P452H/L/Δ	63
1207 G > T	E403X	63	80		1.0	E457G	58
1207 G > T	E403X	NA	NA		NA	E457G	63
1207 G > A	E403K	106	90	33	1.9	E457G	63
1190 insT	F397fs E406X	Low Nl	High Nl	1	Nl	E460K	59

Average for three affected adult members of the family.
del, deletion; ins, insertion; Nl, normal; NA, data not available prior to treatment with L-T4.
The alternative splicing giving rise to TRa2 isoforms occurs at amino acid 372.

with an affected father and unaffected mother. Two-thirds of their infants carried the *THRB* gene mutation, which suggests that nearly all miscarried fetuses had no mutation and thus, a normal response to TH. Furthermore, unaffected infants born to affected mothers had lower birth weights and suppressed serum TSH concentrations when compared to affected infants.[50] These results are in agreement with the finding that infants with excess TH caused by gain-of-function *TSH receptor* gene mutations are born prematurely and have low birth weights.[51] Therefore, management of pregnancies in mothers with RTHβ, carrying unaffected fetuses, may warrant judicious use of antithyroid medication depending on the well-being of the fetus. Further studies are needed before general recommendations can be made.

NonTR-RTH

NonTR-RTH refers to the occurrence of the RTHβ in the absence of a *THRB* gene mutation. It is clinically and biochemically undistinguishable from RTHβ with *THRB* gene mutations. In several of these families, inheritance is autosomal dominant and mutations in *THRB* gene have been excluded by the absence of genetic cosegregation and sequencing, thus ruling out mosaicism. Based on the observations in mice[52,53] and studies in humans[54] mutations of one of the cofactors that interact with the receptors may be responsible for the resistance in these families.[54,55]

RTHα

Background, Incidence, and Prevalence

Thrα gene manipulations in the mouse have long demonstrated that there is an absence of the RTHβ

phenotype and that there may be lack of central hypothyroidism and perturbations in metabolic regulations.[56,57] Our knowledge of RTHα in humans is limited to only 13 subjects belonging to 9 families[58–63] (Table 9.2). The first case was identified by whole-exome sequencing.[58] The proposita, a white female of European descent, was heterozygous for a *de novo* mutation in the *THRA* gene. A second family, of Greek ancestry, had an affected father and daughter. A fourth case was diagnosed at the age of 42 in a woman with epilepsy, growth retardation, constipation, and macrocephaly. Six additional families with *THRA* gene mutations were recently presented. The prevalence remains unknown.

Clinical Features

In general RTHα presents with a range of features that are characterized by growth retardation with skeletal dysplasia resulting in short lower limbs and large head, mental retardation, constipation, bradycardia, and reduced muscle strength. Other associated problems are seizures and red cell macrocytosis. These clinical findings are compatible with the known tissue distribution of the TRα in bone, brain intestine, heart, and muscle.

Laboratory Findings and Differential Diagnosis

Thyroid function tests in RTHα have consistently demonstrated markedly low serum rT_3, slightly low or low normal T_4, relatively high normal T_3, and normal TSH. This constellation of thyroid tests is not explained by a defective feedback regulation at the hypothalamus and pituitary level, as this is mainly the function of RTHβ. It has been demonstrated that TH regulation of deiodinase 3 (DIO3) is dependent on TRα.[64] Therefore, subjects with RTHα may have a reduced DIO3 resulting in a low rT_3. The differential diagnosis would include TH cell transport defects, such as MCT8 deficiency, although the lack of severe psychomotor retardation in RTHα and the lack of bone and growth defects in the MCT8 deficient patients are distinguishing features.

Genetic Pathophysiology

The inheritance is autosomal dominant. As is the case of TRβ defects, the mutant TRα exerts dominant negative effect on the wild-type TRα1 that binds T_3. The first three families reported had mutations that truncated the TRα; these were M338X, E403X, and E406X. Functionally, they corresponded to the following mutations in the TRβ molecule: M442X, E457X, and E460X, one of which, C446X has been reported. The latter produce a very severe form of RTHβ phenotype.[65] More recently missense mutation that involve both the TRα1 and TRα2 isoforms have been reported, A263V and N359Y.[61,62] The mutation affecting the TRα2 isoform does not seem to contribute to the phenotype.[62] While A382PfsX7 and N359Y both have dominant negative effects on TRα1 and TRα2, the A382PfsX7 mutant retains constitutive corepressor binding and there is an absence of coactivator recruitment. The reason for the unusual manifestations and somatic defects present in the subject with the THRA N359Y[61] remains unexplained.

Treatment

Given the limited experience with RTHα, there is no established therapy. Affected subjects have received trials of L-T_4 therapy that alleviated the constipation. Unless treatment is instituted in early life, it is unclear whether there will be improvement on mental function.

THYROID HORMONE CELL TRANSPORTER DEFECT

Background, Incidence, and Prevalence

One form of THCTD was identified in 2004.[10,11] It is caused by *monocarboxylate transporter 8 (MCT8 or SLC16A2)* gene mutations, manifesting a X-chromosome linked syndromic defect that combines severe psychomotor deficiency and abnormal thyroid function tests. The early hallmarks are hypotonia and high serumT_3 levels.

TH is transported across cell membranes by several molecules with different kinetics and substrate preferences.[66] These proteins belong to different families of solute carriers, organic anions, amino acids, and MCT. Differences in cell distribution and kinetics, as well as transport of other ligands, provide them with distinctive roles in the cell-specific delivery of TH.[11] Presumably, defects in each molecule would result in distinct phenotype, the nature of which may be predicted by the generation of mice deficient in a specific transporter.

The incidence is not known. A sex-linked form of mental retardation with motor abnormalities was described in 1944[67] and subsequently named the Allan Herndon Dudley syndrome. Its etiology was recognized in 2004, when the same defect, associated with characteristic thyroid function test abnormalities, was found to be caused by *MCT8* gene mutations.[10,11] In the ensuing 11 years, more than 250 individuals belonging to more than 130 families with *MCT8* gene defects were identified[68–71] (and personal observations), suggesting that this syndrome is more common than initially suspected. Individuals of all races and diverse ethnic origin harbor more than 80 different mutations. These mutations can be maintained in the population because carrier females are asymptomatic, thus no negative selection takes place.

Clinical Features

Male patients present during infancy or early childhood with neurodevelopmental abnormalities. A single female patient with a phenotype identical to that of males, had a *de novo* balanced translocation at the *MCT8* gene site together with a nonrandom unfavorable X-inactivation.[68] Early signs, manifesting in the first few weeks of life, are hypotonia and feeding difficulties. Though variable in severity, the clinical presentation is very similar, with consistent thyroid test abnormalities. Gestation and delivery are normal and infants have normal length, weight, and head circumference.

With advancing age, weight gain lags behind, in most cases microcephaly develops, while linear growth proceeds normally. Although truncal hypotonia persists, there is progressive development of limb rigidity often resulting in spastic quadriplegia. Muscle mass is diminished and there is generalized muscle weakness with typical poor head control. Purposeless movements and characteristic paroxysms of kinesigenic dyskinesias are common. These are usually triggered by somatosensory stimuli. The attacks, lasting a few minutes or less, consist of body extension, opening of the mouth, and stretching or flexing of the limbs.[72] True seizures can also occur. Most affected children are never able to sit by themselves or walk, and have no speech.

Affected individuals present stigmata of TH deficiency as well as excess. In fact, the psychomotor abnormalities are due to hormone deficiency and the hypermetabolic state and difficulty to maintain weight, due to hormone excess. As demonstrated in Mct8 deficient mice,[73–75] this coexistence of TH deficiency and excess in the same individual is due cell specific differences in the expression of the various TH transporters.

Although early death has occurred in some families, some individuals live beyond the age of 70 years.

TABLE 9.3 Types of MCT8 Gene Mutations Reported

Types		Number of different mutations	Number of families	Effects on MCT8 protein
Substitution	Single nucleotide	37	61	Single aa substitution (27 mutations, 45 families).
				Premature stop (10 mutations, 16 families)
Deletion	Single nucleotide	7	8	FrSh with premature termination (6) and extension with 64 aa (1)
	Trinucleotide	3	7	Single aa deletion (F229 Δ, F501 Δ, F554 Δ)
	Eight nucleotides	1	1	FrSh with premature termination
	Fourteen nucleotides	1	1	FrSh with premature termination
	Large	11	11	Lacking part of the gene
Insertion	Single nucleotide	6	9	FrSh with premature termination
	Four nucleotides	1	1	FrSh with premature termination
	InDel	1	1	FrSh with premature termination (c.1678 insA delCC)
Duplication	Three nucleotides	2	2	Single aa insertion (189I, 236V)
	Four nucleotides	2	2	FrSh with premature termination
	Six nucleotides	1	1	Two aa insertion (c.127 insGGCAGC → p.43insGS)
	Eight nucleotides	1	1	FrSh with premature termination
Splice site mutation		1	1	IVS3as -1 G- > C, alternative splicing and in frame deletion of 94 aa
Chr translocation involving MCT8		1	1	Balanced translocation 46,X,t(X;9)(q13.2;p24)
Mutations	At CpG dinucleotides	8	26	42.6% Of 61 families with single nucleotide substitution
	In C repeats	5	8	Missense (2), 1nc insertion (1), 1nc deletion (1), InDel (1)
	In A repeats	1	2	FrSh with premature termination (c.629insA)
De novo mutations	Total	12*	13*	Mutation F229 Δ occurred *de novo* in two different families
	in CpGs	3	3	25% Of the *de novo* mutation

* Might be underestimated, as publications do not always include parental genotype.
aa, amino acid; FrSh, frame shift; nc, nucleotide.

Female carriers do not manifest any of the psychomotor abnormalities described above. However, intellectual delay and frank mental retardation have been reported.[10,76]

Laboratory Findings

All have the characteristic high serum T_3 and low rT_3 concentrations. T_4 has the tendency to be low and TSH levels are normal or slightly elevated. Heterozygous female carriers have serum TH concentrations intermediate between affected males and unaffected family members. Diagnosis is confirmed by sequencing the MCT8 gene in the affected individuals. Table 9.3 summarizes the types of mutations identified in the *MCT8* gene.

Though brain magnetic resonance imaging has been reported as normal, atrophy of the cerebrum, thalamus, and basal ganglia has been occasionally noted.[69,70,77,78] A more common feature is dysmyelination and increase brain levels of choline and myoinositol.[79] For comparison of the other two syndromes of reduced sensitivity to TH, refer to Table 9.4.

Genetic Pathophysiology

MCT8 is a specific transporter of T_4 and T_3 into cells.[81] It is believed that TH deprivation, particularly of the brain during embryonic and early postnatal life, is responsible for the clinical manifestations. In fact, there appears to be a correlation between the degree of functional impairment of the mutant transporter and severity of the psychomotor retardation.[82] Loss of function is the consequence of reduced protein expression, impaired trafficking to plasma membrane or reduced substrate affinity.

Given the variability in the severity of the disease we searched for correlations between phenotypes and

TABLE 9.4   Comparison of Phenotypes and Diagnostic Genetic Testing in Syndromes of Reduced Sensitivity to TH

	Other major phenotypic manifestations	Genes	Number of reported mutations	Prenatal testing reported	Relevant lab contacts (phone/fax)	Should family members be tested
RTHβ	ADD, tachycardia, goiter	*THRB*	132	Yes[80]	Quest, authors	Yes
NonTR RTH	ADD, tachycardia, goiter	??	??	??	Authors	Yes
THCTD*	Severe psychomotor impairment	*MCT8*	76	Yes (personal data)	Authors	Yes
THMD	Growth delay	*SBP2*	16	Yes (personal data)	Authors	Yes
RTHα	Skeletal abnl; mental retardation; constipation; or diarrhea	*THRA*	5	??	Authors	Yes

* Males.
THCTD, thyroid hormone transport defect; THMD, thyroid hormone metabolism defect.

genotypes. A comparison of the clinical picture in the families with identical mutations would have been helpful in determining if such correlations exist. Unfortunately, detailed clinical information in one of each family is not available. However, early deaths were reported in the two families with a truncated MCT8 molecule (S448X). In one family two affected males died at age 13 and 39 years, and in the other, deaths occurred at 20, 22, and 30 years.

Early death was reported in subjects harboring the following mutations: S448X, P537L, 404 frameshift 416X, F230D, S194F, and frameshift with 64 amino acid carboxyl-terminal extension. The cause of death in 4 of 14 was aspiration pneumonia.

Functional analysis, in terms of T_3 uptake, examined in 20 different mutations,[83] revealed no activity in 10 mutations all producing truncated molecules due to deletions or stop codons. In 4, also caused by deletions, uptake was from 2.4% to 5%. In 5 of the mutation tested, namely the missense mutations S194F, V235M, R271H, L434W, and L598P, T_3 uptake ranged from 8.6% to 33% that of the wild-type MCT8.

Using available clinical, chemical and *in vitro* information, there is no clear relation between the degree of impairment of T_3 transport by the mutant MCT8 molecules and the level of serum T_3. This is probably due to the important role played by perturbations in the metabolism of iodothyronines on the production of T_3.[73] Furthermore, except for early death, no other clinical consequence appears to significantly correlate with the degree of functional or physical disruption of the MCT8 molecule. Genetic factors, variability of tissue expression of MCT8, and other iodothyronine cell membrane transporters could be at the basis of this lack of phenotype/genotype correlation. However, the possibility that MCT8 is involved in the transport of other ligands has not been excluded.

Replication of the thyroid function test abnormalities in recombinant mice deficient in MCT8,[73,74] provided some understanding of the mechanisms responsible for the thyroid phenotype.[84] Depending on redundancy in TH transmembrane transporters, tissues such as the liver had high, and brain low T_3 concentration. The resulting increase in liver type 1 deiodinase and brain type 2 deiodinase, enhances the generation of T_3, has a consumptive effect on T_4 and the impaired T_3 uptake in some tissues increases its accumulation in serum. Additional reasons for the low serum T4 are reduced secretion from the thyroid gland and increased urinary loss.[85,86] These tissue-specific differences in TH content and their effect on TH metabolism are in part responsible for the unusual clinical presentation of mixed TH deprivation and excess.[14]

Treatment

Current treatment options for patients with *MCT8* gene mutations are limited to supportive measures including the use of braces to prevent contractures that may require orthopedic surgery. Diet is adjusted to prevent aspiration and dystonia is treated with anticholinergics, L-DOPA carbamazepine, and lioresol. Drooling might be improved with glycopyrolate or scopolamine. Seizures are treated with standard anticonvulsants.

Treatment of the low serum T_4 concentration with physiological doses of L-T_4 has been ineffective presumably because of the impaired uptake of the hormone in MCT8-dependent tissues. Treatment with higher doses of L-T_4 together with propylthiouracil has been effective in improving nutrition by reducing the amount of T_3 generated.[87] Administration of L-T_4 together with propylthiouracil during pregnancy and the efficacy of several TH analogs, that may bypass the molecular defect by using alternative transporters, have therapeutic potential that have been tested in Mct8-deficient mice[88] and to a limited extent in humans.[89] The analog diiodothyropropionic acid given to children older than 9 months improves normalizes the thyroid test abnormalities and nutrition but not the neuropsychiatric abnormalities.[89] This is likely due to the fact that brain abnormalities develop in affected males in embryonic life.[90]

THYROID HORMONE METABOLISM DEFECT

Background, Incidence, and Prevalence

While acquired changes in TH metabolism, as those producing the "low T_3 syndrome" of nonthyroidal illness are common,[91] until recently, no inherited defect was known. Predicted phenotypes based on gene manipulations of mice vary depending on the type of deiodinase deficiency.[92] A defect of selenocysteine insertion sequence-binding protein 2 (*SECISBP2*, in short *SBP2*), one of the 12 known genes involved in deiodinase synthesis and degradation, was identified in humans.[12] It produces a global deiodination defect which gives rise to a low serum T_3 and high T_4 and rT_3.

Intracellular iodothyronine metabolism satisfies the varying requirements of TH depending on tissue, cell type, and time. Through hormone activation and inactivation by site-specific monodeiodination, it provides the proper amount of active hormone at its site of action. Deiodinases are selenoproteins requiring a selenocysteine at the active center of enzymatic activity. Type 1 and 2 deiodinases (D1 and D2) catalyze the activation of TH by 5′ deiodination that converts T_4 to T_3. Type 3 deiodinase (D3) inactivates T_4 and T_3 by 5-deiodination, leading to the generation of rT_3 and T_2, respectively.

SBP2 gene, located in chromosome 9, is expressed at low level in all tissues and at high level in testis.[93] SBP2 binds to the selenocysteine insertion sequence (SECIS) located in the 3′ noncoding sequence of selenoprotein mRNAs. SBP2 then recruits several proteins, among which an elongation factor specific for selenocysteine (eEF^{Sec}), which in turn brings the tRNA^{Sec} to recode a stop codon (UGA) for the incorporation of selenocysteine in the protein nascent chain.

The incidence of inherited defects of metabolism is unknown The first three mutations in the *SBP2* gene were reported in 2005.[12] Several more families were found subsequently (Table 9.5).[12,94–99] The inheritance is recessive, homozygous, or compound heterozygous. The ethnic origins of the affected individuals are Bedouin from Saudi Arabia, Black African, Irish, Brazilian, English (two families), Turkish, Japanese, and Argentinian.

Clinical Features

Most patients found to have SBP2 deficiency came to medical attention during childhood because of short stature and delayed bone age, which prompted thyroid function testing, leading to identification of abnormalities. The extent of manifestations varies among the known cases. Mild cases present with growth delay, while a complex multisystem phenotype was recognized in more severe cases, with delayed motor and intellectual milestones, and progressive congenital myopathy. Increased visceral fat mass associated with paradoxically increased insulin sensitivity was characterized in some patients.[94,97] Several distinct manifestations, including primary infertility with azoospermia, skin photosensitivity, and severe Raynaud disease were seen in the only adult patient known to have SBP2 deficiency. Sensorineural hearing loss has been documented in four cases.[100]

Laboratory Findings and Differential Diagnosis

Typical laboratory findings are high rT_3 and T_4, low T_3, and normal or slightly elevated serum TSH. No other hormonal abnormalities have been detected and serum IGF1 concentrations are normal despite delayed growth. Affected individuals require relatively larger doses of L-T_4 but not L-T_3 to suppress their serum TSH concentration.[12] For comparison to the other two syndromes of reduced sensitivity to TH see Table 9.4. None of the subjects had an enlarged thyroid gland confirmed by ultrasound examination. Bone age delay detected by X-ray was documented in all subjects.[100]

In agreement with the role of SBP2 in the synthesis of selenoproteins, its deficiency affects multiple selenoproteins. Baseline and cAMP-stimulated D2 enzymatic activity was reduced in cultured skin fibroblasts from affected individuals, compared to those from unaffected family members. However, baseline and cAMP-stimulated D2 mRNA levels were similar between the two groups, supporting a posttranscriptional defect in the synthesis of the active D2 enzyme in the affected.[12] Selenoprotein P (SePP) and glutathione peroxidase (Gpx) activities in serum and/or fibroblasts were also decreased in affected individuals, as were the circulating levels of selenium (Se) reflecting a global deficiency in selenoprotein synthesis.[12]

In patients manifesting a more severe phenotype, defects of other selenoproteins than D2, SePP, and Gpx were detected. Selenoprotein N1 (SEPN1) deficiency results in clinical symptoms resembling SEPN1-related myopathies characterized by axial and adductor muscle hypotrophy as well as spinal rigidity.[94,97] Lack of testis-enriched

TABLE 9.5 Mutations in the *SBP2* Gene

Family	*SBP2* gene	Protein	Comments on putative defect	No. of affected	Defect	References
1	c.1619 G > A	R540Q	Hypomorphic allele	3	Homozygous	12
2	c.1312 A > T	K438X	Missing C terminus	1	Compound heterozygous	12
	IVS8ds + 29 G > A	fs	Abnormal splicing			
3	c.382 C > T	R128X	Smaller isoforms*	1	Homozygous	95
4	c.358 C > T	R120X	Smaller isoforms*	1	Compound heterozygous	94
	c.2308 C > T	R770X	Disrupted C-terminus			
5	c.668delT	F223 fs 255X	Truncation and smaller isoforms*	1	Compound heterozygous	97
	Intron 6-155 delC	fs	Abnormal splicing, missing C-terminus			
6	c.2071 T > C	C691R	Increased proteasomal degradation	1	Compound heterozygous	97
	Intronic SNP	fs	Transcripts lacking exons 2–4, or 3–4			
7	c.1529_1541dup CCAGCGCCCCACT	M515 fs 563X	Missing C terminus	1	Compound heterozygous	96
	c.235 C > T	Q79X	Smaller isoforms*			
8	c.2344 C > T	Q782X	Missing C terminus	1	Compound heterozygous	98
	c.2045-2048 delAACA	K682 fs 683X	Missing C terminus			
	c.2037 G > T	E679D	Disrupted SECIS binding			

* *Generated from downstream ATGs; fs, frame shift.*

selenoproteins, led to failure of the latter stages of spermatogenesis and azoospermia.[97] Cutaneous deficits of antioxidant selenoenzymes caused increased cellular reactive oxygen species, and the decreased selenoproteins in peripheral blood cells resulted in immune deficits.[97]

Genetic Pathophysiology

Although the defect causes generalized deficiency of selenoproteins, the phenotype is mild. This is most likely because the deficiency is not complete and due to the hierarchical preservation of selenoproteins.[101,102] The thyroid phenotype of subjects with SBP2 deficiency is consistent with a defect in TH metabolism due to the deficiency in deiodinases. However, targeted disruption in mice of D1, D2, or D3 alone, or both D1 and D2, only partially replicate the thyroid abnormalities observed in SBP2 deficient humans.[100] Thus, a putative, partial and uneven involvement of the three deiodinases might explain the noted difference in the thyroid tests abnormalities. Of note, delayed growth and bone maturation, and impaired auditory functions are also associated with deficiencies of the deiodinases.

Table 9.5 illustrates the consequences of the *SBP2* mutations on SBP2 protein function. Of note smaller functional isoforms can provide partial SBP2 function in mutations that result in early terminations.

Treatment

No specific treatment is available. Se supplementation in both organic and inorganic forms was administrated in three affected siblings,[103] but no effect was observed on the serum glutathione peroxidase activity and SePP levels, and the abnormalities of serum iodothyronine levels were not improved. L-T$_3$ administration was subsequently attempted in four patients and provided some beneficial effects[94–97] with improved linear growth and/or bone maturation. Other clinical features of SBP2 defects are treated symptomatically.

References

1. Friesema EC, Jansen J, Milici C, Visser TJ. Thyroid hormone transporters. *Vitam Horm* 2005;**70**:137–67.
2. Bianco AC, Salvatore D, Gereben B, Berry MJ, Larsen PR. Biochemistry, cellular and molecular biology, and physiological roles of the iodothyronine selenodeiodinases. *Endocr Rev* 2002;**23**(1):38–89.
3. Bassett JH, Harvey CB, Williams GR. Mechanisms of thyroid hormone receptor-specific nuclear and extra nuclear actions. *Mol Cell Endocrinol* 2003;**213**(1):1–11.
4. Yen PM, Ando S, Feng X, Liu Y, Maruvada P, Xia X. Thyroid hormone action at the cellular, genomic and target gene levels. *Mol Cell Endocrinol* 2006;**246**(1-2):121–7.
5. Zhang J, Lazar MA. The mechanism of action of thyroid hormones. *Annu Rev Physiol* 2000;**62**:439–66.
6. Refetoff S, DeWind LT, DeGroot LJ. Familial syndrome combining deaf-mutism, stippled epiphyses, goiter, and abnormally high

PBI: possible target organ refractoriness to thyroid hormone. *J Clin Endocrinol Metab* 1967;**27**:279–94.

7. Sakurai A, Takeda K, Ain K, et al. Generalized resistance to thyroid hormone associated with a mutation in the ligand-binding domain of the human thyroid hormone receptor b. *Proc Natl Acad Sci USA* 1989;**86**:8977–81.

8. Usala SJ, Tennyson GE, Bale AE, et al. A base mutation of the c-erbAb thyroid hormone receptor in a kindred with generalized thyroid hormone resistance. Molecular heterogeneity in two other kindreds. *J Clin Invest* 1990;**85**:93–100.

9. Refetoff S, Weiss RE, Usala SJ. The syndromes of resistance to thyroid hormone. *Endocr Rev* 1993;**14**:348–99.

10. Dumitrescu AM, Liao XH, Best TB, Brockmann K, Refetoff S. A novel syndrome combining thyroid and neurological abnormalities is associated with mutations in a monocarboxylate transporter gene. *Am J Hum Genet* 2004;**74**(1):168–75.

11. Friesema EC, Grueters A, Biebermann H, et al. Association between mutations in a thyroid hormone transporter and severe X-linked psychomotor retardation. *Lancet* 2004;**364**(9443):1435–7.

12. Dumitrescu AM, Liao X-H, Abdullah SYM, et al. Mutations in SECISBP2 result in abnormal thyroid hormone metabolism. *Nat Genet* 2005;**37**(11):1247–52.

13. Refetoff S, Bassett JH, Beck-Peccoz P, et al. Classification and proposed nomenclature for inherited defects of thyroid hormone action, cell transport, and metabolism. *Thyroid* 2014;**24**(3):407–9 *Eur Thyroid J* 2014;**3**:7–9. *J Clin Endocrinol Metab* 2014;**99**:768–770.

14. Fu J, Refetoff S, Dumitrescu AM. Inherited defects of thyroid hormone-cell-membrane transport: review of recent findings. *Curr Opin Endocrinol Diabetes Obes* 2013;**20**(5):434–40.

15. Schoenmakers N, Moran C, Peeters RP, Visser T, Gurnell M, Chatterjee K. Resistance to thyroid hormone mediated by defective thyroid hormone receptor alpha. *Biochim Biophys Acta* 2013;**1830**(7):4004–8.

16. Koenig RJ. Thyroid hormone receptor coactivators and corepressors. *Thyroid* 1998;**8**:703–13.

17. Feng W, Ribeiro RCJ, Wagner RL, et al. Hormone-dependent coactivator binding to a hydrophobic cleft on nuclear receptors. *Science* 1998;**280**:1747–9.

18. Chen JD, Evans RM. A transcriptional co-repressor that interacts with nuclear hormone receptors. *Nature* 1995;**377**:454–7.

19. Hörlein AJ, Näär AM, Heinzel T, et al. Ligand-independent repression by the thyroid hormone receptor mediated by a nuclear receptor co-repressor. *Nature* 1995;**377**:397–404.

20. Busch K, Martin B, Baniahmad A, Renkawitz R, Muller M. At least three subdomains of v-erbA are involved in its silencing function. *Mol Endocrinol* 1997;**11**:379–89.

21. Au-Fliegner M, Helmer E, Casanova J, Raaka BM, Samuels HT. The conserved ninth C-terminal heptad in thyroid hormone and retinoic acid receptors mediates diverse responses by affecting heterodimer but not homodimer formation. *Mol Cel Endocrinol* 1993;**13**(9):5725–37.

22. Forman BM, Samuels HH. Interactions among a subfamily of nuclear hormone receptors: the regulatory zipper model. *Mol Endocrinol* 1990;**4**:1293–301.

23. Kurokawa R, Yu VC, Naar A, et al. Differential orientations of the DNA-binding domain and carboxy-terminal dimerization interface regulate binding site selection by nuclear receptor heterodimers. *Genes Develop* 1993;**7**:1423–35.

24. Beck-Peccoz P, Chatterjee VKK, Chin WW, et al. Nomenclature of thyroid hormone receptor ß-gene mutations in resistance to thyroid hormone: consensus statement from the first workshop on thyroid hormone resistance, July 10–11, 1993, Cambridge, United Kingdom. *J Clin Endocrinol Metab* 1994;**78**:990–3.

25. Adams M, Matthews C, Collingwood TN, Tone Y, Beck-Peccoz P, Chatterjee KK. Genetic analysis of 29 kindreds with generalized and pituitary resistance to thyroid hormone: identification of thirteen novel mutations in the thyroid hormone receptor ß gene. *J Clin Invest* 1994;**94**:506–15.

26. Baniahmad A, Leng X, Burris TP, Tsai SY, Tsai M-J, O'Malley BW. The tau 4 activation domain of the thyroid hormone receptor is required for release of a putative corepressor(s) necessary for transcriptional silencing. *Mol Cell Biol* 1995;**15**:76–86.

27. Hamy F, Helbeque NJ, Henichart P. Comparison between synthetic nuclear localisation signal peptides from the steroid thyroid hormone receptor superfamily. *Biochem Byophys Res Commun* 1992;**183**:289–93.

28. Wurtz J-M, Bourguet W, Renaud J-P, Vivat V, Chambon P, Moras D. A canonical strusture for the ligand-binding domain of the nuclear receptors. *Nature Struct Biol* 1996;**3**:87–94.

29. Beck-Peccoz P, Chatterjee VKK. The variable clinical phenotype in thyroid hormone resistance syndrome. *Thyroid* 1994;**4**:225–32.

30. LaFranchi SH, Snyder DB, Sesser DE, et al. Follow-up of newborns with elevated screening T4 concentrations. *J Pediatr* 2003;**143**(3):296–301.

31. Weiss RE, Weinberg M, Refetoff S. Identical mutations in unrelated families with generalized resistance to thyroid hormone occur in cytosine-guanine-rich areas of the thyroid hormone receptor beta gene: analysis of 15 families. *J Clin Invest* 1993;**91**:2408–15.

32. Brucker-Davis F, Skarulis MC, Grace MB, et al. Genetic and clinical features of 42 kindreds with resistance to thyroid hormone. The National Institutes of Health Prospective Study. *Ann Intern Med* 1995;**123**:573–83.

33. Hauser P, Zametkin AJ, Martinez P, et al. Attention deficit-hyperactivity disorder in people with generalized resistance to thyroid hormone. *N Engl J Med* 1993;**328**:997–1001.

34. Takeda K, Sakurai A, DeGroot LJ, Refetoff S. Recessive inheritance of thyroid hormone resistance caused by complete deletion of the protein-coding region of the thyroid hormone receptor-ß gene. *J Clin Endocrinol Metab* 1992;**74**(1):49–55.

35. Abel ED, Boers ME, Pazos-Moura C, et al. Divergent roles for thyroid hormone receptor beta isoforms in the endocrine axis and auditory system. *J Clin Invest* 1999;**104**:291–300.

36. Beck-Peccoz P, Brucker-Davis F, Persani L, Smallridge RC, Weintraub BD. Thyrotropin-secreting pituitary tumors. *Endocr Rev* 1996;**17**:610–838.

37. Collingwood TN, Wagner R, Matthews CH, et al. A role for helix 3 of the TRß ligand-binding domain in coactivator recruitment identified by characterization of a third cluster of mutations in resistance to thyroid hormone. *EMBO J* 1998;**17**:4760–70 Aug 17.

38. Weiss RE, Marcocci C, Bruno-Bossio G, Refetoff S. Multiple genetic factors in the heterogeneity of thyroid hormone resistance. *J Clin Endocrinol Metab* 1993;**76**:257–9.

39. Hayashi Y, Sunthornthepvarakul T, Refetoff S. Mutations of CpG dinucleotides located in the triiodothyronine (T_3)-binding domain of the thyroid hormone receptor (TR) ß gene that appears to be devoid of natural mutations may not be detected because they are unlikely to produce the clinical phenotype of resistance to thyroid hormone. *J Clin Invest* 1994;**94**:607–15.

40. Takeda K, Weiss RE, Refetoff S. Rapid localization of mutations in the thyroid hormone receptor-ß gene by denaturing gradient gel electrophoresis in 18 families with thyroid hormone resistance. *J Clin Endocrinol Metab* 1992;**74**:712–9.

41. Liu Y, Takeshita A, Misiti S, Chin WW, Yen PM. Lack of coactivator interaction can be a mechanism for dominant negative activity by mutant thyroid hormone receptors. *Endocrinology* 1998;**139**:4197–204.

42. Safer JD, Cohen RN, Hollenberg AN, Wondisford FE. Defective release of corepressor by hinge mutants of the thyroid hormone receptor found in patients with resistance to thyroid hormone. *J Biol Chem* 1998;**273**:30175–82.

43. Yoh SM, Chatterjee VKK, Privalsky ML. Thyroid hormone resistance syndrome manifests as an aberrant interaction between

mutant T_3 receptor and transcriptional corepressor. *Mol Endocrinol* 1997;**11**:470–80.

44. Wu SY, Cohen RN, Simsek E, et al. A novel thyroid hormone receptor-beta mutation that fails to bind nuclear receptor corepressor in a patient as an apparent cause of severe, predominantly pituitary resistance to thyroid hormone. *J Clin Endocrinol Metab* 2006;**91**(5):1887–95.

45. Lado-Abeal J, Dumitrescu AM, Liao XH, et al. A *de novo* mutation in an already mutant nucleotide of the thyroid hormone receptor beta gene perpetuates resistance to thyroid hormone. *J Clin Endocrinol Metab* 2005;**90**(3):1760–7.

46. Flamant F, Samarut J. Thyroid hormone receptors: lessons from knockout and knock-in mutant mice. *Trends Endocrinol Metab* 2003;**14**(2):85–90.

47. Weiss RE, Murata Y, Cua K, Hayashi Y, Seo H, Refetoff S. Thyroid hormone action on liver, heart and energy expenditure in thyroid hormone receptor ß deficient mice. (Erratum, 141:4767, 2000). *Endocrinology* 1998;**139**:4945–52.

48. Dillmann WH. Biochemical basis of thyroid hormone action in the heart. *Am J Med* 1990;**88**(6):626–30.

49. Izumo S, Mahdavi V. Thyroid hormone receptor alpha isoforms generated by alternative splicing differentially activate myosin HC gene transcription. *Nature* 1988;**334**(6182):539–42.

50. Anselmo J, Cao D, Karrison T, Weiss RE, Refetoff S. Fetal loss associated with excess thyroid hormone exposure. *JAMA* 2004;**292**(6):691–5.

51. Vaidya B, Campbell V, Tripp JH, Spyer G, Hattersley AT, Ellard S. Premature birth and low birth weight associated with nonautoimmune hyperthyroidism due to an activating thyrotropin receptor gene mutation. *Clin Endocrinol (Oxf)* 2004;**60**(6):711–8.

52. Weiss RE, Xu J, Ning G, Pohlenz J, O'Malley BW, Refetoff S. Mice deficient in the steroid receptor coactivator-1 (SRC-1) are resistant to thyroid hormone. *EMBO J* 1999;**18**(7):1900–4.

53. Brown NS, Smart A, Sharma V, et al. Thyroid hormone resistance and increased metabolic rate in the RXR-g-deficient mouse. *J Clin Invest* 2000;**106**:73–9.

54. Weiss RE, Hayashi Y, Nagaya T, et al. Dominant inheritance of resistance to thyroid hormone not linked to defects in the thyroid hormone receptors a or ß genes may be due to a defective co-factor. *J Clin Endocrinol Metab* 1996;**81**(12):4196–203.

55. Sadow P, Reutrakul S, Weiss RE, Refetoff S. Resistnace to thyroid hormone in the absence of mutations in the thyroid hormone receptor genes. *Curr Opin Endocrinol Diabetes* 2000;**7**:253–9.

56. Tinnikov A, Nordstrom K, Thoren P, et al. Retardation of post-natal development caused by a negatively acting thyroid hormone receptor alpha1. *Embo J* 2002;**21**(19):5079–87.

57. Liu YY, Schultz JJ, Brent GA. A thyroid hormone receptor alpha gene mutation (P398H) Is associated with visceral adiposity and impaired catecholamine-stimulated lipolysis in mice. *J Biol Chem* 2003;**278**:38913–20.

58. Bochukova E, Schoenmakers N, Agostini M, et al. A mutation in the thyroid hormone receptor alpha gene. *N Engl J Med* 2012;**366**(3):243–9.

59. van Mullem A, van Heerebeek R, Chrysis D, et al. Clinical phenotype and mutant TRalpha1. *N Engl J Med* 2012;**366**(15):1451–3.

60. Moran C, Schoenmakers N, Agostini M, et al. An adult female with resistance to thyroid hormone mediated by defective thyroid hormone receptor alpha. *J Clin Endocrinol Metab* 2013;**98**(11):4254–61.

61. Espiard S, Savagner F, Flamant F, et al. A novel mutation in THRA gene associated with an atypical phenotype of resistance to thyroid hormone. *J Clin Endocrinol Metab* 2015; jc20151120.

62. Moran C, Agostini M, Visser WE, et al. Resistance to thyroid hormone caused by a mutation in thyroid hormone receptor (TR)alpha1 and TRalpha2: clinical, biochemical, and genetic analyses of three related patients. *Lancet Diabetes Endocrinol* 2014;.

63. Tylki-Szymanska A, Acuna-Hidalgo R, Krajewska-Walasek M, et al. Thyroid hormone resistance syndrome due to mutations in the thyroid hormone receptor alpha gene (THRA). *J Med Genet* 2015;**52**(5):312–6.

64. Barca-Mayo O, Liao XH, Alonso M, et al. Thyroid hormone receptor alpha and regulation of type 3 deiodinase. *Mol Endocrinol* 2011;**25**(4):575–83.

65. Groenhout EG, Dorin RI. Severe generalized thyroid hormone resistance due to a sporadic mutation in the c-erbAß gene producing a truncated T3 receptor protein. Annual Meeting of the Endocrine Society. San Antonio, TX; 1992.

66. Hennemann G, Docter R, Friesema EC, de Jong M, Krenning EP, Visser TJ. Plasma membrane transport of thyroid hormones and its role in thyroid hormone metabolism and bioavailability. *Endocr Rev* 2001;**22**(4):451–76.

67. Allan W, Herndon CN, Dudley FC. Some examples of the inheritance of mental deficiency: apparently sex-linked idiocy and microcephaly. *Am J Ment Defic* 1944;**48**:325–34.

68. Frints SG, Lenzner S, Bauters M, et al. MCT8 mutation analysis and identification of the first female with Allan-Herndon-Dudley syndrome due to loss of MCT8 expression. *Eur J Hum Genet* 2008;**16**(9):1029–37.

69. Kakinuma H, Itoh M, Takahashi H. A novel mutation in the monocarboxylate transporter 8 gene in a boy with putamen lesions and low free T4 levels in cerebrospinal fluid. *J Pediatr* 2005;**147**(4):552–4.

70. Papadimitriou A, Dumitrescu AM, Papavasiliou A, Fretzayas A, Nicolaidou P, Refetoff S. A novel monocarboxylate transporter 8 gene mutation as a cause of severe neonatal hypotonia and developmental delay. *Pediatrics* 2008;**121**(1):e199–202.

71. Visser WE, Friesema EC, Jansen J, Visser TJ. Thyroid hormone transport in and out of cells. *Trends Endocrinol Metab* 2008;**19**(2):50–6.

72. Brockmann K, Dumitrescu AM, Best TT, Hanefeld F, Refetoff S. X-linked paroxysmal dyskinesia and severe global retardation caused by defective MCT8 gene. *J Neurol* 2005;**252**(6):663–6.

73. Dumitrescu AM, Liao XH, Weiss RE, Millen K, Refetoff S. Tissue specific thyroid hormone deprivation and excess in Mct8 deficient mice. *Endocrinology* 2006;(**147**):4036–43.

74. Trajkovic M, Visser TJ, Mittag J, et al. Abnormal thyroid hormone metabolism in mice lacking the monocarboxylate transporter 8. *J Clin Invest* 2007;**117**(3):627–35.

75. Di Cosmo C, Liao XH, Ye H, et al. Mct8-deficient mice have increased energy xpenditure and reduced fat mass that is abrogated by normalization of serum T3 levels. *Endocrinology* 2013;**154**(12):4885–95.

76. Herzovich V, Vaiani E, Marino R, et al. Unexpected peripheral markers of thyroid function in a patient with a novel mutation of the MCT8 thyroid hormone transporter gene. *Horm Res* 2007;**67**(1):1–6.

77. Holden KR, Zuniga OF, May MM, et al. X-linked MCT8 gene mutations: characterization of the pediatric neurologic phenotype. *J Child Neurol* 2005;**20**(10):852–7.

78. Namba N, Etani Y, Kitaoka T, et al. Clinical phenotype and endocrinological investigations in a patient with a mutation in the MCT8 thyroid hormone transporter. *Eur J Pediatr* 2008;**167**(7):785–91.

79. Sijens PE, Rodiger LA, Meiners LC, Lunsing RJ. 1H magnetic resonance spectroscopy in monocarboxylate transporter 8 gene deficiency. *J Clin Endocrinol Metab* 2008;**93**(5):1854–9.

80. Asteria C, Rajanayagam O, Collingwood TN, et al. Prenatal diagnosis of thyroid hormone resistance. *J Clin Endocrinol Metab* 1999;**84**:405–10.

81. Friesema EC, Ganguly S, Abdalla A, Manning Fox JE, Halestrap AP, Visser TJ. Identification of monocarboxylate transporter 8

as a specific thyroid hormone transporter. *J Biol Chem* 2003;**278**: 40128–35.

82. Jansen J, Friesema EC, Kester MH, Schwartz CE, Visser TJ. Genotype-phenotype relationship in patients with mutations in thyroid hormone transporter MCT8. *Endocrinology* 2008;**149**(5):2184–90.

83. Friesema EC, Kuiper GG, Jansen J, Visser TJ, Kester MH. Thyroid hormone transport by the human monocarboxylate transporter 8 and its rate-limiting role in intracellular metabolism. *Mol Endocrinol* 2006;**20**(11):2761–72.

84. Bernal J. Role of monocarboxylate anion transporter 8 (MCT8) in thyroid hormone transport: answers from mice. *Endocrinology* 2006;**147**(9):4034–5.

85. Di Cosmo C, Liao XH, Dumitrescu AM, Philp NJ, Weiss RE, Refetoff S. Mice deficient in MCT8 reveal a mechanism regulating thyroid hormone secretion. *J Clin Invest* 2010;**120**(9):3377–88.

86. Trajkovic-Arsic M, Visser TJ, Darras VM, et al. Consequences of monocarboxylate transporter 8 deficiency for renal transport and metabolism of thyroid hormones in mice. *Endocrinology* 2010;**151**(2):802–9.

87. Wemeau JL, Pigeyre M, Proust-Lemoine E, et al. Beneficial effects of propylthiouracil plus L-thyroxine treatment in a patient with a mutation in MCT8. *J Clin Endocrinol Metab* 2008;**93**(6):2084–8.

88. Di Cosmo C, Liao XH, Dumitrescu AM, Weiss RE, Refetoff S. A thyroid hormone analogue with reduced dependence on the monocarboxylate transporter 8 (MCT8) for tissue transport. *Endocrinology* 2009;**150**(9):4450–8.

89. Verge CF, Konrad D, Cohen M, et al. Diiodothyropropionic acid (DITPA) in the treatment of MCT8 deficiency. *J Clin Endocrinol Metab* 2012;**97**(12):4515–23.

90. Lopez-Espindola D, Morales-Bastos C, Grijota-Martinez C, et al. Mutations of the thyroid hormone transporter MCT8 cause prenatal brain damage and persistent hypomyelination. *J Clin Endocrinol Metab* 2014;**99**(12):E2799–804.

91. Koenig RJ. Regulation of type 1 iodothyronine deiodinase in health and disease. *Thyroid* 2005;**15**(8):835–40.

92. Bianco AC, Kim BW. Deiodinases: implications of the local control of thyroid hormone action. *J Clin Invest* 2006;**116**(10):2571–9.

93. Lescure A, Allmang C, Yamada K, Carbon P, Krol A. cDNA cloning, expression pattern and RNA binding analysis of human selenocysteine insertion sequence (SECIS) binding protein 2. *Gene* 2002;**291**(1–2):279–85.

94. Azevedo MF, Barra GB, Naves LA, et al. Selenoprotein-related disease in a young girl caused by nonsense mutations in the SBP2 gene. *J Clin Endocrinol Metab* 2010;**95**(8):4066–71.

95. Di Cosmo C, McLellan N, Liao XH, et al. Clinical and molecular characterization of a novel selenocysteine insertion sequence-binding protein 2 (SBP2) gene mutation (R128X). *J Clin Endocrinol Metab* 2009;**94**(10):4003–9.

96. Hamajima T, Mushimoto Y, Kobayashi H, Saito Y, Onigata K. Novel compound heterozygous mutations in the SBP2 gene: characteristic clinical manifestations and the implications of GH and triiodothyronine in longitudinal bone growth and maturation. *Eur J Endocrinol* 2012;**166**(4):757–64.

97. Schoenmakers E, Agostini M, Mitchell C, et al. Mutations in the selenocysteine insertion sequence-binding protein 2 gene lead to a multisystem selenoprotein deficiency disorder in humans. *J Clin Invest* 2010;**120**(12):4220–35.

98. Sillers L, Gönç N, Kandemir N, et al. Selenoprotein deficiency syndrome caused by novel compound heterozygous mutations in the SBP2 gene. Ninety-fourth Annual Meeting of the Endocrine Society. Houston, TX; 2012.

99. Fu J, Liao X-H, Menucci MB, Dumitrescu AM, Weiss RE. Thyroid hormone metabolism defect caused by novel compound heterozygous mutations in the SBP2 Gene. Ninety-sixth Annual Meeting of the Endocrine Society. Chicago, IL; 2014.

100. Fu J, Dumitrescu AM. Inherited defects in thyroid hormone cell-membrane transport and metabolism. *Best Pract Res Clin Endocrinol Metab* 2014;**28**(2):189–201.

101. Copeland PR. Regulation of gene expression by stop codon recoding: selenocysteine. *Gene* 2003;**312**:17–25.

102. Köhrle J. Selenium and the control of thyroid hormone metabolism. *Thyroid* 2005;**15**(8):841–53.

103. Schomburg L, Dumitrescu AM, Liao XH, et al. Selenium supplementation fails to correct the selenoprotein synthesis defect in subjects with SBP2 gene mutations. *Thyroid* 2009;**19**(3):277–81.

10

Molecular Diagnosis of Thyroid Cancer

Furio Pacini, Silvia Cantara

Department of Medical, Surgical and Neurological Sciences, University of Siena, Siena, Italy

INTRODUCTION

Thanks to great technological advances, our understanding of the genomics of thyroid cancer has dramatically improved over the last few years.[1,2] A number of mutations have been demonstrated in differentiated thyroid cancers, and some of them are considered driver mutations for specific cancer histotypes, often configuring genotype-phenotype relationships.

Genetic events include somatic missense mutations or rearrangements of oncogenes in the MAPK and phosphatidylinositol-3 kinase (PI3K)/AKT signaling pathways.

ONCOGENE REARRANGEMENTS

The most frequent oncogene rearrangement is RET/PTC. More rarely, TRK rearrangements, ALK rearrangements, and PAX8/PPARγ rearrangements are found.

RET/PTC

The Ret proto-oncogene is located at chromosome 10 (10q11.2) and encodes a tyrosine kinase cell surface receptor for members of the glial cell line-derived neurotrophic factor family.[3] All these factors activate RET via different glycosyl phosphatidylinositol-linked GFRα receptors.[4] Ret gene alternative splicing results in the production of three different isoforms named RET51, RET43, and RET9, which contain 51, 43, and 9 amino acids in their C-terminal tail, respectively.[5] The RET51 and RET9 are the most common isoforms in which RET is found *in vivo*.[6] Ret protein is composed of an N-terminal extracellular domain with four cadherin-like repeats, a cysteine-rich region, nine N-glycosilation sites, a hydrophobic transmembrane domain, and a cytoplasmic tyrosine kinase domain.[7]

Ret activates the Raf/Mek/ERK1/2 cascade, which regulates cell proliferation, and the PI3K/Akt signal transduction pathway, which regulates cellular survival.[8,9] Normally Ret is expressed at very low levels in thyroid follicular cells,[10] but the chimeric forms of Ret are constitutively active due to the fusion of the C-terminal kinase domain of Ret with the promoter and N-terminal domains of unrelated genes. Several rearrangements have been described: inversion inv[10] (q11.2;q21) with H4 gene (D10S170 locus) generates the RET/CCDC6 (PTC1) oncogene; inversion inv[10] (q11.2;q11.2) with ELE1 gene generates the RET/NCOA4 (PTC3) oncogene; translocation t(10;14)(q11;q32) with GOLGA5 generates the RET/GOLGA5 (PTC5) oncogene; translocation t(8;10)(p21.3;q11.2) with PCM1 generates the PCM1/RET fusion; translocation t(6;10)(p21.3;q11.2) with RFP generates the Delta RFP/RET oncogene; translocation t(1;10)(p13;q11) with TRIM33 generates the TRIM33/RET (PTC7) oncogene; and translocation t(7;10)(q32;q11) with TRIM24/TIF1 generates the TRIM24/RET (PTC6) oncogene.[11] The most common rearrangements are RET/PTC1 and RET/PTC3, accounting for 10–20% of all papillary thyroid carcinomas.[12]

TRK

Neurotrophic tyrosine kinase receptor type 1 (NTRK1) gene, also named TRK, is located on the q arm of chromosome 1 (1q21-22) and encodes for a member of the neurotrophic tyrosine kinase receptor (NTKR) family.[13] This kinase is a membrane-bound receptor that, upon neurotrophin binding, phosphorylates itself and members of the MAPK pathway leading to cell differentiation.[14] Genes involved in TRK rearrangements are at least three: TPM3 located at 1q22-23, TGF located at 3q11-12, and TPR located at 1q25. Namely, TRK is a rearranged protein between NTRK1 and TPM3; TRK1 and TRK2 are chimeric proteins between NTRK1 and TPR; and TRK3

Genetic Diagnosis of Endocrine Disorders. http://dx.doi.org/10.1016/B978-0-12-800892-8.00010-5

is the rearranged product of NTKR1 and TGF. TRK rearrangements are found in 1–5% of papillary thyroid carcinomas.[15]

ALK

The ALK rearrangement results from the fusion of the striatin, calmodulin-binding protein (STRN, 2p22.2) gene and the anaplastic lymphoma kinase (ALK, 2p23.1) gene. The first one functions as a signaling protein and appears to be involved in dendrite development, tight junction assembly, and regulation of cell proliferation. The ALK gene is implicated in phosphorylation of several substrates, including the MAP kinases MAPK1/ERK2 and MAPK3/ERK1. ALK also acts as a receptor for the ligands pleiotrophin (PTN), a secreted growth factor, and midkine (MDK), a PTN-related factor, thus participating in PTN and MDK signal transduction.

The ALK rearrangement is present in about 9% of poorly differentiated thyroid cancers, 4% of anaplastic thyroid cancers, and 1% of papillary thyroid cancers.[16]

PAX8/PPARγ

The PAX8/PPARγ rearrangement consists in a fusion of the coding region of the paired box 8 (Pax-8) and of the coding region of the PPARγ (peroxisome proliferator-activated receptor gamma).[17] The PAX8 gene is located on chromosome 2 and is a member of the paired box (PAX) family of transcription factors.[18] The gene encodes for a nuclear protein, which contains a paired box domain, an octapeptide, and a paired-type homeodomain. Five different isoforms of the protein have been described due to alternative splicing: isoform a (450 a.a.); isoform b (387 a.a., lack exon 8); isoform c (398 a.a., lack exons 7, 8); isoform d (321 a.a., lack exon 8); and isoform e (287 a.a., lack exons 8, 9, 10).[19] Pax8 is expressed in embryonal human tissues, in particular in the developing thyroid gland and kidney and only in these tissues does it remain expressed in adults.[20] The PPARγ gene is located on chromosome 3 at position p25. The encoded proteins are a group of nuclear receptor proteins that function as transcription factors regulating expression of genes involved in cellular differentiation, development, and metabolism.[21] The PAX8/PPARγ rearrangement is found in 30–40% of follicular carcinomas and at lower frequency in the follicular variant of papillary thyroid carcinoma.[22] The molecular event found in thyroid cancer is represented by an interchromosomal translocation t(2;3)(q13;p25), which fuses promoter elements of the PAX8 gene with most of the coding sequence of the PPARγ gene. PAX8 can break at the junction between exon 7 and exon 8, exon 8 and exon 9, or exon 7 and 9 (losing exon 8) and fuse with exon 1 of the PPARγ gene.[23]

GENE MUTATIONS

Somatic missense point mutations involve mainly the RAS gene family (H-, K-, and N-RAS), the BRAF gene, and the human telomerase reverse transcriptase (hTERT) promoter.

RAS Isoforms

RAS genes (H-RAS, N-RAS, and K-RAS) encode for four different proteins (one H-Ras, one N-Ras, and two K-Ras) approximately of 21 kDa, which are involved in cellular differentiation, division, and cell death.[24] All proteins have at their C-termini covalently attached lipid tails, composed of farnesyl, palmitol, or geranylgeranyl groups, which enable the Ras proteins to become anchored to the cytoplasmic membrane. When Ras is inactive, it binds a GDP molecule. The Ras signaling cycle starts upon an upstream stimulatory signal. Ras moves from inactive state to an active GTP-binding condition due to guanine nucleotide exchange factor (GEF) stimulation.[25] Then, using its own intrinsic GTPase function, Ras cleaves GTP coming back to its quiescent configuration. Point mutations of Ras block this cycle by inactivating the intrinsic GTPase activity of Ras and blocking the protein in its active state. Studies carried out on the amino acidic sequence of Ras showed that the 12th, 13th, and 61st residues are those mainly affected by point mutations. These amino acidic residues are located around the catalytic GTPase domain of Ras. Consequently, almost all substitutions of these amino acids compromise GTPase function. Activated Ras acts, at the same time, downstream through the PI3K pathway, Raf-MEK-ERK1/2 pathway, and Ral-GEF pathway inducing many of the phenotypic changes involved in neoplastic transformation. RAS mutations are probably an early event in thyroid tumorigenesis and are not tumor specific. They can be found in follicular carcinomas, in the follicular variant of papillary thyroid carcinomas,[26–28] in poorly differentiated thyroid cancers,[29] but also in follicular adenomas.[30]

BRAF Oncogene

The BRAF (v-raf murine sarcoma viral oncogene homolog B1) gene is located on chromosome 7 and encodes for an 18-exon cytoplasmic protein, which is recruited to the membrane upon growth factor stimulation. It is a member of the Raf kinase family of serine/threonine-specific protein kinases.[31] BRAF interacts with other proteins such as HRAS, AKT, CRAF, and it has been demonstrated to play a role in cell growth, differentiation, and survival through ERK1/2 activation.[32,33] More than 30 mutations of the BRAF gene associated with human cancers have been identified, and these mutations are mostly located within the part of the gene encoding the kinase domain, clustered to two regions containing the glycine-rich P-loop of the N-lobe and the activation segment.[34] Among these mutations, the V600E (T1799A), resulting in a T-to-A transversion at position 1799 with a valine-to-glutamate substitution, is the most common BRAF mutation in papillary thyroid cancers (about 45% of the cases),[35,36] and it is less frequently found in poorly differentiated or anaplastic thyroid cancers.[37] BRAF mutations are identified in a prevalence of approximately 60% of classic PTC, 80% of tall cell variant PTC, and only 10% of follicular variant PTC. A multicenter study demonstrated a significant association of BRAF V600E with poor clinic-pathological outcomes of PTC[38] not confirmed in other studies.[39] BRAF mutations have been associated with reduced expression of the sodium iodine symporter,[40] resulting in total or partial loss of iodine uptake.

The K601E point mutation has been also described in thyroid cancer but it is less common.[41]

BRAF protein can be activated also by a paracentric inversion of chromosome 7, resulting in the in-frame fusion between exons 1 and 8 of the A kinase (PRKA) anchor protein 9 (AKAP9) gene and exons 9 and 18 of BRAF.[42,43] Normally, AKAP9 binds to type II regulatory subunits of protein kinase A contributing to maintain the integrity of the Golgi apparatus. In the fusion protein, the autoinhibitory aminoterminal part of BRAF is lost and the chimeric product contains only the BRAF kinase domain, thus resulting in an elevated kinase activity. This rearrangement has been found in approximately 11% of PTC after radiation exposure and only in 1% of PTC with no history of radiation.[44]

hTERT

Telomerase is a reverse transcriptase enzyme that maintains chromosome ends by addition of DNA sequence repeat TTAGGG to the 3′ end of DNA strand. The enzyme is composed of a telomerase RNA component (TERC) that binds to the 3′ overhang of telomeric DNA and serves as a template for the addition of repeats, a protein component called telomerase reverse transcriptase (hTERT) and specific accessory proteins.[45–47] TERC maps on chromosome 3q26.2 and is a 451 nucleotide-long noncoding RNA. hTERT maps on chromosome 5p15.33 and is characterized by eight different transcripts, four of them coding for 1132, 1069, 807, and 1120 a.a. length proteins. According to mutational and bioinformatics studies,[45,48–50] hTERT is formed by three main structural elements: (1) a long N-terminal extension containing conserved DNA and RNA-binding domains; (2) a central catalytic domain with the reverse transcriptase activity; and (3) a short C-terminal domain. In embryos during development, telomerase activity is required from the beginning of meiosis and is strongly activated at the morula/blastocyst transition. Its activity and expression are elevated in all tissues and start to be downregulated when tissues differentiate. On the contrary, normal somatic cells do not express telomerase and have a finite replicative capacity regulated by telomere length.

Mutations of the hTERT promoter, encoding for the catalytic domain of telomerase, have been recently described in differentiated thyroid cancer,[51,52] mainly in poorly differentiated, more aggressive tumors and in anaplastic thyroid carcinoma.

OTHER GENETIC ALTERATIONS

Other rare mutations, such as TP53, β-catenin, PIK3CA, AKT1, isocitrate dehydrogenase 1 (IDH1), and epidermal growth factor receptor (EGFR) mutations, have been recently described, mostly in poorly differentiated and anaplastic cancer.[53]

TP53

This protein, also known as cellular tumor antigen P53, is encoded by a gene located at 17p13.1 and responds to diverse cellular stresses to regulate expression of target genes, thereby inducing cell cycle arrest, apoptosis, senescence, DNA repair, or changes in metabolism.[54] Its regulation of cell cycle is mediated by controlling a set of genes required for this process such as an inhibitor of cyclin-dependent kinases. When cell conditions do not allow DNA damage to repair, TP53 induces apoptosis either by stimulation of BAX and FAS antigen expression, or by repression of Bcl-2 expression.[55] Moreover, cooperation with mitochondrial PPIF is involved in activating oxidative stress-induced necrosis.[56]

β-Catenin

Catenin (cadherin-associated protein), beta 1, 88 kDa, also known as the β-catenin gene, is located at 3p22.1. The encoded protein is part of a complex of proteins that constitute adherens junctions, which are required for the creation and maintenance of epithelial cell layers by regulating cell growth and adhesion between cells. In addition, the protein also anchors the actin cytoskeleton and may be responsible for transmitting the contact inhibition signal that causes cells to stop dividing once the epithelial sheet is complete. Finally, this protein acts as a coactivator for transcription factors of the TCF/LEF family, leading to activate Wnt responsive genes.[57]

PI3KCA

Phosphatidylinositol 4,5-bisphosphate 3-kinase catalytic subunit alpha gene is located at 3q26.3. The encoded protein belongs to a family of lipid kinases capable of phosphorylating the 3'OH of the inositol ring of phosphoinositides. They contribute to coordinate a different range of cell functions including proliferation, cell survival, degranulation, vesicular trafficking, and cell migration through activation of AKT1, RhoA, PDK1, PKC, and other factors.[58]

AKT1

AKT1 (gene at 14q32.33) is one of three closely related serine/threonine-protein kinases (AKT1, AKT2, and AKT3) that regulate many processes including metabolism, proliferation, cell survival, growth, and angiogenesis by phosphorylating a range of downstream substrates in response to growth factor stimulation (i.e., EGF, IGF). AKT is responsible of the regulation of glucose uptake by mediating insulin-induced translocation of the SLC2A4/GLUT4 glucose transporter to the cell surface. It also regulates the storage of glucose in the form of glycogen. AKT promotes cell survival via the phosphorylation of the apoptosis signal-related kinase MAP3K5. Phosphorylation of "Ser-83" decreases MAP3K5 kinase activity stimulated by oxidative stress and thereby prevents apoptosis. AKT has an important role in the regulation of NF-kappa-B-dependent gene transcription and positively regulates the activity of the cyclic AMP (cAMP)-response element binding protein (CREB1). The phosphorylation of CREB1 induces the binding of accessory proteins that are necessary for the transcription of pro-survival genes.[59]

IDH1

Isocitrate dehydrogenases (2q34) catalyze the oxidative decarboxylation of isocitrate to 2-oxoglutarate to produce alpha-ketoglutarate (α-KG), playing a key role in the Krebs cycle. The enzyme activity is dependent on nicotinamide adenine dinucleotide phosphate ($NADP^+$) and the biochemical reaction, catalyzed by IDH, leads to the production of NADPH, which plays an important role in the cellular control of oxidative damage. IDH1 mutations were found to be exclusively in codon 132.[60]

Epidermal Growth Factor Receptor

The EGFR gene (7p11.2) encodes for a transmembrane glycoprotein that functions as a receptor for members of the epidermal growth factor family. Binding of the protein to its ligand induces receptor dimerization and tyrosine autophosphorylation and leads to cell proliferation. Known ligands include EGF, TGF-alpha, amphiregulin, epigen, betacellulin, epiregulin, and heparin-binding EGF. The activated receptor recruits adapter proteins like GRB2,

TABLE 10.1   Most Prevalent Oncogene Mutations and Rearrangements in Thyroid Cancer of the Follicular Epithelium (by ThyroSeq)

Oncogene mutations	Rearrangements
BRAF (V600E)	RET/PTC (1, 3, others)
RAS (N, H, K)	PAX8/PPARγ
hTERT	NTRK
PIK3CA	BRAF/AKAP9
PTEN	ALK
AKT1	
TP53	
CTNNB1	
GNAS	

which, in turn, activates important downstream signaling cascades including the RAS-RAF-MEK-ERK, PI3 kinase-AKT, PLC-gamma-PKC, STATs, and NF-kappa-B. EGFR also directly phosphorylates other proteins like RGS16, triggering its GTPase activity and probably coupling the EGF receptor signaling to the G protein-coupled receptor signaling.[61]

In summary, it is now evident that the most important driver of genetic events of papillary and follicular thyroid cancer, respectively, occur in the MAPK and PI3K/AKT pathways. It is also evident that mutations seen in thyroid cancer are almost always mutually exclusive. In the rare cases where multiple mutations accumulate, the cancer phenotype is usually more aggressive.

A summary of the most frequent mutations found in thyroid cancer is reported in Table 10.1.

APPLICATION OF MOLECULAR FINDINGS TO THE CLINICAL DIAGNOSIS OF THYROID CANCER

The frequency of thyroid nodules in clinical practice has been increasing year by year after the introduction of neck ultrasound in clinical practice. Fortunately, most of the nodules are benign, with only 5–7% being malignant. Thyroid nodules are not a unique disease but are the clinical manifestation of several different thyroid diseases. They may be single or multiple and may be found in the context of a normal gland or a diffuse goiter. In multinodular goiter, one of the nodules may become clinically dominant in terms of growth, dimension, and functional character. The main problem posed by the discovery of a thyroid nodule is the distinction between its benign or malignant nature and, consequently, its appropriate treatment. Several clinical and instrumental findings may suggest the nature of the nodule, but none of them is sufficient for a diagnosis. Up to now, the gold standard for the differential diagnosis of thyroid nodules is represented by fine needle aspiration cytology (FNAC). Compared to pre-FNAC years, in centers that use FNAC results for treatment decision making, there has been a 35–75% reduction in the number of patients sent to surgery, and a two– to threefold increase in the percentage of cancers found among surgical nodule specimens. In addition, the result of FNAC has been better planning of the surgical intervention.[62] However, despite its high specificity and sensitivity, FNAC is not diagnostic in about 20–30% of cases due to the presence of an indeterminate category of nodules, the so-called follicular proliferation, in which the cell features are not distinguishable in benign or malignant. To overcome this limitation, the search of genetic profiles associated to thyroid nodules has been proposed with promising results. So far, two possible approaches have been implemented: search for known oncogene mutations or search for benign gene profiling.

Search of Oncogene Mutations

This approach is based on the rationale that cumulating all the mutations that occur in thyroid cancer (point mutations and rearrangements), at least one mutation is found in nearly 90% of thyroid cancers.[63,64] In addition, some of the most frequent mutations are 100% specific for malignancy with no false positive results. BRAF mutation analysis has high specificity (almost 100%) but low sensitivity, being present in only 40–45% of PTC. To increase the sensitivity, a large panel of oncogenes should be included in the analysis. Several studies have shown the validity of using a panel of oncogenes including at least BRAF, RET/PTC, PAX8/PPAR, and hTERT mutations with the possible addition of TRK rearrangement.[63–65] When one of them is present in the nodule specimen, a diagnosis of malignancy is almost certain at final histology, assuming that the analysis is performed in a clinical laboratory and after validation of detection techniques. On the basis of the high probability of cancer in nodules carrying one of the above mutations, these patients should be treated with total thyroidectomy. This will avoid the repeat of FNA and eliminate the need for a two-step surgery (lobectomy followed by thyroidectomy). At variance, missense mutations of RAS have a 74–87% positive predictive value for cancer,[63–65] but in the remaining cases may be associated with a diagnosis of follicular adenoma. These results are only apparently false positive, if one considers that follicular adenomas are most likely premalignant lesions, and thus an indication for surgery based on positive mutational status might be considered as true positive. When no mutations are found in FNA material, the risk that the nodule is malignant is reduced to 3–5%, but is not completely eliminated. Patients are sent for follow-up, FNA

is repeated, and, if the diagnosis of indeterminate nodule remains, lobectomy can be offered. To measure these mutations numerous "homemade" methods are available from end-point polymerase chain reaction (PCR) (including ASO-PCR for BRAFV600E mutation and nested PCR for rearrangements) to real time or high resolution melting PCR. Screening with denaturing high-pressure liquid chromatography and final Sanger sequencing (or better pyrosequencing) are also recommended. In addition, several commercial kits are now offered by companies to assay all RAS and BRAF point mutations by real-time PCR using a multiplex approach whereas the "AmoyDx™ RET 14 Rearrangements Detection Kit" (developed by AmoyDx, China, with a contract agreement with Astra-Zeneca) is the only commercial assay for RET/PTC rearrangements detection with a limitation for research use only. A commercial panel including BRAF, RAS, PI3KA mutations, and RET/PTC, PAX8/ PPARγ rearrangements available in the United States (except for New York and Florida states) is the ThyGenX by Interpace Diagnostics. Specimen collection kits may be ordered by physicians or other authorized health care professionals and, after sample collection, sent to the factory for analysis.

A more comprehensive analysis is offered by the next-generation sequencing (NGS) approach, which is able to interrogate multiple genes simultaneously in a cost-effective way with high sensitivity and working with low input of starting material (10 ng).[1,66–68] A commercial NGS panel for thyroid cancer is represented by Thyroseq® (UPMC Life Changing Medicine),[1] which works on DNA or RNA isolated from FNA material collected in preservative solution, fixed FNA specimens, or formalin-fixed, paraffin-embedded (FFPE) tumor tissue. The analytical sensitivity is 3–5% of mutant alleles for detection of mutations and 1% for detection of gene fusions. Thyroseq is performed by the MGP Laboratory of the University of Pittsburgh Medical Center upon requirement of dedicated tubes for sample collection.

Recently, driver mutations were uncovered in 95% of PTC, and modern techniques will probably lead to even more accurate diagnoses.[69] By extending the molecular analysis to a panel of point mutations in 13 genes and 42 types of gene fusions using a proprietary sequencer and NGS, the risk of malignancy is further decreased if no mutations are detected.[70]

Based on the current experience, it is apparent that mutational analysis of a nodule is particularly useful in case of indeterminate lesions. The results of FNAC integrated with the mutational screening of the nodules may direct the therapeutic strategy as reported in Table 10.2.

Gene Expression Profiling

The second approach is based on a gene expression classifier including 142 mRNA transcripts, differentially

TABLE 10.2   Proposed Therapeutic Algorithm Based on the Combination of Cytology and Molecular Status

Cytology combined with mutational status	Cancer risk (%)	Therapy
Positive cytology and positive mutation	100	Total Tx
Indeterminate cytology and positive mutation (BRAF-RET/PTC-hTERT)	100	Total Tx
Indeterminate cytology and positive mutation (RAS)	85	Total Tx
Indeterminate cytology and negative mutation	6	Follow-up lobectomy
Negative cytology and negative mutation	3	Follow-up

expressed in benign and malignant nodules.[71] This test is commercially available (Afirma®-Veracyte)[72] and it has been validated in a large series of thyroid nodules with indeterminate FNAC. At variance with the mutational analysis, this test is intended to have high negative predictive value (NPV). The sensitivity in detecting benignity is high (92%) and will permit surgery to be avoided for nodules with negative results. However, the specificity is low, around 50%.[72,73] These results have not been confirmed in another study, which found a sensitivity of only 83% and a specificity of 10%.[74,75] Afirma is available in the United States.

miRNAs IN THYROID LESIONS

MicroRNA (miRNAs) are small, nonprotein encoding RNAs that posttranscriptionally regulate gene expression via suppression of specific target mRNAs.[76] Tumor cells have been shown to release miRNAs into the circulation in a highly stable and cell-free form and, consequently, miRNAs can be detected in plasma and serum.[77] Larger-scale miRNA analysis has proven that miRNA expression enables the distinction of benign tissues from their malignant counterpart.[78,79] The mechanisms of miRNA implication in cancer development are linked to down regulation of tumor suppressor genes or upregulation of oncogenes. Several studies have demonstrated a different miRNA signature between benign and malignant thyroid tissues,[80-83] describing miR374a, miR21, miR222, miR146b, miR221, miR146a, miR155, miR181a, miR181c, miR7, and miR30d as the most effective in differentiating the nature of thyroid nodules in the surgical samples. Even in the cytological material, miRNAs appear useful to refine FNAC diagnosis especially in indeterminate lesions[84-87] with a good sensitivity (0.87). The miRNAs found differentially expressed in FNAC material

are miR7, miR146, miR146b, miR155, miR221, miR222, miR21, miR31, miR187, miR30a-3p, miR30d, miR146b-5p, miR199b-5p, miR328, and miR197. In addition, three studies[88-90] hypothesized that serum miRNA can be used for thyroid cancer diagnosis reaching a good sensitivity (ranging from 61.4% to 94%) and specificity (ranging from 57.9% to 98.7%). These markers are being analyzed only at the research level. In the United States a genetic classifier has been recently launched, which classifies the gene expression of 10 miRNAs in FNAC material called ThyraMir™ (PDI Inc.). When the test is used with Thy-GenX, a genetic mutation panel from PDI's parent company Interpace Diagnostics, it has a NPV of 94% and a positive prediction value of 74%, at a cancer prevalence of 32% with a specificity of 85% and sensitivity of 89%.

MICROARRAY

Microarray measures the expression of a large number of genes simultaneously by hybridization of cDNA to an array of short DNA probes specific for genes of interest. Microarray-based gene expression profiles are available for malignant thyroid tumors (follicular thyroid carcinoma and papillary thyroid carcinoma) and for benign lesions and show good sensitivity and high negative predictive value. The most used platform for microarray analysis is the Affymetrix GeneChips (Affymetrix), which provides comprehensive coverage of the transcribed human genome on a single array. Despite the potential of the method, microarray shows some limitations due to costs, lack of validation of data analysis methods, and a strong association with sample quality (i.e., sample collection, storage condition), which may vary the expression levels of many genes altering the functional status of the cells and, consequently, the results.[91]

mRNA EXPRESSION

At the research level, together with microRNA and microarray analysis, the search of functional mRNA in samples is giving promising results and may become commercially available in the future. Thyroid cancer cells have shown to express thyroid-stimulating hormone receptor (TSHR) mRNA, and its measurement in the circulation may be useful in the diagnosis and management of thyroid cancer. In blood, a cut-off value of preoperative TSHR mRNA >1 ng/μg total RNA had 96% predictive value for thyroid cancer and showed a good specificity and sensitivity.[92] Combining TSHR mRNA expression with the FNAC category, the sensitivity reached 90% and specificity 80%[93-95] both for papillary and follicular lesions, indicating a possible use of this marker for refining the diagnosis in indeterminate cases. Moreover, using TSHR mRNA together with ultrasound features, the sensitivity reached 100%, avoiding unnecessary thyroidectomy in patients with benign diseases.[93,95] TSHR mRNA seems to have also a prognostic value, as it shows a short life in the circulation, and elevated TSHR mRNA on postoperative day 1 correlated with disease-free status.[93,95] Another example of a protein whose expression is correlated with malignant phenotype in thyroid cancer is the high mobility group A2 (HMGA2). This protein belongs to the family of the HMGA proteins, which are architectural chromatin proteins. They bind the minor groove of AT-rich DNA sequences through three short basic repeats, called "AT-hooks," located at the N-terminal region of the proteins. By interacting with the transcription machinery, HMGA proteins alter the chromatin structure and, thereby, regulate the transcriptional activity of several genes either enhancing or suppressing the ability of more usual transcriptional activators and repressors. HMGA2 expression was found elevated both in papillary and follicular thyroid carcinoma,[96,97] indicating a new possible marker for thyroid cancer diagnosis.

CONCLUSION

With the advent of newer technologies, the genetic profiling of thyroid nodules in material obtained by FNAC is becoming a complementary diagnostic test for the differential diagnoses of thyroid nodules. Several laboratories have included this analysis as a routine procedure in their diagnostic algorithm. It is very likely that in the near future the diagnostic accuracy will be further refined and, hopefully, will become the new gold standard.

For the moment, we can say that both the mutational analysis and the expression gene classifier are good starting points. Both tests have a low, but still significant, incidence of false positive and false negative results. In addition, the gene expression classifier has the disadvantage of being centralized in the United States, thus requiring shipment of the samples and being very expensive. Mutational analysis can be done in any laboratory with good equipment at lower costs. New technologies, such as NGS, will further improve the accuracy of cancer diagnosis in thyroid nodules with indeterminate cytology being able to simultaneously sequencing a large panel of genes. Generally, molecular testing has opened the way to individualized management of patients.

References

1. Nikiforov YE, Carty SE, Chiosea SI, Coyne C, Duvvuri U, Ferris RL, et al. Highly accurate diagnosis of cancer in thyroid nodules with follicular neoplasm/suspicious for a follicular neoplasm cytology by ThyroSeq v2 next-generation sequencing assay. *Cancer* 2014;**120**:3627–34.

2. National Cancer Institute. Cancer Genome Atlas Research Network. Integrated genomic characterization of papillary thyroid carcinoma. *Cell* 2014;**159**:676–90.

3. Ceccherini I, Bocciardi R, Luo Y, Pasini B, Hofstra R, Takahashi M, Romeo G. Exon structure and flanking intronic sequences of the human RET proto-oncogene. *Biochem Biophys Res Commun* 1993;**96**:1288–95.

4. Airaksinen MS, Titievsky A, Saarma M. GDNF family neurotrophic factor signaling: four masters, one servant? *Mol Cell Neurosci* 1999;**13**:313–25.

5. Myers SM, Eng C, Ponder BA, Mulligan LM. Characterization of RET proto-oncogene 3' splicing variants and polyadenylation sites: a novel C-terminus for RET. *Oncogene* 1995;**11**:2039–45.

6. Hickey JG, Myers SM, Tian X, Zhu SJ, V Shaw JL, Andrew SD, Richardson DS, Brettschneider J, Mulligan LM. RET-mediated gene expression pattern is affected by isoform but not oncogenic mutation. *Genes Chromosomes Cancer* 2009;**48**:429–40.

7. Knowles PP, Murray-Rust J, Kjaer S, Scott RP, Hanrahan S, Santoro M, Ibáñez CF, McDonald NQ. Structure and chemical inhibition of the RET tyrosine kinase domain. *J Biol Chem* 2006;**281**: 33577–87.

8. Zbuk KM, Eng C. Cancer phenomics: RET and PTEN as illustrative models. *Nat Rev Cancer* 2007;**7**:35–45.

9. Liu R, Liu D, Trink E, Bojdani E, Ning G, Xing M. The Akt-specific inhibitor MK2206 selectively inhibits thyroid cancer cells harboring mutations that can activate the PI3K/Akt pathway. *J Clin Endocrinol Metab* 2011;**96**:E577–85.

10. Powers JF, Brachold JM, Tischler AS. Ret protein expression in adrenal medullary hyperplasia and pheochromocytoma. *Endocr Pathol* 2003;**14**:351–61.

11. Zitzelsberger H, Bauer V, Thomas G, Unger K. Molecular rearrangements in papillary thyroid carcinomas. *Clinica Chimica Acta* 2010;**411**:301–8.

12. Nikiforov YE. Radiation-induced thyroid cancer: what we have learned from chernobyl. *Endocr Pathol* 2006;**17**:307–17.

13. Pierotti MA, Greco A. Oncogenic rearrangements of the NTRK1/NGF receptor. *Cancer Lett* 2006;**232**:90–8.

14. Greco A, Miranda C, Pierotti MA. Rearrangements of NTRK1 gene in papillary thyroid carcinoma. *Mol Cell Endocrinol* 2010;**321**: 44–9.

15. Bongarzone I, Vigneri P, Mariani L, Collini P, Pilotti S, Pierotti MA. RET/NTRK1 rearrangements in thyroid gland tumors of the papillary carcinoma family: correlation with clinicopathological features. *Clin Cancer Res* 1998;**4**:223–8.

16. Kelly LM, Barila G, Liu P, Evdokimova VN, Trivedi S, Panebianco F, et al. Identification of the transforming STRN-ALK fusion as a potential therapeutic target in the aggressive forms of thyroid cancer. *Proc Natl Acad Sci USA* 2014;**111**:4233–8.

17. Kroll TG, Sarraf P, Pecciarini L, Chen CJ, Mueller E, Spiegelman BM, et al. PAX8-PPARgamma1 fusion oncogene in human thyroid carcinoma [corrected]. *Science* 2000;**289**:1357–60.

18. Stapleton P, Weith A, Urbánek P, Kozmik Z, Busslinger M. Chromosomal localization of seven PAX genes and cloning of a novel family member, PAX-9. *Nat Genet* 1993;**3**:292–8.

19. Kozmik Z, Kurzbauer R, Dorfler P, Busslinger M. Alternative splicing of Pax-8 gene transcripts is developmentally regulated and generates isoforms with different transactivation properties. *Mol Cell Biol* 1993;**13**:6024–35.

20. Mansouri A, Hallonet M, Gruss P. Pax genes and their roles in cell differentiation and development. *Curr Opin Cell Biol* 1996;**8**: 851–7.

21. Michalik L, Auwerx J, Berger JP, Chatterjee VK, Glass CK, Gonzalez FJ, Grimaldi PA, Kadowaki T, Lazar MA, O'Rahilly S, Palmer CN, Plutzky J, Reddy JK, Spiegelman BM, Staels B, Wahli W. International Union of Pharmacology. LXI. Peroxisome proliferator-activated receptors. *Pharmacol Rev* 2006;**58**:726–41.

22. Dwight T, Thoppe SR, Foukakis T, Lui WO, Wallin G, Hoog A, et al. Involvement of the PAX8/ peroxisome proliferator-activated receptor gamma rearrangement in follicular thyroid tumors. *J Clin Endocrinol Metab* 2003;**88**:4440–5.

23. Hibi Y, Nagaya T, Kambe F, Imai T, Funahashi H, Nakao A, Seo H. Is thyroid follicular cancer in Japanese caused by a specific t(2; 3)(q13; p25) translocation generating Pax8-PPAR gamma fusion mRNA? *Endocr J* 2004;**51**:361–6.

24. Barbacid M. Ras genes. *Annu Rev Biochem* 1987;**56**:779–827.

25. Vetter IR, Wittinghofer A. The guanine nucleotide-binding switch in three dimensions. *Science* 2001;**294**:1299–304.

26. Zhu Z, Gandhi M, Nikiforova MN, Fischer AH, Nikiforov YE. Molecular profile and clinical-pathologic features of the follicular variant of papillary thyroid carcinoma. An unusually high prevalence of ras mutations. *Am J Clin Pathol* 2003;**120**:71–7.

27. Suarez HG, Du Villard JA, Severino M, Caillou B, Schlumberger M, Tubiana M, et al. Presence of mutations in all three ras genes in human thyroid tumors. *Oncogene* 1990;**5**:565–70.

28. Radkay LA, Chiosea SI, Seethala RR, Hodak SP, LeBeau SO, Yip L, McCoy KL, Carty SE, Schoedel KE, Nikiforova MN, Nikiforov YE, Ohori NP. Thyroid nodules with KRAS mutations are different from nodules with NRAS and HRAS mutations with regard to cytopathologic and histopathologic outcome characteristics. *Cancer Cytopathol* 2014;**122**:873–82.

29. Volante M, Rapa I, Gandhi M, Bussolati G, Giachino D, Papotti M, et al. RAS mutations are the predominant molecular alteration in poorly differentiated thyroid carcinomas and bear prognostic impact. *J Clin Endocrinol Metab* 2009;**94**:4735–41.

30. Namba H, Rubin SA, Fagin JA. Point mutations of ras oncogenes are an early event in thyroid tumorigenesis. *Mol Endocrinol* 1990;**4**:1474–9.

31. Sithanandam G, Druck T, Cannizzaro LA, Leuzzi G, Huebner K, Rapp UR. B-raf and a B-raf pseudogene are located on 7q in man. *Oncogene* 1992;**7**:795–9.

32. Mitsutake N, Miyagishi M, Mitsutake S, Akeno N, Mesa Jr C, Knauf JA, Zhang L, Taira K, Fagin JA. BRAF mediates RET/PTC-induced mitogen-activated protein kinase activation in thyroid cells: functional support for requirement of the RET/PTC-RAS-BRAF pathway in papillary thyroid carcinogenesis. *Endocrinology* 2006;**147**:1014–9.

33. Brummer T. B-Raf signalling. In: Schwab M, editor. *Essay for the second edition of the Encyclopedia of Cancer*. Springer-Verlag: Berlin Heidelberg, New York; 2008.

34. Davies H, Bignell GR, Cox C, Stephens P, Edkins S, Clegg S, Teague J, Woffendin H, Garnett MJ, Bottomley W, Davis N, Dicks E, Ewing R, Floyd Y, Gray K, Hall S, Hawes R, Hughes J, Kosmidou V, Menzies A, Mould C, Parker A, Stevens C, Watt S, Hooper S, Wilson R, Jayatilake H, Gusterson BA, Cooper C, Shipley J, Hargrave D, Pritchard-Jones K, Maitland N, Chenevix-Trench G, Riggins GJ, Bigner DD, Palmieri G, Cossu A, Flanagan A, Nicholson A, Ho JW, Leung SY, Yuen ST, Weber BL, Seigler HF, Darrow TL, Paterson H, Marais R, Marshall CJ, Wooster R, Stratton MR, Futreal PA. Mutation of the BRAF gene in human cancer. *Nature* 2002;**417**:949–54.

35. Cohen Y, Xing M, Mambo E, Guo Z, Wu G, Trink B, et al. BRAF mutation in papillary thyroid carcinoma. *J Natl Cancer Inst* 2003;**95**:625–7.

36. Kimura ET, Nikiforova MN, Zhu Z, Knauf JA, Nikiforov YE, Fagin JA. High prevalence of BRAF mutations in thyroid cancer: genetic evidence for constitutive activation of the RET/PTC-RAS-BRAF signaling pathway in papillary thyroid carcinoma. *Cancer Res* 2003;**63**:1454–7.

37. Ghossein RA, Katabi N, Fagin JA. Immunohistochemical detection of mutated BRAF V600E supports the clonal origin of BRAF-induced thyroid cancers along the spectrum of disease progression. *J Clin Endocrinol Metab* 2013;**98**:E1414–21.

38. Xing M, Alzahrani AS, Carson KA, Viola D, Elisei R, Bendlova B, et al. Association between BRAF V600E mutation and mortality in patients with papillary thyroid cancer. *JAMA* 2013;**309**:1493–501.

39. Kim TY, Kim WB, Song JY, Rhee YS, Gong G, Cho YM, et al. The BRAF mutation is not associated with poor prognostic factors in Korean patients with conventional papillary thyroid microcarcinoma. *Clin Endocrinol (Oxf)* 2005;**63**:588–93.

40. Durante C, Puxeddu E, Ferretti E, Morisi R, Moretti S, Bruno R, et al. BRAF mutations in papillary thyroid carcinomas inhibit genes involved in iodine metabolism. *J Clin Endocrinol Metab* 2007;**92**:2840–3.

41. Xing M. BRAF mutation in papillary thyroid cancer: pathogenic role, molecular bases, and clinical implications. *Endocr Rev* 2007;**28**:742–62.

42. Ciampi R, Knauf JA, Kerler R, Gandhi M, Zhu Z, Nikiforova MN, Rabes HM, Fagin JA, Nikiforov YE. *J Clin Invest* 2005;**115**:94–101.

43. Gandhi M, Evdokimova V, Nikiforov YE. Mechanisms of chromosomal rearrangements in solid tumors: the model of papillary thyroid carcinoma. *Mol Cell Endocrinol* 2010;**321**:36–43.

44. Lee JH, Lee ES, Kim YS, Won NH, Chae YS. BRAF mutation and AKAP9 expression in sporadic papillary thyroid carcinomas. *Pathology* 2006;**38**:201–4.

45. Watson JD. Origin of concatemeric T7 DNA. *Nat New Biol* 1972;**239**:197–201.

46. Olovnikov AM. A theory of marginotomy. The incomplete copying of template margin in enzymic synthesis of polynucleotides and biological significance of the phenomenon. *J Theor Biol* 1973;**41**:181–90.

47. Blackburn EH, Gall JG. A tandemly repeated sequence at the termini of the extrachromosomal ribosomal RNA genes in *Tetrahymena*. *J Mol Biol* 1978;**120**:33–53.

48. Fu D, Collins K. Purification of human telomerase complexes identifies factors involved in telomerase biogenesis and telomere length regulation. *Mol Cell* 2007;**28**:773–85.

49. Gillis AJ, Schuller AP, Skordalakes E. Structure of the *Tribolium castaneum* telomerase catalytic subunit TERT. *Nature* 2008;**455**:633–7.

50. Schaetzlein S, Rudolph KL. Telomere length regulation during cloning, embryogenesis and ageing. *Reprod Fertil Dev* 2005;**17**:85–96.

51. Landa I, Ganly I, Chan TA, Mitsutake N, Matsuse M, Ibrahimpasic T, et al. Frequent somatic TERT promoter mutations in thyroid cancer: higher prevalence in advanced forms of the disease. *J Clin Endocrinol Metab* 2013;**98**:E1562–6.

52. Liu X, Bishop J, Shan Y, Pai S, Liu D, Murugan AK, et al. Highly prevalent TERT promoter mutations in aggressive thyroid cancers. *Endocr Relat Cancer* 2013;**20**:603–10.

53. Garcia-Rostan G, Costa AM, Pereira-Castro I, Salvatore G, Hernandez R, Hermsem MJ, et al. Mutation of the PIK3CA gene in anaplastic thyroid cancer. *Cancer Res* 2005;**65**:10199–207.

54. Bode AM, Dong Z. Post-translational modification of p53 in tumorigenesis. *Nat Rev Cancer* 2004;**4**:793–805.

55. Marcel V, Dichtel-Danjoy ML, Sagne C, Hafsi H, Ma D, Ortiz-Cuaran S, Olivier M, Hall J, Mollereau B, Hainaut P, Bourdon JC. Biological functions of p53 isoforms through evolution: lessons from animal and cellular models. *Cell Death Differ* 2011;**18**:1815–24.

56. Wu L, Zhou N, Sun R, Chen XD, Feng SC, Zhang B, Bao JK. Network-based identification of key proteins involved in apoptosis and cell cycle regulation. *Cell Prolif* 2014;**47**:356–68.

57. Abbosh PH, Nephew KP. Multiple signaling pathways converge on beta-catenin in thyroid cancer. *Thyroid* 2005;**15**:551–61.

58. Xing M. Genetic alterations in the phosphatidylinositol-3 kinase/Akt pathway in thyroid cancer. *Thyroid* 2010;**20**:697–706.

59. Kada F, Saji M, Ringel MD. Akt: a potential target for thyroid cancer therapy. *Curr Drug Targets Immune Endocr Metabol Disord* 2004;**4**:181–5.

60. Murugan AK, Bojdani E, Xing M. Identification and functional characterization of isocitrate dehydrogenase 1 (IDH1) mutations in thyroid cancer. *Biochem Biophys Res Commun* 2010;**393**:555–9.

61. Lote H, Bhosle J, Thway K, Newbold K, O'Brien M. Epidermal growth factor mutation as a diagnostic and therapeutic target in metastatic poorly differentiated thyroid carcinoma: a case report and review of the literature. *Case Rep Oncol* 2014;**7**:393–400.

62. Gharib H, Goellner JR. Fine-needle aspiration biopsy of the thyroid: an appraisal. *Ann Intern Med* 1993;**118**:282–9.

63. Nikiforov YE, Steward DL, Robinson-Smith TM, Haugen BR, Klopper JP, Zhu Z, et al. Molecular testing for mutations in improving the fine-needle aspiration diagnosis of thyroid nodules. *J Clin Endocrinol Metab* 2009;**94**:2092–8.

64. Cantara S, Capezzone M, Marchisotta S, Capuano S, Busonero G, Toti P, et al. Impact of proto-oncogene mutation detection in cytological specimens from thyroid nodules improves the diagnostic accuracy of cytology. *J Clin Endocrinol Metab* 2010;**95**: 1365–9.

65. Nikiforov YE, Ohori NP, Hodak SP, Carty SE, LeBeau SO, Ferris RL, et al. Impact of mutational testing on the diagnosis and management of patients with cytologically indeterminate thyroid nodules: a prospective analysis of 1056 FNA samples. *J Clin Endocrinol Metab* 2011;**96**:3390–7.

66. Le Mercier M, D'Haene N, De Nève N, Blanchard O, Degand C, Rorive S, Salmon I. Next-generation sequencing improves the diagnosis of thyroid FNA specimens with indeterminate cytology. *Histopathology* 2015;**66**:215–24.

67. Hadd AG, Houghton J, Choudhary A, Sah S, Chen L, Marko AC, Sanford T, Buddavarapu K, Krosting J, Garmire L, Wylie D, Shinde R, Beaudenon S, Alexander EK, Mambo E, Adai AT, Latham GJ. Targeted, high-depth, next-generation sequencing of cancer genes in formalin-fixed, paraffin-embedded and fine-needle aspiration tumor specimens. *J Mol Diagn* 2013;**15**:234–47.

68. Kanagal-Shamanna R, Portier BP, Singh RR, Routbort MJ, Aldape KD, Handal BA, Rahimi H, Reddy NG, Barkoh BA, Mishra BM, Paladugu AV, Manekia JH, Kalhor N, Chowdhuri SR, Staerkel GA, Medeiros LJ, Luthra R, Patel KP. Next-generation sequencing-based multi-gene mutation profiling of solid tumors using fine needle aspiration samples: promises and challenges for routine clinical diagnostics. *Mod Pathol* 2014;**27**:314–27.

69. Cancer Genome Atlas Research Network. Integrated genomic characterization of papillary thyroid carcinoma. *Cell* 2014;**159**: 676–90.

70. Nikiforov YE, Carty SE, Chiosea SI, Coyne C, Duvvuri U, Ferris RL, et al. Highly accurate diagnosis of cancer in thyroid nodules with follicular neoplasm/suspicious for a follicular neoplasm cytology by ThyroSeq v2 next-generation sequencing assay. *Cancer* 2014;**120**:3627–34.

71. Chudova D, Wilde JI, Wang ET, Wang H, Rabbee N, Egidio CM, et al. Molecular classification of thyroid nodules using high-dimensionality genomic data. *J Clin Endocrinol Metab* 2010;**95**: 5296–304.

72. Alexander EK, Kennedy GC, Baloch ZW, Cibas ES, Chudova D, Diggans J, et al. Preoperative diagnosis of benign thyroid nodules with indeterminate cytology. *N Engl J Med* 2012;**367**:705–15.

73. Alexander EK, Schorr M, Klopper J, Kim C, Sipos J, Nabhan F, et al. Multicenter clinical experience with the Afirma gene expression classifier. *J Clin Endocrinol Metab* 2014;**99**:119–25.

74. McIver B, Castro MR, Morris JC, Bernet V, Smallridge R, Henry M, Kosok L, Reddi H. An independent study of a gene expression classifier (Afirma) in the evaluation of cytologically indeterminate thyroid nodules. *J Clin Endocrinol Metab* 2014;**99**:4069–77.

75. Harrell RM, Bimston DN. Surgical utility of Afirma: effects of high cancer prevalence and oncocytic cell types in patients with indeterminate thyroid cytology. *Endocr Pract* 2014;**20**:364–9.

76. Carthew RW, Sontheimer EJ. Origins and mechanisms of miRNAs and siRNAs. *Cell* 2009;**136**:642–55.

77. Mitchell PS, Parkin RK, Kroh EM, Fritz BR, Wyman SK, Pogosova-Agadjanyan EL, Peterson A, Noteboom J, O'Briant KC, Allen A, Lin DW, Urban N, Drescher CW, Knudsen BS, Stirewalt DL, Gentleman R, Vessella RL, Nelson PS, Martin DB, Tewari M. Circulating microRNAs as stable blood-based markers for cancer detection. *PNAS* 2008;**105**:10513–8.

78. Lu J, Getz G, Miska EA, Alvarez-Saavedra E, Lamb J, Peck D, Sweet-Cordero A, Ebert BL, Mak RH, Ferrando AA, Downing JR, Jacks T, Horvitz HR, Golub TR. MicroRNA expression profiles classify human cancers. *Nature* 2005;**435**:834–8.

79. Volinia S, Calin GA, Liu CG, Ambs S, Cimmino A, Petrocca F, Visone R, Iorio M, Roldo C, Ferracin M, Prueitt RL, Yanaihara N, Lanza G, Scarpa A, Vecchione A, Negrini M, Harris CC, Croce CM. A microRNA expression signature of human solid tumors defines cancer gene targets. *PNAS* 2006;**103**:2257–61.

80. Tetzlaff MT, Liu A, Xu X, Master SR, Baldwin DA, Tobias JW, Livolsi VA, Baloch ZW. Differential expression of miRNAs in papillary thyroid carcinoma compared to multinodular goiter using formalin fixed paraffin embedded tissues. *Endocr Pathol* 2007;**18**: 163–73.

81. Chen YT, Kitabayashi N, Zhou XK, Fahey III TJ, Scognamiglio T. MicroRNA analysis as a potential diagnostic tool for papillary thyroid carcinoma. *Mod Pathol* 2008;**21**:1139–46.

82. He H, Jazdzewski K, Li W, Liyanarachchi S, Nagy R, Volinia S, Calin GA, Liu CG, Franssila K, Suster S, Kloos RT, Croce CM, de la Chapelle A. The role of microRNA genes in papillary thyroid carcinoma. *PNAS* 2005;**102**:19075–80.

83. Swierniak M, Wojcicka A, Czetwertynska M, Stachlewska E, Maciag M, Wiechno W, Gornicka B, Bogdanska M, Koperski L, de la Chapelle A, Jazdzewski K. In-depth characterization of the microRNA transcriptome in normal thyroid and papillary thyroid carcinoma. *J Clin Endocrinol Metab* 2013;**98**:E1401–9.

84. Mazeh H, Levy Y, Mizrahi I, Appelbaum L, Ilyayev N, Halle D, Freund HR, Nissan A. Differentiating benign from malignant thyroid nodules using micro ribonucleic acid amplification in residual cells obtained by fine needle aspiration biopsy. *J Surg Res* 2013;**180**:216–21.

85. Agretti P1, Ferrarini E, Rago T, Candelieri A, De Marco G, Dimida A, Niccolai F, Molinaro A, Di Coscio G, Pinchera A, Vitti P, Tonacchera M. MicroRNA expression profile helps to distinguish benign nodules from papillary thyroid carcinomas starting from cells of fine-needle aspiration. *Eur J Endocrinol* 2012;**167**:393–400.

86. Kitano M, Rahbari R, Patterson EE, Steinberg SM, Prasad NB, Wang Y, Zeiger MA, Kebebew E. Evaluation of candidate diagnostic microRNAs in thyroid fine-needle aspiration biopsy samples. *Thyroid* 2012;**22**:285–91.

87. Keutgen XM, Filicori F, Crowley MJ, Wang Y, Scognamiglio T, Hoda R, Buitrago D, Cooper D, Zeiger MA, Zarnegar R, Elemento O, Fahey III TJ. A panel of four miRNAs accurately differentiates malignant from benign indeterminate thyroid lesions on fine needle aspiration. *Clin Cancer Res* 2012;**18**(7):2032–8.

88. Lee YS, Lim YS, Lee JC, Wang SG, Park HY, Kim SY, Lee BJ. Differential expression levels of plasma-derived miR-146b and miR-155 in papillary thyroid cancer. *Oral Oncol* 2015;**51**:77–83.

89. Yu S, Liu Y, Wang J, Guo Z, Zhang Q, Yu F, Zhang Y, Huang K, Li Y, Song E, Zheng XL, Xiao H. Circulating microRNA profiles as potential biomarkers for diagnosis of papillary thyroid carcinoma. *J Clin Endocrinol Metab* 2012;**97**(6):2084–92.

90. Cantara S, Pilli T, Sebastiani G, Cevenini G, Busonero G, Cardinale S, Dotta F, Pacini F. Circulating miRNA95 and miRNA190 are

sensitive markers for the differential diagnosis of thyroid nodules in a Caucasian population. *Clin Endocrinol Metab* 2014;**99**(11): 4190–8.

91. Eszlinger M, Krohn K, Kukulska A, Jarzab B, Paschke R. Perspectives and limitations of microarray-based gene expression profiling of thyroid tumors. *Endocr Rev* 2007;**28**:322–38.

92. Milas M, Shin J, Gupta M, Novosel T, Nasr C, Brainard J, Mitchell J, Berber E, Siperstein A. Circulating thyrotropin receptor mRNA as a novel marker of thyroid cancer: clinical applications learned from 1758 samples. *Ann Surg* 2010;**252**(4):643–51.

93. Chia SY, Milas M, Reddy SK, Siperstein A, Skugor M, Brainard J, Gupta MK. Thyroid-stimulating hormone receptor messenger ribonucleic acid measurement in blood as a marker for circulating thyroid cancer cells and its role in the preoperative diagnosis of thyroid cancer. *J Clin Endocrinol Metab* 2007;**92**(2):468–75.

94. Wagner K, Arciaga R, Siperstein A, Milas M, Warshawsky I, Sethu S, Reddy K, Gupta MK. Thyrotropin receptor/thyroglobulin messenger ribonucleic acid in peripheral blood and fine-needle aspiration cytology: diagnostic synergy for detecting thyroid cancer. *J Clin Endocrinol Metab* 2005;**90**(4):1921–4.

95. Milas M, Mazzaglia P, Chia SY, Skugor M, Berber E, Reddy S, Gupta M, Siperstein A. The utility of peripheral thyrotropin mRNA in the diagnosis of follicular neoplasms and surveillance of thyroid cancers. *Surgery* 2007;**141**(2):137–46 discussion 146.

96. Chiappetta G, Ferraro A, Vuttariello E, Monaco M, Galdiero F, De Simone V, Califano D, Pallante P, Botti G, Pezzullo L, Pierantoni GM, Santoro M, Fusco A. HMGA2 mRNA expression correlates with the malignant phenotype in human thyroid neoplasias. *Eur J Cancer* 2008;**44**:1015–21.

97. Jang MH, Jung KC, Min HS. The diagnostic usefulness of HMGA2, survivin, CEACAM6, and SFN/14-3-3 δ in follicular thyroid carcinoma. *J Pathol Transl Med* 2015;**49**:112–7.

PART V

PARATHYROID/BONE

11

Genetics of Hyperparathyroidism Including Parathyroid Cancer

Andrew Arnold*, Kelly Lauter**

*Center for Molecular Medicine and Division of Endocrinology and Metabolism, University of Connecticut School of Medicine, Farmington, CT, USA

**Department of Medicine, Endocrine Division, Massachusetts General Hospital, Boston, MA, USA

INTRODUCTION

The parathyroid glands function to maintain calcium homeostasis through secretion of parathyroid hormone (PTH). Parathyroid cells express on their surface a G-protein coupled calcium-sensing receptor, CaSR, which has the ability to detect minute deviations in the extracellular concentration of ionized calcium and trigger a corrective response in the amount of PTH released. For example, in defending against a lowering of serum calcium, PTH secretion is increased, and the hormone acts in concert on multiple target organs to increase serum calcium. PTH increases bone turnover including resorption, releasing calcium into the circulation. PTH acts on the kidney to increase tubular reabsorption of calcium, and to increase phosphate loss. PTH also increases the activity of renal alpha-1-hydroxylase, which hydroxylates 25-OH-vitamin D to form 1-25-$(OH)_2$-vitamin D, the active form of the vitamin. Active vitamin D induces increased calcium uptake in the intestine, and acts in a feedback loop with the parathyroid glands to decrease PTH synthesis and secretion.

In primary hyperparathyroidism, PTH secretion is increased and is no longer appropriately regulated by ambient calcium levels, resulting in hypercalcemia. Patients with this disorder may suffer from symptoms of hypercalcemia and may manifest consequences of excessive PTH action including bone pain, increased bone resorption, osteopenia, fractures, hypercalciuria, and nephrolithiasis. However, in the United States and other parts of the world where serum calcium measurement is part of routine laboratory screening, the diagnosis is often made in individuals who have few symptoms or are asymptomatic.

Primary hyperparathyroidism is typically caused by abnormal proliferation of one or more parathyroid glands. A solitary benign parathyroid adenoma is the most common pathologic basis for the biochemical diagnosis (85%), and less frequently multigland disease, that is, primary parathyroid hyperplasia, is found (10–20%). More rarely, primary hyperparathyroidism is attributed to other causes such as parathyroid carcinoma (1%) or ectopic secretion of PTH from nonparathyroid tumors.

Generally speaking, the large majority of patients with sporadic primary hyperparathyroidism are not known to carry a strong genetic predisposition, and with the exception of sporadic parathyroid carcinoma, genetic diagnosis has no current role in this setting. That said, the discovery that some patients with sporadic parathyroid adenomas have rare germline variants in cyclin-dependent kinase inhibitor genes suggests that such variants, despite their apparently insufficient penetrance to yield clear familial clustering, may importantly contribute to sporadically presenting hyperparathyroidism.[1,2] Further study of these patients and variants, plus others to be discovered, could ultimately define a clinical role for specific genetic information as an example of "personalized medicine." Familial hyperparathyroidism is rare in comparison to sporadic disease; less than 10% of primary hyperparathyroidism is linked to heritable genetic causes.[3,4] However, the genetic basis for several of these familial syndromes has been identified, and in these instances genetic testing can make important (and at times life-saving) contributions to patient/family management. In this chapter, we will focus on the following familial hyperparathyroid syndromes: hyperparathyroidism-jaw tumor syndrome (HPT-JT), familial hypocalciuric hypercalcemia (FHH), neonatal severe hyperparathyroidism (NSHPT), and familial isolated

Genetic Diagnosis of Endocrine Disorders. http://dx.doi.org/10.1016/B978-0-12-800892-8.00011-7

TABLE 11.1 Genetics of Sporadic and Familial Hyperparathyroidism

Parathyroid disorder	Gene(s) involved*
Sporadic parathyroid adenoma	*Cyclin D1* (somatic); *MEN1* (somatic); *CDKN1B* and other *CDKIs* (somatic and germline)
Sporadic parathyroid carcinoma	*HRPT2/CDC73* (somatic and germline)
MEN1 (or MEN4)	*MEN1*; *CDKN1B* (in *MEN4*); and other *CDKIs*
MEN2	*RET*
HPT-JT	*HRPT2/CDC73*
Familial hypocalciuric hypercalcemia (FHH)	*CASR*; *GNA11*; *AP2S1*
Neonatal severe hyperparathyroidism (NSHPT)	*CASR*
Familial isolated hyperparathyroidism (FIHP)	*MEN1*; *HRPT2*; *CASR*; other gene(s), unidentified

* Germline mutations unless otherwise indicated.

hyperparathyroidism (FIHP). Two other important hyperparathyroid syndromes, multiple endocrine neoplasia (MEN) types 1 and 2, are the subjects of separate chapters. A summary of hyperparathyroid disorders and the genes known to be involved, either through acquired or germline mutation, is provided in Table 11.1.

HYPERPARATHYROIDISM-JAW TUMOR SYNDROME (HPT-JT), *HRPT2*, AND PARATHYROID CARCINOMA

Introduction

Hyperparathyroidism-jaw tumor syndrome (HPT-JT) is an autosomal dominant condition that results from mutation of the *HRPT2* (*CDC73*) tumor suppressor gene.[5] HPT-JT includes a predisposition to parathyroid tumors, ossifying fibromas of the jaw, renal lesions such as bilateral cysts, hamartomas, or Wilms' tumors and uterine adenomas.[3,4] Hyperparathyroidism occurs with about 80% penetrance in HPT-JT. All parathyroid cells are at risk for tumor development over the individual's lifespan, and tumors might arise in multiple glands simultaneously, or asynchronously in one gland at a time over many years. Other distinctive features of the hyperparathyroidism in carriers of *HRPT2* germline mutations include a tendency to develop cystic parathyroid adenomas[6] and, importantly, a markedly increased risk of parathyroid carcinoma (10–20%).

Parathyroid carcinoma can be difficult to distinguish from benign adenoma on purely histological grounds, and its unequivocal diagnosis requires the presence of local invasion into surrounding tissues and/or distant metastasis.[7–9] Histologic features frequently present in parathyroid carcinoma, but insufficient for its diagnosis, include cytologic atypia, nuclear pleomorphism, capsular invasion, fibrous bands, and mitoses. Early surgical treatment by *en bloc* resection is necessary to cure parathyroid

carcinoma, since metastatic disease is very difficult to control and usually results in death from hypercalcemia and metabolic complications. As described in the subsequent section, the association of parathyroid carcinoma with HPT-JT also led to the discovery of the role of *HRPT2* mutation in sporadic parathyroid carcinoma.

Genetic Pathophysiology

The *HRPT2* gene, located on chromosome region 1q25-q32, encodes a 531 amino acid nuclear protein termed parafibromin. A nuclear localization signal resides at amino acids 136–139.[10,11] Human parafibromin has homology to the yeast cdc73 protein and has been shown to have an analogous function, interacting with Paf1, Leo1, Ctr9, and RNA polymerase II in the Paf1 complex to play a role in transcript elongation and 3′ end processing.[12] Parafibromin downregulation promotes the S phase of the cell cycle, consistent with its role as a tumor suppressor gene.[13,14] Despite these insights, precisely how the loss of parafibromin function leads to parathyroid cell transformation and the other manifestations of HPT-JT remains to be understood.

Most germline mutations in HPT-JT kindreds are predicted to inactivate or eliminate parafibromin, and examples of inactivating somatic mutations in the remaining allele were found in HPT-JT-associated tumors.[5] This pattern of biallelic inactivating mutation suggested a classical tumor suppressor function for *HRPT2*, consistent with Knudson's two-hit model.[15] *HRPT2* mutations have been detected in about 70% of kindreds with HPT-JT, typical of a tumor suppressor syndrome, the assumption being that most, if not all, of the remaining families have mutations in the same gene but escape detection because the usual sequencing strategy is limited to the coding exons. No HPT-JT families are known to show linkage to loci other than the 1q region in which *HRPT2* lies. In addition, again typical for a tumor suppressor and reflecting the numerous ways in which a gene can become inactivated, mutations are dispersed throughout *HRPT2* without any highly specific hot spot.

The familial predilection to parathyroid carcinoma in HPT-JT led to investigation of the hypothesis that acquired (somatic) mutation of the *HRPT2* gene might participate in the pathogenesis of sporadic, nonfamilial parathyroid carcinomas. Not only were somatic *HRPT2* gene mutations found in sporadic carcinomas,[16,17] indicating that they confer a strong selective advantage leading to parathyroid cell malignancy, but these mutations were detected with an impressively high frequency of 75%.[16–18] Again, because many tumors would be expected to have *HRPT2* gene mutations that would have been missed with the current sequencing strategies, it is likely that virtually all parathyroid carcinomas are driven by *HRPT2* mutation.

Importantly, a subset of patients with apparently sporadic parathyroid carcinoma were unexpectedly found to harbor germline, heritable mutations in the *HRPT2* gene.[16] Such patients may have *de novo* mutations or otherwise represent newly discovered but phenotypically classical HPT-JT, or they may represent a distinct phenotypic variant with unknown penetrance of parathyroid carcinoma. This discovery that relatives of patients with apparently sporadic parathyroid carcinoma can themselves be at risk for parathyroid malignancy by inheriting an *HRPT2* gene mutation provided a new indication for genetic testing.

Diagnosis Genetic Testing and Interpretation

Direct sequencing of the full coding region of the *HRPT2* gene is the standard approach to DNA mutational testing. Typical for a tumor suppressor gene, there is no focused hot spot for detectable mutations within the gene, and therefore the entire gene must be analyzed when the mutation is not already known. As noted earlier, a negative result in a potential proband is not conclusive, given that even among typical HPT-JT kindreds, a false negative rate of about 30% exists. Thus, in the setting of a negative DNA sequencing result, a search for relevant phenotypic traits (e.g., using jaw imaging, renal ultrasound) can be especially valuable, and appropriate surveillance for development of hyperparathyroidism should receive consideration taking pretest likelihood into account. A negative test result provides more valuable information in an asymptomatic relative in a family with a known mutation, since the absence of that specific mutation would indicate that the individual did not inherit the predisposing allele. A list of resources and laboratories that perform *HRPT2/CDC73* DNA testing is available at www.genetests.org or www.ncbi.nlm.nih.gov/gtr.

HRPT2 gene sequencing can be considered in probands with known or suspected familial HPT-JT or familial isolated hyperparathyroidism, as well as patients who present with apparently sporadic carcinoma. Testing of the latter is indicated in most instances because of the known presence of otherwise unsuspected germline mutations in a substantial minority of these individuals.[16] Knowledge that an individual bears a germline mutation of *HRPT2* has important implications. In a phenotypically classical HPT-JT kindred, identification of the proband's mutation enables family members to be easily and less expensively genotyped for the presence or absence of that specific mutation, since only that region must be sequenced to determine if the diseased allele was inherited. Asymptomatic relatives found not to carry the mutation can be spared ongoing surveillance for parathyroid malignancy. Asymptomatic, normocalcemic relatives who do carry the mutation can be monitored in a surveillance program that includes regular testing for the development of biochemical primary hyperparathyroidism, directing early surgical treatment to cure or prevent malignant parathyroid neoplasia. Similar considerations apply in the situation of suspected or clear familial isolated hyperparathyroidism in which an *HRPT2* gene mutation is uncovered, or in relatives of mutation-positive patients with sporadic parathyroid carcinoma. Additionally, in the latter two settings, treatment of a proband already affected with hyperparathyroidism can be altered by a positive genotype as described in the subsequent section, related to the recognition that all parathyroid tissue in such individuals remains at increased risk for new tumor development.

Prenatal or preimplantation testing is possible in the setting of a known familial *HRPT2* gene mutation. However, issues related to this possibility – including the likelihood that mortality from parathyroid cancer in mutation carriers could be largely eliminated by surveillance and early surgery – would need to be factored into these considerations. There are no reports of such testing to date, and its use appears to be rare.

Another potential role for *HRPT2* DNA testing would be in the diagnosis of parathyroid carcinoma, particularly in sporadic presentations. In this instance the idea would be to examine parathyroid tumor tissue for somatic mutations, which are highly specific to parathyroid cancer as opposed to sporadic benign adenomas.[19] In practical terms, the value of such testing would likely be primarily for so-called "atypical parathyroid adenomas," which are diagnostically equivocal, containing some cancer-associated features while failing to meet the strict criteria for diagnosis of carcinoma. This possibility requires further study before its clinical utility can be said to be established. Similarly, studies have been directed at using immunohistochemical assessment of parafibromin expression as a surrogate marker for *HRPT2* gene mutation, for the same diagnostic purpose.[20–22] Again, the clinical utility of this approach remains under investigation.

Treatment

A positive test, which uncovers a germline mutation in the *HRPT2* gene in a normocalcemic, phenotypically unaffected member of an HPT-JT or FIHP kindred, or relative of a patient with sporadic parathyroid carcinoma, would indicate the need for steps to prevent mortality from parathyroid carcinoma. The best ways to accomplish this are under discussion, but most would agree with a recommendation of careful surveillance by monitoring PTH and calcium levels every 6–12 months. It should be emphasized that in the interpretation of such surveillance testing, a low threshold should be maintained for acting upon biochemical results that suggest the development of primary hyperparathyroidism and a low threshold for moving to parathyroid surgery if such results are confirmed. The reported presence of biochemically nonfunctional parathyroid carcinoma in a family with an *HRPT2* gene mutation has led to consideration of whether diagnostic imaging studies should also be part of a surveillance program.[23]

Surgical treatment must be thorough, given the malignant potential of parathyroid neoplasms in individuals with *HRPT2* germline mutation, the fact that all parathyroid tissue is at risk, and that initial complete resection offers the best chance to prevent mortality from metastatic disease. However, the approach should also take into account that most parathyroid tumors in HPT-JT are benign. Therefore, the recommended approach to parathyroid surgery in *HRPT2* gene mutation-positive patients is quite different from that often taken for typical sporadic hyperparathyroidism. For the latter group, in which a solitary benign adenoma is the most common finding, a focused minimally invasive parathyroidectomy is often used. In contrast, mutation-positive patients should have a bilateral neck exploration, with the goal of identifying all parathyroid glands. Any gland that appears abnormal should be removed *en bloc*. Whether to also resect normal-appearing glands, which all carry the genetic potential to give rise to parathyroid cancers in the future, has been the subject of some discussion. At this time, we favor leaving the normal glands *in situ*, possibly tagging them to facilitate identification in a subsequent operation. This suggestion is influenced by: (1) the well-known difficulties faced by patients contending with currently available lifelong replacement therapy for hypoparathyroidism; (2) the fact that the majority of mutation-positive individuals would not be expected to ever develop metastatic parathyroid disease; and (3) the likelihood (still to be proven) that an ongoing program of careful surveillance and, when needed, repeat surgery, would continue to identify tumors early enough to effectively prevent death from parathyroid malignancy. As discussed earlier, a negative DNA mutation result in a proband does not rule out the condition, so the same approach to surveillance and surgery can be appropriate when there is sufficient clinical suspicion of an *HRPT2* gene mutation, even if a mutation was not detected or if DNA testing was not performed.

Genetic counseling for those with *HRPT2* gene mutations should focus on the potential for malignant parathyroid carcinoma, while also reassuring the individual that benign adenoma is the more common pathology in HPT-JT. When appropriate, the importance of continued monitoring of serum calcium and PTH levels to diagnose parathyroid disease promptly and minimize the potential for metastatic spread should be emphasized. Involvement of an experienced endocrinologist and endocrine surgeon is strongly advised.

FAMILIAL HYPOCALCIURIC HYPERCALCEMIA (FHH) AND NEONATAL SEVERE HYPERPARATHYROIDISM (NSHPT)

Introduction

In the parathyroid glands, CaSR detects extracellular calcium, responding to low levels by signaling to increase PTH release. Alternatively, when serum calcium is high, increased CaSR activation signals to diminish PTH secretion. Inactivating mutation of *CASR* is the primary cause of familial hypocalciuric hypercalcemia (FHH) and neonatal severe hyperparathyroidism (NSHPT).[24,25] FHH is an autosomal dominant condition involving elevated serum calcium levels, lower than expected urine calcium levels, and elevated or inappropriately nonsuppressed PTH levels. The fractional excretion of calcium, determined by the ratio of calcium clearance to creatinine clearance, is typically less than 1%. In one population, the prevalence of FHH was measured to be one in 78,000.[26] Typically, individuals with FHH do not have notable symptoms of hypercalcemia. This is probably indicative of the role CaSR plays in multiple organ systems and that its inactivation leads to decreased sensitivity to the hypercalcemic state. Also, the syndrome does not generally share the same pathology, for example, in bone, as found in typical primary hyperparathyroidism. Importantly, parathyroid surgery is not beneficial in most cases of FHH, given the benign nature of the syndrome, combined with the fact that hypercalcemia will generally persist after subtotal parathyroidectomy; in addition, permanent hypocalcemia, itself an undesirable outcome, will occur if all parathyroid tissue is removed.

FHH is a heterogeneous disorder with three genetically different subtypes, FHH1, FHH2, and FHH3. FHH1 comprises 65% of cases, and is due to loss-of-function mutations of *CASR*, located on chromosome 3q21.1. The three genetically defined subtypes appear to have similar clinical manifestations, although the latter two are not as well characterized given limited numbers of reported cases. In FHH2 the mutated protein, Gα11, is involved in downstream signaling from the CaSR.[27] FHH3, resulting from a mutation in the *AP2S1* gene, accounts for over 20% of the FHH patients who do not have a mutation in *CASR*.[28]

Activating mutations of *CASR* or of *GNA11* can be responsible for autosomal dominant hypocalcemia (OMIM #601198), which can be important to distinguish from autosomal dominant hypoparathyroidism; the latter can be caused by dominant mutations of *GCMB*, which encodes a crucial transcription factor in parathyroid gland development, or by mutation in *PTH*, which leads to a processing defect of the PTH protein.

Interestingly, *CASR* gene mutations have been documented as playing a role in a small fraction, on the order of 10%, of familial isolated hyperparathyroidism cases. One such case revealed a mutation in the cytoplasmic tail of CaSR that resulted in a syndrome with hypercalcemia, hypercalciuria, and large parathyroid glands, removal of which effectively treated the hypercalcemia.[29] Due to its major role in FHH and NSHPT, the *CASR* gene has been examined in parathyroid adenomas, carcinomas, and hyperplasias, and the lack of observed somatic mutations implies that it does not commonly drive tumorigenesis in these disorders.[30]

NSHPT is characterized by highly elevated serum calcium and PTH. In marked contrast to FHH, NSHPT is typically lethal or has devastating developmental consequences if treatment with total parathyroidectomy is not performed early in life. NSHPT results from homozygosity for inactivating mutations of the *CASR* gene, or, rarely, mutation of one allele, which plays a dominant negative role.

Genetic Pathophysiology

In the case of FHH1, inactivating mutation of the *CASR* gene causes impairment of the parathyroid glands' ability to sense serum calcium. Therefore, PTH levels are inappropriately normal or slightly elevated at high concentrations of calcium. Similar insensitivity at the level of the kidney leads to increased calcium reabsorption and relative hypocalciuria.

The calcium sensing receptor is a seven-transmembrane domain G-protein coupled receptor, which can sense extracellular calcium levels and signal to influence cellular activity. CaSR has three main domains, a hydrophilic extracellular domain with glycosylation sites, a seven-transmembrane domain, and a cytoplasmic hydrophilic domain at the C-terminus.[25] Acidic amino acids in the extracellular domain are suspected to interact with calcium ions. The cytoplasmic domain contains three apparent protein kinase C phosphorylation sites. CaSR is expressed in many tissues and cell types, including the parathyroid glands, kidney, gastrointestinal tract, placenta, pancreas, brain, osteoblasts, and osteoclasts. Numerous *CASR* mutations have been documented throughout the gene. A list of known *CASR* mutations is maintained at www.casrdb.mcgill.ca. In FHH and NSHPT, *CASR* is inactivated by the respective gene mutations; types of mutations documented include nonsense, insertion, deletion, missense, and splice site changes. Because of the diverse nature of the mutations, genetic testing involves sequencing of all six coding exons and splice sites of the gene. Autosomal dominant hypocalcemia can be conceptualized as the opposite of FHH and NSHPT, since it results from activating mutation in the *CASR* gene, and is briefly addressed in the subsequent section.

A subset of patients presented with the clinical features of FHH, but an absence of *CASR* mutations. Genetic analysis in a group of these individuals revealed linkage to chromosome 19p.[31] Nesbit et al. found these individuals to have loss-of-function mutations of the *GNA11* gene. Activation of the CaSR signals through Gα11, which increases inositol 1,4,5-trisphosphate concentrations and thereby increases the intracellular calcium concentration and decreases PTH release. It is by this impairment of signaling that inactivating mutations of Gα11 result in decreased sensitivity of cells to extracellular calcium.[27] Similar to *CASR*, activating mutations in *GNA11* can result in autosomal dominant hypocalcemia.

FHH3 was originally known as the Oklahoma Variant (FBH$_{OK}$), based on an FHH kindred that showed linkage to chromosome 19q13.3.[32] The specific responsible mutation was then identified in exon 2 of the *AP2S1* gene, which encodes the protein Ap2σ2,[28] the sigma subunit of adaptor protein-2 (AP2). AP2 is a component of clathrin-coated vesicles, which can internalize GPCRs such as CaSR. Sequencing of *AP2S1* in 50 patients in whom *CASR* mutations were excluded demonstrated mutations in 11 of the patients (>20%). Each mutation involved Arg15.[28] Mutations in this codon are thought to disrupt binding to the dileucine motif impairing function of the protein and were shown to decrease sensitivity of cells to extracellular calcium and reduce CaSR endocytosis.[28]

Diagnosis Genetic Testing and Interpretation

DNA testing is often unnecessary since the diagnosis of FHH can usually be made clinically. However, in some unusual presentations, it can be useful to have knowledge of the mutational status of the *CASR* gene or, potentially, *AP2S1* or *GNA11*.

FHH may be diagnosed in a family based on testing for the pattern of hypercalcemia among multiple adults and children. The degree of hypercalcemia in FHH is similar to that observed in typical primary hyperparathyroidism. Diagnosis of FHH is aided by the high penetrance of hypercalcemia, essentially 100%, in all age groups including neonates. Hypercalcemia in the first decade of life is routinely present in FHH, while its presence in other forms of familial hyperparathyroidism is exceedingly rare. Assessment for relative hypocalciuria in hypercalcemic individuals is important in the clinical diagnosis, especially when few family members (and no children) are affected or available for testing. This assessment is best performed by calculating the ratio of calcium clearance to creatinine clearance, which is usually less than 0.01 in FHH but greater than 0.01 in typical primary hyperparathyroidism. However, the discrimination of this cutoff value is imperfect[33,34] and subject to overlap. False-positive values can also result from other causes of hypercalcemia with relative hypocalciuria, such as use of lithium or thiazide diuretics.

Mutation of CASR is the molecular basis in most families with FHH, evidenced by detectable mutations and/or linkage to chromosome 3q.[25] Twenty percent of the patients without detectable CASR mutations were found to have mutations in AP2S1,[28] and mutation in GNA11 can also cause the FHH phenotype, albeit rarely.[27] Even among FHH kindreds linked to 3q, CASR gene mutations are not detected in about 30%. This false-negative rate is substantial but not unexpected, because inactivating mutations would be anticipated to occur occasionally in the noncoding regions, which are not usually included in clinical DNA testing; large DNA deletions encompassing one CASR allele may also escape detection. Thus, the use of CASR gene mutation testing is limited because of its expense, the success of clinical diagnosis in many instances, and its substantial false-negative rate. Settings where CASR gene testing can be particularly useful include individuals with features suggestive of FHH (e.g., young age, hyperparathyroidism with relative hypocalciuria) but without adequate family members available for assessment and familial isolated hyperparathyroidism, in which a CASR gene mutation can sometimes be found to cause atypical presentations of FHH.

The main utility of genetic testing in FHH is that proper diagnosis enables parathyroid surgery to be avoided, but again the diagnosis can often be made without mutation testing. Prenatal testing for CASR gene mutations is technically possible, but its potential utility is quite limited and reports of its use are scarce. A list of laboratories that perform CASR, AP2S1, or GNA11 DNA testing is available at www.genetests.org or www.ncbi.nlm.nih.gov/gtr.

Treatment

As described earlier, the major treatment implication of FHH is that surgical parathyroidectomy should generally be avoided. Even in individuals and families with atypically-presenting CASR gene mutation, such as isolated primary hyperparathyroidism without relative hypocalciuria, the same cautionary note regarding the futility of parathyroid surgery appears to be relevant. Exceptions include rare cases in which the CASR gene mutation exists in the context of familial hyperparathyroidism and hypercalciuria; in such instances, parathyroidectomy can correct the biochemical abnormalities.[29]

NSHPT is a clinical diagnosis that mandates parathyroidectomy in the first few weeks of life, and DNA testing would usually not be central to immediate management.

Genetic counseling regarding CASR gene mutation testing should include the relative value of clinical versus DNA testing, noting the false-negative rate of about 30%. Additionally, genetic counseling should include conversation about NSHPT, and the fact that it can arise from homozygosity for a CASR gene mutation, usually in the setting of consanguinity although compound heterozygotes have been reported.[35]

AUTOSOMAL DOMINANT HYPOPARATHYROIDISM

Although obviously not a form of hyperparathyroidism, autosomal dominant hypoparathyroidism (ADH) is briefly discussed in this chapter because of its conceptual and molecular nature as the inverse of FHH.

ADH is characterized by mild or moderate hypocalcemia, which is often asymptomatic, and serum PTH levels that are inappropriately normal or low. A strong tendency toward hypercalciuria exists, and urinary calcium excretion can be high despite the low filtered load. If the filtered load is increased by treating such individuals with calcium and vitamin D, marked hypercalciuria can result with the risk of nephrolithiasis or nephrocalcinosis. Thus, an important therapeutic corollary is to reserve such treatment for symptomatic ADH and keep dosages to the minimum needed for symptom relief rather than necessarily attempting to normalize serum calcium levels. Genetic testing for CASR mutation is an important consideration in evaluating patients with isolated hypoparathyroidism, especially given the heightened risk of renal complications from calcium/calcitriol therapy in ADH. A negative mutation test does not, however, rule out ADH.

Most individuals with ADH have heterozygous activating mutations of CASR, causing heightened parathyroid cell sensitivity to extracellular calcium and impaired renal calcium reabsorption.[35] Similarly, activating mutations of GNA11 have been identified in what is now referred to as autosomal dominant hypocalcemia type 2.[27,36]

FAMILIAL ISOLATED HYPERPARATHYROIDISM

Introduction

Familial isolated hyperparathyroidism (FIHP) accounts for about 1% of all cases of primary hyperparathyroidism.[3,4] FIHP is applied to familial syndromes of hyperparathyroidism that lack specific characteristics of the currently recognized syndromes, such as MEN1 or 2A, HPT-JT, or FHH. Thus, it can be considered a diagnosis of exclusion. FIHP can occasionally represent a variant presentation of one of these recognized syndromes, but primarily FIHP appears to result from mutation in one or more still unidentified genes, distinct from *MEN1*, *RET*, *HRPT2*, or *CASR*.

Genetic Pathophysiology

FIHP can occasionally be a variant presentation of a recognized familial hyperparathyroid syndrome, evidenced by the detection of mutations in *MEN1*, *CASR*, or (more rarely) *HRPT2* genes in minorities of FIHP kindreds.[37–39] Germline *RET* gene mutation virtually never presents as FIHP. The likelihood that the *HRPT2* gene is responsible for FIHP is markedly increased when parathyroid tumors in the kindred have atypical or malignant features.

Linkage analysis has identified a region of chromosome 2p13.3-14, which appears to be conserved among individuals in multiple Australian FIHP kindreds.[40] Two candidate genes in this region, *PPP3R1* and *PKR1*, were sequenced, but mutations were not uncovered. Therefore, a putative FIHP predisposition gene on 2p remains to be identified. While the existence of genetic heterogeneity in FIHP is already clear, the number of causative genes awaiting identification is unknown.

Diagnosis Genetic Testing and Interpretation

FIHP is a clinical diagnosis with underlying genetic heterogeneity. The development or recognition of a syndrome-associated condition in the kindred, for example, pancreatic endocrine tumor or ossifying fibroma, can change the presumptive diagnosis from FIHP to another syndrome such as MEN1 or HPT-JT. Often, however, there are few or no clinical hints to suggest one of the recognized syndromes.

Because FIHP can be an occult presentation of *MEN1*, *CASR*, or *HRPT2* gene mutation, testing for mutation of these genes can reasonably be considered or recommended given the potential differing management implications for index patient and family. In the large majority of families, however, no mutation will be found, and the expense of testing is substantial. Clinical or biochemical findings in an individual or family that suggest a known syndrome but are insufficient for its diagnosis can sometimes be used to prioritize or inform the order in which these genes are examined for mutations. For individuals or families with isolated hyperparathyroidism, features that might increase the likelihood of finding a syndromal mutation include multigland parathyroid disease, parathyroid cancer, and early age of onset, for example, in the first two decades. When a mutation is found in a proband, one benefit of DNA testing can be the ability to definitively rule out the disorder in clinically unaffected relatives and release them from the potential need for ongoing surveillance.

A role for prenatal testing in FIHP has not been described, and will be quite limited given the factors described earlier.

A list of resources and laboratories that perform genetic testing for the various genes currently recognized as involved in FIHP is available at www.genetests.org or www.ncbi.nlm.nih.gov/gtr.

Treatment

The treatment for FIHP should address the pathogenic parathyroid gland(s), typically through parathyroidectomy, except in most cases of a *CASR* gene mutation. Special issues related to treatment of specific syndromes, variants of which can present as FIHP, are described earlier or in the separate chapter on MEN1. Appropriate follow-up will take into account the fact that all remaining parathyroid tissue bears the causative or predisposing germline mutation. Additionally, patients and physicians should be aware of the nonparathyroid components of the syndromes involving familial parathyroid disease.

SUMMARY

A small but important portion of cases of primary hyperparathyroidism are familial and associated with highly penetrant mutations. The major target genes for these mutations, *HRPT2* (*CDC73*), *CASR*, *MEN1*, and *RET*, have been implicated in causing HPT-JT, sporadic parathyroid carcinoma, FHH, NSHPT, MEN1 (discussed separately), and MEN2A (discussed separately). *HRPT2*, *CASR*, and *MEN1* mutations can occasionally present as FIHP. *RET* gene testing in MEN2 is a crucial and proven modality for preventing death from medullary thyroid carcinoma, although its role in managing hyperparathyroidism is much less important. Mutational testing for *HRPT2*, *CASR* (and potentially *GNA11* and *AP2S1*), and *MEN1* genes can yield information important to clinical management, which may improve morbidity and mortality in specific circumstances.

References

1. Costa-Guda J, Marinoni I, Molatore S, Pellegata NS, Arnold A. Somatic mutation and germline sequence abnormalities in CDKN1B, encoding p27Kip1, in sporadic parathyroid adenomas. *J Clin Endocrinol Metab* 2011;**96**(4):E701–6.

2. Costa-Guda J, Soong CP, Parekh VI, Agarwal SK, Arnold A. Germline and somatic mutations in cyclin-dependent kinase inhibitor genes CDKN1A, CDKN2B, and CDKN2C in sporadic parathyroid adenomas. *Hormones Cancer* 2013;**4**(5):301–7.

3. Arnold A, Marx SJ. Familial primary hyperparathyroidism including MEN, FHH, and HPT-JT. In: Rosen CJ, editor. *Primer on the metabolic bone diseases and disorders of mineral metabolism.* 8th ed. Hoboken, NJ: John Wiley & Sons; 2013. p. 553–61.

4. Marx SJ, Simonds WF, Agarwal SK, et al. Hyperparathyroidism in hereditary syndromes: special expressions and special management. *J Bone Min Res* 2002;**17**(Suppl. 2):N37–43.

5. Carpten JD, Robbins CM, Villablanca A, et al. HRPT2, encoding parafibromin, is mutated in hyperparathyroidism-jaw tumor syndrome. *Nat Genet* 2002;**32**(4):676–80.

6. Mallette LE, Malini S, Rappaport MP, Kirkland JL. Familial cystic parathyroid adenomatosis. *Ann Intern Med* 1987;**107**(1):54–60.

7. Roth SI. Pathology of the parathyroids in hyperparathyroidism. Discussion of recent advances in the anatomy and pathology of the parathyroid glands. *Arch Pathol* 1962;**73**:495–510.

8. Bondeson L, Grimelius L, DeLellis RA. Parathyroid carcinoma. In: DeLellis RA, Lloyd RV, Heitz PU, Eng C, editors. *World health organization classification of tumours pathology and genetics of tumours of endocrine organs.* Lyon: IARC Press; 2004. p. 124–7.

9. Apel R, Asa SL. The parathyroid glands. In: LiVolsi VA, Asa SL, editors. *Endocrine pathology.* Philadelphia: Churchill Livingstone; 2002. p. 103–47.

10. Bradley KJ, Bowl MR, Williams SE, et al. Parafibromin is a nuclear protein with a functional monopartite nuclear localization signal. *Oncogene* 2007;**26**(8):1213–21.

11. Hahn MA, Marsh DJ. Identification of a functional bipartite nuclear localization signal in the tumor suppressor parafibromin. *Oncogene* 2005;**24**(41):6241–8.

12. Rozenblatt-Rosen O, Hughes CM, Nannepaga SJ, et al. The parafibromin tumor suppressor protein is part of a human Paf1 complex. *Mol Cell Biol* 2005;**25**(2):612–20.

13. Yart A, Gstaiger M, Wirbelauer C, et al. The HRPT2 tumor suppressor gene product parafibromin associates with human PAF1 and RNA polymerase II. *Mol Cell Biol* 2005;**25**(12):5052–60.

14. Zhang C, Kong D, Tan MH, et al. Parafibromin inhibits cancer cell growth and causes G1 phase arrest. *Biochem Biophys Res Commun* 2006;**350**(1):17–24.

15. Haber D, Harlow E. Tumour-suppressor genes: evolving definitions in the genomic age. *Nat Genet* 1997;**16**(4):320–2.

16. Shattuck TM, Valimaki S, Obara T, et al. Somatic and germ-line mutations of the HRPT2 gene in sporadic parathyroid carcinoma. *N Engl J Med* 2003;**349**(18):1722–9.

17. Howell VM, Haven CJ, Kahnoski K, et al. HRPT2 mutations are associated with malignancy in sporadic parathyroid tumours. *J Med Genet* 2003;**40**(9):657–63.

18. Cetani F, Pardi E, Borsari S, et al. Genetic analyses of the HRPT2 gene in primary hyperparathyroidism: germline and somatic mutations in familial and sporadic parathyroid tumors. *J Clin Endocrinol Metab* 2004;**89**(11):5583–91.

19. Krebs LJ, Shattuck TM, Arnold A. HRPT2 mutational analysis of typical sporadic parathyroid adenomas. *J Clin Endocrinol Metab* 2005;**90**(9):5015–7.

20. Tan MH, Morrison C, Wang P, et al. Loss of parafibromin immunoreactivity is a distinguishing feature of parathyroid carcinoma. *Clin Cancer Res* 2004;**10**(19):6629–37.

21. Juhlin CC, Villablanca A, Sandelin K, et al. Parafibromin immunoreactivity: its use as an additional diagnostic marker for parathyroid tumor classification. *Endocr Relat Cancer* 2007;**14**(2):501–12.

22. Cetani F, Ambrogini E, Viacava P, et al. Should parafibromin staining replace HRTP2 gene analysis as an additional tool for histologic diagnosis of parathyroid carcinoma? *Eur J Endocrinol* 2007;**156**(5):547–54.

23. Guarnieri V, Scillitani A, Muscarella LA, et al. Diagnosis of parathyroid tumors in familial isolated hyperparathyroidism with HRPT2 mutation: implications for cancer surveillance. *J Clin Endocrinol Metab* 2006;**91**(8):2827–32.

24. Pollak MR, Brown EM, Chou YH, et al. Mutations in the human Ca (2+)-sensing receptor gene cause familial hypocalciuric hypercalcemia and neonatal severe hyperparathyroidism. *Cell* 1993;**75**(7):1297–303.

25. Brown EM. Clinical lessons from the calcium-sensing receptor. *Nat Clin Pract Endocrinol Metab* 2007;**3**(2):122–33.

26. Hinnie J, Bell E, McKillop E, Gallacher S. The prevalence of familial hypocalciuric hypercalcemia. *Calcif Tissue Int* 2001;**68**(4):216–8.

27. Nesbit MA, Hannan FM, Howles SA, et al. Mutations affecting G-protein subunit alpha11 in hypercalcemia and hypocalcemia. *N Engl J Med* 2013;**368**(26):2476–86.

28. Nesbit MA, Hannan FM, Howles SA, et al. Mutations in AP2S1 cause familial hypocalciuric hypercalcemia type 3. *Nat Genet* 2013;**45**(1):93–7.

29. Carling T, Szabo E, Bai M, et al. Familial hypercalcemia and hypercalciuria caused by a novel mutation in the cytoplasmic tail of the calcium receptor. *J Clin Endocrinol Metab* 2000;**85**(5):2042–7.

30. Hosokawa Y, Pollak MR, Brown EM, Arnold A. Mutational analysis of the extracellular Ca (2+)-sensing receptor gene in human parathyroid tumors. *J Clin Endocrinol Metab* 1995;**80**(11):3107–10.

31. Heath III H, Jackson CE, Otterud B, Leppert MF. Genetic linkage analysis in familial benign (hypocalciuric) hypercalcemia: evidence for locus heterogeneity. *Am J Hum Genet* 1993;**53**(1):193–200.

32. Nesbit MA, Hannan FM, Graham U, et al. Identification of a second kindred with familial hypocalciuric hypercalcemia type 3 (FHH3) narrows localization to a <3.5 megabase pair region on chromosome 19q13.3. *J Clin Endocrinol Metab* 2010;**95**(4):1947–54.

33. Marx SJ, Attie MF, Levine MA, Spiegel AM, Downs Jr RW, Lasker RD. The hypocalciuric or benign variant of familial hypercalcemia: clinical and biochemical features in fifteen kindreds. *Medicine* 1981;**60**(6):397–412.

34. Christensen SE, Nissen PH, Vestergaard P, Heickendorff L, Brixen K, Mosekilde L. Discriminative power of three indices of renal calcium excretion for the distinction between familial hypocalciuric hypercalcaemia and primary hyperparathyroidism: a follow-up study on methods. *Clin Endocrinol* 2008;**69**(5):713–20.

35. Egbuna OI, Brown EM. Hypercalcaemic and hypocalcaemic conditions due to calcium-sensing receptor mutations. *Best Pract Res Clin Rheumatol* 2008;**22**(1):129–48.

36. Mannstadt M, Harris M, Bravenboer B, et al. Germline mutations affecting Galpha11 in hypoparathyroidism. *N Engl J Med* 2013;**368**(26):2532–4.

37. Simonds WF, James-Newton LA, Agarwal SK, et al. Familial isolated hyperparathyroidism: clinical and genetic characteristics of 36 kindreds. *Medicine* 2002;**81**(1):1–26.

38. Warner J, Epstein M, Sweet A, et al. Genetic testing in familial isolated hyperparathyroidism: unexpected results and their implications. *J Med Genet* 2004;**41**(3):155–60.

39. Simonds WF, Robbins CM, Agarwal SK, Hendy GN, Carpten JD, Marx SJ. Familial isolated hyperparathyroidism is rarely caused by germline mutation in HRPT2, the gene for the hyperparathyroidism-jaw tumor syndrome. *J Clin Endocrinol Metab* 2004;**89**(1):96–102.

40. Warner JV, Nyholt DR, Busfield F, et al. Familial isolated hyperparathyroidism is linked to a 1.7 Mb region on chromosome 2p13.3-14. *J Med Genet* 2006;**43**(3):e12.

12

Genetic Diagnosis of Skeletal Dysplasias

Murray J. Favus

Section of Endocrinology, Diabetes and Metabolism, Department of Medicine, University of Chicago, Chicago, IL, USA

INTRODUCTION

Genetic skeletal disorders make up one of the largest groups among rare inherited conditions. Referred to as skeletal dysplasias, these hereditary diseases of bone may be classified according to radiologic anatomic appearance, density of bone, and state of mineralization. This chapter examines a number of selected inherited skeletal dysplasias representing each of the three groups: (1) sclerosing bone disorders defined as thickening either of trabecular or cortical bone; (2) disorders of defective mineralization with low bone mass called either rickets in childhood or osteomalacia in adults; and (3) dysplasias of bone and cartilage with normal or low bone mass, called osteopenia or osteoporosis in the absence of a defect in mineralization. In the absence of unique radiographic changes, diseases can be classified by bone histology, serum chemistries, or genetic mutations. This chapter discusses representative disorders (Table 12.1) from each of the three major categories of genetic sclerosing bone disorders, mineralization defects, and other dysplasias of bone and cartilage with normal or low bone mass.

Although cases of most rare bone disorders have been studied in detail, effective therapeutic interventions are lacking for many of these conditions. Given that most of the diseases begin in infancy or childhood and are irreversible, the patients are given the dim prospect of lifelong suffering. It is hoped that the advances in basic science of the past two decades can be used in innovative ways to develop much-needed effective therapeutics.

SCLEROSING BONE DISORDERS

Sclerosing bone dysplasias vary in severity and present with a wide range of radiologic, clinical, and genetic features. Hereditary sclerosing bone dysplasias result from disturbance(s) in pathways involved in osteoblast or osteoclast function, leading to abnormal accumulation of trabecular bone, called osteosclerosis, or increased cortical bone, called hyperostosis. Investigations into the genetic basis of inherited sclerosing bone dysplasias have resulted in identification of gene mutations that may alter critical skeletal developmental or bone cell regulatory functions through activating or inactivating mutations.

Knowledge of the radiologic appearances, distribution, and associated clinical findings with increased bone density are crucial for accurate diagnosis. Acquired disorders with increased bone density may simulate genetic sclerosing bone dysplasias, including hepatitis C-associated osteosclerosis, osteoblastic metastases, Paget's disease of bone, myelofibrosis and sickle cell disease.[1] Selected inherited disorders are discussed in the subsequent section.

Osteopetrosis

Osteopetrosis is a group of closely related inherited disorders due to the loss of osteoclastic bone resorption from either the absence of osteoclasts or functional loss. The central features include fractures, dense bone, low to absent modeling, obliteration of the marrow space, and anemia with extramedullary hematopoiesis. The condition is also known as marble bone disease from the radiologic appearance of the solid, dense skeleton, which was initially described by Albers-Schonberg in 1904. Current terminology uses Albers-Schonberg disease to refer to the more benign adult onset form of osteopetrosis, also called autosomal dominant osteopetrosis type II.[2] The same phenotype is also produced by carbonic anhydrase (CA) II deficiency, which is transmitted by an autosomal recessive pattern of inheritance characterized by a milder course than the severe form of osteopetrosis and accompanied by renal tubular acidosis, mental retardation, and cerebral calcification.[2-8]

Genetic Diagnosis of Endocrine Disorders. http://dx.doi.org/10.1016/B978-0-12-800892-8.00012-9

TABLE 12.1 Skeletal Bone Dysplasias

Sclerosing bone disorders
Inherited sclerosing dysplasias
  Carbonic anhydrase II deficiency
  High bone mass phenotype
  Hyperostosis corticalis generalisata (van Buchem disease, sclerosteosis)
  Juvenile Paget's disease of bone (osteoectasia with hyperphosphatasia)
  Melorheostosis (flowing hyperostosis)
  Osteopathia striata
  Osteopetrosis
  Osteopoikilosis
  Pyknodysostosis
  Progressive diaphyseal dysplasia (Camurati–Engelmann disease)

Bone dysplasias with defective mineralization
  Axial osteomalacia
  Hypophosphatasia

Selected inherited dysplasias of bone and cartilage
  Achondroplasia
  Enchondromatosis
  Multiple exostoses
  Osteogenesis imperfecta
  Pachydermoperiostosis

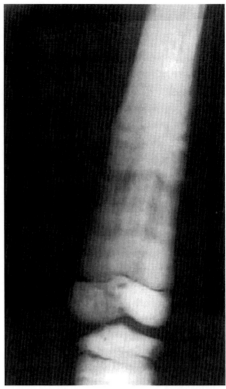

FIGURE 12.1 Osteopetrosis. Intermediate form involving forearm and hand of this boy aged four years, three months, showing characteristic defects in modeling with cortical thickening and obliteration of marrow space. *Adapted from Ref. [116, p. 950, Fig. 41.14].*

The distinctive forms of osteopetrosis differ by the mode of inheritance, age of appearance, and severity of the disease. The autosomal dominant forms are generally more benign with adult onset and minimal disability or symptoms.[2,9–11] The autosomal recessive type appears in infancy and follows a malignant course with high mortality rate.[10,12,13] There is a neuropathic subtype associated with mutations in *OSTM1* or *CLCN7* leading to seizures, developmental delay, hypotonia, and renal atrophy. A rare autosomal recessive form with onset in childhood is intermediate in severity with a more benign course.[14]

Clinical Presentation

The incidence of the severe autosomal recessive malignant osteopetrosis ranges from one in 200,000 to one in 500,000 live births. Hypocalcemia in infants and children is the result of absent osteoclastic bone resorption with loss of bone modeling. Failure of bone modeling may also cause neurologic abnormalities of the cranial nerves, including paralysis due to narrowing of cranial foramina.

Adult benign autosomal recessive osteopetrosis (ARO) is often diagnosed by the presence of typical skeletal changes seen in radiographs taken to evaluate fractures. The prevalence is one in 100,000 to one in 500,000 adults. The course may not be benign, with fractures accompanied by loss of vision, deafness, mandibular osteomyelitis, and other complications often more characteristic of the juvenile or intermediate form of the disease. The pattern of inheritance is not always predictable with

examples of nonpenetrance resulting in skipped generations, or in contrast, severely affected children being born into families with a largely benign form of the disease.

Radiography

The major radiographic change in osteopetrosis is a systemic increase in bone mass with thickening of both trabecular (sclerosis) and cortical bone (hyperostosis) (Fig. 12.1). Metaphyseal areas of bone may contain alternating dense (sclerotic) and radiolucent banding. Metaphyses may widen and create an "Erlenmeyer flask" deformity. Pathologic fractures may be discovered at any site in the skeleton. Changes in the growth plates associated with rickets may be present due to secondary hyperparathyroidism prompted by hypocalcemia. The basal aspects of the skull is thickened, and there is thickening of other skull bone with narrowing of the cranial nerve foramina and minimal or absent development of the mastoid and paranasal sinuses. Nuclear medicine Tc-99m scans reveal fracture sites and complications such as minimal bone marrow space and osteomyelitis.

Laboratory Tests

Serum calcium may be decreased in severe disease and accompanied by increased serum parathyroid hormone and 1,25-dihydroxyvitamin D levels. Serum levels of osteoclastic tartrate-resistant acid phosphatase (TRAP) and the brain isoenzyme creatine kinase may be increased.

Histology

The presence of small areas or "islands" of persistent calcified cartilage within trabecular bone is a unique marker for near absence of osteoclast function during endochondral bone formation.[15] The heterogeneity of osteopetrosis is indicated by the osteoclast number, which may be increased or normal. Rarely is the number of osteoclasts decreased. When nuclear number is increased, the ruffled border and clear zone will be absent. The bone matrix may have multiple areas of immature woven bone.

Genetic Pathophysiology

About 50% of patients with autosomal recessive malignant infantile osteopetrosis have a mutation of the *TCIRG1* gene, which encodes for the osteoclast-specific α3 subunit of the vacuolar proton pump. The pump is responsible for acidification of the microspace of the bone mineral surface enclosed by the osteoclast ruffled border. Autosomal dominant osteopetrosis type II (Albers-Schonberg disease) results from a mutation in the *CLCN7* chloride channel gene that disables osteoclast resorptive function.[6,16] *CLCN7* mutations can also cause autosomal recessive, malignant, or the intermediate osteopetrosis. Mutations of the *CAII* gene cause carbonic anhydrase II deficiency and dysregulation of acidification, normal function of which is required for normal osteoclast resorption.[5,7] Rarely, infants with osteopetrosis have an inactivating mutation in the receptor activator of nuclear factor-*k*B ligand (RANKL) gene.[17,18] Such patients have low osteoclast counts and fail bone marrow transplant therapy.[19,20]

Genetic Diagnostic Testing[21]

Mutation analysis can be used to identify severe cases caused by mutations of the *TCIRG1* and *CLCN7* genes.[21] The genetic basis of the disease is now largely uncovered by identification of mutations in *TCIRG1*, *CLCN7*,[22] *OSTM1*, *SNX10*, *CSF-I*,[23] and *PLEKHM1*,[24] which lead to osteoclast-rich ARO in which osteoclasts are abundant but have severely impaired resorptive function.[25] In contrast, mutations in *TNFSF11* and *TNFRSF11A*[26] lead to osteoclast-poor ARO.[25] In osteoclast-rich ARO, impaired endosomal and lysosomal vesicle trafficking result in defective osteoclast ruffled-border formation and, hence, the inability to resorb bone and mineralized cartilage.[24]

Treatment

Allogeneic human leukocyte antigen-identical bone marrow transplantation has the highest success rate.[27]

Other children may respond to interferon-gamma-1β, 1,25-dihydroxyvitamin D3,[28] methylprednisolone, and a low calcium-high phosphorus diet. Bisphosphonate therapy may increase bone density and decrease fracture rates, but does not correct the underlying genetic defects. Surgical treatment is indicated to correct skeletal deformities and stabilize fractures to optimize healing.

Progressive Diaphyseal Dysplasia

The disorder also known as Camurati–Engelmann disease is an autosomal dominant inherited disease notable for symmetrical thickening and widening of the endosteal and periosteal surfaces of the diaphysis of long bones, primarily the femur and tibia and less frequently the fibula, radius, and ulna.

Clinical

The disorder presents during childhood with lower extremity pain and tenderness of the involved areas, fatigue, muscle wasting, and weakness, often manifest as a waddling gait. On examination, the limbs are thin due to loss of muscle mass while the bones are prominent and tender. Skull involvement may also be present with a large cranium, prominent forehead, proptosis, cranial nerve palsies, and hydrocephalus with increased intracranial pressure. There may be delayed puberty with central hypogonadism. While skeletal radiographic studies show progressive bone changes with time, the clinical manifestations may be quite variable with occasional spontaneous improvement, more likely during the adult years. Some patients with more severe disease may have Raynaud's phenomenon, hepatosplenomegaly, and other changes suggestive of vasculitis.[29]

Radiography

The hyperostosis is progressive and seen as a patchy, irregular thickening of both the endosteal and periosteal surfaces due to new bone formation (Fig. 12.2). The areas spread symmetrically along the diaphysis of long bones with eventual involvement of the metaphysis. Tc-99m bone scan radiotracer uptake is increased in radiographically involved areas. The presence of marked tracer uptake at sites with minimal radiographic changes suggests early skeletal disease.[29]

Laboratory

Metabolic studies show markedly positive calcium balance, which is manifest as a tendency for a mild hypocalcemia and marked hypocalciuria, consistent with increased renal tubule calcium reabsorption and positive calcium balance. Some patients have elevation of serum alkaline phosphatase and urine hydroxyproline excretion, indicating increased bone turnover. Elevated erythrocyte sedimentation rate, leukopenia, and mild anemia reported is an arginine-cysteine amino acid change at codon 218 (R218C) in the latency-associated peptide domain of TGF-β1.[29,30] In one study *in vitro* peripheral blood mononuclear cells obtained from three related CED patients harboring the R218C mutation, TGF-β1 bioactivity, and osteoclast formation *in vitro* were increased and inhibited by addition of the soluble TGF-β1 type II receptor. Increased osteoclast activity is consistent with clinical and biochemical evidence of increased resorption. In addition, the mutation in bone cells activates TGF-β1 resulting in overstimulating of bone matrix formation. Binding of a mutated latency-associated peptide domain promotes the persistence of the TGF-β1 action.

Genetic Diagnostic Testing and Interpretation

Characteristic radiologic changes strongly support the diagnosis, which can be confirmed by the identification of an activating mutation within the *TGF-β1* gene.[29]

Treatment

Small doses of glucocorticoids (prednisone five mg daily) may relieve bone pain and reverse the abnormal histologic changes of bone formation. There is limited experience with bisphosphonates, and no random controlled trials have been done. However, clinical improvement has been reported in a small number of patients. Treatment with the angiotensin II type 1 receptor antagonist Losartan relieved bone pain, increased activity, and increased body composition with increased lean and fat mass during 38 months of therapy in a nine-year-old girl.[31]

Pyknodysostosis

Pyknodysostosis is a rare autosomal recessive disorder caused by an inactivating mutation in the lysosomal cathepsin K (*CTSK*) gene and characterized clinically by osteosclerosis, fragility fractures, short stature, dysmorphic facial features, dental abnormalities, distal phalangeal acroosteolysis, and delayed closure of cranial sutures.[32] There is a range of age at presentation from childhood to later adult years contributing to an underdiagnosis or misdiagnosis of this form of short-limb dwarfism. The condition is thought to be a form of osteosclerosis as both have progressive generalized osteosclerosis and recurrent fractures. Pyknodysostosis may have been responsible for the short stature and other physical features of the French impressionist painter Henri de Toulouse-Lautrec.[33]

Clinical

Features include short stature, kyphoscoliosis, chest deformity, high arched palate, blue sclera, and proptosis. Dysmorphic facial features include small face and chin, pointed nose, large cranium with fronto-occipital prominence, persistence of the open anterior fontanel, and a

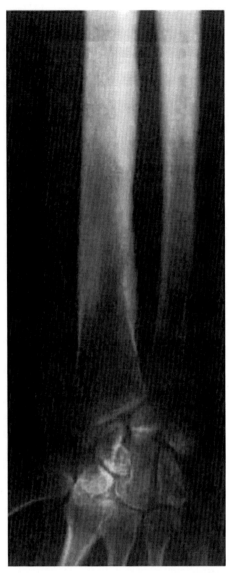

FIGURE 12.2 **Progressive diaphyseal dysplasia.** The diaphyseal shafts of the radius and ulna show characteristic endosteal and periosteal sclerosis. *Adapted from Ref. [117, p. 957, Fig. 41.18].*

in some patients may indicate that the clinical and laboratory changes of inflammation are due to an accompanying systemic connective tissue inflammatory disorder.

Histology

Histologic examination of bone shows new bone formation along the diaphyseal surfaces. Woven bone can be seen undergoing maturation and incorporation into cortical bone. Affected muscle shows both myopathic and vascular changes.

Genetics and Pathophysiology

The genetic defect is caused by various activating mutations at a specific location (chromosome 19q13.2) within one region of the gene that encodes for transforming growth factor (TGF)-β1. The most common mutation large angle of the mandible. Extremities are notable for short upper and lower extremities, small square hands, and hypoplastic nails.

Radiography

Radiographs reveal a generalized increase in bone density with normally shaped long bones, in contrast to the appearance of the long bones in osteopetrosis. Osteosclerosis appears in childhood and increases with age. There are recurrent fractures. Cranial sutures are separated with persistent patency of the anterior fontanel. Hypoplasia of the sinuses, mandible, distal clavicles, ribs, and terminal phalanges are variable findings. Other common cranial features include persistence of deciduous teeth and sclerosis of the calvarium and base of the skull. Long bones have narrow medullary canals.

Laboratory

Bone-related chemistries include normal serum calcium, phosphorus, and alkaline phosphatase. Unlike osteopetrosis, there is no anemia. In a report of six affected children, growth hormone and serum insulin-like growth factor 1 levels were low in five.[34]

Histology

There is normal cortical bone architecture with sclerosis of trabecular bone. Osteoblastic bone formation and osteoclastic bone resorption activities are decreased. Osteoclasts are present in the sclerotic trabecular bone but do not function normally.

Genetic Pathophysiology

The molecular basis for the disorder is a mutation in the *CTSK* gene that encodes cathepsin K, a lysosomal metalloproteinase that is highly expressed in osteoclasts.[32,35] By 2010, some 27 different mutations have been reported in members of 34 unrelated families.[36] The major defect in the disorder appears to be an impaired degradation of matrix collagen, as the cathepsin K enzyme plays an important role in digestion of matrix proteins during osteoclastic bone resorption. Electron microscopy of bone suggests that degradation of bone collagen may be defective, related to the inactive cathepsin K enzymatic digestion. Inclusion bodies have been described in chondrocytes, and virus-like inclusions have been reported in osteoclasts, but their significance remains unknown.

Diagnosis

The diagnosis is usually made during infancy or early childhood based on the major dysmorphic features of short stature, large cranium, and dysplastic facies. Suggestive clinical characteristics can be confirmed by radiographic examination of the entire skeleton and skull. As some features of osteopetrosis may be present, the differential diagnosis includes osteopetrosis, osteoporosis, cleidocranial dysplasia, and idiopathic or localized acroosteolysis of the distal phalanges.[14]

Treatment

There is no known treatment for this disorder, and there are no reports of bone marrow transplantation. Five to 12 years of growth hormone therapy in three children with pyknodysostosis improved longitudinal growth that approached predicted levels and restored body proportion.[36] Management is symptomatic and multidisciplinary, including orthopedic treatment of fractures and maxofacial reconstruction. Long bone fractures tend to be transverse and heal satisfactorily. But delayed union of the long bones may be complicated by large callus formation. The skeletal hardness from the underlying disease makes dental extractions and internal fixation of the long bones difficult. Vertebral surveillance is indicated to detect frequent spondylolisthesis. The prognosis is favorable, and the disease is not progressive.

Hyperostosis Corticalis Generalisata

This autosomal recessive disorder, also known as van Buchem disease, is characterized by endosteal hyperostosis in which osteosclerosis is found in the skull, ribs, and clavicles. A more severe form of endosteal hyperostosis is accompanied by syndactyly and tallness and is known as sclerosteosis.[37] The latter affects predominantly those of Dutch ancestry, including Afrikaners. An autosomal dominant variant of van Buchem disease is called Worth disease, which is a mild form of endosteal hyperostosis.[38]

Clinical

Narrowed cranial foramina are responsible for the major clinical manifestations including optic atrophy, deafness, and facial paralysis. These symptoms may be discovered as early as infancy in the severe form. In the milder form of the disorder, an enlarged mandible may be the initial manifestation and develops during puberty or later in some adults. Long bones may be painful, but fractures are uncommon. In contrast, syndactyly only appears in the more severe sclerosteosis. Sclerosteosis patients are tall and heavy and with time develop an enlarged cranium with facial palsy and disfigurement. Deafness may also develop. They may eventually develop increased intracranial pressure with headache and brainstem compression. Syndactyly may develop either by cutaneous growth or bony fusion. Middle and index fingers are most commonly affected.

Histology

There is uncoupled bone turnover with increased osteoblastic bone formation and low osteoclastic bone

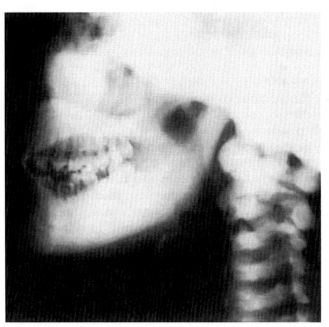

FIGURE 12.3 **Endosteal hyperostosis.** Note the dense sclerosis of the mandible, facial bones, and cervical vertebrae. *Adapted from Ref. [117, p. 960, Fig. 41.21].*

resorption leading to an accumulation of normal bone. The initial report by van Buchem described bone of normal quality.

Bone histomorphometry in one patient with sclerosteosis showed an affected skull with increased bone formation, thickened trabeculae, and accumulation of osteoid. In contrast, osteoclastic bone resorption was reduced.

Radiography

Endosteal thickening creates dense diaphyseal cortical bone with narrow medullary space. Modeling of the long bones is normal.

Osteosclerosis causes mandibular enlargement with density of the base of the skull, facial bones, vertebral bodies, pelvic bones, and ribs (Fig. 12.3). In sclerosteosis, skeletal radiographs appear normal in early childhood except for the syndactyly. With progressive cortical bone growth, there is widening of the mandible and skull and greater density of the vertebral pedicles, ribs, pelvis, and long bones.[39]

Laboratory

In both disorders, serum bone-specific alkaline phosphatase levels may be elevated, reflecting the high bone formation. Serum calcium and phosphorus are normal. Serum sclerostin levels are reduced in both van Buchem disease and sclerosteosis. The lower levels in sclerosteosis may account for the more severe disease with greater osteoblastic activity.[40,41]

Genetic Pathophysiology

Sclerostin is a 213 amino acid product of *SOST* whose physiologic function is to bind LRP5 and LRP6 and BMP2 and other BMPs and antagonize Wnt signaling. As a result, sclerostin inhibits bone formation. Both van Buchem disease and sclerosteosis are autosomal recessive genetic diseases in which defects in the *SOST* gene have been assigned to the same region of the chromosome 17q12-21. However, recent advances show that the two diseases have different genetic defects. Sclerosteosis is caused by a loss-of-function mutation of the *SOST* gene, with increased osteoblastic bone formation and minimal osteoclastic bone resorption.[42] Calcium homeostasis and pituitary function are normal. Van Buchem disease results from a homozygous deletion of a 52-kb regulatory element 35 kb downstream of the SOST gene that diminishes the downstream magnification of *SOST*.[43]

Genetic Diagnostic Testing

The specific diagnosis of hyperostosis corticalis generalisata and sclerosteosis can be obtained by *SOST* gene sequencing, which is available through a select number of academic medical centers and commercial laboratories. The procedures include sequencing the *SOST* gene and analyses for duplication or deletion of regions of *SOST*.[44]

Treatment

There is no specific medical therapy that can change the progression of the endosteal hyperostosis. Losartan, an angiotensin II type 1 receptor antagonist, has been found to downregulate the expression of TGF-β type 1 and 2 receptors. A single case report of a patient who improved level of activity during Losartan therapy suggests that downregulation of TGF-B1 receptor may be an approach to decrease the effects of high levels of TGF-B.[44] Surgical decompression of the narrowed foramina may improve the cranial nerve palsies. In sclerosteosis there is no effective medical therapy. Bony syndactyly is difficult to repair. Neurologic symptoms from cranial nerve root compression and the subsequent complications are managed surgically.

Melorheostosis

Melorheostosis is Greek for "flowing hyperostosis," which is taken from the radiographic appearance of the involved bone that resembles melted wax that has dripped down a candle. The rare disease is an acquired sclerosing bone disorder of unknown etiology.

Clinical

The major manifestation is progressive linear hyperostosis in one or more bones of one limb, usually a lower

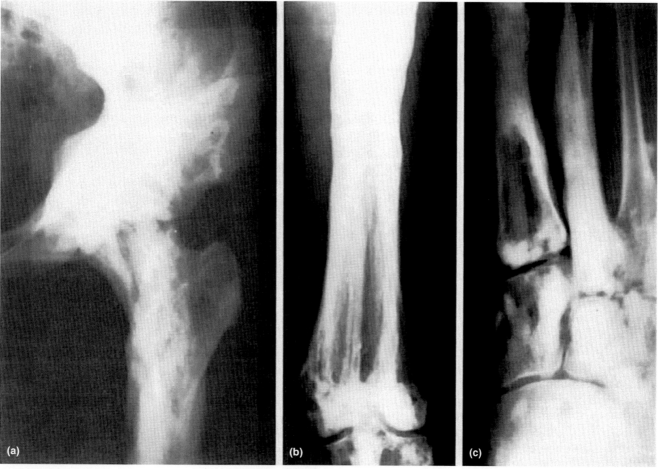

FIGURE 12.4   Melorheostosis. The characteristic radiographic abnormalities of melorheostosis are: (a) evident throughout the left hemipelvis, para-acetabular region, and medial aspect of the proximal femur. (b) There is a linear radiodense pattern in the distal femur that extends across the knee joint. (c) The lesions are also found in the medial aspect of the foot. *Courtesy of R. Freiberger, MD, New York, NY. Adapted from Ref. [117, p. 1216, Fig. 84-8A—C].*

extremity. If bilateral, the disease is usually asymmetric. Symptoms appear in childhood as pain and/or stiffness in the area of the sclerotic bone with the lesions advancing rapidly. Associated soft tissue masses may be present and include scleroderma-like features. Other changes that may overlie the bone lesions include hypertrichosis, ectopic cartilage, bony tissue, fibromas, and hemangiomas. These superficial lesions may become evident before the underlying skeletal changes. In adults, pain and stiffness may persist, and there may be contractures of nearby joints, but the skeletal lesions do not progress.[1]

Histology

The skeletal changes of melorheostosis are endosteal thickening during childhood and periosteal new bone formation during the adult years. There is sclerosis with thick and irregular lamellae. Examination of the skin overlying the skeletal lesions may contain a scleroderma-like lesion, but the composition of collagen is normal and different from collagen in scleroderma lesions.

Radiography

The changes characteristic of melorheostosis involve a single bone or an occasional adjacent bone with irregular, dense hyperostosis involving both periosteal and endosteal surfaces (Fig. 12.4). Bones of the lower extremities are most commonly involved, but the lesions have been found in other bones as well. Ectopic bone may be present in the soft tissue in proximity to the skeletal lesions and near joints. Nuclear medicine bone scans show accumulation of tracer in the lesions indicating high blood flow and can be useful in differentiating melorheostosis from other osteopoikilosis and osteopathic striata.[45,46]

Laboratory

Laboratory tests including serum calcium, phosphorus, and alkaline phosphatase are normal.

Genetic Pathophysiology

No specific etiology is known. A small number of patients with a mutation in the *LEMD3* gene that causes poikilosis have some phenotypic changes suggestive of melorheostosis, but the majority of melorheostosis patients are sporadic and have no *LEMD3* abnormality.[47] A hypothesis that has not been tested is that the disease arises from a somatic mutation confined to the affected bone.

Genetic Diagnostic Testing

There is no available genetic testing available. Radiotracer scanning of the skeleton will demonstrate increased uptake of a single bone consistent with that involved on standard radiography.

Treatment

No specific treatment is available. Surgical repair of contractures is often unsuccessful due to recurrence of the deformities.

High Bone Mass Phenotype

Selective activating mutations of the *LRP5* gene, which encodes for low-density lipoprotein receptor-related protein 5, is inherited as an autosomal dominant trait with the major finding of a systemic increase in skeletal mass composed of normal bone.[48] Serum sclerostin is elevated in men and women with high bone mass with or without an hereditary LRP5 activating mutation.[49] The high bone mass is associated with an increased Wnt signaling and stimulation of osteoblasts. Therefore, these observations indicate there is resistance to sclerostin's inhibitory actions on bone formation. Other features of the phenotype may be observed including torus palatinus, cranial nerve palsies, and oropharyngeal exostoses.[50] Detection of the lesions in other family members is done by finding high DEXA scanning and dense radiographic imaging of young adults at risk.[51]

Juvenile Paget's Disease

The disorder, also known as hereditary hyperphosphatasia or osteoectasia with hyperphosphatasia, is characterized clinically by infants and children with progressive bone deformities, fractures, short stature, and elevated bone alkaline phosphatase.[52] Families have been characterized as having a homozygous deletion of the *TNFRSF11B* gene, which encodes for the circulating receptor activator of nuclear factor-kB (RANKL)-binding protein osteoprotegrin.[53] The familial pattern of disease is consistent with an autosomal dominant pattern of inheritance with variable penetrance.[54] Radiographic findings are the result of uncontrolled, high rates of osteoclastic resorption (Fig. 12.5).

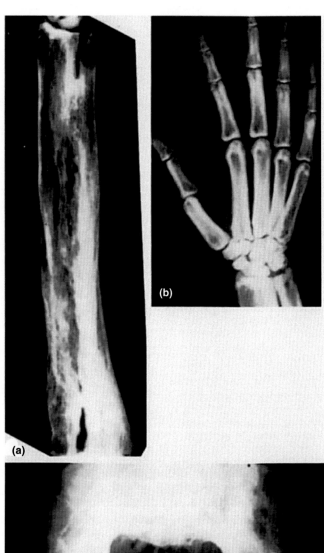

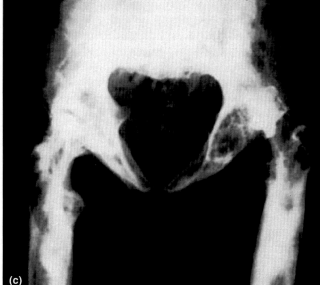

FIGURE 12.5 **Hereditary hyperphosphatasia (Juvenile Paget's disease).** (a) Shows cortical thickening of the radius and ulna associated with loss of definition between the cortical and medullary bone. (b) Shows similar resorptive lesions in the bones of the right hand. (c) The pelvis shows pagetoid appearance with subluxation of the left femoral head resulting from the extensive resorption and weakened new bone. *Adapted from Ref. [118, p. 1145, Fig. 79-39].*

A rare Paget's disease-like disorder called familial expansile osteolysis is an autosomal dominant condition marked by accelerated bone remodeling has been mapped to chromosome 18q21-22.[55] Mutations have been located to exon 1 of the *TNFRSF11A* gene, which encodes for RANK, a pivotal molecule in the activation of osteoclastic bone resorption.[18,54,56] In other families, susceptibility loci have been mapped to a number of chromosomes resulting in mutations of *TNFSF1* (RANKL), *TNFRS11A* (RANK), and *TNFRSF11B* (OPG), which explain the phenotypic similarities with the Paget's disease phenotype.[54,57,26]

Treatment of these related disorders has been antiresorptive agents, including intravenous bisphosphonates and denosumab.[58]

Osteopathia Striata

Osteopathia striata with cranial sclerosis (OSCS) is an X-linked dominant condition marked by linear striations, mainly affecting the metaphyseal region of the long bones including those of the hands, feet, and pelvis.[59] Recently, the disease-causing gene was identified as the *WTX* gene, an inhibitor of WNT signaling. Skeletal changes are usually benign but in some patients may be accompanied by cranial sclerosis due to a mutation of the *WTX* gene.[60] The WTX protein functions as a tumor suppressor and inactivating mutations have been associated with some cancers. However, no increase in cancer among patients with osteopathia striata has been reported.

Females demonstrate sclerotic striations on the long bones, cranial sclerosis, and craniofacial dysmorphism. Males have significant skeletal sclerosis, do not have striations but do display a more severe phenotype commonly associated with gross structural malformations, patterning defects, and significant pre- and postnatal lethality.[61] A more serious variant with high lethality is notable for the presence of focal dermal hypoplasia with bony limb defects in males with *WTX* mutation and transmission as an X-linked recessive trait. Diagnosis is made from the radiographic appearance of the lesions. Genetic testing of a *WTX* mutation would be supportive of the diagnosis, but the absence of a *WTX* mutation does not exclude the disease. There are no specific laboratory tests that are characteristic of the disorder. No specific therapy is known.

Osteopoikilosis

The disease of "spotted bones" is a benign autosomal dominant disorder free of symptoms and discovered by its radiologic changes. Osteopoikilosis can occur either as an isolated anomaly or in association with other abnormalities of skin and bone.[62]

Radiographic Findings

The lesions are characterized by a symmetric but unequal distribution of multiple hyperostotic areas in different parts of the skeleton. Small, numerous, round or oval foci of bony sclerosis are seen in the epiphyses and adjacent metaphyses. The lesions are mainly in long bones and not in the skull, vertebrae, or ribs.

Histology

The lesions represent foci of mature remodeled bone with lamellar structure, either connected to adjacent trabeculae of spongy bone or attached to the subchondral cortex.

Genetic Findings

A genome-wide linkage analysis followed by the identification of a microdeletion in an unrelated individual with these diseases allowed a map of the gene that is mutated in osteopoikilosis.[63] In one series, all the affected individuals investigated were heterozygous with respect to a loss-of-function mutation in *LEMD3* (also called *MAN1*), which encodes an inner nuclear membrane protein. A novel C to T substitution at position 2032 bp (cDNA) in exon 8 of *LEMD3*, resulting in a premature stop codon at amino acid position 678, has been described. The mutation cosegregates with the osteopoikilosis phenotype in two other reported families. Often confused with metastatic lesions, the osteopoikilosis lesions are stable over time and do not take up radiotracer of nuclear bone scanning.[62] No treatment is indicated.[64]

DISORDERS OF DEFECTIVE MINERALIZATION

The following group of disorders presents in childhood or as adults with normal or low bone density and the skeletal signs and symptoms of rickets (children) or osteomalacia (adults). Bone matrix mineralization is decreased with weakened mechanical properties of bone that may permit bowing and fracture of weight-bearing pelvis and lower extremities. Heritable causes of rickets and osteomalacia due to mutations and dysfunction of the vitamin D metabolic pathway or the vitamin D receptor are presented in detail in Chapter 13.

Axial Osteomalacia

This rare sporadic disorder of skeletal mineralization typically involves the axial but not the appendicular skeleton.

Clinical Features

The disorder most commonly affects middle-aged or older men who present with pain distributed along the axial skeleton with the most intense involvement, often affecting the cervical spine.[65,66]

Radiologic Changes

Pathologic coarsening of trabecular bone of the spine and pelvis suggests osteomalacia. The skull and long bones show no coarsening or other specific abnormalities. Changes of ankylosing spondylitis have been described in a number of patients.[65]

Laboratory

Normal serum calcium, phosphorus, 25-hydroxyvitamin D and 1,25-dihydroxyvitamin D levels separate axial osteomalacia from rickets or osteomalacia due to vitamin D deficiency. Bone-specific alkaline phosphatase may be mildly elevated.

Pathogenesis and Histology

The primary defect is a poorly understood loss of function of osteoblasts. Histologic examination shows coarsened trabeculae in the axial but not the appendicular skeleton, defective mineralization with flat, inactive osteoblasts, and disorganized collagen fibrils. Analyses of iliac crest bone biopsy specimens show increased cortical and total bone volume with excessive osteoid and normal staining of the lamellar collagen matrix. Mineralization is decreased and irregularly distributed.[66]

Genetic Pathophysiology and Diagnosis

The disorder can be found in family clusters, including vertical transmission from mother to son, suggesting an inheritable disorder. However, there is no evidence for a genetic mutation or other abnormality that could account for the family clusters or sporadic cases.

Treatment

The course is generally benign and is not benefited by long-term calcium or vitamin D therapy. Treatment is symptomatic.

Hypophosphatasia

Hypophosphatasia (HPP) is a rare inherited disorder in which a loss-of-function mutation of the *tissue-nonspecific isozyme of alkaline phosphatase* (*TNSALP*) gene results in low levels of TNSALP[67,68] and defective mineralization involving both bone and teeth. Serum total ALP is composed of ALP isoenzymes expressed mainly in liver, placenta, bone, and kidney.[69] Despite the presence of a functional BSAP isoenzyme in bone in HPP, rickets and osteomalacia develop.

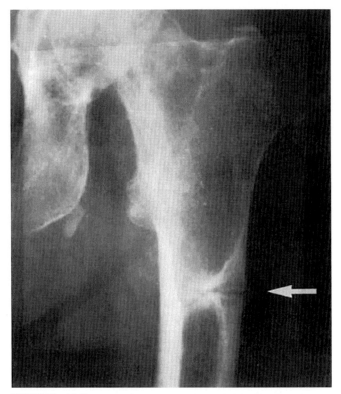

FIGURE 12.6 Hypophosphatasia. An adult with characteristic pseudofracture in the lateral aspect of the left proximal femur (arrow). *Adapted from Ref. [117, p. 942, Fig. 41.5].*

Clinical

HPP may present at any time including *in utero*, in the newborn, early childhood, or adulthood. The clinical signs and symptoms are those of rickets in infants and children and osteomalacia in adults.[70] The severity of the disorder varies greatly from intrauterine polyhydramnios and death with profound unmineralized bone to demineralization and mild rickets in childhood and mild osteomalacia with fractures in adults. Infantile onset presents with symptoms and signs before age six months with failure to thrive, rachitic deformities, flail chest, and functional craniosynostosis with increased intracranial pressure.[71] The hallmark of childhood onset is a premature loss of deciduous teeth and mild rickets.[70–72] HPP often improves during puberty and may recur during the adult years. Adult onset disease activity often presents as metatarsal fractures with poor healing and femoral pseudofractures (Fig. 12.6) with thigh pain a common symptom. In severe cases, the disorder is inherited with an autosomal recessive pattern.[70,71] In milder forms, the inheritance pattern may be autosomal dominant or autosomal recessive.[73,74]

Pathogenesis

Unlike most causes of rickets and osteomalacia in which serum ALK and BSAP are elevated, patients with

HPP have low total ALP.[70] Low TNSALP enzyme activity and protein levels occur while gene expressions of other ALP isoenzymes in germ cell, placental, and intestinal tissues are normal. In the absence of a normal functioning ALP, the naturally occurring ALP substrates phosphoethanolamine (PEA), inorganic pyrophosphate (PPi), and pyridoxal 5'-phosphate (PLP) are not metabolized and accumulate in tissues.[75] In bone, in the absence of TSALP the accumulation of PPi specifically interferes with mineralization of bone through inhibition of hydroxyapatite crystal growth.[76]

Laboratory Tests

Low serum ALP, normal or elevated serum calcium and phosphorus, and clinical and radiologic evidence of rickets or osteomalacia suggest hypophosphatasia. Serum calcium and hypercalciuria may be present. Serum levels of parathyroid hormone, 25-OH-vitamin D, and 1,25-dihydroxyvitamin D are normal. Elevated urine PLP is specific for hypophosphatasia and may be found in asymptomatic parents of affected children.[70]

Genetic Testing

The clinical presentations vary from neonatal to adult and mild to severe and are the result of reduced activity of either one or two pathologic mutations in *ALP*, the gene encoding alkaline phosphatase, and specifically the TNSALP isoenzyme. *ALP* is the only gene known to be associated with HPP. To date, more than 200 distinct pathogenic variants have been described in *ALP* and *TNSALP* in persons from North America, Japan, and Europe. The mutations may be found in any one of the gene's 12 exons. Missense mutations account for about 80% of mutations; the remainder comprise microdeletions/insertions (10%), splicing mutations (5%), nonsense mutations (4%), gross deletions (1%), and a nucleotide substitution affecting the major transcription initiation site. This variety of mutations results in highly variable clinical expression and in a great number of compound heterozygous genotypes.[77,78]

Treatment

Recent advances in enzyme purification have created a recombinant human TNSALP-specific therapy for hypophosphatasia by replacing the missing alkaline phosphatase enzyme activity. This study reported by Whyte et al. represents the first human trial of therapeutic enzyme replacement therapy for hypophosphatasia.[79] Hypophosphatasia often causes premature infant death due to respiratory problems and severe bone disease. In the initial trial of recombinant human enzyme replacement therapy using TNSALP (ENB-0040), 11 young children with severe infantile or perinatal hypophosphatasia were given ENB-0040 treatment for 12 months. Nine patients who completed 12 months of therapy all showed reduced evidence of rickets by six months followed by resolution of the rickets and improved pulmonary function. Growth resumed and developmental milestones were reached. Urine excretion of PPi, PEA, and PLP return to normal. No sign of serious adverse events were reported following the 48-week study.[79]

Effective therapies for the other causes of rickets and osteomalacia, including calcium and vitamin D supplements, are to be avoided due to potential worsening of hypercalcemia and hypercalciuria that are common in hypophosphatasia. Fracture healing is poor, and so insertion of long bone rods for prevention of impending fracture is indicated.

Vitamin D-Dependent Rickets Types I and II

See Chapter 13 for detailed discussions of these two genetic disorders of vitamin D metabolism and function.

DYSPLASIAS OF BONE AND CARTILAGE WITH NORMAL OR LOW BONE MASS

Achondroplasias

The disorder and its variants present as short-limb dwarfism with proximal portions of the long bones primarily affected. There is also a characteristic enlarged skull, saddle nose, and normal trunk with exaggerated lumbar lordosis. Severe spinal deformity may lead to cord compression. The disease occurs in one in 15,000 to one in 40,000 live births, and arises from a gain-of-function mutation of the *fibroblast growth factor receptor 3* (*FGFR3*) gene, resulting in abnormal chondrocyte proliferation at the long bone growth plate. Most patients with achondroplasia (ACH), a milder form called hypochondroplasia (HCH), present as a milder form of achondroplasia, but lack the large head and have less stunting of the long bones. A less common form called thanatophoric dysplasia type I and II (TD I and TD II) present with early onset and high mortality from respiratory failure. All four of these disorders share the same one or two mutations of the same nucleotide in the *FGFR3* gene.[80–83] Sporadic mutations account for 80% of cases, and those with a family history follow an autosomal dominant pattern of inheritance and often may have a more serious expression of the disorder.

Diagnosis

The diagnosis of achondroplasia is largely based on the characteristic findings on physical examination and skeletal radiographs. Genetic testing is available through commercial laboratories and can be done in patients whose diagnosis is uncertain, such as distinction between ACH and HCH or when either ACH or HCH is suspected. Testing for ACH consists of the two point mutations for p.Gly380Arg, whereas testing for HCH is usually limited to p.Asn540Glu. Given the overlapping mutations between ACH and HCH, both p.Gly380Arg and p.Asn540Glu mutations should be tested.

Treatment

There is no effective therapy to reverse the skeletal deformities resulting from the *FGFR3* mutation. Short stature has many causes, but the small number of mutations of the *FGFR3* gene is responsible for two-thirds of all forms of human dwarfism. Yamashita et al.[118] established a human disease-based system for screening potential drugs to treat skeletal growth defects including potential attenuation of activated *FGFR3* in achondroplasia, but to date no agent has been identified that can reverse the activated *FGFR3* mutation.

Multiple Exostoses

The condition is also known as diaphyseal acalsis or osteochondromatosis and is an autosomal dominant pattern of inheritance in which there is displacement of the growth plates that grow through a defect in the perichondrium.[84]

Clinical

The changes are usually asymptomatic; however, the lesions may interfere with joint or tendon functions and may compress peripheral nerves. The growth of the lesions stop when growth ceases and may resume during pregnancy. The risk of malignant transformation into chondrosarcoma is small.[85]

Radiographic Features

A specific change is seen in the growth plates in which a mass forms that is in direct communication with the marrow cavity. As a result, single or multiple lesions are located in the metaphyseal region of the long bones in close proximity to the growth plate. There is also localized resorption of cortical bone.[86,87]

Pathophysiology and Genetic Basis

The characteristic lesions are due to an inactivating mutation of the *EXT1* and *EXT2* genes that normally produce factors that regulate chondrocyte protein processing.[85,88] There is some evidence that the *EXT* gene protein products may function as tumor suppressors.[87] The loss-of-function mutation results in an abnormal proliferation of the growth plate cartilage.

Treatment

There is no known specific therapy. Surgical removal of the lesions may be required when symptoms develop.

Osteogenesis Imperfecta

Osteogenesis imperfecta (OI) known as "brittle bone disease" is a genetic disorder of connective tissue resulting from inactivating mutations or protein abnormalities of type 1 collagen genes and their protein products.[89] OI is characterized by fragile bone that predisposes to spontaneous or low trauma fractures, deformities, and shortened lifespan. OI has a birth prevalence of approximately 6–7/100,000.

Clinical

Recent reviews of clinical characteristics, severity of presentation, and indications for treatment have led to revision of the classification of OI.[90] As many as 11 phenotypes have been described that cover a broad range of severity while sharing the clinical manifestations of fractures and short stature. Recently, the International Nomenclature Committee for Constitutional Disorders of the Skeleton proposed a classification of OI recognizing five types (OI types I–V), based on distinct clinical and radiographic characteristics. Genetic studies of OI types I–IV have yielded more than 1,000 different variants in the *COL1A1* and *COL1A2* genes.[91] Types I–IV are due to autosomal dominant inheritance of mutations of the *COLIA1 and COLIA2* genes and vary in severity from the perinatal lethal (type II),[92,93] severe but not lethal (type III),[93] moderate (type IV),[93] and mild (type I).[94] Autosomal recessive OI is associated with mutations of *CRTAP* and *LEPRE1*[92,95] and PLOD2[96] and FKBP10[97] genes. Three recessive OI types arise from defects in any of the components of the collagen prolyl 3-hydroxylation complex (CRTAP, P3H1, CyPB), which modifies the collagen α1(I)Pro986 residue and leads to abnormal procollagen triple helix folding.[98] A mutation causing deficient BMP1 proteolysis is an autosomal recessive form of OI that generates PICP deficient activity.[99]

The nomenclature has also been expanded to include OI type V as a distinctive subtype.[100] Type V causes mild to moderate disability and is clinically distinguishable from others for the presence of ossification of the interosseous membrane of the forearm with radial head dislocation, hyperplastic callus formation, and no evidence of a type I collagen disorder or abnormal histologic pattern.[101] Phenotypic variability may be found in type V due to a mutation in *IFITM5*.[100]

Laboratory

Serum calcium, phosphorus, parathyroid hormone, and vitamin D levels are normal. High blood or urine levels of bone turnover markers indicate increased bone resorption and formation, with resorption exceeding the rate of formation of the abnormal *COLIA1 and COLIA2* genes.

Radiology and Imaging

The most striking characteristic finding of radiographs of OI patients is the marked reduction in bone density of both the axial and appendicular skeletons. Radiographic changes in the extremities may reveal them to be thin or thick, straight or curving. Infrequently, there may be multiple cysts at the end of long bones with flaring of the metaphyses and a honeycomb composed of thick and coarse trabeculae. Serial radiographs during rapid growth phase may show evolution from one pattern to another. Bowing of the lower extremities may be accompanied by transverse fractures. Horizontal growth plates may be absent with severe growth retardation. Lower extremity fractures that result in deformity of the articular surfaces may result in premature degenerative joint disease of large and small joints. Healing of fractures is generally normal but may generate a more exuberant callus.

In the spine, kyphoscoliosis may be found in almost half of patients, and multiple thoracic anterior wedge compression fractures and biconcave vertebrae frequently accompany the kyphoscoliosis and contribute to reduced respiratory function.

The skull may have several deformities including thickening of both tables, decreased bone density, platybasia with or without basal impression, and enlarged frontal and mastoid sinuses.

Pathophysiology

The most common cause of OI is a mutation in the type I collagen *COL1A1* and *COL1A2* genes or in one of the genes that encode for proteins involved in the several steps of protein assembly, postexpression modification, and osteoblast secretion of type I collagen. Autosomal dominant mutations account for most of the cases with the recessive forms of the disorder being uncommon.[95] One suggested mechanism includes the observation of reduced BMP1 function, leading to bone fragility.[102]

Genetic Testing[103]

Testing is available through a small number of accredited academic research and commercial laboratories. Several applications of genetic testing are useful in OI, including: prenatal counseling for those with OI and affected family members, in patients with severe phenotype including short stature and fractures who test negative for the common *COL1A1* and *COL1A2* mutations, children and adults with multiple fractures who do not have short stature or deformities, to distinguish the autosomal dominant types I–IV from the less common recessive forms, and to confirm the diagnosis of OI when therapeutic intervention with bisphosphonates is being considered. Genes found to have at least one mutation in patients with OI are listed in Table 12.2.

TABLE 12.2 Osteogenesis Imperfecta: Analyzed Genes or DNA/Chromosome Segments

Gene	Name*	Locus**	Inheritance†
BMP1	Bone morphogenetic protein	1 8p21.3	AR
COL1A1	Collagen, type I, alpha 1	17q21.33	AD
COL1A2	Collagen, type I, alpha 2	7q22.3	AD‡
CRTAP	Cartilage-associated protein	3p22.3	AR
FKBP10	FK506-binding protein 10, 65 kDa	17q21.2	AR
IFITM5	Interferon-induced transmembrane protein 5	11p15.5	AD
LEPRE1	Leucine proline-enriched proteoglycan (leprecan)	11p34.2	AR
PLOD2	Procollagen-lysine, 2-oxoglutarate 5-dioxygenase	23q24	AR
PPIB	Peptidylprolyl isomerase B (cyclophilin B)	15q22.31	AR
SERPINF1	Serpin peptidase inhibitor, clade F (alpha-2 antiplasmin, pigment epithelium derived factor), member	117p13.3	AR
SERPINH1	Serpin peptidase inhibitor, clade H (heat shock protein 47) member 1 (collagen binding protein 1)	11q13.5	AR
SP7	Sp7 transcription factor	12q13.13	AR

The gene names are written according to the HUGO Gene Nomenclature Committee website (http://www.genenames.org).

**The loci are written according to the UCSC genome browser (http://genome.ucsc.edu/) using the current genome assembly (GRCh37/hg19).*

†*AD, autosomal dominant; AR, autosomal recessive.*

‡*The vast majority of COL1A2 mutations are autosomal dominantly inherited, but there is one that is nominally recessive: c.3105þ2T4C. Adapted from Ref. [115].*

Treatment

Novel approaches to reduce fractures and deformities now use medical therapy, stem cell transplantation, and surgical procedures to improve the length and quality of life of OI patients.[104,105] For medical therapy, there is evidence that intravenous bisphosphonate (pamidronate) therapy in children may increase bone density and reduce fractures.[106–108] While bisphosphonates do not alter the abnormal collagen or the underlying genetic defect in OI, the well-known effect to reduce bone turnover may contribute to the beneficial bone response.[109] More recently, other effective antiresorptive agents, such as the RANKL inhibitor denosumab, have been shown to potentially have beneficial effects on skeletal growth.[110] The anabolic agent teriparatide in adults with OI increases serum bone turnover markers and bone mineral density in type II, but not for those with types III and IV.[111] In all cases of intervention, DEXA scans offer a quantitative assessment of response.

Transplantation of fetal mesenchymal stem cells for treatment of the severe forms of OI in the pre- and postnatal states is being explored.[112] To improve the quality of life, a variety of surgical procedures have been developed for use in OI patients to stabilize fractures and repair spinal and long bone deformities.[113]

Pachydermoperiostosis (Hypertrophic Osteoarthropathy)

A primary or idiopathic form is an autosomal dominant condition with features of periosteal new bone formation involving the distal aspects of the upper and lower extremities. The primary autosomal recessive disorder is understood as a loss-of-function mutation of the gene that encodes for the 15-hydroxyprostaglandin dehydrogenase.[114]

Clinical

The condition becomes evident in adolescence with the major features of clubbing of the digits of the hands and feet, hyperhidrosis and thickening of the skin, especially of the face and forehead. The features progress over the next decade and then stabilize or resolve. During the active phase, the changes may be accompanied by arthralgias, reduced mobility, and pseudogout. The primary disorder must be differentiated from secondary hypertrophic osteopathy that accompanies serious pulmonary disorders.

Radiographic Features

Standard radiographs can differentiate the two conditions by observing the digits. The primary hypertrophic osteopathy has an irregular periosteal surface in contrast to the secondary pachydermoperiostosis, which has extensive new bone formation with a smooth, undulating surface.

Laboratory Testing

There are no blood, urine, or synovial fluid tests with diagnostic specificity or suggestive of disordered calcium or vitamin D.

Treatment

No effective therapy is known, although nonsteroidal anti-inflammatory agents may reduce arthralgic pain. Surgical intervention is indicated for release of compression of blood vessels or peripheral nerves.

References

1. Ihde LL, Forrester DM, Gottsegen CJ, et al. Sclerosing bone dysplasias: review and differentiation from other causes of osteosclerosis. *Radiographics* 2011;**31**(7):1865–82.
2. Bollerslev J. Autosomal dominant osteopetrosis: bone metabolism and epidemiological, clinical, and hormonal aspects. *Endocr Rev* 1989;**10**(1):45–67.
3. Sly WS, Whyte MP, Sundaram V, et al. Carbonic anhydrase II deficiency in 12 families with the autosomal recessive syndrome of osteopetrosis with renal tubular acidosis and cerebral calcification. *N Engl J Med* 1985;**313**(3):139–45.
4. Sly WS, Sato S, Zhu XL. Evaluation of carbonic anhydrase isozymes in disorders involving osteopetrosis and/or renal tubular acidosis. *Clin Biochem* 1991;**24**(4):311–8.
5. Borthwick KJ, Kandemir N, Topaloglu R, et al. A phenocopy of CAII deficiency: a novel genetic explanation for inherited infantile osteopetrosis with distal renal tubular acidosis. *J Med Genet* 2003;**40**(2):115–21.
6. Whyte MP, Kempa LG, McAlister WH, Zhang F, Mumm S, Wenkert D. Elevated serum lactate dehydrogenase isoenzymes and aspartate transaminase distinguish Albers-Schonberg disease (chloride channel 7 deficiency osteopetrosis) among the sclerosing bone disorders. *J Bone Min Res* 2010;**25**(11):2515–26.
7. Shah GN, Bonapace G, Hu PY, Strisciuglio P, Sly WS. Carbonic anhydrase II deficiency syndrome (osteopetrosis with renal tubular acidosis and brain calcification): novel mutations in CA2 identified by direct sequencing expand the opportunity for genotype-phenotype correlation. *Hum Mutat* 2004;**24**(3):272.
8. Sly WS, Hewett-Emmett D, Whyte MP, Yu YS, Tashian RE. Carbonic anhydrase II deficiency identified as the primary defect in the autosomal recessive syndrome of osteopetrosis with renal tubular acidosis and cerebral calcification. *Proc Natl Acad Sci USA* 1983;**80**(9):2752–6.
9. Henriksen K, Gram J, Hoegh-Andersen P, et al. Osteoclasts from patients with autosomal dominant osteopetrosis type I caused by a T253I mutation in low-density lipoprotein receptor-related protein 5 are normal *in vitro*, but have decreased resorption capacity *in vivo*. *Am J Pathol* 2005;**167**(5):1341–8.
10. Johnston Jr CC, Lavy N, Lord T, Vellios F, Merritt AD, Deiss Jr WP. Osteopetrosis. A clinical, genetic, metabolic, and morphologic study of the dominantly inherited, benign form. *Medicine* 1968;**47**(2):149–67.
11. Kant P, Sharda N, Bhowate RR. Clinical and radiological findings of autosomal dominant osteopetrosis Type II: a case report. *Case Rep Dentist* 2013;**2013**:707343.
12. Pangrazio A, Cassani B, Guerrini MM, et al. RANK-dependent autosomal recessive osteopetrosis: characterization of five new cases with novel mutations. *J Bone Min Res* 2012;**27**(2):342–51.

13. Rashid BM, Rashid NG, Schulz A, Lahr G, Nore BF. A novel missense mutation in the CLCN7 gene linked to benign autosomal dominant osteopetrosis: a case series. *J Med Case Rep* 2013;**7**:7.

14. Pangrazio A, Puddu A, Oppo M, et al. Exome sequencing identifies CTSK mutations in patients originally diagnosed as intermediate osteopetrosis. *Bone* 2014;**59**:122–6.

15. Helfrich MH, Aronson DC, Everts V, et al. Morphologic features of bone in human osteopetrosis. *Bone* 1991;**12**(6):411–9.

16. Henriksen K, Gram J, Schaller S, et al. Characterization of osteoclasts from patients harboring a G215R mutation in ClC-7 causing autosomal dominant osteopetrosis type II. *Am J Pathol* 2004;**164**(5):1537–45.

17. Sobacchi C, Frattini A, Guerrini MM, et al. Osteoclast-poor human osteopetrosis due to mutations in the gene encoding RANKL. *Nat Genet* 2007;**39**(8):960–2.

18. Boyce BF, Xing L. Functions of RANKL/RANK/OPG in bone modeling and remodeling. *Arch Biochem Biophys* 2008;**473**(2):139–46.

19. Lo Iacono N, Blair HC, Poliani PL, et al. Osteopetrosis rescue upon RANKL administration to Rankl (−/−) mice: a new therapy for human RANKL-dependent ARO. *J Bone Min Res* 2012;**27**(12):2501–10.

20. Whyte MP, Wenkert D, McAlister WH, et al. Dysosteosclerosis presents as an "osteoclast-poor" form of osteopetrosis: comprehensive investigation of a 3-year-old girl and literature review. *J Bone Min Res* 2010;**25**(11):2527–39.

21. Sobacchi C, Schulz A, Coxon FP, Villa A, Helfrich MH. Osteopetrosis: genetics, treatment and new insights into osteoclast function. *Nat Rev Endocrinol* 2013;**9**(9):522–36.

22. Waguespack SG, Hui SL, Dimeglio LA, Econs MJ. Autosomal dominant osteopetrosis: clinical severity and natural history of 94 subjects with a chloride channel 7 gene mutation. *J Clin Endocrinol Metab* 2007;**92**(3):771–8.

23. Harris SE, MacDougall M, Horn D, et al. Meox2Cre-mediated disruption of CSF-1 leads to osteopetrosis and osteocyte defects. *Bone* 2012;**50**(1):42–53.

24. Van Wesenbeeck L, Odgren PR, Coxon FP, et al. Involvement of PLEKHM1 in osteoclastic vesicular transport and osteopetrosis in incisors absent rats and humans. *J Clin Invest* 2007;**117**(4):919–30.

25. Taranta A, Migliaccio S, Recchia I, et al. Genotype-phenotype relationship in human ATP6i-dependent autosomal recessive osteopetrosis. *Am J Pathol* 2003;**162**(1):57–68.

26. Hughes AE, Ralston SH, Marken J, et al. Mutations in TNFRSF11A, affecting the signal peptide of RANK, cause familial expansile osteolysis. *Nat Genet* 2000;**24**(1):45–8.

27. Villa A, Pangrazio A, Caldana E, et al. Prognostic potential of precise molecular diagnosis of autosomal recessive osteopetrosis with respect to the outcome of bone marrow transplantation. *Cytotechnology* 2008;**58**(1):57–62.

28. Key L, Carnes D, Cole S, et al. Treatment of congenital osteopetrosis with high-dose calcitriol. *N Engl J Med* 1984;**310**(7):409–15.

29. Janssens K, Vanhoenacker F, Bonduelle M, et al. Camurati-Engelmann disease: review of the clinical, radiological, and molecular data of 24 families and implications for diagnosis and treatment. *J Med Genet* 2006;**43**(1):1–11.

30. Janssens K, Gershoni-Baruch R, Guanabens N, et al. Mutations in the gene encoding the latency-associated peptide of TGF-beta 1 cause Camurati-Engelmann disease. *Nat Genet* 2000;**26**(3):273–5.

31. Ayyavoo A, Derraik JG, Cutfield WS, Hofman PL. Elimination of pain and improvement of exercise capacity in Camurati-Engelmann disease with losartan. *J Clin Endocrinol Metab* 2014;**99**(11):3978–82.

32. Caracas HP, Figueiredo PS, Mestrinho HD, Acevedo AC, Leite AF. Pycnodysostosis with craniosynostosis: case report of the craniofacial and oral features. *Clin Dysmorphol* 2012;**21**(1):19–21.

33. Hodder A, Huntley C, Aronson JK, Ramachandran M. Pycnodysostosis and the making of an artist. *Gene* 2015;**555**(1):59–62.

34. Soliman AT, Rajab A, AlSalmi I, Darwish A, Asfour M. Defective growth hormone secretion in children with pycnodysostosis and improved linear growth after growth hormone treatment. *Arch Dis Child* 1996;**75**(3):242–4.

35. Arman A, Bereket A, Coker A, et al. Cathepsin K analysis in a pycnodysostosis cohort: demographic, genotypic and phenotypic features. *Orphanet J Rare Dis* 2014;**9**:60.

36. Rothenbuhler A, Piquard C, Gueorguieva I, Lahlou N, Linglart A, Bougneres P. Near normalization of adult height and body proportions by growth hormone in pycnodysostosis. *J Clin Endocrinol Metab* 2010;**95**(6):2827–31.

37. Beighton P, Barnard A, Hamersma H, van der Wouden A. The syndromic status of sclerosteosis and van Buchem disease. *Clin Genet* 1984;**25**(2):175–81.

38. Moretti C, D'Osualdo F, Modesto A, Benedetti A, Corsi M. Endosteal hyperostosis with dominant transmission. Description of 8 cases in 3 generations of the same nuclear family. *La Radiologia Medica* 1982;**68**(3):151–8.

39. Wengenroth M, Vasvari G, Federspil PA, Mair J, Schneider P, Stippich C. Case 150: Van Buchem disease (hyperostosis corticalis generalisata). *Radiology* 2009;**253**(1):272–6.

40. van Lierop AH, Hamdy NA, Hamersma H, et al. Patients with sclerosteosis and disease carriers: human models of the effect of sclerostin on bone turnover. *J Bone Min Res* 2011;**26**(12):2804–11.

41. van Lierop AH, Moester MJ, Hamdy NA, Papapoulos SE. Serum Dickkopf 1 levels in sclerostin deficiency. *J Clin Endocrinol Metab* 2014;**99**(2):E252–6.

42. Balemans W, Ebeling M, Patel N, et al. Increased bone density in sclerosteosis is due to the deficiency of a novel secreted protein (SOST). *Hum Mol Genet* 2001;**10**(5):537–43.

43. Van Hul W, Balemans W, Van Hul E, et al. Van Buchem disease (hyperostosis corticalis generalisata) maps to chromosome 17q12-q21. *Am J Hum Genet* 1998;**62**(2):391–9.

44. Simsek-Kiper PO, Dikoglu E, Campos-Xavier B, et al. Positive effects of an angiotensin II type 1 receptor antagonist in Camurati-Engelmann disease: a single case Observation. *Am J Med Genet Part A* 2014;**164A**(10):2667–71.

45. Suresh S, Muthukumar T, Saifuddin A. Classical and unusual imaging appearances of melorheostosis. *Clin Radiol* 2010;**65**(8):593–600.

46. Whyte MP, Murphy WA, Siegel BA. 99mTc-pyrophosphate bone imaging in osteopoikilosis, osteopathia striata, and melorheostosis. *Radiology* 1978;**127**(2):439–43.

47. Mumm S, Wenkert D, Zhang X, McAlister WH, Mier RJ, Whyte MP. Deactivating germline mutations in LEMD3 cause osteopoikilosis and Buschke-Ollendorff syndrome, but not sporadic melorheostosis. *J Bone Min Res* 2007;**22**(2):243–50.

48. Boyden LM, Mao J, Belsky J, et al. High bone density due to a mutation in LDL-receptor-related protein 5. *N Engl J Med* 2002;**346**(20):1513–21.

49. Gregson CL, Poole KE, McCloskey EV, et al. Elevated circulating sclerostin concentrations in individuals with high bone mass, with and without LRP5 mutations. *J Clin Endocrinol Metab* 2014;**99**(8):2897–907.

50. Rickels MR, Zhang X, Mumm S, Whyte MP. Oropharyngeal skeletal disease accompanying high bone mass and novel LRP5 mutation. *J Bone Min Res* 2005;**20**(5):878–85.

51. Frost M, Andersen T, Gossiel F, et al. Levels of serotonin, sclerostin, bone turnover markers as well as bone density and microarchitecture in patients with high-bone-mass phenotype due to a mutation in Lrp5. *J Bone Min Res* 2011;**26**(8):1721–8.

52. Golob DS, McAlister WH, Mills BG, et al. Juvenile Paget disease: life-long features of a mildly affected young woman. *J Bone Min Res* 1996;**11**(1):132–42.

53. Whyte MP, Singhellakis PN, Petersen MB, Davies M, Totty WG, Mumm S. Juvenile Paget's disease: the second reported, oldest patient is homozygous for the TNFRSF11B "Balkan" mutation (966_969delTGACinsCTT), which elevates circulating immunoreactive osteoprotegerin levels. J Bone Min Res 2007;**22**(6):938–46.

54. Whyte MP, Tau C, McAlister WH, et al. Juvenile Paget's disease with heterozygous duplication within TNFRSF11A encoding RANK. Bone 2014;**68**:153–61.

55. Osterberg PH, Wallace RG, Adams DA, et al. Familial expansile osteolysis. A new dysplasia. J Bone Joint Surg Br 1988;**70**(2):255–60.

56. Martin TJ. Paracrine regulation of osteoclast formation and activity: milestones in discovery. J Musculoskel Neuron Interact 2004;**4**(3):243–53.

57. Whyte MP, Obrecht SE, Finnegan PM, et al. Osteoprotegerin deficiency and juvenile Paget's disease. N Engl J Med 2002;**347**(3):175–84.

58. Polyzos SA, Singhellakis PN, Naot D, et al. Denosumab treatment for juvenile Paget's disease: results from two adult patients with osteoprotegerin deficiency ("Balkan" mutation in the TNFRSF11B gene). J Clin Endocrinol Metab 2014;**99**(3):703–7.

59. Bass HN, Weiner JR, Goldman A, Smith LE, Sparkes RS, Crandall BF. Osteopathia striata syndrome. Clinical, genetic and radiologic considerations. Clin Pediatr 1980;**19**(5):369–73.

60. Perdu B, de Freitas F, Frints SG, et al. Osteopathia striata with cranial sclerosis owing to WTX gene defect. J Bone Min Res 2010;**25**(1):82–90.

61. Pellegrino JE, McDonald-McGinn DM, Schneider A, Markowitz RI, Zackai EH. Further clinical delineation and increased morbidity in males with osteopathia striata with cranial sclerosis: an X-linked disorder? Am J Med Genet 1997;**70**(2):159–65.

62. Ng C, Schwartzman L, Moadel R, Haigentz Jr M. Osteopoikilosis: a benign condition with the appearance of metastatic bone disease. J Clin Oncol 2014;**33**(18):e77–8.

63. Korkmaz MF, Elli M, Ozkan MB, et al. Osteopoikilosis: report of a familial case and review of the literature. Rheumatol Int 2014;**35**(5):921–4.

64. Hill CE, McKee L. Osteopoikilosis: an important incidental finding. Injury 2015;**46**(7):1403–5.

65. Nelson AM, Riggs BL, Jowsey JO. Atypical axial osteomalacia. Report of four cases with two having features of ankylosing spondylitis. Arthritis Rheum 1978;**21**(6):715–22.

66. Whyte MP, Fallon MD, Murphy WA, Teitelbaum SL. Axial osteomalacia. Clinical, laboratory and genetic investigation of an affected mother and son. Am J Med 1981;**71**(6):1041–9.

67. Fedde KN, Cole DE, Whyte MP. Pseudohypophosphatasia: aberrant localization and substrate specificity of alkaline phosphatase in cultured skin fibroblasts. Am J Hum Genet 1990;**47**(5):776–83.

68. Fedde KN, Whyte MP. Alkaline phosphatase (tissue-nonspecific isoenzyme) is a phosphoethanolamine and pyridoxal-5'-phosphate ectophosphatase: normal and hypophosphatasia fibroblast study. Am J Hum Genet 1990;**47**(5):767–75.

69. Buchet R, Millan JL, Magne D. Multisystemic functions of alkaline phosphatases. Methods Mol Biol (Clifton, NJ) 2013;**1053**:27–51.

70. Whyte MP, Teitelbaum SL, Murphy WA, Bergfeld MA, Avioli LV. Adult hypophosphatasia. Clinical, laboratory, and genetic investigation of a large kindred with review of the literature. Medicine 1979;**58**(5):329–47.

71. Whyte MP, Zhang F, Wenkert D, et al. Hypophosphatasia: validation and expansion of the clinical nosology for children from 25 years experience with 173 pediatric patients. Bone 2015;**75**:229–39.

72. Berkseth KE, Tebben PJ, Drake MT, Hefferan TE, Jewison DE, Wermers RA. Clinical spectrum of hypophosphatasia diagnosed in adults. Bone 2013;**54**(1):21–7.

73. Henthorn PS, Raducha M, Fedde KN, Lafferty MA, Whyte MP. Different missense mutations at the tissue-nonspecific alkaline phosphatase gene locus in autosomal recessively inherited forms of mild and severe hypophosphatasia. Proc Natl Acad Sci USA 1992;**89**(20):9924–8.

74. Weinstein RS, Whyte MP. Heterogeneity of adult hypophosphatasia. Report of severe and mild cases. Arch Intern Med 1981;**141**(6):727–31.

75. Whyte MP, Mahuren JD, Vrabel LA, Coburn SP. Markedly increased circulating pyridoxal-5'-phosphate levels in hypophosphatasia. Alkaline phosphatase acts in vitamin B6 metabolism. J Clin Invest 1985;**76**(2):752–6.

76. Register TC, Wuthier RE. Effect of pyrophosphate and two diphosphonates on 45Ca and 32Pi uptake and mineralization by matrix vesicle-enriched fractions and by hydroxyapatite. Bone 1985;**6**(5):307–12.

77. Taillandier A, Sallinen SL, Brun-Heath I, De Mazancourt P, Serre JL, Mornet E. Childhood hypophosphatasia due to a de novo missense mutation in the tissue-nonspecific alkaline phosphatase gene. J Clin Endocrinol Metab 2005;**90**(4):2436–9.

78. Watanabe H, Goseki-Sone M, Orimo H, Hamatani R, Takinami H, Ishikawa I. Function of mutant (G1144A) tissue-nonspecific ALP gene from hypophosphatasia. J Bone Min Res 2002;**17**(11):1945–8.

79. Whyte MP, Greenberg CR, Salman NJ, et al. Enzyme-replacement therapy in life-threatening hypophosphatasia. N Engl J Med 2012;**366**(10):904–13.

80. Bellus GA, Hefferon TW, Ortiz de Luna RI, et al. Achondroplasia is defined by recurrent G380R mutations of FGFR3. Am J Hum Genet 1995;**56**(2):368–73.

81. Bellus GA, McIntosh I, Smith EA, et al. A recurrent mutation in the tyrosine kinase domain of fibroblast growth factor receptor 3 causes hypochondroplasia. Nat Genet 1995;**10**(3):357–9.

82. Bellus GA, Spector EB, Speiser PW, et al. Distinct missense mutations of the FGFR3 lys650 codon modulate receptor kinase activation and the severity of the skeletal dysplasia phenotype. Am J Hum Genet 2000;**67**(6):1411–21.

83. Park WJ, Bellus GA, Jabs EW. Mutations in fibroblast growth factor receptors: phenotypic consequences during eukaryotic development. Am J Hum Genet 1995;**57**(4):748–54.

84. Francannet C, Cohen-Tanugi A, Le Merrer M, Munnich A, Bonaventure J, Legeai-Mallet L. Genotype-phenotype correlation in hereditary multiple exostoses. J Med Genet 2001;**38**(7):430–4.

85. Hecht JT, Hogue D, Wang Y, et al. Hereditary multiple exostoses (EXT): mutational studies of familial EXT1 cases and EXT-associated malignancies. Am J Hum Genet 1997;**60**(1):80–6.

86. Kok HK, Fitzgerald L, Campbell N, et al. Multimodality imaging features of hereditary multiple exostoses. Brit J Radiol 2013;**86**(1030):20130398.

87. Duncan G, McCormick C, Tufaro F. The link between heparan sulfate and hereditary bone disease: finding a function for the EXT family of putative tumor suppressor proteins. J Clin Invest 2001;**108**(4):511–6.

88. Jamsheer A, Socha M, Sowinska-Seidler A, Telega K, Trzeciak T, Latos-Bielenska A. Mutational screening of EXT1 and EXT2 genes in Polish patients with hereditary multiple exostoses. J Appl Genet 2014;**55**(2):183–8.

89. Deyle DR, Khan IF, Ren G, et al. Normal collagen and bone production by gene-targeted human osteogenesis imperfecta iPSCs. Mol Ther 2012;**20**(1):204–13.

90. Van Dijk FS, Sillence DO. Osteogenesis imperfecta: clinical diagnosis, nomenclature and severity assessment. Am J Med Genet Part A 2014;**164A**(6):1470–81.

91. Warman ML, Cormier-Daire V, Hall C, et al. Nosology and classification of genetic skeletal disorders: 2010 revision. Am J Med Genet Part A 2011;**155A**(5):943–68.

92. Cabral WA, Barnes AM, Adeyemo A, et al. A founder mutation in LEPRE1 carried by 1.5% of West Africans and 0. 4% of African Americans causes lethal recessive osteogenesis imperfecta. Genet Med 2012;**14**(5):543–51.

93. Pyott SM, Schwarze U, Christiansen HE, et al. Mutations in PPIB (cyclophilin B) delay type I procollagen chain association and result in perinatal lethal to moderate osteogenesis imperfecta phenotypes. *Hum Mol Genet* 2011;**20**(8):1595–609.

94. Forlino A, Cabral WA, Barnes AM, Marini JC. New perspectives on osteogenesis imperfecta. *Nat Rev Endocrinol* 2011;**7**(9):540–57.

95. Baldridge D, Schwarze U, Morello R, et al. CRTAP and LEPRE1 mutations in recessive osteogenesis imperfecta. *Hum Mutat* 2008;**29**(12):1435–42.

96. Puig-Hervas MT, Temtamy S, Aglan M, et al. Mutations in PLOD2 cause autosomal-recessive connective tissue disorders within the Bruck syndrome – osteogenesis imperfecta phenotypic spectrum. *Hum Mutat* 2012;**33**(10):1444–9.

97. Kelley BP, Malfait F, Bonafe L, et al. Mutations in FKBP10 cause recessive osteogenesis imperfecta and Bruck syndrome. *J Bone Min Res* 2011;**26**(3):666–72.

98. Marini JC, Blissett AR. New genes in bone development: what's new in osteogenesis imperfecta. *J Clin Endocrinol Metab* 2013;**98**(8):3095–103.

99. Martinez-Glez V, Valencia M, Caparros-Martin JA, et al. Identification of a mutation causing deficient BMP1/mTLD proteolytic activity in autosomal recessive osteogenesis imperfecta. *Hum Mutat* 2012;**33**(2):343–50.

100. Shapiro JR, Lietman C, Grover M, et al. Phenotypic variability of osteogenesis imperfecta type V caused by an IFITM5 mutation. *J Bone Min Res* 2013;**28**(7):1523–30.

101. Lee DY, Cho TJ, Choi IH, et al. Clinical and radiological manifestations of osteogenesis imperfecta type V. *J Korean Med Sci* 2006;**21**(4):709–14.

102. Asharani PV, Keupp K, Semler O, et al. Attenuated BMP1 function compromises osteogenesis, leading to bone fragility in humans and zebrafish. *Am J Hum Genet* 2012;**90**(4):661–74.

103. Byers PH, Krakow D, Nunes ME, Pepin M. American College of Medical Genetics. Genetic evaluation of suspected osteogenesis imperfecta (OI). *Genet Med* 2006;**8**(6):383–8.

104. Harrington J, Sochett E, Howard A. Update on the evaluation and treatment of osteogenesis imperfecta. *Pediatr Clin N Am* 2014;**61**(6):1243–57.

105. Biggin A, Munns CF. Osteogenesis imperfecta: diagnosis and treatment. *Curr Osteoporosis Rep* 2014;**12**(3):279–88.

106. Glorieux FH, Bishop NJ, Plotkin H, Chabot G, Lanoue G, Travers R. Cyclic administration of pamidronate in children with severe osteogenesis imperfecta. *N Engl J Med* 1998;**339**(14):947–52.

107. Chan JK, Gotherstrom C. Prenatal transplantation of mesenchymal stem cells to treat osteogenesis imperfecta. *Front Pharmacol* 2014;**5**:223.

108. Falk MJ, Heeger S, Lynch KA, et al. Intravenous bisphosphonate therapy in children with osteogenesis imperfecta. *Pediatrics* 2003;**111**(3):573–8.

109. Hald JD, Evangelou E, Langdahl BL, Ralston SH. Bisphosphonates for the prevention of fractures in osteogenesis imperfecta: meta-analysis of placebo-controlled trials. *J Bone Min Res* 2015;**30**(5):929–33.

110. Hoyer-Kuhn H, Semler O, Schoenau E. Effect of denosumab on the growing skeleton in osteogenesis imperfecta. *J Clin Endocrinol Metab* 2014;**99**(11):3954–5.

111. Orwoll ES, Shapiro J, Veith S, et al. Evaluation of teriparatide treatment in adults with osteogenesis imperfecta. *J Clin Invest* 2014;**124**(2):491–8.

112. Gotherstrom C, Westgren M, Shaw SW, et al. Pre- and postnatal transplantation of fetal mesenchymal stem cells in osteogenesis imperfecta: a two-center experience. *Stem Cells Transl Med* 2014;**3**(2):255–64.

113. Georgescu I, Vlad C, Gavriliu TS, Dan S, Parvan AA. Surgical treatment in osteogenesis imperfecta – 10 years experience. *J Med Life* 2013;**6**(2):205–13.

114. Uppal S, Diggle CP, Carr IM, et al. Mutations in 15-hydroxyprostaglandin dehydrogenase cause primary hypertrophic osteoarthropathy. *Nat Genet* 2008;**40**(6):789–93.

115. van Dijk FS, Dalgleish R, Malfait F, Maugeri A, Rusinska A, Semler O, Symoens S, Pals G. Clinical utility gene card for: osteogenesis imperfecta. *Eur J Hum Gent.* 2013;**21**. doi:10.1038/ejhg.2012.210; published online September 26, 2012.

116. Whyte MP. Hereditary metabolic and dysplastic skeletal disorders. In: Coe FL, Favus MJ, editors. *Disorders of Bone and Mineral Metabolism.* 2nd ed. Philadelphia: Lippincott Williams and Wilkins, Wolters Kluwer Co; 2002.

117. Resnick D., editor. *Bone and Joint Imaging.* 2nd ed. Philadelphia: W.B. Saunders Co; 1996.

118. Yamashita A, Morioka M, Kishi H, Kimura T, Yahara Y, Okada M, Fujita K, Sawai H, Ikegawa S, Tsumaki N. Statin treatment rescues FGFR3 skeletal dysplasia phenotypes. *Nature* 2014;**513**:507–11.

13

Vitamin D Disorders

Michael F. Holick

Department of Medicine, Section of Endocrinology, Nutrition, and Diabetes, Vitamin D, Skin and Bone Research Laboratory, Boston University Medical Center, Boston, MA, USA

INTRODUCTION

Vitamin D is essential for bone health. It is responsible for maintaining calcium and phosphorus metabolism that results in the development and maintenance of a healthy skeleton. Genetic defects in the metabolism and recognition of vitamin D lead to alterations in calcium, phosphorus, and bone metabolism. The severity of the bone disease associated with these genetic defects depends on the severity of the consequences of the genetic defect on calcium and phosphorus metabolism. In children there is often poor mineralization of the bone leading to classic rachitic changes in the skeleton. The skeletal deformities often process throughout adulthood. The goal of this chapter is to provide a perspective on the causes and consequences of these genetic defects on calcium, phosphorus, and bone metabolism and to provide treatment strategies.

CALCIUM, PHOSPHORUS, AND VITAMIN D METABOLISM

Vitamin D is recognized as the sunshine vitamin because it is made in the skin during exposure to sunlight.[1] During exposure to sunlight, the ultraviolet B photons are absorbed by 7-dehydrocholesterol in the epidermis and dermis, resulting in the production of previtamin D_3 (Fig. 13.1). Once formed in the plasma membrane of the skin cell it rapidly isomerizes to vitamin D_3. Vitamin D_3 is ejected out of the plasma membrane into the extracellular space, whereby diffusion enters the derma capillary bed for transport on the vitamin D binding protein (DBP) to the liver. Vitamin D coming from the diet is incorporated into the chylomicrons and absorbed into the lymphatic system, which then enters into the venous blood supply. There are two forms of vitamin D. Vitamin D_2 comes from the UVB irradiation of yeast. Vitamin D_3 is

made from 7-dehydrocholesterol obtained from sheep's lanolin. Physiologic doses of vitamin D_2 and vitamin D_3 are equally effective in maintaining vitamin D status (D represents either D_2 or D_3).[2]

Vitamin D is biologically inert and requires a hydroxylation on carbon 25 by vitamin D-25-hydroxylase to form the major circulating form of vitamin D 25-hydroxyvitamin D (25(OH)D) (Fig. 13.1). This metabolite is also biologically inert and requires further activation in the kidneys on carbon 1 by the 25-hydroxyvitamin D-1-hydroxylase (cyp27B1;1-OHase) to form 1,25-dihydroxyvitamin D (1,25(OH)$_2$D). Vitamin D_2 and vitamin D_3 are metabolized in the liver and kidneys in a similar manner.

1,25(OH)$_2$D interacts with its nuclear receptor, the vitamin D receptor (VDR), in the small intestine to increase the efficiency of intestinal calcium absorption. In a vitamin D deficient state, the small intestine is able to absorb passively about 10–15% of dietary calcium. Vitamin D sufficiency enhances the absorption of calcium to 30–40%.[1,3] 1,25(OH)$_2$D also plays an important role in phosphorus metabolism by enhancing the efficiency of phosphorus absorption in the jejunum and ilium. The small intestine passively absorbs about 60% of dietary phosphate, and vitamin D sufficiency increases it to about 80%. When there is adequate calcium and phosphorus in the diet, vitamin D is able to maintain the serum calcium in a normal range of 8.6–10.2 mg/dL and the serum phosphorus level at 2.5–4.5 mg/dL. 1,25(OH)$_2$D interacts with its receptor in the kidneys to enhance calcium and phosphate reabsorption. It is the calcium phosphate product in the circulation and in the extravascular space that is critically important for the mineralization of osteoid laid down by osteoblasts.

When dietary calcium is inadequate to maintain extracellular calcium concentrations, 1,25(OH)$_2$D interacts with its VDR in osteoblasts to increase the expression of receptor activator of NFκB ligand (RANKL). This sets into motion the ability of the osteoclast to dock with the RANK

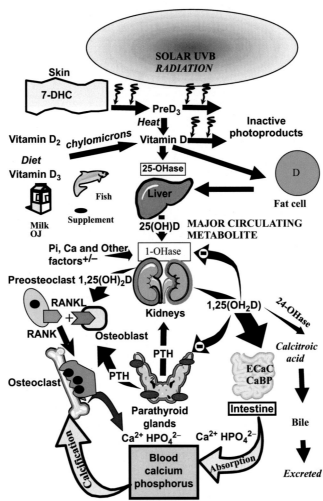

FIGURE 13.1 **Schematic representation of the synthesis and metabolism of vitamin D for regulating calcium, phosphorus, and bone metabolism.** During exposure to sunlight, 7-dehydrocholesterol in the skin is converted to previtamin D_3. Previtamin D_3 immediately converts by a heat-dependent process to vitamin D_3. Excessive exposure to sunlight degrades previtamin D_3 and vitamin D_3 into inactive photoproducts. Vitamin D_2 and vitamin D_3 from dietary sources is incorporated into chylomicrons, transported by the lymphatic system into the venous circulation. Vitamin D (D represents D_2 or D_3) made in the skin or ingested in the diet can be stored in and then released from fat cells. Vitamin D in the circulation is bound to the vitamin D binding protein, which transports it to the liver where vitamin D is converted by the vitamin D-25-hydroxylase to 25-hydroxyvitamin D [25(OH)D]. This is the major circulating form of vitamin D that is used by clinicians to measure vitamin D status (although most reference laboratories report the normal range to be 20–100 ng/mL, the preferred healthful range is 30–60 ng/mL). It is biologically inactive and must be converted in the kidneys by the 25-hydroxyvitamin D-1α-hydroxylase (1-OHase) to its biologically active form 1,25-dihydroxyvitamin D [1,25(OH)$_2$D]. Serum phosphorus, calcium fibroblast growth factors (FGF 23) and other factors can either increase (+) or decrease (−) the renal production of 1,25(OH)$_2$D. 1,25(OH)$_2$D feedback regulates its own synthesis and decreases the synthesis and secretion of parathyroid hormone (PTH) in the parathyroid glands. 1,25(OH)$_2$D increases the expression of the 25-hydroxyvitamin D-24-hydroxylase (24-OHase) to catabolize 1,25(OH)$_2$D to the water soluble biologically inactive calcitroic acid, which is excreted in the bile. 1,25(OH)$_2$D enhances intestinal calcium absorption in the small intestine by stimulating the expression of the epithelial calcium channel (ECaC) and the calbindin 9K (calcium binding protein; CaBP). 1,25(OH)$_2$D is recognized by its receptor in osteoblasts causing an increase in the expression of receptor activator of NFκB ligand (RANKL). Its receptor RANK on the preosteoclast binds RANKL, which induces the preosteoclast to become a mature osteoclast. The mature osteoclast removes calcium and phosphorus from the bone to maintain blood calcium and phosphorus levels. Adequate calcium and phosphorus levels promote the mineralization of the skeleton.

receptor on the preosteoclast, resulting in the formation of a multinucleated mature osteoclast (Fig. 13.2). The mature osteoclast releases HCl and collagenases resulting in the destruction of the mineral and matrix releasing calcium and phosphorus into the circulation.[1,4]

The interaction of 1,25(OH)$_2$D with its VDR in osteoblasts also results in the expression of osteocalcin and

alkaline phosphatase, which play an important role in bone metabolism. However, 1,25(OH)$_2$D does not appear to play a direct role in the mineralization process. It is known in both rodents and humans that vitamin D is not necessary for the mineralization of osteoid matrix. This was demonstrated when vitamin D deficient rats were either infused with calcium and phosphorus to maintain

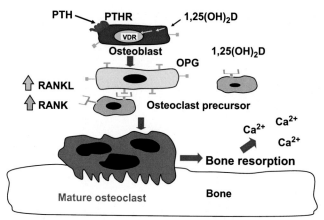

FIGURE 13.2 Both 1,25(OH)₂D and PTH stimulate the mobilization of calcium from the skeleton by interacting with their respective receptors on osteoblasts, which induces expression of receptor activator of NFκB (RANK) ligand (RANKL). The RANK on the immature osteoclast's plasma membrane binds to RANKL, causing it to mature and coalesce with other osteoclast precursors to become mature multinuclear osteoclasts. Osteoprotogerin (OPG) is a soluble decoy of RANK that binds to RANKL to inhibit osteoclastogenesis. *Reproduced with permission from Holick (2004).*

a normal calcium–phosphorus product in the circulation or when they received a high calcium lactose, high phosphorus diet that maintained a normal serum calcium–phosphorus product.[5,6] In both studies, bone development and mineralization occurred in a normal manner similar to animals that were vitamin D sufficient.

VITAMIN D DEFICIENCY AND RICKETS

Vitamin D deficiency is the most common cause of rickets and osteomalacia. Because vitamin D plays such an important role in maintaining serum calcium, it was assumed that the mechanism by which vitamin D deficiency causes a mineralization defect of the osteoid is due to hypocalcemia. However, the body maintains its serum calcium in the normal range in order to maintain a wide variety of metabolic functions and neuromuscular activity, and thus most patients who are vitamin D deficient have a normal serum calcium due to the increased bone calcium mobilization by 1,25(OH)₂D and parathyroid hormone (PTH) (Figs 13.1 and 13.2). In a vitamin D deficient state, ionized calcium declines, causing an immediate stimulus to the calcium sensor in the parathyroid glands, resulting in an increase in the expression and production of PTH.[7] PTH increases tubular reabsorption of calcium in the kidneys and increases the expression of RANKL in osteoblasts, resulting in an increased number of osteoclasts, which remove precious calcium stores from the skeleton and increases the kidneys' production of 1,25(OH)₂D. PTH also causes the internalization of the sodium-dependent phosphate cotransporter (NaPi-2A),

resulting in loss of phosphate into the urine. This results in a lowering of the serum phosphorus level, causing an inadequate calcium–phosphorus product needed for matrix mineralization. Thus, the major cause of osteomalacia and rickets is due to secondary hyperparathyroidism and low–normal or low serum phosphorus level.

GENETIC CAUSES OF RICKETS – OSTEOMALACIA: DISORDERS IN VITAMIN D METABOLISM AND RECOGNITION

Vitamin D-25 Hydroxylase Deficiency

There are at least four different *hepatic* enzymes found in the mitochondria and microsomes that are capable of metabolizing vitamin D to 25(OH)D.[8] This is the likely reason why there have only been rare reports of a 25-hydroxylase deficiency causing rickets in children.[9]

Vitamin D-Dependent Rickets Type I: Pseudovitamin D Deficiency Rickets (PDDR)

Before the discovery that vitamin D needed to be metabolized in the liver and kidneys before it could carry out its biologic actions on calcium and phosphorus metabolism, it had been reported that there were children who had rickets and who did not respond to physiologic doses of vitamin D. The children had hypocalcemia, hypophosphatemia, elevated alkaline phosphatase, elevated PTH, and severe rachitic changes on X-ray. However, many of these children responded to very large pharmacologic doses of vitamin D. As a result, these children received the diagnosis of vitamin D-dependent rickets because of their need for much higher amounts of vitamin D to treat their rickets. With the revelation that vitamin D needed to be metabolized in the kidneys to its active form, it was speculated that vitamin D-dependent rickets was caused by a mutation of the 1-OHase, resulting in either inadequate production of 1,25(OH)₂D or the lack of production of 1,25(OH)₂D. The first insight into the cause of this disorder was when 1,25(OH)₂D₃ was chemically synthesized and provided to these patients.[10] Within several months there was a dramatic improvement in their serum calcium and phosphorus level, with a decrease in alkaline phosphatase and PTH. Thus, it was concluded that vitamin D-dependent rickets was caused by a hereditary defect in the 1-OHase.

The cloning of the 1-OHase gene led to the identification of inactivating mutations that confirmed the hypothesis for the cause of this rare genetic disorder.[11] Patients with PDDR present in their first year of life with severe hypocalcemia, which can cause seizures and carpal pedal spasms, hypophosphatemia, elevated alkaline

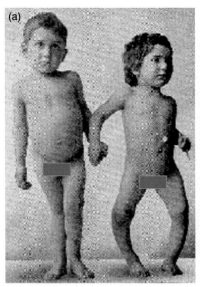 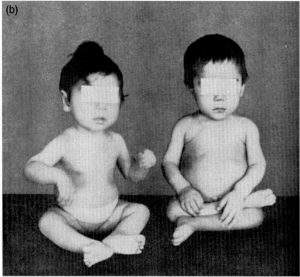

FIGURE 13.3 (a) Sister (left) and brother (right), ages 10 months, and 2.5 years with the enlargement of the ends of the bones at the wrist, carpopedal spasm and a typical "Taylorwise" posture of rickets. (b) Children with classic rickets with inward or outward bowing of the legs. *Reproduced with permission from Holick (2006).*

phosphatase, and PTH. Their blood level of 25(OH)D is usually normal, and treating them with physiologic doses of vitamin D had little effect on correcting their abnormal biochemistries. The hallmark for making the diagnosis is a low or undetectable blood level of 1,25(OH)$_2$D.

If these patients are not appropriately treated with replacement doses of 1,25(OH)$_2$D$_3$, they will have the same skeletal deformities seen in children with severe vitamin D deficiency (Fig. 13.3). These include global poor mineralization of the skeleton and skeletal manifestations in areas of rapid bone growth, including the long bones, epiphyses, and costochondral junctions. Classic clinical manifestation of rickets is usually observed between the ages of 4–12 months. The first skeletal deformities that are observed are the classic rachitic rosary due to hypertrophy of the costochondral junctions leading to beading along both sides of the sternum. There is also involution of the ribs, protrusion of the sternum (pigeon chest), recession of the costochondral junctions, and traverse depressions causing Harrison's groove. Chondrocyte hypertrophy is disrupted, and as a result, there is uncontrolled chondrocyte hypertrophy, leading to expansion of the epiphyseal plates, which appears as widening of the wrists and ends of the other long bones. Once the child begins to stand, gravity pushing on the lower limbs results in an inward (genu valgum) or outward (genu varum) bowing of the tibia and femur. Poor mineralization of the skull in the occipital region causes rachitic craniotabes, enlarged sutures and fontanelles,

delayed closing of fontanelles, and occipital or parietal flattening, causing squaring of the cranium and frontal bossing. Tooth development and eruption is impaired and delayed, and the hypoplasia of the enamel leads to early dental caries.[13]

Extraskeletal manifestations associated with hypocalcemia lead to tetany, seizures, laryngospasm and hypocalcemic myocardiopathy, and death. These children have a delay in motor development, causing severe muscle weakness, especially of the proximal muscles and thoracic muscles. The combination of poor thoracic muscular function and softening of the rib cage leads to poor ventilation, increasing the risk of upper respiratory tract infections and respiratory obstruction.[13,14]

These patients respond well to replacement doses of 1–2 µg of 1,25(OH)$_2$D$_3$ (calcitriol) along with adequate calcium intake. The serum calcium levels begin to rise within 24 h, and radiologic healing is observed by 3 months.[12,13]

Vitamin D-Dependent Rickets Type II: Hereditary Vitamin D Resistant Rickets (HVDRR)

Children who had severe biochemical abnormalities and skeletal abnormalities associated with vitamin D deficiency rickets and who did not respond to physiologic doses of vitamin D and only rarely responded to pharmacologic doses of vitamin D were considered to have vitamin D-dependent rickets. However, unlike children with PDDR who had a low or undetectable blood level of 1,25(OH)₂D, these children had a marked-ly elevated blood level of 1,25(OH)₂D.[13,15] It was assumed that these children must have a genetic defect causing a lack of responsiveness to the calcium and bone metabo-lism effects of 1,25(OH)₂D. The most likely cause was a defective or absent recognition of 1,25(OH)₂D due to a mutation of the VDR. These patients had variable re-sponses to replacement doses of 1,25(OH)₂D₃. There has been a multitude of point mutations of the VDR iden-tified, causing disruption of hormone binding or DNA binding, leading to partial resistance to 1,25(OH)₂D₃ replacement therapy.[13,17] In some cases, these patients respond to pharmacologic doses of 1,25(OH)₂D₃. How-ever, some patients have a point mutation that prevents the production of VDR or prevents the VDR from either binding to 1,25(OH)₂D or permitting the VDR-RXR-1,25(OH)₂D complex from binding to its responsive element within the DNA. These patients often are re-sistant to both physiologic and pharmacologic doses of 1,25(OH)₂D₃.[12,17] (See Table 13.1.)

These patients have all the biochemical and clinical manifestations of PDDR with the exception of often having a low blood level of 25(OH)D and a markedly elevated level of 1,25(OH)₂D. Another clinical manifes-tation that is not seen in PDDR patients is that some patients develop progressive alopecia beginning in the first year of life and progressing to alopecia totalis[12] (Fig. 13.4).

Vitamin D-Dependent Rickets Type III

There has been one reported case of vitamin D re-sistant rickets that is caused by the abnormal expres-sion of hormone responsive element-binding protein that binds to the vitamin D responsive element, thus preventing the VDR-RXR-1,25(OH)₂D complex from binding to its responsive element. This patient had a normal VDR expression and was completely resistant to 1,25(OH)₂D₃.[18]

Treatment Strategies for Vitamin D Resistant Rickets

Children with these vitamin D resistant syndromes often suffer from severe bony deformities and more marked hypocalcemia than children with vitamin D deficiency rickets. Treatment depends on the cause and severity of the vitamin D resistance. Children have re-sponded to pharmacologic doses of vitamin D, physi-ologic and pharmacologic doses of 1,25(OH)₂D₃, and its analog 1-hydroxyvitamin D₃. Children who have a complete resistance to vitamin D and 1,25(OH)₂D₃ can respond to intravenous infusions of calcium and phosphorus.[1,12,19]

GENETIC CAUSES OF RICKETS: HYPOPHOSPHATEMIC DISORDERS

These disorders are caused by a decrease in tubular reabsorption of phosphorus in the kidneys, resulting in hypophosphatemia. These patients usually have normal serum 25(OH)D, PTH, and calcium with an elevated alkaline phosphatase and very low serum phosphorus levels. The blood level of 1,25(OH)₂D is usually low or low–normal, which is considered to be pathologic, since normal patients with low serum phosphorus have a high normal or elevated level of 1,25(OH)₂D. Thus, low or low–normal 1,25(OH)₂D is considered to be both inap-propriate for the degree of hypophosphatemia and a key biochemical feature of these disorders.[13,20–23] Despite the fact that these patients often have a normal serum cal-cium level, they often have very severe rachitic skeletal deformities (Fig. 13.5).

X-Linked Hypophosphatemic Rickets (XLH)

Children with severe hypophosphatemia were thought to have a genetic mutation of the renal phosphate trans-porter. However, it is now recognized that a major factor(s) that controls phosphorus metabolism in the kid-neys is phosphatonins, including fibroblast growth factor 23(FGF 23), matrix extracellular phosphoglycoprotein, and frizzled-related proteins 4.[24] FGF 23 and the other phosphatonins are made in osteoblasts and osteocytes. FGF 23 causes an internalization of the sodium phosphate cotransporter in both the kidneys and intestine, caus-ing phosphate loss in the urine and reducing intestinal phosphate absorption. FGF 23 also inhibits renal 1-OHase (Fig. 13.6).

Originally it was thought that these patients had a genetic defect that caused an increase in the produc-tion of FGF 23 and other phosphatonins, which could easily explain all of the biochemical abnormalities seen in this disorder. However, studies in mice that have the same biochemical abnormalities suggested an alternative explanation, that is, that there was a de-fect in the enzymatic destruction of FGF 23 and other phosphatonins. These studies have led to the identifi-cation of the PHEX (phosphate-regulating gene with homologies to endopeptidases on the X-chromosome) located at XP 22.1–22.2. This gene encodes a membrane-bound endopeptidase, primarily expressed in osteo-blasts and osteocytes. Thus, it is believed that many patients with XLH have a defect in the metabolism of FGF 23 and other phosphatonins because of a muta-tion of the *PHEX* gene. Many of these patients have elevated blood levels of FGF 23 but some do not, suggesting that there may be other causes for this disorder.[24]

TABLE 13.1 Properties of Mutant VDRs Causing HVDRR

Mutation	Ligand binding	DNA binding	RXR binding	Coactivator binding	Alopecia
Arg30stop	−	−	−	−	Yes
Gly33Asp	+++	−	+++	+++	Yes
His35Gln	+++	−	+++	+++	Yes
Cys41Tyr	+++	−	+++	+++	Yes
Lys45Glu	+++	−	+++	+++	Yes
Gly46Asp	+++	−	+++	+++	Yes
Phe47Ile	+++	−	+++	+++	Yes
Phe48frameshift Splice site defect	−	−	−	−	Yes
Arg50Gln	+++	−	+++	+++	Yes
Arg73stop	−	−	−	−	Yes
Arg80Gln	+++	−	+++	+++	Yes
Arg80Gln	+++	−	+++	+++	Yes
Leu141Trp142Ala143 Insertion/substitution	+	+	+	+	No
Gln152stop	−	−	−	−	Yes
Glu92frameshift Splice site defect	−	−	−	−	No
Arg121frameshift 366delC	−	−	−	−	Yes
Leu233frameshift Splice site defect	−	−	−	−	Yes
ΔLys246	+++	−	−	−	Yes
Phe251Cys	+	+	+	+	Yes
Gln259Pro	+++	+	−	+++	Yes
Leu263Arg	−	−	−	−	Yes
Ile268Thr	++	+	+	+	No
Arg274Leu	+	+	+	+	No
Trp286Arg	−	−	−	−	No
Tyr295stop	−	−	−	−	Yes
His305Glu	++	++	++	++	No
Ile314Ser	+++	+	+	+++	No
Gln317stop	−	−	−	−	Yes
Glu329Lys	+++	−	−	−	Yes
Val346Met	+++	+	+	+	No
Arg391Cys	+++	+	+	+	Yes
Arg391Ser	+++	+	+	+	Yes
Tyr401stop Insertion/ duplication	−	+	+	−	No
Glu420Lys	+++	+++	+++	−	No

(+) = present, graded + to +++; (−) = absent.
Adapted from Ref. [16].

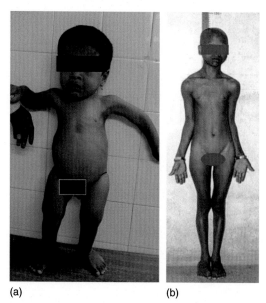

(a)   (b)

FIGURE 13.4 (a) and (b) Both of these patients have biochemical abnormalities and skeletal deformities for vitamin D resistant rickets. The only difference is that the child (a) does not have alopecia totalis as is seen in patient (b).

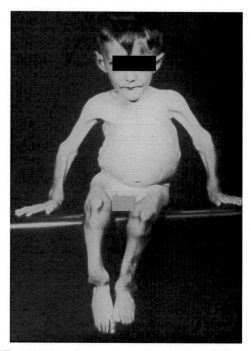

FIGURE 13.5 Child with X-linked hypophospatemic rickets.

Autosomal Dominant Hypophosphatemic Rickets (ADHR)

Patients with autosomal dominant hypophosphatemic rickets have a mutation in the *FGF 23* gene that prevents or reduces FGF 23 metabolic breakdown, leading to elevated FGF 23 levels. These patients present

both biochemically and clinically, similar to patients with XLH.

Treatment Strategies

The major cause of the skeletal deformities in these hypophosphatemic syndromes is the inadequate calcium–phosphorus product, causing a mineralization defect of the skeleton. Thus, correcting hypophosphatemia with phosphate replacement has been the mainstay of treatment, along with physiologic or pharmacologic doses of 1,25(OH)$_2$D$_3$. Phosphate supplementation can be provided either in liquid form or in tablet form. High doses of phosphate can cause diarrhea and also cause one of the complications of the treatment of this disease, which is secondary and tertiary hyperparathyroidism due to the chronic lowering of ionized calcium as a result of the phosphate treatment. K-phosphate preparations are preferred since the sodium phosphate provides a high sodium load causing calciuria. Another complication of giving high doses of phosphate is nephrocalcinosis. Curiously, these patients often have normal or even elevated bone mineral density, which may be due to the chronic elevation in PTH. This can lead to the formation of osteophytes on vertebral bodies, causing encroachment of nerves exiting the spinal cord resulting in severe peripheral neuropathies. Thus, treatment needs to be monitored carefully with routine measurements of kidney functions, serum PTH levels, and urine calcium/creatinine ratio.

GENETIC CAUSES OF HYPERCALCEMIA ASSOCIATED WITH ALTERATIONS IN VITAMIN D METABOLISM

Idiopathic Infantile Hypercalcemia

It is well documented that Williams syndrome, which affects one in 10,000 people worldwide, is associated with elfin faces, learning disabilities, supravalvular aortic stenosis, and sometimes hypercalcemia.[25] Although the exact cause for the hypercalcemia is not well documented, it is known that those affected have a hypersensitivity to vitamin D, thought to be due to the lack of regulation in the renal production of 1,25(OH)$_2$D. There have been sporadic reports of unexplained hypercalcemia in infants and adults that have been associated with suppressed PTH levels and normal or elevated 25(OH)D levels with no evidence of ingesting excessive amounts of vitamin D. The 25-hydroxyvitamin D-24-hydroxylase (cyp24A1;24-OHase) is responsible for the degradation of 1,25(OH)$_2$D and is present in cells that have a VDR. 1,25(OH)$_2$D markedly upregulates this enzyme at the same time

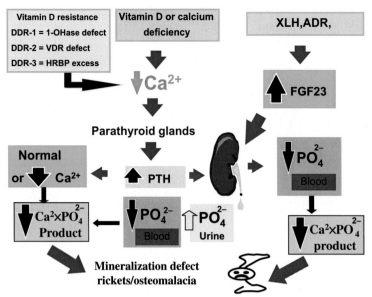

FIGURE 13.6 Biochemical changes in calcium and phosphorus metabolism due to vitamin D or calcium deficiency, vitamin D resistant syndromes or hypophosphatemic syndromes that cause rickets or osteomelacia. Vitamin D and/or calcium deficiency leads to a decrease in the level of ionized calcium (Ca^{2+}), resulting in an increase in PTH. PTH increases tubular reabsorption of calcium to correct the serum calcium into the normal range. However, in severe calcium and vitamin D deficiency, the serum calcium is below normal. In addition, PTH causes phosphorus loss via the urine, resulting in a decrease in serum HPO_4^{2-}. Inadequate calcium–phosphorus product ($Ca^{+2} \times HPO_4^{2-}$) leads to a defect in bone mineralization that causes rickets in children and osteomalacia in adults. There are various inherited and acquired disorders that can disrupt calcium and phosphorus metabolism that can also result in defective mineralization of the skeleton. There are three inherited disorders that cause vitamin D resistance. Vitamin D-dependent rickets type 1 (DDR-1) is due to a mutation of the 1-OHase. A mutation of the VDR gene results in an ineffective recognition of $1,25(OH)_2D$, causing DDR-2. A genetic defect that results in the overproduction of hormone response element-binding protein (HRBP) eliminates the interaction of $1,25(OH)_2D$ with its VDR, resulting in DDR-3. There are also inherited and acquired disorders that cause severe hypophosphatemia and decrease renal production of $1,25(OH)_2D$. The acquired disorders X-linked hypophosphatemic rickets (XLH) and autosomal dominant hypophosphatemic rickets (ADHR) are caused by the increased production or decreased destruction, respectively, of phosphatins that include FGF 23.

as it carries out its other physiologically important genomic activities. Therefore, an inactivation mutation of the 24-OHase would predictably result in an increase in intestinal calcium absorption, an inappropriately normal or elevated $1,25(OH)_2D$ level, hypercalcemia, hyperphosphatemia, nephrocalcinosis, and nephrolithiasis. Schlingmann et al.[26] reported in a cohort of familial cases of idiopathic infantile hypercalcemia, recessive mutations in their 24-OHase. Some of the children were reported to have elevated $1,25(OH)_2D$ levels with associated hypercalcemia, hypercalciuria, medullary nephrocalcinosis, and suppressed PTH levels. A 10-month-old patient with a mutation of the 24-OHase had a serum calcium of 16.3 mg%, fractional absorption of calcium >90% with a normal circulating level of 25(OH)D, $1,25(OH)_2D$, suppressed PTH level, and an undetectable serum $24,25(OH)_2D$ level.[27]

Treatment Strategies

This syndrome is usually caused by vitamin D supplementation at birth or a bolus dose of vitamin D at birth. The hypercalcemia is treated short term by intravenous hydration, furosemide, glucocorticoids, calcitonin, and/or pamidronate. Once identified, these patients should be cautiously given relatively small amounts vitamin D to prevent vitamin D deficiency, but at the same time they should be carefully monitor for hypercalciuria and hypercalcemia.

References

1. Holick MF. Vitamin D deficiency. *N Engl J Med* 2007;**357**:266–81.
2. Holick MF, Biancuzzo RM, Chen TC, Klein EK, Young A, Bibuld D, Reitz R, Salameh W, Ameri A, Tannenbaum AD. Vitamin D_2 is as effective as Vitamin D_3 in maintaining circulating concentrations of 25-hydroxyvitamin D. *J Clin Endocr Metab* 2008;**93**:677–81.
3. Heaney RP, Dowell MS, Hale CA, Bendich A. Calcium absorption varies within the reference range for serum 25-hydroxyvitamin D. *J Am Coll Nutr* 2003;**22**(2):142–6.
4. Khosla S. The OPG/RANKL/RANK system. *Endocrinology* 2001; **142**(12):5050–5.
5. DeLuca H. Overview of general physiologic features and functions of vitamin D. *Am J Clin Nutr* 2004;**80**(Suppl.):1689S–96S.
6. Holtrop ME, Cox KA, Carnes DL, Holick MF. Effects of serum calcium and phosphorus on skeletal mineralization in vitamin-D-deficient rats. *Am J Physiol* 1986;**251**:E234–40.
7. Brown EM, Gamba G, Riccardl D, Lombardi M, Butters R, Klfor O, Sun A, Hedlger MA, Lytton J, Hebert SC. Cloning and characterization of an extracellular Ca^{2+}-sensing receptor from bovine parathyroid. *Nature* 1993;**366**:575–80.

8. Jones G. Expanding role for vitamin D in chronic kidney disease: importance of blood 25-OH-D levels and extra-renal 1α-hydroxylase in the classical and nonclassical actions of 1α,25-dihydroxyvitamin D₃. *Semin Dialysis* 2007;**20**(4):316–24.

9. Casella SJ, Reiner BJ, Chen TC, Holick MF. A possible defect in 25-hydroxylation as a cause of rickets. *J Pediatr* 1994;**124**:929–32.

10. Fraser D, Kooh SW, Kind HP, Holick MF, Tanaka Y, DeLuca HF. Pathogenesis of hereditary vitamin-D-dependent rickets. An inborn error of vitamin D metabolism involving defective conversion of 25-hydroxyvitamin D to 1 alpha,25-dihydroxyvitamin D. *N Engl J Med* 1973;**289**:817–22.

11. Kitanaka S, Takeyama KI, Murayama A, Sato T, Okumura K, Nogami M, Hasegawa Y, Nimi H, Yanagisawa J, Tanaka T, et al. Inactivating mutations in the human 25-hydroxyvitamin D₃ 1α-hydroxylase gene in patients with pseudovitamin D-deficient rickets. *N Engl J Med* 1998;**338**:653–61.

12. Demay MB. Rickets caused by impaired vitamin D activation and hormone resistance: pseudovitamin D deficiency rickets and hereditary vitamin D-resistant rickets. In: Favus MJ, editor. *Primer on the metabolic bone diseases and disorders of mineral metabolism*. 6th ed. Washington, DC: American Society for Bone and Mineral Research; 2006. p. 338–41.

13. Holick MF. Resurrection of vitamin D deficiency and rickets. *J Clin Invest* 2006;**116**:2062–72.

14. Hess AF. *Collected writings, volume I*. Illinois: Charles C. Thomas; 1936, p. 669–719.

15. Brooks MH, Bell NH, Love L, Stern PH, Orfei E, Queene S, Hamstra A, DeLuca H. Vitamin-D-dependent rickets type II: resistance of target organs to 1,25-dihydroxyvitamin D. *N Engl J Med* 1978;**298**: 996–9.

16. Malloy PJ, Pike JW, Feldman D. Hereditary 1,25 dihydroxyvitamin D resistant rickets. *Vitamin D*. 2nd ed. Boston: Elsevier Academic Press; 2005. p. 1207–37.

17. Feldman D, Chen T, Cone C, Hirst M, Shani S, Benderli A, Hochberg Z. Vitamin-D resistant rickets with alopecia: cultured skin fibroblasts exhibit defective cytoplasmic receptors and unresponsiveness to 1,25 (OH)₂D₃. *J Clin Endocrinol Metab* 1982;**55**:1020–2.

18. Chen H, Hewison M, Hu B, Adams JS. Heterogeneous nuclear ribonucleoprotein (hnRNP) binding to hormone response elements: a cause of vitamin D resistance. *PNAS* 2003;**100**:6109–14.

19. Balsan S, Garabedian M, Larchet M, et al. Long-term nocturnal calcium infusions can cure rickets and promote normal mineralization in hereditary resistance to 1,25-dihydroxyvitamin D. *J Clin Invest* 1986;**77**:1661–7.

20. Drezner MK. Clinical disorders of phosphate homeostasis. *Vitamin D*. 2nd ed. Boston: Elsevier Academic Press; 2005. p. 1159–87.

21. Econs MJ. Disorders of phosphate metabolism: autosomal dominant hypophosphatemic rickets, tumor induced osteomalacia, fibrous dysplasia, and the pathophysiological relevance of FGF23. *Vitamin D*. 2nd ed. Boston: Elsevier Academic Press; 2005. p. 1189–95.

22. Portale AA, Halloran BP, Murphy MM, Morris RC. Oral intake phosphorus can determine the serum concentration of 1,25-dihyroxyvitamin D by determining its production rate in humans. *J Clin Invest* 1986;**77**:7–12.

23. Shimada T, Mizutani S, Muto T, Yoneya T, Hino R, Takeda S, Takeuchi Y, Fujita T, Fukumoto S, Yamashita T. Cloning and characterization of FGF23 as a causative factor of tumor-induced osteomalacia. *PNAS* 2001;**98**:6500–5.

24. Hruska KA. Hyperphosphatemia and hypophosphatemia. In: Favus MJ, editor. *Primer on the metabolic bone diseases and disorders of mineral metabolism*. 6th ed. Washington, DC: American Society for Bone and Mineral Research; 2006. p. 233–42.

25. Wacker M, Holick MF. Sunlight and vitamin D: a global perspective for health. *Dermato Endocrinol* 2013;**5**(1):51–108.

26. Schlingmann KP, Kaufmann M, Weber S, et al. Mutation in CYP24A1 and idiopathic infantile hypercalcemia. *N Engl J Med* 2011;**365**: 410–21.

27. Dauber A, Wguyen TT, Sochett E, et al. Genetic defect in CYP24A1, the vitamin D 24-hydroxylase gene, in a patient with severe infantile hypercalcemia. *J Clin Endocrinol Met* 2012;**97**:E268–74.

P A R T   VI

ADRENAL

14

Congenital Adrenal Hyperplasia

Mabel Yau*, Saroj Nimkarn**, Maria I. New*

Mount Sinai School of Medicine, Department of Pediatrics, New York, NY, USA
**Bumrungrad International Hospital, Department of Pediatrics, Bangkok, Thailand

INTRODUCTION

Background, Incidence, Prevalence

Congenital adrenal hyperplasia (CAH) refers to a group of autosomal recessive disorders in which genetic enzyme deficiencies impair normal steroid synthesis. The production of cortisol in the zona fasciculata of the adrenal cortex occurs in five major enzyme-mediated steps, and deficiency in any one of these leads to CAH. The most common form is 21-hydroxylase deficiency (21-OHD), which accounts for over 90% of CAH cases. Impaired cortisol synthesis, the common feature to all forms of CAH, leads to chronic elevation of adrenocorticotropic hormone (ACTH) and overstimulation of the adrenal cortex, resulting in adrenal hyperplasia. Impaired enzyme function at each step of adrenal cortisol biosynthesis leads to an increase in precursors and deficient products. The clinical and hormonal findings of five forms of CAH are summarized in Table 14.1.[1] This chapter will focus on CAH owing to 21-OHD, the most frequent form of CAH.

Data from nearly 6.5 million newborn screens worldwide for 21-hydroxylase deficiency indicate that classical CAH occurs in 1:13,000 to 1:15,000 live births.[2] Nonclassical 21-OHD CAH (NC21-OHD) is more common. The incidence of NC21-OHD in the heterogeneous population of New York City is one in 100, making NC21-OHD one of the most frequent recessive disorders in humans.[3] NC21-OHD is particularly frequent in Ashkenazi Jews, in whom one in three are carriers of the allele, and one in 27 are affected with NC21-OHD.[3-5] Steroid 11 deficiency (11-OHD) is the second most common cause of CAH, accounting for 5–8% of all cases.[6] It occurs in one in 10,000 live births in the general population[7,8] and is more common in some populations of North African origin.[9] In Moroccan Jews, for example, the disease incidence of 11-OHD was initially estimated to be one in

5000 live births;[10] subsequently, it was shown to occur less frequently,[11] but remains more common in Moroccan Jews than in other populations. 17α-Hydroxylase deficiency is found to be common in Mennonite descendants of Dutch Frieslanders and the Brazilian population.[12,13] The disease incidence of congenital lipoid adrenal hyperplasia is shown to be common in Japan and Korea.[14] The other forms are considered rare diseases, and the incidence is unknown in the general population.

A very rare form of CAH not included in this table is cytochrome P450 oxidoreductase deficiency. It is characterized by partial, combined deficiencies of P450C17 (17α-hydroxylase), P450C21 (21-hydroxylase), and aromatase. Affected girls are born with ambiguous genitalia, indicating intrauterine androgen excess. However, virilization does not progress after birth. The 17-hydroxyprogesterone levels are elevated, as in 21-hydroxylase deficiency, while androgen levels are low; cortisol may be normal but is poorly responsive to ACTH. Conversely, affected boys may be born undervirilized, owing to 17α-hydroxylase deficiency. Boys and girls can also present with bone malformations, resembling a pattern seen in patients with Antley–Bixler syndrome.

Steroid 21 Hydroxylase Deficiency (21-OHD)

CAH owing to 21-OHD is divided into classical and nonclassical forms. Classical CAH is further divided into salt-wasting (SW) and simple-virilizing (SV) forms. The SW form is the most severe form of CAH and is characterized by inadequate aldosterone synthesis, which places patients at risk for SW crises, whereas the SV form is distinguished by the ability to conserve salt.[15] Both the SW and SV forms result in genital ambiguity in the female. Females affected with classical forms of CAH are born with genital ambiguity termed "46XX, disorder of sex development."[16] In 21-OHD, the precursors to the

Genetic Diagnosis of Endocrine Disorders. http://dx.doi.org/10.1016/B978-0-12-800892-8.00014-2

TABLE 14.1 Enzyme Deficiencies Resulting in CAH[1]

Enzyme deficiency	Substrate	Product	Androgen	Mineralocorticoid
Steroidogenic acute regulatory protein	Cholesterol	Mediates cholesterol transport across mitochondrial membrane	Deficiency*	Deficiency**
3β-dehydrogenase	Pregnenolone, 17-OH pregnenolone, DHEA	Progesterone, 17-OHP, 4-androstenedione	Deficiency*	Deficiency**
17α-hydroxylase	Pregnenolone, progesterone	17-OH pregnenolone, 17-OH progesterone (17-OHP)	Deficiency*	Excess†
21-hydroxylase	Progesterone, 17-OH progesterone	Deoxycorticosterone (DOC), 11-deoxycortisol	Excess‡	Deficiency**
11β-hydroxylase	Deoxycorticosterone	Corticosterone	Excess‡	Excess†

Males are undervirilized at birth.
**Associated with salt wasting.*
†*Associated with hypertension.*
‡*Females are virilized at birth or later.*

TABLE 14.2 Clinical Features in Individuals with Classical and Nonclassical 21-Hydroxylase Deficiency[1]

Feature	Salt-wasting (classical)	Simple-virilizing (classical)	Nonclassical
Prenatal genital virilization	Present in females	Present in females	Absent
Postnatal hyperandrogenic signs	Males and females	Males and females	Variable presentation
Salt-wasting	Present	None	Absent
Cortisol deficiency	Present	Present	Absent

21-hydroxylase enzyme are shunted into the androgen pathway. Virilization of the female genitalia begins *in utero* as a result of fetal adrenal androgen production. Males do not exhibit genital ambiguity because the major source of androgens in males is the testes, not the adrenals. The nonclassical form of CAH causes postnatal symptoms of hyperandrogenism that may present any time after birth, including adulthood.[17] Table 14.2 summarizes the main clinical characteristics of all forms of 21-OHD CAH.

In all three forms of 21-OHD CAH, both affected males and females experience the effects of postnatal hyperandrogenism unless treatment intervenes. These hyperandrogenic symptoms include premature development of pubic and axillary hair, frontal hair loss, severe acne, advanced somatic and epiphyseal maturation and induced central precocious puberty (a condition in which puberty develops unusually early or rapidly). Early epiphyseal maturation may lead to compromised final adult height.[18] Milder symptoms and/or later presentation are typical in nonclassical 21-OHD CAH owing to genetic mutations associated with less severe enzymatic defects. A synthetic ACTH stimulation test can differentiate 21-OHD from other forms of CAH. A logarithmic nomogram was developed to provide hormonal standards for diagnosis and further assignment of the 21-OHD type by relating baseline to ACTH-stimulated serum concentrations of 17-OHP[19] (see Fig. 14.1).[20]

GENETIC PATHOPHYSIOLOGY

Hormonally and clinically defined forms of 21-OHD CAH are associated with distinct genotypes characterized by varying enzyme activity, demonstrated through *in vitro* expression studies. The gene encoding 21-hydroxylase is a microsomal cytochrome P450 termed cytochrome P450, family 21, subfamily A, polypeptide 21 (*CYP21A2*[21]; previously called *P450c21B*, *CYP21B*, or *CYP21*) (Online Mendelian Inheritance in Man [OMIM] database number 201910) mapped to the short arm of chromosome 6, at locus 6p21 within the human leukocyte antigen (HLA) complex (see Fig. 14.2).[22] *CYP21A2* and its homolog, the pseudogene *CYP21A1P*, alternate with two genes, *C4B* and *C4A*, that encode the two isoforms of the fourth component of the serum complement system.[23,24]

More than 140 mutations have been described, including point mutations, small deletions, small insertions, and complex rearrangements of the gene.[25] About 20% of the mutant alleles are meiotic recombinations deleting a 30-kb gene segment[26] that encompasses the

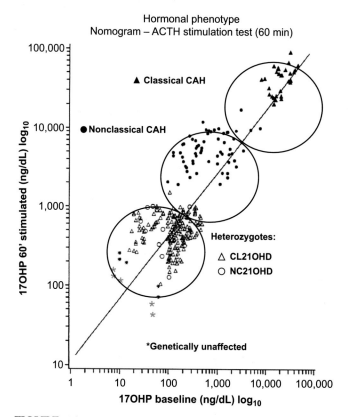

FIGURE 14.1 17-OHP nomogram for the diagnosis of steroid 21-hydroxylase deficiency (60-min cotrosyn stimulation test). The data for this nomogram were collected between 1982 and 1991 at the Department of Pediatrics, the New York Hospital-Cornell Medical Center, New York.[20]

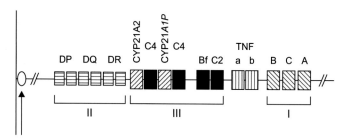

FIGURE 14.2 Gene map of *CYP21A2* on human chromosome 6p. Class I, II, III refer to the HLA genes. C4 refers to the complement genes (2). TNF, tumor necrosis factor.[24]

3′ end of the *CYP21A1P* pseudogene, all of the adjacent C4B complement gene, and the 5′ end of *CYP21A2*, producing a nonfunctional chimera. Gene conversion causes mutations through transfer of deleterious mutations from the pseudogene to the active gene. *De novo* mutations in the active gene can also occur, but account for a small percentage of mutations in *CYP21A2*.[25] These mutations result in complete or partial insufficiency of 21-hydroxylase activity based on *in vitro* expression studies.[27] The common mutations in CYP21A2 are

summarized in Table 14.3.[17] Approximately 95–98% of the mutations causing 21-OHD have been identified through molecular genetic studies of gene rearrangement and point mutations arrays.[28–30] Some mutations are particularly prevalent in certain ethnic groups, indicating a high ethnic specificity.[31]

Pathophysiology of Mutations

In recessive disorders, the less severe mutation of the two alleles is expressed, which results in the phenotype. Classical 21-OHD is most often caused by two severe allelic mutations. In contrast to the classical form, patients with nonclassical 21-OHD are predicted to have mild mutations on both alleles or one severe and one mild mutation (compound heterozygosity) of *CYP21A2*. The severity of each mutation is characterized by the percentage of the remaining enzyme activity found by *in vitro* expression studies. Table 14.3 demonstrates degree of impairment of 21-hydroxylase activity in the common mutations.

DIAGNOSIS: GENETIC TESTING AND INTERPRETATION

CYP21A2 analysis in the probands provides confirmation of clinical and hormonal diagnosis. In the majority of cases, phenotypic disease severity can be predicted from genotypic findings. This phenotype–genotype correlation can provide a significant guide for short- and long-term management of the patients. Clinical uses for family genetic testing include determining carrier status, prenatal diagnosis and treatment, and the future use of preimplantation diagnosis. The available laboratories providing clinical testing are listed in Table 14.4.

Molecular genetic testing of the *CYP21A2* gene for a panel of common mutations (target mutation analysis) and gene deletions detects about 80–98% of disease-causing alleles in affected individuals. Now, molecular genetic testing is widely commercially available. The mutations arise as a result of gene conversion or unequal crossing over between *CYP21A2* and its adjacent pseudogene *CYP21A1P*, which contains multiple deleterious mutations. Entire gene sequencing may detect rarer alleles in affected individuals in whom the panel of nine common mutations and Southern blot analysis identify large gene (30 kb) deletion or gene conversion. A minisequencing technique to detect the most common mutations was recently developed and has become available for clinical use.[32] The majority of individuals from heterogeneous populations with 21-OHD are compound heterozygotes.

TABLE 14.3 Common Gene Mutations of the 21-Hydroxylase Gene CYP21A2[17,*]

Exon/intron	Mutation type	Mutation	Phenotype	Severity of enzyme defect (% enzyme activity)	References
1. Nonclassical mutations					
Exon 1	Missense mutation	P30L	NC	Mild (30–60%)	Tusie-Luna, 1991
Exon 7	Missense mutation	V281L	NC	Mild (20–50%)	Speiser, 1988
Exon 8	Missense mutation	R339H	NC	Mild (20–50%)	Helmberg, 1992
Exon 10	Missense mutation	P453S	NC	Mild (20–50%)	Helmberg, 1992; Owerbach, 1992
2. Classical mutations					
Deletion	30 kb deletion	–	SW	Severe (0%)	White, 1984
Intron 2	Aberrant splicing of intron 2	656 A/C-G	SW, SV	Severe (ND)	Higashi, 1988
Exon 3	Eight-base deletion	G110 8nt	SW	Severe (0%)	White, 1994
Exon 4	Missense mutation	I172N	SV	Severe (1%)	Amor, 1988; Tusie-Luna, 1990
Exon 6	Cluster mutations	I236N, V237E, M239K	SW	Severe (0%)	Amor, 1988; Tusie-Luna, 1990
Exon 8	Nonsense mutation	Q318X	SW	Severe (0%)	Globerman, 1988
Exon 8	Missense mutation	R356W	SW, SV	Severe (0%)	Chiou, 1990
Exon 10	Missense mutation	R483P*	SW	Severe (1–2%)	Wedell, 1993

Full references for this table are listed in Ref. [17].

TABLE 14.4 List of Laboratories for CYP21A2 Analysis in the United States[1,*]

Laboratory	Director	Genetic counselor	Contact information
Center for Genetics at Saint Francis	Nancy Carpenter, PhD		lhwhetsell@saintfrancis.com
Quest Diagnostics, Nichols Institute SJC, San Juan Capistrano, CA	Charles Strom, MD, PhD	Wendy Conlon, MS	Wendy.a.conlon@questdiagnostics.com Phone 800 642 4657 ext 4077 Fax 949 728 4874
Esoterix Molecular Endocrinology, Calabasas, Hills, CA	Samuel H. Pepkowitz, MD	–	Toni R. Prezant, PhD prezant@labcorp.com Phone 818 880 8040 Fax 818 880 1048
Athena Diagnostic Inc., Reference Lab, Worcester, MA	Sat Dev Batish, PhD	Khalida Liaquat, MS	khalida.x.liaquat@athenadiagnostics.com
Emory Molecular Genetics Laboratory, Atlanta, GA	Madhuri Hegde, PhD	Alice Tanner, PhD	atanner@geneticsemory.edu Phone 404 778 8469 Fax 404 778 8559
Mayo Clinic	Edward Highsmith, PhD		

The information was obtained from a public domain, www.genetests.org, which contains a full list of laboratories and their details.[32]

Interpretation of Test Results

Complexity of the Active Genes and Their Nearby Pseudogenes

A large-scale gene conversion can transfer sequences containing more than one mutation from the pseudogene to the active gene. In this instance, targeted mutation analysis that aims to screen for common mutations may detect multiple mutations. The method will not distinguish whether the mutations lie on the same allele, representing only one mutant allele, or the mutations lie on both alleles. To avoid such potential errors, studying both parents as well as the proband is recommended to confirm the mutations and to determine if they are in the *cis* configuration or the *trans* configuration.

Another potential cause of misdiagnosis is duplication of the *CYP21A2* gene. This could affect the screening of individuals who are not known carriers. A person carrying a functional gene and a copy with a mutation on the same chromosome may be incorrectly labeled a carrier.[1]

Phenotype–Genotype Correlation

Mutations in the CYP21A2 gene cause varying degrees of 21-hydroxylase deficiency, resulting in differences in clinical severity. In a study of 1507 subjects with CAH, a direct genotype–phenotype correlation was noted in less than 50% of the genotypes studied.[33] In SW and nonclassical CAH, a phenotype was strongly correlated to a genotype.[33] In SV CAH, there is a high degree of genotype–phenotype nonconcordance. A recent study used computational modeling to correlate disease severity with 113 known mutations on the basis of the extent to which the enzyme is disrupted *in silico*. Mutations that affect critical enzyme functions or alter enzyme stability result in a complete loss of functionality and SW disease.[34]

In the context of prenatal diagnosis, it is important to distinguish classical and nonclassical genotypes in order to determine the necessity of prenatal treatment. In our series of publications on prenatal diagnosis,[35,36] we divided the genotypes into mutation identical groups. The Exon 7 V281L mutation is known to cause a nonclassical phenotype, and therefore should not indicate prenatal dexamethasone treatment. However, in approximately 5% of our prenatal diagnoses, the homozygous Exon 7 V281L mutation resulted in a newborn with classical disease. We have also found the P30L mutation to cause nonconcordance, but less frequently than the V281L mutation.[37–39] In our recent review of 723 patients from mixed ethnic backgrounds,[31] the classical versus nonclassical phenotype could be predicted from genotypes in most cases. However, rare exceptions existed when patients carried the V281L and P30L mutations. These mutations conferred the classical phenotype in less than 3% of the patients when a nonclassical phenotype was expected.[40]

Prenatal Diagnosis and Treatment of 21-OHD CAH

In utero virilization, owing to excessive fetal adrenal androgen production, leads to ambiguous genitalia in affected females with classical 21-OHD CAH, placing them at risk for misassignment to male sex. Females may have to undergo genital surgical reconstruction. Prenatal treatment for 21-OHD CAH reduces female genital ambiguity, and its consequences include sex misassignment and gender identity confusion. Further, the controversies about the appropriate time and technique for feminizing surgery in virilized females[41,42] are avoided.

CAH is one of the few genetic disorders for which prenatal treatment is able to improve postnatal phenotypic outcome. An algorithm has been developed for prenatal diagnosis and treatment for fetuses at risk for classic 21-OHD as demonstrated in Fig. 14.3.[43]

Prenatal diagnosis in pregnancies at risk for classic 21-OHD has been performed for several decades, and prenatal treatment with dexamethasone to reduce the virilization of affected females and thus to reduce their need for clitoroplasty and/or vaginoplasty has been used successfully since 1984.[44,45] Dexamethasone administration begins as early as the 8th week of gestation; the treatment is blind to the disease status and sex of the fetus. If the fetus is later determined upon karyotype to be a male or an unaffected female upon DNA analysis on amniocentesis or chorionic villus sampling, treatment is discontinued. Because an affected male fetus does not carry a risk of ambiguous genitalia in this disorder, prenatal dexamethasone treatment is not indicated. Both males and females receive partial and unnecessary treatment until fetal genetic diagnosis is established. Treatment is continued to term in the affected female fetus. The optimal dosage and timing is 20 g/kg/day of dexamethasone per maternal prepregnancy body weight, in three divided doses, starting as soon as pregnancy is confirmed. Treatment must begin before the 9th week of gestation for prevention of genital ambiguity in the female.[43,46,47]

In general, prenatal diagnosis and treatment of subsequent pregnancy is unnecessary if the proband in the family has the nonclassical form of the disease. Situations do, however, exist in which a nonclassical index case may inherit one severe mutation and one mild mutation. The parent who passed on the mild mutation may be an asymptomatic, unrecognized nonclassical patient who is a compound heterozygote with a mild and a severe mutation. The affectation status of the nonclassical parent in this case is recognized only by genetic analysis during family evaluation, as symptoms of androgen excess tend to wax and wane in this mild form of 21-hydroxylase defect. If both parents pass on severe mutations to their children, these children are at risk of the classical disease, despite the proband having the nonclassical form.

TREATMENT

The Effect of Genetic Information in Treatment Decision

In most of the cases, genotypic characterization provides information to confirm or exclude diagnosis of 21-OHD CAH. This has a significant impact

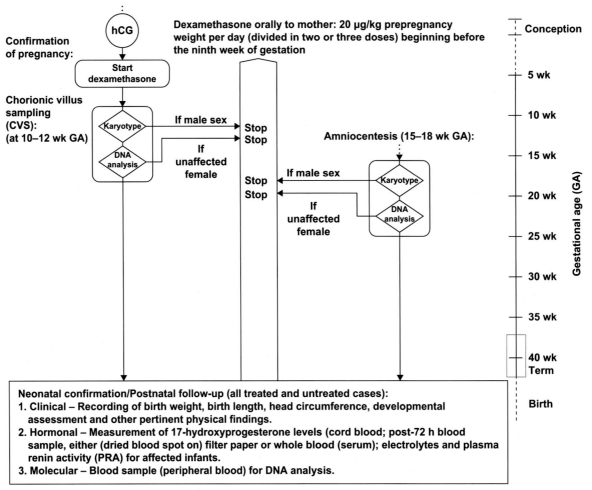

FIGURE 14.3 Algorithm of treatment, diagnosis, and decision-making for prenatal treatment of fetuses at risk for 21-hydroxylase deficiency CAH.[43]

on the clinical decision of lifetime glucocorticoid replacement, and in the SW form of CAH, mineralocorticoid replacement. Unnecessary long-term exposure to medications that are not clearly indicated can therefore be avoided if the patients prove not to have 21-OHD CAH. This benefit of confirming diagnosis becomes more evident when a universal neonatal screening program for CAH is considered. Newborn screening uses hormonal assays of a blood spot on filter paper. The test only detects the classical form of CAH. Thus, nonclassical 21-OHD is not detected, though the child may require treatment for hyperandrogenism. In many circumstances, temporal hormonal confirmation cannot practically be performed before initiation of glucocorticoid replacement. Once the patient, especially newborns, received glucocorticoid, hormonal testing with or without synthetic ACTH can give a misleading result. DNA analysis for *CYP21A2* in a reliable laboratory is the test of choice and can be conducted after and during treatment to confirm the diagnosis.

In prenatal diagnosis and treatment of 21-OHD CAH, genetic analysis of the fetus, his/her parents, and affected siblings is crucial in determining sex and affectation status of the fetus. This in turn dictates the necessity of prenatal treatment.

Genetic Counseling

CAH is transmitted as an autosomal recessive trait. Usually, the parents of a proband are obligate heterozygotes with one normal allele and one mutated allele. Heterozygotes are asymptomatic but may have slightly elevated 17-OHP levels when stimulated with ACTH, as compared to individuals with two normal alleles. A very small percentage of mutations occur *de novo* (approximately 1%). In some instances, a parent who was previously not known to be affected may be found to have the nonclassic form of 21-OHD. It is appropriate to evaluate both parents of a proband with molecular genetic testing and hormonal profiling to determine if either has two allelic mutations and is therefore an undiagnosed patient. Among other immediate family members, each sibling of a proband has a 25% chance of inheriting both mutant alleles and being affected, a 50% chance of inheriting one altered allele and being an unaffected carrier, and a 25% chance of inheriting both normal alleles and being unaffected. If the family study revealed that one parent of a proband is heterozygous and the other has 21-OHD, each sibling has a 50% chance of inheriting both mutant alleles and being affected, and a 50% chance of inheriting one mutated allele and being a carrier.

Risk for the Offspring of a Proband

Given the high carrier rate for 21-OHD, it is appropriate to offer molecular genetic testing of the *CYP21A2* gene to the reproductive partner of a proband. If the reproductive partner is determined to be heterozygous for an identified mutation, the risk to each child of being affected is 50%.

In the process of counseling, it is important first to know that the ability to predict the phenotype based on genotype is imperfect, but is more reliable within the same family. Second, mutation analysis does not detect 100% of mutant alleles; there is a slight residual risk that the reproductive partner may carry a mutant allele that is not detected. This risk becomes significantly less if the entire gene is sequenced. Finally, the optimal time for determination of genetic risk, clarification of carrier status, and discussion of the availability of prenatal testing and treatment is before pregnancy so that the risk of having a child with CAH can be explained.

The other family members of the probands can also be carriers. Siblings of the proband's obligate heterozygous parents are at 50% risk of also being carriers. Carrier testing using molecular genetic testing of the *CYP21A2* gene is recommended to at-risk relatives when one or both disease-causing mutations have been identified in the proband.

Recommendation for Diagnosis and Treatment

As patients with both classical and nonclassical CAH have normal intelligence and normal development when properly treated, early correct diagnosis and treatment should be carried out. Continued care by an experienced pediatric endocrinologist is recommended. Some children with CAH may suffer from a low final height (when the average of the parents' height is considered). Those with a poor height prediction (usually owing to an advanced bone age) may need treatment with growth hormone and a luteinizing hormone releasing hormone (LHRH) analog or an aromatase inhibitor to delay puberty and thus permit a longer duration of growth before epiphyseal fusion.[48]

Resources

CARES (Congenital Adrenal Hyperplasia Research, Education and Support) Foundation, Inc.
2414 Morris Avenue
Suite 110, Union, NJ 07083, USA
Phone: 866-227-3737
Email: contact@caresfoundation.org
www.caresfoundation.org
Congenital Adrenal Hyperplasia Due to 21-Hydroxylase Deficiency: A Guide for Patients and Their Families
CAH printable booklet
National Library of Medicine Genetics Home Reference
21-Hydroxylase Deficiency
NCBI Genes and Disease
Congenital Adrenal Hyperplasia
Genereviews
21-Hydroxylase-Deficient Congenital Adrenal Hyperplasia –
GeneReviews – NCBI Bookshelf

References

1. New MI, Lekarev O, Parsa A, Yuen T, O'Malley B, Hammer G, editors. *Genetic Steroid Disorders*. San Diego, CA: Elsevier; 2013.
2. Pang SY, Wallace MA, Hofman L, et al. Worldwide experience in newborn screening for classical congenital adrenal hyperplasia due to 21-hydroxylase deficiency. *Pediatrics* 1988;**81**:866–74.
3. Speiser PW, Dupont B, Rubinstein P, et al. High frequency of nonclassical steroid 21-hydroxylase deficiency. *Am J Hum Genet* 1985;**37**:650–67.
4. Sherman SL, Aston CE, Morton NE, et al. A segregation and linkage study of classical and nonclassical 21-hydroxylase deficiency. *Am J Hum Genet* 1988;**42**:830–8.
5. Zerah M, Ueshiba H, Wood E, et al. Prevalence of nonclassical steroid 21-hydroxylase deficiency based on a morning salivary 17-hydroxyprogesterone screening test: a small sample study. *J Clin Endocrinol Metab* 1990;**70**:1662–7.
6. Zachmann M, Tassinari D, Prader A. Clinical and biochemical variability of congenital adrenal hyperplasia due to 11 beta-hydroxylase deficiency. A study of 25 patients. *J Endocrinol Metab* 1983;**56**:222–9.
7. Curnow KM, Slutsker L, Vitek J. Mutations in the CYP11B1 gene causing congenital adrenal hyperplasia and hypertension cluster in exons 6, 7, and 8. *Proc Natl Acad Sci USA* 1993;**90**:4552–6.
8. White PC, Obeid J, Agarwal AK, et al. Genetic analysis of 11 beta-hydroxysteroid dehydrogenase. *Steroids* 1994;**59**:111–5.
9. Khemiri M, Ridane H, Bou YO, et al. 11 beta hydroxylase deficiency: a clinical study of seven cases. *Tunis Med* 2006;**84**:106–13.
10. Rosler A, Leiberman E, Cohen T. High frequency of congenital adrenal hyperplasia (classic 11 beta-hydroxylase deficiency) among Jews from Morocco. *Am J Med Genet* 1992;**42**:827–34.
11. Paperna T, Gershoni-Baruch R, Badarneh K, et al. Mutations in CYP11B1 and congenital adrenal hyperplasia in Moroccan Jews. *J Clin Endocrinol Metab* 2005;**90**:5463–5.
12. Costa-Santos M, Kater CE, Auchus RJ. Two intronic mutations cause 17-hydroxylase deficiency by disrupting splice acceptor sites: direct demonstration of aberrant splicing and absent enzyme activity by expression of the entire CYP17 gene in HEK-293 cells. *J Clin Endocrinol Metab* 2004;**89**:49–60.
13. Imai T, Yanase T, Waterman M, et al. Canadian Mennonites and individuals residing in the Friesland region of The Netherlands share the same molecular basis of 17 alpha-hydroxylase deficiency. *Hum Genet* 1992;**89**:95–6.

14. Bose H, Sugawara T, Strauss JR, et al. The pathophysiology and genetics of congenital lipoid adrenal hyperplasia. *N Engl J Med* 1996;**335**:1870–8.

15. Nimkarn S, Lin-Su K, Berglind N, et al. Aldosterone-to-renin ratio as a marker for disease severity in 21-hydroxylase deficiency congenital adrenal hyperplasia. *J Clin Endocrinol Metab* 2007;**92**:137–42.

16. Hughes IA, Houk C, Ahmed SF, Lee PA. LWPES Consensus Group and ESPE Consensus Group. Consensus statement on management of intersex disorders. *Arch Dis Child* 2006;**91**:554–63.

17. New MI. Extensive clinical experience: nonclassical 21-hydroxylase deficiency. *J Clin Endocrinol Metab* 2006;**91**:4205–14.

18. Lin-Su K, Harbison MD, Lekarev O, Vogiatzi MG, New MI. Final adult height in children with congenital adrenal hyperplasia treated with growth hormone. *J Clin Endocrinol Metab* 2011;**96**(6):1710–7.

19. New MI, Wilson RC. Steroid disorders in children: congenital adrenal hyperplasia and apparent mineralocorticoid excess. *Proc Natl Acad Sci USA* 1999;**96**:12790–7.

20. New M, Lorenzen F, Lerner A, et al. Genotyping steroid 21-hydroxylase deficiency: hormonal reference data. *J Clin Endocrinol Metab* 1983;**57**:320–6.

21. Nebert DW, Nelson DR, Coon MJ. The P450 superfamily: update on new sequences, gene mapping, and recommended nomenclature. *DNA Cell Biol* 1991;**10**:1–14.

22. Dupont B, Oberfield S, Smithwick E, et al. Close genetic linkage between HLA and congenital adrenal hyperplasia (21-hydroxylase deficiency). *Lancet* 1977;**2**:1309–12.

23. Donohoue P, van Dop C, McLean R, et al. Gene conversion in salt-losing congenital adrenal hyperplasia with absent complement C4B protein. *J Clin Endocrinol Metab* 1986;**62**:995–1002.

24. Speiser PW, Dupont J, Zhu D, et al. Disease expression and molecular genotype in congenital adrenal hyperplasia due to 21-hydroxylase deficiency. *J Clin Invest* 1992;**90**:584–95.

25. Stenson PD, Ball EV, Mort M, et al. Human Gene Mutation Database (HGMD): 2003 update. *Hum Mutat* 2003;**21**:577–81.

26. White P, Vitek A, Dupont B, et al. Characterization of frequent deletions causing steroid 21-hydroxylase deficiency. *Proc Natl Acad Sci USA* 1998;**85**:4436–40.

27. White P, Speiser P. Congenital adrenal hyperplasia due to 21-hydroxylase deficiency. *Endocr Rev* 2000;**21**:245–91.

28. Lee HH, Lee YJ, Chan P, et al. Use of PCR-based amplification analysis as a substitute for the southern blot method for CYP21 deletion detection in congenital adrenal hyperplasia. *Clin Chem* 2004;**50**:1074–6.

29. Nimkarn S, Weerakulwattana L, Chaichanwatanakul K, et al. In: *Eleventh Asian congress of pediatrics*. Bangkok, Thailand; 2003, p. 80.

30. Tukel T, Uyguner O, Wei JQ, et al. A novel semiquantitative polymerase chain reaction/enzyme digestion-based method for detection of large scale deletions/conversions of the CYP21 gene and mutation screening in Turkish families with 21-hydroxylase deficiency. *J Clin Endocrinol Metab* 2003;**88**:5893–7.

31. Wilson RC, Nimkarn S, Dumic M, et al. Ethnic-specific distribution of mutations in 716 patients with congenital adrenal hyperplasia owing to 21-hydroxylase deficiency. *Mol Genet Metab* 2007;**90**:414–21.

32. Keen-Kim D, Redman JB, Alanes RU, et al. Validation and clinical application of a locus-specific polymerase chain reaction- and

minisequencing-based assay for congenital adrenal hyperplasia (21-hydroxylase deficiency). *J Mol Diagn* 2005;**7**:236–46.

33. New MI, Abraham M, Gonzalez B, Dumic M, Razzaghy-Azar M, Chitayat D, Sun L, Zaidi M, Wilson RC, Yuen T. Genotype-phenotype correlation in 1,507 families with congenital adrenal hyperplasia owing to 21-hydroxylase deficiency. *Proc Natl Acad Sci USA* 2013;**110**:2611–6.

34. Haider S, Islam B, D'Atri V, Sgobba M, Poojari C, Sun L, Yuen T, Zaidi M, New MI. Structure-phenotype correlations of human CYP21A2 mutations in congenital adrenal hyperplasia. *Proc Natl Acad Sci USA* 2013;**110**:2605–10.

35. New MI. Prenatal treatment of congenital adrenal hyperplasia. The United States experience. *Endocrinol Metab Clin North Am* 2001;**30**:1–13.

36. New M, Carlson A, Obeid J, et al. Prenatal diagnosis for congenital adrenal hyperplasia in 532 pregnancies. *J Clin Endocrinol Metab* 2001;**86**:5651–7.

37. Krone N, Braun A, Roscher AA, et al. Predicting phenotype in steroid 21-hydroxylase deficiency? Comprehensive genotyping in 155 unrelated, well defined patients from southern Germany. *J Clin Endocrinol Metab* 2000;**85**:1059–65.

38. Dolzan V, Solyom J, Fekete G, et al. Mutational spectrum of steroid 21-hydroxylase and the genotype-phenotype association in Middle European patients with congenital adrenal hyperplasia. *Eur J Endocrinol* 2005;**153**:99–106.

39. Friaes A, Rego AT, Aragues JM, et al. CYP21A2 mutations in Portuguese patients with congenital adrenal hyperplasia: identification of two novel mutations and characterization of four different partial gene conversions. *Mol Genet Metab* 2006;**88**:58–65.

40. Stikkelbroeck NM, Hoefsloot LH, de Wijs IJ, et al. CYP21 gene mutation analysis in 198 patients with 21-hydroxylase deficiency in The Netherlands: six novel mutations and a specific cluster of four mutations. *J Clin Endocrinol Metab* 2003;**88**:3852–9.

41. Hughes IA, Houk C, Ahmed SF, et al. Consensus statement on management of intersex disorders. *Arch Dis Child* 2006;**91**:554–63.

42. Crouch NS, Minto CL, Liao L-M, et al. Genital sensation after feminizing genitoplasty for congenital adrenal hyperplasia: a pilot study. *BJU Internat* 2004;**93**:135–8.

43. Mercado AB, Wilson RC, Cheng KC, et al. Prenatal treatment and diagnosis of congenital adrenal hyperplasia owing to steroid 21-hydroxylase deficiency. *J Clin Endocrinol Metab* 1995;**80**:2014–20.

44. Nimkarn S, New MI. Prenatal diagnosis and treatment of congenital adrenal hyperplasia. *Horm Res* 2007;**67**:53–60.

45. New MI, Abraham M, Yuen T, Lekarev O. An update on prenatal diagnosis and treatment of congenital adrenal hyperplasia. *Semin Reprod Med* 2012;**30**:396–9.

46. Forest M, Betuel H, David M. Prenatal treatment in congenital adrenal hyperplasia due to 21-hydroxylase deficiency: up-date 88 of the French multicentric study. *Endocr Res* 1989;**15**:277–301.

47. New MI, Carlson A, Obeid J, et al. Prenatal diagnosis for congenital adrenal hyperplasia in 532 pregnancies. *J Clin Endocrinol Metab* 2001;**86**:5651–7.

48. Lin-Su K, Vogiatzi MG, Marshall I, et al. Treatment with growth hormone and luteinizing hormone releasing hormone analog improves final adult height in children with congenital adrenal hyperplasia. *J Clin Endocrinol Metab* 2005;**90**:3318–25.

15

Genetics of Adrenocortical Tumors (ACT) and Hypersecretory Syndromes

Rossella Libé, Guillaume Assié, Lionel Groussin,
Jérôme Bertherat, Xavier Bertagna

Département Hospitalo-Universitaire, Faculté de Médecine Paris Descartes, Université Paris, Paris, France;
Service des Maladies Endocriniennes et Métaboliques, Centre de Recherche des Maladies Rares de la
Surrénale, Hôpital Cochin, India

The identification of germline molecular defects in hereditary syndromes responsible for adrenocortical tumors (ACTs) has unraveled new pathophysiological mechanisms. Indeed, similar molecular defects have since been identified as somatic alterations in sporadic tumors. The familial diseases are Li–Fraumeni syndrome, which may be due to germline mutation of the tumor suppressor gene *TP53*; Beckwith–Wiedemann syndrome, which is caused by dysregulation of the imprinted *IGF-II* locus at 11p15; type 1 multiple endocrine neoplasia (MEN1), which is characterized by a germline mutation of the *menin* gene; and Cushing's syndrome due to primary pigmented nodular adrenocortical disease (PPNAD), which has been observed in Carney complex patients presenting inactivating germline *PRKAR1A* mutations or inactivating germline PDE11A mutations. Interestingly, allelic losses at 17p13 and 11p15 have been demonstrated in sporadic adrenocortical cancer, and somatic *PRKAR1A* mutations have been found in secreting adrenocortical adenomas. Genetic alterations of β-catenin, a Wnt pathway key component, have been described in ACT. More rarely, mutations in *gsp* and *ACTH-receptor* have been observed in ACT. The genetics of another group of adrenal diseases that can lead to adrenal nodular hyperplasia – congenital adrenal hyperplasia (CAH) and glucocorticoid-remediable aldosteronism (GRA) – have also been extensively studied. Aberrantly expressed G-protein-coupled receptors in the adrenal cortex have been reported in the ACTH independent macronodular adrenal hyperplasia (AIMAH).

Most recent achievements have unraveled the integrated genomic landscape of adrenal cortical carcinomas (ACCs) and revealed new driver genes, identified the role of mutated PKACA subunits in the pathogenesis of more than a third of sporadic adrenal cortical adenomas (ACAs) with overt Cushing's syndrome, and discovered a new and nonanticipated gene, *ARMC5* (*armadillo repeat containing 5*), whose germline mutations cause a familial form of AIMAH.

This chapter summarizes recent advances in the genetics of ACT, highlighting both improvements in our understanding of the pathophysiology and the diagnosis of these tumors.

The particular, aldosterone-producing ACT, responsible for Conn's adenomas, are discussed in another, specific Chapter.

INTRODUCTION

Considerable advances toward understanding the molecular mechanisms of ACT development have recently been made. The study of rare genetic syndromes associated with ACT has greatly facilitated progress and has increased our understanding of sporadic adrenal tumors. Furthermore, several observations have demonstrated that genetic alterations are frequent in both benign and malignant ACT.

The Li–Fraumeni Syndrome: *TP53* and Locus 17p13

Germline mutations in tumor suppressor gene *TP53* are present in 70% of families with Li–Fraumeni syndrome (LFS). This syndrome displays dominant inheritance and confers susceptibility to breast carcinoma,

Genetic Diagnosis of Endocrine Disorders. http://dx.doi.org/10.1016/B978-0-12-800892-8.00015-4

soft tissue sarcoma, brain tumors, osteosarcoma, leukemia, and adrenocortical carcinoma (ACC).[1] Other possible tumors include melanoma, gonadal germ cells tumors, and carcinoma of the lung, pancreas, and prostate. These tumors have an early onset, affecting mostly children and young adults. Germline mutations in *TP53* have been observed in 50–80% of children with apparently sporadic ACC in North America and Europe.[2] The incidence of pediatric ACC is about 10 times higher in southern Brazil than in the rest of the world, and a specific germline mutation has been identified in exon 10 of the *TP53* gene (R337H) in almost all cases.[3] In sporadic ACC in adults, somatic mutations of *TP53* are found in only 25% of ACC cases and are located in four "hot spot regions" within exons 5 and 8, as reported in a recent review.[4] More recently, Libé et al.,[5] by sequencing the entire *TP53* coding region, reported *TP53* mutation in 33% of ACT, which is associated with a more aggressive and advanced staging. Loss of heterozygosity (LOH) at 17p13 has been consistently demonstrated in ACC but not in adrenocortical adenoma (ACA). LOH at 17p13 was recently reported to occur in 85% of ACC and only in 30% of ACA. Increased prevalence of LOH at 17p13 is correlated with higher Weiss score – an index of pathological alterations used to determine the malignancy of ACT. It has therefore been suggested that 17p13 LOH could be used as a molecular marker of malignancy in ACT: in a large prospective study of ACT patients, 17p13 LOH was demonstrated to be an independent variable predictive of recurrence after complete surgical removal of localized ACT.[6] The discrepancy between the frequencies of *TP53* mutation and 17p13 LOH may be accounted for by the existence of another tumor suppressor gene in this region. More recently, a common minimal region of loss on 17p13 in ACCs is identified, whereas no minimal region of loss in ACAs could be demonstrated. A significant downregulation of two genes that map in this region, *ACADVL* and *ALOX15B*, in ACCs compared with ACAs was demonstrated, suggesting that the ACADVL and ALOX15B expression are good discriminators between ACCs and ACAs.[7]

The Beckwith–Wiedemann Syndrome: IGF-II (Insulin-Like Growth Factor II) and 11p15 Alterations

The 11p15 region is organized into two different clusters: a telomeric domain including the *IGF-II* gene, *H19*, and a centromeric domain including *CDKNIC* (*p57kip2*).[8] The *IGF-II* gene encodes an important fetal growth factor, is maternally imprinted, and is therefore expressed only from the paternal allele.[8] The *H19* mRNA is not translated and this gene may modulate *IGF-II* expression. The *p57kip2* gene encodes a cyclin-dependent kinase inhibitor involved in the G1/S phase of the cell cycle. The *H19* and *p57kip2* genes are paternally imprinted and are therefore

expressed from the maternal allele only. Genetic or epigenetic changes in the imprinted 11p15 region, resulting in increases in *IGF-II* expression, and these alterations have been implicated in Beckwith–Wiedemann syndrome (BWS). This overgrowth disorder is characterized by macrosomia, macroglossia, organomegaly, and developmental abnormalities (in particular abdominal wall defects with exomphalos). It predisposes patients to the development of embryonal tumors – such as Wilms' tumor – ACC,[9] neuroblastoma, and hepatoblastoma. *IGF-II* mRNA is efficiently translated and malignant tumors contain large amounts of IGF-II protein, some of which is in the prohormone form.

The insulin-like growth factors system is involved in the development and maintenance of differentiated adrenocortical functions, and its role has been largely documented in ACTs.[6] Many studies have demonstrated that *IGF-II* is often strongly overexpressed in malignant adrenocortical tumors, with such overexpression observed in approximately 90% of ACC.[10,11] Transcriptome analysis of ACT has demonstrated that *IGF-II* is the gene most overexpressed in ACC by comparison with ACA or normal adrenal glands.[12,13] The mechanism underlying *IGF-II* overexpression is paternal isodisomy (loss of the maternal allele and duplication of the paternal allele) or, less frequently, loss of imprinting with maintenance of both parental alleles but a paternal-like *IGF-II* gene expression pattern.[11] The mitogenic effect of *IGF-II* is dependent on the IGF-I receptor, as reported by Logié et al.,[14] who demonstrated that IGF-II is involved in NCI H295R cell line proliferation and acts via the IGF-I receptor. IGF-II effects are restricted to tumors, and plasma IGF-II concentrations are usually in the normal range. In ACC, only the maternal *H19* allele is expressed, so expression of this gene is abolished in most ACC displaying paternal isodisomy.[11] Expression of *p57kip2* is also abolished in ACC,[15] but the precise role of the product encoded by this gene in the cell cycle machinery and tumorigenesis requires confirmation. Like 17p13 LOH, 11p15 LOH is associated with a higher risk of tumor recurrence, is more frequent in ACC than in ACA (78.5% vs. 9.5%), and correlates with Weiss score.[6] These genetic abnormalities generate a mosaic-like pattern in some tumors, suggesting that the tumor is made up of different subpopulations of cells. Thus, 11p15 alterations could be used as a biological marker for predicting ACC malignancy after surgical removal of the tumor.[6] However, 11p15 LOH seems to have a lower predictive value than 17p13 LOH.

Multiple Endocrine Neoplasia Type 1: The *Menin* Gene and Locus 11q13

The *menin* gene, located at the 11q13 locus, is thought to be a tumor suppressor gene. A heterozygous inactivating germline mutation of *menin* is found in about 90% of families affected by multiple endocrine neoplasia type 1 (MEN1). This is an autosomal dominant syndrome with high penetrance and an equal sex distribution. The principal clinical features include parathyroid (95%), endocrine pancreas (45%), and pituitary (45%) tumors, thymic carcinoids, and thyroid adenomas.[16] Adrenocortical tumors and/or hyperplasia are observed in 25–40% of MEN1 patients.[17,18] In most cases, they are nonfunctional ACA that can be managed conservatively with radiological/hormonal follow-up. ACC has rarely been reported in MEN1 patients.

Somatic mutation of the *menin* gene is very rare: as reported in a recent series,[4] one mutation was identified in a series of 41 ACA in one study, and one mutation in a series of ACC was found in another. By contrast, LOH at 11q13 was identified in more than 90% of informative ACC, whereas it has been reported in less than 20% of informative adenomas.[17] However, LOH in ACC involves almost the entire 11q domain, suggesting that an as yet unidentified tumor suppressor gene located on the long arm of the chromosome is involved in ACC formation. For more details, see Chapter 26.

The Carney Complex: *PRKAR1A* Gene and Locus 17q22-24

The regulatory R1A subunit of protein kinase A (PRKAR1A) is a key component of the cAMP signaling pathway that has been implicated in endocrine tumorigenesis.[4] This gene maps to 17q22-24, a locus that has been implicated, by linkage analysis in a dominantly multiple neoplasia inherited syndrome with many clinical and pathological manifestations, the Carney complex (CNC).[19] Heterozygous inactivating germline mutations of *PRKAR1A* have been demonstrated in about 45–65 % of CNC families.[20,21] LOH at 17q22-24 is observed in tumors from CNC patients, suggesting that *PRKAR1A* is a tumor suppressor gene. The main features of CNC include spotty skin pigmentation (lentiginosis), endocrine overactivity, and cardiac myxomas.[22] The tumors observed in CNC patients include growth-hormone (GH)-secreting pituitary adenoma, thyroid adenomas or carcinomas, testicular tumors (large-cell calcifying Sertoli cell tumors), ovarian cysts, melanocytic schwannomas, breast ductal adenomas, and adrenocortical lesions. ACTH-independent Cushing's syndrome caused by primary pigmented nodular adrenocortical disease (PPNAD) is observed in 25–30% of patients with CNC. PPNAD is caused by a primary bilateral adrenal defect and may occur in patients with no other CNC features and no family history of CNC. ACTH-independent Cushing's syndrome is often atypical in PPNAD: it may be cyclic, associated with a paradoxical increase in cortisol concentration after dexamethasone administration,

and may be found in patients with normal CT scans. The frequency of *PRKAR1A* mutations is about 80% in CNC patients with Cushing's syndrome, suggesting 17q22-24 defects are more likely to be found in families with PPNAD.[23] Moreover, patients with isolated PPNAD and no family history of CNC may present *de novo* germline mutation of *PRKAR1A*.[24] Somatic mutation of *PRKAR1A* can occur in PPNAD, inactivating the wild-type allele in a patient already presenting a germline mutation.[24]

Somatic *PRKAR1A* mutations have been also demonstrated in sporadic secreting ACA, with clinical, hormonal, and pathological characteristics similar to those of PPNAD.[25] LOH at 17q22-24 has been also observed in sporadic ACA and seems to be restricted to the *PRKAR1A* locus, suggesting the possible involvement of this tumor suppressor gene. By contrast, LOH seems to affect a large part of 17q in ACC, suggesting that *PRKAR1A* alteration may play only a minor role in malignant ACT growth.

Recently, a whole-genome association study led to the discovery of a new actor of the cAMP pathway involved in adrenal cortex tumorigenesis, the *phosphodiesterase 11A4 (PDE11A)* gene. *PDE11A* inactivating mutations have been reported in a subgroup of patients with Cushing's syndrome due to micronodular adrenal hyperplasia (MAH).[26] Moreover, two functional missense substitutions were identified, suggesting a role of PDE11A as genetic predisposition factor. These results has been confirmed by Libé et al.[27] in a large cohort of ACT (ACA, ACC, and AIMAH): in fact, a higher frequency of missense variants were found in particular in the AIMAH group and a higher frequency of synonymous polymorphisms were found in ACC.

Familial Adenomatous Polyposis (FAP) and the β-Catenin Gene; and the Lynch Syndrome

Genetic alterations of the Wnt signaling pathway were initially identified in familial adenomatous polyposis coli (APC) and have been extended to a variety of cancers.[28] ACT have been observed in some case reports of patients with familial APC.[29] Furthermore, familial APC patients with germline mutations of the *APC* gene that lead to an activation of the Wnt signaling pathway may develop ACTs.[30] Molecular studies have suggested that somatic mutations of *APC* could occur in these tumors in patients already having a germline defect.

The Wnt signaling pathway is normally activated during embryonic development. β-Catenin is a key component of this signaling pathway. It has a structural role in cell–cell adhesion, and is a transcription cofactor with T-cell factor/lymphoid enhancer factor (TCF/LEF) mediating transcriptional activation of target genes of the Wnt signaling pathway.

Interestingly, gene profiling studies in various types of ACTs have shown the frequent activation of Wnt signaling target genes: in ACC a microarray approach has shown that ectodermal-neural cortex one (ECN-1) was upregulated.[12] In both benign and malignant ACT, β-catenin accumulation can be observed. These alterations seem very frequent in ACC, consistent with an abnormal activation of the Wnt-signaling pathway. This is explained in a subset of ACT by somatic mutations of the β-catenin gene altering the glycogen synthase kinase 3-β (GSK3-β) phosphorylation site.[31] GSK3-β is implicated in the regulation of β-catenin. In the absence of Wnt signaling, the level of β-catenin is low: β-catenin is phosphorylated at critical NH2-terminal residues by the GSK3-β bound to a scaffolding complex of axin and adenomatous polyposis proteins (APC), and subsequently the phosphorylated protein is degraded by the ubiquitin–proteasome system. Wnt stimulation leads to the inactivation of GSK3-β and thereby the stabilization of β-catenin in the cytoplasm. In the Lynch syndrome, DNA instability is associated with an array of tumors, primarily in the gut. Yet evidence has been provided that rare ACTs may also be part of the syndrome.[32]

The McCune–Albright Syndrome: GNAS1 Gene

The trimeric G protein (α, β, γ subunit) is responsible for transmembrane signal transduction following ligand activation of a G-protein-coupled seven-transmembrane domain adrenocorticotropin receptor (ACTH-R). Somatic activating mutations of the GNAS1 gene (mutant Gs protein, termed gsp) responsible for excess activity of the cyclic AMP (cAMP) signaling pathway have been reported in McCune–Albright syndrome (MAS).[33] This disease is characterized by polyostotic fibrous dysplasia, café au lait spots, precocious puberty, and hyperfunction of multiple endocrine glands (thyroid, adrenal glands, and pituitary gland). Hypercortisolism occurs in 5% of patients and is due to ACTH-independent macronodular adrenocortical hyperplasia (AIMAH).[33] In MAS, the gsp mutation occurs during embryonic development, as demonstrated by its mosaic pattern of distribution in various tissues. Few somatic GNAS mutations have been found in ACTs, as reported in a recent review.[34]

Two different gsp mutations have been reported in three patients with Cushing's syndrome due to AIMAH without MAS features.[35] The authors speculated as to whether these patients presented a disease in the spectrum of MAS, with a late somatic mutation leading to a single defect, or whether they were the first reported cases of isolated AIMAH with gsp mutations involved in molecular pathogenesis.[35]

Congenital Adrenal Hyperplasia

CAH is one of the most frequent genetic endocrine diseases, inherited as an autosomal recessive trait. It is caused by the loss of or severe decrease in activity of one of the steroidogenic enzymes involved in cortisol biosynthesis (mostly 21-hydroxylase (21-OH), 11β-hydroxylase (11β-OH), and 3β-hydroxysteroid dehydrogenase). Deficiencies in 21-hydroxylase (CYP21) are the most common causes of CAH, accounting for 90–95% of cases. All the known biochemical defects impair cortisol secretion, resulting in the stimulation of pituitary corticotrophs, leading to compensatory hypersecretion of ACTH, resulting in hyperplasia of the adrenal cortex. In the past, both homozygous and heterozygous patients with CAH have been reported to have substantially enlarged adrenal glands and a prevalence of adrenal incidentalomas.[36] No mutation in CYP21 was detected in a study of leukocyte DNA from a series of 27 patients, whereas two heterozygous CYP21 mutations were found in adrenal tumor DNA.[37] By contrast, in another series, a higher frequency of classic CAH carriers (16%) and of manifest CAH (2%) was reported among patients with adrenal adenomas than in the general population.[38] For more details, see Chapter 18.

Glucocorticoid-Remediable Aldosteronism

GRA was the first described familial form of hyperaldosteronism. This disorder is characterized by the chronic regulation of aldosterone secretory function by ACTH. Aldosterone hypersecretion can therefore be chronically blocked by exogenous glucocorticoids, such as dexamethasone. This autosomal dominant disorder has been shown to be caused by a hybrid gene formed by crossover between the ACTH-responsive regulatory portion of the 11β-hydroxylase (CYP11B1) gene and the coding region of the aldosterone synthase gene (CYP11B2). Adrenal tumors, together with micronodular and homogeneous hyperplasia of the adrenal cortex, have been observed in the familial cases.[39]

ACTH Receptor (ACTH-R) Gene

ACTH-R belongs to a subgroup of five receptors of the G-protein-coupled receptors superfamily. This subgroup consists of ACTH-R (or MCR-2), MSH-R (MCR-1), and three other receptors (MCR3 to 5). It is encoded by an intron-less gene on chromosome 18p11.2. Inactivating mutations in ACTH-R have been identified in several families with hereditary isolated glucocorticoid deficiency.[40] Screening for ACTH-R mutations in a variety of adrenal tumors has identified no somatic activating mutations to date.[41] Swords et al.[42] reported the functional characterization of a potential activating germline mutation of ACTH-R and demonstrated that it displays high levels of basal activity due to a defect in receptor desensitization. ACTH-R LOH has also been investigated in ACA and ACT: it was observed in two of four informative cancers, but not in 15 hyperfunctioning adenomas, suggesting a role for ACTH-R in cellular differentiation.[43]

ACTH-Independent Macronodular Adrenal Hyperplasia (AIMAH)

ACTH-Independent macronodular adrenal hyperplasia (AIMAH) is a rare cause of endogenous Cushing's syndrome, in which clinical features usually become apparent only after several decades of life. The pathophysiology of this entity is heterogeneous and has been intensely explored in recent years. Several G-protein coupled receptors aberrantly expressed in the adrenal cortex have been implicated in the regulation of steroidogenesis and in the initial cell proliferation in AIMAH,[44] such as the gastric inhibitory polypeptide receptor (GIP-R), β-adrenergic receptors, vasopressin (V2–V3) receptors, serotonin (5-HT7) receptor, and probably angiotensin II receptor (AT1R). Increased expression or over activity of eutopic receptors such as those for vasopressin (V1) receptor, luteinizing hormone/human chorionic gonadotropin (LH/hCG-R), serotonin (5-HT4) receptor, and leptin receptor.[44]

The molecular mechanisms leading to the abnormal expression of eutopic and ectopic receptors in the adrenal glands of patients with AIMAH and, less commonly in ACAs, remains incompletely understood. The GIP-R is the more extensively characterized ectopic receptor in the adrenal CS.[45–47] It was unclear whether aberrant hormone receptors are a primary phenomenon responsible for the pathogenesis of AIMAH or adenomas, or an epiphenomenon resulting from cell proliferation and dedifferentiation; there are now several evidences in favor of the former hypothesis.

Initially reported cases of AIMAH appeared to be sporadic; more recently, first-degree relatives, screenings identified several familial cases with an autosomal dominant pattern of transmission. Up to now, the genetic basis of these familial forms has not been extensively studied. Moreover, the potential presence of aberrant receptors was evaluated only in the recently studied families. As reported in a recent review,[48] some aberrant receptors have been identified so far: V1-vasopressin and β-adrenergic in one family; β-adrenergic in a second one; V1–V2 and V3-vasopressin in another family, and combined 5HT4 and V1–V2-vasopressin in another one. A systematic clinical screening of a family with hereditary cortisol-secreting β-adrenergic responsive AIMAH revealed unsuspected subclinical Cushing's syndrome and aberrant β-adrenergic regulation of cortisol in all familial cases studied with subclinical Cushing's syndrome.

In a most recent paper describing the very thoughtful approach of Hervé Lefebvre's team,[49] the pathophysiological role of ACTH locally produced by the tumoral adrenocortical cells themselves has been convincingly demonstrated. This intracrine action of ACTH should logically modify the name of these tumors, which are indeed "ACTH-dependent" if not "pituitary-dependent," Primary macronodular adrenocortical hyperplasian (PMAH) seems therefore more appropriate.

Assié G. et al.[50] have established that germline mutations of ARMC5 (armadillo repeat containing 5) was responsible for a familial form of benign primary macronodular adrenal hyperplasia with Cushing's syndrome. This unanticipated new actor is a tumor suppressor gene, the normal allele being functionally lost, by different ways, in each of the many adrenal nodules in the same patient. Mutations of ARMC5 have been confirmed by other groups.[51–54] Familial screening now allows early diagnosis of mutated carriers who can be closely and prospectively followed clinically. It remains to be established whether mutated ARMC5 also acts, in some way, through the pivotal same cAMP pathway.

Sporadic Adrenocortical Adenomas With Cushing's Syndrome

Beuschlein's group[55] published the molecular mechanism responsible for more than a third of sporadic adrenocortical adenomas with clinically overt Cushing's syndrome: somatic, activating mutations of the gene coding for the PKA catalytic subunit (PKACα). This mutation was primarily found by exome sequencing in eight of 10 such tumors, and subsequently screened in more than 200 adrenal cortical tumors of various types, within the ENS@T (European Network for the Study of @drenal Tumors) collection of stored tumor specimens. Remarkably one single hot spot mutation, the c.617A>C single nucleotide variation, inducing the p.Leu206Arg substitution, was reported by far as the most common alteration in the cortisol-secreting adenomas (95%). There are many lessons from this discovery: the PKACα mutations are strictly related to adrenocortical adenomas with overt Cushing's syndrome; they are not found in ACC, nor in aldosterone producing adenomas; quite unexpectedly they are not found either in patients with subclinical hypercortisolism, highly suggesting that these rather frequent adenomas, often diagnosed as "incidentalomas," are a different pathophysiological entity and not merely an early form of overt hypercortisolism; there remain other molecular causes to be found, in the two-thirds of "nonmutated" tumors with overt Cushing's syndrome, and in all the others; it is remarkable that few mutations were present in adenomas (an average of five by tumor), quite different from the high numbers in ACCs (see further): these benign, homogeneous, tumors are ideal biological specimens for genetic discoveries. In the same paper, the authors report on five patients with cortisol-secreting bilateral adrenocortical hyperplasia who had copy number gains on the PKACα in their constitutive DNA. *In vitro* studies demonstrated the allelic dosage effect of PKACα as an alternative mechanism to activate the cAMP pathway. Indeed, these data further emphasize the pivotal role of this pathway, which, through different and exclusive actors, can provoke a variety of benign – most often with cortisol hypersecretion – ACT.

Beuschlein's paper was "rapidly" – within weeks! – confirmed by three other independent teams, in high level journals.[56–58]

Integrated Genomic Organization of ACCs

Assié et al.[59] reported the first "integrated genomic organization of adrenocortical carcinoma." This exhaustive work was accomplished by the combination of a well-organized and rather exceptional collection of these rare tumors, more than 122 specimens collected within ENS@T, and the use of the most actual genomic approaches: exome sequencing, genome, miRnome, transcriptome, and methylome analyses. These long-time and European-wide efforts ultimately allowed drawing the "genomic landscape" of these rare but devastating endocrine tumors. ACCs have a high rate of mutations (median of 24 per tumor, excluding two tumors accounting together for 1881 mutations!); interestingly, these mutations point to few definite pathways: the Wnt/βcat pathway activation is the most frequently involved, and, among its actors, a new tumor suppressor gene was identified, *ZNRF3*, which is altered in up to 21% of the tumors. Along the same pathway, *CTNNB1* is also frequently mutated (16%), and the *ZNRF3/CTNNB1* alterations are mutually exclusive, stressing both the high prevalence and the pathophysiological consequences of altered Wnt/βcat pathway in ACCs; gene alterations in other known driver genes were confirmed (*TP53, CDKN2A, RB1*, and *MEN1*), and new ones unraveled (*DAXX, TERT*, and *MED12*), each being found in more than 5% of ACCs. Altogether these recurrent gene alterations were present in ca. 60% of the tumors. The integrated genomic approach also studied the miRNA profiles and the methylation patterns of the tumors. Remarkably, it separated different groups of tumors, with different prognoses, which also correlated nicely with previous transcriptomic data obtained 5 years earlier in the same tumors.[60] These data convincingly show that ACCs can now be classified on a molecular basis, which impact on the prognosis and ultimately on personalized treatment.[61]

CONCLUSIONS

Studies of hereditary neoplasia syndromes have led to the identification of various loci or chromosomal regions and genes responsible for ACT development.

The same molecular defects are observed in the germline DNA in cases of hereditary disease and as somatic defects in tumor DNA in cases of sporadic ACT. For a given genetic defect, the tumor phenotype observed in sporadic tumors displays some similarities to the tumor phenotype observed in familial diseases. This may have important clinical implications, as the molecular study of tumor DNA could provide important information for diagnostic and/or prognostic purposes.

Interestingly, in sporadic tumors, there is almost no overlap between the genetic alterations observed in cancers and those found in adrenal adenomas. For instance, LOH at 17p13 or 11q15, mutations of p53 or *PRKAR1A*, or loss of CREB expression are observed in cancers but not in adenomas.[62] Conversely, *GNAS* mutations or ectopic expression of GIP-R have been identified to date only in secreting adrenocortical adenomas and ACTH-independent macronodular adrenal hyperplasia.

No molecular defect has been identified that would be consistently present in both benign and malignant tumors. The development of tumors in other tissues, such as the digestive tract, is thought to be based on the accumulation of numerous molecular defects, resulting in progression from benign polyp to colon cancer. Some rare tumors in which a malignant and a benign zone are associated within the same adrenal gland are consistent with this model. However, from what we have learned so far from the genetics of ACT, it would seem premature to suggest that such a model could be applied to the adrenal cortex. However, it is tempting to speculate that genetic defects might stimulate the growth of some benign cortisol-secreting tumors with such a level of cellular differentiation that progression toward a malignant dedifferentiated tumor would be prevented. This is illustrated by the various cellular and molecular defects activating the cAMP signaling pathway that have been observed in benign hyperplasia or tumors causing Cushing's syndrome. Nevertheless, this hypothesis is consistent with an apparently benign adenoma with a lower level of differentiation, not responsible for overt cortisol secretion, being able to progress toward a malignant tumor. However, the high frequency of such adenomas, which are usually discovered by chance, contrasts with the rarity of adrenal cancer, suggesting that this multistep progression from benign to malignant tumors might be very rare. Clearly, despite progress in studies of the genetics of ACT, much remains to be done if we are to identify the many molecular alterations involved.

Acknowledgments

This work was supported in part by the *Plan Hospitalier de Recherche Clinique* to the COMETE network (AOM 02068), the *Ligue Nationale Contre le Cancer* (04-7571), the *GIS-INSERM Institut des Maladies Rares* for the Carney Complex network, and the European Network for the Study of Adrenal Tumors (ENSAT).

References

1. Hisada M, Garber J, Fung C, Fraumeni JJ, Li F. Multiple primary cancers in families with Li-Fraumeni syndrome. *J Natl Cancer Inst* 1998;**90**:606–11.

2. Wagner J, Portwine C, Rabin K, Leclerc J, Narod S, Malkin D. High frequency of germline p53 mutations in childhood adrenocortical cancer. *J Natl Cancer Inst* 1994;**86**:1707–10.

3. Ribeiro R, Sandrini F, Figueiredo B, Zambetti G, Michalkiewicz E, Lafferty A, DeLacerda L, Rabin M, Cadwell C, Sampaio G, Cat I, Stratakis C, Sandrini R. An inherited p53 mutation that contributes in a tissue-specific manner to pediatric adrenal cortical carcinoma. *Proc Natl Acad Sci USA* 2001;**98**:9330–5.

4. Libè R, Fratticci A, Bertherat J. Adrenocortical cancer: pathophysiology and clinical management. *Endocr Relat Cancer* 2007;**14**:13–28.

5. Libè R, Groussin L, Tissier F, Elie C, René-Corail F, Fratticci A, Jullian E, Beck-Peccoz P, Bertagna X, Gicquel C, Bertherat J. Somatic TP53 mutations are relatively rare among adrenocortical cancers with the frequent 17p13 loss of heterozygosity. *Clin Cancer Res* 2007;**13**:844–50.

6. Gicquel C, Bertagna X, Gaston V, Coste J, Louvel A, Baudin E, Bertherat J, Chapuis Y, Duclos J, Schlumberger M, Plouin P, Luton J, Le Bouc Y. Molecular markers and long-term recurrences in a large cohort of patients with sporadic adrenocortical tumors. *Cancer Res* 2001;**61**:6762–7.

7. Soon P, Libe R, Benn D, Gill A, Shaw J, Sywak M, Groussin L, Bertagna X, Gicquel C, Bertherat J, McDonald K, Sidhu S, Robinson B. Loss of heterozygosity of 17p13, with possible involvement of ACADVL and ALOX15B, in the pathogenesis of adrenocortical tumors. *Ann Surg* 2008;**247**:157–64.

8. DeChiara T, Robertson E, Efstratiadis A. Parental imprinting of the mouse insulin-like growth factor II gene. *Cell* 1991;**64**:849–59.

9. Wiedemann H, Burgio G, Aldenhoff P, Kunze J, Kaufmann H, Schirg E. The Proteus syndrome. Partial gigantism of the hands and/or feet, nevi, hemihypertrophy, subcutaneous tumors, macrocephaly or other skull anomalies and possible accelerated growth and visceral affections. *Eur J Pediatr* 1983;**140**:5–12.

10. Boulle N, Logié A, Gicquel C, Perin L, Le Bouc Y. Increased levels of insulin-like growth factor II (IGF-II) and IGF-binding protein-2 are associated with malignancy in sporadic adrenocortical tumors. *J Clin Endocrinol Metab* 1998;**83**:1713–20.

11. Gicquel C, Raffin-Sanson M, Gaston V, Bertagna X, Plouin P, Schlumberger M, Louvel A, Luton J, Le Bouc Y. Structural and functional abnormalities at 11p15 are associated with the malignant phenotype in sporadic adrenocortical tumors: study on a series of 82 tumors. *J Clin Endocrinol Metab* 1997;**82**:2559–65.

12. Giordano TJ, Thomas DG, Kuick R, Lizyness M, Misek DE, Smith AL, Sanders D, Aljundi RT, Gauger PG, Thompson NW, Taylor JM, Hanash SM. Distinct transcriptional profiles of adrenocortical tumors uncovered by DNA microarray analysis. *Am J Pathol* 2003;**162**:521–31.

13. de Fraipont F, El Atifi M, Cherradi N, Le Moigne G, Defaye G, Houlgatte R, Bertherat J, Bertagna X, Plouin P, Baudin E, Berger F, Gicquel C, Chabre O, Feige J. Gene expression profiling of human adrenocortical tumors using complementary deoxyribonucleic acid microarrays identifies several candidate genes as markers of malignancy. *J Clin Endocrinol Metab* 2005;**90**:1819–29.

14. Logié A, Boulle N, Gaston V, Perin L, Boudou P, Le Bouc Y, Gicquel C. Autocrine role of IGF-II in proliferation of human adrenocortical carcinoma NCI H295R cell line. *J Mol Endocrinol* 1999;**23**:23–32.

15. Liu J, Voutilainen R, Heikkilä P, Kahri A. Ribonucleic acid expression of the CLA-1 gene, a human homolog to mouse high density lipoprotein receptor SR-BI, in human adrenal tumors and cultured adrenal cells. *J Clin Endocrinol Metab* 1997;**82**:2522–7.

16. Thakker R. Multiple endocrine neoplasia – syndromes of the twentieth century. *J Clin Endocrinol Metab* 1998;**83**:2617–20.

17. Kjellman M, Roshani L, Teh B, Kallioniemi O, Höög A, Gray S, Farnebo L, Holst M, Bäckdahl M, Larsson C. Genotyping of adrenocortical tumors: very frequent deletions of the MEN1 locus in 11q13 and of a 1-centimorgan region in 2p16. *J Clin Endocrinol Metab* 1999;**84**:730–5.

18. Schulte K, Mengel M, Heinze M, Simon D, Scheuring S, Köhrer K, Röher H. Complete sequencing and messenger ribonucleic acid expression analysis of the MEN I gene in adrenal cancer. *J Clin Endocrinol Metab* 2000;**85**:441–8.

19. Kirschner L, Carney J, Pack S, Taymans S, Giatzakis C, Cho Y, Cho-Chung Y, Stratakis C. Mutations of the gene encoding the protein kinase A type I-alpha regulatory subunit in patients with the Carney complex. *Nat Genet* 2000;**26**:89–92.

20. Kirschner L, Sandrini F, Monbo J, Lin J, Carney J, Stratakis C. Genetic heterogeneity and spectrum of mutations of the PRKAR1A gene in patients with the carney complex. *Hum Mol Genet* 2000;**9**: 3037–46.

21. Veugelers M, Wilkes D, Burton K, McDermott D, Song Y, Goldstein M, La Perle K, Vaughan C, O'Hagan A, Bennett K, Meyer B, Legius E, Karttunen M, Norio R, Kaariainen H, Lavyne M, Neau J, Richter G, Kirali K, Farnsworth A, Stapleton K, Morelli P, Takanashi Y, Bamforth J, Eitelberger F, Noszian I, Manfroi W, Powers J, Mochizuki Y, Imai T, Ko G, Driscoll D, Goldmuntz E, Edelberg J, Collins A, Eccles D, Irvine A, McKnight G, Basson C. Comparative PRKAR1A genotype-phenotype analyses in humans with Carney complex and prkar1a haploinsufficient mice. *Proc Natl Acad Sci USA* 2004;**101**:14222–7.

22. Carney J, Gordon H, Carpenter P, Shenoy B, Go V. The complex of myxomas, spotty pigmentation, and endocrine overactivity. *Medicine (Baltimore)* 1985;**64**:270–83.

23. Groussin L, Kirschner L, Vincent-Dejean C, Perlemoine K, Jullian E, Delemer B, Zacharieva S, Pignatelli D, Carney J, Luton J, Bertagna X, Stratakis C, Bertherat J. Molecular analysis of the cyclic AMP-dependent protein kinase A (PKA) regulatory subunit 1A (PRKAR1A) gene in patients with Carney complex and primary pigmented nodular adrenocortical disease (PPNAD) reveals novel mutations and clues for pathophysiology: augmented PKA signaling is associated with adrenal tumorigenesis in PPNAD. *Am J Hum Genet* 2002;**71**:1433–42.

24. Groussin L, Jullian E, Perlemoine K, Louvel A, Leheup B, Luton J, Bertagna X, Bertherat J. Mutations of the PRKAR1A gene in Cushing's syndrome due to sporadic primary pigmented nodular adrenocortical disease. *J Clin Endocrinol Metab* 2002;**87**:4324–9.

25. Bertherat J, Groussin L, Sandrini F, Matyakhina L, Bei T, Stergiopoulos S, Papageorgiou T, Bourdeau I, Kirschner L, Vincent-Dejean C, Perlemoine K, Gicquel C, Bertagna X, Stratakis C. Molecular and functional analysis of PRKAR1A and its locus (17q22-24) in sporadic adrenocortical tumors: 17q losses, somatic mutations, and protein kinase A expression and activity. *Cancer Res* 2003;**63**:5308–19.

26. Horvath A, Boikos S, Giatzakis C, Robinson-White A, Groussin L, Griffin K, Stein E, Levine E, Delimpasi G, Hsiao H, Keil M, Heyerdahl S, Matyakhina L, Libè R, Fratticci A, Kirschner L, Cramer K, Gaillard R, Bertagna X, Carney J, Bertherat J, Bossis I, Stratakis C. A genome-wide scan identifies mutations in the gene encoding phosphodiesterase 11A4 (PDE11A) in individuals with adrenocortical hyperplasia. *Nat Genet* 2006;**38**:794–800.

27. Libé R, Fratticci A, Coste J, Tissier F, Horvath A, Ragazzon B, Rene-Corail F, Groussin L, Bertagna X, Raffin-Sanson M, Stratakis C, Bertherat J. Phosphodiesterase 11A (PDE11A) and genetic predisposition to adrenocortical tumors. *Clin Cancer Res* 2008;**14**:4016–24.

28. Kikuchi A. Tumor formation by genetic mutations in the components of the Wnt signaling pathway. *Cancer Sci* 2003;**94**:225–9.

29. Naylor E, Gardner E. Adrenal adenomas in a patient with Gardner's syndrome. *Clin Genet* 1981;**20**:67–73.

30. Bläker H, Sutter C, Kadmon M, Otto H, Von Knebel-Doeberitz M, Gebert J, Helmke B. Analysis of somatic APC mutations in rare extracolonic tumors of patients with familial adenomatous polyposis coli. *Genes Chromosomes Cancer* 2004;**41**:93–8.

31. Tissier F, Cavard C, Groussin L, Perlemoine K, Fumey G, Hagneré A, René-Corail F, Jullian E, Gicquel C, Bertagna X, Vacher-Lavenu M, Perret C, Bertherat J. Mutations of beta-catenin in adrenocortical tumors: activation of the Wnt signaling pathway is a frequent event in both benign and malignant adrenocortical tumors. *Cancer Res* 2005;**65**:7622–7.

32. Raymond VM, Everett JN, Furtado LV, Gustafson SL, Jungbluth CR, Gruber SB, Hammer GD, Stoffel EM, Greenson JK, Giordano TJ, Else T. Adrenocortical carcinoma is a lynch syndrome-associated cancer. *J Clin Oncol* 2013;**31**(24):3012–8.

33. Weinstein L, Shenker A, Gejman P, Merino M, Friedman E, Spiegel A. Activating mutations of the stimulatory G protein in the McCune-Albright syndrome. *N Engl J Med* 1991;**325**:1688–95.

34. Libé R, Bertherat J. Molecular genetics of adrenocortical tumours, from familial to sporadic diseases. *Eur J Endocrinol* 2005;**153**:477–87.

35. Fragoso M, Domenice S, Latronico A, Martin R, Pereira M, Zerbini M, Lucon A, Mendonca B. Cushing's syndrome secondary to adrenocorticotropin-independent macronodular adrenocortical hyperplasia due to activating mutations of GNAS1 gene. *J Clin Endocrinol Metab* 2003;**88**:2147–51.

36. Falke T, van Seters A, Schaberg A, Moolenaar A. Computed tomography in untreated adults with virilizing congenital adrenal cortical hyperplasia. *Clin Radiol* 1986;**37**:155–60.

37. Kjellman M, Holst M, Bäckdahl M, Larsson C, Farnebo L, Wedell A. No overrepresentation of congenital adrenal hyperplasia in patients with adrenocortical tumours. *Clin Endocrinol (Oxf)* 1999;**50**:343–6.

38. Baumgartner-Parzer S, Pauschenwein S, Waldhäusl W, Pölzler K, Nowotny P, Vierhapper H. Increased prevalence of heterozygous 21-OH germline mutations in patients with adrenal incidentalomas. *Clin Endocrinol (Oxf)* 2002;**56**:811–6.

39. Jeunemaitre X, Charru A, Pascoe L, Guyene T, Aupetit-Faisant B, Shackleton C, Schambelan M, Plouin P, Corvol P. Hyperaldosteronism sensitive to dexamethasone with adrenal adenoma. Clinical, biological and genetic study. *Presse Med* 1995;**24**:1243–8.

40. Vamvakopoulos N, Rojas K, Overhauser J, Durkin A, Nierman W, Chrousos G. Mapping the human melanocortin 2 receptor (adrenocorticotropic hormone receptor; ACTHR) gene (MC2R) to the small arm of chromosome 18 (18p11.21-pter). *Genomics* 1993;**18**:454–5.

41. Latronico A, Reincke M, Mendonça B, Arai K, Mora P, Allolio B, Wajchenberg B, Chrousos G, Tsigos C. No evidence for oncogenic mutations in the adrenocorticotropin receptor gene in human adrenocortical neoplasms. *J Clin Endocrinol Metab* 1995;**80**:875–7.

42. Swords F, Baig A, Malchoff D, Malchoff C, Thorner M, King P, Hunyady L, Clark A. Impaired desensitization of a mutant adrenocorticotropin receptor associated with apparent constitutive activity. *Mol Endocrinol* 2002;**16**:2746–53.

43. Reincke M, Mora P, Beuschlein F, Arlt W, Chrousos G, Allolio B. Deletion of the adrenocorticotropin receptor gene in human adrenocortical tumors: implications for tumorigenesis. *J Clin Endocrinol Metab* 1997;**82**:3054–8.

44. Lacroix A, Ndiaye N, Tremblay J, Hamet P. Ectopic and abnormal hormone receptors in adrenal Cushing's syndrome. *Endocr Rev* 2001;**22**:75–110.

45. Reznik Y, Allali-Zerah V, Chayvialle J, Leroyer R, Leymarie P, Travert G, Lebrethon M, Budi I, Balliere A, Mahoudeau J. Food-dependent Cushing's syndrome mediated by aberrant adrenal sensitivity to gastric inhibitory polypeptide. *N Engl J Med* 1992;**327**:981–6.

46. Lacroix A, Bolté E, Tremblay J, Dupré J, Poitras P, Fournier H, Garon J, Garrel D, Bayard F, Taillefer R. Gastric inhibitory polypeptide-dependent cortisol hypersecretion – a new cause of Cushing's syndrome. *N Engl J Med* 1992;**327**:974–80.

47. Groussin L, Perlemoine K, Contesse V, Lefebvre H, Tabarin A, Thieblot P, Schlienger J, Luton J, Bertagna X, Bertherat J. The ectopic expression of the gastric inhibitory polypeptide receptor is frequent in adrenocorticotropin-independent bilateral macronodular adrenal hyperplasia, but rare in unilateral tumors. *J Clin Endocrinol Metab* 2002;**87**:1980–5.

48. Costa M, Lacroix A. Cushing's syndrome secondary to ACTH-independent macronodular adrenal hyperplasia. *Arq Bras Endocrinol Metabol* 2007;**51**:1226–37.

49. Louiset E, Duparc C, Young J, Renouf S, Tetsi Nomigni M, Boutelet I, Libé R, Bram Z, Groussin L, Caron P, Tabarin A, Grunenberger F, Christin-Maitre S, Bertagna X, Kuhn JM, Anouar Y, Bertherat J, Lefebvre H. Intraadrenal corticotropin in bilateral macronodular adrenal hyperplasia. *N Engl J Med* 2013;**369**(22):2115–25.

50. Assié G, Libé R, Espiard S, Rizk-Rabin M, Guimier A, Luscap W, Barreau O, Lefèvre L, Sibony M, Guignat L, Rodriguez S, Perlemoine K, René-Corail F, Letourneur F, Trabulsi B, Poussier A, Chabbert-Buffet N, Borson-Chazot F, Groussin L, Bertagna X, Stratakis CA, Ragazzon B, Bertherat J. ARMC5 mutations in macronodular adrenal hyperplasia with Cushing's syndrome. *N Engl J Med* 2013;**369**(22):2105–14.

51. Elbelt U, Trovato A, Kloth M, Gentz E, Finke R, Spranger J, Galas D, Weber S, Wolf C, König K, Arlt W, Büttner R, May P, Allolio B, Schneider JG. Molecular and clinical evidence for an ARMC5 tumor syndrome: concurrent inactivating germline and somatic mutations are associated with both primary macronodular adrenal hyperplasia and meningioma. *J Clin Endocrinol Metab* 2015;**100**(1): E119–28.

52. Gagliardi L, Schreiber AW, Hahn CN, Feng J, Cranston T, Boon H, Hotu C, Oftedal BE, Cutfield R, Adelson DL, Braund WJ, Gordon RD, Rees DA, Grossman AB, Torpy DJ, Scott HS. ARMC5 mutations are common in familial bilateral macronodular adrenal hyperplasia. *J Clin Endocrinol Metab* 2014;**99**(9):E1784–92.

53. Alencar GA, Lerario AM, Nishi MY, Mariani BM, Almeida MQ, Tremblay J, Hamet P, Bourdeau I, Zerbini MC, Pereira MA, Gomes GC, Rocha Mde S, Chambo JL, Lacroix A, Mendonca BB, Fragoso MC. ARMC5 mutations are a frequent cause of primary macronodular adrenal hyperplasia. *J Clin Endocrinol Metab* 2014;**99**(8): E1501–9.

54. Faucz FR, Zilbermint M, Lodish MB, Szarek E, Trivellin G, Sinaii N, Berthon A, Libé R, Assié G, Espiard S, Drougat L, Ragazzon B, Bertherat J, Stratakis CA. Macronodular adrenal hyperplasia due to mutations in an armadillo repeat containing 5 (ARMC5) gene: a clinical and genetic investigation. *J Clin Endocrinol Metab* 2014;**99**(6):E1113–9.

55. Beuschlein F, Fassnacht M, Assié G, Calebiro D, Stratakis CA, Osswald A, Ronchi CL, Wieland T, Sbiera S, Faucz FR, Schaak K, Schmittfull A, Schwarzmayr T, Barreau O, Vezzosi D, Rizk-Rabin M, Zabel U, Szarek E, Salpea P, Forlino A, Vetro A, Zuffardi O, Kisker C, Diener S, Meitinger T, Lohse MJ, Reincke M, Bertherat J, Strom TM, Allolio B. Constitutive activation of PKA catalytic subunit in adrenal Cushing's syndrome. *N Engl J Med* 2014;**370**(11): 1019–28.

56. Goh G, Scholl UI, Healy JM, Choi M, Prasad ML, Nelson-Williams C, Kunstman JW, Korah R, Suttorp AC, Dietrich D, Haase M, Willenberg HS, Stålberg P, Hellman P, Akerström G, Björklund P, Carling T, Lifton RP. Recurrent activating mutation in PRKACA in cortisol-producing adrenal tumors. *Nat Genet* 2014;**46**(6):613–7.

57. Sato Y, Maekawa S, Ishii R, Sanada M, Morikawa T, Shiraishi Y, Yoshida K, Nagata Y, Sato-Otsubo A, Yoshizato T, Suzuki H, Shiozawa Y, Kataoka K, Kon A, Aoki K, Chiba K, Tanaka H, Kume H, Miyano S, Fukayama M, Nureki O, Homma Y, Ogawa S. Recurrent somatic mutations underlie corticotropin-independent Cushing's syndrome. *Science* 2014;**344**(6186):917–20.

58. Cao Y, He M, Gao Z, Peng Y, Li Y, Li L, Zhou W, Li X, Zhong X, Lei Y, Su T, Wang H, Jiang Y, Yang L, Wei W, Yang X, Jiang X, Liu L, He J, Ye J, Wei Q, Li Y, Wang W, Wang J, Ning G. Activating hotspot L205R mutation in PRKACA and adrenal Cushing's syndrome. *Science* 2014;**344**(6186):913–59.

59. Assié G, Letouzé E, Fassnacht M, Jouinot A, Luscap W, Barreau O, Omeiri H, Rodriguez S, Perlemoine K, René-Corail F, Elarouci N, Sbiera S, Kroiss M, Allolio B, Waldmann J, Quinkler M, Mannelli M, Mantero F, Papathomas T, De Krijger R, Tabarin A, Kerlan V, Baudin E, Tissier F, Dousset B, Groussin L, Amar L, Clauser E, Bertagna X, Ragazzon B, Beuschlein F, Libé R, de Reyniès A, Bertherat J. Integrated genomic characterization of adrenocortical carcinoma. *Nat Genet* 2014;**46**(6):607–12.

60. de Reyniès A, Assié G, Rickman DS, Tissier F, Groussin L, René-Corail F, Dousset B, Bertagna X, Clauser E, Bertherat J. Gene expression profiling reveals a new classification of adrenocortical tumors and identifies molecular predictors of malignancy and survival. *J Clin Oncol* 2009;**27**(7):1108–15.

61. Assié G, Jouinot A, Bertherat J. The "omics" of adrenocortical tumours for personalized medicine. *Nat Rev Endocrinol* 2014;**10**(4): 215–28.

62. Groussin L, Massias J, Bertagna X, Bertherat J. Loss of expression of the ubiquitous transcription factor cAMP response element-binding protein (CREB) and compensatory overexpression of the activator CREMtau in the human adrenocortical cancer cell line H295R. *J Clin Endocrinol Metab* 2000;**85**:345–54.

16

Hereditary Syndromes Involving Pheochromocytoma and Paraganglioma

Jennifer L. Geurts*, Thereasa A. Rich**, Douglas B. Evans*, Tracy S. Wang*

*Medical College of Wisconsin, Milwaukee, WI, USA
**Myriad Genetic Laboratories, Denver, Co, USA

Pheochromocytomas are rare neural crest-derived tumors of the adrenal medulla. By definition, a pheochromocytoma results in hypersecretion of catecholamines, which can cause headache, palpitations, elevated blood pressure (hypertension), and excessive perspiration, as well as other nonspecific symptoms. Pheochromocytomas have an incidence of approximately one in 100,000–300,000 in the general population. Although most pheochromocytomas are benign, they can cause major morbidity and death due to uncontrolled hypertension precipitated by stressful events such as anesthesia or pregnancy.

Extra-adrenal pheochromocytomas are referred to as paragangliomas. They are histologically identical, but occur outside of the adrenal medulla, anywhere in the sympathetic or parasympathetic paraganglia. Paragangliomas arising from the sympathetic paraganglia commonly hypersecrete catecholamines and are located mainly in the chest, abdomen, and pelvis. Parasympathetic paragangliomas are typically located within the head and neck (particularly the carotid body) and usually do not secrete excess catecholamines. Therefore, parasympathetic paragangliomas, if symptomatic, may be associated with a mass effect causing a visible or palpable neck mass, headaches, vocal cord disturbance, or cranial nerve deficit such as tongue weakness, shoulder drop, hearing loss, tinnitus, or problems with balance. The terms pheochromocytoma and paraganglioma are often used interchangeably in the medical literature; however, genetic risk assessment is different for pheochromocytoma and paraganglioma. In addition, terms such as ectopic pheochromocytoma, chemodectoma, nonchromaffin tumor, and glomus tumors are being phased out as these are anatomically nonspecific. The preferred terminology, which is used in this chapter, is pheochromocytoma for intra-adrenal tumors, and paraganglioma for extra-adrenal tumors, in addition to specifying the anatomic location (e.g., carotid body paraganglioma).

Pheochromocytomas have historically been referred to as "the 10% tumor," given previous estimates that the rates of extra-adrenal location, malignancy, and heredity were each 10%. However, recent evidence suggests that the rate of an underlying hereditary condition may be closer to 30–40% if one includes patients with a familial presentation as well as those with an occult germline mutation with an apparently sporadic (nonfamilial) presentation, which may account for up to half of all hereditary cases.[1–5] Therefore, lack of family history should not preclude referral for genetic assessment. The hereditary causes of pheochromocytoma and paraganglioma include multiple endocrine neoplasia types 2A and 2B (MEN2A/2B), von Hippel–Lindau syndrome (VHL), the familial paraganglioma/pheochromocytoma syndromes (SDHA, SDHB, SDHC, SDHD, SDHAF2, TMEM127, MAX), and neurofibromatosis type 1 (NF1). Pheochromocytoma has rarely been reported in patients with multiple endocrine neoplasia type 1 (MEN1).[5] In addition, a proportion of familial cases of pheochromocytoma and paraganglioma have no known underlying genetic basis, suggesting the existence of additional susceptibility loci and/or limitations in current molecular genetic testing techniques.[6] Currently, family history is the strongest independent predictor of a germline mutation in individuals with pheochromocytoma or paraganglioma; approximately 90% with a family history have an identifiable gene mutation.[1]

Genetic Diagnosis of Endocrine Disorders. http://dx.doi.org/10.1016/B978-0-12-800892-8.00016-6

MULTIPLE ENDOCRINE NEOPLASIA TYPE 2

Multiple endocrine neoplasia type 2 (MEN2) is characterized by a very high lifetime risk of developing medullary thyroid carcinoma (MTC), given a penetrance of >95% in untreated patients. Pheochromocytoma is associated with two of the clinical subtypes of hereditary MTC, MEN2A, and MEN2B. Familial MTC (FMTC) is now thought to be a phenotypic variant of MEN2A rather than a separate subtype. The genotype-phenotype association of *RET* gene mutations has increasingly been recognized as the main stimulus of penetrance disease estimates. It is important to recognize that patients previously classified as having FMTC may still have a risk for developing additional endocrinopathies.[7,8] In addition to MTC and pheochromocytoma, MEN2A is also associated with risk for primary hyperparathyroidism, and rarely individuals can develop a pruritic skin lesion (cutaneous lichen amyloidosis) or Hirschsprung disease. MEN2B is associated with a distinct physical appearance, which includes enlarged lips, a "bumpy" tongue, and eversion of the eyelids resulting from mucosal neuromas, as well as long, flexible limbs and joints, often referred to as a "Marfanoid" body habitus. The prevalence of MEN2 has been estimated at one in 35,000 individuals.[9]

MEN2 is inherited in an autosomal dominant manner, meaning that those affected with MEN2 have a 50% chance of passing the condition onto their children with each pregnancy, with males and females having an equally likely chance to be affected. Most cases of MEN2A are inherited (approximately 10% of patients are the first affected person in their family, i.e., *de novo*), whereas the majority of MEN2B patients do not have an affected parent and represent spontaneous or *de novo* mutations.

A definitive diagnosis of MEN2 in cases of apparently sporadic MTC or pheochromocytoma and in patients with an equivocal family history usually depends on the identification of a germline mutation in the rearranged during transfection (*RET*) proto-oncogene. *RET* is a 21-exon proto-oncogene located on chromosome 10q11.2 and encodes a receptor tyrosine kinase. The RET receptor interacts with the glial-derived neurotrophic factor family of ligands. Ligand binding induces receptor dimerization, autophosphorylation of intracellular tyrosine kinase residues, and activation of signaling cascades that ultimately lead to promotion of cell growth and survival. Virtually all MEN2-associated *RET* mutations are missense mutations, which cause constitutive activation of the RET receptor leading to unregulated cell growth and survival.[10] MEN2-associated mutations are almost always located in exons 10, 11, or 13–16 of the *RET* proto-oncogene, although mutations in exons 5 and 8 have been reported on rare occasions.[11,12] Strong genotype-phenotype correlations exist such that the codon in which the mutation occurs can be used to

predict MEN2 subtype, risk for pheochromocytoma and hyperparathyroidism, and age at onset and aggressiveness of MTC.[13] Mutations in the extracellular, ligand-binding domain (exons 5 and 8) and the first intracellular tyrosine kinase domain (exons 13 and 14) are mainly associated with low risk for pheochromocytoma and lower penetrant and less aggressive medullary thyroid cancer. Mutations in the extracellular cysteine-rich domain, which mediates dimerization (exons 10 and 11), are mainly associated with MEN2A and have a moderate aggressiveness of medullary thyroid cancer. Finally, the two specific mutations causing MEN2B, A883F, and M918T are located in the second intracellular tyrosine kinase domain and cause the earliest onset and most aggressive form of MTC (Table 16.1).

Pheochromocytomas occur in up to 50% of individuals with MEN2A and MEN2B. Pheochromocytomas occur most commonly with codon 634 (MEN2A) and codon 918 (MEN2B) *RET* proto-oncogene mutations, and with lesser frequency in kindreds with mutations of codons 532–534, 630, 777, and 912. Pheochromocytomas appear to be more commonly seen in association with the high-risk mutations for MTC.

Pheochromocytomas tend to develop after MTC is identified; however, there are well-documented examples of MEN2-related pheochromocytomas being responsible for the initial manifestations of this syndrome. Even though fewer than 3% of cases of apparently sporadic pheochromocytoma occurring before age 50 years are due to germline mutations of the *RET* proto-oncogene, *RET* testing of certain patients with apparently sporadic pheochromocytoma should still be done given the important clinical implications of making an accurate diagnosis.[2,5]

The pattern of catecholamine production in MEN2 differs from that seen in other hereditary forms of pheochromocytoma. MEN2-related pheochromocytomas usually secrete epinephrine/metanephrines and may or may not secrete noradrenergic catecholamines.[14] This biochemical phenotype results in an early clinical presentation characterized by attacks of palpitations, nervousness, anxiety, and headaches, rather than the more common pattern of hypertension seen with sporadic or other hereditary tumors. Some symptoms may be so subtle that the patient will be unaware of them and not bring any symptoms to medical attention. Bilateral pheochromocytomas occur in approximately half of patients with MEN2 with high risk *RET* mutations; their development is frequently asynchronous, with separation by as much as 15 years. Malignant pheochromocytomas and paragangliomas are extremely rare in MEN2. Malignant pheochromocytomas are generally associated with large infiltrating tumors. There have been reports of sympathetic paragangliomas in MEN2, although most of these have been found in the adrenal region and may actually represent a tumor that has developed in an adrenal rest, recurrence of a previously excised adrenal medullary tumor, or seeding from a

TABLE 16.1   Genotype–Phenotype Correlations in MEN2[19,21]

Level	*RET* codon	Associated clinical subtypes	Relative risk for pheochromocytoma	Relative risk of hyperparathyroidism	Age of prophylactic thyroidectomy
MOD	768 790 791 804 891	Usually FMTC	Low	Low	Consider by age five, but may be delayed*
	609 611 630	Usually FMTC	Low	Low	
	618 620	MEN2A or FMTC	Moderate		
H	634	MEN2A	High	High	By three to five years
HST	883 918	MEN2B	High	No increased risk	<12 months

* Thyroidectomy can be delayed as long as:
- *Annual calcitonin level is not elevated*
- *Annual thyroid ultrasound is normal*
- *MTC is not aggressive in the family*
- *Patient preference*

malignant pheochromocytoma. Therefore, young patients presenting with apparently sporadic adrenergic pheochromocytoma should be offered testing for *RET* mutations, whereas patients with entirely noradrenergic pheochromocytomas, paragangliomas, or malignant tumors are unlikely to benefit from *RET* testing.

MEN2-associated pheochromocytomas are often detected by routine biochemical screening or present symptomatically due to palpitations, headache, tachycardia, or sweating. Plasma-free metanephrines have the highest sensitivity and specificity for detecting pheochromocytoma, particularly for those with hereditary syndromes, and are the preferred method of surveillance. However, measurement of 24-h urine catecholamines and metanephrines and serum catecholamines are also frequently used.[15,16] Importantly, biochemical screening for pheochromocytoma should occur before any type of surgical intervention or pregnancy in patients with MEN2; the consequences of an undiagnosed pheochromocytoma in these situations can be life threatening. For MEN2 patients with biochemical evidence or convincing symptoms of pheochromocytoma, imaging studies such as CT or MRI are useful for determining whether one or both adrenal glands are involved. Patients with a pheochromocytoma should be treated with alpha- and sometimes beta-blockade before undergoing surgery.[17] There is really no role for adrenal biopsy in patients with suspected pheochromocytoma, and biopsy carries the risk of adrenergic crisis.

The typical age at onset of MEN2-associated pheochromocytomas is in the third decade of life, which is approximately 10–20 years younger than the typical age of sporadic pheochromocytoma development. Surgery is the treatment modality of choice, and due to the high frequency of bilateral tumors, bilateral adrenal surgery is often required at some point. Bilateral total adrenalectomy renders the patient dependent on replacement doses of corticosteroid drugs for life and at risk for acute adrenal insufficiency (Addisonian crisis), and therefore a cortex-sparing adrenalectomy should be considered whenever technically possible in all patients. *In situ* preservation of the intact functional adrenal cortex, while removing most of the adrenal medulla and all of the associated pheochromocytoma (cortical-sparing adrenalectomy) would be the ideal operation. While associated with higher rates of recurrence (because it is impossible to remove all medullary tissue when preserving some cortex), this approach minimizes the risk of adrenal insufficiency, which is an appropriate balance with regard to risk for recurrence given the very low malignant potential for MEN2-related pheochromocytomas. In one of the largest studies of patients with hereditary pheochromocytomas, 96 patients underwent adrenalectomy, including 39 patients who underwent intentional cortical-sparing adrenalectomy. In these patients, only one (3%) patient developed acute adrenal insufficiency, two (7%) patients developed recurrent disease in the adrenal remnant, and 21 of 27 patients with ≥3 years of follow-up were steroid independent.[18] Adrenalectomy, total or cortical-sparing, can be safely performed via the laparoscopic (transabdominal or retroperitoneoscopic) or open approach, ideally by high-volume adrenal surgeons, given the association between surgeon volume and patient outcomes.[19,20] Patients who do undergo bilateral total adrenalectomy should wear medical alert bracelets indicating that hydrocortisone could be life-saving in emergency situations.

Because nearly all MEN2 patients eventually develop MTC, patients with a germline RET mutation are generally recommended to undergo early intervention for MTC. For the highest risk mutations, early detection of MTC is difficult and the treatment options for locally advanced and metastatic disease are limited. Given the acceptably low morbidity and mortality associated with thyroidectomy, early thyroidectomy has become widely accepted as standard of care. Guidelines are available to determine the earliest age at which prophylactic thyroidectomy would be appropriate in mutation carriers.[19] Mutations associated with the least aggressive forms of MTC are "moderate risk" (MOD) mutations and occur in RET codons 609, 611, 618, 620, 630, 768, 790, 791, 804, or 891; these were previously described as Level A and B mutations. MTC associated with these codons is typically later onset, less aggressive with rare reports of MTC-related death, and may not be fully penetrant.[13] However, there are rare cases of more aggressive MTC occurring with mutations in these codons. Genetic testing and initiation of MTC screening for RET mutation carriers is still recommended by three to five years of age; however, thyroidectomy can be delayed as long as: thyroid ultrasound and measurement of serum levels of calcitonin, performed on at least a yearly basis, remain without evidence of MTC; the MTC in other family members is not aggressive; and, perhaps most importantly, the patient prefers to delay surgery.[19] "High risk" (H) mutations (codons 634 and 883) are associated with moderately aggressive MTC and thyroidectomy is recommended by age five years, based on serum calcitonin levels. The "highest risk" (HST) mutation is associated with the most aggressive form of MTC and includes the MEN2B-related mutations on codon 918. Individuals with an HST mutation should undergo prophylactic thyroidectomy by twelve months of age, with some experts advocating for even earlier surgery (Table 17.1). However, this rarely occurs, as newborns do not have the phenotypic appearance of MEN2B, and the mutation is spontaneous in almost all affected infants. These recommendations are based on consensus guidelines published by expert panels, and they have evolved over time

and are likely to continue to evolve as additional data on long-term outcomes are collected.[19,21]

VON HIPPEL–LINDAU SYNDROME (VHL)

In addition to MEN2, VHL is a common cause of hereditary pheochromocytoma and should be one of the first syndromes to consider, particularly if the patient is very young (VHL accounts for nearly half of pheochromocytomas presenting before age 20 years), or if the tumor has noradrenergic catecholamine secretion. Overall, VHL accounts for approximately 11% of apparently sporadic pheochromocytomas.[2] Most VHL-associated pheochromocytomas are benign and intra-adrenal, however (in contrast to MEN2), extra-adrenal paragangliomas and malignant tumors may occur beyond that of a truly anecdotal patient.[14,22,23] VHL is also inherited in an autosomal dominant manner and has a prevalence of about one in 36,000 persons, with an estimated 20% of cases being *de novo* mutations. As such, a significant proportion of patients with VHL do not have a family history of disease.

In addition to pheochromocytoma, VHL is characterized by hemangioblastomas in the retina and central nervous system, renal cysts and renal cell carcinoma, pancreatic cysts and pancreatic endocrine tumors, endolymphatic sac tumors, and papillary cystadenomas of the epididymis and broad ligament. Some families with VHL have been reported to have developed only pheochromocytomas and/or paragangliomas; however, it is not clear whether these families were simply ascertained before the onset of other VHL-related manifestations. VHL is classified into two main types, based on the risk for pheochromocytoma; VHL type 1 is characterized by a low risk of pheochromocytoma, whereas VHL type 2 is characterized by a high risk of pheochromocytoma. VHL type 2 is further divided based on risk for renal cell carcinoma; type 2A is associated with a low risk of renal cell carcinoma, type 2B is associated with a high risk of renal cell carcinoma, and type 2C is defined as pheochromocytoma occurring without other manifestations of VHL disease (Table 16.2).[24,25] VHL is highly variable, both within and

TABLE 16.2   VHL Subtypes and Genotype/Phenotype Correlations

| VHL subtype | Associated tumor types | | | Most common gene mutation type |
	HB	RCC	PGL/PCC	
Type 1	+	+	−	Deletions, truncations
Type 2A	+	−	+	Missense (Y98H, Y112H)
Type 2B	+	+	+	Missense
Type 2C	−	−	+	Missense (L188V, V84L, S80L)

HB, hemangioblastoma (retinal or central nervous system); RCC, renal cell carcinoma (clear cell); PGL/PCC, paraganglioma/pheochromocytoma.

between different families, in terms of the age at onset, number and types of tumors, and severity of the disease. VHL is thought to be fully penetrant by age 60, although the most common age of onset is in the second to third decade of life.[26] Like MEN2, VHL-associated pheochromocytomas are often detected on routine biochemical screening (preferably through measurement of plasma-free metanephrines and normetanephrines) or due to patient-reported symptoms, and they are often bilateral. Diagnosis and treatment of VHL-associated pheochromocytoma is similar to that of MEN2-associated pheochromocytomas; however, the possibility of extra-adrenal tumors (i.e., paragangliomas) and malignancy must be considered in patients with VHL.

VHL is caused by germline inactivating mutations of the *VHL* tumor suppressor gene, which is comprised of three exons and is located on chromosome 3p25. The most well-defined role of the VHL protein is in its participation in the regulation of the hypoxia-inducible factor (HIF) pathway. Normally, HIF is active in hypoxic states to promote oxygen-independent ATP synthesis and angiogenesis and its alpha subunit is degraded in normoxic states. VHL, in complex with elongins B and C, is involved in proteolytic degradation, ultimately resulting in reduced or abnormal function of the VHL protein, thus permitting increased HIF1 and HIF2 activity. VHL's involvement in the HIF pathway likely contributes to the highly vascular nature of VHL-associated tumors. However, failure to properly degrade HIF does not seem to be an important mechanism in the development of VHL-associated pheochromocytomas.[27,28]

Sequencing and large deletion/duplication testing detects nearly 100% of VHL-associated mutations, making genetic testing the most definitive and cost-effective method to diagnose or rule out VHL syndrome in patients presenting with VHL-related diseases such as pheochromocytoma.[25] Rare cases of VHL with an undetectable mutation are documented and are likely due to mutations acquired during embryogenesis resulting in somatic mosaicism. Unlike MEN2, in which a few recurring mutations account for the majority of the disease, over 500 *VHL* gene mutations have been described, and there is considerable variation in the number, location, and types of mutations involved. Despite this, there are genotype-phenotype correlations in VHL and there are a limited number of recurring mutations that occur either in hypermutable sequences (delPhe76, N78S/h/T, P86L, R161Ter, C162Y/F/W, R167Q/W, and L178P) or as founder mutations, such as the Black Forest Y98H mutation in German families.[29] The majority (96%) of VHL patients with pheochromocytoma (i.e., VHL type 2) have missense mutations, whereas 95% of null or truncating mutations do not cause VHL-associated pheochromocytoma (i.e., VHL type 1).[25,30,31] Mutations at codon 167 are responsible for approximately 46% of VHL type 2

families.[25] Four mutations, L188V, V84L, S80L, and F161Q, have been associated with the pheochromocytoma-only subtype (type 2C), and three mutations, L128F, S111C, and R161Q, have been associated with a risk of pheochromocytoma without renal cell carcinoma (i.e., type 2A) (Table 16.2).[25,29] Importantly, not all laboratories offer both sequencing and large deletion testing for the *VHL* gene. Large deletion testing is necessary to achieve a near 100% mutation detection rate.

If a diagnosis of VHL is confirmed, the patient should be referred for appropriate tumor surveillance studies. The recommended surveillance strategy varies depending on patient age and whether there are any symptomatic complaints, but typically involves: (1) ophthalmologic evaluation, at least yearly and biannually during puberty; (2) annual abdominal ultrasound of the kidneys starting around 10 years of age and switching to MRI abdominal surveillance around age 20; (3) baseline MRI of the brain and spine starting around puberty and repeating every one to three years; and (4) annual audiology exam to monitor hearing. This monitoring should be implemented in all *VHL* gene mutation carriers, even if they appear to have a pheochromocytoma-only phenotype. Surveillance for pheochromocytoma in type 2 families involves annual measurement of plasma-free metanephrines and normetanephrines. Families who appear to be type 1 based on family history and genetic test results may not need to undergo as rigorous screening for pheochromocytoma as type 2 families; however, screening should be performed prior to any surgical procedure or pregnancy/delivery.

NEUROFIBROMATOSIS TYPE 1

NF1 is also associated with hereditary pheochromocytoma; however, genetic testing is generally not necessary to establish a diagnosis of NF1. Patients with NF1 will have manifestations that are obvious on physical examination, most commonly including café au lait macules, neurofibromas, and axillary and inguinal freckling by the time they are at risk for the development of pheochromocytoma.[32] The National Institutes of Health diagnostic criteria for NF1 include a patient with two or more of the following clinical features: (1) six or more café au lait macules measuring at least five mm in prepubertal individuals and at least 15 mm in postpubertal individuals; (2) two or more neurofibromas; (3) a plexiform neurofibroma; (4) axillary or inguinal freckling; (5) optic glioma; (6) more than one Lisch nodule; (7) sphenoid dysplasia or tibial pseudoarthrosis; (8) a parent, sibling, or child with NF1.[33] By eight years of age, 97% of NF1 patients meet these criteria, and all do so by 20 years of age.[34]

Pheochromocytomas rarely develop in NF1 (approximately 1% risk) and tend to behave in a similar manner as sporadic pheochromocytoma.[35] The average age at onset is in the fourth decade, but they can occur in childhood. There are rare examples of multigenerational pheochromocytomas. Most pheochromocytomas in NF1 produce mainly norepinephrine and therefore most commonly present with hypertension and noradrenergic symptomatology. However, 22% of pheochromocytomas have no symptoms related to excessive catecholamine secretion. Approximately 11–12% of tumors are malignant, 10% are bilateral, and over 94% are located within the adrenal gland.[35,36]

HEREDITARY PARAGANGLIOMA/ PHEOCHROMOCYTOMA SYNDROMES

Germline mutations in the succinate dehydrogenase complex genes (abbreviated *SDHx*), along with *TMEM127* and *MAX* genes, are known to cause hereditary paraganglioma-pheochromocytoma (PGL/PCC) syndromes. Hereditary PGL/PCC syndromes are characterized by susceptibility to multiple paragangliomas of the head and neck, thoracic, and abdominal paraganglia, as well as pheochromocytoma. Most patients with hereditary PGL/PCC and 8–50% of apparently sporadic, young onset paragangliomas are caused by a mutation in the *SDHB*, *SDHD*, or *SDHC* genes.[37,38]

Of the *SDHx* genes, *SDHB* and *SDHD* are the most common genes responsible for hereditary PGL/PCC syndromes, followed by *SDHC*, whereas *SDHA* and *SDHAF2* gene mutations have only been observed in a handful of families. The typical age at tumor development in patients with *SDHx*-related hereditary PGL/PCC syndrome is in the late 20s to early 30s; however, a wide range of ages of onset have been reported and penetrance can be incomplete, particularly for *SDHB* and *SDHC* mutation carriers.[39] Each of the *SDHx* genes is distinguished by the most common location of tumor development, risk for malignancy and the inheritance pattern. The *SDHB* and *SDHD* genes were the first to be discovered, are much more commonly mutated than the others, and therefore are the most well characterized of all the *SDHx* deficiencies.

In *SDHB* gene mutation carriers, paragangliomas develop most frequently in the abdomen, followed by head and neck locations, and are less common in the chest or adrenal gland.[39,40] *SDHB*-related paragangliomas are also associated with a high rate of malignancy reported to approach 38%.[40] The risk for malignancy appears to be highest in those patients with extra-adrenal abdominal tumors (notice the consistent theme here regardless of the underlying disease), risk for malignant biologic behavior (distant metastases) of head/neck paragangliomas or pheochromocytoma is unknown, but overall, uncommon. Malignant behavior may be evident at the time of diagnosis, or may not be apparent for months and sometimes years after initial diagnosis. Similar to sporadic paragangliomas, *SDHB*-associated paragangliomas that develop in abdominal locations tend to secrete catecholamines, whereas those that develop in the head and neck are usually nonfunctioning. Patients present with either the characteristic symptoms associated with catecholamine excess such as hypertension, headache, excessive perspiration, and palpitations, or they may be asymptomatic due to a nonfunctional tumor. An important management consideration for *SDHB* carriers is that approximately 10% of abdominal/pelvic tumors are biochemically nonfunctional and only cause symptoms related to mass effect, such as abdominal or back pain, abdominal distention, urinary abnormalities, deep vein thrombosis, or weight loss. The lack of symptoms of catecholamine excess is likely why these *SDHB*-associated tumors tend to be larger than sporadic tumors.[3] The majority of functional tumors secrete noraderenergic catecholamines and frequently secrete dopamine.[41] Initial estimates of the age-related penetrance of tumor development in *SDHB* mutation carriers was 30–50% by age 30–35 years, 45% by age 40 years, and 77% by age 50 years.[42] However, these estimates were largely based on affected individuals from high risk families initially ascertained for gene discovery studies and may overestimate the true frequency of tumor development due to selection bias. More recently, studies conducted on mutation carriers identified through predictive genetic testing have estimated the lifetime risk to be approximately 30%, which seems like a more accurate estimate based on clinical experience with these families.[43] The earliest reported diagnosis of a paraganglioma in an *SDHB* mutation carrier was age 10 years.[39,41] Many (approximately 72%) *SDHB* mutation carriers develop only a single tumor and the majority have an apparently sporadic presentation, although the percentage of patients with a *de novo* mutation has not yet been well established.[41] *SDHB* mutations are inherited in an autosomal dominant manner.

Germline *SDHB* mutations have been found to be an independent predictor of poor prognosis in individuals with a malignant pheochromocytoma or paraganglioma. *SDHB* mutation carriers have a significantly lower five-year survival rate and a shorter metanephrine excretion doubling time than patients with malignant sporadic tumors.[44] In fact, an *SDHB* germline mutation is the only independent predictor of malignancy in pheochromocytomas and paragangliomas at this time. The high rate of malignancy in *SDHB* mutation carriers warrants close and aggressive tumor surveillance and treatment. At present, evidence-based guidelines on the diagnosis, surveillance, and management of *SDHB*-associated tumors are not available. Based on available natural history data, it is reasonable to recommend annual plasma or urinary fractionated metanephrine measurement in addition to cross-sectional imaging studies every 1–2 years in all mutation carriers. In asymptomatic patients, MRI is the preferred screening method due to the minimal radiation exposure associated with it. We favor MRI (as a screening study) for any patient at risk for an abdominal tumor/recurrence when it is anticipated that they will require more than 10 imaging studies and have an expected survival greater than 10 years. Whole body rapid MRI protocols have been developed at some centers, appear to be a promising screening modality, and are less expensive than ordering MRI of each system separately.[45] If a tumor is suspected based on screening studies or symptoms, further diagnostic imaging studies such as targeted CT or MRI can be utilized and have a high sensitivity for detecting pheochromocytoma and paraganglioma. In general, if the tumor is large enough to cause symptoms, it will be seen on cross-sectional imaging. Functional imaging modalities such as ^{123}I-labeled meta-iodobenzylguanide scintigraphy, which has a lower sensitivity but higher specificity, or studies such as 6-[^{18}F]-fluorodopamine, [^{18}F]-dihydroxyphenylalanine, [^{11}C]-hydroxyephedrine, or [^{11}C]-epinephrine PET scans may also be helpful.[46] The preferred imaging modality may be influenced by genotype; one recent study determined that [^{18}F]-FDG-PET had a sensitivity of almost 100% for the detection of SDHB-related paraganglioma metastases and may also be useful for preoperative staging as well as surveillance after tumor resection.[46,47] However, functional imaging is not recommended as a routine surveillance test in asymptomatic patients.

Sequencing of the coding region (exons 1–8) and intron-exon junctions as well as large deletion/rearrangement analysis are clinically available for SDHB. The detection rate using these testing methodologies is currently unknown. Large deletions are now believed to occur in a significant proportion of cases. For example, one study found three (12.5%) SDHB deletions in 24 patients who had either familial, multifocal, extra-adrenal, or young age of onset (age 35 years and younger) disease and who had previously tested negative for point mutations in RET, VHL, SDHB, SDHD, and SDHC.[48] Therefore, we advocate that large deletion/duplication testing be utilized in all patients who undergo SDHB testing.

Paragangliomas in SDHD gene mutation carriers tend to develop most often in the head and neck; however, abdominal and thoracic paragangliomas and adrenal pheochromocytomas are also observed at a lower frequency.[39,40] The vast majority of paragangliomas in SDHD mutation carriers demonstrate benign biologic behavior, although some can be locally invasive and rarely can metastasize.[49] The age-related penetrance of tumors in carriers of a paternally inherited SDHD mutation is estimated at 50% by age 31 years and 86% by age 50 years.[42] Most (approximately 75%) have multifocal tumors such

as bilateral carotid body tumors. The earliest reported age of paraganglioma development in an SDHD mutation carrier was 5 years.

SDHD mutations are inherited in an autosomal dominant pattern with parent of origin effects.[50] While any person with an SDHD mutation is at 50% risk of passing the mutation on to each of their children, the gender of the transmitting parent determines whether the child is at risk for tumor development. Individuals who inherit an SDHD gene mutation from their father are at high risk for paraganglioma development; those that inherit a mutation from their mother are not at risk for paraganglioma development. While there are reports of finding a paraganglioma in patients who inherited an SDHD mutation from their mother, these appear to be exceptional, and do not justify routine screening for patients with a maternally transmitted SDHD mutation.[51] The rate of de novo mutations has not yet been established.

There are no consensus guidelines for surveillance of SDHD mutation carriers. As the majority of tumors are nonfunctional and develop primarily in head and neck locations, cross-sectional imaging studies will usually be necessary for tumor detection. However, because functioning abdominal pheochromocytoma and paraganglioma can also occur, annual biochemical screening using urinary or plasma fractionated metanephrine levels is also recommended. Whether abdominal imaging in the absence of biochemical evidence of a functioning pheochromocytoma or paraganglioma is warranted is unclear; however, the utilization of serial whole body MRI for screening asymptomatic patients should be considered in these patients as well.[45]

Most SDHC mutation carriers present with benign tumors of the head and neck, although abdominal, thoracic, intra-adrenal, and malignant tumors have been observed on multiple occasions.[52–57] In one large published cohort of unselected patients diagnosed with pheochromocytoma/paragangliomas, SDHC mutations were found in 1.85% of index cases, which comprised 8.3% of all the mutations identified.[58] SDHC mutations are inherited in an autosomal dominant manner and appear to be incompletely penetrant, though precise lifetime risk estimates have not yet been established. The rate of de novo mutations is currently unknown.

SDHB, SDHC, SDHD, and SDHA encode proteins that comprise the four subunits that make up the succinate dehydrogenase complex, an enzyme involved in the primary mechanisms of ATP synthesis within the mitochondria, the electron transport chain, and the Krebs cycle. Succinate dehydrogenase is responsible for oxidizing succinate to fumerate within the Krebs cycle. The SDHC and SDHD subunits are responsible for anchoring the succinate dehydrogenase enzyme to the mitochondrial membrane, while SDHB and SDHA form the enzyme's catalytic core. Germline mutations of SDHB, SDHD,

SDHC, or SDHA are thought to result in disassembly of the succinate dehydrogenase enzyme, causing reduced enzymatic function leading to accumulation of succinate and oxaloglutarate within the cell. In addition to being key intermediaries of the Krebs cycle, succinate and oxaloglutarate are also cofactors of the polyhydroxylase enzymes involved in HIF degradation. Therefore, the downstream effect of abnormal SDHB, SDHD, SDHC, or SDHA subunit function is an increase in available HIF, thus leading to a similar mechanism of tumorigenesis as is seen in VHL.[4]

While SDHA has long been known for its role in Leigh syndrome, a severe neurodegenerative disorder caused by biallelic germline SDHA mutations, heterozygous SDHA mutations have only recently been implicated in the development of pheochromocytoma/paragangliomas. To date, only a handful of patients have been described; and given that pheochromocytomas and paragangliomas have not been a reported feature in family members of patients with Leigh syndrome, heterozygous SDHA mutations appear to confer low-penetrant susceptibility.

The SDHAF2 gene (also known as SDH5) was implicated in hereditary PGL/PCC syndromes in 2009. Mutations in this gene have only been identified in two families, which have exclusively developed benign head/neck paragangliomas.[59] The SDHAF2 gene produces a protein that facilitates posttranslational modification of the SDHA protein.

Shortly after the discovery of the SDHAF2 gene, mutations in TMEM127 were identified in a cohort of sporadic and autosomal dominantly inherited pheochromocytomas.[60] Subsequently, additional studies have identified TMEM127 mutations in extra-adrenal paragangliomas, bilateral pheochromocytomas, and rarely in patients with malignant pheochromocytomas or paragangliomas.[61,62] The gene is thought to be a negative regulator of mTOR, with a kinase receptor signaling transcription signature similar to that of NF1 and RET gene mutations. This is in contrast to the expression profiles seen in VHL or SDHx gene mutations, which are characteristic of hypoxia response.[63] Most recently, whole-exome sequencing (WES) identified mutations in the MAX gene in individuals with pheochromocytoma, including those with bilateral and malignant disease.[64] The MAX (MYC-associated factor X) gene encodes a protein involved in the regulation of cell proliferation, differentiation, and apoptosis.[65] TMEM127 and MAX mutations appear to be quite rare.[66,67]

The utility of WES was once again proven successful in the recent identification of mutations in the FH gene predisposing to pheochromocytomas and paragangliomas.[68] Mutations in FH are associated with hereditary leiomyomatosis and renal cell cancer, but no associations with pheochromocytomas or paragangliomas were previously appreciated. FH mutations have been identified in patients with malignant pheochromocytomas or paragangliomas. The authors proposed, based on immunohistochemistry staining patterns in both the SDHB-deficient tumors and FH-deficient tumors, that the mechanism for tumorigenesis involved a hypermethylation phenotype leading to genome-wide alterations in DNA methylation.[68]

FH, TMEM127, and SDHA mutations are inherited in a classic autosomal dominant pattern. All of the families reported to date with mutations in SDHAF2 and MAX have exhibited an inheritance pattern consistent with maternal imprinting (e.g., similar to SDHD, tumors develop only in carriers of a paternally inherited mutation). However, phenotype characterization of additional families with mutations in these genes is required before making any definitive conclusions. Importantly, further research on large population-based cohorts needs to be performed before penetrance estimates, malignant potential, and multifocality can be accurately predicted in carriers of these newly described gene mutations. Studies of TMEM127, MAX, and FH mutations in unselected patients with pheochromocytomas/paragangliomas suggest a mutation frequency in the range of 1% for each gene. Newly emerging associations with the KIF1B, EGLN1, and HIF2A genes are currently being investigated.[69–71]

It is now well established that there is a small but real increased risk for gastrointestinal stromal tumors (GISTs) in patients with SDHx mutations, and renal carcinoma with SDHB mutations.[72] Germline SDHB, SDHC, and SDHD mutations have been identified in families with both GISTs and paragangliomas (referred to as the Carney–Stratakis dyad) and germline mutations in all four SDH genes have been found in patients with early-onset apparently sporadic GIST.[73–75] GISTs in SDHx carriers tend to lack KIT and PDGFRA mutations (wild type), to be wild type (i.e., not associated with an acquired KIT mutation), show loss of heterozygosity for the corresponding SDHx gene as expected according to the two-hit hypothesis, and also demonstrate loss of immunostaining for SDHB.[73] SDHB mutations have been found in 4.4% of patients with apparently isolated familial, early-onset, or bilateral renal cell carcinoma, and renal tumors in SDHB mutation carriers show biallelic inactivation of the SDHB locus and loss of immunostaining of the SDHB protein, consistent with SDHB having a causative mechanism in these tumors.[76,77] Similarly, there is growing evidence for a role of SDHx in pituitary tumors, which have been reported in a few cases.[77,78] There are several case reports of papillary thyroid cancer in SDHx carriers, but SDHx does not seem to be causative, but rather an incidental finding.[72] Several other neoplasms have been reported in SDHx mutation carriers; however, it is not clear that the reported incidence exceeds that of the frequency in the general population.

GENETIC RISK ASSESSMENT IN PATIENTS WITH APPARENTLY SPORADIC PHEOCHROMOCYTOMA

In patients presenting with an apparently sporadic pheochromocytoma or paraganglioma, defined as a patient with a single tumor without a family history or overt clinical evidence suggestive of a particular syndrome, one should consider the known inherited endocrinopathies including VHL, MEN2, and SDH-related syndromes. Several retrospective studies have assessed the frequency of germline mutations associated with these conditions in patients with apparently sporadic pheochromocytoma and paraganglioma, the largest of which identified a germline mutation in 24–27% of patients.[1-3] However, when one considers the age of diagnosis, tumor location and focality, and tumor biology (malignant vs. benign), the mutation prevalence ranged from less than 2–70%.[2,5,38] The patients most likely to have an underlying germline mutation were those with multifocal tumors (approximately 81% for RET, VHL, and SDHD mutations combined), age of onset 18 years and younger (approximately 56% have a mutation in VHL, SDHB, or SDHD), and malignant extra-adrenal paraganglioma (almost 50% have a mutation in SDHB).[2,37,79] Patients presenting with a single benign pheochromocytoma after age 20 have a mutation prevalence of 10–20% with the prevalence dropping below 2% after 50 years of age. Mutation rates overall are higher for paragangliomas.[2,38]

Recommendations from the First International Symposium on Pheochromocytoma were published in 2007; it was recommended that all individuals with a young onset pheochromocytoma or paraganglioma (age 50 and under), more than one tumor, a malignant tumor, a family history of a pheochromocytoma or paraganglioma, or with suggestive features for a syndromic condition, should be offered genetic assessment.[80] Several authors have recently suggested that all individuals with pheochromocytomas or paragangliomas, regardless of age or family history, should be referred for genetic counseling and testing.[1,2] The genetic risk assessment can provide crucial information on the patient's future risk for additional tumors and the assessed risk for their family members. In addition, as with most other neuroendocrine tumors, the presence or absence of a malignant pheochromocytoma cannot be predicted reliably based on tumor histology alone, and is generally identified only by the presence/development of metastatic disease. Therefore, the presence or absence of a particular gene mutation can provide information regarding the risk of metachronous metastases. For example, a patient with an SDHB mutation has a higher risk of having synchronous or metachronous metastases than does a patient with an SDHD mutation, a VHL mutation, or a sporadic pheochromocytoma.

Genetic testing is complicated by the fact that there are multiple candidate genes with overlapping clinical phenotypes. Testing each gene individually in a sequential method is costly and time consuming. As above, a careful clinical assessment in addition to considering the age at onset, tumor location (adrenal, extra-adrenal head/neck, or abdominal), presence or absence of malignancy, and the tumor biochemical phenotype can help target genetic testing (Table 16.3). It is relevant to consider mutation likelihood; SDHB, SDHD, and VHL are the major genes whereas SDHC, SDHA, SDHAF2, TMEM127, MAX, RET, and NF1 genes are rarely mutated in presumed nonsyndromic patients. In addition, the clinical utility of genetic testing should also be considered; for example, if there is an approximately equal likelihood of detecting an SDHB mutation as an SDHD mutation, SDHB should be considered first as the consequences of a missed SDHB mutation may be more clinically significant than a missed SDHD mutation. Similarly, the clinical utility of testing for MEN2 and VHL is also very high, given that individuals with these syndromes are at risk for multiple tumor types and benefit from surveillance and early detection, including the need for prophylactic thyroid surgery in the case of MEN2. Most recently, gene panel testing has emerged as a comprehensive approach to maximize mutation detection, particularly in patients that lack overt syndromic features.[81] This strategy capitalizes on the high throughput, sensitivity, cost effectiveness, and rapid data turnaround of massively parallel next-generation sequencing (NGS) technology when compared to conventional Sanger sequencing.[82] Regardless of the sequencing technology utilized, analysis for larger structural abnormalities with multiplex ligation-dependent probe amplification or microarray technology should also be deployed as deletions, duplications, inversions, and translocations of the various genes have been described.

Ideally all first-degree relatives of SDHx mutation positive individuals should be referred for genetic counseling and considered for genetic testing. The utility of screening asymptomatic SDHx mutation positive family members has been demonstrated by the detection of a significant number of occult paragangliomas in this population.[83] Genetic counseling is warranted, and genetic testing should be strongly considered for many reasons including the inherent anxiety and stress of the at-risk individuals, the potential for genetic discrimination, and the elevated costs of surveillance (imaging, blood tests, etc.).

TABLE 16.3 Characteristics of Genetic Syndromes Associated with Paraganglioma or Pheochromocytoma

Gene	Primary location	Biochemistries	Malignancy risk	Involved signaling pathway	Inheritance	PGL/PCC penetrance (%)	De novo rate (%)
RET (MEN2)	PCC/bPCC	Metanephrine	+	MAP kinase/AKT/mTOR	AD	50	5% MEN2A; 50% MEN2B
VHL (VHL type 2)	bPCC>PCC>PGL	Normetanephrine	++	Pseudohypoxia/HIF	AD	40–60	20%
NF1	PCC	Metanephrine	+	MAP kinase/AKT/mTOR	AD	1	50%
SDHB	TAPGL>HNPGL>PCC>bPCC	Dopamine +/− normetanephrine	+++	Pseudohypoxia/HIF	AD	30–50	UNK
SDHD	HNPGL>TAPGL>PCC	Dopamine +/− normetanephrine	+	Pseudohypoxia/HIF	AD-PI	70–90*	UNK
SDHC	HNPGL>TAPGL>PCC	Dopamine +/− normetanephrine	+	Pseudohypoxia/HIF	AD	UNK	UNK
SDHA	PGL	UNK	UNK	Pseudohypoxia/HIF	AD	UNK	UNK
SDHAF2	HNPGL	UNK	−	Pseudohypoxia/HIF	AD-PI	UNK	UNK
MAX	PCC>bPCC>PGL	Norepinepherin +/− epinephrine	+	MAP kinase/AKT/mTOR	AD-PI	UNK	UNK
TMEM127	PCC>bPCC>PGL	Metanephrine	−	MAP kinase/AKT/mTOR	AD	UNK	UNK
FH	PCC/PGL	Normetanephrine	+++	Pseudohypoxia/HIF	AD	UNK	UNK

Only for paternally inherited mutation.
TAPGL, thoracic or abdominal paraganglioma; HNPGL, head and neck paraganglioma; PCC, pheochromocytoma; bPCC, bilateral pheochromocytoma; PGL, paraganglioma; AD-PI, paternal inheritance, parent-of-origin effect.

Authors' Recommended Genetic Testing Strategy for Patients with Apparently Sporadic Pheochromocytoma/Paraganglioma

A list of commercial laboratories offering clinical genetic testing for pheochromocytoma/paraganglioma syndromes can be found through the Genetic Testing Registry website administered by the National Center for Biotechnology Information: http://www.ncbi.nlm.nih.gov/gtr/. Various institutions have proposed complex genetic testing algorithms based on tumor phenotype and clinical history in an attempt to balance mutation detection rate and the cost of testing.[1,53,84–86]

- *Syndromic features*: The most important first step in assessing risk for hereditary pheochromocytoma in an individual patient is to collect an accurate family and medical history to determine whether there are any obvious indicators of MEN2, VHL, and NF1. For example, with MEN2, it is important to consider whether there is a history of thyroid cancer or a thyroid nodule, an elevated calcium, sudden death (a complication of pheochromocytoma), or visible signs of MEN2B. The clinician should assess whether there are any obvious features of VHL in a patient or their family, such as retinal tumors/sudden onset of blindness, central nervous system tumors, or kidney cancer. In addition, most patients with a pheochromocytoma will have undergone abdominal CT or MRI imaging, which allows for evaluation of renal or pancreatic cysts. Importantly, some patients with VHL develop only a pheochromocytoma or paraganglioma and not the other features of VHL; thus, the absence of extra-adrenal features of VHL, even in an older patient, cannot by itself exclude this diagnosis.
- *Age at onset*: All patients younger than 50 years of age should be offered genetic testing. VHL should be the first diagnosis to be considered in patients with pheochromocytomas diagnosed before the age of 20, followed by *SDHB*. *SDHD* may also be considered, although as above, the utility of *SDHD* testing is less than that of VHL and *SDHB*. Pheochromocytomas are rare in children with MEN2 (usually develops a bit later in life) and therefore *RET* testing should be considered when patients under the age of 20 are found to have intra-adrenal, nonmetastatic tumors that produce epinephrine/metanephrines.
- *Tumor location*:
  - Pheochromocytomas: *RET* should be considered first for patients with epinephrine/metanephrine-producing tumors. *VHL* should be considered first for patients with noradrenergic tumors, particularly those occurring at a young age. *SDHB* followed by *SDHD* testing should then be considered in that order.
  - Abdominal, thoracic, and pelvic sympathetic paragangliomas: *SDHB* should be considered first followed by *VHL* (especially if the tumor produces noradrenaline). *SDHD* could also be considered if the above testing is negative.
  - Head and neck paragangliomas: For individuals with multifocal tumors, *SDHD* should be considered first as this is the most likely gene to be involved, followed by *SDHB* and *SDHC*. For individuals with a single tumor, the likelihood of an *SDHB* mutation is approximately equal to that of an *SDHD* mutation, and therefore *SDHB* testing should take preference given the higher clinical utility.[6,56] *SDHC* genetic testing can also be considered if a mutation in the above genes is not detected.
- *Malignancy*: The presence of a malignant tumor should prompt *SDHB* testing. *VHL* can also be considered in patients who test negative for *SDHB* and who have a norepinephrine-producing tumor.

Patients with a single, apparently benign pheochromocytoma presenting after 50 years of age and a negative family history are unlikely to have a mutation, and genetic testing is generally not indicated.

Tumor immunohistochemistry of the SDH subunits is currently available in some pathology laboratories. The *SDHx* genes act as tumor suppressors, with biallelic inactivation in the tumor tissue. It is proposed that a germline mutation in any of the *SDHx* subunit genes results in instability of the mitochondrial complex II leading to the degradation of the SDHB subunit.[87,88] Therefore, staining of SDHB is absent whenever *SDHA*, *SDHB*, *SDHD*, or *SDHAF2* is completely inactivated, due to loss of heterozygosity.[89] The staining may help to identify tumors with an *SDHx* gene mutation, but does not predict the precise gene involved.

Debate exists as to whether the future of NGS will replace or simplify these genetic testing strategies. As the list of candidate genes expands, the cost effectiveness and proficiency of NGS will become apparent. In addition, it is likely that as the understanding of phenotypic variability and penetrance of these gene mutations expands, researchers will uncover an even greater overlapping phenotype between genes, further complicating the clinical testing algorithms.

SUMMARY

Recent studies of the genetics of pheochromocytomas and paragangliomas have identified a high rate of underlying hereditary conditions, even in patients with seemingly sporadic tumors. Clinicians should recognize the need for a genetic evaluation in such patients to allow for optimal medical management. The most important inherited genetic conditions to consider include VHL and MEN2 (given the risk for additional tumors and the proven reduction of morbidity and mortality from surveillance protocols) and *SDHB*-associated mutations given their high risk of malignancy. In addition to the healthcare management of the proband, the risk of disease in relatives must also be adequately addressed.

References

1. Fishbein L, Merrill S, Fraker DL, Cohen DL, Nathanson KL. Inherited mutations in pheochromocytoma and paraganglioma: why all patients should be offered genetic testing. *Ann Surg Oncol* 2013;**20**(5):1444–50.

2. Neumann HP, Bausch B, McWhinney SR, et al. Germ-line mutations in nonsyndromic pheochromocytoma. *N Engl J Med* 2002;**346**(19): 1459–66.

3. Amar L, Bertherat J, Baudin E, et al. Genetic testing in pheochromocytoma or functional paraganglioma. *J Clin Oncol* 2005;**23**(34): 8812–8.

4. Gimenez-Roqueplo AP, Burnichon N, Amar L, Favier J, Jeunemaitre X, Plouin PF. Recent advances in the genetics of phaeochromocytoma and functional paraganglioma. *Clin Exp Pharmacol Physiol* 2008;**35**(4):376–9.

5. Jimenez C, Cote G, Arnold A, Gagel RF. Review: should patients with apparently sporadic pheochromocytomas or paragangliomas be screened for hereditary syndromes? *J Clin Endocrinol Metab* 2006;**91**(8):2851-L2858.

6. Baysal BE, Willett-Brozick JE, Lawrence EC, et al. Prevalence of SDHB, SDHC, and SDHD germline mutations in clinic patients with head and neck paragangliomas. *J Med Genet* 2002;**39**(3): 178–83.

7. Mian C, Sartorato P, Barollo S, Zane M, Opocher G. RET codon 609 mutations: a contribution for better clinical managing. *Clinics (Sao Paulo, Brazil)* 2012;**67**(Suppl 1):33–6.

8. Eng C, Clayton D, Schuffenecker I, et al. The relationship between specific RET proto-oncogene mutations and disease phenotype in multiple endocrine neoplasia type 2. International RET mutation consortium analysis. *J Am Med Assoc* 1996;**276**(19):1575–9.

9. DeLellis R, Lloyd R, Heitz P, Eng C. *Pathology and genetics of tumours of the endocrine organs*. Lyon: IARC Press; 2004.

10. Santoro M, Carlomagno F, Melillo RM, et al. Dysfunction of the RET receptor in human cancer. *Cell Mol Life Sci* 2004;**61**(23): 2954–64.

11. Dvorakova S, Vaclavikova E, Duskova J, Vlcek P, Ryska A, Bendlova B. Exon 5 of the RET proto-oncogene: a newly detected risk exon for familial medullary thyroid carcinoma, a novel germline mutation Gly321Arg. *J Endocrl Invest* 2005;**28**(10):905–9.

12. Da Silva AM, Maciel RM, Da Silva MR, Toledo SR, De Carvalho MB, Cerutti JM. A novel germ-line point mutation in RET exon 8 (Gly(533)Cys) in a large kindred with familial medullary thyroid carcinoma. *J Clin Endocrinol Metab* 2003;**88**(11):5438–43.

13. Rich TA, Feng L, Busaidy N, et al. Prevalence by age and predictors of medullary thyroid cancer in patients with lower risk germline RET proto-oncogene mutations. *Thyroid* 2014;**24**(7): 1096–106.

14. Eisenhofer G, Walther MM, Huynh TT, et al. Pheochromocytomas in von Hippel–Lindau syndrome and multiple endocrine neoplasia type 2 display distinct biochemical and clinical phenotypes. *J Clin Endocrinol Metab* 2001;**86**(5):1999–2008.

15. Lenders JW, Pacak K, Walther MM, et al. Biochemical diagnosis of pheochromocytoma: which test is best? *J Am Med Assoc* 2002;**287**(11): 1427–34.

16. Sawka AM, Jaeschke R, Singh RJ, Young Jr WF. A comparison of biochemical tests for pheochromocytoma: measurement of fractionated plasma metanephrines compared with the combination of 24-hour urinary metanephrines and catecholamines. *J Clin Endocrinol Metab* 2003;**88**(2):553–8.

17. Pacak K. Preoperative management of the pheochromocytoma patient. *J Clin Endocrinol Metab* 2007;**92**(11):4069–79.

18. Grubbs EG, Rich TA, Ng C, et al. Long-term outcomes of surgical treatment for hereditary pheochromocytoma. *J Am Coll Surg* 2013;**216**(2):280–9.

19. Wells SA, Asa SL, Dralle H, et al. Revised American Thyroid Association Guidelines for the Management of Medullary Thyroid Carcinoma. *Thyroid* 2015;**25**(6):567–610.

20. Dickson PV, Alex GC, Grubbs EG, et al. Posterior retroperitoneoscopic adrenalectomy is a safe and effective alternative to transabdominal laparoscopic adrenalectomy for pheochromocytoma. *Surgery* 2011;**150**(3):452–8.

21. Brandi ML, Gagel RF, Angeli A, et al. Guidelines for diagnosis and therapy of MEN type 1 and type 2. *J Clin Endocrinol Metab* 2001;**86**(12):5658–71.

22. Hull MT, Roth LM, Glover JL, Walker PD. Metastatic carotid body paraganglioma in von Hippel–Lindau disease. An electron microscopic study. *Arch Pathol Lab Med* 1982;**106**(5):235–9.

23. Walther MM, Linehan WM. Von Hippel-Lindau disease and pheochromocytoma. *J Am Med Assoc* 1996;**275**(11):839–40.

24. Neumann HP, Wiestler OD. Clustering of features of von Hippel–Lindau syndrome: evidence for a complex genetic locus. *Lancet* 1991;**337**(8749):1052–4.

25. Stolle C, Glenn G, Zbar B, et al. Improved detection of germline mutations in the von Hippel–Lindau disease tumor suppressor gene. *Hum Mutat* 1998;**12**(6):417–23.

26. Maher ER, Iselius L, Yates JR, et al. Von Hippel–Lindau disease: a genetic study. *J Med Genet* 1991;**28**(7):443–7.

27. Clifford SC, Cockman ME, Smallwood AC, et al. Contrasting effects on HIF-1alpha regulation by disease-causing pVHL mutations correlate with patterns of tumourigenesis in von Hippel–Lindau disease. *Hum Mol Genet* 2001;**10**(10):1029–38.

28. Hoffman MA, Ohh M, Yang H, Klco JM, Ivan M, Kaelin Jr WG. von Hippel–Lindau protein mutants linked to type 2C VHL disease preserve the ability to downregulate HIF. *Hum Mol Genet* 2001;**10**(10):1019–27.

29. Richards FM. Molecular pathology of von Hippel–Lindau disease and the VHL tumour suppressor gene. *Expert Rev Mol Med* 2001;**2001**:1–27.

30. Chen F, Kishida T, Yao M, et al. Germline mutations in the von Hippel–Lindau disease tumor suppressor gene: correlations with phenotype. *Hum Mutat* 1995;**5**(1):66–75.

31. Crossey PA, Richards FM, Foster K, et al. Identification of intragenic mutations in the von Hippel–Lindau disease tumour suppressor gene and correlation with disease phenotype. *Hum Mol Genet* 1994;**3**(8):1303–8.

32. Gutmann DH, Aylsworth A, Carey JC, et al. The diagnostic evaluation and multidisciplinary management of neurofibromatosis 1 and neurofibromatosis 2. *J Am Med Assoc* 1997;**278**(1): 51–7.

33. Neurofibromatosis. Conference statement. National Institutes of Health Consensus Development Conference. *Arch Neurol* 1988;**45**(5): 575–8.

34. DeBella K, Szudek J, Friedman JM. Use of the national institutes of health criteria for diagnosis of neurofibromatosis 1 in children. *Pediatrics* 2000;**105**(3 Pt 1):608–14.

35. Bausch B, Borozdin W, Neumann HP. European–American pheochromocytoma study G. Clinical and genetic characteristics of patients with neurofibromatosis type 1 and pheochromocytoma. *N Engl J Med* 2006;**354**(25):2729–31.

36. Opocher G, Conton P, Schiavi F, Macino B, Mantero F. Pheochromocytoma in von Hippel–Lindau disease and neurofibromatosis type 1. *Fam Cancer* 2005;**4**(1):13–6.

37. Brouwers FM, Eisenhofer G, Tao JJ, et al. High frequency of SDHB germline mutations in patients with malignant catecholamine-producing paragangliomas: implications for genetic testing. *J Clin Endocrinol Metab* 2006;**91**(11):4505–9.

38. Gimenez-Roqueplo AP, Lehnert H, Mannelli M, et al. Phaeochromocytoma, new genes and screening strategies. *Clin Endocrinol* 2006;**65**(6):699–705.

39. Neumann HP, Pawlu C, Peczkowska M, et al. Distinct clinical features of paraganglioma syndromes associated with SDHB and SDHD gene mutations. *J Am Med Assoc* 2004;**292**(8):943–51.

40. Benn DE, Gimenez-Roqueplo AP, Reilly JR, et al. Clinical presentation and penetrance of pheochromocytoma/paraganglioma syndromes. *J Clin Endocrinol Metab* 2006;**91**(3):827–36.

41. Timmers HJ, Kozupa A, Eisenhofer G, et al. Clinical presentations, biochemical phenotypes, and genotype-phenotype correlations in patients with succinate dehydrogenase subunit B-associated pheochromocytomas and paragangliomas. *J Clin Endocrinol Metab* 2007;**92**(3):779–86.

42. Ricketts CJ, Forman JR, Rattenberry E, et al. Tumor risks and genotype-phenotype-proteotype analysis in 358 patients with germline mutations in SDHB and SDHD. *Hum Mutat* 2010;**31**(1): 41–51.

43. Schiavi F, Milne RL, Anda E, et al. Are we overestimating the penetrance of mutations in SDHB? *Hum Mutat* 2010;**31**(6):761–2.

44. Amar L, Baudin E, Burnichon N, et al. Succinate dehydrogenase B gene mutations predict survival in patients with malignant pheochromocytomas or paragangliomas. *J Clin Endocrinol Metab* 2007;**92**(10):3822–8.

45. Jasperson KW, Kohlmann W, Gammon A, et al. Role of rapid sequence whole-body MRI screening in SDH-associated hereditary paraganglioma families. *Fam Cancer* 2014;**13**(2):257–65.

46. Timmers HJ, Chen CC, Carrasquillo JA, et al. Staging and functional characterization of pheochromocytoma and paraganglioma by 18F-fluorodeoxyglucose (18F-FDG) positron emission tomography. *J Natl Cancer Inst* 2012;**104**(9):700–8.

47. Venkatesan AM, Trivedi H, Adams KT, Kebebew E, Pacak K, Hughes MS. Comparison of clinical and imaging features in succinate dehydrogenase-positive versus sporadic paragangliomas. *Surgery* 2011;**150**(6):1186–93.

48. Cascon A, Montero-Conde C, Ruiz-Llorente S, et al. Gross SDHB deletions in patients with paraganglioma detected by multiplex PCR: a possible hot spot? *Genes Chromosomes Cancer* 2006;**45**(3): 213–9.

49. Havekes B, Corssmit EP, Jansen JC, van der Mey AG, Vriends AH, Romijn JA. Malignant paragangliomas associated with mutations in the succinate dehydrogenase D gene. *J Clin Endocrinol Metab* 2007;**92**(4):1245–8.

50. Baysal BE. Genomic imprinting and environment in hereditary paraganglioma. *American J Med Genet Part C, Semin Med Genet* 2004;**129C**(1):85–90.

51. Pigny P, Vincent A, Cardot Bauters C, et al. Paraganglioma after maternal transmission of a succinate dehydrogenase gene mutation. *J Clin Endocrinol Metab* 2008;**93**(5):1609–15.

52. Fishbein L, Nathanson KL. Pheochromocytoma and paraganglioma: understanding the complexities of the genetic background. *Cancer Genet* 2012;**205**(1–2):1–11.

53. Jafri M, Maher ER. The genetics of phaeochromocytoma: using clinical features to guide genetic testing. *Eur J Endocrinol/Eur Fed Endocr Soc* 2012;**166**(2):151–8.

54. Pasini B, Stratakis CA. SDH mutations in tumorigenesis and inherited endocrine tumours: lesson from the phaeochromocytoma-paraganglioma syndromes. *J Intern Med* 2009;**266**(1):19–42.

55. Gimenez-Roqueplo AP, Dahia PL, Robledo M. An update on the genetics of paraganglioma, pheochromocytoma, and associated hereditary syndromes. *Horm Metab Res* 2012;**44**(5):328–33.

56. Schiavi F, Boedeker CC, Bausch B, et al. Predictors and prevalence of paraganglioma syndrome associated with mutations of the SDHC gene. *J Am Med Assoc* 2005;**294**(16):2057–63.

57. Peczkowska M, Cascon A, Prejbisz A, et al. Extra-adrenal and adrenal pheochromocytomas associated with a germline SDHC mutation. *Nat Clin Prac Endocr Metab* 2008;**4**(2):111–5.

58. Buffet A, Venisse A, Nau V, et al. A decade (2001–2010) of genetic testing for pheochromocytoma and paraganglioma. *Horm Metab Res* 2012;**44**(5):359–66.

59. Hao HX, Khalimonchuk O, Schraders M, et al. SDH5, a gene required for flavination of succinate dehydrogenase, is mutated in paraganglioma. *Science* 2009;**325**(5944):1139–42.

60. Qin Y, Yao L, King EE, et al. Germline mutations in TMEM127 confer susceptibility to pheochromocytoma. *Nat Genet* 2010;**42**(3):229–33.

61. Burnichon N, Lepoutre-Lussey C, Laffaire J, et al. A novel TMEM127 mutation in a patient with familial bilateral pheochromocytoma. *Eur J Endocrinol/Eur Fed Endocr Soc* 2011;**164**(1):141–5.

62. Neumann HP, Sullivan M, Winter A, et al. Germline mutations of the TMEM127 gene in patients with paraganglioma of head and neck and extraadrenal abdominal sites. *J Clin Endocrinol Metab* 2011;**96**(8):E1279–82.

63. Dahia PL, Ross KN, Wright ME, et al. A HIF1alpha regulatory loop links hypoxia and mitochondrial signals in pheochromocytomas. *PLoS Genet* 2005;**1**(1):72–80.

64. Comino-Mendez I, Gracia-Aznarez FJ, Schiavi F, et al. Exome sequencing identifies MAX mutations as a cause of hereditary pheochromocytoma. *Nat Genet* 2011;**43**(7):663–7.

65. Atchley WR, Fitch WM. Myc and Max: molecular evolution of a family of proto-oncogene products and their dimerization partner. *Proc Natl AcadSciUSA* 1995;**92**(22):10217–21.

66. Yao L, Schiavi F, Cascon A, et al. Spectrum and prevalence of FP/TMEM127 gene mutations in pheochromocytomas and paragangliomas. *J Am Med Assoc* 2010;**304**(23):2611–9.

67. Burnichon N, Cascon A, Schiavi F, et al. MAX mutations cause hereditary and sporadic pheochromocytoma and paraganglioma. *Clin Cancer Res* 2012;**18**(10):2828–37.

68. Castro-Vega LJ, Buffet A, De Cubas AA, et al. Germline mutations in FH confer predisposition to malignant pheochromocytomas and paragangliomas. *Hum Mol Genet* 2014;**23**(9):2440–6.

69. Schlisio S, Kenchappa RS, Vredeveld LC, et al. The kinesin KIF1Bbeta acts downstream from EglN3 to induce apoptosis and is a potential 1p36 tumor suppressor. *Genes Dev* 2008;**22**(7):884–93.

70. Ladroue C, Carcenac R, Leporrier M, et al. PHD2 mutation and congenital erythrocytosis with paraganglioma. *N Engl J Med* 2008;**359**(25):2685–92.

71. Zhuang Z, Yang C, Lorenzo F, et al. Somatic HIF2A gain-of-function mutations in paraganglioma with polycythemia. *N Engl J Med* 2012;**367**(10):922–30.

72. Papathomas TG, Gaal J, Corssmit EP, et al. Non-pheochromocytoma (PCC)/paraganglioma (PGL) tumors in patients with succinate dehydrogenase-related PCC-PGL syndromes: a clinicopathological and molecular analysis. *Eur J Endocrinol/Eur Fed Endocr Soc* 2014;**170**(1):1–12.

73. Miettinen M, Wang ZF, Sarlomo-Rikala M, Osuch C, Rutkowski P, Lasota J. Succinate dehydrogenase-deficient GISTs: a clinicopathologic, immunohistochemical, and molecular genetic study of 66 gastric GISTs with predilection to young age. *Am J Surg Pathol* 2011;**35**(11):1712–21.

74. Miettinen M, Killian JK, Wang ZF, et al. Immunohistochemical loss of succinate dehydrogenase subunit A (SDHA) in gastrointestinal stromal tumors (GISTs) signals SDHA germline mutation. *Am J Surg Pathol* 2013;**37**(2):234–2340.

75. Pasini B, McWhinney SR, Bei T, et al. Clinical and molecular genetics of patients with the Carney-Stratakis syndrome and germline mutations of the genes coding for the succinate dehydrogenase subunits SDHB, SDHC, and SDHD. *Eur J Hum Genet* 2008;**16**(1):79–88.

76. Ricketts C, Woodward ER, Killick P, et al. Germline SDHB mutations and familial renal cell carcinoma. *J Natl Cancer Inst* 2008;**100**(17):1260–2.

77. Vanharanta S, Buchta M, McWhinney SR, et al. Early-onset renal cell carcinoma as a novel extraparaganglial component of SDHB-associated heritable paraganglioma. *Am J Hum Genet* 2004;**74**(1):153–9.

78. Xekouki P, Pacak K, Almeida M, et al. Succinate dehydrogenase (SDH) D subunit (SDHD) inactivation in a growth-hormone-producing pituitary tumor: a new association for SDH? *J Clin Endocrinol Metab* 2012;**97**(3):E357–66.

79. Boedeker CC, Neumann HP, Maier W, Bausch B, Schipper J, Ridder GJ. Malignant head and neck paragangliomas in SDHB mutation carriers. *Otolaryngol – Head Neck Surg* 2007;**137**(1):126–9.

80. Pacak K, Eisenhofer G, Ahlman H, et al. Pheochromocytoma: recommendations for clinical practice from the First International Symposium. October 2005. *Nat Clin Pract Endocrinol Metab* 2007;**3**(2):92–102.

81. Rattenberry E, Vialard L, Yeung A, et al. A comprehensive next generation sequencing-based genetic testing strategy to improve diagnosis of inherited pheochromocytoma and paraganglioma. *J Clin Endocrinol Metab* 2013;**98**(7):E1248–56.

82. Meldrum C, Doyle MA, Tothill RW. Next-generation sequencing for cancer diagnostics: a practical perspective. *Clin Biochem Rev/Austr Assoc Clin Biochem* 2011;**32**(4):177–95.

83. Heesterman BL, Bayley JP, Tops CM, et al. High prevalence of occult paragangliomas in asymptomatic carriers of SDHD and SDHB gene mutations. *Eur J Hum Genet* 2013;**21**(4):469–70.

84. Burnichon N, Rohmer V, Amar L, et al. The succinate dehydrogenase genetic testing in a large prospective series of patients with paragangliomas. *J Clin Endocrinol Metab* 2009;**94**(8):2817–27.

85. Neumann HP, Erlic Z, Boedeker CC, et al. Clinical predictors for germline mutations in head and neck paraganglioma patients: cost reduction strategy in genetic diagnostic process as fall-out. *Cancer Res* 2009;**69**(8):3650–6.

86. Erlic Z, Rybicki L, Peczkowska M, et al. Clinical predictors and algorithm for the genetic diagnosis of pheochromocytoma patients. *Clin Cancer Res: an official journal of the American Association for Cancer Research* 2009;**15**(20):6378–85.

87. Gimenez-Roqueplo AP, Favier J, Rustin P, et al. The R22X mutation of the SDHD gene in hereditary paraganglioma abolishes the enzymatic activity of complex II in the mitochondrial respiratory chain and activates the hypoxia pathway. *Am J Hum Genet* 2001;**69**(6):1186–97.

88. Gimm O, Armanios M, Dziema H, Neumann HP, Eng C. Somatic and occult germ-line mutations in SDHD, a mitochondrial complex II gene, in nonfamilial pheochromocytoma. *Cancer Res* 2000;**60**(24):6822–5.

89. Gill AJ, Benn DE, Chou A, et al. Immunohistochemistry for SDHB triages genetic testing of SDHB, SDHC, and SDHD in paraganglioma-pheochromocytoma syndromes. *Hum Pathol* 2010;**41**(6):805–14.

17

Genetic Conditions Associated with Congenital Adrenocortical Insufficiency or Glucocorticoid and/or Mineralocorticoid Resistance

Margarita Raygada, Constantine A. Stratakis

Section on Endocrinology & Genetics, Program on Developmental Endocrinology & Genetics, (PDEGEN), *Eunice Kenney Shriver* National Institute of Child Health and Human Development (NICHD), National Institutes of Health (NIH), Bethesda, MD, USA

INTRODUCTION

Reference to the existence of the adrenal glands dates back to the times of Galen (130–201 AD); he was believed to be the first to describe the left adrenal vein.[1] There are also some accounts that the phenotype of the biblical monozygotic twins, Esau and Jacob, in the Book of Genesis was consistent with congenital adrenal hyperplasia (CAH).[2] But it was not until 1552 that a professor at the *Collegio della Sapienza* in Rome, Batholemeus Eustachius, gave the first clear description of the adrenal glands.[3] In the last 30 years, an unprecedented production of new knowledge about the adrenal glands has led to subspecialization in the field: enzymatic deficiencies of steroidogenesis are mostly associated with CAH; in this chapter, we focus on congenital causes of adrenal insufficiency (AI) associated with *hypoplasia* of the adrenal glands. Disorders that affect aspects of adrenal function at a young age have also been recently molecularly elucidated; these diseases do not affect embryonic or early (infantile) development of the adrenal glands. We also briefly discuss genetic causes of autoimmunity that affect the adrenal gland, as these patients often present with AI at an early age.[4] A brief outline of adrenal embryonic development precedes nosology, as the genetics of the latter can only be understood in the context of the former.

GENETICS OF EMBRYOLOGY AND FUNCTION OF THE ADRENAL GLANDS

The human adrenal cortex forms at the fourth week of embryonic development: on the 25th gestational day (embryonic day 10 in the mouse), a blastema of undifferentiated cells of mesodermal origin forms from the medial part of the urogenital ridge as a condensation of coelomic epithelial cells or from mesoderm that is intermediate between the mesonephros and the coelomic cavity.[5] During the fifth embryonic week, the adrenogonadal primordium cells (which express steroidogenic factor 1, SF1) proliferate and invade the underlying mesenchyme; the adrenocortical primordium is separate from the gonads by 33 days postconception. A second wave of mesodermal cells then penetrates and surrounds the original cell mass by the sixth to eighth week, forming the first evidence for zonation. The latter cells are smaller than those of the first migration and form what has been called the *definitive* zone (DZ), whereas the former mesothelial cells develop into the *fetal* zone (FZ). These processes are to some extent under the control of fetal adrenocorticotropic hormone (ACTH, or corticotrophin) and cortisol is being made for the first time at about the sixth week reaching a peak between the eighth and ninth week of development; cortisol secretion at this early stage is from the rapidly growing FZ and is necessary to protect female

Genetic Diagnosis of Endocrine Disorders. http://dx.doi.org/10.1016/B978-0-12-800892-8.00017-8

sexual development from excess ACTH-driven adrenal androgen production.[6] An additional cell type takes origin from the mesonephron and seems to arise from the region of Bowman's capsule. Transitional zone (TZ) is a thin layer of cells between the small outer DZ and the much bigger FZ; although cortisol synthesis at the early phase is from the FZ,[6] gradually aldosterone and cortisol are being made from the DZ and TZ cells, whereas the FZ produces primarily dehydroepiandrosterone (DHEA) and dehydroepiandrosterone-sulfate (DHEAS) that support estrogen production through the fetal-placental unit.[7]

Thus, adrenal cortex derives from three embryologically distinct mesodermal cells, two from the celomic epithelium and one from the mesonephron.[5,7] The adrenal capsule and outer regions of the adrenal cortex have long been hypothesized to contain progenitor cells that mediate adrenocortical maintenance. Recent studies have elucidated the cellular origin of the capsule[8] and have identified two mutually exclusive capsular progenitor populations of the adult adrenal cortex, expressing Nr5a1 and Gli1.[8] By the ninth week, the now encapsulated adrenal accepts migrating neural crest cells that will form the adrenal medulla. It is not until after birth and after the involution of the FZ that a clear separation exists at the corticomedullary junction between steroid hormone-producing and catecholamine-secreting cells.[5] During fetal life, medullary cells are intermingled with cortical cells of the FZ; examples of that intermingling between neuroendocrine and cortical cells remain visible even in the adult adrenal gland, and are important regulators of corticomedullary coordination of adrenal function.[9]

Between the ninth and 12th embryonic week, sinusoidal vascularization of the glands forms the framework for the zonation of the adult cortex.[5] By the fourth month, the fetal adrenal attains its maximum size and becomes actually larger than the kidney; most of its volume occupied by the FZ. After the fourth month, the FZ starts receding, very gradually, at least at the beginning (despite some original evidence for massive collapse and hemorrhagic necrosis), while the thin DZ and TZ start expanding and will gradually give rise to the adult zona glomeruloza (ZG) and zona fasciculata (ZF), respectively.[5,7] At birth, a dramatic reduction in adrenal mass occurs as a result of rapid degeneration of the FZ. It is not clear what initiates the process of degeneration of the fetal zone; it happens in anencephalic infants, although in the absence of proopiomelanocortin (POMC) or ACTH, regression of the FZ is even faster.[7,10]

By the end of the first year postpartum, a TZ between adult and FZ is still present, FZ at this time being primarily fibrous tissue between the adult cortex and the medulla. The first evidence for an anatomically distinct zona reticularis (ZR) appears by the end of the second

year of life, but unequivocal immunohistochemical evidence of steroidogenic activity is not present until the age of five years,[8] concomitant with the onset of adrenarche. It is not clear when the adult adrenal cortex reaches maturity: estimates vary from as early as eight years of age to as late as after mid-puberty (11 years of age).[7,11–13]

GENETIC DEFECTS CAUSING CAI: AN OVERVIEW AND A COMMENT ON TREATMENT

As one would expect from its complex developmental biology, several genes work together in parallel as well as vertical networks orchestrating adrenocortical organogenesis.[5] They were discovered mostly over the last 20 years from mouse models, as well as the cloning of genes responsible for unique human disorders affecting the adrenal gland.[5] In a Canadian cohort with experience of treating children with CAI defects over 20 years (1981–2001), CAH was the most frequent etiology (71.8%) and non-CAH etiologies accounted for 28.2%, of which 55% were nonautoimmune in etiology.[4]

Many of the genetic disorders affecting adrenocortical development and causing CAI have not been molecularly elucidated. One of the recently elucidated ones is IMAGe (intrauterine growth retardation, metaphyseal dysplasia, adrenal hypoplasia, genitourinary anomalies), which is now known to be caused by mutations in the CDKN1C gene in chromosome 11p15.[14] Cases of chromosomal anomalies, such as one with a 5p defect and CAI are useful potential indicators of new genetic loci involved in adrenal development and/or function.[15] Finally, rare associations, such as the one of CAI and CHARGE syndrome (coloboma of the eye; heart anomaly; atresia, choanal; retardation of mental and somatic development; microphallus; ear abnormalities and/or deafness) that is caused by mutations in the chromodomain helicase DNA-binding protein-7 (CHD7) gene may point to new functions of already identified genes.[16,17]

CAI Associated with Hypothalamic-Pituitary-Adrenal (HPA) Axis Defects

As presented earlier, sufficient levels and function of ACTH and its receptor (MC2R) are essential for normal adrenocortical growth and differentiation.[18] Genetic diseases affecting the pituitary (mutations of the HESX1, LHX3, LHX4, SOX3, and TPIT genes), homozygote or compound heterozygote genetic defects of the proopiomelanocortin (POMC) gene (the ACTH precursor), its processing enzyme (prohormone convertase 1 or PC1), the product of the proprotein convertase subtilisin/kexin type 1 (PCSK1) gene, and the MC2R gene or the genes providing its accessory molecules (AAAS, MRAP for definition of the abbreviation see later) all lead to CAI.[4,19–22] Defects that lead to hypopituitarism (ACTH deficiency in combination with other defects),[18] and POMC and PC1 gene mutations that cause, among other symptoms, pediatric obesity[20,21] are discussed elsewhere in this book, so in this chapter we discuss further CAI due to TPIT (isolated ACTH deficiency), MC2R, AAAS, and MRAP (ACTH resistance) gene defects.

CAI Associated with Primary Adrenocortical Development Defects

The development of the adrenogonadal primordium from the urogenital ridge is dependent on normal function and interplay between the transcriptional factors SF1 and DAX-1 (for definition of the abbreviation see later).[5,23] Mice knockouts (KO) for Sf1 have complete absence of the adrenal glands, whereas mice KO for Dax1 have developmental defects in their adrenal glands (but they are not adrenally insufficient).[5]

Humans with heterozygous SF1 (coded by the NR5A1 gene) mutations have AI and gonadal abnormalities but this defect is a rare cause of CAI,[24] although recently more patients with AI and heterozygous NR5A1 mutations have been described. On the other hand, X-linked DAX-1 (the NR0B1 gene) defects cause the most common form of CAI in humans, adrenal hypoplasia congenita (AHC):[25,26] out of 117 patients with mostly CAI (and some adults with AI), DAX-1 gene mutations were found in 37 (58%) 46,XY phenotypic boys referred with adrenal hypoplasia and in eight boys with hypogonadotropic hypogonadism and a family history of male-only CAI; SF1 gene mutations were found in only two patients who also had 46,XY gonadal dysgenesis.[27] Mutations in the NR0B1 (DAX1) gene are now known to associate growth hormone deficiency with CAI.[28] More than 200 mutations in the NR0B1 (DAX1) gene have been identified to date;[29] the phenotype varies with the type of mutation including hypogonadotropic hypogonadism, CAI, and impaired synthesis and release of gonadotropins.

Since adrenocortical development is tightly linked to that of the gonads and other structures derived from the urogenital ridge, genes encoding proteins such as Wilms' tumor 1 (WT1), a transcriptional regulator that is mutated in Denys–Drash and Frasier syndromes,[30,31] and WNT4 and WNT11, both involved in signaling via the frizzled receptor family, are important for normal adrenocortical formation. Mice that are KO for Wt1 and Wnt4 have hypoplastic adrenals[32] but the adrenocortical function of patients with Denys–Drash and Frasier syndromes has not been studied extensively; mild functional defects in adrenarche and pubertal anomalies may be present but not CAI.[31] A female with 46,XY karyotype associated with a 1p31-p35 duplication including the WNT4

gene was described;[33] recently, a missense substitution of the WNT4 gene[34] was found in an autosomal recessive syndrome designated SERKAL: female-to-male sex reversal and kidney, adrenal, and lung dysgenesis. The defect reduced WNT4 mRNA levels and thus downregulated WNT4-dependent inhibition of the degradation of beta-catenin, a process essential for adrenal and other organ development.

SF1 (NR5A1) mutations have rarely been identified in patients with CAI without any evidence of gonadal defects; one of the first patients to be described was a phenotypically normal girl who presented at age 14 months with adrenal insufficiency, normal female chromosomal pattern (46,XX), normal luteinizing hormone, and follicle-stimulating hormone. The defect was a heterozygous G-to-T transversion in exon 4 of the NR5A1 (SF1) gene, leading to an arg255-to-leu mutation in the hinge region of the NR5A1 protein.[35]

See more discussion on DAX1 and SF1 defects, under the listing of adrenal hypoplasia congenital (AHC).

CAI Associated with Metabolic Disorders

There are a few metabolic disorders that affect adrenocortical function in early life and can, thus, cause CAI. Wolman's disease (familial xanthomatosis) may cause CAI and is occasionally associated with adrenal calcification, too.[36,37] This rare metabolic disorder is caused by lysosomal acid cholesteryl ester hydrolase defects.[37] Smith–Lemli–Opitz syndrome (SLOS), a disorder of cholesterol biosynthesis associated with developmental delay, dysmorphic features, male undervirilization, and/or hypogonadism, is often cited as one frequently associated with CAI.[5,38] Rare cases of SLOS may present with adrenal crisis,[39] but in our experience, most SLOS patients have mostly compensated adrenocortical dysfunction and not CAI, consistent with prior reports of functional abnormalities of the fetal-placental unit in this condition.[40] X-linked adrenoleukodystrophy (X-ALD)[41] is by far the most common metabolic disorder causing CAI and is discussed further in the subsequent section.

CAI Associated with Adrenal Calcifications and/or Hemorrhage: Genetic Factors

Adrenal hemorrhage, a well-recognized obstetrical complication of the newborn, is now a rare cause of CAI due to the improvements in care during labor and delivery. The adrenal glands often become calcified and the classic presentation is that of an infant in adrenal crisis and calcifications in the adrenal fields, usually bilaterally.[42] Risk factors, besides difficulties in labor and delivery, include genetic disorders of the newborn,[36] such as macrosomia and related syndromes, coagulation defects, metabolic disorders, congenital masses including hemorrhagic and teratomatous cysts, ganglioneuromas, pheochromocytomas, neuroblastomas, and adrenocortical cancer.[42]

A newly described rare disease is hereditary cystatin C amyloid angiopathy; it is a genetic amyloid disease that occurs frequently in Iceland. The condition is caused by a mutation in cystatin C that causes amyloid deposition, predominantly in brain arteries and arterioles, but also in tissues outside the brain including the adrenal cortex. The abnormal deposits lead to occlusion or rupture resulting in hemorrhage in the brain, adrenal and rarely in other tissues.[43]

CAI Associated with Genetic Defects Leading to Autoimmunity

Autoimmune polyglandular failure, involving the adrenal glands, usually presents after the first two years of life along with other autoimmune diseases in the form of two genetic syndromes: multiple endocrine abnormalities types 1 and 2 (MEA-1 and MEA-2).[44] In the Canadian cohort of primary AI, the most frequently encountered autoimmune CAI was the autoimmune polyendocrinopathy-candidiasis-ectodermal dysplasia (APECED) syndrome. Patients with APECED presented with AI four years earlier than those with nonautoimmune disease.[2] AI in APECED occurs in association with chronic mucocutaneous candidiasis and/or acquired hypoparathyroidism, chronic active hepatitis, malabsorption, juvenile onset pernicious anemia, alopecia and primary hypogonadism, whereas insulin-dependent diabetes and/or autoimmune thyroid disease are infrequent. AI in the context of MEA-2 is associated with insulin-dependent diabetes and/or autoimmune thyroid disease, has a later but variable age of onset, and occurs predominantly in females. The association of HLA-B8, -DR3, and -DR4 with MEA-2, and more recently the cytotoxic T-lymphocyte antigen (CTLA-4) gene, and the protein tyrosine phosphatase nonreceptor type 22 (PTPN22) genes on chromosomes 6, 2, and 1, respectively,[45] but not with MEA-1, further confirms the different clinical and genetic nature of these two syndromes, which accounts for almost all pediatric and approximately 50% of adult cases of autoimmune AI.[40] In addition, 100% concordance of disease in identical twins has not been shown.[45]

MEA-1 or APECED is a genetic disorder inherited in an autosomal recessive manner and caused by mutations in the *autoimmune regulator* (*AIRE*) gene.[47] Although technically not a congenital form of AI, we discuss it briefly in the subsequent section because if its frequency among young pediatric patients with AI[2] (for details see Chapter 27). We do not discuss here MEA-2, a syndrome that affects mostly older children and adults and has no known single gene as a cause (see Chapter 27).

Treatment

All disorders leading to AI, whether primary (peripheral or due to adrenal defects) or secondary (central, due to hypothalamic or pituitary defects) require treatment with adequate glucocorticoid coverage, most commonly with hydrocortisone at the usually recommended dose of 10–12 mg/m^2/day. The dose schedule varies, but most patients do well with a twice-a-day administration of two-thirds of the total dose in the morning and one-third in the early evening hours. The usual stress coverage dosing recommendations also apply. It is important to note that the ACTH levels of patients with primary AI vary widely and should not be routinely monitored, as this often leads to increased glucocorticoid administration with detrimental long-term effects. Titration of coverage should be done by careful clinical examination and consideration of the patient's overall well-being and not aim at suppressing ACTH levels.

Mineralocorticoid replacement is usually not necessary in disorders associated with secondary (central) AI. It is also rarely needed in patients with ACTH resistance syndromes. Most patients with primary AI due to an adrenal developmental defect require mineralocorticoid replacement: the usual dose is 0.1 mg fludrocortisone per day; salt may also be given in these patients. For detailed guidelines, please refer to the newly published consensus statement on the diagnosis, treatment, and follow-up of patients with primary adrenal insufficiency.[48]

SPECIFIC GENETIC CONDITIONS ASSOCIATED WITH CAI

Congenital Isolated ACTH Deficiency

TPIT (or *TBX19*) is a transcription factor, member of the T-box gene family that is required for expression of the *POMC* gene in the differentiating pituitary corticotrophs. All members of the T-box gene family encode an N-terminal DNA-binding domain (the T-box) and are important for development of several, mostly mesodermal, tissues in the human and mouse embryo. Holt–Oram ulnar-mammary and DiGeorge syndromes are all caused by mutations in a member of this family of transcription factors.[49]

TPIT gene mutations are associated with autosomal recessive CAI.[50] Most patients with CAI and *TPIT* mutations presented with neonatal hypoglycemia, seizures, and occasionally death; all patients had very low (but not necessarily undetectable) ACTH and cortisol levels. Patients with isolated ACTH deficiency and CAI have a high likelihood of having *TPIT* mutations,[4] but patients that have either late onset AI or other pituitary and/or other developmental defects should generally not be tested for defects of this gene.[4,51,52] Studies of congenital isolated ACTH cases found that about 65% of patients with CAI (neonatal onset) had *TPIT* mutations.[4,52] A recent study identified mutations that led to loss of function by different mechanisms, such as nonsense-mediated mRNA decay, abnormal mRNA splicing, loss of *TPIT* DNA binding, or protein-protein interaction defects.[53]

ACTHR Defects (Familial Glucocorticoid Deficiency 1, FGD-1)

Hereditary resistance to ACTH action (RACTH) or familial glucocorticoid deficiency (FGD) type 1 (FGD-1) is an autosomal recessive disorder[22] that is caused by defects of the ACTH receptor (ACTHR) (the *MC2R* gene).[54] The *MC2R* gene, with only one coding exon, is on chromosome 18 (18p11). The presentation of the disease is quite variable with little genotype-phenotype correlation, although certain mutations may have a more specific phenotype, such as the homozygous insertion of a cytosine nucleotide between codons 153 and 154 (c.459_460insC) that was associated with thyroid dysfunction, growth hormone secretion abnormalities, and hyperferritinemia.[55] Patients can present in infancy with severe AI, hypoglycemia, and seizures, or later in childhood with a milder form of AI. The occasional patient may require mineralocorticoid replacement,[56] but most require only glucocorticoid replacement; ACTH levels can vary from slightly above the normal range to several-fold higher. Finally, patients with *MC2R* defects tend to be taller than their expected (by genetic target) height, a phenotype that is largely unexplained.[57]

Inactivating *MC2R* gene mutations in FGD-1 are responsible for 25–40% of FGD cases.[58] Thus, for genetic counseling purposes, one needs to know that the chance of *MC2R* gene mutation(s) in a patient with isolated ACTHR defects (with no other health or developmental problems) appears to be significantly less than 50%, with only few of the remaining patients bearing mutations in a known gene (e.g., *MRAP*).

Recently, disruption of *Mc2r* in mice led to neonatal lethality in most but not all animals, in agreement with the variability of the phenotype in humans.[59] Mice that survived had small adrenal glands with atrophic ZF confirming the role of ACTH and ACTHR in the development of the mature (adult) adrenal cortex.[59]

Other Genetic Defects for Familial Glucocorticoid Deficiency Type-2 (FGD-2)

Recently, it was discovered that at least some of the non-ACTHR-mutant FGD cases, FGD-2, have mutations in the gene encoding an accessory protein required for ACTH signaling, the MC2 receptor accessory protein (MRAP).[60] Lack of functional *MRAP* gene defects, like

MC2R gene defects, causes RACTH and severe glucocorticoid deficiency that can be fatal if the condition is not recognized and treated. MRAP exists as a homodimer and is essential for the cell surface expression of the MC2R. In the absence of MRAP, MC2R is trapped in the endoplasmic reticulum, but with MRAP, MC2R is glycosylated and can be present in the plasma membrane for proper ACTH signaling.[60]

MRAP gene mutations are an overall rare cause of RACTH; other genes are continuously sought, including genes that confer an AI phenotype in mice: a gene (*acd*) responsible for the urogenital and caudal dysgenesis and adrenocortical dysplasia in mice was investigated in patients with FGD-2 and no mutations were found.[61]

New gene defects in FGD have recently been recognized in the mini-chromosome maintenance-deficient 4 homologue (*MCM4*) and nicotinamide nucleotide transhydrogenase (*NNT*) genes.[62] These findings highlight new pathogenetic mechanisms involved in replicative and oxidative stress and set the basis for future studies to expand the spectrum of pathways for FGD.

Allgrove or Triple-A Syndrome

Allgrove or triple A syndrome (AAAS) is an autosomal recessive condition characterized by RACTH AI, reduced or absent tearing (*a*lacrima), and *a*chalasia.[63] Patients can also exhibit other signs of autonomic dysfunction such as ocular abnormalities, an abnormal reaction to intradermal histamine, abnormal sweating, orthostatic hypotension, and heart rate disturbances, leading some to suggest the name "4A syndrome."[63–65] Both static and progressive neurological abnormalities – including microcephaly, mental retardation/learning disabilities, bulbospinal amyotrophy, dysarthria/nasal speech, optic atrophy, ataxia, muscle weakness, dementia, hyperreflexia, Parkinsonian features, and sensory impairment – have been reported. Palmar and plantar keratosis is a frequent dermatologic manifestation.[63]

Thus, AAAS is a multisystem disorder with endocrine, gastrointestinal, ocular, and neurologic manifestations. The "classic" presentation reported in the literature is that of a child born healthy who does not make tears when crying, develops ACTH-resistant adrenal insufficiency in the first decade of life, and goes on to have achalasia some time in the first or second decade of life. Our experience, however, is that AAAS is clinically a heterogeneous disorder with highly variable severity of presentation.[65] For example, it is well recognized that while most children are born healthy and normal, a small percentage appears to have developmental abnormalities such as microcephaly and mental retardation/developmental delay. AI due to RACTH may present in a dramatic fashion (e.g., hypoglycemic seizures, adrenal crisis) or may be mild and not require pharmacologic replacement of glucocorticoids until teenage years or later. For example, we encountered a case who did not present with AI until age 13, despite being a compound heterozygote for two AAAS mutations. The pituitary may appear hypoplastic on magnetic resonance imaging, despite clear evidence of RACTH being the cause of adrenal insufficiency. While adrenal failure is generally limited to glucocorticoid deficiency, mineralocorticoid deficiency has also been reported. However, in at least one case, an intact ZG was reported on autopsy consistent with the low incidence of mineralocorticoid deficiency.[63] Achalasia may also have a variable – often insidious – presentation. Dysphagia may be present for years before a diagnosis of achalasia is made and may present as gastroesophageal reflux. In our experience, surgical intervention is usually successful. However, anesthesiologists should be warned about the possible proclivity of peripheral neuropathy in these patients and should take appropriate intraoperative precautions. The autopsy case presented by Allgrove showed muscular hypertrophy, loss of ganglion cells, and a paucity of small nerves in the distal esophagus of an affected patient.

Ocular abnormalities may be the most invariable and present the earliest.[63,64] The loss of basal and reflex tearing (which is under parasympathetic control) likely stems from disease of the autonomic nervous system and may result in corneal punctate epitheliopathy, melting, and/or scarring. The main lacrimal gland appears small to absent on neuroimaging, either because of primary hypoplasia or secondary denervation atrophy. Autonomic dysfunction may also result in pupillary abnormalities and/or accommodative spasms. Optic nerve pallor, generally described in the literature as "optic nerve atrophy" with either delayed timing or reduced amplitude on visual-evoked potential testing, has been reported is several cases. It is unclear from most reports whether these optic nerve changes are secondary to episodes of hypoglycemia from adrenal insufficiency or represent a primary, progressive neurodegenerative process.

The most concerning and least treatable manifestations of AAAS are related to central and peripheral neurodegeneration.[65] While symptoms may present early in life, they are not invariably present. Polyneuropathy (sensory, motor, and autonomic components), Parkinsonian features, dementia, bulbospinal amyotrophy, and selective ulnar nerve involvement in patients with peripheral motor neuropathy and amyotrophy have all been reported in separate case reports. Counseling presymptomatic, mutation-proven patients regarding neurologic prognosis is difficult, as it is currently unclear if all patients with AAAS go on to develop significant neurologic deterioration.

Mutations in the AAAS gene on 12q13,[65] which codes for a 546 amino acid protein called ALADIN (for *a*lacrima-*a*chalasia-a*d*renal *i*nsufficiency *n*eurologic disorder), have been described in several individuals.[66–68] A splicing mutation, the IVS14+1G>A change, is the most common AAAS gene mutation.[65] The frequent presence of this mutation appears to be the result of a founder effect. New mutations have also been found, however, in ethnically diverse or mixed populations. There is little, if any, genotype-phenotype correlation.

The results of a recent international literature search revealed a delayed onset of symptoms in patients with missense mutations in one allele as compared to those with truncation mutations in both alleles; for example, alacrima, adrenal insufficiency, and motor impairment developed significantly later in patients with missense mutations than in those with only truncation mutations.[69] The frequency of adrenal insufficiency was also significantly higher in patients with only truncation mutations; in contrast, the frequencies of occurrence of other major symptoms (achalasia, alacrima, and motor impairment) did not differ significantly.[69] ALADIN, the AAAS protein, belongs to the WD-repeat family of regulatory proteins that have functions ranging from transmembrane signaling and transcription to cell division and intracellular trafficking.[68,70] While the precise function of ALADIN is unknown, it appears to be a protein in the nuclear core complex of cells; a variety of missense, nonsense, and splicing mutations in ALADIN cause the protein to mislocalize to the cytoplasm. Because microscopic analysis of cells from an AAAS patient showed no morphologic abnormalities in the nuclei, nuclear envelope, or nuclear pore complexes, it has been suggested that most missense AAAS mutations result in a functional, rather than a structural, abnormality in the nuclear pore complex.[68] Interestingly, *Aaas* deficient mice largely failed to replicate the human phenotype.[70]

Is AAAS a genetically homogeneous disorder? Some of the studied patients, although they clinically had unequivocally AAAS, did not have mutations in both alleles of the AAAS gene.[65,71] In addition, AAAS gene mutations are completely absent in a number of patients with the triple A syndrome. These findings raise the possibility of mutations in regulatory or deeper intronic sequences of the AAAS gene, and/or of genetic heterogeneity with mutations expected to be found in molecules that are functional partners of the ALADIN protein.

Adrenal Hypoplasia Congenita (AHC)

AHC is a disease that has at least two major histologic types: the "cytomegalic" form and the "miniature" form. The cytomegalic form typically presents in infancy with adrenal insufficiency; these children who are typically male ultimately fail to undergo puberty because of hypogonadotropic hypogonadism. The inheritance is X-linked recessive, although genetic heterogeneity may exist. The gene for the X-linked cytomegalic AHC was cloned in 1994: it was called *DAX-1*, for *d*osage-sensitive sex-reversal [and] *a*drenal *h*ypoplasia *c*ongenital [on the] *X* chromosome.[25–27] DAX-1 is an orphan nuclear receptor that is expressed in the adrenal gland, gonads, ventromedial hypothalamus (VMH), and the pituitary gonadotropes.[19] Most *DAX1* gene mutations introduce frameshifts and/or cause premature termination of the protein. Relatively few missense mutations have been described and most are located within the carboxy-terminal half of the protein.[19] Functional assays show that AHC-associated *DAX-1* mutations abrogate the ability of this gene to act as a transcriptional repressor of SF-1. There are "mild" missense *DAX-1* gene mutations but, in general, genotype-phenotype correlation is not as strong as one might expect. Hypogonadropic hypogonadism, for example, appears to be particularly variable among patients both in its severity and age of onset.[19] The *Dax-1* KO mice show an important role of this gene in testis development and spermatogenesis, abnormalities that are also present in humans.[5,19,23]

The miniature form of AHC appears to be autosomal recessive (it has been reported in both males and females) and is much less common than the cytomegalic form. It is often associated with other developmental defects including abnormalities of the pituitary gland and the central nervous system. When these defects are present, the clinical course is ominous and diagnosis is often made in infancy. The disease may be genetically heterogeneous.

Targeted mutagenesis of *Sf1* in mice (the *FtzF1* gene) prevents gonadal and adrenal development and causes male-to-female sex reversal.[5,19,23] Several naturally occurring mutations have been previously reported and reviewed.[24,27] At first, a heterozygous loss-of-function human *SF1* gene mutation (G35E) in the first zinc finger of the DNA binding domain was described in a patient with adrenal insufficiency and 46,XY sex reversal. A second patient, a 46,XX prepubertal female with adrenal insufficiency was otherwise phenotypically normal; she had a heterozygous mutation of the *SF1* gene in the hinge region corresponding to the AF-1 domain. Another mutation caused sex reversal but had no effect on adrenal function, whereas the newest SF1 mutation, in an infant with homozygosity for the R92Q genetic *SF1* change, both adrenal insufficiency and sex reversal were present. In functional assays, this mutant SF1 protein exhibited partial loss of DNA binding and transcriptional activity when compared with the more severe G35E P-box mutant The parents of the patient and a heterozygote sibling were all unaffected, suggesting that overall *SF1* gene mutations can present with a recessive or a dominant inheritance, depending on the defect.[19,27]

X-Linked Adrenoleukodystrophy (X-ALD)

X-ALD is the most common inherited peroxisomal disorder, affecting 1/15,000–20,000 males in the Caucasian population.[41] The disease is characterized by progressive demyelination of the white matter and by AI, although the latter is rarely congenital. In X-ALD, peroxisomal 5-oxidation of unbranched saturated very-long-chain fatty acids is defective or absent, depending on the molecular defect. The phenotype can vary widely from severe ALD and CAI (less than half of the patients) to only adrenomyeloneuropathy (one-quarter of the patients) or only AI (in a tenth of the patients). The remaining male carriers of pathogenic defects have no detectable disease until later in life or none at all, indicating that there are other, perhaps some environmental, factors that determine disease expression. The gene for X-ALD, ATP-binding cassette transporter 1 (*ABCD1*) encodes a peroxisomal membrane transporter that belongs to the "ATP-binding cassette" superfamily of membrane proteins.[41] To date, a total of 591 mutations have been identified in the *ABCD1* gene in 1,260 kindred worldwide.[72] The phenotypic variability is extremely broad, ranging from cerebral inflammatory demyelination of childhood onset, and early death, to adults remaining presymptomatic through more than five decades. There is no general genotype-phenotype correlation in X-ALD.

APECED (MEA-1)

APECED is a rare autosomal recessive disorder[46,47] that is also extensively discussed in Chapter 27. APECED patients usually have at least two out of three main symptoms: Addison's disease, hypoparathyroidism, and chronic mucocutaneous candidiasis. Patients may also develop other organ-specific autoimmune disorders leading to gonadal failure, pancreatic (β-cell) deficiency, gastric (parietal cells) dysfunction, hepatitis, and thyroiditis. Other, less common, clinical manifestations include ectodermal dystrophy, affecting the dental enamel and nails, alopecia, vitiligo, and corneal disease (keratopathy). MEA-1 usually occurs in early childhood, but new, tissue-specific symptoms may appear throughout life.[46] Immunologically, the main finding in APECED patients is the presence of autoantibodies against the affected organs, including those against steroidogenic enzymes (P450scc, P450c17, and P450c21) in patients with Addison's disease, glutamic acid decarboxylase in patients with diabetes, and the enzymes aromatic L-amino acid decarboxylase and P4501A2 in patients with liver disease.[46] Mucocutaneous candidiasis, hypoparathyroidism, and AI usually present in this order in pediatric patients with APECED. As with the other manifestations of the syndrome, there is a wide variability of age of onset, from six months to 41 years, with a peak around 13 years of age. AI is usually one of the reasons these patients end up being diagnosed with MEA-I; it develops in 60–100% of patients with APECED and may be preceded by months or years of detectable adrenal cortex autoantibodies.[40] APECED is more common in certain genetically isolated populations. In Finland, the incidence has been estimated to be 1:25,000 and in Iranian Jews, 1:9,000.[47] APECED is also relatively common among Sardinians (1:14,400) and in northern Italy. Based on linkage analysis in Finnish APECED families, the locus for the *APECED* gene was mapped to chromosome 21q22.3, and recently, the gene responsible for this disease was cloned. It encodes a predicted 545 amino acid protein, which was named *AIRE* (*autoimmune regulator*). It contains two plant homeodomain (PHD)-type zinc finger motifs and a newly described putative DNA-binding domain SAND, a proline rich region, and three LXXLL motifs, all suggestive of a transcription regulator.[47] AIRE is expressed in thymus, lymph node, and fetal liver, tissues that have important roles in the maturation of immune system and development of immune tolerance. These findings, together with the immunologic deficiency in APECED patients, suggest that AIRE may have an important role in the control of immune recognition and may function as a transcription factor or as transcriptional coactivator. To date, several mutations in the *AIRE* gene have been described in APECED patients: a common Finnish mutation, R257X, was shown to be responsible for 82% of Finnish APECED cases, whereas a deletion of 13 nucleotides (1094–1106del) has been detected in several patients of different ethnic origin, and another nonsense mutation, R139X, is the major mutation among Sardinian APECED patients.[47]

GENETIC CONDITIONS ASSOCIATED WITH RESISTANCE TO GLUCOCORTICOIDS OR MINERALOCORTICOIDS

These disorders do not affect the development of the adrenal glands but affect adrenal function at a young age, sometimes as early as in infancy. The prototype of these disorders is glucocorticoid resistance, but the field has expanded in the last two decades with description of defects in the mineralocorticoid receptor (MR, the *NR3C2* gene) and epithelial sodium channel (ENaC) α, β, and γ subunit genes (*SCNN1A*, *SCNN1B*, and *SCNN1G*, respectively) in the autosomal dominant and recessive forms of pseudohypoaldosteronism (PHA) type-I.

Pseudohypoaldosteronism type-II, or Gordon syndrome (and its variants), is not an aldosterone resistance syndrome, but rather a familial form of hyperkalemic hypertension caused by mutations of the *WNK1*, *WNK4*, *OSR1*, *SPAK*, *KLHL3*, *Cullin3*, and *Nedd4-2* and possibly additional genes.

Primary Familial or Sporadic Generalized Glucocorticoid Resistance

Primary familial or sporadic generalized glucocorticoid resistance (FGR or GGR, respectively) is a rare hereditary disorder characterized by hypercortisolism and the absence of stigmata of Cushing's syndrome.[73–77] Glucocorticoids are crucial regulators of affect and behavior, metabolism, cardiovascular function, inflammation, and immunity and, thus, complete inability of glucocorticoids to exert their effects on target tissues would be incompatible with life in primates.[78] Thus, only syndromes of partial or incomplete glucocorticoid resistance exist and are caused by defects of the ubiquitous, classic glucocorticoid receptor (GR, the *NR3C1* gene), which mediates most of glucocorticoid actions. GGR was the penultimate nuclear hormone resistance syndrome to be described, as late as 1976,[73] and to date has been reported in several unrelated kindreds and in individual subjects.[74–7,79,80]

The daily production of glucocorticoids is tightly controlled by an elaborate feedback system in which glucocorticoids exert negative feedback on hypothalamic secretion of corticotropin-releasing hormone and arginine vasopressin and on pituitary secretion of ACTH.[77] In states of resistance, this complex system is insensitive to concentrations of cortisol considered normal for the general population, and the HPA axis is reset to a higher level because of compensatory increases of ACTH and cortisol secretion.[77,79] Since the defect of the receptor is partial, adequate compensation for the end-organ insensitivity appears to be achieved by the elevated circulating cortisol; however, the excess ACTH secretion results in increased production of adrenal steroids with salt-retaining or androgenic activity.[77,78] Because in GGR the peripheral tissues are presumably normally sensitive to mineralocorticoids and androgens, the clinical characteristics of the condition reflect the increased production of these hormones. Corticosterone, deoxycorticosterone, and even cortisol, through their interaction with the mineralocorticoid receptor, cause symptoms and signs of mineralocorticoid excess, such as hypertension and/or hypokalemic alkalosis.[73–76] The excess androgens, such as Δ4-androstenedione, DHEA, and DHEA-S, on the other hand, lead to signs and symptoms of hyperandrogenism.[76] The latter in women include cystic acne, hirsutism, male pattern baldness, menstrual irregularities, oligo- and amenorrhea, anovulation, and infertility.[80] In children, the excessive and early prepubertal adrenal androgen secretion can cause sexual precocity.[76] Finally, the interference of adrenal androgens with the regulation of pituitary secretion of follicle stimulating hormone and/or ACTH-induced intratesticular growth of adrenal rests appear to be responsible for abnormal spermatogenesis and infertility in men with GGR. These clinical manifestations were not reported in all patients with FGR or GGR, and presentation varied even within families, including asymptomatic members. The diagnosis of GGR is made by demonstrating a high cortisol production rate, that is, high plasma total and free cortisol, elevated 24-h urinary free corticol, or 17-OH-corticosteroid excretion, resistance to dexamethasone suppression, and maintenance of the circadian and stress-induced patterns of glucocorticoid secretion, albeit at higher levels, along with absence of Cushing's syndrome stigmata.[73–80] The presentation varies from hypertension and hyperandrogenism in both adult males and females to isosexual precocious puberty in young boys with GR mutations. Patients with FGR are easily managed with high doses of synthetic glucocorticoids with no mineralocorticoid activity, such as dexamethasone, which should be carefully titrated per individual, obtaining a "graded" dexamethasone suppression test.[81] Thereafter, monitoring of clinical and biochemical parameters is essential to avoid glucocorticoid excess; patients with FGR or GGR may be at increased risk for forming ACTH-producing adenomas of the pituitary gland.[82]

Pseudohypoaldosteronism (Type-I)

Cheek and Perry first reported PHA in an infant with a severe salt-wasting syndrome in 1958.[83] The syndrome was subsequently reported in several patients.[84,85] Approximately one-fifth of these cases were familial, and although the disease was inherited mostly in an autosomal recessive manner, dominant inheritance was also demonstrated in several pedigrees, pointing to at least two clinically distinct forms of the disease.[84,85] Both these forms of the syndrome are referred to as PHA type-I (OMIM 264350) in order to be distinguished from Gordon syndrome or hyporeninemic hypoaldosteronism (PHA type-II, OMIM 145260), which is characterized by renal tubular unresponsiveness to the kaliuretic, but not sodium and chloride reabsortive effect of aldosterone. PHAII is also inherited in an autosomal dominant manner.

Type-I PHA is manifested early in infancy with salt wasting, vomiting, and feeding difficulties that lead to severe dehydration and death if untreated.[83–86] Hyponatremia, hyperkalemia, and acidosis are present despite marked renin excess and hyperaldosteronemia, and normal 17-ketosteroid excretion. Unresponsiveness to aldosterone may be limited to the renal tubule or generalized, in which case sodium excretion is increased not only in the urine, but also in sweat, saliva, and stool. The sweat chloride test is positive.[87] The symptoms improve with age, and older patients with PHA require smaller doses of salt supplements and salt-retaining steroids.[84]

The human mineralocorticoid receptor (MR) was cloned in 1987, and its chromosomal locus was mapped to chromosome 4q31.1-2.[88] A number of reports in the 1980s and early 1990s suggested that quantitative and/or qualitative defects of the MR might be responsible for PHA, based on studies that reported absent or decreased [3H]-aldosterone binding to circulating leucocytes from patients with the disorder.[89] However, mutations of the coding and 5'-regulatory regions of the MR gene were not identified until relatively recently[90] and do not appear to be responsible for the majority of cases of type-I PHA; to date about 50 MR gene mutations have been reported.[91,92] Most patients with this disease appear to be carriers of epithelial sodium channel (ENaC) α, β, and γ subunit gene (SCNN1A, SCNN1B, and SCNN1G, respectively) mutations.[93–96]

Patients with type-I PHA are aggressively treated in early infancy with salt supplementation (0.5–1.5 g/kg/day in four to six divided doses). Close monitoring of the sodium balance is needed especially during infancy, when plasma renin activity is not useful because of the large variation of this parameter in the first year of life. After the age of two years, the reduction of sodium supplement is often possible with gradual improvement of the symptoms and decrease of the dehydration crises. Carbenoxolone, an inhibitor of 11-β-hydroxysteroid dehydrogenase activity, has also been used successfully in the treatment of type-I PHA.[97] Patients with systemic disease are more difficult to treat due to respiratory, renal, and other complications.[98]

PHAII is a rare heterogeneous syndrome; symptoms include hypertension, hyperkalemia, and metabolic acidosis. PHAII is caused by gain-of-function mutations in NaCl cotransporters. In the past 10 years, several studies have identified novel proteins involved in the pathogenesis of PHAII, namely, WNK1, WNK4, OSR1, SPAK, KLHL3, Cullin3, and Nedd4-2.[99] This condition is not extensively reviewed here since it is covered elsewhere in this book.

Acknowledgment

This work is supported by the US National Institutes of Health, National Institute of Child Health and Human Development intramural program (PI: Dr C.A. Stratakis).

References

1. Carmichael SW. *The adrenal medulla*. New York: Oxford University Press; 1986.

2. Greenblatt RB. *Search the scriptures*. Philadelphia, USA: J.P. Lippincott; 1977. p. 45–48.

3. Eustachius B. 1563 Tabular Anatomica. Chapter VL, Vemica and Eustachius B. Opuscula Anatomica de Renum structura, Efficio et Administratione. Venice: V.V. Lucchino; 1564. Cited by: Hiatt JR, Hiatt N. The conquest of Addison's disease. Am J Surg 1997;174:280.

4. Perry R, Kecha O, Paquette J, Huot C, Van Vliet G, Deal C. Primary adrenal insufficiency in children: twenty years experience at the Sainte-Justine Hospital, Montreal. *J Clin Endocrinol Metab* 2005;**90**(6): 3243–50.

5. Else T, Hammer GD. Genetic analysis of adrenal absence: agenesis and aplasia. *Trends Endocrinol Metab* 2005;**16**(10):458–68.

6. Goto M, Piper Hanley K, Marcos J, Wood PJ, Wright S, Postle AD, Cameron IT, Mason JI, Wilson DI, Hanley NA. In humans, early cortisol biosynthesis provides a mechanism to safeguard female sexual development. *J Clin Invest* 2006;**116**(4):953–60.

7. Nguyen AD, Conley AJ. Adrenal androgens in humans and non-human primates: production, zonation and regulation. *Endocr Dev* 2008;**13**:33–54.

8. Woods MA, Acharya A, Finco I, Swonger JM, Elston MJ, Tallquist MD, Hammer GD. Fetal adrenal capsular cells serve as progenitor cells for steroidogenic and stromal adrenocortical cell lineages in *M. musculus*. *Development* 2013;**140**(22):4522–32.

9. Merke DP, Chrousos GP, Eisenhofer G, Weise M, Keil MF, Rogol AD, Van Wyk JJ, Bornstein SR. Adrenomedullary dysplasia and hypofunction in patients with classic 21-hydroxylase deficiency. *N Engl J Med* 2000;**343**(19):1362–8.

10. Young MC, Laurence KM, Hughes IA. Relationship between fetal adrenal morphology and anterior pituitary function. *Horm Res* 1989;**32**(4):130–5.

11. Suzuki T, Sasano H, Takeyama J, Kaneko C, Freije WA, Carr BR, Rainey WE. Developmental changes in steroidogenic enzymes in human postnatal adrenal cortex: immunohistochemical studies. *Clin Endocrinol (Oxf)* 2000;**53**(6):739–47.

12. Merke DP, Stratakis CA. The adrenal life cycle: the fetal and adult cortex and the remaining questions. *J Pediatr Endocrinol Metab* 2006;**19**(11):1299–302.

13. Sucheston ME, Cannon MS. Development of zonular patterns in the human adrenal gland. *J Morphol* 1968;**126**(4):477–91.

14. Arboleda VA, Lee H, Parnaik R, Fleming A, Banerjee A, Ferraz-de-Souza B, Délot EC, Rodriguez-Fernandez IA, Braslavsky D, Bergadá I, Dell'Angelica EC, Nelson SF, Martinez-Agosto JA, Achermann JC, Vilain E. IMAGe mutations in the PCNA-binding domain of CDKN1C cause IMAGe syndrome. *Nat Genet* 2012;**44**(7):788–92.

15. Chen H, Hoffman WH, Kusyk CJ, Tuck-Muller CM, Hoffman MG, Davis LS. De novo dup (5p) in a patient with congenital hypoplasia of the adrenal gland. *Am J Med Genet* 1995;**55**(4):489–93.

16. James PA, Aftimos S, Hofman P. CHARGE association and secondary hypoadrenalism. *Am J Med Genet A* 2003;**117A**(2):177–80.

17. Jongmans MCJ, Admiraal RJ, van der Donk KP, Vissers LELM, Baas BF, Kapusta L, van Hagen JM, Donnai D, de Ravel TJ, Veltman JA, Geurts van Kessel A, De Vries BBA, Brunner HG, Hoefsloot LH, van Ravenswaaij CMA. CHARGE syndrome: the phenotypic spectrum of mutations in the CHD7 gene. *J Med Genet* 2006;**43**:306–14.

18. Karpac J, Kern A, Hochgeschwender U. Pro-opiomelanocortin peptides and the adrenal gland. *Mol Cell Endocrinol* 2007;**265–266**: 29–33.

19. Ferraz-de-Souza B, Achermann JC. Disorders of adrenal development. *Endocr Dev* 2008;**13**:19–32.

20. Krude H, Biebermann H, Schnabel D, Zerjav Tansek M, Theunissen P, Mullis PE, Gruters A. Obesity due to proopiomelanocortin deficiency: three new cases and treatment trials with thyroid hormone and ACTH4-10. *J Clin Endocr Metab* 2003;**88**:4633–40.

21. Jackson RS, Creemers JWM, Ohagi S, Raffin-Sanson M-L, Sanders L, Montague CT, Hutton JC, O'Rahilly S. Obesity and impaired prohormone processing associated with mutations in the human prohormone convertase 1 gene. *Nature Genet* 1997;**16**:303–6.

22. Metherell LA, Chan LF, Clark AJ. The genetics of ACTH resistance syndromes. *Best Pract Res Clin Endocrinol Metab* 2006;**20**(4): 547–60.

23. Hanley NA, Arlt W. The human fetal adrenal cortex and the window of sexual differentiation. *Trends Endocrinol Metab* 2006;**17**(10): 391–7.

24. Achermann JC, Ito M, Ito M, Hindmarsh PC, Jameson JL. A mutation in the gene encoding steroidogenic factor-1 causes XY sex reversal and adrenal failure in humans. *Nature Genet* 1999;**22**: 125–6.

25. Zanaria E, Muscatelli F, Bardoni B, Strom TM, Guioli S, Guo W, Lalli E, Moser C, Walker AP, McCabe ER, et al. An unusual member of the nuclear hormone receptor superfamily responsible for X-linked adrenal hypoplasia congenita. *Nature* 1994;**372**(6507): 635–41.

26. Muscatelli F, Strom TM, Walker AP, Zanaria E, Recan D, Meindl A, Bardoni B, Guioli S, Zehetner G, Rabl W, Schwarz HP, Kaplan J-C, Camerino G, Meitinger T, Monaco AP. Mutations in the DAX-1 gene give rise to both X-linked adrenal hypoplasia congenita and hypogonadotropic hypogonadism. *Nature* 1994;**372**:672–6.

27. Lin L, Gu W-X, Ozisik G, To WS, Owen CJ, Jameson JL, Achermann JC. Analysis of DAX1 (NR0B1) and steroidogenic factor-1 (NR5A1) in children and adults with primary adrenal failure: ten years' experience. *J Clin Endocr Metab* 2006;**91**:3048–54.

28. Rojek A, Obara-Moszynska M, Malecka E, Slomko-Jozwiak M, Niedziela M. NR0B1 (DAX1) mutations in patients affected by congenital adrenal hypoplasia with growth hormone deficiency as a new finding. *J Appl Genet* 2013;**54**(2):225–30.

29. The Human Gene Mutation Database – data for 10.08.11. http://www.hgmd.org.

30. Kreidberg JA, Sariola H, Loring JM, Maeda M, Pelletier J, Housman D, Jaenisch R. WT-1 is required for early kidney development. *Cell* 1993;**74**:679–91.

31. Melo KF, Martin RM, Costa EM, Carvalho FM, Jorge AA, Arnhold IJ, Mendonca BB. An unusual phenotype of Frasier syndrome due to IVS9+4C>T mutation in the WT1 gene: predominantly male ambiguous genitalia and absence of gonadal dysgenesis. *J Clin Endocrinol Metab* 2002;**87**(6):2500–5.

32. Vainio S, Heikkila M, Kispert A, Chin N, McMahon AP. Female development in mammals is regulated by Wnt-4 signalling. *Nature* 1999;**397**:405–9.

33. Jordan BK, Mohammed M, Ching ST, Delot E, Chen XN, Dewing P, Swain A, Rao PN, Elejalde BR, Vilain E. Up-regulation of WNT-4 signaling and dosage-sensitive sex reversal in humans. *Am J Hum Genet* 2001;**68**:1102–9.

34. Mandel H, Shemer R, Borochowitz ZU, Okopnik M, Knopf C, Indelman M, Drugan A, Tiosano D, Gershoni-Baruch R, Choder M, Sprecher E. SERKAL syndrome: an autosomal-recessive disorder caused by a loss-of-function mutation in WNT4. *Am J Hum Genet* 2008;**82**:39–47.

35. Biason-Lauber A, Schoenle EJ. Apparently normal ovarian differentiation in a prepubertal girl with transcriptionally inactive steroidogenic factor 1 (NR5A1/SF-1) and adrenocortical insufficiency. *Am J Hum Genet* 2000;**67**:1563–8.

36. Kahana D, Berant M, Wolman M. Primary familial xanthomatosis with adrenal involvement (Wolman's disease): report of a further case with nervous system involvement and pathogenetic considerations. *Pediatrics* 1968;**42**:70–6.

37. Anderson RA, Byrum RS, Coates PM, Sando GN. Mutations at the lysosomal acid cholesteryl ester hydrolase gene locus in Wolman disease. *Proc Nat Acad Sci* 1994;**91**:2718–22.

38. Chemaitilly W, Goldenberg A, Baujat G, Thibaud E, Cormier-Daire V, Abadie V. Adrenal insufficiency and abnormal genitalia in a 46XX female with Smith-Lemli-Opitz syndrome. *Horm Res* 2003;**59**(5):254–6.

39. Nowaczyk MJ, Siu VM, Krakowiak PA, Porter FD. Adrenal insufficiency and hypertension in a newborn infant with Smith-Lemli-Opitz syndrome. *Am J Med Genet* 2001;**103**(3):223–5.

40. Bionconi SE, Conley SK, Keil MF, Sinaii N, Rother Kl, Porter FD, Stratakis CA. Adrenal function in Smith-Lemli-Opitz syndrome. *Am J Med Genet A* 2011;**155A**(11):2732–8.

41. Feigenbaum V, Lombard-Platet G, Guidoux S, Sarde CO, Mandel JL, Aubourg P. Mutational and protein analysis of patients and heterozygous women with X-linked adrenoleukodystrophy. *Am J Hum Genet* 1996;**58**(6):1135–44.

42. Perl S, Kotz L, Keil M, Patronas NJ, Stratakis CA. Calcified adrenals associated with perinatal adrenal hemorrhage and adrenal insufficiency. *J Clin Endocrinol Metab* 2007;**92**(3):754.

43. Palsdottir A, Snorradottir AO, Thorsteinsson L. Hereditary cystatin C amyloid angiopathy: genetic, clinical, and pathological aspects. *Brain Pathol* 2006;**16**(1):55–9.

44. Stratakis CA. Cushing syndrome and Addison disease. Hughes IA, Clark AJL, editors. *Endocrine development. Adrenal disease in childhood. Clinical and molecular aspects*, vol. 2. Basel: Karger; 2000. p. 163–173.

45. Kahaly GJ. Polyglandular autoimmune syndrome type II. *Presse Med* 2012;**41**:663–70.

46. Dittmar M, Kahaly GJ. Polyglandular autoimmune syndromes: immunogenetics and long-term follow-up. *J Clin Endocrinol Metab* 2003;**88**:2983–92.

47. Heino M, Scott HS, Chen Q, Peterson P, Mäebpää U, Papasavvas MP, Mittaz L, Barras C, Rossier C, Chrousos GP, Stratakis CA, Nagamine K, Kudoh J, Shimizu N, Maclaren N, Antonarakis SE, Krohn K. Mutation analyses of North American APS-1 patients. *Hum Mutat* 1999;**13**(1):69–74.

48. Husebye ES, Allolio B, Arlt W, Badenhoop H, Bensing S, Betterle C, Falorni A, Gan EH, Hulting AL, Kasperlik-Zaluska A, Kämpe O, Løvås K, Meyer G, Pearce SH. Consensus statement on the diagnosis, treatment and follow-up of patients with primary adrenal insufficiency. *J Intern Med* 2014;**275**(2):104–15.

49. Packham EA, Brook JD. T-box genes in human disorders. *Hum Mol Genet* 2003;**12**(R1):R37–44.

50. Lamolet B, Pulichino A-M, Lamonerie T, Gauthier Y, Brue T, Enjalbert A, Drouin J. A pituitary cell-restricted T box factor, Tpit, activates POMC transcription in cooperation with Pitx homeoproteins. *Cell* 2001;**104**:849–59.

51. Fujieda K, Tajima T. Molecular basis of adrenal insufficiency. *Pediatr Res* 2005;**57**(5 Pt. 2):62R–9R.

52. Metherell LA, Savage MO, Dattani M, Walker J, Clayton PE, Farooqi IS, Clark AJL. TPIT mutations are associated with early-onset, but not late-onset isolated ACTH deficiency. *Eur J Endocr* 2004;**151**:463–5.

53. Couture C, Saveanu A, Barlier A, Carel JC, Fassnacht M, Flück CE, Houang M, Maes M, Phan-Hug F, Enjalbert A, Drouin J, Brue T, Vallette S. Phenotypic homogeneity and genotypic variability in a large series of congenital isolated ACTH-deficiency patients with TPIT gene mutations. *J Clin Endocrinol Metab* 2012;**97**(3):E486–95.

54. Tsigos C, Arai K, Hung W, Chrousos GP. Hereditary isolated glucocorticoid deficiency is associated with abnormalities of the adrenocorticotropin receptor gene. *J Clin Invest* 1993;**92**:2458–61.

55. Al Kandari HM, Katsumata N, al Alwan I, al Balwi M, Rasoul MS. Familial glucocorticoid deficiency in five Arab kindreds with homozygous point mutations of the ACTH receptor (MC2R): genotype and phenotype correlations. *Horm Res Paediatr* 2011;**76**(3):165–71.

56. Lin L, Hindmarsh PC, Metherell LA, Alzyoud M, Al-Ali M, Brain CE, Clark AJ, Dattani MT, Achermann JC. Severe loss-of-function mutations in the adrenocorticotropin receptor (ACTHR, MC2R) can be found in patients diagnosed with salt-losing adrenal hypoplasia. *Clin Endocrinol (Oxf)* 2007;**66**(2):205–10.

57. Elias LL, Huebner A, Metherell LA, Canas A, Warne GL, Bitti ML, Cianfarani S, Clayton PE, Savage MO, Clark AJ. Tall stature in familial glucocorticoid deficiency. *Clin Endocrinol (Oxf)* 2000;**53**(4):423–30.

58. Wu SM, Stratakis CA, Chan CH, Hallermeier KM, Bourdony CJ, Rennert OM, Chan WY. Genetic heterogeneity of adrenocorticotropin (ACTH) resistance syndromes: identification of a novel mutation of the ACTH receptor gene in hereditary glucocorticoid deficiency. *Mol Genet Metab* 1998;**64**(4):256–65.

59. Chida D, Nakagawa S, Nagai S, Sagara H, Katsumata H, Imaki T, Suzuki H, Mitani F, Ogishima T, Shimizu C, Kotaki H, Kakuta S, Sudo K, Koike T, Kubo M, Iwakura Y. Melanocortin 2 receptor is required for adrenal gland development, steroidogenesis, and neonatal gluconeogenesis. *Proc Nat Acad Sci* 2007;**104**:18205–10.

60. Metherell LA, Chapple JP, Cooray S, David A, Becker C, Rüschendorf F, Naville D, Begeot M, Khoo B, Nürnberg P, Huebner A, Cheetham ME, Clark AJ. Mutations in MRAP, encoding a new interacting partner of the ACTH receptor, cause familial glucocorticoid deficiency type 2. *Nat Genet* 2005;**37**(2):166–70.

61. Keegan CE, Hutz JE, Krause AS, Koehler K, Metherell LA, Boikos S, Stergiopoulos S, Clark AJ, Stratakis CA, Huebner A, Hammer GD. Novel polymorphisms and lack of mutations in the ACD gene in patients with ACTH resistance syndromes. *Clin Endocrinol (Oxf)* 2007;**67**(2):168–74.

62. Meimaridou E, Hughes CR, Kowalczyk J, Chan LF, Clark AJ, Metherell LA. ACTH resistance: genes and mechanisms. *Endocr Dev* 2013;**24**:57–66.

63. Brooks BP, Kleta R, Caruso RC, Stuart C, Ludlow J, Stratakis CA. Triple-A syndrome with prominent ophthalmic features and a novel mutation in the AAAS gene: a case report. *BMC Ophthalmol* 2004;**4**:7.

64. Tsilou E, Stratakis CA, Rubin BI, Hay BN, Patronas N, Kaiser-Kupfer MI. Ophthalmic manifestations of Allgrove syndrome: report of a case. *Clin Dysmorphol* 2001;**10**(3):231–3.

65. Brooks BP, Kleta R, Stuart C, Tuchman M, Jeong A, Stergiopoulos SG, Bei T, Bjornson B, Russell L, Chanoine JP, Tsagarakis S, Kalsner L, Stratakis C. Genotypic heterogeneity and clinical phenotype in triple A syndrome: a review of the NIH experience 2000-2005. *Clin Genet* 2005;**68**(3):215–21.

66. Tullio-Pelet A, Salomon R, Hadj-Rabia S, Mugnier C, de Laet MH, Chaouachi B, Bakiri F, Brottier P, Cattolico L, Penet C, Bégeot M, Naville D, Nicolino M, Chaussain JL, Weissenbach J, Munnich A, Lyonnet S. Mutant WD-repeat protein in triple-A syndrome. *Nat Genet* 2000;**26**(3):332–5.

67. Handschug K, Sperling S, Yoon SJ, Hennig S, Clark AJ, Huebner A. Triple A syndrome is caused by mutations in AAAS, a new WD-repeat protein gene. *Hum Mol Genet* 2001;**10**(3):283–90.

68. Cronshaw JM, Matunis MJ. The nuclear pore complex protein ALADIN is mislocalized in triple A syndrome. *Proc Natl Acad Sci USA* 2003;**100**(10):5823–7.

69. Ikeda M, Hirano M, Shinoda K, Katsumata N, Furutama D, Nakamura K, Ikeda S, Tanaka T, Hanafusa T, Kitajima H, Kohno H, Nakagawa M, Nakamura Y, Ueno S. Triple A syndrome in Japan. *Muscle Nerve* 2013;**48**(3):381–6.

70. Huebner A, Mann P, Rohde E, Kaindl AM, Witt M, Verkade P, Jakubiczka S, Menschikowski M, Stoltenburg-Didinger G, Koehler K. Mice lacking the nuclear pore complex protein ALADIN show female infertility but fail to develop a phenotype resembling human triple A syndrome. *Mol Cell Biol* 2006;**26**(5):1879–87.

71. Sandrini F, Farmakidis C, Kirschner LS, Wu SM, Tullio-Pelet A, Lyonnet S, Metzger DL, Bourdony CJ, Tiosano D, Chan WY, Stratakis CA. Spectrum of mutations of the AAAS gene in Allgrove syndrome: lack of mutations in six kindreds with isolated resistance to corticotropin. *J Clin Endocrinol Metab* 2001;**86**(11): 5433–7.

72. Hung KL, Wang JS, Keng WT, Chen HJ, Liang JS, Ngu LH, Lu JF. Mutational analyses on X-linked adrenoleukodystrophy reveal a novel cryptic splicing and three missense mutations in the ABCD1 gene. *Pediatr Neurol* 2013;**49**(3):185–90.

73. Vingerhoeds AC, Thijssen JH, Schwarz F. Spontaneous hypercortisolism without Cushing's syndrome. *J Clin Endocrinol Metab* 1976;**43**:1128–33.

74. Hurley DM, Accili D, Stratakis CA, Karl M, Vamvakopoulos N, Rorer E, Constantine K, Taylor SI, Chrousos GP. Point mutation causing a single amino acid substitution in the hormone binding domain of the glucocorticoid receptor in familial glucocorticoid resistance. *J Clin Invest* 1991;**87**:680–6.

75. Karl M, Lamberts SW, Detera-Wadleigh SD, Encio IJ, Stratakis CA, Hurley DM, Accili D, Chrousos GP. Familial glucocorticoid resistance caused by a splice site deletion in the human glucocorticoid receptor gene. *J Clin Endocrinol Metab* 1993;**76**:683–9.

76. Malchoff DM, Brufsky A, Reardon G, McDermott P, Javier EC, Bergh CH, Rowe D, Malchoff CD. A mutation of the glucocorticoid receptor in primary cortisol resistance. *J Clin Invest* 1993;**91**: 1918–25.

77. Stratakis CA, Karl M, Schulte HM, Chrousos GP. Glucocorticoid resistance in humans: elucidation of the molecular mechanisms and implications for pathophysiology. *Ann NY Acad Sci* 1994;**746**: 362–76.

78. Schaaf MJ, Cidlowski JA. Molecular mechanisms of glucocorticoid action and resistance. *J Steroid Biochem Mol Biol* 2002;**83**:37–48.

79. Charmandari E, Kino T, Ichijo T, Jubiz W, Mejia L, Zachman K, Chrousos GP. A novel point mutation in helix 11 of the ligand-binding domain of the human glucocorticoid receptor gene causing generalized glucocorticoid resistance. *J Clin Endocrinol Metab* 2007;**92**: 3986–90.

80. Mendonca BB, Leite MV, de Castro M, Kino T, Elias LL, Bachega TA, Arnhold IJ, Chrousos GP, Latronico AC. Female pseudohermaphroditism caused by a novel homozygous missense mutation of the GR gene. *J Clin Endocrinol Metab* 2002;**87**:1805–9.

81. Chrousos GP, Vingerhoeds A, Brandon D, Eil C, Pugeat M, DeVroede M, Loriaux DL, Lipsett MB. Primary cortisol resistance in man. A glucocorticoid receptor-mediated disease. *J Clin Invest* 1982;**69**:1261–9.

82. Karl M, Lamberts SW, Koper JW, Katz DA, Huizenga NE, Kino T, Haddad BR, Hughes MR, Chrousos GP. Cushing's disease preceded by generalized glucocorticoid resistance: clinical consequences of a novel, dominant-negative glucocorticoid receptor mutation. *Proc Assoc Am Physicians* 1996;**108**:296–307.

83. Cheek DB, Perry JW. A salt wasting syndrome in infancy. *Arch Dis Child* 1958;**33**:252–6.

84. Geller DS. Mineralocorticoid resistance. *Clin Endocrinol (Oxf)* 2005;**62**: 513–20.

85. Hanukoglu A. Type I pseudohypoaldosteronism includes two clinically and genetically distinct entities with either renal or multiple target organ defects. *J Clin Endocrinol Metab* 1991;**73**: 936–44.

86. Oberfield SE, Levine LS, Carey RM, Bejar R, New MI. Pseudohypoaldosteronism: multiple target organ unresponsiveness to mineralocorticoid hormones. *J Clin Endocrinol Metab* 1979;**48**: 228–34.

87. Hanukoglu A, Bistritzer T, Rakover Y, Mandelberg A. Pseudohypoaldosteronism with increased sweat and saliva electrolyte values and frequent lower respiratory tract infections mimicking cystic fibrosis. *J Pediatr* 1994;**125**:752–5.

88. Pearce D, Bhargava A, Cole TJ. Aldosterone: its receptor, target genes, and actions. *Vitam Horm* 2003;**66**:29–76.

89. Arai K, Zachman K, Shibasaki T, Chrousos GP. Polymorphisms of amiloride-sensitive sodium channel subunits in five sporadic cases of pseudohypoaldosteronism: do they have pathologic potential? *J Clin Endocrinol Metab* 1999;**84**:2434–7.

90. Geller DS, Rodriguez-Soriano J, Vallo Boado A, Schifter S, Bayer M, Chang SS. Mutations in the mineralocorticoid receptor gene cause autosomal dominant pseudohypoaldosteronism type I. *Nat Genet* 1998;**19**:279–81.

91. Sartorato P, Cluzeaud F, Fagart J, Viengchareun S, Lombes M, Zennaro MC. New naturally occurring missense mutations of the human mineralocorticoid receptor disclose important residues involved in dynamic interactions with deoxyribonucleic acid, intracellular trafficking, and ligand binding. *Mol Endocrinol* 2004;**18**: 2151–65.

92. Fernandes-Rosa FL, de Castro M, Latronico AC, Sippell W, Riepe FG, Antonini SR. Recurrence of the R947X mutation in unrelated families with autosomal dominant pseudohypoaldosteronism type 1: evidence for a mutational hot spot in the mineralocorticoid receptor gene. *J Clin Endocrinol Metab* 2006;**91**:3671–5.

93. Zennaro MC, Lombes M. Mineralocorticoid resistance. *Trends Endocrinol Metab* 2004;**15**:264–70.

94. Chang SS, Grunder S, Hanukoglu A, Rosler A, Mathew PM, Hanukoglu I. Mutations in subunits of the epithelial sodium channel cause salt wasting with hyperkalaemic acidosis, pseudohypoaldosteronism type 1. *Nature Genet* 1996;**12**:248–53.

95. Strautnieks SS, Thompson RJ, Gardiner RM, Chung E. A novel splice-site mutation in the gamma subunit of the epithelial sodium channel gene in three pseudohypoaldosteronism type 1 families. *Nature Genet* 1996;**13**:248–50.

96. Belot A, Ranchin B, Fichtner C, Pujo L, Rossier BC, Liutkus A, Morlat C, Nicolino M, Zennaro MC, Cochat P. Pseudohypoaldosteronisms, report on a 10-patient series. *Nephrol Dial Transplant* 2008;**23**:1636–41.

97. Arai K, Tsigos C, Suzuki Y, Irony I, Karl M, Listwak S, Chrousos GP. Physiological and molecular aspects of mineralocorticoid receptor action in pseudohypoaldosteronism: a responsiveness test and therapy. *J Clin Endocrinol Metab* 1994;**79**:1019–23.

98. Kerem E, Bistritzer T, Hanukoglu A, Hofmann T, Zhou Z, Bennett W. Pulmonary epithelial sodium-channel dysfunction and excess airway liquid in pseudohypoaldosteronism. *N Engl J Med* 1999;**341**:156–62.

99. Pathare G, Hoenderop JG, Bindels RJ, San-Cristobal P. A molecular update on pseudohypoaldosteronism type II. *Am J Physiol Renal Physiol* 2013;**305**(11):F1513–20.

REPRODUCTIVE

18

Genetic Considerations in the Evaluation of Menstrual Cycle Irregularities

Milad Abusag*, David A. Ehrmann**, Leslie Hoffman***

*University of Pittsburgh Medical Center – Horizon (UPMC-Horizon), Pittsburgh, PA, USA
**Section of Endocrinology, Diabetes, and Metabolism, The University of Chicago,
Maryland, Chicago, IL, USA
***Columbus Endocrinology, Columbus, OH, USA

INTRODUCTION

Normal menstrual cycling results from a complex and integrated process that requires the temporal coordination of hormonal secretion and signaling within the hypothalamic-pituitary-ovarian axis. Disruption of these signals at any point in the axis can manifest clinically as amenorrhea (the absence or abnormal cessation of menstrual cycles), or as oligomenorrhea (less than nine menstrual cycles a year). Primary amenorrhea refers to the failure to menstruate by age 15 years in the presence of normal secondary sexual development or within five years of breast development, if that occurs before age 10 years.[1] Secondary amenorrhea refers to amenorrhea that occurs after menarche.

The causes of oligomenorrhea and both primary and secondary amenorrhea are similar, and the differential diagnosis is extremely broad. Causes can be generally divided into genetic (covered here) and nongenetic. The latter include psychosocial disorders (e.g., anorexia nervosa) as well as infectious, autoimmune, and metabolic conditions. Chromosomal and genetic causes are becoming increasingly appreciated as important and complex contributors to menstrual cycle abnormalities. Genetic disorders leading to defects in hypothalamic-pituitary regulation of the menstrual cycle, such as hypogonadotropic hypogonadism and Kallman's syndrome, as well as mutations of the luteinizing hormone (LH) and follicle-stimulating hormone (FSH) receptors, are covered elsewhere in this book (see Chapter XX). This chapter will therefore focus on genetic causes of oligomenorrhea that primarily affect ovarian and/or adrenal function. Specific attention will be given to the most common etiologies: primary ovarian insufficiency (POI), polycystic ovary syndrome (PCOS), and nonclassic congenital adrenal hyperplasia (CAH).

OVARIAN DISORDERS

Primary Ovarian Insufficiency

POI refers to the cessation of ovarian function (i.e., ovulation and menstrual cycling) before the age of 40 years. The term primary ovarian insufficiency emphasizes that these disorders are intrinsic to the ovary and not the result of defects elsewhere in the hypothalamic-pituitary-ovarian axis. The diagnosis of POI is typically based on the presence of oligo- or amenorrhea in conjunction with circulating FSH levels that are in the menopausal range (usually above 40 IU/L) detected on at least two occasions a few weeks apart. The condition affects approximately 10–28% of women with primary amenorrhea and 4–18% of those with secondary amenorrhea.[2,3] Although the majority of cases of POI are idiopathic, several familial cases have been identified. The overall prevalence of familial POI ranges from 4% to 31%. This wide range is likely due to variation regarding inclusion criteria and the detail sought in assessment of families.[4] Pedigree analyses in affected families have identified global X-chromosome abnormalities, as well as specific mutations with autosomal dominant sex-linked inheritance or X-linked inheritance with incomplete penetrance.[5]

Genetic Diagnosis of Endocrine Disorders. http://dx.doi.org/10.1016/B978-0-12-800892-8.00018-X

Anti-Müllerian hormone (AMH) is produced by the granulosa cells of growing follicles and appears to inhibit the growth of primordial follicles. Clinically, serum AMH may be a useful biomarker of ovarian reserve. In women and mice, serum AMH declines with advancing age. Although it is difficult to establish a direct link between serum AMH and the primordial follicle pool in humans, antral follicle number is positively correlated with AMH levels.[6]

X-Chromosome Abnormalities

Both familial and sporadic X-chromosome abnormalities have been described in women with POI. These include the complete absence of one X chromosome, as seen in Turner's syndrome, trisomy X, as well as partial X-chromosome defects including deletions and balanced autosome translocations.[7]

Turner's Syndrome

Classic Turner's syndrome is characterized by the absence of one X chromosome. Clinical features are variable, and at birth newborns may present with low birth weight, lymphedema of the upper and lower extremities (in 30% of Turner babies), and a webbed neck (pterygium colli). Additional dysmorphic features may include low-set prominent ears, a low posterior hairline, micrognathia, high-arched palate, epicanthal folds, hypoplastic nail beds, and/or hypoplastic fourth and fifth metacarpals. At adolescence, the most common presentation is one of short stature, amenorrhea, and lack of secondary sex characteristics, although the latter is variable and depends upon the extent of gonadal dysgenesis. The diagnosis of Turner's syndrome should prompt a search for renal anomalies (incidence between 30% and 50%), as well as coarctation of the aorta and hearing loss.

Thirty percent of patients with Turner's syndrome have the classic 45, X karyotype, and the remaining have a mosaic form where a 45, X cell line is associated with another cell line such as 46, XX or 46, XY. Of note, patients with mosaic forms of Turner's syndrome can lack the classic phenotypic features, and usually are not diagnosed until presenting with POI later in life.

While one X chromosome in females is inactivated for dosage compensation, several genes escape inactivation and are essential for normal function of the X chromosome.[7] The phenotypic traits of Turner's syndrome, including POI, have been mapped to a critical region in Xp11.2-p22.1 which escapes X inactivation, and contains the 18 candidate genes that have been reported.[8,9]

The loss of one X chromosome leads to an accelerated loss of ovarian follicles prior to birth. Many patients with Turner's syndrome lose all of their follicles prenatally, and some lose their remaining germ cells during childhood. Accordingly, such patients will present with primary amenorrhea. Less than 15% of Turner's patients will lose their germ cells after puberty and may even have enough germ cells for regular cycles before presenting with secondary amenorrhea. Rarely, some patients with Turner's syndrome will even achieve pregnancy prior to developing ovarian insufficiency.[10]

Once the germ cells are depleted from the ovaries, the residual gonads consisting only of connective stroma are called "streak gonads." Patients with a Y cell line Turner's syndrome have an increased risk of development of gonadoblastoma and malignant germ cell tumors. Consequently, it is recommended that streak gonads be surgically removed.

Trisomy X

Trisomy X affects one in 900 women and is rarely associated with POI. In one series, two of 52 (3.8%) patients with POI had trisomy X.[11] The underlying mechanism is not clear, but may be similar to that observed among patients with Klinefelter's syndrome.[12]

Deletions

Deletions on an X chromosome leave a portion of the normal X unpaired and can lead to oocyte atresia. Although deletions more often involve the short arm of the X chromosome (Xp), deletions associated with POI more commonly involve the Xq13-25 region.[13] Deletions at Xp11 result in both primary and secondary amenorrhea, whereas deletions at Xq13 usually lead to primary amenorrhea. It has also been noted that more proximal deletions are associated with ovarian failure while distal deletions appear to be innocuous.[10] However, it remains unclear why large deletions that remove the whole critical region for POI, in Xq21, are not associated with ovarian failure.[14]

Translocations

Balanced X/autosomal translocations, while rare, often lead to POI and have been reported in more than 100 postpubertal women. Breakpoints that fall between Xq13 and Xq26 are associated with ovarian failure. Thus, it has been proposed that this is a critical region for normal ovarian function.[15] The most frequent breakpoints involve two specific POI regions: POI1 Xq26-qter,[16] and POI2 Xq13.3-Xq21.1.[17] The chromosome in this region may be particularly sensitive to structural changes that ultimately lead to oocyte apoptosis. Distal deletions involving the POI1 locus are typically associated with POI at age 24–39 years, whereas translocations involving the POI2 locus cause POI at an earlier age of 16–21 years.

POI Genes on the X Chromosome

Multiple candidate genes for POI have been proposed, although actual mutations have been identified in less than 10% of cases.[18] Because the functions of many of these genes are not known, they are not used as genetic markers for POI.[10]

Fragile X Metal Retardation (*FMR1*) Gene

The most common form of inherited mental retardation, the fragile X syndrome, is associated with expansion of a CGG triplet repeat in the 5' untranslated region of the *FMR1* gene located in Xq27.3, outside the Xq POI critical region. Four types of alleles have been identified based on the number of repeats: normal (6–40), gray zone (41–60), premutated (61–200), and fully mutated (>200). The overall prevalence of the fragile X premutation ranges from 1/259 in a nonselected population of over 10,000 French-Canadian women[19] to 1/100 in a population of Israeli women.[20] Women who carry the fragile X premutation (61–200 repeats) are at risk for having a child with fragile X syndrome. In addition, the association between fragile X premutation carrier status and POI has been well documented. The results of an international study examining premature menopause in 760 women from families with fragile X syndrome showed that 16% of the 395 premutation carriers experienced menopause prior to the age of 40, compared with none of the 238 full mutation carriers and one (0.4%) of the 237 controls.[21] Early menopause, however, is only one end of a spectrum of the ovarian dysfunction that occurs, since some women may have elevated serum FSH levels but maintain regular menstrual cycles. Collectively, premutation carriers experience earlier menopause by approximately 5 years.[22]

The mechanism underlying the association between the fragile X premutation and POI, however, remains unclear. The *FMR1* gene is expressed in oocytes and encodes an RNA-binding protein involved in translation. Few functional studies of the FMR1 protein have been performed with POI in view. Rather, theories are based on studies using neurons and lymphocytes from patients with a late-onset neurodegenerative disorder also associated with the fragile X premutation, termed fragile X tremor/ataxia syndrome. Premutation carriers are known to produce transcripts with large CGG repeat tracks. In the presence of the full fragile X mutation, however, the *FMR1* gene is methylated, and its transcription is blocked. Since ovarian failure has not been demonstrated among women with the full mutation, it cannot be attributed to a lack of the FMR protein. Rather, it has been proposed that the large tracks of rCGGs act in a toxic manner through increased mRNA levels or the unusual structural features of the large repeat tracks in the transcripts. Such structures could bind other important proteins in the cell, rendering them inactive or targeting them for degradation.[23] Alternatively, proteins or transcription factors that bind to the CGG repeats may be unable to perform their normal functions.[24]

Since the fragile X premutation is one of the few genetic causes of premature ovarian failure that can be tested for, it is of particular importance to document the mutation for diagnostic purposes and for assessing the risk of bearing offspring of fragile X mental retardation.

In most people, the CGG segment is repeated in the gene approximately 5–44 times. Increased expression of the CGG segment on the *FMR1* gene is associated with impaired cognitive and reproductive function.

Bone Morphogenetic Protein 15 Gene (*BMP15*)

Bone morphogenetic proteins (BMPs) are extracellular signaling proteins belonging to transforming growth factor-β superfamily that also includes growth/differentiation factors (GDFs) encoded by the *BMP15* gene. BMP15 is an oocyte-specific GDF that stimulates folliculogenesis and granulosa cell growth and is expressed in oocytes during early folliculogenesis. It is thought that this protein may be involved in oocyte maturation and follicular development as a *homodimer* or by forming *heterodimers* with a related protein, Gdf9.[25] The *BMP15* gene is located within the POI critical region on Xp11.2.[25]

In 2004, Di Pasquale et al. reported a heterozygous mutation in the *BMP15* gene in two sisters with primary amenorrhea. The mutation involved an A to G transition at nucleotide 704 of the *BMP15* gene produce Y235C. The father was unaffected as a hemizygous carrier and the mother had only wild-type *BMP15* coding sequence.[26] The mutation was not found in 210 alleles from 120 ethnically matched controls.

Autosomal Genes

Blepharophimosis-Ptosis-Epicanthus Inversus Syndrome (BPES)

BPES is an autosomal dominant condition characterized by eyelid malformations, low nasal bridge, and ptosis of the eyelids with (type I) or without (type II) POI. Both types map to 3q22-q23[27] and are associated with a mutation in the *FOXL2* gene, which encodes a winged helix/forkhead transcription factor that plays an important role in development, present mostly in the craniofacial region, the pituitary gland, and ovary.[28] Gene mutations leading to a truncated protein tend to produce BPES type I, whereas elongated proteins typically lead to BPES type II. In the human ovary, *FOXL2* is expressed initially in cells of the female genital ridge prior to sex determination, and through adulthood is highly expressed in granulosa cells and to a lesser extent in theca and stromal cells.[29] Ovarian failure may be caused by either a decreased number of follicles forming during development or an increased rate of follicle loss. In addition, since FOXL2 is also expressed during pituitary organogenesis, mutations that affect the hypothalamic-pituitary-gonadal axis may also play a role in ovarian failure.

Because the *FOXL2* gene is the first human autosomal gene in which dominant mutations have been linked to ovarian function, its potential role in nonsyndromic POI has also been a subject of investigation. However, sequence analysis of the *FOXL2* gene in a total of 290 nonsyndromic POI patients from five different studies found only one mutation potentially responsible for the POI.[30] Thus, the role *FOXL2* gene mutations in nonsyndromic POI remains unclear.

Inhibin (*INH*) Gene

Inhibin is a gonadal protein that inhibits the synthesis and secretion of FSH from the pituitary, and has been proposed as a strong candidate gene in POI. Low serum inhibin levels have been documented in women with POI.[31] Inhibins have also been implicated in regulating numerous cellular processes including cell proliferation, apoptosis, and tumor suppressor activity.[32] One variation of the *INH alpha* gene, G769A, has been associated with POI, the prevalence of which may vary in different populations from 0% to 11%.[33,34]

Galactose-1-Phosphate Uridyltransferase (*GALT*) Gene

Galactosemia is a rare autosomal recessive disorder associated with an impairment in GALT metabolism and consequent excessive galactose levels. GALT deficiency is the most common type of galactosemia. Intracellular accumulation of galactose or its metabolites may cause follicular damage or decrease the number of oogonia formed initially, therefore leading to POI. The *GALT* gene maps to chromosome 9p13, and female patients with galactosemia have a 60–70% prevalence of POI.[35] Women can present with primary amenorrhea, although the majority develop POI shortly after puberty.[36]

Folliculogenesis Specific Basic Helix–Loop–Helix Gene Mutation (*FIGLA* gene)

This gene encodes a protein that functions in postnatal oocyte-specific gene expression that regulates multiple oocyte-specific genes, including genes involved in folliculogenesis and those that encode the zona pellucida. Mutations in this gene cause POI. It seems likely that genetic hierarchies expressed within oocytes complement these somatic signals to maintain appropriate germ cell identity by activating oocyte-associated genes and repressing sperm-associated genes during postnatal oogenesis.[37]

Zhao et al. analyzed the FIGLA gene in 100 Chinese women with POI and identified a 22-bp deletion (*608697.0001*) and a 3-bp in-frame deletion (140delN; *608697.0002*) in two patients, respectively. The mutations were not found in 340 female controls with regular menstrual cycles and no history of infertility.[38]

NOG Gene Mutations Causing Noggin Deficiency

Proximal symphalangism (SYM1) is an autosomal dominant disorder characterized by ankylosis of the proximal interphalangeal joint, fusion of the carpal and tarsal bones, brachydactyly, and conductive deafness.[39] It is caused by haploinsufficiency of the *NOG* gene on 17q22 encoding Noggin. NOG is also expressed in the ovary and acts as an antagonist for bone morphogenic proteins (BMPs), including BMP 4 and BMP 7, which play an important role in ovarian function.[39] POI has been reported in one female with SYM1 and a documented *NOG* gene, suggesting increase in the susceptibility to POI by disrupting the function of BMPs.[40,41]

The Estrogen Receptor (ER)

The estrogen receptor (ER) exists primarily in two isoforms (ERalpha and ERbeta) with specific tissue and cell patterns of expression. The ER gene has been mapped to 6q25.1 and its product is a ligand-activated transcription factor. Variants in the ER have been associated with differences in the clinical expression of risk for breast carcinoma,[42] bone mineral density[43] (primarily in men), and age at menopause. In one large study of 900 postmenopausal women[44] genetic variation in the *ESR* gene contributed to the variability in the onset of menopause. The ESR genotypes (PP, Pp, and pp) were assessed by polymerase chain reaction using the PvuII endonuclease to detect an anonymous intronic RFLP. Compared with women carrying the pp genotype, homozygous PP women had a 1.1 year ($P < 0.02$) earlier onset of menopause. Furthermore, an allele dose effect was observed, corresponding to a 0.5 year ($P < 0.02$) earlier onset of menopause per copy of the P allele.

Down Syndrome

Down syndrome is a genetic disorder caused by the presence of all or part of a third copy of chromosome 21 (Hsa21), which was first described in 1866. Depending on the maternal age structure of the population and the utilization of paternal testing, the incidence of trisomy 21 ranges from 1/600 to 1/1000 live births, making it one of the most common chromosomal abnormalities in live-born individuals. Like most trisomies, the incidence of trisomy 21 is highly correlated with maternal age, increasing from about 1/1500 live births for women 20 years of age to 1/30 for women ≥ 45 years.[45,46]

Down syndrome is associated with intellectual disability, a characteristic facial appearance, and hypotonia. All affected individuals experience cognitive delays, but the intellectual disability is usually mild to moderate. Down syndrome is also associated with an increased risk of hypothyroidism (15%), core binding factor acute myeloid leukemia, and the development of Alzheimer disease. Approximately half of adults with Down syndrome develop Alzheimer disease, and they usually develop this condition in their fifties or sixties.[47]

Most cases of Down syndrome result from trisomy 21; less commonly, Down syndrome occurs when part of chromosome 21 becomes translocated to another chromosome in a parent or very early in fetal development. A very small percentage of people with Down syndrome have an extra copy of chromosome 21 in only some cells (mosaic Down syndrome).

Most cases of Down syndrome are not inherited. When the condition is caused by trisomy 21, the chromosomal abnormality occurs as a random event during the formation of reproductive cells in a parent. Nondisjunction results in a reproductive cell with an abnormal number of chromosomes. If one of these atypical reproductive cells contributes to the genetic make-up of a child, the child will have an extra chromosome 21 in each of the body's cells.

Individuals with translocation Down syndrome can inherit the condition from an unaffected parent. The parent carries a rearrangement of genetic material between chromosome 21 and another chromosome. This rearrangement is called a balanced translocation. No genetic material is gained or lost in a balanced translocation, so these chromosomal changes usually do not cause any health problems. However, as this translocation is passed to the next generation, it can become unbalanced. Individuals who inherit an unbalanced translocation-involving chromosome 21 may have extra genetic material from chromosome 21, which causes Down syndrome.

Early studies suggested that a limited region of Hsa21, termed the Down syndrome critical region (DSCR), might contain one or more dosage-sensitive genes that contribute to many of the Down syndrome phenotypes.[48,49] However, further studies that included larger numbers of partial trisomy cases and more detailed genetic mapping have shown that different regions of Hsa21 contribute to different phenotypes, arguing against a single DSCR.[50]

Polycystic Ovary Syndrome

In contrast to the relatively rare genetic disorders discussed above, polycystic ovary syndrome (PCOS) is among the most common hormonal disorders in women and accounts for a large majority of women presenting with menstrual irregularities. PCOS is characterized by irregular menstrual cycles and androgen excess, with symptoms typically emerging late in puberty or shortly following. After exclusion of other causes (hyperprolactinemia, nonclassic congenital adrenal hyperplasia (see below), Cushing's syndrome) diagnosis requires at least two of the following to be present:[51] oligoovulation or anovulation, typically presenting as oligomenorrhea or amenorrhea,[52] hyperandrogenemia without or with clinical manifestations of androgen excess including acne, hirsutism, and/or hair thinning in a male pattern, and[53] polycystic ovaries on ultrasonography. Although not included in the formal definition, PCOS is also associated with a variety of metabolic abnormalities including obesity, impaired glucose tolerance, type 2 diabetes, hyperlipidemia, and obstructive sleep apnea.[54]

Multiple etiologic factors, at all levels of the hypothalamic-pituitary-ovarian axis, have been implicated in the pathogenesis of PCOS that ultimately lead to increased ovarian androgen biosynthesis. For instance, there is an increased frequency of hypothalamic gonadotropin-releasing hormone (GnRH) pulses that favor the production of LH over FSH. The relative increase in pituitary secretion of LH leads to an increase in androgen production by ovarian theca cells.[55]

Insulin also plays an important role in the pathogenesis of PCOS by acting synergistically with LH to enhance androgen production by theca cells. In addition, insulin increases the proportion of biologically active testosterone by inhibiting the hepatic synthesis of sex hormone-binding globulin, the key circulating protein that binds testosterone.

In PCOS, menstrual dysfunction reflects chronic anovulation and is characterized by irregular and infrequent bleeding. There is a broad spectrum of menstrual patterns seen among women with PCOS: 5–10% may even have normal or near-normal menstrual function, whereas approximately 20% will have the complete absence of menses.[56]

Multiple studies suggest that PCOS is heritable. For instance, a prospective study of the relatives of 195 consecutive PCOS patients found that 35% of mothers and 40% of sisters of patients with PCOS had PCOS themselves.[57] In addition, a Dutch twin study demonstrated that concordance of PCOS was greater among monozygotic than dizygotic twins.[58,59] The metabolic complications associated with PCOS are also familial. Women with PCOS have an increased likelihood of having at least one first degree relative with type 2 diabetes,[60] and mothers of women with PCOS have increased total cholesterol and LDL cholesterol levels, as well as increased prevalence of the metabolic syndrome.[61] The mode of inheritance for PCOS is difficult to define due to clinical heterogeneity and lack of a male phenotype, but our current understanding suggests that PCOS is a complex multigenic disorder.

Over the past decade, a large number of case–control and family-based association studies have been performed to search for candidate genes in PCOS. The most obvious genes to consider include those that may regulate the hypothalamic-pituitary-ovarian axis, as well as those responsible for insulin resistance and its sequelae. A recent review by Urbanek documented over 60 studies on the genetic contribution of more than 70 genes to the etiology of PCOS.[62] With a few exceptions, the results of these studies have been ambiguous or lacked replication, a finding attributed to the clinical heterogeneity of PCOS and limited sample sizes.

In some instances, single gene mutations can give rise to the PCOS phenotype, as has been documented in three patients with cortisone reductase deficiency, in whom mutations of the genes for 11β-hydroxysteroid dehydrogenase type 1 and hexose-6-phosphate dehydrogenase were found.[63] The inability to convert cortisone to cortisol, as seen with cortisone reductase deficiency, leads to the accumulation of adrenocorticotrophin-mediated androgens and mimics the phenotype of PCOS.

CYP11A Gene

One ideal functional candidate gene for PCOS is *CYP11A*, since it encodes the enzyme that cleaves the side chain of cholesterol p450, the rate-limiting step in androgen biosynthesis. In 1997, Gharani et al. showed evidence for linkage between the *CYP11A* gene and PCOS.[64] Although additional studies demonstrated further evidence for an association between a polymorphism at the CYP11A promoter and PCOS,[65] a subsequent large case-control study found no associations between the promoter polymorphisms in *CYP11A* and either polycystic ovary morphology or testosterone levels.[66] Similarly conflicting results have been found in studies of other genes, where larger follow-up studies failed to confirm initially reported associations with the insulin *VNTR* gene (a regulator of transcription of the insulin) and the *CAPN10* gene (involved in insulin secretion and action).

Chromosome 19p13.2 PCOS Susceptibility Locus (D19S884)

One locus that has been consistently associated with PCOS is on chromosome 19p13.2, at the dinucleotide repeat marker D19S884, which maps 800 kilobases centromeric to the *insulin receptor* (*INSR*) gene and is thought to be in the region of the *fibrillin 3* (*FBN-3*) gene.[67] In addition to a structural role in connective tissue, fibrillins are also believed to regulate the activity of members of the TGFβ family. TGFh maps 800 kilobases centromeric to 19p13.2 D19S884 allele 8 (A8), within intron 55 of the *FBN-3* gene.[68] A8 was also associated with higher levels of fasting insulin homeostasis model assessment (HOMA index) for insulin resistance in women with PCOS. The association of D19S884 with markers of insulin resistance and pancreatic beta-cell dysfunction suggests that the same variant contributes to the reproductive and metabolic abnormalities of PCOS in affected women. The roles of both fibrillin 3 and its potential interactions with INSR remain an area of active research.[69]

The Androgen Receptor Gene

In humans, the androgen receptor is encoded by the *AR* gene located on the X chromosome at Xq11-12. The *androgen receptor* (*AR*) gene and the role of X-inactivation have also received increasing attention as a possible epigenetic contributor to the PCOS phenotype. The *AR* gene contains a CAG repeat encoding the polyglutamine tract in the N-terminal transactivation domain. *In vitro* studies have demonstrated an inverse relationship between the length of the CAG repeat and receptor activity, and suggest that the length of the CAG repeat could affect androgen sensitivity.[70] Genetic studies of the role variation in the CAG repeat, however, have not been consistent.[71,72]

Since the *AR* gene is X-linked, other studies have found that the pattern of X-inactivation could influence AR activity and PCOS. One study looking at the pattern of X-inactivation in 88 sisters of women with PCOS found that sisters with the same *AR* CAG repeat genotype and the same clinical presentation (both unaffected or both with PCOS) more frequently showed the same pattern of X-inactivation than did sisters with different clinical presentations (85% vs. 16%).[72] Thus, this study adds further support to the hypothesis that the *AR* CAG repeat number has an effect on PCOS phenotype.

Thus, clinical studies point to a strong heritable factor in the predisposition to PCOS, the genetic details remain poorly understood. Although the search for candidate genes will continue, the key to better understanding of complex multigenic disorders may be in expensive and large genome-wide association studies, such as those applied in type 2 diabetes.

ADRENAL DISORDERS

Nonclassic Congenital Adrenal Hyperplasia

Another genetic disorder that can lead to oligo- or amenorrhea is nonclassic congenital adrenal hyperplasia (CAH) due to 21-hydroxylase (P450c21) deficiency. The *CYP21A2* gene encodes P450c21 and is responsible for the conversion of 17-hydroxyprogesterone to 11-deoxycortisol. P450c21 deficiency results in decreased cortisol synthesis and therefore increased ACTH secretion and consequent increased production of androgens. Different mutations compromise enzyme activity to varying degrees. In patients with the nonclassic form, enzymatic activity is reduced but sufficient to maintain normal glucocorticoid and mineralocorticoid production, at the expense of excessive androgen production.

The most severely affected individuals with classic CAH present during the neonatal period and early infancy with adrenal insufficiency (with or without salt wasting), and in females, genital ambiguity. Nonclassic or "late-onset" P450c21 deficiency, however, often presents in adolescent and adult females with acne, hirsutism, and menstrual irregularity. A study of 220 adolescent and adult females examined the presenting clinical features in this disorder and demonstrated primary amenorrhea in 4% and oligomenorrhea in 54%.[73] Most patients with the nonclassic form will not be identified by the neonatal screening studies, which rely on detection of very high levels of 17-hydroxyprogesterone.[74]

TABLE 18.1 Genetic Causes of Oligomenorrhea

Clinical disorder	Historical features	Physical examination findings	Laboratory findings	Genetic testing
Turner's syndrome		At birth: low birth weight, lymphedema of the upper and lower extremities, webbed neck, low-set prominent ears, low posterior hairline, micrognatia, high-arched palate, epicanthal folds, hypoplastic nail beds, and/or hypoplastic fourth and fifth metacarpals At adolescence: short stature, amenorrhea, and variable lack of secondary sex characteristics. Renal anomalies, coarctation of the aorta, and hearing loss may be present	FSH↑, E2↓	Abnormal karyotype (45,X or mosaic)
X-chromosome deletions/translocations			FSH↑, E2↓	Karyotype
FMR1	Family history of mental retardation or fragile X tremor/ataxia syndrome		FSH↑, E2↓	FMR1 gene premutation
Bone morphogenetic protein 15			FSH↑, E2↓	BMP15
Blepharophimosis, ptosis, epicanthus inversus syndrome (BPES)		Eyelid malformations, short nasal bridge, ptosis	FSH↑, E2↓	FOXL2
Inhibin			FSH↑, E2↓	INH alpha, G769A
Galactosemia		Hepatomegaly, jaundice	FSH↑, E2↓	GALT
Folliculogenesis-specific basic helix–loop–helix		Primary ovarian insufficiency	FSH↑, E2↓	*FIGLA* gene
Proximal symphalangism		Ankylosis of the proximal interphalangeal joint, fusion of the carpal and tarsal bones, brachydactyly, conductive deafness	FSH↑, E2↓	NOG
Polycystic ovary syndrome	Oligomenorrhea hyperandrogenism, family or personal history of diabetes, impaired glucose tolerance, dyslipidemia	Hirsutism, acne, male pattern alopecia, obesity	FSH↔, E2 ↔/↓, androgens↑	
Nonclassic congenital adrenal hyperplasia	Mediterranean, Hispanic, Yuogoslavian, Ashkenazi Jewish ethnicity	Acne, hirsutism	FSH↔, E2↔, androgens↑, 17-hydroxyprogesterone↑	CYP21A2
Kallman's syndrome*	Anosmia	Synkinesis, high arched palate, renal agenesis	FSH↓, LH↓, E2↓	KAL1
FSH receptor mutation*			FSH↑, E2↓	FSHR

* Not covered in this chapter.

The nonclassic form of CAH is one of the most common autosomal recessive diseases, and the frequency is ethnicity specific. Among whites, the prevalence may be as high as one in 1,000 to one in 100, with the prevalence being even higher among Mediterraneans, Hispanics, and Eastern European Jews.[75]

P450c21 deficiency is an autosomal recessive disorder, and accordingly, both *CYP21A2* alleles must be affected to produce a clinical phenotype. Humans have two *CYP21A* genes, a nonfunctional pseudogene (*CYP21A1*) and the active gene (*CYP21A2*), both located in a 25-kb region of chromosome 6p21.3 within the major histocompatibility locus. The two *CYP21A* genes are more than 90% homologous, which facilitates recombination events during meiosis. Large or unequal crossover exchanges can result in a large deletion of the active gene leading to severe enzyme deficiency (0–5% of native enzyme activity); patients who are homozygous for such mutations present with the classic form of CAH. If smaller amounts of material are exchanged, hybrid *CYP21A1/CYP21A2* gene products can result with reduced but not absent enzyme activity (preserving about 20% of native enzyme activity), and can lead to the nonclassic form of CAH with either mild–mild combinations or mild–severe combinations. Women who are compound heterozygotes for two different *CYP21A2* gene mutations usually have the phenotype associated with the less severe of the two genetic defects.[76] Obligate heterozygote carriers (with one normal allele) may have mild biochemical abnormalities with no important clinical sequelae.[77] The diagnosis of 21-OH deficient nonclassic CAH is basal or an ACTH stimulated 17-hydroxyprogesterone level greater than 10 ng/mL. The diagnosis can be confirmed by genotyping, although it should be noted that not all mutations can be readily identified if they are new and unknown or rare and not included in screening assays.

CLINICAL AND LABORATORY EVALUATION

The differential diagnosis for both genetic and nongenetic causes of oligomenorrhea is extremely broad. Patients presenting with menstrual cycle abnormalities require a thorough evaluation with attention to the many potential causes. However, the history and physical can provide clues that a genetic etiology is involved, and a detailed family history for menstrual disorders is a critical component of the initial evaluation. A family history of metabolic disorders such as type 2 diabetes or dyslipidemia is also relevant, as this may be indicative of a predisposition to PCOS. Ethnic background is also important, and can help narrow the diagnostic considerations. For instance, nonclassic CAH is rarely found among African-American women.

Physical examination should include assessment for signs of hyperandrogenism including acne, hirsutism, temporal balding, or even frank virilization. A thorough evaluation of external and internal genitalia is particularly important in the evaluation of primary amenorrhea. A blind or absent vagina with breast development can indicate Müllerian agenesis or androgen insensitivity syndrome. An abdominal ultrasound may be useful to confirm the presence of a uterus.

Initial laboratory investigation should include tests to exclude pregnancy along with measurement of appropriate sex steroids, prolactin, TSH, and FSH concentrations. If FSH concentration is elevated, indicating gonadal failure, and the patient is less than 40 years of age, a karyotype should be done to identify chromosomal abnormalities. A screen for the fragile X premutation is also important, and is the only mutation that currently can be screened for in routine clinical practice. If serum FSH concentration is normal or low, the most likely cause is PCOS, nonclassic CAH, or hypothalamic amenorrhea. Details of appropriate testing are outlined in Table 18.1.

CONCLUSIONS

A variety of disorders with a genetic basis deserve consideration in the evaluation of oligomenorrhea. It is remarkable that diseases with defects that range from a single gene mutation to entire chromosome abnormalities to complex multigenic disorders can present with a common phenotype. We are clearly still in the early phases of understanding these disorders. As genetic techniques become more sophisticated and efficient, so that larger numbers of individuals can be included in research protocols, our understanding of the genetic basis of menstrual abnormalities is sure to expand and allow for more careful diagnosis, evaluation, and treatment of these disorders.

References

1. Hermann-Giddens ME, Slora EJ, Wasserman RC, et al. Secondary sexual characteristics and menses in young girls seen in office practice: a study from the Pediatric Research in Office Settings Network. *Pediatrics* 1997;**99**:505–12.
2. Coulam CB, et al. Incidence of premature ovarian failure. *Obstet Gynecol* 1986;**67**:604–6.
3. Anasti JN. Premature ovarian failure: an update. *Fertil Steril* 1998;**75**:438–9.
4. Davis CJ, Davison RM, Payne NN, Rodeck CH, Coneay GS. Female sex preponderance for idiopathic familial premature ovarian failure suggests an X chromosome defect: opinion. *Hum Reprod* 2000;**15**:2418–22.
5. Vegetti W, Marozzi A, Manfredini, Testa G, Alagna F, Nicolosi A, Caliari I, Taborelli M, Tibiletti MG, Dalpra L, et al. Premature ovarian failure. *Mol Cell Endocrinol* 2000;**161**:53–7.
6. Edson MA, Nagaraja AK, Matzuk MM. The mammalian ovary from genesis to revelation. *Endocr Rev* 2009;**30**(6):624–712.

7. Zinn AR. The X xhromosome and the ovary. *J Soc Gynecol Investig* 2001;**8**:S34–6.

8. Rappold GA. The pseudoautosomal regions of the human sex chromosomes. *Hum Genet* 1993;**92**:315–24.

9. Lahn BT, Page DC. Functional coherence of the human Y chromosome. *Science* 1997;**278**:675–80.

10. Reindollar RH, Byrd JR, McDonough PG. Delayed sexual development: a study of 252 patients. *Am J Obstet Gynecol* 1981;**140**(4): 372–80.

11. Goswami R, Goswami D, Kabra M, Gupta N, Dubey S, Dadwahl V. Prevalence of the triple X syndrome in phenotypically normal women with premature ovarian failure and its association with autoimmune thyroid disorders. *Fertil Steril* 2003;**80**:1052–4.

12. Goswami, Conway GS. Premature ovarian failure. *Hum Reprod Update* 2005;**11**:391–410.

13. Simpson JL, Rajkovic. Ovarian differentiation and gonadal failure. *Am J Genet* 1999;**89**:186–200.

14. Merry DE, Lesko JG, Sosnoski DM, Lewis RA, Lubinsky RA, Lubinsky M, Trask B, van den Engh G, Collins FS, Nussbaum RL. Choroideremia and deafness with stapes fixation: a contiguous gene deletion syndrome in Xq21. *Am J Hum Gene* 1989;**45**:530–40.

15. Therman E, Laxova R, Susman B. The critical region on the human Xq. *Hum Genet* 1990;**85**:455–61.

16. Tharapel AT, Anderson KP, Simposn JL, Martens PR, Wilroy Jr RS, Llerena Jr JC, Schwartz CE. Deletion (X)(q26.1 → q28) in a proband and her mother: molecular characterization and phenotypickaryotypic deductions. *Am J Hum Genet* 1993;**52**:463–71.

17. Powell CM, Taggart RT, Drumheller TC, Wangsa D, Qian C, Nelson LM, White BJ. Molecular and cytogenetic studies of an X-autosome translocation in a patient with premature ovarian failure and review of the literature. *Am J Med Genet* 1994;**52**:19–26.

18. Harris SE, Chand AL, Winship IM, Gersak K, Aittomaki K, Shelling AN. Identification of novel mutations in FOXL2 associated with premature ovarian failure. *Mol Hum Reprod* 2002;**8**:729–33.

19. Rousseau F, Rouillard P, Morel ML, Khandjian EW, Morgan K. Prevalence of carriers of premutation-size alleles of the FMRI gene – and implications for the population genetics of the fragile X syndrome. *Am J Hum Genet* 1995;**57**:1006–18.

20. Pesso R, Berkenstadt M, Cuckle H, et al. Screening for fragile X syndrome in women of reproductive age. *Prenat Diagn* 2000;**20**:611–4.

21. Allingham-Hawkins DJ, Babul-Hirji R, Chitayat D, Holden JJ, Yang KT, Lee C, Hudson R, Gorwill H, Nolin SL, Glickman A, et al. Fragile X premutation is a significant risk factor for premature ovarian failure: the International Collaborative POI in Fragile X study – preliminary data. *Am J Med Genet* 1999;**83**:322–5.

22. Murray A, Ennis S, MacSwiney F, Webb J, Morton NE. Reproductive and menstrual history of females with fragile X expansions. *Eur J Hum Genet* 2000;**8**:247–52.

23. Jin P, Zarnescu DC, Zhang F, et al. RNA-mediated neurodegeneration caused by the fragile X premutation rCGG repeats in *Drosophila*. *Neuron* 2003;**39**:739–47.

24. Ranum LP, Day JW. Dominantly inherited, non-coding microsatellite expansion disorders. *Curr Opin Genet Dev* 2002;**12**:266–71.

25. Dube JL, Wang P, Elvin J, Lyons KM, Celeste AJ, Matzuk MM. The bone morphogenetic protein 15 gene is X-linked and expressed in oocytes. *Mol Endocrinol* 1998;**12**:1809–17.

26. Di Pasquale E, Back-Peccoz P, Persani L. Hypergonadotropic ovarian failure associated with an inherited mutation of the human bone morphogenetic protein-15 (BMP15) gene. *Am J Hum Genet* 2004;**75**:106–11.

27. Amati P, Gasparini P, Zelante L, Chomel JC, Kitzis A, Kaplan J, Bonneau D. A gene for premature ovarian failure associated with eyelid malformation maps to chromosome 3q22-q23. *Am J Hum Genet* 1996;**58**:1089–92.

28. Crisponi L, Deiana M, Loi A, Chiappe F, Uda M, Amati P, Bisceglia L, Zelante L, Nagaraja R, Porcu S, Ristaldi MS, Marzella R, Rocchi M,

Nicolino M, Lienhardt-Roussie A, Nivelon A, Verloes A, Schlessinger D, Gasparini P, Bonneau D, Cao A, Pilia G. The putative forkhead transcription factor FOXL2 is mutated in blepharophimosis/ ptosis/epicanthus inversus syndrome. *Nat Genet* 2001;**27**(2): 159–66.

29. Cocquet J, Da Baere E, Gareil M, Pannetier M, Xia X, Fellous M, Veitia RS. Structure, evolution, and expression of the FOXL2. *J Med Genet* 2003;**39**:916–21.

30. Laissue P, Vinci G, Veitia RA, Fellous M. Recent advances in the study of genes involved in non-syndromic premature ovarian failure. *Mol Cell Endocrinol* 2008;**282**:101–11.

31. Petraglia F, Hartmann B, Luisis S, Florio P, Kichengast S, Santuz M, Genazzani AD, Genazzani AR. Low levels of serum inhibin A and inhibin B in women with hypergonadotropic amenorrhea and evidence of high levels of activin A in women with hypothalamic amenorrhea. *Fertil Steril* 1998;**70**:907–12.

32. Kim YI, Shim J, Kim BH, Lee SJ, Lee HK, Cho C, Cho BN. Transcriptional silencing of the inhibin-α gene in human gastric carcinoma cells. *Int J Oncol* 2012;**41**(2):690–700.

33. Dixit H, Deendayal M, Singh L. Mutational analysis of the mature eptide region of inhibin genes in Indian women with ovarian failure. *Hum Reprod* 2004;**19**:1760–4.

34. Jeong HJ, Cho SW, Kim HA, Lee SH, Cho JH, Choi DH, Kwon H, Cha WT, Han JE, Cha KY. G769A variation of inhibin alpha gene in Korean women with premature ovarian failure. *Yonsei Med J* 2004;**45**:479–82.

35. Laml T, Preyer O, Umek W, Hengstschlager M, Hanazal H. Genetic disorders in premature ovarian failure. *Hum Reprod Update* 2002;**8**:483–91.

36. Waggoner DD, Buist NR, Donnell GN. Long-term prognosis in galactosemia, results of a survey of 350 cases. *J Inherit Metab Dis* 1990;**13**:802–18.

37. Hu W, Gauthier L, Baibakov B, Jimenez-Movilla M, Jurrien D. FIGLA, a basic helix–loop–helix transcription factor, balances sexually dimorphic gene expression in postnatal oocytes. *Mol Cell Biol* 2010;**30**(14):3661–71.

38. Zhao H, Chen Z-J, Qin Y, Shi Y, Wang S, Choi Y, Simpson JL, Rajkovic A. Transcription factor FIGLA is mutated in patients with premature ovarian failure. *Am J Hum Genet* 2008;**82**: 1342–8.

39. Gong Y, Krakow D, Marcelino J, Wilkin D, Chitayat D, Babul-Hirji R, Hodgkins L, Cremers CW, Cremers FP, Brunner HG, et al. Heterogeneous mutations in the gene encoding noggin affect human joint morphogenesis. *Nat Genet* 1999;**21**:302–4.

40. Shimasaki S, Moore RE, Erikson GF, Otsuka F. The role of bone morphogenetic proteins in ovarian function. *Reprod Suppl* 2003;**61**: 323–37.

41. Kosaki K, Sato S, Hasegawa T, Matuso N, Suzuki T, Ogata T. Premature ovarian failure in a female with proximal symphalangism and Noggin mutation. *Fertil Steril* 2004;**81**:1137–9.

42. Holst F, Stahl PR, Ruiz C, Hellwinkel O, Jehan Z, Wendland M, Lebeau A, Terracciano L, Al-Kuraya K, Janicke F, Sauter G, Simon R. Estrogen receptor alpha (ESR1) gene amplification is frequent in breast cancer. *Nature Genet* 2007;**39**:655–60.

43. Khosla S, Riggs BL, Atkinson EJ, Oberg AL, Mavilia C, Del Monte F, Melton III LJ, Brandi ML. Relationship of estrogen receptor genotypes to bone mineral density and to rates of bone loss in men. *J Clin Endocr Metab* 2004;**89**:1808–16.

44. Weel AEAM, Uitterlinden AG, Westendorp ICD, Burger H, Schuit SCE, Hofman A, Helmerhorst TJM, van Leeuwen JPTM, Pols HAP. Estrogen receptor polymorphism predicts the onset of natural and surgical menopause. *J Clin Endocr Metab* 1999;**84**: 3146–50.

45. Penrose LS. The relative effects of paternal and maternal age in mongolism. *J Genet* 1933;**27**:219.

46. Roizen NJ, Patterson D. Down's syndrome. *Lancet* 2003;**361**:1281–9.

47. Pennington BF, Moon J, Edgin J, Stedron J, Nadel L. The neuropsychology of Down syndrome: evidence for hippocampal dysfunction. *Child Dev* 2003;**74**:75–93.

48. McCormick MK, Schinzel A, Petersen MB, Stetten G, Driscoll DJ, Cantu ES, Tranebjaerg L, Mikkelsen M, Watkins PC, Antonarakis SE. Molecular genetic approach to the characterization of the "Down syndrome region" of chromosome 21. *Genomics* 1989;**5**: 325–31.

49. Sinet PM, Theophile D, Rahmani Z, Chettouh Z, Blouin JL, Prieur M, Noel B, Delabar JM. Mapping of the Down syndrome phenotype on chromosome 21 at the molecular level. *Biomed Pharmacother* 1994;**48**:247–52.

50. Lyle R, Bena F, Gagos S, Gehrig C, Lopez G, Schinzel A, Lespinasse J, Bottani A, Dahoun S, Taine L, et al. Genotype-phenotype correlations in Down syndrome identified by array CGH in 30 cases of partial trisomy and partial monosomy chromosome 21. *Eur J Hum Genet* 2009;**17**:454–66.

51. Casarini L, Brigante G. The polycystic ovary syndrome evolutionary paradox: a GWAS-based, *in silico*, evolutionary explanation. *J Clin Endocrinol Metab* 2014;**99**:E2220–412.

52. McAllister JM, Modi B, Miller BA, Biegler J, Bruggeman R, Legro RS, Strauss III JF. Overexpression of a DENND1A isoform produces a polycystic ovary syndrome theca phenotype. *Proc Natl Acad Sci USA* 2014;**111**(15):E1519–27.

53. Wang P, Zhao H, Li T, Zhang W, Wu K, Li M, Bian Y, Liu H, Ning Y, Li G, Ch ZJ. Hypomethylation of the LH/choriogonadotropin receptor promoter region is a potential mechanism underlying susceptibility to polycystic ovary syndrome. *Endocrinology* 2014;**155**(4): 1445–52.

54. Rotterdam ESHRE/ASRM-Sponsored PCOS Consensus Workshop Group. Revised 2003 consensus on diagnostic criteria and long-term health risks related to polycystic ovary syndrome (PCOS). *Hum Reprod* 2004;**19**:41–7.

55. Ehrmann DA. Polycystic ovary syndrome. *New Eng J Med* 2005;**352**: 1223–36.

56. Nelson VL, Qin KN, Rosenfield RL, et al. The biochemical basis for increased testosterone production in theca cells propagated from patients with polycystic ovary syndrome. *J Clin Endocrinol Metab* 2001;**86**:5925–33.

57. Chang JR. The reproductive phenotype in polycystic ovary syndrome. *Nat Clin Pract Endocrin Metab* 2007;**3**(10):688–95.

58. Azziz R, Kashar-Miller MD. Family history as a risk factor for the polycystic ovary syndrome. *J Pediatr Endocrinol Metab* 2000; **13**(Suppl. 5):1303–6.

59. Vink JM, Sadrzadeh S, Lambalk CB, Boonsma DI. Heritability of polycystic ovary syndrome in a Dutch twin-family study. *J Lin Endocrinol Metab* 2006;**91**:2100–4.

60. Fox R. Prevalence of a positive family history of type 2 diabetes in women with polycystic ovarian disease. *Gynecol Endocrinol* 1999;**13**:390–3.

61. Sam S, et al. Evidence for metabolic and reproductive phenotypes in mothers of women with polycystic ovary syndrome. *Proc Natl Acad Sci USA* 2006;**103**:7030–5.

62. Urbanek M. The genetics of the polycystic ovary syndrome. *Nat Clin Pract Endocrinol Metab* 2007;**3**(2):103–11.

63. Draper N, Walker EA, Bujalska IJ, Tomlinson JW, Chalder SM, Arlt W, Lavery GG, Bedendo O, Ray DW, Laing I, Malunowicz E, White PC, Hewison M, Mason PJ, Connell JM, Shackelton CH, Stewart PM. Mutations in the genes encoding 11beta-hydroxysteroid dehydrogenase type 1 and hexose-6-phosphate dehydrogenase interact to cause cortisone reductase deficiency. *Nat Genet* 2003;**34**(4):434–9.

64. Gharani N, et al. Association of the steroid synthesis gene CYP11a with polycystic ovary syndrome and hyperandrogenism. *Hum Mol Genet* 1997;**6**:397–402.

65. Diamanti-Kandarakis E, et al. Microsatellite polymorphism (ttta) (n) at-528 base pairs of gene CYP11a influences hyperandrogenemia in patients with polycytic ovary syndrome. *Fertil Steril* 2000;**73**: 735–41.

66. Gasenbeek M, Powell BL, Sovio U, Haddad L, Gharani N, Bennett A, Groves CJ, Rush K, Goh MJ, Conway GS, et al. Large-scale analysis of the relationship between CYP11A promoter variation, polycytic ovarian syndrome, and serum testosterone. *J Clin Endocrinol Metab* 2004;**89**:2408–13.

67. Stewart DR, Dombroski BA, Urbanek M, Ankener W, Ewns KG, Wood JR, Legro RS, Strauss III JF, Dunaif A, Spielman RS. Fine mapping of genetic susceptibility to polycystic ovary syndrome on chromosome 19p13.2 and tests for regulatory activity. *J Clin Endocrinol Metab* 2006;**91**:4112–7.

68. Urbanek M, Sam S, Legro RS, Dunaif A. Identification of a polycystic ovary syndrome susceptibility variant in fibrillin-3 and association with a metabolic phenotype. *J Clin Endocrin Metab* 2007;**92**(11):4191–8.

69. Tut TG, et al. Long polyglutamine tracts in the androgen receptor are associated with reduced transactivation, impaired sperm production, and male infertility. *J Clin Endocrinol Metab* 1997;**82**: 3777–82.

70. Jaaskelainen J, et al. Androgen receptor gene CAG length polymorprphism in women with polycystic ovary syndrome. *Fertil Steril* 2005;**83**:1724–8.

71. Mifsud A, et al. Androgen receptor gene CAG trinucleotide repeats in anovulatory infertility and polycystic ovaries. *J Clin Endocrinol Metab* 2000;**8**:3483–8.

72. Mohlig M, et al. The androgen receptor CAG repeat modifies the impact of testosterone in insulin resistance in women with polycystic ovary syndrome. *Eur J Endocrinol* 2006;**155**:127–30.

73. Azziz R, Sanchez LA, Knochenhauer ES, Moran C, Lazenby J, et al. Androgen excess in women: experience with over 1000 consecutive patients. *J Clin Endocrinol Metab* 2004;**89**:453.

74. Votava F, Torok D, Kovacs J, et al. Estimation of the false-negative rate in newborn screening for congenital adrenal hyperplasia. *Eur J Endocrinol* 2005;**152**:869.

75. White PC, Speiser PW. Congenital adrenal hyperplasia due to 21-hydroxylase deficiency. *Endocr Rev* 2001;**21**:245.

76. Lajic S, Clauin S, Robins T, et al. Novel mutations ion CYP21 detected individuals with hyperandrogenism. *J Clin Endocrinol Metab* 2002;**85**:4562.

77. Charmandari E, Merke DP, Negro PJ, et al. Endocrinologic and psychologic evaluation of 21-hydroxylase deficiency carriers and matched normal subjects; evidence for physical and/or psychologic vulnerability to stress. *J Clin Endocrinol Metab* 2004;**89**:2228.

19

Disorders of Sex Development

Valerie Arboleda*, Eric Vilain**

*Department of Pathology and Human Genetics, David Geffen School of Medicine,
University of California, Los Angeles, CA, USA
**Department of Human Genetics, Pediatrics, and Medical Genetics, David Geffen School of Medicine,
University of California, Los Angeles, CA, USA

DISORDERS OF SEX DEVELOPMENT

"Is it a boy or a girl?" is probably the most frequently asked question to a new mother. To this apparently simple question, the answers can become rather complicated. First, one could try to define precisely what "boy or girl" entails. In other words, how do we define sex? Although common sense would dictate that the appearance of the external genitalia should define what sex really is, biological complexity suggests otherwise. Many biological parameters are crucial to precisely delineate the sex of an individual: chromosomal constitution (46,XX or 46,XY), sex determining genes (presence or absence of Sry), gonadal histology (testis or ovary), hormonal output (testosterone or estradiol), sex of internal reproductive organs (uterus, fallopian tubes, or prostate, epipidymis, vas deferens). Each of these parameters can be disrupted in disorders of sex development (DSD), defined as "congenital conditions in which development of chromosomal, gonadal or anatomical sex is atypical."[1]

DSD are typically categorized in disorders of sex determination and disorders of sex differentiation (see Fig. 19.1). In the former, the development of the gonads is disrupted, and leads to either gonadal dysgenesis (GD), or the development of ovotestis. In the latter, the development of the gonads is normal, but the fetus develops either more virilized than typical for a 46,XX individual (in the case of congenital adrenal hyperplasia (CAH) for instance, see below), or more feminized than a typical 46,XY individual (e.g., in the case of androgen insensitivity syndrome (AIS)). The goal of this chapter is to help the physician navigate the complexities of atypical sex development and identify rapid tools to diagnose patients with DSD.

DISORDERS OF SEX DETERMINATION

46,XY Disorders

Gonadal Dysgenesis

In 46,XY GD, testes fail to undergo normal development. 46,XY GD can be isolated or syndromic (see Table 19.1), and can be pure, partial, or mixed (Table 19.2).

Pure 46,XY GD is characterized by intra-abdominal bilateral fibrous streak gonads that fail to produce anti-Müllerian hormone (AMH) and testosterone, resulting in an unambiguous female phenotype (formerly Swyer syndrome), with an occasionally hypoplastic but typically well-formed uterus and fallopian tubes, and female external genitalia. Partial 46,XY GD entails varying amounts of testicular dysgenesis and ambiguous genitalia. Approximately 15% of pure and partial 46,XY GD is attributable to mutations in SRY,[2–5] a transcription factor that targets another male-determining gene, SOX9.[6] Over 50 mutations of the SRY open reading frame (ORF) have been identified in 46,XY GD, and deletions of Yp, a region of the Y chromosome containing SRY, have also been implicated.[3,4,7] Most known mutations in SRY disrupt the high mobility group (HMG) box, which results in reduced nuclear import of SRY protein[8] or impaired binding or bending of target gene DNA by SRY.[5,9] Another 15% of pure and partial 46,XY GD is due to mutations in SF1 (NR5A1), an orphan nuclear receptor required for testis and adrenal development.[10] SF-1 interacts with transcription factors GATA-4 and FOG-2 to regulate SRY expression in developing testes.[11,12] Human mutations in GATA4[13] and FOG2/ZFPM2[14] have recently been described in rare cases of 46,XY DSD. Most recently, MAP3K1 has been identified in a subset of patients with isolated 46,XY GD[15,16] and knockdown of other

Genetic Diagnosis of Endocrine Disorders. http://dx.doi.org/10.1016/B978-0-12-800892-8.00019-1

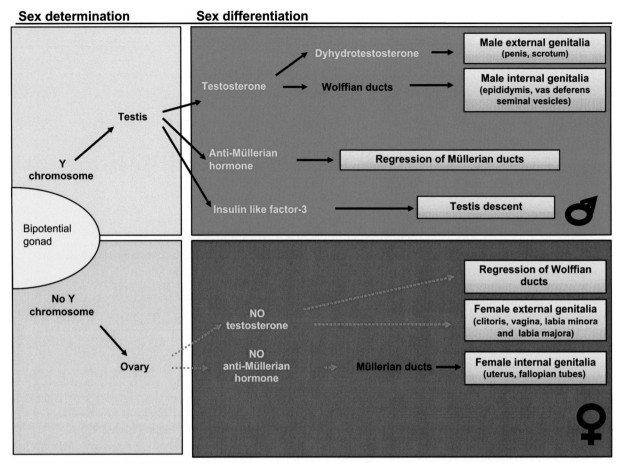

FIGURE 19.1 **Sex development.** Sex determination is the process by which genetic sex determines gonadal sex, the testis, or ovary. Sex differentiation follows and is the process by which the now determined gonads and their respective hormones direct the differentiation of the internal and external genitalia.

Map-kinase factor mouse models have been shown to act upstream of *Sry* to promote its upregulation (Warr, 2012[17]; Bogani, 2009[18]). 46,XY GD due to SF1 haploinsufficiency can be isolated[10] or syndromic (Table 19.1). Very rarely, pure and partial 46,XY GD result from duplications of putative "antitestis" genes (*Rspondin1*, *WNT4*, *DAX1*), or mutations in genes necessary for testes organogenesis (*XH2, SOX9, WT1, DHH*)[19–21] (Table 19.1). Together, these genetic factors account for only about 40% of pure and partial 46,XY GD, indicating that additional, as yet unknown, genes are required for human testis determination.

Mixed 46,XY GD refers to asymmetric gonadal development resulting in asymmetric virilization of the external and/or internal genitalia and unilateral cryptorchidism. One gonad may be an abdominally located, fully dysgenic streak with no ipsilateral virilization. The contralateral testis appears normal to partially dysgenic, and its level of activity determines its location and the extent of virilization on that side. Typically, the etiology of mixed 46,XY GD is 45,X/46,XY mosaicism,[22] causing a range of asymmetric patterns of male-determining gene expression and

consequent phenotype. The proportion of Y chromosomal material in the gonad appears to correlate with the amount of testicular tissue and of phenotypic maleness.[23,24]

Presentation and Diagnosis

Pure 46,XY GD should be considered when an adolescent presents as a phenotypic female with delayed puberty and primary amenorrhea. Pure 46,XY GD patients may be of normal to tall stature and have normal to small Müllerian structures, bilateral streak gonads, and no Turner stigmata. Rarely, they may present with a detectable abdominal or pelvic gonadoblastomal mass. Patients with partial or mixed 46,XY GD typically present much earlier than those with pure 46,XY GD – as infants or in early childhood – with ambiguity of internal and/or external genitalia. Patients with 45,X/46/46,XY mosaicism, in particular, can present with a wide range of phenotypes.[25]

When the presumptive diagnosis suggests GD, and there are no indications of syndromic involvement (Table 19.1), further criteria may strengthen the diagnosis of isolated 46,XY GD (Table 19.2 and Fig. 19.2a). The karyotype of peripheral leukocytes will show 46,XY

TABLE 19.1 Syndromes Associated with Ambiguous Genitalia

	Syndromes associated with ambiguous genitalia	Phenotype	Gene	Locus	OMIM #
Syndromes associated with GD	Gonadal dysgenesis with adrenal hypoplasia	XY GD with adrenal hypoplasia. Gonadal dysgenesis due to SF1 mutations can also be isolated	SF1/NR5A1	9q33	184757
	Denys–Drash	XY GD, early diffuse mesangial sclerosis of kidneys, Wilms' tumor	WT-1	11p13	194080
	Frasier	XY GD, early adolescent development of focal segmental glomerulosclerosis, gonadoblastoma	WT-1	11p13	136680
	Campomelic dysplasia	Ambiguous genitalia, congenital bowing of long bones, hypoplastic scapulae, and thoracic vertebrae pedicles	SOX9	17q24.3-q25.1	114290
	GD with neuropathy	XY GD with associated minifascicular neuropathy (one of four cases)	DHH	12q12-13.1	607080
	X-linked alpha-thalassemia/mental retardation	Hemoglobin H disease, mental retardation, dysmorphic facies, genital abnormalities	XH2	Xq13.3	301040
	Palmoplantar hyperkeratosis with squamous cell carcinoma and XX sex reversal	Variable degrees of XX sex reversal, palmoplantar keratoderma, squamous cell carcinoma, congenital bilateral corneal opacities, onychodystrophy, and hearing impairment	RSPO1	1p34.3	610644
	Blepharophimosis-ptosis-epicanthus inversus syndrome type I	Blepharophimosis, ptosis, and epicanthus inversus syndrome, either with or without POF	FOXL2	3q23	110100
Syndromes associated with small penis and/or cryptorchidism	VACTERL/VATER	Vertebral defects, anal atresia, tracheoesophageal fistula with esophageal atresia, and radial dysplasia, limb anomalies	Unknown	Unknown	192350
	Goldenhar syndrome (hemifacial microsomia)	Unilateral deformity of the external ear and small ipsilateral half of the face with epibulbar dermoid and vertebral anomalies	Unknown	14q32	164210
	Smith–Lemli–Opitz syndrome (SLOS)	Multiple congenital malformation and mental retardation syndrome	DHCR7	11q12-q13	270400
	Pallister–Hall syndrome (PHS)	Hypothalamic hamartoblastoma, postaxial polydactyly, and imperforate anus	GLI3	7p13	146510
	Robinow syndrome	Mesomelic limb shortening associated with facial and genital abnormalities	ROR2	9q22	180700
	Prader–Willi syndrome	Obesity, muscular hypotonia, mental retardation, short stature, hypogonadotropic hypogonadism	SNRPN	15q12, 15q11-q13	17620
	Kallmann syndrome	Hypogonadotropic hypogonadism and anosmia	FGFR-1	8p11.2-p11.1	147950
	Holoprosencephaly	Craniofacial dysmorphology	Many	21q22.3, 2q37.1-q37.3	236100
	Malpeuch facial clefting syndrome	Short stature, hypertelorism, eye anomalies, facial clefting, hearing loss, urogenital abnormalities, mental retardation	Unknown	Unknown	248340
	Najjar syndrome	Genital anomaly, mental retardation, and cardiomyopathy	Unknown	Unknown	212120
	Varadi–Papp syndrome	Big toes, hexadactyly, cleft lip/palate or lingual nodule, and somatic and psychomotor retardation. Some showed absent olfactory bulbs and tracts, cryptorchidism	Unknown	Unknown	277170

(Continued)

TABLE 19.1 Syndromes Associated with Ambiguous Genitalia (cont.)

	Syndromes associated with ambiguous genitalia	Phenotype	Gene	Locus	OMIM #
	Juberg–Marsidi syndrome	Cleft lip/palate with abnormal thumbs and microcephaly	Unknown	Unknown	216100
	Johanson–Blizzard syndrome	Aplasia or hypoplasia of the nasal alae, congenital deafness, hypothyroidism, postnatal growth retardation, malabsorption, mental retardation, midline ectodermal scalp defects, and absent permanent teeth	Unknown	15q15-q21.1	243800
	Borjeson–Forssman–Lehmann syndrome	Severe mental defect, epilepsy, hypogonadism, hypometabolism, marked obesity	Unknown	Xq26.3	301900
	Torticollis, keloids, cryptorchidism, renal dysplasia	Torticollis, keloids, cryptorchidism, renal dysplasia	Unknown	Xq28	314300
	Hypertelorism with esophageal abnormality and hypospadias	Laryngotracheoesophageal cleft; clefts of lip, palate, and uvula; swallowing difficulty and hoarse cry; genitourinary defects; mental retardation; congenital heart defects	Unknown	22q11.2	145410
	Faciogenitopopliteal syndrome	Cleft palate and webbing intercrural pterygium	Unknown	1q32-q41	119500
	Dubowitz syndrome	Short stature, microcephaly, mild mental retardation with behavior problems, eczema, and distinctive facies	Unknown	Unknown	223370
	Noonan syndrome	Hypertelorism, a downward eye slant, and low-set posteriorly rotated ears, short stature, webbed neck, cardiac anomalies	PTPN11	12q24.1	163950
	Aarskog syndrome (faciogenital dysplasia)	Embryonic ocular hypertelorism, anteverted nostrils, broad upper lip, and peculiar penoscrotal relation	FGD1	Xp11.21	305400
	Cornelia de Lange syndrome	Low anterior hairline, anteverted nares, maxillary prognathism, long philtrum, "carp" mouth) in association with prenatal and postnatal growth retardation, mental retardation	NIPBL	5p13.1	122470
	Rubinstein–Taybi syndrome	Mental retardation, broad thumbs and toes, and facial abnormalities	CREBBP	16p13.3, 22q13	180849
	Seckel syndrome	Growth retardation, microcephaly with mental retardation, and a characteristic "bird-headed" facial appearance	SCKL1	3q22-q24	210600
	Miller–Dieker syndrome	Microcephaly and a thickened cortex with 4/6 layers	LIS1	17p13.3	247200
	Lenz–Majewski hyperostosis syndrome	High palate, short, yellow, carious teeth, progeroid appearance, short, increased venous pattern of the forehead and thorax	Unknown	Unknown	151050
	Lowe syndrome	Ophthalmic, cataract, mental retardation, vitamin D-resistant rickets, amino aciduria	OCRL1	Xq26.1	309000
Syndromes associated with Müllerian malformation	MURCs	Müllerian duct aplasia, renal aplasia, and cervicothoracic somite dysplasia	Unknown	Unknown	601076
	Mayer–Rokitansky–Kuster–Hauser syndrome (MRKH)	Müllerian duct aplasia	WNT4 in subset with hyperandrogenism	Unknown; 1p35	277000
	McKusick–Kaufman syndrome (MKKS)	Hydrometrocolpos, congenital heart malformations, postaxial polydactyly	Unknown	20p12	236700

TABLE 19.2 Diagnostic Criteria for DSD

DSD		Biochemical changes	Differentiating features	Genetic diagnostic criteria
DISORDERS OF SEX DETERMINATION				
46,XY disorders	46,XY pure/partial/mixed GD	Pure/partial/mixed: (+) FSH, LH Nml to (−) AMH No (+) with hCG Pure: (−−) T, DHT, E2 (−−) AMH Partial/Mixed: (−) T, DHT, E2	Presence of partial testicular function (T, AMH) points towards partial or mixed GD. However, histological examination of testes after prophylactic gonadectomy can differentiate between pure, partial, and mixed GD. There is little genotype–phenotype correlation, but the presence of mosaicism 45,X/46,XY is often associated with mixed GD	Sequencing for *SF1* and *SRY* mutations; if no mutation is found, CGH should be performed to look for copy number variants as a source for XY GD. If partial/mixed GD suspected, patient should be tested for mosaicism
46,XX disorders	XX testicular/ovotesticular DSD	(+) FSH, LH (−) T, DHT Nml AMH No (+) with hCG	Testicular and ovotesticular DSD are at different ends of a phenotypic spectrum. The only way to definitively differentiate between 46,XX testicular and ovotesticular DSD is through complete gonadal histological examination looking for the presence of ovarian tissue	Testicular DSD: FISH for presence of *SRY* or *SOX9*. If clinical phenotype appropriate, sequence *RSPO1* gene. If no molecular diagnostic is successful, CGH should be performed. Ovotesticular DSD: Search for XX/XY mosaicism and sequencing for *SRY* mutation
	XX GD	(+) FSH, LH No (+) with hCG No AMH	Full female phenotype with amenorrhea and lack of secondary sex development	For cases of isolated 46,XX GD, sequencing for FSH mutations can be performed. If clinical presentation matches, sequence *FOXL2* mutation. CGH can also be done to identify causative duplications or mutations
DISORDERS OF SEX DIFFERENTIATION				
Disorders of androgen synthesis and action	StAR deficiency (*StAR*); P450scc deficiency (*CYP11A1*)	(+) Renin (−) Aldo (+) K (−) Na (−) All adrenal hormones (−) 17OH-P	Presence of lipid vacuoles in adrenals on histology. P450scc deficiency does present with enlarged adrenals. No HTN and hyperkalemia differentiates from CYP 17 deficiency	Presence of lipid filled vacuoles on histology; sequencing of the *StaR* or *CYP11A1* genes and presence of a mutation in either one gives definitive diagnosis
	3βHSD type II (*HSD3B2*)	(+) Renin; (−) Aldo, F (+) K, (−) Na (+) Ratio Δ^5-17-pregnenolone: F (−) A, T	Baseline and ACTH-stimulated ratios of Δ^5-17-pregnenolone:cortisol consistently distinguished between patients affected and nonaffected patients	Sequencing of the *HSD3B2* gene
	17α-Hydroxylase/17,20-lyase (*CYP17*)	(−) Renin (−) Aldo, F (−)17OH-P (+) LH, FSH (+) Progesterone, DOC, B (−) Na (−) K (−) DHEA-S, A, T; No response to hCG stim	Hypertension and hypokalemic alkalosis in the presence of low 17-OH progesterone	Sequencing of the *CYP17* gene
	POR	(+) 17OH-P (+) Progesterone (−) F, DHEA-S (−) A, T	Hypertension in the presence of elevated 17OH-P. Occasionally presence of Antley–Bixler skeletal malformations	Sequencing of *CYP450 oxidoreductase*

(Continued)

TABLE 19.2 Diagnostic Criteria for DSD (cont.)

DSD	Biochemical changes	Differentiating features	Genetic diagnostic criteria
17βHSD type 3 (HSD17B3)	Nml to ↑ A; (+) Ratio A/T (>15); (−) T, DHT	Differentiate from 5α-reductase type 2 by levels and ratios of serum androgens	Sequencing of the HSD17B3 gene for deletions, insertions, and point mutations
LCH	(+) LH Nml FSH; (+) AMH; (−) T, DHT, E_2; (−) hCG response; Nml A/T ratio	To differentiate LCH from GD, AMH is used as a marker of testicular function	Sequencing of the LHCHR gene for deletions, insertions, and point mutations
5α-Reductase type 2	Nml FSH, LH; Nml T, E_2; (−) DHT; (+) Ratio T/DHT (>30)	Development of male secondary sex characteristics in puberty with fine and sparse facial hair. Unlike HSD17B3 and AIS, no gynecomastia during puberty	Sequencing of the SRD5A2 gene
AR	Nml FSH, LH (PAIS); (−) FSH, LH (CAIS); Nml to (+) AMH; Nml A, T, DHT; (+ +) hCG response	Female phenotype with breast development at puberty, with sparse pubic and axilla hair	Sequencing of AR gene looking for single amino acid substitutions, which account for 90% of reported cases
AMH/AMHR	Nml hormonal profile	Presence of both Müllerian and Wolffian derivatives usually discovered incidentally	
Disorders of androgen excess	21α-Hydroxylase (CYP21) — (+) Renin (−) Aldo, DOC, F; (+) K; (−) Na; (+) 17OH-P; (+) DHEA-S, A, T	17-OH progesterone elevated in the absence of hypertension, which is typical in HSD11B1 deficiency	Sequencing of CYP21 gene gives definitive diagnosis and can inform parents about genetic counseling in future pregnancies
	11βHSD1 (HSD11B1) — (−) Aldo, F, Renin; (+) DOC; (+) K, (−) Na; (+) 17OH-P; (+) DHEA-S, A, T	Differentiate from CYP21 deficiency by presences of hypertension and hypokalemic alkalosis	Sequencing of the HSD11B1 gene
	P450 aromatase (CYP19) — (+) 16OH-A (maternal); (+) FSH, LH, A, T; (−) Estrone, E_2	The presence of maternal virilization during pregnancy and XX virilization, which stops after delivery	Sequencing of CYP19 gives definitive diagnosis

FSH, follicle stimulating hormone; Aldo, aldosterone; 17OH-P, 17OH-progesterone; DOC, deoxycorticosterone; A, androsteindione; LH, lutenizing hormone; T, testosterone; hCG, human chorionic gonadotropin; AMH, anti-Müllerian hormone; DHT, dihydrotestosterone; F, cortisol; Nml, normal; DHEA, dehydroepiandrosterone.

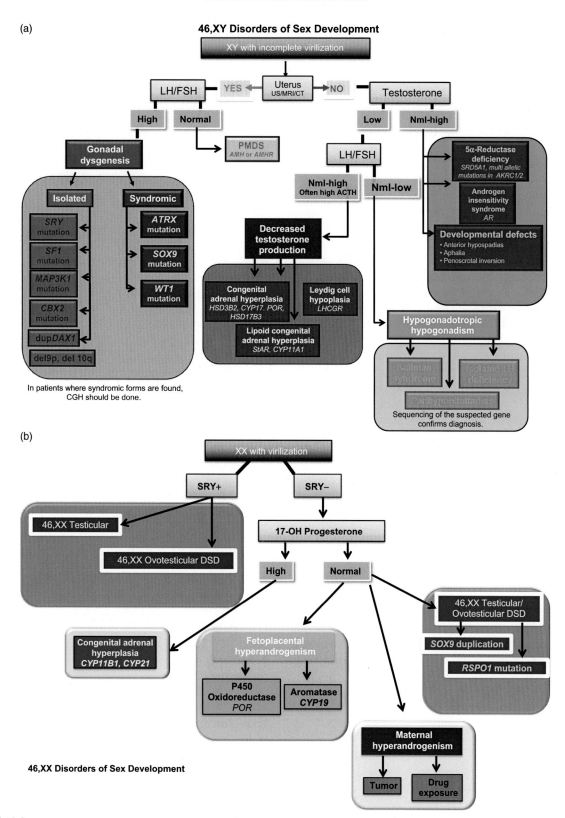

FIGURE 19.2 Characterization of the subsets of patients within disorders of sex development based on functional data (above) and genetics (below). 46,XY (a) and 46,XX (b). *Adapted from Fleming et al.*[30]

for pure and partial GD, and, frequently, 45,X/46,XY mosaicism for mixed GD. Pelvic ultrasound may show bilateral atrophic gonads with a normal to hypoplastic uterus in the case of pure to partial 46,XY GD, but a unilateral streak gonad with asymmetric Müllerian and Wolffian structures in mixed GD. Diagnostic biochemical criteria are listed in Table 19.2, and include elevated follicle stimulating hormone (FSH) and lutenizing hormone (LH). Fluorescent *in situ* hybridization (FISH) can be used to detect *SRY* or Yp, and sequencing the ORF is typically performed in all patients with 46,XY GD. However, given the advent of exome sequencing for clinical genetic diagnosis and the ever-increasing number of genes associated with 46,XY GD, we believe that single gene sequencing is no longer efficient or cost effective (discussed in the Section "High Throughput Sequencing in the Diagnosis of Disorders of Sex Determination"). In addition, whole-exome sequencing can also detect deletions and duplications at the level of exons, although the gold standard for larger deletions/duplications remains comparative genomic hybridization (CGH). CGH can alternatively be used to detect relatively small deletions or duplications in or around any of the known sex-determining genes. As such, noncoding duplications and deletions have also been identified surrounding *SOX9*,[26,27] *GATA-4*,[28] and *SOX3*.[29] Therefore, in patients with 46,XY GD we recommend whole-exome sequencing to identify the causative genetic diagnosis for patients with DSD.

Ultimately, the definitive diagnosis to determine whether one has complete, partial, or mixed GD is through biopsy of the gonads. While this is a useful endeavor, it may not provide significant insight into the biological or genetic basis of disease or any predictive value outside of fertility and gonadoblastoma risk. Bilateral streak gonads are seen in pure GD, bilateral dysgenic testes in partial GD, and a unilateral streak gonad with contralateral normal to dysgenic testis in mixed 46,XY GD. The risk of otherwise undetected gonadoblastoma in 46,XY GD patients is high, and prophylactic or therapeutic gonadectomy is therefore often indicated when 46,XY GD is diagnosed.

46,XX Disorders

Testicular and Ovotesticular DSD

There is a range of 46,XX individuals in which the gonads are testicular or ovotesticular. Patients with 46,XX testicular or ovotesticular DSD are likely to be part of the same phenotypic spectrum. Ultimately, differentiating the two can only be done with gonadal biopsy, which is not always performed; therefore, an exact diagnosis cannot always be made.

In 46,XX testicular DSD, with an approximate incidence of 1:20,000, varying amounts of testicular (but

no ovarian) tissue are present. About 85% of 46,XX testicular DSD patients are phenotypic males with unambiguous male genitalia at birth, and are not diagnosed until puberty fails to proceed normally.[31,32] There can be similarities with Klinefelter (46,XXY), including diminished secondary sex development, gynecomastia, small testes, and azoospermia.[33] However, in contrast, cognitive ability is normal, stature is normal to short, and ejaculation and sexual function are normal. Approximately 85% of these patients carry an Xp:Yp translocation that contains *SRY*. The remaining 15% of 46,XX testicular DSD individuals have ambiguous genitalia, and ectopic presence of SRY accounts for only a minority of these cases.[32,34] In addition, disruption of R-spondin1 (*RSPO1*), a female-determining gene, results in a recessive syndrome that includes complete 46,XX sex reversal[35] (Table 19.1), but the etiology of many 46,XX testicular DSD patients remains unclear.

In 46,XX ovotesticular DSD (formerly true hermaphroditism), both ovarian and testicular tissues (defined by the presence of follicles and seminiferous tubules, respectively) are present. Approximately 50% of cases have an ovary and an ovotestis; 30% have two ovotestes; and 20% have an ovary and a testis,[36] resulting in variable development of internal and external genitalia. The ipsilateral predominance of ovarian or testicular tissue typically correlates with the location of each gonad and the degree of Müllerian- and/or Wolffian-derived development on that side. About 7–10% of ovotesticular DSD are actually 46,XY; mutations in SRY have been found in two such patients.[4,36,37] An estimated 30–33% of ovotesticular DSD are mosaics – including 46,XX/46,XY – with some Yp involvement.[36,37] The majority – approximately 60% – of ovotesticular DSD are 46,XX. Of these, a minority are SRY-positive,[4,38–41] placing these patients on a phenotypic continuum with 46,XX testicular DSD. As with some 46,XX testicular DSD patients, the etiology can be an Xp:Yp translocation.[33,34,42] There is evidence based on genotyping of peripheral lymphocytes that translocation of Yp onto the active X in a majority of cells can result in 46,XX testicular DSD, while translocation onto the inactive X in most cells may cause inactivation to spread into the Yp region and result in the more ambiguous phenotype of 46,XX ovotesticular DSD.[33,34,37] Syndromic 46,XX ovotesticular, as well as testicular, DSD has been attributed to a mutation in *RSPO1*.[35]

A duplication containing the male-determining gene Sox9 has been implicated in 46,XX testicular and ovotesticular DSD in one patient[43] and, more recently, duplications and deletions in the regulatory regions of SOX9 and other SOX genes leading to aberrant upregulation of *SOX9* or other SOX genes is sufficient to promote testis formation in an 46,XX individual.[30,44]

Presentation and Diagnosis

Most patients with 46,XX testicular DSD have "de la Chapelle syndrome" with an unambiguous male phenotype that is unassociated with other disease diagnoses, and present in adolescence with delayed puberty or infertility. Although patients do occasionally present in childhood with undervirilized genitalia, SRY (found in the majority of 46,XX testicular DSD) is often correlated with relatively high virilization of the genitals, and fewer than 20% of 46,XX testicular DSD patients present as preadolescents. In 46,XX testicular DSD, analysis of the semen typically shows a normal volume with azoospermia. Karyotyping of peripheral blood cells shows 46,XX in the majority of cases. FISH for SRY is positive in 90% of patients; if negative, FISH for SOX9 microduplications can be performed, and CGH can be done if molecular etiology is still needed. 46,XX testicular DSD patients may have hypergonadotropic hypogonadism with elevated FSH and LH, decreased T, dihydrotestosterone (DHT), less than twofold increase in response to the human chorionic gonadotropin (hCG) stimulation test, and no uterus, as determined by pelvic ultrasound (refer to Table 19.2 and Fig. 19.2b).

The majority of patients with ovotesticular DSD present with ambiguous genitalia in infancy or early childhood. Unlike patients with 46,XX testicular DSD, those with ovotesticular DSD often have some ovarian function. Hormonal levels (Table 19.2) correlate with the relative amounts of testicular and ovarian tissue; FSH, LH, E_2, T, DHT, and Δ4A can be in the normal female range. Hormonal levels in turn correlate with the degree of genital ambiguity. The karyotype for the majority of patients with ovotesticular DSD will be 46,XX, but approximately one-third will be mosaic (including 46,XX/46,XY), or, rarely, 46,XY.

Syndromic effects of SOX9 duplications or RSPO1 mutations (Table 19.1) will suggest appropriate tests for these criteria, but results will not differentiate between 46,XX testicular and ovotesticular DSD. The biochemical marker for Sertoli cells is a serum AMH level greater than 75 nmol/L, which is unequivocal evidence of functional testicular tissue, and suggests either 46,XX testicular or ovotesticular DSD,[45] but again does not distinguish between the two. Ultimately, a thorough biopsy of both gonads remains the basis for a definitive differential diagnosis: the presence of any ovarian tissue distinguishes ovotesticular from 46,XX testicular DSD. In addition, as SRY expression has been detected in gonads of patients who are otherwise SRY negative,[31,46,47] tests for karyotype, cryptic mosaicism, and SRY can be performed in gonadal tissue, if indicated.

Sex Chromosome Disorders

Sex chromosome DSD results from having an abnormal number of sex chromosomes. Errors in paternal meiosis are a cause in a majority of these disorders.

Turner Syndrome

Turner syndrome is a disorder affecting females in which all or a part of one X chromosome is missing. The majority of Turner syndrome patients are 45,X; the rest are 46,XX with X chromosome deletions or mosaics with various combinations of sex chromosome number or content. 45,X Turner syndrome is relatively common in the population, and while less than 3% of Turner zygotes survive to term,[48] approximately one in 2000 newborn phenotypic females are 45,X.[49]

Most Turner females are short in stature due to lack of one copy of the homeobox gene SHOX, which is located in the pseudoautosomal region. They possess streak gonads, with fewer and poorly developed follicles *in utero*.[50] Given the reduced number of follicles, Turner patients have less estrogen secretion, resulting in delayed puberty or primary amenorrhea.

Presentation and Diagnosis

Turner syndrome should be considered if prenatal ultrasound reveals short femur, total body lymphangiectasia, large septate cystic hygromas, nuchal thickening, and/or cardiac defects. Turner newborns may present with low birth weight, lymphedema of the upper and lower extremities (in 30% of Turner babies), a webbed neck (pterygium colli), and dysmorphic features: low-set prominent ears, low posterior hairline, micrognatia, high-arched palate, epicanthal folds, hypoplastic nail beds, and/or hypoplastic fourth and fifth metacarpals.[51] Turner adolescents present most frequently with short stature, amenorrhea, and lack of secondary sex characteristics, although (depending on the amount of GD) approximately 30% do have some spontaneous puberty.[52]

Additional characteristics of Turner syndrome may include: renal anomalies (incidence between 30% and 50%), increased frequency of cardiovascular disease such as coarctation of the aorta, hearing loss, shield-like chest, and higher carrying angle of the arms (cubitus valgus).

If a patient's presentation suggests Turner, karyotyping should be done to detect 45,X Turner. If this assay is normal, FISH or CGH should be done to detect possible cryptic deletions in the SHOX-containing pseudoautosomal region of an X chromosome in a 46,XX patient.

Klinefelter Syndrome

Klinefelter syndrome, or 47,XXY, males have a normal number of primordial germ cells *in utero*, which degenerate through childhood probably due to a fault of communication between Sertoli and germ cells.[53] Roughly 50% of Klinefelter syndrome occurrence is of paternal origin[54] with a possible increase in 46,XY sperm frequency as paternal age increases.[55]

Presentation and Diagnosis

The phenotype of Klinefelter syndrome is often not obvious and as such remains primarily underdiagnosed in the general population.[56] Behavioral disorders, abnormally small testes and legs disproportionately long compared to upper extremities, may be seen in Klinefelter boys. A patient's IQ may be somewhat lower than siblings, though typically still in the normal range.[57] In adolescence, most Klinefelter patients present with small, firm testes and hypogonadism with some degree of androgen deficiency.[58] Later, males may present at infertility centers with azoospermia.

Diagnosis of Klinefelter syndrome is performed by karyotype of lymphocytes. Some mosaic cases will only be detected by karyotype of skin fibroblasts and occasionally of testicular biopsy.

DISORDERS OF SEX DIFFERENTIATION

The presence of a functional testis causes the development of male internal and external genitalia. Therefore, the testis actively governs *sexual differentiation* (see Fig. 19.1). Hormones secreted from testes are essential to the development of male internal and external genitalia. Normally developed testes have both Sertoli cells and testicular cords. Sertoli cells secrete AMH causing Müllerian (paramesonephric) duct regression. At the same time, Leydig cells, the steroidogenic cells of the testes, secrete testosterone and INSL3 (insulin-like factor 3), to promote development of Wolffian structures (epididymis, vas deferens, and seminal vesicles) and mediate transabdominal descent of the testes to the internal inguinal ring, respectively.[59]

To mediate the development of the male external genitalia, testosterone is converted to DHT, a more potent androgen, by the enzyme 5α-reductase (Fig. 19.1). Except for phallic growth and inguinoscrotal descent in the third trimester, male sexual differentiation is essentially complete by week 14. After this point, defects in labioscrotal fusion and urogenital sinus growth cannot be affected by high doses of androgens.

In the absence of testes-secreted hormones, such as testosterone, DHT, and AMH, Wolffian ducts regress and the Müllerian ducts develop into female genitalia. Defects in steroidogenic enzymes and in androgen receptor (AR) action can result in conditions ranging from full 46,XY sex reversal, virilized 46,XX individuals, and undervirilized 46,XY individuals.

Disorders in Testosterone Biosynthesis

Congenital Adrenal Hyperplasia

Depending on the enzyme implicated, CAH can cause either 46,XY DSD or 46,XX DSD. Mutations in 21-hydroxylase (*CYP21*) deficiency account for over 90% of CAH and represent the most common etiology of ambiguous genitalia in the 46,XX newborn. The disease frequency in the general population is 1:15,000 people, but is higher in certain ethnic groups, including Hispanics and Ashkenazy Jews.[60] The mechanism of excess androgen production is through blockage of both aldosterone and cortisol biosynthetic pathways. Therefore, early precursors are shunted towards androgen biosynthesis, resulting in virilization in some 46,XX fetuses and newborns. Mutations in 11β-hydroxylase (*CYP11B1*) account for another 5% of virilizing CAH.[61,62]

Rarely, CAH with ambiguous genitalia occurs in 46,XY individuals. Mutations in 3β-hydroxysteroid dehydrogenase 2 (*HSD3B2*) or 17α-hydroxylase/17,20-lyase deficiency (*CYP17*) can cause CAH and ambiguous genitalia in 46,XY individuals. Mutations in *HSD3B2* affect the biosynthesis of three major adrenal steroid hormones: cortisol, aldosterone, and testosterone. The resulting phenotype is adrenal insufficiency in 46,XX and 46,XY patients and genital ambiguity only in 46,XY patients. Mutations in *CYP17* result in decreased cortisol and androgen biosynthesis, which shunts steroidogenic precursors to a mineralocorticoid precursor causing hypertension with adrenal insufficiency and ambiguous genitalia.

Cytochrome P450 oxidoreductase (*POR*) deficiency is a recently characterized cause of CAH that affects the enzymatic activity of all microsomal P450 enzymes, including the steroidogenic enzymes *CYP17*, *CYP21*, and *CYP19*.[63–66] *POR* causes partial enzymatic activity in multiple enzymes, presents with a wide range of phenotypes, and is particularly difficult to diagnose using serum hormone levels. Additionally, it can cause ambiguous genitalia in both 46,XX and 46,XY individuals. In mothers carrying a fetus with a *POR* mutation, virilization of the mother can occur. This virilization is due to decreased fetal aromatase activity, which causes excess testosterone to be present in bloodstream of both the fetus and mother.

Presentation and Diagnosis

Among both 46,XX and 46,XY fetuses, there is a wide range of potential phenotypes resulting from the multitude of genes that can be involved in CAH. For *CYP21A2* and *CYP11B1* mutations, 46,XX fetuses can present at birth with ambiguous genitalia while 46,XY fetuses do not exhibit any degree of undervirilization. CAH can present in classical salt wasting, simple virilizing, or in the nonclassical form, which presents with hyperandrogenism and precocious puberty with normal genitalia. The numerous genes, some of the biochemical variability, make CAH cases ideal for high throughput sequencing. The biochemical diagnoses for the different genes involved in CAH are outlined in Table 19.2.

CYP21A2 deficiencies are often not detected in the newborn period outside of newborn screening programs, unless accompanied by a life-threatening salt-wasting crisis or with the presence of ambiguous genitalia. 46,XY patients often present with premature masculinization and accelerated physical development during childhood. If left untreated, there is premature fusion of epiphyses resulting in short stature.

Diagnosis of 21-hydroxylase deficiency can be ascertained by elevated 17-OH progesterone. Biochemical changes are outlined in Table 19.2. Clinically CYP11B1 deficiency can be differentiated from CYP21A2 deficiency by the presence of hypertension with hypokalemic alkalosis. Molecular diagnosis should be obtained through targeted sequencing approaches provided by a center that specializes in DSD. However, if there is a family history of CAH and a gene mutation is already known, it can also be diagnosed through single gene sequencing of the genetic defect.

Congenital Lipoid Adrenal Hyperplasia

Congenital lipoid adrenal hyperplasia (CLAH) is a rare form of CAH in which lipoid vacuoles accumulate in steroidogenic tissues, specifically the adrenal glands. CLAH results from mutations in enzymes involved in early steroidogenesis, primarily steroidogenic acute regulatory (StAR) protein, which transports cholesterol across the inner and outer mitochondrial membranes of steroidogenic cells, and cytochrome P450 side chain cleavage (CYP11A1).[67,68] Mutations in these genes result in global silencing of adrenal and gonadal steroidogenesis. On histology, patients with mutations in StAR protein or CYP11A1 have lipid vacuoles in adrenals due to the accumulation of cholesterol in the adrenals and gonads. This lipid accumulation results in primary adrenal and gonadal failure.[69]

Presentation and Diagnosis

The majority of newborns with mutations in StAR or CYP11A1 present with a salt-wasting adrenal crisis at birth that is usually fatal if not immediately diagnosed and treated. All 46,XY patients present with a genital phenotype with adrenal insufficiency, while 46,XX patients only exhibit the features of adrenal insufficiency at birth.

46,XY patients have testes, no Müllerian structures due to the presence of AMH, partial or absent Wolffian derivatives due to a lack of testosterone biosynthesis, and a blind vaginal pouch. The spectrum of external genital phenotype ranges from ambiguous to fully feminized. Testes are not completely descended and can be located in the abdomen, inguinal canal, or labia. Clinically, infants with intrauterine glucocorticoid deficiency can present with generalized hyperpigmentation at birth as a result of elevated ACTH levels.

46,XX patients with partial deficiency of StAR protein experience spontaneous puberty, menarche, and anovulatory menses because their ovaries are able to produce estrogen through StAR-independent pathways.[69-71] Since the ovaries do not produce steroids until puberty, they are spared from the cholesterol-induced damage that occurs from birth in the adrenals.[71] At puberty, however, 46,XX females develop multiple cysts in their ovaries, possibly from anovulation.

Definitive diagnosis of CLAH is done by sequencing of StAR or CYP11A1. Characteristic biochemical abnormalities are outlined in Table 19.2. Another distinguishing characteristic is adrenal enlargement, which has only been found in patients with StAR mutations.

Defects in 5α-Reductase Type 2

Steroid 5α-reductase deficiency results from a defect in the enzyme converting testosterone to DHT. DHT is responsible for the differentiation of male external genitalia. Peripheral conversion of testosterone to DHT is an irreversible reaction catalyzed by the two isoenzymes of 5α-reductase, SRD5A1 and SRD5A2. During puberty, the SRD5A1 isozyme is active in skin fibroblasts. During fetal development, SRD5A2 isozyme is expressed in the genital skin tissue and male accessory sex organs. Mutations in SRD5A2 result in undervirilized external genitalia, such as micropenis and perineal hypospadias. Internal genitalia consist of fully differentiated Wolffian ducts, no Müllerian structures, and a blind vaginal pouch.[72,73] Adrenal and gonadal steroid biosynthesis remains normal.

Presentation and Diagnosis

In 46,XY infants with recessive mutations in SRD5A2, the degree of ambiguity ranges from isolated hypospadias to severe undermasculinization, complete with perineal hypospadias, micropenis, bifid scrotum, and hypoplastic prostate.[74] At puberty, 46,XY patients experience a surge of testosterone, which can cause development of male secondary sex characteristics, such as male body habitus, deepening of the voice, and penile enlargement. However, these individuals have unusually fine upper lip hair. Due to the underdevelopment of the prostate and seminal vesicles, their semen is highly viscous and their ejaculate volume is extremely low (<0.5–1 mL). Despite the low ejaculate volume, sperm counts are normal. Patients who are unambiguously female at birth often present in puberty with complaints of amenorrhea, a deepening voice, clitoromegaly, and increased musculature. In contrast to patients with AIS, mutations in SRD5A2 do not develop gynecomastia during puberty. 46,XX females with SRD5A2 deficiency have normal sexual differentiation and fertility, but delayed puberty and sparse pubic hair.

Diagnosis of SRD5A2 deficiency can be made during early infancy and puberty based on an elevated ratio of testosterone to DHT (normal <30:1), either with or without hCG stimulation.[72,75] DHT is low in infants but can reach near-normal levels during adolescence without treatment, presumably by peripheral SRD5A1 activity. If the diagnosis is suspected during prepubertal childhood or during adolescence, hCG stimulation should be used to obtain diagnostic hormone levels. Affected females are phenotypically normal, but have the same biochemical abnormalities (Table 19.2) as affected males.

Mutational analysis of the *SR5A2* gene can be performed, and mutations have been reported in all five exons. To date, over 50 mutations in *SR5A2* have been found, ranging from point mutations to deletions but the genotype-phenotype correlations are poor.

3α-Reductase, AKR1C2 (Backdoor Pathway of DHT Synthesis)

The major enzyme in DHT synthesis from testosterone (5-α-reductase-2) is found primarily in the genital skin and promotes labioscrotal fusion of the external genitals *in utero* and male puberty. However, a secondary pathway, which bypasses testosterone formation, exists in marsupial animals, and is partially conserved in humans and reviewed in Refs 76 and 77. Recent studies have provided convincing evidence that these pathways also play a role in pre- and postnatal masculinization in 46,XY individuals. While the complete role of the backdoor pathway for DHT synthesis remains unclear, emerging evidence indicates it may play a role in pathologic virilization in CAH.[78]

Presentation and Diagnosis

While this has only been described in two families in the literature, the phenotypic variability within these families is particularly striking as affected individuals took on varying gender roles. Patients are 46,XY with ambiguous genitalia, normal testes on histology, and cryptorchidism. All patients had a urinary steroid profile significant for elevated pregnanetriol and pregnanetriolone, which were hyperresponsive to both corticotropin and hCG stimulation tests. Urinary dehydroepiandrosterone (DHEA) was unmeasurable after hCG stimulation.

Human mutations in two genes along the same backdoor pathways were described in patients with 46,XY DSD indicating that multigenic inheritance occurs in families with 46,XY DSD with phenotypic variability.[79] In one family, coinheritance of multiple mutations on both alleles of *AKR1C2* and a heterozygous missense mutation on closely linked *AKR1C4* resulted in a 46,XY DSD and female phenotype, while inheritance of biallelic mutations on *AKR1C2* alone resulted in an undervirilization 46,XY phenotype. In a second family, a heterozygous mutation in *AKR1C2* coupled with a complex genomic rearrangement between the maternal and paternal on the other allele was sufficient to result in a 46,XY DSD phenotype with normal testis.

These types of cases are particularly difficult to diagnose with traditional methods. It requires sequencing of multiple genes involved in sex differentiation to identify multiallelic inheritance.

Defects in Androgen Activity

Complete and Partial AIS

In the presence of normal gonads and steroid biosynthesis, mutations in the steroid hormone receptors can mute the effects of circulating steroid on specific tissues. Mutations in the AR, which is located on the X chromosome, result in AIS. Both testosterone and DHT can activate AR, and when AR is bound to either steroid hormone, it can activate transcription of specific downstream genes. The AR plays important roles in the differentiation of male internal and external genitalia and in the maintenance of spermatogenesis. Currently, there is an estimated incidence of 1 per 20,400 liveborn 46,XY individuals in which more than 300 mutations have been identified.

AIS is characterized by a range of phenotypes in 46,XY individuals, from unambiguous females, termed complete AIS (CAIS) to phenotypic males, termed minimal AIS (MAIS). In CAIS, the only manifestations are infertility and amenorrhea, which present in puberty.[80,81] At the other extreme is MAIS in which phenotypically male patients often present with infertility, gynecomastia, and/or hypospadias. Patients who exhibit various degrees of 46,XY ambiguity are referred to as partial AIS (PAIS). All known forms of AIS are caused by disruption of AR activity. CAIS is generally associated with a complete absence of androgen binding and AR activation. Beyond this, there is almost no correlation between residual AR activity and genital phenotype.[82] This suggests that there exist other genetic modifiers that modulate the phenotype. Even within families, the phenotype resulting from a given mutation can vary between PAIS and CAIS.[83,84]

The majority of mutations resulting in AIS are due to single amino acid substitutions, which account for approximately 90% of the cases. Mutations are spread throughout the gene, and there is no single mutation that appears to be prevalent over others.[85] Interestingly, exon 1 rarely possesses a causative mutation.[85] The majority of cases are inherited, but 30% of all AIS cases are *de novo* mutations, in which the risk of recurrence in future offspring is low.

In familial PAIS, and in both familial and sporadic CAIS, mutations in *AR* exonic sequences are found in 85–90% of cases.[84,85] In contrast, detectable mutations in *AR* account for only 10–15% of sporadic cases with hormonal profile and clinical presentation suggestive of *de novo* PAIS,[84] and the molecular defect in these cases is unknown. While the genotypes causing CAIS are consistent in phenotypic presentation, in PAIS there is phenotypic variability among affected individuals carrying the same mutation.

Presentation and Diagnosis

The phenotype of patients with PAIS is extremely heterogeneous. Patients present in infancy or childhood with variable degrees of virilization such as micropenis, cryptorchidism, and perineoscrotal hypospadias. Infants and children may present with unilateral or bilateral inguinal hernias. Testes are present and functional, producing high levels of testosterone and AMH resulting in variable development of Wolffian derivatives and regression of Müllerian structures in most patients.[86] In patients presenting at puberty, breast development and sparse pubic hair are suggestive of PAIS, and help differentiate it from SRD5A2.

CAIS usually presents at puberty with primary amenorrhea. Physical exam reveals a short, blind vagina, absent uterus, and sometimes palpable inguinal or labial testes. Testosterone-dependent Wolffian derivatives and prostate are absent or vestigial. Additionally, height, bone maturation, and breast development are normal, but pubic and axillary hair, an androgen-mediated feature, is absent or sparse. Patients' identity and behavior are feminine without gender dysphoria.[87] Less commonly, CAIS may present in infancy with phenotypic female genitalia and inguinal or labial masses representing testes.

Diagnostic criteria for serum hormone levels are outlined in Table 19.2. In prepubertal children, basal LH and testosterone may be normal, but hCG stimulation elicits an exaggerated androgen response (a tripling instead of a doubling of testosterone and DHT). Exam and pelvic ultrasound reveals abdominal testes and the absence of Müllerian structures.

Molecular genetic testing of the *AR* gene detects mutations in more than 95% of probands with CAIS. Molecular testing has been shown to be more consistent than biochemical assays of AR function, which has been discredited based on a high degree of variation due to biopsy site and testing laboratory.[88] PCR-based sequencing of *AR* exons 2–10 can be routinely performed,[89] as well as sequencing of the much longer exon 1 and some intronic and promoter regions. Prenatal testing by mutation analysis is available for families in which the AIS-causing allele has been identified in an affected family member.[84]

Gonadotropin and Gonadotropin Receptor Defects

Leydig Cell Hypoplasia/Agenesis

hCG and LH activate a shared G protein-coupled receptor, luteinizing hormone chorionic gonadotropin receptor (LHCGR). *In utero*, placental hCG stimulates Leydig cells to produce testosterone, resulting in male internal and external genitalia. LH takes over during the third trimester of gestation and neonatal life to complete Leydig development and continue testosterone production.

The importance of the LHCGR in male testicular development is underscored by mutations in this receptor, which leads to Leydig cell hypoplasia (LCH) in 46,XY males. Inactivating mutations of *LHCGR* result in impaired Leydig cell differentiation and testosterone production. Over 20 inactivating mutations of *LHCGR* have been identified scattered throughout the gene,[90,91] and cause variable degrees of loss of receptor activity. More severe mutations resulting in truncation, decreased surface expression, or decreased coupling efficiency are usually associated with an unambiguous female phenotype. Partial inactivating mutations often result in an undervirilized phenotype such as micropenis or hypospadias.[92] One specific mutation highlights the differential binding sites of hCG and LH to the same receptor. A splicing variant resulting in the absence of exon 10 was described in an 18-year-old male who presented with a normal male phenotype, pubertal delay, small testicles, and delayed bone age.[93] The receptors' response to hCG was normal, inferred from normal sex differentiation. However, LHCGR was not activated by LH, resulting in delayed puberty and bone age (see Fig. 19.2).

In females, the LHCGR is important in maintenance of the menstrual cycle and pregnancy.

Presentation and Diagnosis

Severe inactivating mutations in LHCGR in 46,XY patients are often missed at birth and present in puberty with amenorrhea. Partially inactivating mutations present at birth, with undervirilization of the external genitals (e.g., micropenis, hypospadias, and cryptorchidism). In 46,XX individuals, pelvic ultrasound demonstrates absence of uterus and fallopian tubes. To distinguish LCH from patients with 46,XY GD, AMH is used as a marker of testicular Sertoli cell function and is normal-to-high in LCH patients and low-to-undetectable in 46,XY GD patients.[45] Histological analysis of the testis shows normal Sertoli cells, hyalinized seminiferous tubules, without mature Leydig cells or spermatogenesis. In rare patients with partially inactivating mutations, there may be early signs of spermatogenesis but no viable sperm produced.

Therefore, patients have potential for fertility using assisted reproductive technology.

Biochemical changes are outlined in Table 19.2. Definitive diagnosis requires sequencing of the *LHCGR* gene for deletions, insertions, and point mutations.

LHCGR Mutations in 46,XY Precocious Puberty

Constitutively active mutations of LHCGR can be due to amino acid substitution, and cause a 12-fold increase in the basal level of cAMP. The increased levels of cAMP cause transduction of the signal without binding of LH or hCG to the receptor.[94,95]

Presentation and Diagnosis

46,XY patients with precocious puberty present before 8 years old. Formal evaluation requires a complete medical history and physical examination to stage physical development, reviewed in Ref. 96. Furthermore, a bone-age test should be done to confirm the diagnosis. Specific diagnosis of LHCGR mutations can be done through sequencing of the gene.

LHCGR Mutations in 46,XX Patients

Presentation and Diagnosis

Inactivating LHCGR mutations result in hypergonadotropic hypogonadism, or primary hypogonadism. The major symptoms are primary amenorrhea or oligoamenorrhea, cystic ovaries, and infertility.[97–99] However, 46,XX females will undergo spontaneous breast and pubic hair development. There has been one reported case of a splicing mutation in LHCGR, in which a 46,XX female experienced regular menstrual cycles but was infertile.[100] Inactivating mutations in LHCGR and FSHR are reviewed in Ref. 101.

FSHR Mutations in Ovarian Hyperstimulation Syndrome (OHSS)

Constitutively active mutations in the FSH receptor occur in the regions that are responsible for specificity to FSH. These mutations change the receptor-ligand binding site, thereby decreasing receptor specificity. During pregnancy, the FSH receptor is stimulated by endogenous hCG resulting in OHSS.[102–104]

Presentation and Diagnosis

These patients typically present during a spontaneous pregnancy, with multiple ovarian cysts due to multiple follicular development. More serious complications include acute fluid shifts out of the intravascular space, resulting in hypovolemia and hemoconcentration. OHSS is an acute condition that can ultimately result in renal failure, acute respiratory distress syndrome, and even death. Diagnosis of an FSHR mutation can be done through sequencing of the entire gene.

FSHR Mutations in 46,XX Ovarian Dysgenesis

A rare inactivating mutation in FSHR decreases the action of the FSH receptor by limiting its translocation to the cell surface,[105] resulting in decreased folliculogenesis. The ala189val mutation is found at a frequency of 0.96% in the Finnish population, which has a frequency of 1:8,300 of ovarian dysgenesis.[105] However, this same mutation has not been identified in North American or French populations.[106] In affected patients, the follicles are able to develop until the antral stage, but do not progress any further, indicating the essential role for FSH in terminal maturation of the follicle. Other genetic causes of premature ovarian failure (POF) and ovarian dysgenesis are reviewed in Ref. 107.

Presentation and Diagnosis

Patients with FSHR mutations present with primary amenorrhea. This distinguishes these patients from the majority of POF patients, who often experience regular menstrual cycles before the diagnosis of POF. Diagnosis of an FSH mutation can be done through sequencing of the entire gene.

FSH Inactivating Mutations

FSH is a required protein for ovarian folliculogenesis and for Sertoli cell proliferation in the testis. Inactivating mutations in FSH result in delayed puberty and hypogonadism in 46,XX and 46,XY individuals.

Presentation and Diagnosis

In 46,XY males, FSH mutations result in delayed puberty and hypogonadism. In 46,XX females, FSH mutations present with primary amenorrhea with normal LH levels.[108] However, in these patients, ovulation and normal pregnancy can be attained with exogenous FSH administration. Mutations can be detected through direct sequencing of the FSH gene.

Disorders of AMH or AMH Receptor

Persistent Müllerian Duct Syndrome

During the critical time in male sex differentiation, between weeks 10 and 12 of gestation, testicular AMH causes the regression of the Müllerian derivatives. With persistent Müllerian duct syndrome (PMDS), inactivating mutations in primarily AMH and AMH receptor 2 (AMHR2) can cause failure of this regression, resulting in both male and female internal genitalia, but normal male external genitalia.

Presentation and Diagnosis

The diagnosis of PMDS is commonly made incidentally during abdominal imaging studies or surgical exploration of the abdomen.[109] 46,XY patients with PMDS are phenotypically male. Aside from having a uterus and fallopian tubes in addition to the male external genitalia, they often have no clinical abnormalities. Therefore the true prevalence of PMDS is difficult to ascertain. The vast majority of cases of PMDS have found mutations in either the *AMH* or *AMHR2*, and in approximately equal proportions.[110–112] Inheritance is autosomal recessive for both loci.

Boys present with cryptorchidism (20%) or with an inguinal hernia[112] containing Müllerian structures, but normal virilization.[111,113] The increased likelihood of abdominal testis in PMDS causes an increased incidence of gonadoblastoma.[114]

Androgen Excess: Fetoplacental Causes

P450 Aromatase Deficiency

Aromatase (*CYP19A1*) catalyzes the conversion of androgens to estrogen. Mutations that affect aromatase activity have effects in both male and females. Estrogen is required for spermatogenesis in males and for the development of secondary sex characteristics in females.

Presentation and Diagnosis

The initial manifestation of aromatase deficiency is *in utero*, where the mother carrying a child with a *CYP19A1* mutation presents with maternal virilization.[115–117] The elevated levels of androgenic precursors in the maternal and fetal blood result in virilization of the mother and 46,XX fetuses. Biochemical findings are outlined in Table 19.2. CAH is the major differential diagnosis considered and can be ruled out with ACTH-stimulation test showing increase in adrenal hormones but not androgens.

46,XX patients fail to undergo spontaneous puberty, and often have polycystic ovaries, increased virilization, and amenorrhea, without breast development.[115,118,119] 46,XY males experience normal puberty but are infertile due to the lack of estrogen.[117,120,121]

Neither 46,XX nor 46,XY patients experience a growth spurt during puberty but are considered tall as a result of linear growth. Skeletal maturity is delayed and osteoporosis develops early.[117,122] Patients often complain of bone pain.[117]

P450 Oxidoreductase Deficiency

POR is a rare cause of ambiguous genitalia in 46,XX and 46,XY infants. Pathophysiology and presentation are described in detail in the Section "46,XY DSD."

Androgen Excess: Maternal Etiologies

Maternal ingestion of androgens is a potential cause of virilization of 46,XX newborns. However, present-day progestin-containing oral contraceptives are not highly androgenic, and this cause of 46,XX virilization is therefore rare.[123–125] Hormonally active tumors, such as luteomas, are another uncommon causes of maternal virilization and ambiguous genitalia in a 46,XX baby.[126] Rare cases of hyperreactio luteinalis, a cystic ovarian condition associated with virilization of the mother and fetus, are associated with 46,XX virilization in 15% of cases.[126] While CAH causes the majority of cases of 46,XX virilization, one should always consider oncologic and iatrogenic causes of maternal and fetal virilization.

HIGH-THROUGHPUT SEQUENCING IN THE DIAGNOSIS OF DISORDERS OF SEX DETERMINATION

As the technology around high-throughput sequencing has penetrated the clinical diagnostics arena, there is much debate as to its clinical utility relative to the perceived expense. Only since 2012 has whole exome sequencing of all the coding regions within the genome been both financially and bioinformatically feasible for clinical diagnostics, particularly for rare Mendelian disorders.[127] Major academic centers that offer whole-exome sequencing, such as the University of California, Los Angeles, and Baylor College of Medicine, report a genetic diagnosis rate for all patients of approximately 30%, which is significantly higher than the proportion of patients diagnosed with a disorder of sex development who ultimately receive a genetic diagnosis. Similarly, targeted approaches, which sequence only genes known to be important in sex determination, have a similar utility and diagnostic rate[128] but are less powered to identify atypical presentations of genetic diseases.

The argument that next generation sequencing is neither cost effective nor timely in the diagnosis of DSD is largely untrue. Currently, the cost is similar, and will likely continue to drop. Additionally, it comes with the added benefit of sequencing multiple genes at once. This will increase the likelihood of finding a causative mutation and identifying atypical phenotypes that exist along the same spectrum of disease. Turnaround time for comprehensive sequencing approaches is virtually the same for single or multiple gene Sanger sequencing. More importantly, these technologies are most likely to decrease the diagnostic odyssey for patients and their families,[129] provide closure, and provide predictive utility for what the family can expect in the future with regards to recurrence risk in future siblings, fertility, gender identity, cancer risk, and other associated conditions in the affected child.[130] Furthermore, understanding the genetic underpinnings of DSD and how these alter endocrine function can further guide the functional testing of steroidogenic pathways to determine the effect of specific gene mutations. Genetic testing would not replace endocrine tests

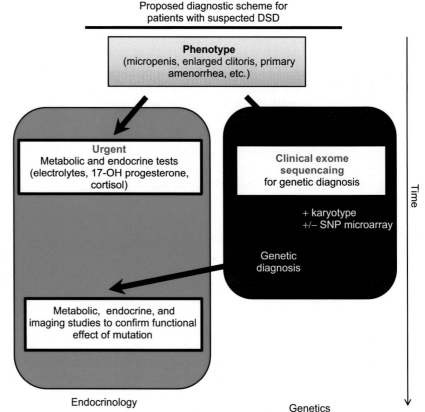

FIGURE 19.3 **Proposed diagnostic flow chart for disorders of sex development.** Upon recognition of a DSD, patients should undergo urgent metabolic and endocrine testing to rule out life-threatening disorders. Additionally, genetic testing should be ordered as one of the primary tests in order to quickly identify a causative genetic mutation. Upon identification of the variant, the patient should undergo targeted functional testing to confirm the functionality of the mutated gene and that the identified mutation is the cause of DSD.

to rule out life-threatening conditions, such as CAH in cases of a newborn 46,XX female with ambiguous genitalia.[131] As more and more genes involved in DSD are identified, single gene sequencing has been cumbersome, ineffective, expensive, and in many cases not feasible on a clinical basis, as no clinical labs offer the testing. Given the major progress in next generation sequencing, either targeted or whole-exome sequencing approaches are appropriate.

We propose that physicians prioritize the genetic diagnosis earlier in the diagnostic odyssey, as one of the primary tools to obtain a diagnosis for DSD (Fig. 19.3). Once a genetic diagnosis is made, further follow-up testing should be performed to assess the degree of function of the mutated gene and limited to testing that confirms the genetic diagnosis. We believe this will alleviate both unnecessary endocrine testing and provide a more definitive diagnosis for more patients affected by DSD.

CONCLUSIONS

Although DSD are complex to decipher, advances in genetics have made the genetic diagnosis easier and, most importantly, quite rapid. Despite advances in diagnosis that have made considerable progress in the past decade, management strategies have remained

controversial.[130] In particular, the question of early genital surgery remains in dispute, with little outcome data to support either early surgical intervention or later decision by the patient him or herself. The question of disclosure – that is, whether to inform the child or adolescent about his or her diagnosis and genital surgery, if applicable – also remains contentious, with many physicians (and parents) supporting a model of nondisclosure in order to prevent gender identity confusion. The challenging view is that nondisclosure reinforces secrecy and shame, and full information adjusted to the cognitive and psychological development of the child will enhance psychosocial adaptation.

References

1. Lee PA, Houk CP, Ahmed SF, Hughes IA. Consensus statement on management of intersex disorders. International Consensus Conference on Intersex. *Pediatrics* 2006;**118**(2):e488–500.
2. Knower KC, Kelly S, Harley VR. Turning on the male – SRY, SOX9 and sex determination in mammals. *Cytogenet Genome Res* 2003;**101**(3–4):185–98.
3. McElreavey K, Vilain E, Abbas N, Costa JM, Souleyreau N, Kucheria K, et al. XY sex reversal associated with a deletion 5′ to the SRY "HMG box" in the testis-determining region. *Proc Natl Acad Sci USA* 1992;**89**(22):11016–20.
4. Uehara S, Hashiyada M, Sato K, Nata M, Funato T, Okamura K. Complete XY gonadal dysgenesis and aspects of the SRY genotype and gonadal tumor formation. *J Hum Genet* 2002;**47**(6):279–84.

5. Assumpcao JG, Benedetti CE, Maciel-Guerra AT, Guerra Jr G, Baptista MT, Scolfaro MR, et al. Novel mutations affecting SRY DNA-binding activity: the HMG box N65H associated with 46,XY pure gonadal dysgenesis and the familial non-HMG box R30I associated with variable phenotypes. *J Mol Med* 2002;**80**(12): 782–90.

6. Aittomaki K. The genetics of XX gonadal dysgenesis. *Am J Hum Genet* 1994;**54**(5):844–51.

7. Uehara S, Funato T, Yaegashi N, Suziki H, Sato J, Sasaki T, et al. SRY mutation and tumor formation on the gonads of XY pure gonadal dysgenesis patients. *Cancer Genet Cytogenet* 1999;**113**(1):78–84.

8. Li B, Zhang W, Chan G, Jancso-Radek J, Liu S, Weiss MA. Human sex reversal due to impaired nuclear localization of SRY. *J Biol Chem* 2001;**276**(49):46480–4.

9. Mitchell CL, Harley VR. Biochemical defects in eight SRY missense mutations causing XY gonadal dysgenesis. *Mol Genet Metab* 2002;**77**(3):217–25.

10. Kohler B, Lin L, Ferraz-de-Souza B, Wieacker P, Heidemann P, Schroder V, et al. Five novel mutations in steroidogenic factor 1 (SF1, NR5A1) in 46,XY patients with severe underandrogenization but without adrenal insufficiency. *Hum Mutat* 2008;**29**(1):59–64.

11. Ozisik G, Achermann JC, Jameson JL. The role of SF1 in adrenal and reproductive function: insight from naturally occurring mutations in humans. *Mol Genet Metab* 2002;**76**(2):85–91.

12. Tremblay JJ, Viger RS. A mutated form of steroidogenic factor 1 (SF-1 G35E) that causes sex reversal in humans fails to synergize with transcription factor GATA-4. *J Biol Chem* 2003;**278**(43): 42637–42.

13. Lourenco D, Brauner R, Rybczynska M, Nihoul-Fekete C, McElreavey K, Bashamboo A. Loss-of-function mutation in GATA4 causes anomalies of human testicular development. *Proc Natl Acad Sci USA* 2011;**108**(4):1597–602.

14. Bashamboo A, Brauner R, Bignon-Topalovic J, Lortat-Jacob S, Karageorgou V, Lourenco D, et al. Mutations in the FOG2/ZFPM2 gene are associated with anomalies of human testis determination. *Hum Mol Genet* 2014;**23**(14):3657–65.

15. Pearlman A, Loke J, Le Caignec C, White S, Chin L, Friedman A, et al. Mutations in MAP3K1 cause 46,XY disorders of sex development and implicate a common signal transduction pathway in human testis determination. *Am J Hum Genet* 2010;**87**(6): 898–904.

16. Das DK, Rahate SG, Mehta BP, Gawde HM, Tamhankar PM. Mutation analysis of mitogen activated protein kinase 1 gene in Indian cases of 46,XY disorder of sex development. *Indian J Hum Genet* 2013;**19**(4):437–42.

17. Warr Nick, et al. Gadd45γ and Map3k4 interactions regulate mouse testis determination via p38 MAPK-mediated control of Sry expression. *Dev Cell* 2012;**23**:1020–31.

18. Debora Bogani, et al. Loss of mitogen-activated protein kinase kinase kinase 4 (MAP3K4) reveals a requirement for MAPK signalling in mouse sex determination. *PLoS Biol* 2009;**7**(9):e1000196.

19. Capel B. Sex in the 90s: SRY and the switch to the male pathway. *Annu Rev Physiol* 1998;**60**:497–523.

20. Saylam K, Simon P. WT1 gene mutation responsible for male sex reversal and renal failure: the Frasier syndrome. *Eur J Obstet Gynecol Reprod Biol* 2003;**110**(1):111–3.

21. Domenice S, Correa RV, Costa EM, Nishi MY, Vilain E, Arnhold IJ, et al. Mutations in the SRY, DAX1, SF1 and WNT4 genes in Brazilian sex-reversed patients. *Braz J Med Biol Res* 2004;**37**(1):145–50.

22. Migeon CJ, Wisniewski AB. Human sex differentiation and its abnormalities. *Best Pract Res Clin Obstet Gynaecol* 2003;**17**(1):1–18.

23. Reddy KS, Sulcova V. Pathogenetics of 45,X/46,XY gonadal mosaicism. *Cytogenet Cell Genet* 1998;**82**(1–2):52–7.

24. Rosenberg C, Frota-Pessoa O, Vianna-Morgante AM, Chu TH. Phenotypic spectrum of 45,X/46,XY individuals. *Am J Med Genet* 1987;**27**(3):553–9.

25. Knudtzon J, Aarskog D. 45, X/46,XY mosaicism. A clinical review and report of ten cases. *Eur J Pediatr* 1987;**146**(3):266–71.

26. Xiao B, Ji X, Xing Y, Chen YW, Tao J. A rare case of 46, XX SRY-negative male with approximately 74-kb duplication in a region upstream of SOX9. *Eur J Med Genet* 2013;**56**(12):695–8.

27. Benko S, Gordon CT, Mallet D, Sreenivasan R, Thauvin-Robinet C, Brendehaug A, et al. Disruption of a long distance regulatory region upstream of SOX9 in isolated disorders of sex development. *J Med Genet* 2011;**48**(12):825–30.

28. White S, Ohnesorg T, Notini A, Roeszler K, Hewitt J, Daggag H, et al. Copy number variation in patients with disorders of sex development due to 46,XY gonadal dysgenesis. *PLoS One* 2011;**6**(3):e17793.

29. Sutton E, Hughes J, White S, Sekido R, Tan J, Arboleda V, et al. Identification of SOX3 as an XX male sex reversal gene in mice and humans. *J Clin Invest* 2011;**121**(1):328–41.

30. Fleming A, Eric V. The endless quest for sex determination genes. *Clin Genet* 2005;**67**(1):15–25.

31. Queipo G, Zenteno JC, Pena R, Nieto K, Radillo A, Dorantes LM, et al. Molecular analysis in true hermaphroditism: demonstration of low-level hidden mosaicism for Y-derived sequences in 46,XX cases. *Hum Genet* 2002;**111**(3):278–83.

32. Zenteno-Ruiz JC, Kofman-Alfaro S, Mendez JP. 46, XX sex reversal. *Arch Med Res* 2001;**32**(6):559–66.

33. Bouayed Abdelmoula N, Portnoi MF, Keskes L, Recan D, Bahloul A, Boudawara T, et al. Skewed X-chromosome inactivation pattern in SRY positive XX maleness: a case report and review of literature. *Ann Genet* 2003;**46**(1):11–8.

34. Kusz K, Kotecki M, Wojda A, Szarras-Czapnik M, Latos-Bielenska A, Warenik-Szymankiewicz A, et al. Incomplete masculinisation of XX subjects carrying the SRY gene on an inactive X chromosome. *J Med Genet* 1999;**36**(6):452–6.

35. Tomaselli S, Megiorni F, De Bernardo C, Felici A, Marrocco G, Maggiulli G, et al. Syndromic true hermaphroditism due to an R-spondin1 (RSPO1) homozygous mutation. *Hum Mutat* 2008;**29**(2): 220–6.

36. Krob G, Braun A, Kuhnle U. True hermaphroditism: geographical distribution, clinical findings, chromosomes and gonadal histology. *Eur J Pediatr* 1994;**153**(1):2–10.

37. Hadjiathanasiou CG, Brauner R, Lortat-Jacob S, Nivot S, Jaubert F, Fellous M, et al. True hermaphroditism: genetic variants and clinical management. *J Pediatr* 1994;**125**(5 Pt 1):738–44.

38. Boucekkine C, Toublanc JE, Abbas N, Chaabouni S, Ouahid S, Semrouni M, et al. Clinical and anatomical spectrum in XX sex reversed patients. Relationship to the presence of Y specific DNA-sequences. *Clin Endocrinol (Oxf)* 1994;**40**(6):733–42.

39. McElreavey K, Rappaport R, Vilain E, Abbas N, Richaud F, Lortat-Jacob S, et al. A minority of 46,XX true hermaphrodites are positive for the Y-DNA sequence including SRY. *Hum Genet* 1992;**90**(1-2):121–5.

40. Berkovitz GD, Fechner PY, Marcantonio SM, Bland G, Stetten G, Goodfellow PN, et al. The role of the sex-determining region of the Y chromosome (SRY) in the etiology of 46,XX true hermaphroditism. *Hum Genet* 1992;**88**(4):411–6.

41. Damiani D, Fellous M, McElreavey K, Barbaux S, Barreto ES, Dichtchekenian V, et al. True hermaphroditism: clinical aspects and molecular studies in 16 cases. *Eur J Endocrinol* 1997;**136**(2):201–4.

42. Abbas N, McElreavey K, Leconiat M, Vilain E, Jaubert F, Berger R, et al. Familial case of 46,XX male and 46,XX true hermaphrodite associated with a paternal-derived SRY-bearing X chromosome. *C R Acad Sci III* 1993;**316**(4):375–83.

43. Huang B, Wang S, Ning Y, Lamb AN, Bartley J. Autosomal XX sex reversal caused by duplication of SOX9. *Am J Med Genet* 1999;**87**(4):349–53.

44. Cox JJ, Willatt L, Homfray T, Woods CG. A SOX9 duplication and familial 46,XX developmental testicular disorder. *N Engl J Med* 2011;**364**(1):91–3.

45. Rey RA, Belville C, Nihoul-Fekete C, Michel-Calemard L, Forest MG, Lahlou N, et al. Evaluation of gonadal function in 107 intersex patients by means of serum antimullerian hormone measurement. *J Clin Endocrinol Metab* 1999;**84**(2):627–31.

46. Jimenez AL, Kofman-Alfaro S, Berumen J, Hernandez E, Canto P, Mendez JP, et al. Partially deleted SRY gene confined to testicular tissue in a 46,XX true hermaphrodite without SRY in leukocytic DNA. *Am J Med Genet* 2000;**93**(5):417–20.

47. Ortenberg J, Oddoux C, Craver R, McElreavey K, Salas-Cortes L, Guillen-Navarro E, et al. SRY gene expression in the ovotestes of XX true hermaphrodites. *J Urol* 2002;**167**(4):1828–31.

48. Eiben B, Borgmann S, Schubbe I, Hansmann I. A cytogenetic study directly from chorionic villi of 140 spontaneous abortions. *Hum Genet* 1987;**77**(2):137–41.

49. Stochholm K, Juul S, Juel K, Naeraa RW, Gravholt CH. Prevalence, incidence, diagnostic delay, and mortality in Turner syndrome. *J Clin Endocrinol Metab* 2006;**91**(10):3897–902.

50. Reynaud K, Cortvrindt R, Verlinde F, De Schepper J, Bourgain C, Smitz J. Number of ovarian follicles in human fetuses with the 45,X karyotype. *Fertil Steril* 2004;**81**(4):1112–9.

51. Ferguson-Smith MA. Karyotype-phenotype correlations in gonadal dysgenesis and their bearing on the pathogenesis of malformations. *J Med Genet* 1965;**39**:142–55.

52. Pasquino AM, Passeri F, Pucarelli I, Segni M, Municchi G. Spontaneous pubertal development in Turner's syndrome. Italian Study Group for Turner's Syndrome. *J Clin Endocrinol Metab* 1997;**82**(6): 1810–3.

53. Lue Y, Rao PN, Sinha Hikim AP, Im M, Salameh WA, Yen PH, et al. XXY male mice: an experimental model for Klinefelter syndrome. *Endocrinology* 2001;**142**(4):1461–70.

54. Thomas NS, Hassold TJ. Aberrant recombination and the origin of Klinefelter syndrome. *Hum Reprod Update* 2003;**9**(4):309–17.

55. Eskenazi B, Wyrobek AJ, Kidd SA, Lowe X, Moore 2nd D, Weisiger K, et al. Sperm aneuploidy in fathers of children with paternally and maternally inherited Klinefelter syndrome. *Hum Reprod* 2002;**17**(3):576–83.

56. Bojesen A, Juul S, Gravholt CH. Prenatal and postnatal prevalence of Klinefelter syndrome: a national registry study. *J Clin Endocrinol Metab* 2003;**88**(2):622–6.

57. Robinson A, de la Chapelle A. Sex chromosome abnormalities. In: Rimoin's DL, editor. *Emery and Rimoin's Principles and Practice of Medical Genetics.* 3rd ed. Churchill Livingstone; 1996.

58. Lanfranco F, Kamischke A, Zitzmann M, Nieschlag E. Klinefelter's syndrome. *Lancet* 2004;**364**(9430):273–83.

59. Bogatcheva NV, Agoulnik AI. INSL3/LGR8 role in testicular descent and cryptorchidism. *Reprod Biomed Online* 2005;**10**(1): 49–54.

60. New MI, Rapaport R. The adrenal cortex. *Pediatric endocrinology.* Philadelphia: Saunders; 1996:281–314.

61. New MI. Inborn errors of adrenal steroidogenesis. *Mol Cell Endocrinol* 2003;**211**(1-2):75–83.

62. Peter M. Congenital adrenal hyperplasia: 11beta-hydroxylase deficiency. *Semin Reprod Med* 2002;**20**(3):249–54.

63. Peterson RE, Imperato-McGinley J, Gautier T, Shackleton C. Male pseudohermaphroditism due to multiple defects in steroid-biosynthetic microsomal mixed-function oxidases. A new variant of congenital adrenal hyperplasia. *N Engl J Med* 1985;**313**(19): 1182–91.

64. Shackleton C, Marcos J, Arlt W, Hauffa BP. Prenatal diagnosis of P450 oxidoreductase deficiency (ORD): a disorder causing low pregnancy estriol, maternal and fetal virilization, and the Antley–Bixler syndrome phenotype. *Am J Med Genet* 2004;**129A**(2): 105–12.

65. Arlt W, Walker EA, Draper N, Ivison HE, Ride JP, Hammer F, et al. Congenital adrenal hyperplasia caused by mutant P450 oxidoreductase and human androgen synthesis: analytical study. *Lancet* 2004;**363**(9427):2128–35.

66. Scott RR, Miller WL. Genetic and clinical features of P450 oxidoreductase deficiency. *Horm Res* 2008;**69**(5):266–75.

67. Tajima T, Fujieda K, Kouda N, Nakae J, Miller WL. Heterozygous mutation in the cholesterol side chain cleavage enzyme (p450scc) gene in a patient with 46,XY sex reversal and adrenal insufficiency. *J Clin Endocrinol Metab* 2001;**86**(8):3820–5.

68. Katsumata N, Ohtake M, Hojo T, Ogawa E, Hara T, Sato N, et al. Compound heterozygous mutations in the cholesterol side-chain cleavage enzyme gene (CYP11A) cause congenital adrenal insufficiency in humans. *J Clin Endocrinol Metab* 2002;**87**(8):3808–13.

69. Stocco DM. Clinical disorders associated with abnormal cholesterol transport: mutations in the steroidogenic acute regulatory protein. *Mol Cell Endocrinol* 2002;**191**(1):19–25.

70. Fujieda K, Okuhara K, Abe S, Tajima T, Mukai T, Nakae J. Molecular pathogenesis of lipoid adrenal hyperplasia and adrenal hypoplasia congenita. *J Steroid Biochem Mol Biol* 2003;**85**(2-5):483–9.

71. Fujieda K, Tajima T, Nakae J, Sageshima S, Tachibana K, Suwa S, et al. Spontaneous puberty in 46,XX subjects with congenital lipoid adrenal hyperplasia. Ovarian steroidogenesis is spared to some extent despite inactivating mutations in the steroidogenic acute regulatory protein (StAR) gene. *J Clin Invest* 1997;**99**(6): 1265–71.

72. Peterson RE, Imperato-McGinley J, Gautier T, Sturla E. Male pseudohermaphroditism due to steroid 5-alpha-reductase deficiency. *Am J Med* 1977;**62**(2):170–91.

73. Imperato-McGinley J, Gautier T, Zirinsky K, Hom T, Palomo O, Stein E, et al. Prostate visualization studies in males homozygous and heterozygous for 5 alpha-reductase deficiency. *J Clin Endocrinol Metab* 1992;**75**(4):1022–6.

74. Imperato-McGinley J. 5alpha-reductase-2 deficiency and complete androgen insensitivity: lessons from nature. *Adv Exp Med Biol* 2002;**511**:121–31 discussion 31–34.

75. Saenger P, Goldman AS, Levine LS, Korth-Schutz S, Muecke EC, Katsumata M, et al. Prepubertal diagnosis of steroid 5 alpha-reductase deficiency. *J Clin Endocrinol Metab* 1978;**46**(4):627–34.

76. Biason-Lauber A, Miller WL, Pandey AV, Fluck CE. Of marsupials and men: "backdoor" dihydrotestosterone synthesis in male sexual differentiation. *Mol Cell Endocrinol* 2013;**371**(1–2):124–32.

77. Fukami M, Homma K, Hasegawa T, Ogata T. Backdoor pathway for dihydrotestosterone biosynthesis: implications for normal and abnormal human sex development. *Develop Dynam: an official publication of the American Association of Anatomists* 2013;**242**(4): 320–9.

78. Kamrath C, Hartmann MF, Wudy SA. Androgen synthesis in patients with congenital adrenal hyperplasia due to 21-hydroxylase deficiency. *Horm Metab Res* 2013;**45**(2):86–91.

79. Fluck CE, Meyer-Boni M, Pandey AV, Kempna P, Miller WL, Schoenle EJ, et al. Why boys will be boys: two pathways of fetal testicular androgen biosynthesis are needed for male sexual differentiation. *Am J Hum Genet* 2011;**89**(2):201–18.

80. Galli-Tsinopoulou A, Hiort O, Schuster T, Messer G, Kuhnle U. A novel point mutation in the hormone binding domain of the androgen receptor associated with partial and minimal androgen insensitivity syndrome. *J Pediatr Endocrinol Metab* 2003;**16**(2): 149–54.

81. Pinsky L, Kaufman M, Killinger DW, Burko B, Shatz D, Volpe R. Human minimal androgen insensitivity with normal dihydrotestosterone-binding capacity in cultured genital skin fibroblasts: evidence for an androgen-selective qualitative abnormality of the receptor. *Am J Hum Genet* 1984;**36**(5):965–78.

82. Bevan CL, Hughes IA, Patterson MN. Wide variation in androgen receptor dysfunction in complete androgen insensitivity syndrome. *J Steroid Biochem Mol Biol* 1997;**61**(1-2):19–26.

83. Brinkmann AO. Molecular basis of androgen insensitivity. *Mol Cell Endocrinol* 2001;**179**(1-2):105–9.

84. Sultan C, Lumbroso S, Paris F, Jeandel C, Terouanne B, Belon C, et al. Disorders of androgen action. *Semin Reprod Med* 2002;**20**(3):217–28.

85. Gottlieb B, Beitel LK, Wu JH, Trifiro M. The androgen receptor gene mutations database (ARDB): 2004 update. *Hum Mutat* 2004;**23**(6):527–33.

86. Rutgers JL, Scully RE. The androgen insensitivity syndrome (testicular feminization): a clinicopathologic study of 43 cases. *Int J Gynecol Pathol* 1991;**10**(2):126–44.

87. Mazur T. Gender dysphoria and gender change in androgen insensitivity or micropenis. *Arch Sex Behav* 2005;**34**(4):411–21.

88. Brown TR, Migeon CJ. Cultured human skin fibroblasts: a model for the study of androgen action. *Mol Cell Biochem* 1981;**36**(1):3–22.

89. Ris-Stalpers C, Hoogenboezem T, Sleddens HF, Verleun-Mooijman MC, Degenhart HJ, Drop SL, et al. A practical approach to the detection of androgen receptor gene mutations and pedigree analysis in families with x-linked androgen insensitivity. *Pediatr Res* 1994;**36**(2):227–34.

90. Richter-Unruh A, Martens JW, Verhoef-Post M, Wessels HT, Kors WA, Sinnecker GH, et al. Leydig cell hypoplasia: cases with new mutations, new polymorphisms and cases without mutations in the luteinizing hormone receptor gene. *Clin Endocrinol (Oxf)* 2002;**56**(1):103–12.

91. Wu SM, Leschek EW, Rennert OM, Chan WY. Luteinizing hormone receptor mutations in disorders of sexual development and cancer. *Front Biosci* 2000;**5**:D343–52.

92. Martens JW, Verhoef-Post M, Abelin N, Ezabella M, Toledo SP, Brunner HG, et al. A homozygous mutation in the luteinizing hormone receptor causes partial Leydig cell hypoplasia: correlation between receptor activity and phenotype. *Mol Endocrinol* 1998;**12**(6):775–84.

93. Gromoll J, Eiholzer U, Nieschlag E, Simoni M. Male hypogonadism caused by homozygous deletion of exon 10 of the luteinizing hormone (LH) receptor: differential action of human chorionic gonadotropin and LH. *J Clin Endocrinol Metab* 2000;**85**(6):2281–6.

94. Latronico AC, Shinozaki H, Guerra Jr G, Pereira MA, Lemos Marini SH, Baptista MT, et al. Gonadotropin-independent precocious puberty due to luteinizing hormone receptor mutations in Brazilian boys: a novel constitutively activating mutation in the first transmembrane helix. *J Clin Endocrinol Metab* 2000;**85**(12):4799–805.

95. Latronico AC. Naturally occurring mutations of the luteinizing hormone receptor gene affecting reproduction. *Semin Reprod Med* 2000;**18**(1):17–20.

96. Carel JC, Leger J. Clinical practice. Precocious puberty. *N Engl J Med* 2008;**358**(22):2366–77.

97. Latronico AC, Anasti J, Arnhold IJ, Rapaport R, Mendonca BB, Bloise W, et al. Brief report: testicular and ovarian resistance to luteinizing hormone caused by inactivating mutations of the luteinizing hormone-receptor gene. *N Engl J Med* 1996;**334**(8):507–12.

98. Latronico AC, Chai Y, Arnhold IJ, Liu X, Mendonca BB, Segaloff DL. A homozygous microdeletion in helix 7 of the luteinizing hormone receptor associated with familial testicular and ovarian resistance is due to both decreased cell surface expression and impaired effector activation by the cell surface receptor. *Mol Endocrinol* 1998;**12**(3):442–50.

99. Toledo SP, Brunner HG, Kraaij R, Post M, Dahia PL, Hayashida CY, et al. An inactivating mutation of the luteinizing hormone receptor causes amenorrhea in a 46,XX female. *J Clin Endocrinol Metab* 1996;**81**(11):3850–4.

100. Bruysters M, Christin-Maitre S, Verhoef-Post M, Sultan C, Auger J, Faugeron I, et al. A new LH receptor splice mutation responsible for male hypogonadism with subnormal sperm production in the propositus, and infertility with regular cycles in an affected sister. *Hum Reprod* 2008;**23**(8):1917–23.

101. Latronico AC, Arnhold IJ. Inactivating mutations of LH and FSH receptors – from genotype to phenotype. *Pediatr Endocrinol Rev* 2006;**4**(1):28–31.

102. Chae HD, Park EJ, Kim SH, Kim CH, Kang BM, Chang YS. Ovarian hyperstimulation syndrome complicating a spontaneous singleton pregnancy: a case report. *J Assist Reprod Genet* 2001;**18**(2):120–3.

103. Smits G, Olatunbosun O, Delbaere A, Pierson R, Vassart G, Costagliola S. Ovarian hyperstimulation syndrome due to a mutation in the follicle-stimulating hormone receptor. *N Engl J Med* 2003;**349**(8):760–6.

104. Vasseur C, Rodien P, Beau I, Desroches A, Gerard C, de Poncheville L, et al. A chorionic gonadotropin-sensitive mutation in the follicle-stimulating hormone receptor as a cause of familial gestational spontaneous ovarian hyperstimulation syndrome. *N Engl J Med* 2003;**349**(8):753–9.

105. Aittomaki K, Lucena JL, Pakarinen P, Sistonen P, Tapanainen J, Gromoll J, et al. Mutation in the follicle-stimulating hormone receptor gene causes hereditary hypergonadotropic ovarian failure. *Cell* 1995;**82**(6):959–68.

106. Layman LC, Amde S, Cohen DP, Jin M, Xie J. The Finnish follicle-stimulating hormone receptor gene mutation is rare in North American women with 46,XX ovarian failure. *Fertil Steril* 1998;**69**(2):300–2.

107. Christin-Maitre S, Vasseur C, Portnoi MF, Bouchard P. Genes and premature ovarian failure. *Mol Cell Endocrinol* 1998;**145**(1–2):75–80.

108. Matthews CH, Borgato S, Beck-Peccoz P, Adams M, Tone Y, Gambino G, et al. Primary amenorrhoea and infertility due to a mutation in the beta-subunit of follicle-stimulating hormone. *Nat Genet* 1993;**5**(1):83–6.

109. Beyribey S, Cetinkaya M, Adsan O, Memis A, Ozturk B. Persistent Mullerian duct syndrome. *Scand J Urol Nephrol* 1993;**27**(4):563–5.

110. Josso N, Belville C, di Clemente N, Picard JY. AMH and AMH receptor defects in persistent Mullerian duct syndrome. *Hum Reprod Update* 2005;**11**(4):351–6.

111. Josso N, Picard JY, Imbeaud S, di Clemente N, Rey R. Clinical aspects and molecular genetics of the persistent Mullerian duct syndrome. *Clin Endocrinol (Oxf)* 1997;**47**(2):137–44.

112. Koksal S, Tokmak H, Tibet HB, Olgun E. Persistent Mullerian duct syndrome. *Br J Clin Pract* 1995;**49**(5):276–7.

113. Lang-Muritano M, Biason-Lauber A, Gitzelmann C, Belville C, Picard Y, Schoenle EJ. A novel mutation in the anti-Mullerian hormone gene as cause of persistent Mullerian duct syndrome. *Eur J Pediatr* 2001;**160**(11):652–4.

114. Berkmen F. Persistent Mullerian duct syndrome with or without transverse testicular ectopia and testis tumours. *Br J Urol* 1997;**79**(1):122–6.

115. Ito Y, Fisher CR, Conte FA, Grumbach MM, Simpson ER. Molecular basis of aromatase deficiency in an adult female with sexual infantilism and polycystic ovaries. *Proc Natl Acad Sci USA* 1993;**90**(24):11673–7.

116. Shozu M, Akasofu K, Harada T, Kubota Y. A new cause of female pseudohermaphroditism: placental aromatase deficiency. *J Clin Endocrinol Metab* 1991;**72**(3):560–6.

117. Herrmann BL, Saller B, Janssen OE, Gocke P, Bockisch A, Sperling H, et al. Impact of estrogen replacement therapy in a male with congenital aromatase deficiency caused by a novel mutation in the CYP19 gene. *J Clin Endocrinol Metab* 2002;**87**(12):5476–84.

118. Mullis PE, Yoshimura N, Kuhlmann B, Lippuner K, Jaeger P, Harada H. Aromatase deficiency in a female who is compound heterozygote for two new point mutations in the P450arom gene: impact of estrogens on hypergonadotropic hypogonadism, multicystic ovaries, and bone densitometry in childhood. *J Clin Endocrinol Metab* 1997;**82**(6):1739–45.

119. Conte FA, Grumbach MM, Ito Y, Fisher CR, Simpson ER. A syndrome of female pseudohermaphrodism, hypergonadotropic hypogonadism, and multicystic ovaries associated with missense mutations in the gene encoding aromatase (P450arom). *J Clin Endocrinol Metab* 1994;**78**(6):1287–92.
120. Carreau S, Lambard S, Delalande C, Denis-Galeraud I, Bilinska B, Bourguiba S. Aromatase expression and role of estrogens in male gonad: a review. *Reprod Biol Endocrinol* 2003;**1**(1):35.
121. Rochira V, Balestrieri A, Madeo B, Spaggiari A, Carani C. Congenital estrogen deficiency in men: a new syndrome with different phenotypes; clinical and therapeutic implications in men. *Mol Cell Endocrinol* 2002;**193**(1–2):19–28.
122. Belgorosky A, Pepe C, Marino R, Guercio G, Saraco N, Vaiani E, et al. Hypothalamic-pituitary-ovarian axis during infancy, early and late prepuberty in an aromatase-deficient girl who is a compound heterocygote for two new point mutations of the CYP19 gene. *J Clin Endocrinol Metab* 2003;**88**(11):5127–31.
123. Nelson KG, Goldenberg RL. Sex hormones and congenital malformations: a review. *J Med Assoc State Ala* 1977;**46**(11):31.
124. Coenen CM, Thomas CM, Borm GF, Rolland R. Comparative evaluation of the androgenicity of four low-dose, fixed-combination oral contraceptives. *Int J Fertil Menopausal Stud* 1995;**40**(Suppl 2): 92–7.

125. Ziaei S, Rajaei L, Faghihzadeh S, Lamyian M. Comparative study and evaluation of side effects of low-dose contraceptive pills administered by the oral and vaginal route. *Contraception* 2002;**65**(5):329–31.
126. Joshi R, Dunaif A. Ovarian disorders of pregnancy. *Endocrinol Metab Clin North Am* 1995;**24**(1):153–69.
127. Yang Y, Muzny DM, Reid JG, Bainbridge MN, Willis A, Ward PA, et al. Clinical whole-exome sequencing for the diagnosis of mendelian disorders. *N Engl J Med* 2013;**369**(16):1502–11.
128. Arboleda VA, Lee H, Sanchez FJ, Delot EC, Sandberg DE, Grody WW, et al. Targeted massively parallel sequencing provides comprehensive genetic diagnosis for patients with disorders of sex development. *Clin Genet* 2013;**83**(1):35–43.
129. Crissman HP, Warner L, Gardner M, Carr M, Schast A, Quittner AL, et al. Children with disorders of sex development: a qualitative study of early parental experience. *Int J Pediatr Endocrinol* 2011;**2011**(1):10.
130. Sandberg DE, Mazur T. A noncategorical approach to the psychosocial care of persons with DSD and their families. Gender Dysphoria and Disorders of Sex Development. US: Springer; 2014. p. 93–114.
131. Arboleda VA, Vilain E. Introduction to human sex development. Yen and Jaffe's Reproductive Endocrinology. Philadelphia: Saunders/Elsevier; 2013. p. 351.

20

Androgen Insensitivity Due to Mutations of the Androgen Receptor

Michael J. McPhaul

Division of Endocrinology, University of Texas Southwestern Medical Center, Dallas, TX

Androgens are critical to the normal developmental processes in vertebrates, affecting the development of the normal male phenotype during embryogenesis and regulating a range of processes in adults.[1] In normal development, the actions of two related steroids, testosterone and 5-alpha-dihydrotestosterone (DHT), are required for the development of the normal male phenotype. These steroid hormones bind to and modulate the activity of the androgen receptor (AR), a member of the steroid hormone receptor family that is encoded on the X-chromosome. The cloning of the human AR has permitted a wealth of information about the types of alterations of the *AR* gene that result in alterations of AR function.

MALE PHENOTYPIC DEVELOPMENT IS CONTROLLED BY ANDROGENS

The pioneering studies of Jost established the endocrine basis of male phenotypic development,[2] and subsequent experiments have confirmed and refined this model (Fig. 20.1). In mammals, the determination of whether the indifferent gonad will differentiate as a testis or an ovary is determined by the complement of sex chromosomes that are present. This fundamental developmental step determines the pattern of sex-specific secretion of gonadal hormones. In male embryos, testosterone is secreted by the Leydig cells of the testes beginning at approximately nine weeks of gestation.[3,4] This steroid hormone acts to virilize the fetal Wolffian duct structures. Although the synthesis of testosterone is required for male sexual development, it is not sufficient for full male phenotypic development. The synthesis of DHT from testosterone by the 5-alpha reductase 2 enzyme is required for virilization of the male external genitalia and development of the prostate,[5] conclusions supported by both human studies of patients deficient in the 5-alpha reductase 2 enzyme and the experimental use of inhibitors of the 5-alpha reductase 2 enzyme.

While this chapter is focused on the molecular defects of the AR, it is important to note that the Sertoli cells of the fetal testes also secrete a second hormone essential for normal male phenotypic development: Müllerian inhibiting substance. This polypeptide hormone, a member of the TGF-beta hormone family, acts to effect an involution of the embryonic Müllerian ducts, structures that persist and develop in the female embryo, giving rise to the uterus and fallopian tubes. Abnormalities of MIS synthesis or action underlie forms of male pseudohermaphroditism in which male phenotypic development has occurred (virilization of the internal and external genitalia has occurred normally), but in whom the uterus and fallopian tubes are present.[1,4]

THE ANDROGEN RECEPTOR

The AR was one of the first receptor proteins to be characterized following the synthesis of high specific activity androgen ligands.[6,7] The cloning of cDNAs encoding the receptor protein[8–10] permitted its identification as a member of a family of structurally related transcriptionally regulated transcription factors, the nuclear receptor (NR) family.[11] Each family member contains specific structural motifs: an amino terminus of varying length, a DNA-binding domain (DBD) containing two "zinc fingers," and a carboxyl-terminal ligand-binding domain (LBD) (Fig. 20.2).

The central DBD is the most highly conserved element within each of the NR proteins. The sequence similarities of this distinctive motif were sufficiently high

Genetic Diagnosis of Endocrine Disorders. http://dx.doi.org/10.1016/B978-0-12-800892-8.00020-8

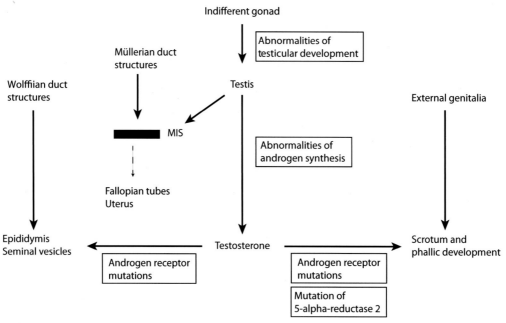

FIGURE 20.1 **Male phenotypic development.** Genotypic sex dictates the subsequent developmental pathway that occurs within the indifferent gonad. If the gonad differentiates into a testis, Müllerian inhibiting substance (MIS) and testosterone are secreted. In selected tissues, the synthesis of 5-alpha-DHT is required for virilization to occur.

among the different members that nucleotide sequence homology between them permitted the cloning of several members of the NR family, including the AR.[11] This region of each of the NR proteins encodes two short segments that contain highly conserved, specifically spaced cysteine residues that each coordinate a zinc atom (zinc fingers). In a number of naturally occurring mutations and in site-directed mutagenesis studies, these highly structured elements have been demonstrated to define the components of each receptor protein that are able to directly recognize and bind to specific DNA sequences within target DNA sequences within and surrounding target genes.

The region of the NR proteins responsible for the high-affinity binding of ligands is the LBD. This region is also highly conserved, albeit not to the degree observed between the DBD regions of the NRs. The overall structural homology of this segment is present, even among members of the family not known to bind a ligand. The LBD of the AR has been demonstrated to extend from approximately amino acid residue 690 to its carboxyl terminus, a size consistent with this segment in other members of the NR family.[12] Structural studies have established the crystal structures of the AR and shown that its structure is similar to the organization of this region in other members of this family. The primary structure of the AR constitutes 12 alpha helices that are arranged in a three-layered sandwich around a central binding pocket. The position of the terminal helix relative to the remainder of the LBD is modulated by the binding of agonist

ligands. The position of this region determines whether critical surfaces needed for the recruitment of receptor coactivators are exposed.[13,14]

In contrast to the DBD and LBD of the AR, the functional role of the amino terminus of the receptor protein is less easily defined. This region is large, encompassing more than half of the primary amino acid sequence of the AR. In deletion experiments using androgen responsive genes to measure the function of the AR, it was observed that deletions of large segments of the AR amino terminus were required to substantially perturb AR function;[12] no individual amino acids or small segments were shown to be critical for alter AR function. This may well indicate that the functional elements of the AR amino terminus are spread throughout this region in a redundant fashion. Alternatively, the requirement of large portions of the amino terminus for AR function may reflect its participation in facilitating the binding of DNA, ligand, and coactivators needed for full function. It is interesting to note that several lines of investigation have concluded that interactions between regions of the amino and carboxyl termini of the AR can influence one another and may exert discernible effects on AR function.[15–17]

The amino acid sequence of the human AR contains three segments composed of repeated amino acid residues: one of glutamine, one of glycine, and one of proline. Similar repeat elements have been identified in other transcription factors, including other members of the steroid receptor family. The repeated proline motif does not appear to be constant, and variations in

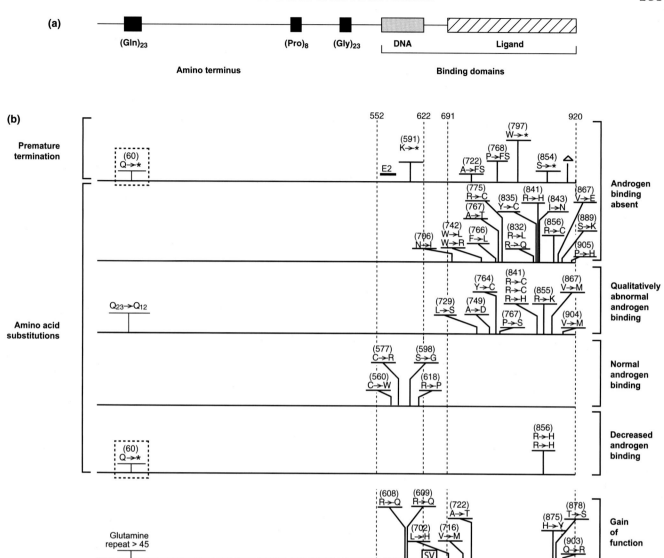

FIGURE 20.2   Schematic of the human AR and overview of differ types of AR mutations. (a) Diagram of the predicted amino acid sequence of the human AR, including a depiction of its functional domains. (b) The loss of function mutations depicted here is largely derived from a single center where such analyses were routinely performed (1). The representative gain-of-function mutations are from published reports. The approximate position within the AR open reading frame, where the constitutive AR variants identified in prostate cancer deviate from the sequence of the normal AR sequence, is indicated within the boxed "SV." *The coordinates employed in this figure are those employed in the Androgen Receptor Database (NCBI reference sequence (NM_000044.2)).*

its length have not been reported. Similarly, polymorphisms of the glycine repeat appear to be relatively infrequent and of little biological consequence. These circumstances stand in contrast with the frequent polymorphisms of the glutamine repeat in the human population.[15,18,19] Of note, differences in the length of this segment have been associated with subtle differences in AR function.[12] Shortening of this same region has been associated with biological attributes such as infertility and a predisposition to prostate and breast cancer. Importantly, expansion of this region has been shown to cause spinal and bulbar muscular atrophy, the first of the "triplet expansion" diseases to be identified.[20]

MEASUREMENTS OF AR AND ITS FUNCTION

Assays to detect the AR were first developed using high specific activity radiolabeled ligands to measure the high affinity binding of these ligands in target tissues and cell lines. These assays were used by several groups to characterize the physical properties of the binding protein is tissue extracts and in samples from cell lines established from normal subjects and from patients with known or suspected abnormalities of AR function.[1] Monolayer binding assays permitted many such subjects to be identified as having abnormalities of the AR on the basis of alterations in the absolute amounts of AR measured or qualitative abnormalities of ligand binding. Although these assays were often informative, a significant number of patients displayed no differences of ligand binding, many times in circumstances where androgen insensitivity was highly likely (e.g., on the basis of family history).

Following the cloning of AR, the predicted amino acid sequence was used to generate antibodies to different regions of the AR protein. The most useful of these were those directed at the first 20 amino acids of the receptor proteins. These epitopes are highly conserved and are present in a number of mammalian species. While useful at the level of the research laboratory to categorize AR expression in patient samples,[21] such assays have not been widely used clinically.

As noted previously, the AR is highly related to other members of the steroid receptor family, particularly the progesterone and glucocorticoid receptors. The definition of model genes responsive to the progesterone and glucocorticoid receptors in cell transfection assays was adapted to the study of the function of the AR. This provided methods to define the functional domains of the AR and to assess the impact of mutations of the AR on its function, independent of the ability to bind ligand. The delivery of such reporter genes directly into patient fibroblasts permitted the assay of the activity of the natively expressed receptor protein. In a subset of such samples, the activities were suggestive of defects in non-AR proteins important for AR function.[22]

THE MODULATION OF GENE EXPRESSION BY THE AR

The human AR exerts its principal effects in the nucleus and the sequences within the AR responsible for nuclear localization sequences within and adjacent to the second zinc finger have been defined.[23,24] In the absence of ligand, the AR exists in association with a complex of chaperone proteins. Following binding of agonist by the receptor, it undergoes extensive conformational changes, particularly the position of the terminal helix of the LBD.[13] This change permits the binding of target DNA sequences and the recruitment of ancillary proteins that facilitate and stabilize interactions with complexes of proteins, including receptor coactivators, that interact with the basal transcription machinery. These alterations result in the binding of the activated AR to specific sites within the genome, alterations of chromatic structure, and changes in activity of androgen responsive genes.[25] Distinct conformations of the AR LBD result when complexed to AR antagonists.[13] These distinctive conformations recruit different classes of proteins (corepressors) that recruit distinct complexes that result in decreases in androgen responsive genes.

ANDROGEN INSENSITIVITY: A SPECTRUM OF ABNORMALITIES CAUSED BY DEFECTS OF THE AR

Genetic defects of the AR are relatively common, a fact that can be explained by several features of the AR and of male phenotypic development. First, the AR gene is located on the human X chromosome, a chromosomal location that ensures that in genetic males, only a single copy of the AR gene is present and that a significant alteration of AR function will be translated into a discernible alteration of male development. Second, the detection of abnormalities of sexual development during the routine careful examination of neonates is often a trigger for extensive evaluations. Finally, while defects of AR function may have substantial effects on the development of the male phenotype, they appear to have limited impact on viability.

In early descriptions, it was not clear how the different phenotypes associated with abnormalities of male sexual development caused by defects of the AR related to one another. It was not clear whether patients with apparently different phenotypes such as complete androgen insensitivity (complete testicular feminization) and the Reifenstein syndrome were due to specific classes of mutations within the AR. It is now evident that androgen insensitivity is a spectrum, and the phenotype that is observed in an individual patient represents the degree to which the function of the AR proteins has been disrupted or the level of the AR expressed has been altered. In many instances these attributes (level of expression and function) are two intertwined aspects of the underlying genetic change that dictates the extent that the effects of androgen in male phenotypic development have been disturbed.

At one extreme of the phenotypic spectrum are individuals in whom the AR possesses no remaining function. Such alterations may include partial or complete gene deletions, premature termination, or an amino acid substitution in a critical domain of the receptor protein. As a result of this absence of AR function, none of the normal effects of testosterone and 5-alpha-DHT occur: phallic growth and scrotal formation do not occur despite the secretion of normal amounts of androgen. Instead of virilization, such individuals with complete androgen insensitivity display a feminized phenotype. The circulating levels of testosterone – although unable to effect virilization – are aromatized to estrogen, which stimulates normal breast development. The normal testes present in these individuals secrete MIS, which mediates the involution of the structures derived from the Müllerian ducts. As a result, these individuals do not have fallopian tubes or a uterus, the vagina is blind-ending, and the testes are either located within the abdomen or within the inguinal canals. A large number of diverse mutations of the AR have been reported in association with the phenotype of complete androgen insensitivity.

At the other end of the phenotypic spectrum are individuals in whom the male phenotype reflects a nearly normal level of virilization. Such affected individuals (often referred to as mild androgen insensitivity syndrome, MAIS) display normal phallic and scrotal development, but exhibit subtle signs of undervirilization, such as gynecomastia, poor beard growth, and infertility. Only a limited number of mutations in the AR have been described in such individuals. In all instances in which this has been examined, the mutations result in limited impairment of AR function in functional assays.

Between these two extremes of phenotypic development are individuals in whom substantial alterations of virilization are evident. Historically, these syndromes have been referred to using a variety of terms, including Reifenstein's syndrome, partial androgen insensitivity syndrome (PAIS), and incomplete testicular feminization.[1,26] In the most severely affected individuals, only minor amounts of virilization are evident. Such individuals may have a primarily female phenotype, but with only minor degrees of clitoromegaly. In individuals in whom a greater degree of AR function is preserved, significant virilization is present, but in the context of substantial defects of urogenital development, such as perineal hypospadias (the Reifenstein phenotype). Quigley et al. have employed a more detailed system with which to categorize patients with partial androgen insensitivity.[26]

THE GENETIC BASIS OF ANDROGEN INSENSITIVITY

More than 500 different mutations have been described within the AR gene associated with androgen insensitivity, both within and adjacent to the coding region of the receptor protein.[27]

The following summary groups the types of mutations that have been identified and presents generalizations regarding effects on receptor function and resulting phenotype for each of the categories. The locations of representative examples of the different mutation types are depicted in Fig. 20.2.

DISRUPTION OF THE PRIMARY AMINO ACID SEQUENCE

Many mutations have been described that interrupt the AR open reading frame. These include deletions and insertions (both small and large scale) and mutations that result in changes in the splicing of the mRNA encoding the AR protein. As noted previously, the regions of the AR that are responsible for mediating the recognition of androgen-responsive DNA sequences and for the binding of ligand by the AR are located in the center and carboxyl terminal regions of the protein, respectively (Fig. 20.2). For this reason, termination of the receptor within the open reading frame will result in the synthesis of an AR protein that is lacking part or all of one of these critical domains. In general, small insertions or deletions within the critical DBD and LBD that maintain the integrity of the open reading frame completely disrupt AR function.

ALTERATIONS OF THE DBD

In studies performed in genital skin fibroblasts from patients with clinical features, endocrine testing, and family history consistent with androgen insensitivity, approximately one-fifth displayed normal levels of qualitatively normal androgen ligand binding.[1]

The AR genes from a number of such individuals have been analyzed. In virtually all instances, alterations of the structure of the DBD of the receptor have been identified. In one of the first series examining this class of AR defect,[28] investigators found that when recreated in heterologous cells, the amino acid substitutions identified within the DBD of receptor protein had no effect on ligand binding, When examined in functional assays, however, these mutant receptors were impaired in their ability to bind to target DNA sequences and to stimulate a model reporter gene, findings consistent with results from a number of other laboratories.[29–33] These findings are also relevant to alterations of the DBD caused by deletions and insertions and by alterations of mRNA splicing.

ALTERATIONS OF LBD STRUCTURE

Mutations that result in replacement of single amino acids are the most frequent mutation identified in patients with all forms of androgen insensitivity. Owing to the wide range of effects that such mutations have on the function of the AR, such mutations have been identified in patients across the entire spectrum of androgen insensitivity. Such mutations may also have markedly different effects on the binding of ligand. These differences reflect the extent to which the mutations alter the highly ordered structure of the LBD.

A number of amino acid substitutions have been identified in patients in whom no binding of ligand was detectable in assays performed in monolayer binding assays. When these mutations were created and studied in heterologous cells, they have been found to exert a range of effects on ligand binding by the receptor protein. In virtually all instances, these mutant ARs can be shown to bind specifically ligand. When examined carefully, the ligand binding is of lower affinity and the ligand complexes formed are unstable. In functional assays, these mutant ARs are impaired in their ability to stimulate androgen-responsive genes.[21,34] In rare instances, the amino acid substitutions disrupt the structure of the LBD so markedly that ligand binding by the mutant receptor cannot be detected.[35]

When characterized in monolayer binding assays, a subset of patients was found to have normal levels of AR binding, but to display clear qualitative disturbances of binding. Such differences included both differences in affinity, instability of the receptor-ligand complexes to temperature elevation, or accelerated rates of ligand dissociation.

The mutations causing androgen insensitivity that result in qualitative abnormalities of ligand binding have invariably been traced to single amino acid changes within the LBD of the receptor. The positions of the substituted amino residues are similar to the distribution observed in subjects with mutations that cause absent ligand binding.[36] Experiments examining the effects of different amino acid substitutions at the identical residue within the AR coding sequence have demonstrated that either absent ligand binding or qualitative defects of ligand binding are observed, depending on the nature of the amino acid substitution: more disruptive substitutions leading to absent binding and less disruptive substitutions leading to qualitative abnormalities of ligand binding.[37–40]

The analysis of ARs with amino acid substitutions in the LBD has provided some interesting insights into the critical nature of the stability of the LBD bound to ligand in AR function. When analyzed in heterologous cells with active oxidation of androgens at the 17-hydroxyl group, different responses were observed when stimulated by testosterone, DHT, and mibolerone (a potent androgen agonist, resistant to metabolism).[35] The mutant receptors showed the greatest degree of deficiency when stimulated with testosterone, lesser degrees of deficient function in response to DHT, and even less deficit of function was observed when stimulated with mibolerone. Repeated pulses of DHT and testosterone were also shown to stimulate the activity of many mutant ARs of this type. These experiments demonstrate that ARs that carry mutations of the LBD that result in qualitative abnormalities of binding can be stimulated pharmacologically.[35,41] These conclusions have been reinforced by instances in which patients treated with supraphysiologic levels of androgen have demonstrated discernible clinical effects.[42,43]

MUTATIONS WITHIN THE AMINO TERMINUS

A significant number of mutations have been identified within the amino terminus of the AR. These alterations have been reported in association with a wide range of phenotypes, including partial and minimal androgen insensitivity. As noted previously, a significant portion of the AR amino terminus must be altered before alterations of AR function can be measured in functional assays. Thus, only those mutations resulting in frameshift or termination are readily supported by functional studies. Reports of clinical androgen insensitivity due to missense mutations within the amino terminus of the AR are infrequent and deserving of additional study.[44–46] One report has suggested that specific reporter gene assays are needed in order to characterize the alterations of AR function in these instances.[47]

MUTATIONS THAT CAUSE DECREASED LEVELS OF LIGAND BINDING

This category of alteration of AR function was defined using ligand-binding assays to define patients with androgen with measureable ligand binding in fibroblasts, but in which all measurable qualitative aspects of ligand and DNA binding appeared normal. The only difference discernible was the absolute quantity of AR present in the individual fibroblast strains. The types of mutations causing decreased levels of ligand binding have proven to be heterogeneous. Mutations that have been identified to cause androgen insensitivity in this context appear to principally alter the amount of AR that is synthesized.

The first example of a mutation of this type was uncovered in the analysis of a pedigree with complete androgen insensitivity. Binding studies in genital skin fibroblasts demonstrated a qualitatively normal androgen protein, albeit present in reduced amounts.[48] Immunoblots of extracts prepared for these fibroblasts detected a shortened AR that did not react with antibodies to the amino terminus. Sequence analysis of the *AR* gene revealed a missense mutation that introduced a premature termination codon at position 188. Subsequent studies demonstrated that this shortened form of the AR (AR-A) results from internal initiation at methionine 188 and is present in normal fibroblasts and tissues at low levels.[49,50] Studies performed in transfected cells demonstrated that the AR-A isoform displays subtle differences of function on selected response elements, suggesting that phenotype of complete androgen insensitivity that was observed in the original pedigree was due to the expression of a reduced level of AR that possesses a reduced functional capacity.[51]

An even more direct example of androgen insensitivity caused by a reduced amount of AR is that described by Choong et al.[52] In this pedigree, the basis for a form of partial androgen insensitivity was traced to a single amino acid substitution at the second amino acid residue (E2K; lysine residue in place of the normal glutamic acid residue) of the AR open reading frame. Analysis of the mutant AR in transfection studies supported the conclusion that reduced efficiency of translation resulted from the missense mutation and its effect on the Kozak consensus sequence surrounding the initiator methionine.

PHENOTYPE AND GENOTYPE IN PATIENTS WITH VARIOUS FORMS OF ANDROGEN INSENSITIVITY

As noted, any disturbance of gonadal development or androgen synthesis can produce an abnormality of the male phenotype. In instances where strict definitions are employed to identify individuals likely to have complete or partial forms of androgen insensitivity (on the basis of family history and endocrine testing), a high proportion of individuals will be found to harbor mutations of the *AR* gene. Consistently, genetic mutations are identified with highest frequency in patients with complete androgen insensitivity and at a lower frequency in patients with partial forms of androgen insensitivity.[53–55]

A large number of mutations of the AR have been characterized at the genetic level and many have been examined in functional studies. These studies permit several generalizations to be made.

Mutations that result in termination of the primary structure of the AR are associated with complete androgen insensitivity. The function of the AR in the virilization mediated by androgen requires an intact LBD. Interruption of the receptor protein sequence, whether caused by a frameshift mutation, aberrant splicing, deletion, or insertion of a premature termination codon, will lead to the production of a receptor that lacks all or part of at least one critical functional domain. For this reason, such receptor proteins cannot be activated by ligand and are associated with complete androgen insensitivity.

Amino acid substitutions within the AR coding segment have been associated with the entire spectrum of androgen resistance. This is because the range of impairment of AR function can be much subtler than the complete loss of function associated with premature termination of the AR, such as slight alterations of ligand binding or partial impairments of DNA binding. In these instances, the phenotype appears to correlate with the level of mutant AR protein and its residual function.

In a limited number of families, significant differences in phenotype have been observed between individuals in different families that harbor the same mutation. In some instances, the differences are relatively small, but in other instances, a significant difference in phenotype has been reported.[27] The mechanism for this mismatch in phenotype has not been clarified for all instances, but has been postulated to reflect either differences in the level of the expressed mutant protein or difference in the levels of 5-alpha reductase expressed in the affected patients in the different pedigrees.[56] In a small number of cases,

the AR mutation that was identified appeared seemingly inconsistent with the phenotype that is present. In a number of instances, this discordance was traced to the impact of somatic mosaicism.[57–59]

SPINAL AND BULBAR MUSCULAR ATROPHY AND PROSTATE CANCER

Classic forms of androgen insensitivity are caused by varying degrees of impairments of AR function. Two other diseases have been traced to mutations of the AR.

Spinal and bulbar muscular atrophy was shown to be caused by an expansion of the glutamine repeat within the AR amino terminus.[20] This disease, which results in the degeneration of neurons within selected spinal and bulbar nuclei, is believe to be caused by the acquisition of toxic effects of the glutamine expanded AR. SBMA was the first example of a class of neurodegenerative diseases caused by glutamine expansion within the coding sequence of different proteins.[60] The mild androgen resistance that is observed clinically may reflect a modest decrease in AR function or abundance that is caused by the glutamine expansion.

Gain-of-function mutations have been identified in advanced forms of prostate cancer. These mutations have been shown to broaden the ligand specificity of the AR, conferring responsiveness to ligands not observed in the normal, nonmutant ARs.[61–64]

Following the cloning of cDNAs encoding the different members of the steroid receptor family, it was demonstrated that removal of the hormone-binding domain in the receptor proteins resulted in truncated receptors that possessed constitutive activity in functional assays. This was demonstrated specifically for the AR.[65] These observations did not appear to be clinically relevant until experiments demonstrated the presence of splice variants that removed the LBD and which exhibited constitutive activity in functional assays.[66] More recent findings have indicated that these receptor forms have importance in the progression of prostate cancer to an androgen-independent state.[67,68]

DIAGNOSTIC RESOURCES

Sequencing of the AR gene is performed by a number of laboratories in the United States and worldwide.

As with most compilations, the list of laboratories performing AR gene sequencing may not always be completely up to date. A broad list of testing laboratories can be found on GeneTests: https://www.genetests.org/disorders/?disid=2575.

Laboratories providing sequencing of the AR gene in the United States are listed as follows.

Institution	Location	Website	Phone
All Children's Hospital	St. Petersburg, FL	http://www.allkids.org	(727)-767-8985
Baylor Miraca Genetics Laboratory	Houston, TX	http://www.bmgl.com/BMGL/Default.aspx	800-411-GENE (4363)
Denver Genetics Laboratories	Denver, CO	http://www.ucdenver.edu/academics/colleges/ medicalschool/programs/genetics/DiagnosticTests/ DNATests/Pages/DNAtests.aspx	(720) 777-0510
Center for Genetics at St. Francis Genetics Laboratory	Tulsa, OK	http://www.saintfrancis.com/genetics/Pages/ default.aspx	(918) 502-1725
Gene DX	Gaithersburg, MD	https://www.genedx.com/	(301)-519-2100
Prevention Genetics	Marshfield, WI	https://www.preventiongenetics.com/	(715)-387-0484

References

1. Griffin JE, McPhaul MJ, Russell DW, et al. The androgen resistance syndromes: steroid 5α-reductase 2 deficiency, testicular feminization, and related disorders. In: Scriver CR, Beaudet AL, Sly WS, editors. *The metabolic and molecular bases of inherited disease.* 8th ed. New York: McGraw-Hill; 2001. p. 4117–46.

2. Jost A. Hormonal factors in the sex differentiation of the mammalian foetus. *Philos Trans R Soc Lond B Biol Sci* 1970;**259**:119–30.

3. Siiteri PK, Wilson JD. Testosterone formation and metabolism during male sexual differentiation in the human embryo. *J Clin Endocrinol Metab* 1974;**38**:113–25.

4. Rey R, Picard JY. Embryology and endocrinology of genital development. *Baillieres Clin Endocrinol Metab* 1998;**12**:17–33.

5. Wilson JD. Role of dihydrotestosterone in androgen action. *Prostate Suppl* 1996;**6**:88–92.

6. Bruchovsky N, Wilson JD. The intranuclear binding of testosterone and 5-alpha-androstan-17-beta-ol-3-one by rat prostate. *J Biol Chem* 1968;**243**:5953–60.

7. Fang S, Anderson KM, Liao S. Receptor proteins for androgens. On the role of specific proteins in selective retention of 17-beta-hydroxy-5-alpha-androstan-3-one by rat ventral prostate *in vivo* and *in vitro*. *J Biol Chem* 1969;**244**:6584–95.

8. Lubahn DB, Joseph DR, Sullivan PM, Willard HF, French FS, Wilson EM. Cloning of human androgen receptor complementary DNA and localization to the X chromosome. *Science* 1988;**240**:327–30.

9. Chang CS, Kokontis J, Liao ST. Molecular cloning of human and rat complementary DNA encoding androgen receptors. *Science* 1988;**240**:324–6.

10. Tilley WD, Marcelli M, Wilson JD, McPhaul MJ. Characterization and expression of a cDNA encoding the human androgen receptor. *Proc Natl Acad Sci USA* 1989;**86**:327–31.

11. Mangelsdorf DJ, Thummel C, Beato M, et al. The nuclear receptor superfamily: the second decade. *Cell* 1995;**83**:835–9.

12. Gao TS, Marcelli M, McPhaul MJ. Transcriptional activation and transient expression of the human androgen receptor. *J Steroid Biochem Mol Biol* 1996;**59**:9–20.

13. Jin L, Li Y. Structural and functional insights into nuclear receptor signaling. *Adv Drug Deliv Rev* 2010;**62**:1218–26.

14. Lonard DM, O'Malley BW. Nuclear receptor coregulators: modulators of pathology and therapeutic targets. *Nat Rev Endocrinol* 2012;**8**:598–604.

15. McPhaul MJ, Marcelli M, Tilley WD, Griffin JE, Isidro-Gutierrez RF, Wilson JD. Molecular basis of androgen resistance in a family with a qualitative abnormality of the androgen receptor and responsive to high-dose androgen therapy. *J Clin Invest* 1991;**87**:1413–21.

16. Wilson Langley E, Kemppainen JA, Wilson EM. Intermolecular NH2-/carboxyl-terminal interactions in androgen receptor dimerization revealed by mutations that cause androgen insensitivity. *J Biol Chem* 1998;**273**:92–101.

17. Doesburg P, Kuil CW, Berrevoets CA, Steketee K, Faber PW, Mulder E, Brinkmann AO, Trapman J. Functional *in vivo* interaction between the amino-terminal, transactivation domain and the ligand binding domain of the androgen receptor. *Biochemistry* 1997;**36**:1052–64.

18. Edwards A, Hammond HA, Jin L, Caskey CT, Chakraborty R. Genetic variation at five trimeric and tetrameric tandem repeat loci in four human population groups. *Genomics* 1992;**12**:241–53.

19. Sleddens HF, Oostra BA, Brinkmann AO, Trapman J. Trinucleotide repeat polymorphism in the androgen receptor gene (AR). *Nucleic Acids Res* 1992;**20**:1427.

20. La Spada AR, Wilson EM, Lubahn DB, Harding AE, Fischbeck KH. Androgen receptor gene mutations in X-linked spinal and bulbar muscular atrophy. *Nature* 1991;**352**:77–9.

21. Avila DM, Wilson CM, Nandi N, Griffin JE, Wilson JD, McPhaul MJ. Immunoreactive androgen receptor (AR) in genital skin fibroblasts from subjects with androgen resistance and undetectable levels of AR in ligand binding assays. *J Clin Endocrinol Metab* 2002;**87**:182–8.

22. McPhaul MJ, Schweikert H-U, Allman DR. Assessment of androgen receptor function in genital skin fibroblasts using a recombinant adenovirus to deliver an androgen-responsive reporter gene. *J Clin Endocrinol Metab* 1997;**82**:1944–8.

23. Zhou ZX, Sar M, Simental JA, Lane MV, Wilson EM. A ligand-dependent bipartite nuclear targeting signal in the human androgen receptor. Requirement for the DNA-binding domain and modulation. *J Biol Chem* 1994;**269**:13115–23.

24. Jenster G, Trapman J, Brinkmann AO. Nuclear import of the human androgen receptor. *Biochem J* 1993;**293**:761–8.

25. Cai C, Yuan X, Balk SP. Androgen receptor epigenetics. *Transl Androl Urol* 2013;**2**:148–57.

26. Quigley CA, Quigley CA, De Bellis A, Marschke KB, el-Awady MK, Wilson EM, French FS. Androgen receptor defects: historical, clinical, and molecular perspectives. *Endocr Rev* 1995;**16**:271–321.

27. Gottlieb B, Beitel LK, Nadarajah A, Palioura M, Trifiro M. The Androgen Receptor Gene Mutations Database (ARDB): 2012 Update. *Hum Mutat* 2012;**33**:887–94.

28. Zoppi S, Marcelli JP, Griffin JE, Wilson JD, McPhaul MJ. Amino acid substitutions in the DNA-binding domain of the human androgen receptor are a frequent cause of receptor-binding positive androgen resistance. *Mol Endocrinol* 1992;**6**:409–15.

29. Beitel LK, Prior L, Vasiliou DM, Gottlieb B, Kaufman M, Lumbroso R, Alvarado C, McGillivray B, Trifiro M, Pinsky L. Complete androgen insensitivity due to mutations in the probable alpha-helical segments of the DNA-binding domain in the human androgen receptor. *Hum Mol Genet* 1994;**3**:21–7.

30. De Bellis A, Quigley CA, Marschke KB, el-Awady MK, Lane MV, Smith EP, Sar M, Wilson EM. Characterization of mutant androgen receptors causing partial androgen insensitivity syndrome. *J Clin Endocrinol Metab* 1994;**78**:513–22.

31. Sultan C, Lumbroso S, Poujol N, Belon C, Boudon C, Lobaccaro JM. Mutations of androgen receptor gene in androgen insensitivity syndromes. *J Steroid Biochem Mol Biol* 1993;**46**:519–30.

32. Lumbroso S, Lobaccaro JM, Belon C, Martin D, Chaussain JL, Sultan C. A new mutation within the deoxyribonucleic acid-binding domain of the androgen receptor gene in a family with complete androgen insensitivity syndrome. *Fertil Steril* 1993;**60**:814–9.

33. Mowszowicz I, Lee HJ, Chen HT, Mestayer C, Portois MC, Cabrol S, Mauvais-Jarvis P, Chang C. A point mutation in the second zinc finger of the DNA-binding domain of the androgen receptor gene causes complete androgen insensitivity in two siblings with receptor-positive androgen resistance. *Mol Endocrinol* 1993;**7**: 861–9.

34. Marcelli M, Tilley WD, Zoppi S, Griffin JE, Wilson JD, McPhaul MJ. Androgen resistance associated with a mutation of the androgen receptor at amino acid 772 (Arg Cys) results from a combination of decreased messenger ribonucleic acid levels and impairment of receptor function. *J Clin Endocrinol Metab* 1991;**73**:318–25.

35. Marcelli M, Zoppi S, Wilson CM, Griffin JE, McPhaul MJ. Amino acid substitutions in the hormone-binding domain of the human androgen receptor alter the stability of the hormone receptor complex. *J Clin Invest* 1994;**94**:1642–50.

36. McPhaul MJ, Marcelli M, Zoppi S, Wilson CM, Griffin JE, Wilson JD. Mutations in the ligand-binding domain of the androgen receptor gene cluster in two regions of the gene. *J Clin Invest* 1992;**90**: 2097–101.

37. Prior L, Bordet S, Trifiro MA, Mhatre A, Kaufman M, Pinsky L, Wrogeman K, Belsham DD, Pereira F, Greenberg C, et al. Replacement of arginine 773 by cysteine or histidine in the human androgen receptor causes complete androgen insensitivity with different receptor phenotypes. *Am J Hum Genet* 1992;**51**:143–55.

38. Beitel LK, Kazemi-Esfarjani P, Kaufman M, Lumbroso R, DiGeorge AM, Killinger DW, Trifiro MA, Pinsky L. Substitution of arginine-839 by cysteine or histidine in the androgen receptor causes different receptor phenotypes in cultured cells and coordinate degrees of clinical androgen resistance. *J Clin Invest* 1994;**94**:546–54.

39. Kazemi-Esfarjani P, Beitel LK, Trifiro M, Kaufman M, Rennie P, Sheppard P, Matusik R, Pinsky L. Substitution of valine-865 by methionine or leucine in the human androgen receptor causes complete or partial androgen insensitivity, respectively with distinct androgen receptor phenotypes. *Mol Endocrinol* 1993;**7**:37–46.

40. Ris-Stalpers C, Trifiro MA, Kuiper GG, Jenster G, Romalo G, Sai T, van Rooij HC, Kaufman M, Rosenfield RL, Liao S, et al. Substitution of aspartic acid-686 by histidine or asparagine in the human androgen receptor leads to a functionally inactive protein with altered hormone-binding characteristics. *Mol Endocrinol* 1991;**5**: 1562–9.

41. Szafran AT, Sun H, Hartig S, Shen Y, Mediwala SN, Bell J, McPhaul MJ, Mancini MA, Marcelli M. Androgen receptor mutations associated with androgen insensitivity syndrome: a high content analysis approach leading to personalized medicine. *Adv Exp Med Biol* 2011;**707**:63–5.

42. Grino PB, Isidro-Gutierrez RF, Griffin JE. Androgen resistance associated with a qualitative abnormality of the androgen receptor and responsive to high dose androgen therapy. *J Clin Endocrinol Metab* 1989;**68**:578–84.

43. Tincello DG, Saunders PT, Hodgins MB, Simpson NB, Edwards CR, Hargreaves TB, Wu FC. Correlation of clinical, endocrine and molecular abnormalities with *in vivo* responses to high-dose testosterone in patients with partial androgen insensitivity syndrome. *Clin Endocrinol* 1997;**46**:497–506.

44. Holterhus PM, Werner R, Struve D, Hauffa BP, Schroeder C, Hiort O. Mutations in the amino terminal domain of the human androgen receptor may be associated with partial androgen insensitivity and impaired transactivation *in vitro*. *Exp Clinical Endocrinol Diabetes* 2005;**113**:457–63.

45. Ferlin A, Vinanzi C, Garolla A, Selice R, Zuccarello D, Cazzadore C, Foresta C. Male infertility and androgen receptor gene mutations: clinical features and identification of seven novel mutations. *Clin Endocrinol* 2006;**65**:606–10.

46. Yong EL, Ng SC, Roy AC, Yun G, Ratnam SS. Pregnancy after hormonal correction of severe spermatogenic defect due to mutation in androgen receptor gene. *Lancet* 1994;**344**:826–7.

47. Zuccarello D, Ferlin A, Vinanzi C, Prana E, Garolla A, Callewaert L, Claessens F, Brinkmann AO, Foresta C. Detailed functional studies on androgen receptor mild mutations demonstrate their association with male infertility. *Clin Endocrinol (Oxf)* 2008;**68**: 580–8.

48. Zoppi S, Wilson CM, Harbison MD, Griffin JE, Wilson JD, McPhaul MJ, Marcelli M. Complete testicular feminization caused by an amino-terminal truncation of the androgen receptor with downstream initiation. *J Clin Invest* 1993;**91**:1105–12.

49. Wilson CM, McPhaul MJ. A and B forms of the androgen receptor are present in human genital skin fibroblasts. *Proc Natl Acad Sci USA* 1994;**91**:1234–8.

50. Wilson CM, McPhaul MJ. A and B forms of the androgen receptor are expressed in a variety of human tissues. *Mol Cell Endocrinol* 1996;**120**:51–7.

51. Gao TS, McPhaul MJ. Functional activities of the A- and B-forms of the human androgen receptor in response to androgen receptor agonists and antagonists. *Mol Endocrinol* 1998;**12**:654–63.

52. Choong CS, Quigley CA, French FS, Wilson EM. A novel missense mutation in the amino-terminal domain of the human androgen receptor gene in a family with partial androgen insensitivity syndrome causes reduced efficiency of protein translation. *J Clin Invest* 1996;**98**:1423–31.

53. Ahmed SF, Cheng A, Dovey L, Hawkins JR, Martin H, Rowland J, Shimura N, Tait AD, Hughes IA. Phenotypic features, androgen receptor binding, and mutational analysis in 278 clinical cases reported as androgen insensitivity syndrome. *J Clin Endocrinol Metab* 2000;**85**:658–65.

54. Melo KF, Mendonca BB, Billerbeck AE, Costa EM, Inacio M, Silva FA, Leal AM, Latronico AC, Arnhold IJ. Clinical, hormonal, behavioral, and genetic characteristics of androgen insensitivity syndrome in a Brazilian cohort: five novel mutations in the androgen receptor gene. *J Clin Endocrinol Metab* 2003;**88**:3241–50.

55. Audi L, Fernandez-Cancio M, Carrascosa A, Andaluz P, Toran N, Piro C, Vilaro E, Vicens-Calvet E, Gussinye M, Albisu MA, Yeste D, Clemente M, Hernandez de la Calle I, Del Campo M, Vendrell T, Blanco A, Martinez-Mora J, Granada ML, Salinas I, Forn J, Calaf J, Angerri O, Martinez-Sopena MJ, Del Valle J, Garcia E, Gracia-Bouthelier R, Lapunzina P, Mayayo E, Labarta JI, Lledo G, Sanchez Del Pozo J, Arroyo J, Perez-Aytes A, Beneyto M, Segura A, Borras V, Gabau E, Caimari M, Rodriguez A, Martinez-Aedo MJ, Carrera M, Castano L, Andrade M, Bermudez de la Vega JA. Grupo de Apoyo al Sindrome de Insensibilidad a los Androgenos (GrApSIA). Novel (60%) and recurrent (40%) androgen receptor gene mutations in a series of 59 patients with a 46,XY disorder of sex development. *J Clin Endocrinol Metab* 2010;**95**: 1876–88.

56. Boehmer AL, Brinkmann AO, Nijman RM, Verleun-Mooijman MC, de Ruiter P, Niermeijer MF, Drop SL. Phenotypic variation in a family with partial androgen insensitivity syndrome explained by differences in 5α dihydrotestosterone availability. *J Clin Endocrinol Metab* 2001;**86**:1240–6.

57. Holterhus PM, Bruggenwirth HT, Hiort O, Kleinkauf-Houcken A, Kruse K, Sinnecker GH, Brinkmann AO. Mosaicism due to a somatic mutation of the androgen receptor gene determines phenotype in androgen insensitivity syndrome. *J Clin Endocrinol Metab* 1997;**82**: 3584–9.

58. Gottlieb B, Beitel LK, Trifiro MA. Somatic mosaicism and variable expressivity. *Trends Genet* 2001;**17**:79–82.

59. Kohler B, Lumbroso S, Leger J, et al. Androgen insensitivity syndrome: somatic mosaicism of the androgen receptor in seven families and consequences for sex assignment and genetic counseling. *J Clin Endocrinol Metab* 2005;**90**:106–11.

60. Fan HC, Ho LI, Chi CS, Chen SJ, Peng GS, Chan TM, Lin SZ, Harn HJ. Polyglutamine (polyQ) diseases: genetics to treatments. *Cell Transplant* 2014;**23**:441–58.

61. Veldscholte J, Ris-Stalpers C, Kuiper G, Jenster G, Berrevoets C, Claassen E, van Rooij HC, Trapman J, Brinkmann AO, Mulder E. A mutation in the ligand binding domain of the androgen receptor of human LNCaP cells affects steroid binding characteristics and response to anti-androgens. *Biochem Biophys Res Commun* 1990;**173**: 534–40.

62. Taplin ME, Bubley GJ, Shuster TD, Frantz ME, Spooner AE, Ogata GK, Keer HN, Balk SP. Mutation of the androgen-receptor gene in metastatic androgen-independent prostate cancer. *N Engl J Med* 1995;**332**:1393–8.

63. Tilley WD, Buchanan G, Hickey TE, Bentel JM. Mutations in the androgen receptor gene are associated with progression of human prostate cancer to androgen independence. *Clin Cancer Res* 1997;**2**: 1277–85.

64. Taplin ME, Bubley GJ, Ko YJ, Small EJ, Upton M, Rajeshkumar B, Balk SP. Selection for androgen receptor mutations in prostate cancers treated with androgen antagonist. *Cancer Res* 1999;**59**: 2511–5.

65. Marcelli M, Tilley WD, Wilson CM, Griffin JE, Wilson JD, McPhaul MJ. Definition of the human androgen receptor gene structure permits the identification of mutations that cause androgen resistance: premature termination of the receptor protein at amino acid residue 588 causes complete androgen resistance. *Mol Endocrinol* 1990;**4**: 1105–16.

66. Dehm SM, Tindall DJ. Alternatively spliced androgen receptor variants. *Endocr Relat Cancer* 2011;**18**:R183–96.

67. Ware KE, Garcia-Blanco MA, Armstrong AJ, Dehm SM. Biologic and clinical significance of androgen receptor variants in castration resistant prostate cancer. *Endocr Relat Cancer* 2014;**21**: T87–T103.

68. Lu J, Lonergan PE, Nacusi LP, Wang L, Schmidt LJ, Sun Z, Van der Steen T, Boorjian SA, Kosari F, Vasmatzis G, Klee GG, Balk SP, Huang H, Wang C, Tindall DJ. The cistrome and gene signature of androgen receptor splice variants in castration resistant prostate cancer cells. *J Urol* 2015;**193**:690–8.

ADIPOCYTE

21

Obesity

Hélène Huvenne*,**, Patrick Tounian**,†, Karine Clément**,‡,
Béatrice Dubern**,†

*Department of Pediatrics, Saint-Vincent de Paul Hospital, GHICL, Lille, France
**Institute of Cardiometabolism and Nutrition (ICAN), Nutriomique, University Pierre et Marie
Curie-Paris, Pitie-Salpêtrière Hospital, Paris, France
†Nutrition and Gastroenterology Department, Armand-Trousseau Hospital, Assistance Publique
Hôpitaux de Paris, Paris, France
‡Nutrition Department, Pitié-Salpêtrière Hospital, Assistance Publique Hôpitaux de Paris, Paris, France

INTRODUCTION

Background, Incidence, Prevalence

According to the World Health Organization (http://www.who.int), there are an estimated 1.5 billion adults who are overweight (body mass index (BMI) $\geq$25 kg/m^2), and 500 million of these are considered clinically obese (BMI $\geq$30 kg/m^2). The worldwide prevalence of childhood overweight and obesity also increased from 4.2% in 1990 to 6.7% in 2010, but has stabilized in recent years.[1] Obesity, defined by an excess fat mass having an impact on health, is characterized by a wide phenotype heterogeneity. There are also individual differences in progression of weight (i.e., different trajectories) and risk of associated comorbidities. It is now well accepted that the development of obesity stems from the interaction of multiple environmental factors (such as overeating and/or reduction in physical activity) with genetic factors. Numerous epidemiological and intervention studies carried out in different cohorts (twins brought up together or separately, adopted children, nuclear families, etc.) have recognized the role of individual genetic and biological susceptibilities in response to the current weight-gain promoting environment.[2] The severity of obesity will be thus determined by the impact of the environment on the genetic background of each individual.

While obesity was first thought to be a disease that obeys the rules of Mendelian inheritance, new technologies paint a far more complicated picture of this complex disorder. Obesity due to a single, naturally occurring dysfunctional gene (i.e., monogenic obesity) is both severe

and rare (from <1% to 2–3% depending on the gene) when compared to the more common forms in which numerous genes make minor contributions in determining phenotypic traits in large populations (i.e., polygenic obesities).[3]

Clinical Presentation

Several clinical presentations are described in obesity depending on the involved genes:

1. *Monogenic obesity,* described as rare and severe early-onset obesity associated with endocrine disorders. This is mainly due to mutations in genes of the leptin-melanocortin axis involved in food intake regulation (genes of leptin (*LEP*) and its receptor (*LEPR*), proopiomelanocortin (*POMC*), proconvertase 1(*PC1*), etc.). New mutations were identified in recent years.
2. *Melanocortin-4 receptor (MC4R)-linked obesity,* characterized by a variable severity of obesity and the absence of very noticible phenotype. This is responsible for 2–3% of obesity in adults and children.
3. *Syndromic obesity,* corresponding to obesity associated with other genetic syndromes, as reviewed elsewhere.[4] Patients are clinically obese and additionally distinguished by mental retardation, dysmorphic features, and organ-specific developmental abnormalities.
4. *Polygenic obesity,* which is the more common clinical situation (>95% of cases). Here each susceptibility gene, taken individually, would only have a slight effect on weight. The cumulative contribution

Genetic Diagnosis of Endocrine Disorders. http://dx.doi.org/10.1016/B978-0-12-800892-8.00021-X

of these genes would become significant only in an "obesogenic lifestyle" (such as overfeeding, sedentariness, stress, and others). This part will not be developed in this chapter.

GENETIC PATHOPHYSIOLOGY

Monogenic Obesity

At least 200 cases of human obesity have been associated with a single gene mutation. These gene mutations all lie in an increased number of genes, but the monogenic forms of obesity remain rare in the obese population.[5] Unlike syndromic obesity, the reason why excess body fat mass develops in these subjects is quite well understood since the genetic anomalies affect key factors related to the leptin and the melanocortin pathways well-known to be pivotal in energy balance regulation (Fig. 21.1). Most of the severe forms of childhood obesity are due to single gene mutations; new ones are regularly discovered in different families.

The hypothalamic leptin/melanocortin pathway is activated following the systemic release of the adipokine LEP and its subsequent interaction with its receptor LEPR, located on the surface of neurons of the arcuate nucleus. The downstream signals that regulate satiety and energy homeostasis are then propagated via POMC, cocaine-and-amphetamine-related transcript (CART), and the melanocortin system.[6] While POMC/CART neurons synthesize the anorectic peptide α-melanocyte stimulating hormone (α-MSH), a separate group of neurons express the orexigenic neuropeptide Y (NPY) and the agouti-related protein (AGRP), which acts as a potent

inhibitor of melanocortin 3 (MC3R) and MC4R receptors. The nature of the POMC-derived peptides depends on the type of endoproteolytic enzyme present in the specific brain region. In the anterior pituitary, the presence of the PC1 enzyme produces adrenocorticotropic hormone (ACTH) and β-lipotropin peptides, while the combined presence of PC1 and PC2 in the hypothalamus controls the production of α-, β-, γ-MSH, and β-endorphins.

Mutations in human genes coding for LEP,[7–9] LEPR,[10] POMC,[11] and PC1[12,13] lead to severe obesity occurring soon after birth, with generally complete penetrance and autosomal-recessive transmission (Table 21.1). Patients carrying mutations show a rapid and dramatic increase in weight, as illustrated by the weight curve of LEPR deficient subjects (Fig. 21.2). Evaluating body composition in some *LEPR* mutation carriers shows a large amount of total body fat mass (>50%) but resting energy expenditure remains related to the level of body corpulence. Feeding behavior is characterized by major hyperphagia and ravenous hunger.[14] Surprisingly, a LEP-deficient Austrian girl has been recently described with more moderate obesity (BMI 31.5 kg/m[2]), despite an increased consumption of calories in a test meal.[15] The phenotype was explained by extremely low daily calorie intake. Even if one takes into account a substantial underreporting, this observation might suggest that despite leptin deficiency, it was possible to control energy intake and thus to prevent extreme obesity. In that specific case the parents' role was described as determinant. They provided a favorable environment by vigorously controlling the patient's eating behavior from early infancy onward.[16] A further explanation might be related to the different genetic backgrounds of different subjects with LEP or LEPR

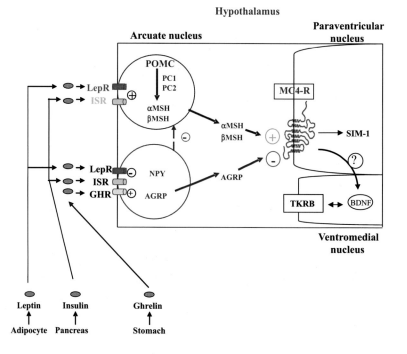

FIGURE 21.1 **The leptin/melanocortin pathway.** Neuronal populations propagate the signaling of various molecules (leptin, insulin, ghrelin) to control food intake and satiety. POMC neurons in the arcuate nucleus are activated by leptin and insulin and produce the α-MSH, which then activates the MC4R receptor in the paraventricular nucleus, resulting in a satiety signal. The downstream roles of SIM1, BDNF, and TKRB are currently being explored. A separate group of neurons expressing NPY and AGRP produce molecules that act as potent inhibitors of MC4R signaling. Several mutations of those genes involved in the leptin/melanocortin pathway are responsible for early-onset and severe obesity. POMC, proopiomelanocortin; LepR, leptin receptor; ISR, insulin receptor; GHR, ghrelin receptor; NPY, neuropeptide Y; AGRP, agouti-related protein; SIM1, single-minded 1; BDNF, brain-derived neurotropic factor; TRKB, tyrosine kinase receptor; PC1 and 2, proconvertase 1 and 2.

TABLE 21.1 Rare Monogenic Forms of Human Obesity

Gene (references)	Mutation type	Obesity	Associated phenotypes
Leptin[7,8,9,15,16]	Homozygous mutation	Severe, from the first days of life	Gonadotropic and thyrotropic insufficiency
Leptin receptor[10,17,18,19]	Homozygous mutation	Severe, from the first days of life	Gonadotropic, thyrotropic, and somatotropic insufficiency
Proopiomelanocortin (POMC)[11]	Homozygous or compound heterozygous	Severe, from the first months of life	ACTH insufficiency. Mild hypothyroidism and ginger hairs if the mutation leads to the absence of POMC production
POMC but in the β-MSH coding region[21-23]	Heterozygous non synonymous mutations	Severe obesity occurring in childhood	Rapid size growth
Proprotein convertase subtilisin/kexin type 1 (PCSK1 or PC1)[12,13,24,25]	Homozygous or compound heterozygous	Severe obesity occurring in childhood	Adrenal, gonadotropic, somatotropic, and thyrotropic insufficiency. Postprandial hypoglycemic malaises. Central diabetes insipidus
Single-minded 1 (SIM1)[26]	Translocation between chr 1p22.1 and 6q16.2 in the SIM 1 gene	Severe obesity occurring in childhood	Inconstantly neurobehavioral abnormalities (including emotional lability or autism-like behavior)
Neurotrophic tyrosine kinase receptor type 2 (NTRK2)[35]	De novo heterozygous mutation	Severe obesity from the first months of life	Developmental delay. Behavioral disturbance. Blunted response to pain
Dedicator of cytokinesis 5 (DOCK5)[38]	Variable number tandem repeats (VNTRs)	Childhood and adult severe obesity	–
Kinase suppressor of Ras2 (KSR2)[91]	Heterozygous frameshift, nonsense, or missense variants	Severe obesity	Hyperphagia in childhood. Low heart rate. Reduced basal metabolic rate. Severe insulin resistance
Tubby-like protein (TUB)[92]	Homozygous frameshift mutation	Early-onset obesity	Night blindness, decreased visual acuity, and electrophysiological features of a rod cone dystrophy

deficiency. However, despite this particular case, severe early-onset obesity with major hyperphagia is recognized as a main clinical presentation of LEP or LEPR deficiency.

Associated to severe early-onset obesity with major hyperphagia, hypogonadotrophic hypogonadism, and thyrotropic insufficiency complete the patient phenotype. Insufficient somatotrophic secretion, leading to moderate growth delay, is also described in some patients with a *LEPR* mutation. In LEP deficient subjects, it was described as a high rate of infection, particularly recurrent respiratory tract infections, associated with a deficiency in T-cell number and function.[17-19] In individuals with leptin deficiency, either due to *LEP* or *LEPR* mutations, no pubertal development was observed, while in others, there is evidence of spontaneous pubertal development, suggesting a recovery of hormonal functions with time. For example, the follow-up of the initially described LEPR deficient sisters revealed the normalization of thyroid mild dysfunction at adult age and a normal spontaneous pregnancy.[20]

Measurement of circulating leptin may help the diagnosis in some cases; it is undetectable in *LEP* mutation carriers. However, leptin levels can be correlated to fat mass or extremely elevated in *LEPR* mutation carriers.[8,10,18] Thus, *LEPR* gene screening might be considered

in subjects with the association of severe obesity with endocrine dysfunctions such as hypogonadism but with leptin related to corpulence level.[18]

Obese children with a complete POMC deficiency have ACTH deficiency, which can lead to acute adrenal insufficiency from birth. These children display a mild central hypothyroidism that necessitates hormonal replacement[11] and sometimes alterations in the somatotropic and gonadotropic axes.[21] The reason for these endocrine anomalies is unknown, even if the role of melanocortin peptides in influencing the hypothalamic pituitary axis has been proposed. Children have inconstantly ginger hair due to the absence of α-MSH, which activates the peripheral melanocortin receptor type 1 (MC1R) (involved in pigmentation). Several observations suggest that the skin and hair phenotype might vary according to the ethnic origin of *POMC* mutation carriers.[21-23] The modifications in color hair, adrenal function, and body weight are consistent with the lack of POMC-derived ligands for the melanocortin receptors MC1R, MC2R, and MC4R, respectively.

Patient carriers of a rare mutation in the proprotein convertase subtilisin/kexin type 1 (*PCSK1*) gene leading to PC1 deficiency have, in addition to severe obesity,

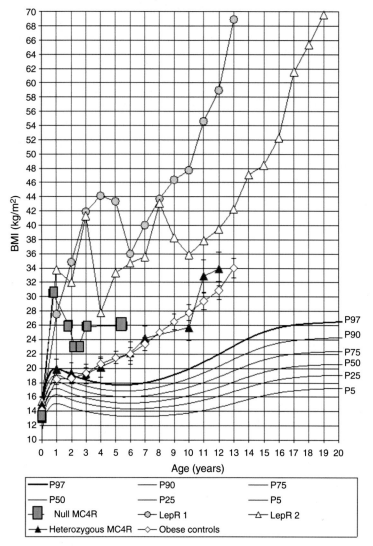

FIGURE 21.2   **BMI curves of two homozygous null LEPR mutants (LEPR 1 and 2), one homozygous null MC4R patient, six heterozygous MC4R carriers, and 40 nonmutated obese controls.**[41] The reference curves are the standard French/Institut National de la Santé et de la Recherche Médicale percentile curves.

postprandial hypoglycemic malaises, fertility disorders due to hypogonadotrophic hypogonadism, central hypothyroidism, and adrenal insufficiency secondary to lack of POMC maturation. The delayed postprandial malaises are explained by the accumulation of proinsulin through lack of PC1, which is involved in the synthesis of mature insulin from proinsulin. The absence of POMC maturation due to *PCSK1* mutation causes a dysfunction in the melanocortin pathway that explains the obese phenotype.[12] In patients suffering from a congenital PC1 deficiency, severe and rebel diarrhea due to a defect in intestinal absorption is described, secondary to lack in mature GLP-1 (glucagon-like peptide-1).[13] The processing of prohormones, progastrin and proglucagon, is altered, explaining, at least in part, the intestinal phenotype and also suggesting a role for PC1 in absorptive functions in

the intestine. Recently, in a proband who was compound heterozygous for a maternally-inherited frameshift mutation and a paternally-inherited 474-kb (kilobase) deletion that encompasses *PCSK1* and in 13 children with total PC1 deficiency, persistent polydipsia and polyuria were noted, as the disease progressed, due to the development of central diabetes insipidus improved by oral desmopressin. These observations suggest that PCSK1 may be involved in the full functioning or central sensing of osmolality in humans.[24,25] Growth hormone deficiency was also noted in the 13 children with complete PC1 deficiency.[25]

These studies have played an important part in confirming the critical role of the leptin and melanocortin pathways in controlling food intake and energy expenditure, as well as their strong implication in controlling endocrine pathways. Furthermore, these studies encouraged the pursuit of screens for genes encoding proteins acting both upstream and downstream of the G-protein coupled receptor MC4R (Table 21.1 and Fig. 21.1). Several additional genes, implicated in the development of the hypothalamus and the central nervous system, have been found to cause monogenic obesity in humans. A deletion of the *SIM1* (*single-minded homolog 1*) gene, located on the 6q chromosome, secondary to a *de novo* translocation between 1p22.1 and 6q16.2 chromosomes, was identified in a girl with early-onset obesity (since first months of life) associated with hyperphagia and food impulsivity.[26] She had a rate of early weight gain comparable to the weight curve of *LEP* and *LEPR*-deficient children. Izumi et al. identify an interstitial 6q deletion, including the *SIM1* gene in a subject with Prader–Willi-like features (neonatal hypotonia, dysmorphy, developmental delay, early-onset obesity, short stature, hypopituitarism).[27] *SIM1* encodes a transcriptional factor implicated in the development of the hypothalamic paraventricular nucleus. It plays a role in the melanocortin signaling pathway and appears to regulate feeding rather than energy expenditure.[28,29] The sequencing of the coding region of *SIM1*, in 2,100 unrelated patients with severe, early-onset obesity and in 1,680 unrelated population-based controls, identified 13 different heterozygous variants in 28 severely obese patients. Variants carriers exhibited severe obesity, increased *ad libitum* food intake at a test meal, normal basal metabolic rate, and inconstantly neurobehavorial phenotype (impaired concentration, memory deficit, emotional lability, or autistic spectrum behavior). Nine of the 13 variants significantly reduced the ability of *SIM1* to activate a *SIM1*-responsive reporter gene. These mutations cosegregated with obesity in extended family studies with variable penetrance. Rare variants in *SIM1* should be considered in patients with hyperphagic early-onset obesity, associated or not to Prader–Willi-like syndrome features (including severe obesity that starts between 1 and 2 years of age) and/or in patients with associated neurobehavorial abnormalities (including emotional lability or autism-like behavior).[30–32]

A decreased expression of the brain-derived neurotropic factor (BDNF) was found to regulate eating behavior.[33] BDNF, encoded by the *NTRK2* gene (*neurotrophic tyrosine kinase receptor type 2*) and its associated tyrosine kinase receptor (TRKB) are both expressed in the ventromedial hypothalamus. They have been attributed a role downstream of MC4R signaling implicated in feeding regulation.[34] A *de novo* heterozygous mutation in the *NTRK2* gene was described in an eight-year-old boy with severe early-onset obesity, mental retardation, developmental delay, and anomalies of higher neurological functions like the impairment of early memory, learning, and nociception.[35] Other mutations in *NTRK2* were found in patients with early-onset obesity and

developmental delay, but their functional consequences and their implication in obesity are yet to demonstrate. *In vitro* studies of some but not all mutations have suggested that these mutations could impair hypothalamic-signaling processes.[36]

Finally, the contribution of copy number variants (CNVs) to complex disease susceptibility, such as severe obesity, has been the subject of debates in recent years. Variable number tandem repeats (VNTRs) constitute a relatively underexamined class of genomic variants in the context of complex diseases. Rare CNVs have been shown to be responsible for severe highly penetrant forms of obesity, initially observed in 31 subjects who were heterozygous for deletions of at least 593 kb at 16p11.2 locus. In addition to obesity, these carriers presented with congenital malformations and/or developmental delay. The phenotype may result from the haploinsufficiency of multiple genes that impact on pathways central to the development of obesity.[37] An investigation of a complex region on chromosome 8p21.2 encompassing the *DOCK5* (*dedicator of cytokinesis 5*) gene has shown a significant association of the *DOCK5* VNTRs with childhood and adult severe obesity.[38] More systematic investigation of the role of VNTRs in obesity had to be performed to study their relatively unexplored contribution to this disease and their potential link with the leptin/melanocortin pathway.

Melanocortin 4 and 3 Receptor-Linked Obesities

The obesities linked to *MC4R* and *MC3R* mutations can be placed between the rare forms of monogenic obesity with complete penetrance and the polygenic forms of common obesity.

Considering the pivotal role of the melanocortin pathway in the control of food intake, the *MC4R* gene is a major candidate gene in human obesity (Fig. 21.1). Since 1998, its genetic evaluation revealed that *MC4R*-linked obesity is the most prevalent form of genetic obesity identified to date. It represents approximately 2–3% of childhood and adult obesity with currently 166 different mutations described in different populations (European, North American, and Asian).[39,40] They include frameshift, inframe deletion nonsense, and missense mutations located throughout the *MC4R* gene (Fig. 21.3). The frequency of heterozygosity for these mutations in (extremely) obese individuals cumulates to approximately 2–5%.[40] In addition, the frequency of such heterozygous carriers in nonobese controls or in the general population is about 10-fold lower than in the cohorts of obese patients.[41,42]

In contrast with rare monogenic obesities, even a meticulous clinical analysis does not easily detect obesity stemming from *MC4R* mutations because of the lack of additional obvious phenotypes. In families with

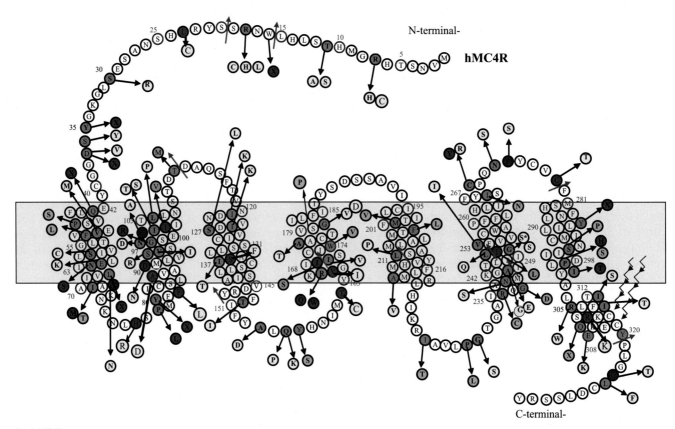

FIGURE 21.3 Localization of human MC4R mutations, including polymorphisms described in adults and children. Red circles: mutations responsible for intracellular retention of the mutated receptor or frameshift. Orange circles: polymorphisms. Yellow circles: mutation with functional consequence (altered α-MSH production, impaired basal activity, etc.). Green circles: any known or described functional effect.

MC4R-linked obesity, obesity tends to have an autosomal dominant mode of transmission, but the penetrance of the disease can be incomplete and the clinical expression variable (moderate to severe obesity), underlying the role of the environment and other potentially modulating genetic factors.[41,42] Homozygous or compound heterozygous carriers of MC4R mutations are very rare. Four carriers of homozygous null mutations in the MC4R gene have been detected,[43–45] and as expected from a dominant condition, obesity is developed earlier in life and is more severe than for heterozygous carriers, but it does not display any additional unrelated phenotypes. In heterozygous MC4R mutations carriers, the onset and severity of obesity vary and are related to the severity of the functional alteration caused by the mutation.

Many authors agree that MC4R mutations facilitate early-onset obesity. MC4R mutations carriers display increased linear growth, in particular in the first five years of life[43] but appear to be taller as adults only in rare cases.[41,46,47] This trend is often observed in overweight and obese children. Assessment of body composition in these patients demonstrates increase in both fat and lean mass.[48]

One study performed in English children with MC4R mutations has suggested that bone mineral density and size increase.[43] This potential increase of bone density may be explained, at least in part, by a decrease in bone resorption, as illustrated by decreases in bone resorption markers in the serum of patients with MC4R homozygous and heterozygous mutations.[49,50] Obese children carrying MC4R mutations have a marked hyperphagia that decreases with age when compared to their siblings,[48] while in both children and adults, no evidence has been found for a decreased metabolic rate in these patients. Meanwhile, the association between "binge eating" disorder and MC4R gene sequence changes[51] has not been confirmed.[41,52] Adult MC4R mutation carriers do not have an increased prevalence of diabetes or other obesity complications.[41] In UK children, fasting insulinemia was found to be significantly elevated in MC4R mutation carriers, particularly before the age of 10 years when compared to age, sex, and BMI matched control.[43] This hyperinsulinemia has not been consistently observed in children[44,45] and in adults.[41,53] Finally, with respect to endocrine function, hypothalamo-pituitary axis and reproductive axis,[43–46] as well as thyroid function, are normal in MC4R mutation carriers.

The role of *MC4R* mutations in cases of human obesity relies on two main arguments based on the frequency of *MC4R* mutations in different populations and their *in vitro* functional consequences. Firstly, *MC4R* mutations are more abundant in obese populations. Functional mutations have also been reported in nonobese subjects but to a significantly lesser frequency (<1%).[42] Secondly, investigating the molecular mechanisms by which loss-of-function mutations in *MC4R* cause obesity have suggested a panel of functional anomalies: abnormal MC4R membrane expression, defect to the agonist response, and a disruption in the intracellular transport of this protein. Normally, after ligand binding, MC4R activation stimulates Gs protein, leading to a subsequent increase in cAMP levels; however, the production of intracellular cAMP in response to α-MSH peptides demonstrated a broad heterogeneity in the activation of the different MC4R mutants in response to α-MSH, ranging from normal or partial activation to a total absence in activation.[39,46] The intracellular transport defect of the mutated receptor, by intracytoplasmic retention, has been described for the majority of *MC4R* mutations found in childhood obesity,[54] but also in adults.[41,42] This mechanism explains the impaired response to agonists. In addition, MC4R has a constitutive activity, meaning a basal activity not necessitating the presence of a ligand, for which agouti-related peptide (AGRP) acts as an inverse agonist.[55] In the absence of the ligand, MC4R has an inhibitory action on food intake. The systematic study of basal activity of some mutations has shown that an alteration in this activity may be the only functional anomaly found, in particular for mutations located in the N-terminal extracytoplasmic part of the receptor.[56] A tonic satiety signal, provided by the constitutive activity of MC4R, could be required in the long-term regulation of energy balance. It is accepted that *MC4R* mutations cause obesity by a haploinsufficiency mechanism rather than a dominant negative activity. While the roles of homo- and heterodimerization in G protein synthesis and maturation are emphasized, some dominant negative effects of *MC4R* mutations might not be excluded.

The MC3R has also been involved in obesity. *MC3R* deficient (*MC3R⁻/⁻*) mice have increased fat mass, reduced lean mass, and higher feed efficiency than wild-type littermates, despite being hypophagic and maintaining normal metabolic rates.[57] In humans, strong evidence of a causative role for *MC3R* mutations is still lacking. Some mutations, leading to amino acid changes in the receptor, have been described in a group of 290 obese subjects,[58] but the total prevalence of rare *MC3R* variants was not significantly different between cohorts of severely obese subjects and lean controls.[59] No specific phenotype of *MC3R* mutations has been identified.

Obesity Syndromes

There are between 20 and 30 Mendelian disorders in which patients are clinically obese with such additional phenotypes as mental retardation, dysmorphic features, and organ-specific developmental abnormalities. These syndromes arise from discrete genetic defects or chromosomal abnormalities and are both autosomal and X-linked disorders. The most common and well-known disorders are Prader–Willi and Bardet–Biedl syndromes, but many others have been reported (Table 21.2).[4] The Online Mendelian Inheritance in Man database provides access to their clinical descriptions (OMIM; http://www.ncbi.nlm.nih.gov/omim/).

The Prader–Willi syndrome (PWS) (one in 20,000–25,000 births) is characterized by obesity, hyperphagia, neonatal hypotonia, mental retardation, and hypogonadism. It is due to physical or functional absence of the paternal chromosomal segment 15q11-q13. Three genetic mechanisms are described: *de novo* microdeletion (70% of cases), uniparental maternal disomy (29%), imprinting defect, or very rare reciprocal translocation (1%).[60] At least five genes, located in the PWS chromosomal region and expressed in the hypothalamus, have been identified. They encode for different proteins whose functions are incompletely understood: *MRKN3* (makorin 3), *MAGEL2* (MAGE-like 2), *NDN* (necdin), *NPAP1* (nuclear pore associated protein 1), and *SNURF-SNRPN* (SNRPN upstream reading frame – small nuclear ribonucleoprotein polypeptide N).[61,62] The *SNURF-SNRPN* region encodes for several proteins implicated in the imprinting center or in alternative splicing. These genes are expressed in the hypothalamus and may be involved in common obesity. The *necdin* gene is implicated in axonal outgrowth. The *MAGEL2* gene encodes a ubiquitin ligase enhancer, which is required for endosomal protein recycling. Several other genes are probably affected and could explain the variability of the PWS phenotype. But the exact genetic mechanisms leading to obesity are still to be defined.[63] The genetic basis of hyperphagia also remains undefined in part due to the fact that none of the currently available PWS mouse models have an obese phenotype.[64] One genetic candidate that may mediate the obese phenotype and disrupt the control of food intake is the gastric hormone ghrelin,[65] via its regulation of hunger and stimulating growth factor hormone (GH) secretion.[17] Patients suffering from PWS have high ghrelin levels.[65] Ghrelin's implication in PWS is additionally reinforced by the positive findings that GH supplementation is capable of reversing several dysfunctional processes associated with PWS[66,67] however, in the absence of a suitable experimental model, identifying the genetic components of this syndrome is challenging. Even if the specific genes responsible for the PWS phenotype are not still identified, understanding the molecular mechanism

TABLE 21.2 Principal Obesity Syndromes

Syndrome	Clinical features in addition to obesity	Locus	Gene
AUTOSOMAL DOMINANT			
Prader–Willi syndrome (PWS)	Neonatal hypotonia, mental retardation, facial dysmorphy, hypogonadotrophic hypogonadism, short stature	Lack of the paternal segment 15q11.2-q12 (microdeletion or maternal disomy)	Unknown SRNPN region
Albright hereditary osteodystrophy	Short stature, skeletal defects, facial dysmorphy, endocrine anomalies	20q13.2	GNAS1
AUTOSOMAL RECESSIVE			
Bardet–Biedl syndrome (BBS)	Mental retardation, dysmorphic extremities, retinal dystrophy or pigmentary retinopathy, hypogonadism, kidney anomalies (structural abnormalities or functional renal impairment)	1q13 (BBS1), 16q21 (BBS2), 3p13 (BBS3), 15q22 (BBS4), 2q31 (BBS5), 20p12 (BBS6), 4q27 (BBS7), 14q32 (BBS8)....	BBS 1–19
Alström syndrome	Retinal dystrophy, neurosensory deafness, diabetes	2p13	ALMS1
Cohen syndrome	Prominent central incisors, dysmorphic extremities, ophthalmopathy, microcephaly, cyclic neutropenia	8q22	COH
X-LINKED			
Borjeson–Forsmman–Lehman syndrome	Mental retardation, hypogonadism, facial dysomoprhy with large ears	Xq26	PHF6
Fragile X syndrome	Mental retardation, hyperkinetic behavior, macroorchidism, large ears, prominent jaw	Xq27.3	FMR1

has improved the management of those children and their families. Schaaf et al. report four individuals with truncating mutations on the paternal allele of *MAGEL2*, a gene within the PWS domain. All four subjects had autism spectrum disorder, developmental delay or intellectual disability, and a varying degree of clinical and behavioral features of PWS (neonatal hypotonia, severe early-onset obesity, hypogonadism, speech articulation defects). It suggests that *MAGEL2* may be a novel gene causing complex autism spectrum disorder and that *MAGEL2* loss of function can contribute to several aspects of the PWS phenotype.[62]

Bardet–Biedl syndrome (BBS) is characterized by obesity, retinal dystrophy (pigmentary retinopathy), malformed extremities (syndactyly, polydactyly), kidney diseases, and eventually mental disabilities. It concerns one in 100,000 births, with an increased prevalence in Arab and Bedouin populations (one in 13,500 births). BBS was first considered as a monogenic disease; however, large-scale molecular screening in families revealed that BBS is associated with at least 19 different chromosomal locations with mutations identified within some of these locations in more than 80% of the affected families (*BBS1* on 11q13; *BBS2* on 16q21; *BBS3* on 3p12-p13; *BBS4* on 15q22.3-q23; *BBS5* on 2q31; *BBS6* on 20p12; *BBS7* on 4q27; *BBS8* on 14q32.11; *BBS9* on 7p14; *BBS10* on 12q; *BBS11* on 9q33.1; *BBS12* on 4q27; *BBS 13* on 17q23; *BBS14* on 12q21.3; *BBS15* on 2p15; *BBS16* on 1q43, *BBS17* on 3p21, *BBS18* on 10q25 and *BBS19* on 22q12).[68–72] While BBS was considered to be autosomal recessive, it has been hypothesized that the clinical symptoms of certain forms of BBS are due to recessive mutations on one of the *BBS* loci associated with one or several heterozygous mutations on another locus; this finding prompting for the first time the hypothesis of a tri- or tetra-allelic mode of transmission.[68,73,74] However, it seems that the tri- or tetra-allelic transmission is only present in a few families. *BBS* genes encode for proteins that are involved in primary cilium function. For the *BBS6* loci, positional cloning identified the *MKKS* (*McKusick–Kaufman syndrome*) gene, which codes for a chaperone protein located in the basal body of primary cilium. Mutations identified in *MKKS* result in a shortened chaperone protein and represent 5–7% of all BBS cases; however, the links between *MKKS*, its eventual target proteins, and the BBS clinical traits are largely unknown. Recently mutations in the locus *BBS10* encoding for *C12orf58*, a vertebrate-specific chaperone-like protein, have been described in 20% of subjects from various ethnic backgrounds.[69] Unlike BBS6 and BBS10, the genes implicated in BBS1, BBS2, and BBS4 are different from *MKKS* and *C12orf58* genes, but it is conceivable that they are encoding for protein substrates of these chaperones.[75] The E3 ubiquitin ligase *TRIM32* (*tripartite motif-containing protein-32*) was identified as the 11th loci associated with BBS, suggesting that the list of genetic components for this syndrome may yet be incomplete.[71] For BBS13, mutations in *MKS1* (*Meckel syndrome-1*) gene have been implicated. Fascinating functional studies performed in single cell organisms have shown that certain *BBS* genes are specific to ciliated cells.[76] These cells have a role in mammalian development, where they contribute to right/left symmetry and enable the organs (heart, liver, lungs) to be correctly positioned within the biological system. Such dysfunction in the processes affecting the ciliated cells may contribute to alterations in pigmentary epithelia and to structural anomalies noted with certain organs like the kidney.[77] BBS is now defined as a ciliopathy (primary cilium dysfunction). But the precise mechanisms leading to obesity in BBS are still to be understood.[72] However, obesity in BBS mice is associated with hyperleptinemia and leptin resistance. BBS proteins seem also to be required for LEPR signaling in the hypothalamus. *BBS2*[-/-], *BBS4*[-/-], and *BBS6*[-/-] mice are resistant to the action of leptin, regardless of serum leptin levels and obesity. In addition, the activation of the hypothalamic STAT3 by leptin is significantly decreased in these models.[78] Furthermore, BBS1 protein was found to physically interact with the LEPR and loss of BBS proteins perturbs LEPR trafficking, suggesting that BBS proteins mediate LEPR trafficking and that impaired LEPR signaling underlies energy imbalance in BBS.[78] Finally, a recent study indicates that leptin resistance is present only when BBS mutant mice are obese and that a defect in the leptin signaling axis may not be the initiating event leading to hyperphagia and obesity in BBS.[79]

Recent data highlighted a significant difference in obesity rates in young fragile X males (31%) compared to age-matched controls (18%). Fragile X syndrome (FXS) is the most common cause of inherited intellectual disability in males and is also frequently implicated in autism. Common features of FXS include mild to severe cognitive impairment in males (cognitive impairment is milder in females), attention deficit hyperactivity disorder, elongated face, prominent ears, severe obesity (onset in the first decade of life) with hyperphagia and lack of satiation after meals, hypogonadism or delayed puberty, and postpubertal macroorchidism. This X-linked disorder is caused by an expansion of a trinucleotide CGG repeat (>200) on the promotor region of *FMR1* (*fragile X mental retardation 1 gene*). Variable methylation of this promotor region leads to a deficiency or absence of the FMR1 protein.[80] An Xq27.3-q28 duplication, including the *FMR1* gene, was identified in a male propositus with profound intellectual disability, hypogonadotropic hypogonadism, short stature, morbid obesity, and gynecomastia.[81]

Alström syndrome is a very rare autosomal recessive disease that, apart from severe early onset obesity, associates several phenotypes reminiscent of BSS-like retinal cone dystrophy. Patients with Alström disease also develop dilated cardiomyopathy and sensorineural deafness but no polydactyly. The *ALMS1* gene encoding for a protein with ubiquitous expression but unknown function has been implicated. Because of its particular localization in the centrosome and basal bodies, which resembles the pattern of protein expression for some BBS-linked genes, it has been suggested that Alström syndrome could also belong to a class of ciliopathy.[77,82]

Cohen syndrome is characterized by late-onset obesity (after the age of 10 years) associated to moderate mental retardation, typical craniofacial features (downwards slanting and wave-shaped palpebral fissures, short philtrum, heavy eyebrows, thick hair, and prominent nasal base), progressive pigmentary retinopathy appearing at mid-childhood, early onset and severe myopia, and intermittent neutropenia. It is an autosomal recessive disorder with variable clinical manifestations. Mutations in the gene *COH1* or *VPS13B* (*vacuolar protein sorting 13B*) located in chromosome 8q22-q23 encoding a transmembrane protein presumably involved in intracellular protein transport are described.[83–85]

The OMIM database (http://www.ncbi.nlm.nih.gov/omim) provides access to clinical descriptions and genetic anomalies for all these obesity syndromes.

DIAGNOSIS GENETIC TESTING AND INTERPRETATION

Monogenic Obesity

In case of a clinical situation suggesting a monogenic obesity (severe early-onset obesity associated with endocrine anomalies and consanguineous parents), direct sequencing of the candidate gene (*LEP*, *LEPR* or *POMC*, etc.) is necessary for diagnosis confirmation. It will detect homozygous or compound heterozygous mutations responsible for an interruption of the leptin-melanocortin axis. Family members are needed to be tested for segregation analysis and to evaluate the risk of recurrence.

A few genetics laboratories routinely perform those analyses, which usually are part of research programs. For example, in Europe, this genetic testing can be found at:

- UF Nutrigénétique (Pitié Salpêtrière Hospital). Contact: B. Dubern or K. Clément. Address: Endocrinology and Nutrition Department, Pitié-Salpêtrière Hospital, Boulevard de l'Hôpital, 75013 Paris or Pediatric Nutrition Department, Trousseau Hospital, 75012 Paris. Email: beatrice.dubern @trs.aphp.fr; karine.clement@psl.aphp.fr; Tel: 33(0) 142177928; Website: http://www.cgmc-psl.fr/
- S. O'Rahilly's team (University of Cambridge). Contact: S. Farooqi or S. O'Rahilly. Address: Department of Clinical Biochemistry, University of Cambridge, Addenbrooke's Hospital, Hills Road, Cambridge, CB2 2QR, UK. Email: isf20@cam.ac.uk; Tel: +44 (0) 1223 336792; Fax: +44 (0) 1223 330598; Website: http://www.mrl.ims.cam.ac.uk

Melanocortin 4 Receptor-Linked Obesity

Direct sequencing of the *MC4R* gene (1 exon) leads to the detection of *MC4R* mutations. To date, it is questionable to perform a routine systematic detection of *MC4R* mutations in obese subjects with a strong obesity familial history. Although knowing the biological reason (i.e., altered melanocortin pathway) leading to an increased susceptibility to obesity might be of interest in some individuals, no specific therapeutic is currently available and the severity of the phenotype is highly variable within *MC4R* mutation carrier families. It may, however, become necessary in few years in case of development of specifics drugs such as MC4R agonists in order to detect patients that may be eligible for such treatments.[86]

Obesity Syndromes

In PWS, genetic analysis is crucial to assess the absence of the paternal chromosomal segment 15q11.2-q12 and to predict the risk of recurrence even if the vast majority of PWS cases occur sporadically. The investigation of DNA methylation of *SNRPN* is now the gold standard and will be completed to detect the exact mechanism (microdeletion, maternal uniparental disomy, translocations, etc.).

For the other few genes or loci identified in obesity syndromes, direct sequencing of the candidate gene (example.g., the *COH1* gene in Cohen syndrome) or specific genetic tests (FISH to detect microdeletions, etc.) are indicated in order to diagnose those syndromes.

Whole-exome sequencing showed its power to identify mutations responsible for rare diseases in a small number of unrelated affected individuals. This tool could probably help to identify new molecular abnormalities in patients with syndromic obesity. It was tested and validated in a study for the molecular diagnosis of 43 forms of monogenic diabetes or obesity. Forty patients (19 with a monogenic form of diabetes and 21 with a monogenic form of obesity) carrying a known causal mutation for those subtypes according to diagnostic laboratories were blindly reanalyzed. Except for one variant, all causal mutations in each patient were reidentified, associated with an almost perfect sequencing of the targets (mean of 98.6%). Moreover, in three individuals, other mutations were detected with a putatively deleterious effect, in addition to those previously reported by the genetic diagnosis laboratories.[87] In another study based on 39 unrelated severely obese Pakistani children, the whole-exome sequencing revealed two novel homozygous *LEPR* mutations in two probands who were phenotypically indistinguishable from age-matched leptin deficient subjects from the same population.[88] The whole-exome sequencing could also be a highly efficient and practical method for genetic diagnosis of BBS, it successfully revealed six pathological mutations, including five novel mutations, in *BBS* genes in five typical BBS patients and family members.[89] For another example, this method of whole-exome sequencing led to identify a paternal 2pter deletion, encompassing the *ACP1* (*acid phosphatase 1*), *TMEM18* (*transmembrane protein 18*), and *MYT1L* (*myelin transcription factor 1-like*) genes, in five unrelated patients presenting with severe early-onset obesity, hyperphagia, intellectual deficiency, and severe behavioural difficulties.[90] It also allowed the identification of multiple rare variants in the *KSR2* (*kinase supressor of Ras 2*) gene in 45 unrelated severely obese individuals with impaired cellular fatty acid oxidation and glucose oxidation. In addition to severe obesity, mutations carriers exhibit hyperphagia in childhood, low heart rate, reduced basal metabolic rate, and severe insulin resistance.[91] A homozygous frameshift mutation in the *TUB* (tubby-like protein) gene was also identified by whole-exome sequencing in a proband who presented with obesity, night blindness, decreased visual acuity, and electrophysiological features of a rod cone dystrophy.[92] So, the whole-exome sequencing seems to be helpful to identify new molecular abnormalities in patients with syndromic obesity.

TREATMENT

Monogenic Obesity

Leptin deficient children and adults benefit from the subcutaneous injection of leptin, resulting in weight loss, mainly of fat mass, with a major effect on reducing food intake and on other dysfunctions including immunity as described previously.[93] A detailed microanalysis of eating behavior of three leptin deficient adults, before and after leptin treatment, revealed reduced overall food consumption, a slower rate of eating, and diminished duration of eating of every meal in the three subjects after leptin therapy. Leptin treatment also induces features of puberty even in adults.[14] This study supports a role of leptin in influencing the motivation to eat before each meal.[94] Another study shows that leptin treatment had a major effect on food intake, *ad libitum* energy intake at a test meal was reduced, hunger ratings in the fasted state decreased, and satiety following a meal increased. Leptin acts on neural circuits governing food intake to diminish perception of food reward while enhancing the response to satiety signals generated during food consumption.[95] In a separate study, hormonal and metabolic changes were evaluated before and after leptin treatment.[14] Leptin treatment was able to induce aspects of puberty even in adults, as illustrated by the effect of leptin treatment in one 27-year-old adult male with hypogonadism. In two women between 35 and 40 years, leptin treatment led to regular menstrual periods and hormonal peaks of progesterone evoking a pattern of ovulation. Although cortisol deficiency was not initially found in leptin deficient patients, 8 months of leptin treatment modified the pulsatility of cortisol with a greater morning rise of cortisol. Leptin could have a previously unsuspected impact on human hypothalamic-pituitary-adrenal function in humans. Metabolic parameters of leptin deficient patients improved in parallel with weight loss.

Because of a nonfunctional LEPR, leptin treatment is useless in the LEPR deficient subjects. Factors that could possibly bypass normal leptin delivery systems are being developed but are not yet currently available. The ciliary neurotrophic factor (CNTF) was one of the candidate molecules. CNTF activates downstream signaling molecules such as STAT-3 in the hypothalamus area that regulate food intake, even when administered systemically. Treatment with CNTF in humans and animals, including *db/db* mice, induced substantial loss of fat mass.[96] The neurotrophic factor, Axokine, an agonist for CNTF receptor, was under development by the Regeneron Company, for the potential treatment of obesity and its metabolic associated complications. But the phase III clinical trials were stopped due to development of antibodies against Axokine in nearly 70% of the tested subjects after approximately 3 months of treatment. In addition, the effect of Axokine had a small positive effect.[97] It is also possible that side effects of CNTF, a molecule possibly acting in the immune function, might be expected.[98]

In children with a complete POMC deficiency, a 3-month trial using an MC4R agonist with a low affinity was inefficient on weight or food intake.[11] POMC, PC1, or LEPR deficient families might benefit from the development of new MC4R agonists if such drugs become available, in order to restore the melanocortin signal. Likewise, deep brain stimulation trials with an electrode inserted in the anterior third ventricle contiguous to the ventromedial hypothalamus are actually performed. In monkeys, this chronic 8-week stimulation induces a significant decrease in corpulence with reduction of 8% of body weight and 18% of fat mass, without side effects.[99]

Melanocortin 4 Receptor-Linked Obesity

To date, no specific management has been proposed for *MC4R* mutation obese patients except for a well-balanced diet and physical activity. However, interestingly, physical activity may have a specific role for modulating the obese phenotype in the case of MC4R anomalies. In *MC4R*[-/-] mice, regular physical activity is described to be more efficient for limiting weight gain during life when compared to wild type.[100] Those findings are also observed in some patients with *MC4R*

mutation in clinical practice. It suggests its specific role in the management of patients with *MC4R* mutations and possibly the prevention of weight gain or regain after intervention. In addition, due to its important role in obesity, MC4R is becoming an attractive candidate drug target suggesting that identification and design of ligands or peptides may rescue the phenotype of the particular molecular mechanistic defect. Several synthetic ligands from the classical NDP-MSH peptides to the multiple tetrapeptides and small molecule MC4R agonists have been *in vitro* tested with variable results.[101,102] But they have to face normal concerns of targeting GPCRs and specific difficulties of possible side effects due to the widespread expression of MC4R in the brain and the already demonstrated role of MC4R in erectile function.[101,102] In the long term, this type of treatment should be evaluated in heterozygous patients for *MC4R* mutations with impaired α-MSH activity, in specific clinical investigation protocols in order to provide effective antiobesity treatment probably in combination with other approaches, such as diet and physical activity. Actually, novel pharmacological MC4R agonists have been tested *in vitro* and can restore normal activity in a mutated receptor.[84] Preclinical trials are now performed.[103] So, treatment with a highly selective novel MC4R agonist in an obese nonhuman primate model resulted in decreased food intake (35%), increased total energy expenditure (14%), and weight loss after 8 weeks of treatment (13.5%). No side effect, in particular in blood pressure or heart rate, was observed in this study.[104]

Obesity Syndromes

No specific therapeutic is described in obesity syndromes except for their global management (diet and physical activity, psychomotricity, hormone substitution, etc.). In PWS, treatment by GH must be discussed. Studies had shown that GH therapy with doses of GH typically used for childhood growth, started before 1 year old, improves growth, body composition, physical strength and agility, fat utilization, and lipid metabolism in children and adults with PWS.[105–108] In addition, in case of hypogonadotrophic hypogonadism, hormonal substitution is necessary.

Bariatric Surgery

Today, bariatric surgery is the only long-term efficient treatment for severe obesity[109] using several operative methods (laparoscopic gastric bypass, gastric banding, or sleeve gastrectomy). The question of such treatment and its potential efficiency is crucial in patients with genetic abnormalities previously described in this chapter. Currently, data on bariatric surgery in patients with genetic obesity are limited and controversial.

In four patients with heterozygous *MC4R* mutations, weight loss after Roux-en-Y gastric bypass surgery was identical to controls without *MC4R* mutations, suggesting that heterozygous *MC4R* mutation status should not influence the decision to perform surgery.[110] Valette et al. confirmed these findings in a group of obese adults.[111] In contrast, in a teenager with complete MC4R loss of function, laparoscopic adjustable gastric banding resulted in the absence of long-term weight loss (12 months postoperatively) suggesting that the full interruption of the melanocortin pathway may not be counteracted by bariatric surgery.[112] Other studies on the effect of bariatric surgery in *MC4R* mutated patients are needed.

In one LEPR-deficient patient, vertical gastroplasty was beneficial and sufficient to induce and maintain a 40-kg weight loss (−20% of the initial weight) over eight years of regular follow-up, whereas the patient remained obese.[113] In contrast, a relative failure of surgical therapy was illustrated in a report of rapid weight regain one year after bypass in another *LEPR*-deficient morbidly obese woman. But this patient of low socioeconomic status had extreme difficulties after postsurgical counseling. She was noncompliant with the recommendations provided in this type of purely restrictive surgery and her medical follow-up was very irregular. This report illustrated the important role of environment on the benefits of bariatric surgery, especially in the case of monogenic obesities but also underlined the poor efficiency of bariatric surgery in these patients.[113]

In three patients with genetic diagnosis of PWS, bariatric surgery was beneficial. Two patients underwent laparoscopic sleeve gastrectomy and one patient underwent laparoscopic gastric bypass. After a median follow-up of 33 months, mean weight loss and percentage of excessive weight loss at 2 years were 32.5 kg and 63.2%, respectively. No major complication was observed.[114] These observations were in contrast with a retrospective review including 60 cases of PWS patients who underwent bariatric surgery. Various bariatric procedures have been used, with poor results in PWS patients in comparison with normal obese individuals. The average weight loss results at 5 years after gastric bypass were only a 2.4% and 3.5% gain above preoperative weight after gastric banding. In addition to poor results on weight, a variety of postoperative issues were reported, including deaths, pulmonary embolus, postoperative wound infection, and gastric perforation.[115] So, due to the limited number of cases, the long-term efficacy and safety of bariatric surgery needs further evaluation. A multidisciplinary team approach should always be adopted in order to establish the correct indication and realistic explanation after surgical treatment of obese patients.

References

1. de Onis M, Blössner M, Borghi E. Global prevalence and trends of overweight and obesity among preschool children. *Am J Clin Nutr* 2010;**92**:1257–64.
2. Sorensen TI. The genetics of obesity. *Metabolism* 1995;**44**:4–6.
3. Farooqi IS. Genetic aspects of severe childhood obesity. *Pediatr Endocrinol Rev* 2006;**3**(Suppl 4):528–36.
4. Chung WK, Leibel RL. Molecular physiology of syndromic obesities in humans. *Trends Endocrinol Metab* 2005;**16**:267–72.
5. Farooqi IS, O'Rahilly S. Monogenic obesity in humans. *Annu Rev Med* 2005;**56**:443–58.
6. Harrold JA, Williams G. Melanocortin-4 receptors, beta-MSH and leptin: key elements in the satiety pathway. *Peptides* 2006;**27**:365–71.
7. Strosberg AD, Issad T. The involvement of leptin in humans revealed by mutations in leptin and leptin receptor genes. *Trends Pharmacol Sci* 1999;**20**:227–30.
8. Montague CT, Farooqi IS, Whitehead JP, Soos MA, Rau H, Wareham NJ, et al. Congenital leptin deficiency is associated with severe early-onset obesity in humans. *Nature* 1997;**387**:903–8.
9. Fatima W, Shahid A, Imran M, Manzoor J, Hasnain S, Rana S, et al. Leptin deficiency and *leptin* gene mutations in obese children from Pakistan. *Int J Pediatr Obes* 2011;**6**:419–27.
10. Clément K, Vaisse C, Lahlou N, Cabrol S, Pelloux V, Cassuto D, et al. A mutation in the human leptin receptor gene causes obesity and pituitary dysfunction. *Nature* 1998;**392**:398–401.
11. Krude H, Biebermann H, Schnabel D, Tansek MZ, Theunissen P, Mullis PE, et al. Obesity due to proopiomelanocortin deficiency: three new cases and treatment trials with thyroid hormone and ACTH4-10. *J Clin Endocrinol Metab* 2003;**88**:4633–40.
12. Jackson RS, Creemers JW, Ohagi S, Raffin-Sanson ML, Sanders L, Montague CT, et al. Obesity and impaired prohormone processing associated with mutations in the human prohormone convertase 1 gene. *Nat Genet* 1997;**16**:303–6.
13. Jackson RS, Creemers JW, Farooqi IS, Raffin-Sanson ML, Varro A, Dockray GJ, et al. Small-intestinal dysfunction accompanies the complex endocrinopathy of human proprotein convertase 1 deficiency. *J Clin Invest* 2003;**112**:1550–60.
14. Licinio J, Caglayan S, Ozata M, Yildiz BO, de Miranda PB, O'Kirwan F, et al. Phenotypic effects of leptin replacement on morbid obesity, diabetes mellitus, hypogonadism, and behavior in leptin-deficient adults. *Proc Natl Acad Sci USA* 2004;**101**:4531–6.
15. Fischer-Posovszky P, von Schnurbein J, Moepps B, Lahr G, Strauss G, Barth TF, et al. A new missense mutation in the leptin gene causes mild obesity and hypogonadism without affecting T cell responsiveness. *J Clin Endocrinol Metab* 2010;**95**:2836–40.
16. Birch LL. Development of food acceptance patterns in the first years of life. *Proc Nutr Soc* 1998;**57**:617–24.
17. Farooqi IS. Genetic and hereditary aspects of childhood obesity. *Best Pract Res Clin Endocrinol Metab* 2005;**19**:359–74.
18. Farooqi IS, Wangensteen T, Collins S, Kimber W, Matarese G, Keogh JM, et al. Clinical and molecular genetic spectrum of congenital deficiency of the leptin receptor. *N Engl J Med* 2007;**356**:237–47.
19. Mazen I, El-Gammal M, Abdel-Hamid M, Farooqi IS, Amr K. Homozygosity for a novel missense mutation in the leptin receptor gene (P316T) in two Egyptian cousins with severe early onset obesity. *Mol Genet Metab* 2011;**102**:461–4.
20. Nizard J, Dommergues M, Clément K. Pregnancy in a woman with a leptin-receptor mutation. *N Engl J Med* 2012;**366**:1064–5.
21. Clément K, Dubern B, Mencarelli M, Czernichow P, Ito S, Wakamatsu K, et al. Unexpected endocrine features and normal pigmentation in a young adult patient carrying a novel homozygous mutation in the POMC gene. *J Clin Endocrinol Metab* 2008;**93**:4955–62.

22. Farooqi IS, Drop S, Clements A, Keogh JM, Biernacka J, Lowenbein S, et al. Heterozygosity for a POMC-null mutation and increased obesity risk in humans. *Diabetes* 2006;**55**:2549–53.

23. Carroll L, Voisey J, van Daal A. Gene polymorphisms and their effects in the melanocortin system. *Peptides* 2005;**26**:1871–85.

24. Frank GR, Fox J, Candela N, Jovanovic Z, Bochukova E, Levine J, et al. Severe obesity and diabetes insipidus in a patient with PCSK1 deficiency. *Mol Genet Metab* 2013;**110**:191–4.

25. Martin MG, Lindberg I, Solorzano-Vargas RS, Wang J, Avitzur Y, Bandsma R, et al. Congenital proprotein convertase 1/3 deficiency causes malabsorptive diarrhea and other endocrinopathies in a pediatric cohort. *Gastroenterology* 2013;**145**:138–48.

26. Holder JLJr, Butte NF, Zinn AR. Profound obesity associated with a balanced translocation that disrupts the SIM1 gene. *Hum Mol Genet* 2000;**9**:101–8.

27. Izumi K, Housam R, Kapadia C, Stallings VA, Medne L, Shaikh TH, et al. Endocrine phenotype of 6q16.1-q21 deletion involving SIM1 and Prader–Willi syndrome-like features. *Am J Med Genet* 2013;**9999**:1–7.

28. Michaud JL, Boucher F, Melnyk A, Gauthier F, Goshu E, Levy E, et al. Sim1 haploinsufficiency causes hyperphagia, obesity and reduction of the paraventricular nucleus of the hypothalamus. *Hum Mol Genet* 2001;**10**:1465–73.

29. Kublaoui BM, Holder JLJr, Gemelli T. Sim1 haploinsufficiency impairs melanocortin-mediated anorexia and activation of paraventricular nucleus neurons. *Mol Endocrinol* 2006;**20**:2483–92.

30. Ramachandrappa S, Raimondo A, Cali AMG, Keogh JM, Henning E, Saeed S, et al. Rare variants in single-minded 1 (SIM1) are associated with severe obesity. *J Clin Invest* 2013;**123**:3042–50.

31. Bonnefond A, Raimondo A, Stutzmann F, Ghoussaini M, Ramachandrappa S, Bersten DC, et al. Loss-of-function mutations in *SIM1* contribute to obesity and Prader–Willi-like features. *J Clin Invest* 2013;**123**:3037–41.

32. Zegers D, Beckers S, Hendrickx R, Van Camp JK, de Craemer V, Verrijken A, et al. Mutation screen of the SIM1 gene in pediatric patients with early-onset obesity. *Int J Obes* 2013;1–5.

33. Kernie SG, Liebl DJ, Parada LF. BDNF regulates eating behavior and locomotor activity in mice. *Embo J* 2000;**19**:1290–300.

34. Xu B, Goulding EH, Zang K, Cepoi D, Cone RD, Jones KR, et al. Brain-derived neurotrophic factor regulates energy balance downstream of melanocortin-4 receptor. *Nat Neurosci* 2003;**6**:736–42.

35. Yeo GS, Connie Hung CC, Rochford J, Keogh J, Gray J, Sivaramakrishnan S, et al. A *de novo* mutation affecting human TrkB associated with severe obesity and developmental delay. *Nat Neurosci* 2004;**7**:1187–9.

36. Gray J, Yeo G, Hung C, Keogh J, Clayton P, Banerjee K, et al. Functional characterization of human NTRK2 mutations identified in patients with severe early-onset obesity. *Int J Obes (Lond)* 2007;**31**:359–64.

37. Walters RG, Jacquemont S, Valsesia A, de Smith AJ, Martinet D, Andersson J, et al. A novel highly-penetrant form of obesity due to microdeletions on chromosome 16p11.2. *Nature* 2010;**463**:671–5.

38. El-Sayed Moustapha JS, Eleftherohorinou H, de Smith AJ, Andersson-Assarsson JC, Couto Alves A, Hadjigeorgiou E, et al. Novel association approach for variable number tandem repeats (VNTRs) identifies *DOCK5* as a susceptibility gene for severe obesity. *Hum Mol Genet* 2012;**21**:3727–38.

39. Govaerts C, Srinivasan S, Shapiro A, Zhang S, Picard F, Clement K, et al. Obesity-associated mutations in the melanocortin 4 receptor provide novel insights into its function. *Peptides* 2005;**26**:1909–19.

40. Hinney A, Volckmar AL, Knoll N. Melanocortin-4 receptor in energy homeostasis and obesity pathogenesis. *Prog Mol Biol Transl Sci* 2013;**114**:147–91.

41. Lubrano-Berthelier C, Dubern B, Lacorte JM, Picard F, Shapiro A, Zhang S, et al. Melanocortin 4 receptor mutations in a large cohort of severely obese adults: prevalence, functional classification, genotype-phenotype relationship, and lack of association with binge eating. *J Clin Endocrinol Metab* 2006;**91**:1811–8.

42. Hinney A, Bettecken T, Tarnow P, Brumm H, Reichwald K, Lichtner P, et al. Prevalence, spectrum, and functional characterization of melanocortin-4 receptor gene mutations in a representative population-based sample and obese adults from Germany. *J Clin Endocrinol Metab* 2006;**91**:1761–9.

43. Farooqi IS, Keogh JM, Yeo GS, Lank EJ, Cheetham T, O'Rahilly S. Clinical spectrum of obesity and mutations in the melanocortin 4 receptor gene. *N Engl J Med* 2003;**348**:1085–95.

44. Lubrano-Berthelier C, Le Stunff C, Bougnères P, Vaisse C. A homozygous null mutation delineates the role of the melanocortin-4 receptor in humans. *J Clin Endocrinol Metab* 2004;**89**:2028–32.

45. Dubern B, Bisbis S, Talbaoui H, Le Beyec J, Tounian P, Lacorte JM, et al. Homozygous null mutation of the melanocortin-4 receptor and severe early-onset obesity. *J Pediatr* 2007;**150**:613–7.

46. Vaisse C, Clément K, Durand E, Hercberg S, Guy-Grand B, Froguel P. Melanocortin-4 receptor mutations are a frequent and heterogeneous cause of morbid obesity. *J Clin Invest* 2000;**106**:253–62.

47. Martinelli CE, Keogh JM, Greenfield JR, Henning E, van der Klaauw AA, Blackwood A, et al. Obesity due to melanocortin 4 receptor (MC4R) deficiency is associated with increased linear growth and final height, fasting hyperinsulinemia, and incompletely suppressed growth hormone secretion. *J Clin Endocrinol Metab* 2011;**96**:181–8.

48. MacKenzie RG. Obesity-associated mutations in the human melanocortin-4 receptor gene. *Peptides* 2006;**27**:395–403.

49. Elefteriou F, Ahn JD, Takeda S, Starbuck M, Yang X, Liu X. Leptin regulation of bone resorption by the sympathetic nervous system and CART. *Nature* 2005;**434**:514–20.

50. Ahn JD, Dubern B, Lubrano-Berthelier C, Clément K, Karsenty G. Cart overexpression is the only identifiable cause of high bone mass in melanocortin 4 receptor deficiency. *Endocrinology* 2006;**147**:3196–202.

51. Branson R, Potoczna N, Kral JG, Lentes KU, Hoehe MR, Horber FF. Binge eating as a major phenotype of melanocortin 4 receptor gene mutations. *N Engl J Med* 2003;**348**:1096–103.

52. Valette M, Poitou C, Kesse-Guyot E, Bellisle F, Carette C, Le Beyec J, et al. Association between melanocortin-4 receptor mutations and eating behaviors in obese patients: a case-control study. *Int J Obes* 2013;1–3.

53. Mergen M, Mergen H, Ozata M, Oner R, Oner C. A novel melanocortin 4 receptor (MC4R) gene mutation associated with morbid obesity. *J Clin Endocrinol Metab* 2001;**86**:3448.

54. Lubrano-Berthelier C, Durand E, Dubern B, Shapiro A, Dazin P, Weill J, et al. Intracellular retention is a common characteristic of childhood obesity-associated MC4R mutations. *Hum Mol Genet* 2003;**12**:145–53.

55. Nijenhuis WA, Oosterom J, Adan RA. AgRP(83-132) acts as an inverse agonist on the human-melanocortin-4 receptor. *Mol Endocrinol* 2001;**15**:164–71.

56. Srinivasan S, Lubrano-Berthelier C, Govaerts C, Picard F, Santiago P, Conklin BR, et al. Constitutive activity of the melanocortin-4 receptor is maintained by its N-terminal domain and plays a role in energy homeostasis in humans. *J Clin Invest* 2004;**114**:1158–64.

57. Chen AS, Marsh DJ, Trumbauer ME, Frazier EG, Guan XM, Yu H, et al. Inactivation of the mouse melanocortin-3 receptor results in increased fat mass and reduced lean body mass. *Nat Genet* 2000;**26**:97–102.

58. Mencarelli M, Walker GE, Maestrini S, Alberti L, Verti B, Brunani A, et al. Sporadic mutations in melanocortin receptor 3 in morbid obese individuals. *Eur J Hum Genet* 2008;**16**:581–6.

59. Calton MA, Ersoy BA, Zhang S, Kane JP, Malloy MJ, Pullinger CR, et al. Association of functionally significant melanocortin-4 but not melanocortin-3 receptor mutations with severe adult obesity in a large North American case-control study. *Hum Mol Genet* 2009;**18**:1140–7.

60. Horsthemke B, Wagstaff J. Mechanisms of imprinting of the Prader-Willi/Angelman region. *Am J Med Genet* 2008;**146**:2041–52.

61. Horsthemke B, Buiting K. Imprinting defects on human chromosome 15. *Cytogenet Genome Res* 2006;**113**:292–9.

62. Schaaf CP, Gonzalez-Garay ML, Xia F, Potocki L, Gripp KW, Zhang B, et al. Truncating mutations of *MAGEL2* cause Prader–Willi phenotypes and autism. *Nat Genet* 2013;**45**:1405–9.

63. Kousta E, Hadjiathanasiou CG, Tolis G, Papathanasiou A. Pleiotropic genetic syndromes with developmental abnormalities associated with obesity. *J Pediatr Endocrinol Metab* 2009;**22**:581–92.

64. Goldstone AP. Prader-Willi syndrome: advances in genetics, pathophysiology and treatment. *Trends Endocrinol Metab* 2004;**15**:12–20.

65. Cummings DE, Clément K, Purnell JQ, Vaisse C, Foster KE, Frayo RS, et al. Elevated plasma ghrelin levels in Prader Willi syndrome. *Nat Med* 2002;**8**:643–4.

66. Franzese A, Romano A, Spagnuolo MI, Ruju F, Valerio G. Growth hormone therapy in children with Prader-Willi syndrome. *J Pediatr* 2006;**148**:846 author reply 846-847.

67. Carrel AL, Moerchen V, Myers SE, Bekx MT, Whitman BY, Allen DB. Growth hormone improves mobility and body composition in infants and toddlers with Prader-Willi syndrome. *J Pediatr* 2004;**145**:744–9.

68. Katsanis N, Lupski JR, Beales PL. Exploring the molecular basis of Bardet–Biedl syndrome. *Hum Mol Genet* 2001;**10**:2293–9.

69. Stoetzel C, Laurier V, Davis EE, Muller J, Rix S, Badano JL, et al. BBS10 encodes a vertebrate-specific chaperonin-like protein and is a major BBS locus. *Nat Genet* 2006;**38**:521–4.

70. Nishimura DY, Swiderski RE, Searby CC, Berg EM, Ferguson AL, Hennekam R, et al. Comparative genomics and gene expression analysis identifies BBS9, a new Bardet–Biedl syndrome gene. *Am J Hum Genet* 2005;**77**:1021–33.

71. Chiang AP, Beck JS, Yen HJ, Tayeh MK, Scheetz TE, Swiderski RE, et al. Homozygosity mapping with SNP arrays identifies TRIM32, an E3 ubiquitin ligase, as a Bardet–Biedl syndrome gene (BBS11). *Proc Natl Acad Sci USA* 2006;**103**:6287–92.

72. Zaghloul NA, Katsanis N. Mechanistic insights into Bardet-Biedl syndrome, a model ciliopathy. *J Clin Invest* 2009;**119**:428–37.

73. Eichers ER, Lewis RA, Katsanis N, Lupski JR. Triallelic inheritance: a bridge between Mendelian and multifactorial traits. *Ann Med* 2004;**36**:262–72.

74. Abu-Safieh L, Al-Anazi S, Al-Abdi L, Hashem M, Alkuraya H, Alamr M, et al. In search of triallelism in Bardet-Biedl syndrome. *Eur J Hum Genet* 2012;**20**:420–7.

75. Slavotinek AM, Searby C, Al-Gazali L, Hennekam RC, Schrander-Stumpel C, Orcana-Losa M, et al. Mutation analysis of the MKKS gene in McKusick-Kaufman syndrome and selected Bardet–Biedl syndrome patients. *Hum Genet* 2002;**110**:561–7.

76. Fan Y, Esmail MA, Ansley SJ, Blacque OE, Boroevich K, Ross AJ, et al. Mutations in a member of the Ras superfamily of small GTP-binding proteins causes Bardet–Biedl syndrome. *Nat Genet* 2004;**36**:989–93.

77. Badano JL, Mitsuma N, Beales PL, Katsanis N. The ciliopathies: an emerging class of human genetic disorders. *Annu Rev Genomics Hum Genet* 2006;**7**:125–48.

78. Seo S, Guo DF, Bugge K, Morgan DA, Rahmouni K, Sheffield VC. Requirement of Bardet-Biedl syndrome proteins for leptin receptor signaling. *Hum Mol Genet* 2009;**18**:1323–31.

79. Berbari NF, Pasek RC, Malarkey EB, Yazdi SMZ, McNair AD, Lewis WR, et al. Leptin resistance is a secondary consequence of the obesity in ciliopathy mutant mice. *Proc Natl Acad Sci* 2013;**110**:7796–801.

80. McLennan Y, Polussa J, Tassone F, Hagerman R. Fragile X syndrome. *Curr Genomics* 2011;**12**:216–24.

81. Hickey SE, Walters-Sen L, Mihalic Mosher T, Pfau RB, Pyatt R, Snyder PJ, et al. Duplication of the Xq27.3-q28 region, including the *FMR1* gene, in an X-linked hypogonadism, gynecomastia, intellectual disability, short stature, and obesity syndrome. *Am J Med Genet* 2013;**161**:2294–9.

82. Marshall JD, Hinman EG, Collin GB, Beck S, Cerqueira R, Maffei P, et al. Spectrum of ALMS1 variants and evaluation of genotype-phenotype correlations in Alström syndrome. *Hum Mutat* 2007;**28**:1114–23.

83. Kolehmainen J, Black GC, Saarinen A, Chandler K, Clayton-Smith J, Traskelin AL, et al. Cohen syndrome is caused by mutations in a novel gene, COH1, encoding a transmembrane protein with a presumed role in vesicle-mediated sorting and intracellular protein transport. *Am J Hum Genet* 2003;**72**:1359–69.

84. Seifert W, Holder-Espinasse M, Spranger S, Hoeltzenbein M, Rossier E, Dollfus H, et al. Mutational spectrum of COH1 and clinical heterogeneity in Cohen syndrome. *J Med Genet* 2006;**43**:e22.

85. Balikova I, Lehesjoki AE, de Ravel TJL, Thienpont B, Chandler KE, Clayton-Smith J, et al. Deletions in the VPS13B (COH1) gene as a cause of Cohen syndrome. *Hum Mutat* 2009;**30**:845–54.

86. Roubert P, Dubern B, Plas P, Lubrano-Berthelier C, Alihi R, Auger F, et al. Novel pharmacological MC4R agonists can efficiently activate mutated MC4R from obese patient with impaired endogenous agonist response. *J Endocrinol* 2010;**207**:177–83.

87. Bonnefond A, Philippe J, Durand E, Muller J, Saeed S, Arslan M, et al. Highly sensitive diagnosis of 43 monogenic forms of diabetes or obesity, through one step PCR-based enrichment in combination with next-generation sequencing. *Diabetes Care* 2013;**37**:1–26.

88. Saeed S, Bonnefond A, Manzoor J, Philippe J, Durand E, Arshad M, et al. Novel LEPR mutations in obese Pakistani children identified by PCR-based enrichment and next generation sequencing. *Obesity* 2013;**22**:1112–7.

89. Xing DJ, Zhang HX, Huang N, Wu KC, Huang XF, Huang F, et al. Comprehensive molecular diagnosis of Bardet-Biedl syndrome by high-throughput targeted exome sequencing. *PloS One* 2014;**9**:90599.

90. Doco-Fenzi M, Leroy C, Schneider A, Petit F, Delrue MA, Andrieux J, et al. Early-onset obesity and paternal 2pter deletion encompassing the *ACP1*, *TMEM18*, and *MYT1L* genes. *Eur J Hum Genet* 2013;1–9.

91. Pearce LR, Atanassova N, Banton MC, Bottomley B, van der Klaauw AA, Revelli JP, et al. KSR2 mutations are associated with obesity, insulin resistance, and impaired cellular fuel oxidation. *Cell* 2013;**155**:765–77.

92. Borman AD, Pearce LR, Mackay DS, Nagel-Wolfrum K, Davidson AE, Henderson R, et al. A homozygous mutation in the *TUB* gene associated with retinal dystrophy and obesity. *Hum Mutat* 2014;**35**:289–93.

93. Farooqi IS, Matarese G, Lord GM, Keogh JM, Lawrence E, Agwu C, et al. Beneficial effects of leptin on obesity, T cell hyporesponsiveness, and neuroendocrine/metabolic dysfunction of human congenital leptin deficiency. *J Clin Invest* 2002;**110**:1093–103.

94. Williamson DA, Ravussin E, Wong ML, Wagner A, Dipaoli A, Caglayan S, et al. Microanalysis of eating behavior of three lean and obese leptin deficient adults treated with leptin therapy. *Appetite* 2005;**45**:75–80.

95. Farooqi IS, Bullmore E, Keogh J, Gillard J, O'Rahilly S, Fletcher PC. Leptin regulates striatal regions and human eating behavior. *Science* 2007;**317**:1355.

96. Sleeman MW, Anderson KD, Lambert PD, Yancopoulos GD, Wiegand SJ. The ciliary neurotrophic factor and its receptor, CNTFR alpha. *Pharm Acta Helv* 2000;**74**:265–72.

97. Preti A. Axokine (Regeneron). *IDrugs* 2003;**6**:696–701.

98. Sariola H. The neurotrophic factors in non-neuronal tissues. *Cell Mol Life Sci* 2001;**58**:1061–6.

99. Torres N, Chabardes S, Piallat B, Devergnas A, Benabid AL. Body fat and body weight reduction following hypothalamic deep brain stimulation in monkeys: an intraventricular approach. *Int J Obes* 2012;**36**:1537–44.

100. Irani BG, Xiang Z, Moore MC, Mandel RJ, Haskell-Luevano C. Voluntary exercise delays monogenetic obesity and overcomes reproductive dysfunction of the melanocortin-4 receptor knockout mouse. *Biochem Biophys Res Commun* 2005;**326**:638–44.

101. Xiang Z, Pogozheva ID, Sorenson NB, Wilczynski AM, Holder JR, Litherland SA, et al. Peptide and small molecules rescue the functional activity and agonist potency of dysfunctional human melanocortin-4 receptor polymorphisms. *Biochemistry* 2007;**46**: 8273–87.

102. Ujjainwalla F, Sebhat IK. Small molecule ligands of the human melanocortin-4 receptor. *Curr Top Med Chem* 2007;**7**:1068–84.

103. Fani L, Bak S, Delhanty P, van Rossum EFC, van den Akker ELT. The melanocortin-4 receptor as target for obesity treatment: a systematic review of emerging pharmacological therapeutic options. *Int J Obes* 2013;**38**:1–7.

104. Kievit P, Halem H, Marks DL, Dong JZ, Glavas MM, Sinnayah P, et al. Chronic treatment with a melanocortin-4 receptor agonist causes weight loss, reduces insulin resistance, and improves cardiovascular function in diet-induced obese rhesus macaques. *Diabetes* 2012;**62**:490–7.

105. Carrel AL, Myers SE, Whitman BY, Allen DB. Growth hormone improves body composition, fat utilization, physical strength and agility, and growth in Prader-Willi syndrome: a controlled study. *J Pediatr* 1999;**134**:215–21.

106. Festen DA, de Lind van Wijngaarden R, van Eekelen M, Otten BJ, Wit JM, Duivenvoorden HJ, et al. Randomized controlled growth hormone trial: effects on anthropometry, body composition, and body proportions in a large group of children with Prader-Willi syndrome. *Clin Endocrinol (Oxf)* 2008;**69**:443–51.

107. Carrel AL, Myers SE, Whitman BY, Eickhoff J, Allen DB. Long-term growth hormone therapy changes the natural history of body composition and motor function in children with Prader-Willi syndrome. *J Clin Endocrinol Metab* 2010;**95**:1131–6.

108. Sanchez-Ortiga R, Klibanski A, Tritos NA. Effects of recombinant human growth hormone therapy in adults with Prader-Willi syndrome: a meta-analysis. *Clin Endocrinol* 2012;**77**: 86–93.

109. Sjöström L. Review of the key results from the Swedish Obes Subjects (SOS) trial – a prospective controlled intervention study of bariatric surgery. *J Intern Med* 2013;**273**:219–34.

110. Aslan IR, Campos GM, Calton MA, Evans DS, Merriman RB, Vaisse C. Weight loss after Roux-en-Y gastric bypass in obese patients heterozygous for MC4R mutations. *Obes Surg* 2011;**21**: 930–4.

111. Valette M, Poitou C, Le Beyec J, Bouillot JL, Clement K, Czernichow S. Melanocortin-4 receptor mutations and polymorphisms do not affect weight loss after bariatric surgery. *PloS One* 2012;**7**: 48221.

112. Aslan IR, Ranadive SA, Ersoy BA, Rogers SJ, Lustig RH, Vaisse C. Bariatric surgery in a patient with complete MC4R deficiency. *Int J Obes* 2011;**35**:457–61.

113. Le Beyec J, Cugnet-Anceau C, Pépin D, Alili R, Cotillard A, Lacorte JM, et al. Homozygous leptin receptor mutation due to uniparental disomy of chromosome 1: response to bariatric surgery. *J Clin Endocrinol Metab* 2013;**98**:397–402.

114. Fong AKW, Wong SKH, Lam CCH, Ng EKW. Ghrelin level and weight loss after laparoscopic sleeve gastrectomy and gastric mini-bypass for Prader-Willi syndrome in Chinese. *Obes Surg* 2012;**22**:1742–5.

115. Scheimann AO, Butler MG, Gourash L, Cuffari C, Kish W. Critical analysis of bariatric procedures in Prader-Willi syndrome. *JPGN* 2008;**46**:80–3.

22

Syndromes of Severe Insulin Resistance and/or Lipodystrophy

Robert K. Semple, David B. Savage, Gemma V. Brierley,
Stephen O'Rahilly

Metabolic Research Laboratories, University of Cambridge Institute of Metabolic Science,
Addenbrooke's Hospital, Cambridge, UK

INTRODUCTION

Background, Incidence, Prevalence

Prospective studies have established that systemic insulin resistance (IR) is among the earliest detectable abnormalities in those who go on to develop type 2 diabetes,[1] and consequently there is major interest in teasing out its genetics and molecular pathology. In the face of the burgeoning prevalence of obesity, IR is also increasing, and this may sometimes be severe. However, a small minority of patients have severe IR without obesity or any other obvious acquired precipitant. This is sometimes seen together with a failure of adipose tissue development, with abnormalities in adipose tissue distribution or function, or in the context of more complex recessive syndromes. For many of these conditions, a single genetic defect has now been defined. Primary forms of IR of known genetic basis are thus the focus of this chapter.

IR only produces diabetes in conjunction with beta cell decompensation, which may take decades to occur. Thus, although it is commonly brought to clinical attention as a failure to adequately control hyperglycemia despite large doses of insulin, the majority of patients are unrecognized in the prediabetic phase. Coupled to a common failure to appreciate the significance of the clinical signs of IR even once, hyperglycemia has supervened; this means that the syndrome is significantly underdiagnosed. This problem is compounded by the arbitrary biochemical thresholds often taken to denote severe IR. In some syndromes, characteristic clinical features permit diagnosis without reference to biochemical criteria, while in others the degree of hyperinsulinemia is critical. One set of operational diagnostic criteria is shown in Box 22.1. Because of these complexities, no accurate prevalence figures for severe insulin resistance exist, though cumulative experience in one center suggests that it affects of the order of 0.5% of patients with type 2 diabetes.

Clinical Presentation

Some clinical features of severe IR are generic to all known genetic forms of the condition, while some are particular only to some genetic subgroups. Although IR is normally thought of in terms of reduced ability of insulin to lower blood glucose, insulin is a highly pleiotropic hormone with a plethora of different metabolic and mitogenic effects, and this complexity is reflected in the range of its clinical manifestations. At least some of these appear to be due to preserved or enhanced signaling stimulated by very high insulin levels, either through the type 1 insulin-like growth factor (IGF) receptor, which has some ability to bind insulin, or via the insulin receptor through preserved arms of the intracellular insulin signaling network. The cardinal feature of severe IR is acanthosis nigricans, which is nearly a *sine qua non* for the condition, and is often associated with flexural skin tags (acrochordons). By far the commonest reason for seeking medical advice, however, is cosmetically distressing hirsutes and/or oligo- or amenorrhoea, driven by the effects of high levels of insulin, in tandem with gonadotropins, on the ovary. Indeed, ovarian hyperandrogenism may be severe, with serum testosterone concentration sometimes in the normal male range, and imaging usually reveals enlarged, polycystic ovaries.

Genetic Diagnosis of Endocrine Disorders. http://dx.doi.org/10.1016/B978-0-12-800892-8.00022-1

BOX 22.1

SUGGESTED CRITERIA FOR THE DIAGNOSIS OF SEVERE INSULIN RESISTANCE

1. Nondiabetic and BMI <30 kg/m^2
   Fasting insulin above 150 pmol/L or peak insulin on oral glucose tolerance testing above 1500 pmol/L.
2. Absolute Insulin Deficiency and BMI <30 kg/m^2
   Exogenous insulin requirement >3 U/kg/day.
3. Partial Beta Cell Decompensation and/or BMI >30 kg/m^2
   Insulin levels are difficult to interpret in the context of obesity, although comparison with

sex and BMI-adjusted normal ranges is of use. Furthermore, in diabetes glucotoxicity, impairing islet function, and mixtures of endogenous and exogenous insulin in the circulation confuses the biochemical picture. In these settings the clinical history and features such as acanthosis nigricans assume particular importance in making a diagnosis of likely monogenic severe IR.

A common clinical label used in cases of severe IR is "type A insulin resistance syndrome," a term coined in the 1970s to denote lean women with the previously mentioned clinical features of severe IR, but without evidence of a causative circulating antibody indicative of "type B" extreme IR.[2] "HAIR-AN" syndrome, another commonly used term,[3] denotes "hyperandrogenism, IR and acanthosis nigricans," and is thus essentially identical to the type A IR syndrome except that it has often come to be used only in women with BMI >30 kg/m^2. This distinction is of some use, as there is a great enrichment of monogenic disease in lean, very insulin resistant patients; however, both descriptors in essence simply capture the general clinical features of severe IR described previously, and therefore, in isolation, do not imply any specific molecular defect.

Although monogenic severe IR most commonly presents as an aggressive form of polycystic ovary syndrome in young, lean women associated with acanthosis nigricans, diabetes is often diagnosed on oral glucose tolerance testing during the diagnostic work-up on the basis of hyperglycemia after the glucose challenge; however, not uncommonly fasting glucose may be low or normal. Indeed, a frequent but underrecognized early feature of severe IR is spontaneous and symptomatic postprandial hypoglycemia, which may be severe and require medical intervention, and which sometimes is reported to be exacerbated by metformin therapy. Only later in the natural history of severe IR does refractory hyperglycemia become the dominant problem, usually manifest as poor metabolic control despite very large doses of exogenous insulin. Because it often takes many years for beta cells to decompensate even in the face of severe IR, this may not occur until the fourth or fifth decade of life. Men are more likely to present at this stage than women.

The degree of IR in an individual with a monogenic defect is not invariant, and physiological or pathological

influences that lead to IR often synergize with the inherited defect to exaggerate the clinical problem. Thus, puberty and the later stages of pregnancy, as well as intercurrent infection or illness, may in some cases lead to an increase in acanthosis nigricans and hyperandrogenism, and/or hyperglycemia, which is resistant even to huge doses of exogenous insulin.

The previous observations are true for all forms of severe IR. However, in some cases specific syndromic or biochemical features are present, which give a strong clue to the underlying single gene defect.

Primary Insulin Signaling Defects

Inherited defects in the insulin receptor most commonly present as type A IR, without obviously discriminating clinical features, although the lack of fatty liver or metabolic dyslipidemia, and preserved or elevated serum adiponectin concentration, are highly characteristic.[4–6] In contrast, the most severe genetic defects in the insulin receptor present in early childhood or infancy. By virtue of their historical descriptions, which long antedated identification of the insulin receptor, the resulting constellations of clinical features are often classified either as Donohue syndrome (formerly leprechaunism; OMIM #246200) or Rabson–Mendenhall syndrome (OMIM#262190), though in truth there is a spectrum of clinical defects of which these two descriptions are arbitrary snapshots. In fact the clinical features of Donohue and Rabson–Mendenhall syndromes are remarkably similar, differing essentially only in the relative prominence of particular components of the syndromes. Features of the syndromes are summarized in Table 22.1.

The consequences of severe loss of insulin receptor function can be grouped into metabolic and growth defects. Metabolism is characterized by severe hyperinsulinemia (one to three orders of magnitude higher

TABLE 22.1 Clinical Features of Donohue and Rabson–Mendenhall Syndromes

	Donohue	Rabson–Mendenhall
Prognosis	Death in infancy	Death at 5–20 years
Metabolic abnormalities	Postprandial hyperglycemia; fasting hypoglycemia; extreme hyperinsulinemia; no ketoacidosis	Postprandial hyperglycemia; fasting hypoglycemia; later refractory hyperglycemia; extreme hyperinsulinemia; late ketoacidosis
Linear growth impairment	Low birth weight; postnatal growth retardation; severe failure to thrive	Low birth weight; postnatal growth retardation; short stature, low weight
Impaired development of tissues with high insulin-dependent glucose uptake	Paucity of adipose tissue; low muscle mass	Paucity of adipose tissue; low muscle mass
Soft tissue overgrowth	"Elfin" facies; large, low-set ears; prominent eyes; wide nostrils; thick lips; gingival hyperplasia; large mouth; acanthosis nigricans; large hands and feet; dysplastic nails; hypertrichosis	Coarse facies; prognathism; large, fissured tongue; gingival hyperplasia; dental dysplasia; premature eruption of teeth; acanthosis nigricans; dry, lichenified skin; onychauxis; hypertrichosis; pineal hypertrophy
Visceral abnormalities	Abdominal distention; nephromegaly; hepatomegaly; Cholestasis; hepatic fibrosis nephrocalcinosis; islet of Langerhans hyperplasia	Abdominal distention; nephromegaly; hepatomegaly; nephrocalcinosis; islet of Langerhans hyperplasia
Overgrowth of sex hormone-dependent tissues	Breast hyperplasia (female); prominent nipples; cystic ovaries; juvenile ovarian granulosa cell tumor; Leydig cell hyperplasia; large penis; large clitoris	Large penis; large clitoris; cystic ovaries
Miscellaneous	Frequent infections; decreased lymphatic tissue; delayed bone age	Frequent infections; motor developmental delay; precocious puberty

than the reference range) with initial preprandial hypoglycemia and postprandial hyperglycemia eventually degenerating into sustained hyperglycemia. Infants appear to be protected from ketoacidosis, although this becomes a major and refractory problem in older children. Leptin levels are usually low or undetectable, while adiponectin is paradoxically elevated after infancy.[5] Growth defects include impaired linear growth and poor development of adipose and muscle tissue, which heavily rely on insulin-stimulated glucose uptake, contrasting with pseudoacromegaloid overgrowth of many other soft tissues, with additional features such as hypertrichosis. Particularly prominent is exaggerated growth of androgen-dependent tissues, which is a consequence of the ability of extreme hyperinsulinemia to synergize with gonadotropin actions on the gonads even in the absence of the insulin receptor.

Genetic defects affecting a more distal insulin signaling component have proved to be extremely rare. A single family with a loss of function mutation in the serine threonine kinase encoded by AKT2, a key effector of many of the metabolic actions of insulin, showed severe insulin resistance and hyperandrogenisim, but unlike receptor defects they also showed femorogluteal lipodystrophy, fatty liver, and dyslipidemia.[6,7] More recently, mutations in the C-terminal of the p85α regulatory subunit of phosphatidylinositol-3-kinase, a signaling enzyme that functions downstream from the insulin receptor as well as many other tyrosine kinases, were found to underlie SHORT syndrome (short stature, hyperextensibility, ocular abnormalities, rieger's anomaly, and teething delay). Insulin resistance, sometimes severe, was described in many of the patients with PIK3R1 mutations, and lipodystrophy has also been reported to be common;[8–10] however, their penetrance and natural history have yet to be fully defined, and this condition will not be discussed further in this chapter.

Lipodystrophy

Lipodystrophic syndromes encompass a heterogeneous group of conditions characterized by partial or complete absence of adipose tissue.[11] They may be genetic or acquired, and are further classified according to the anatomical distribution of the lipodystrophy. IR is a feature of most, but not all, of these disorders and may be severe. Where it complicates lipodystrophy it presents as acanthosis nigricans and dysglycemia in prepubertal children, and has the same clinical features as described for the type A IR syndrome in postpubertal patients. As with all forms of IR, the clinical expression is more pronounced in women and may be both more subtle and delayed in men. As nonalcoholic fatty

TABLE 22.2   Clinical and Biochemical Features of the Most Prevalent Inherited Lipodystrophies

	CGL		Familial partial lipodystrophy	
Subtype	BSCL1	BSCL2	FPLD2	FPLD3
Defective gene	*AGPAT2*	*BSCL2*	*LMNA*	*PPARG*
Clinical onset	Soon after birth	Soon after birth	Puberty	Usually puberty, but may present in younger children
Fat distribution	**Generalized absence**	**Generalized absence**	**Loss of limb and gluteal fat; typically excess facial and nuchal fat; trunk fat often lost**	**Loss of limb and gluteal fat; preserved facial and trunk fat**
Cutaneous features	Acanthosis nigricans and skin tags; hirsutism common in women	Acanthosis nigricans and skin tags; hirsutism common in women	Acanthosis nigricans and skin tags; hirsutism common in women	Acanthosis nigricans and skin tags; hirsutism common in women
Musculoskeletal	Acromegaloid features common	Acromegaloid features common	Frequent muscle hypertrophy; some have overlap features of muscular dystrophy	Nil specific
NAFLD	Severe	Severe	Yes	Yes
Dyslipidemia	Severe associated with pancreatitis	Severe associated with pancreatitis	Yes, may be severe	Yes, may be severe
Insulin resistance	Severe early onset	Severe early onset	Severe	Severe; early onset in some
Diabetes onset	<20 years	<20 years	Variable; generally later in men than women	Variable; generally later in men than women
Hypertension	Common	Common	Common	Very common
Other		Mild mental retardation possible		

liver and dyslipidemia are present in the vast majority (perhaps all) of patients with insulin resistance related to lipodystrophy, their presence in an insulin-resistant patient should prompt careful assessment of fat distribution and fat mass. Clinical and biochemical features of the most prevalent subtypes of congenital lipodystrophy are summarized in Table 22.2.

Congenital generalized lipodystrophy (CGL), also known as Berardinelli–Seip congenital lipodystrophy (BSCL),[12,13] is characterized by a generalized absence of adipose tissue from birth. Children with the condition have increased appetite due to leptin deficiency,[14] accelerated growth, and advanced bone age. Skeletal muscles, peripheral veins, and the thyroid gland are particularly prominent, due to the paucity of subcutaneous fat. Hyperinsulinemia is present from early childhood and leads to organomegaly and acromegaloid features as well as acanthosis nigricans. Diabetes tends to develop in the second decade. Hepatomegaly is often prominent and caused by severe nonalcoholic fatty liver disease (NAFLD), which generally progresses to nonalcoholic steatohepatitis (NASH) and even cirrhosis (believed to be the most common cause of death).

In addition to the biochemical features of severe insulin resistance, these disorders are frequently characterized by severe hypertriglyceridemia, though this may be less striking in young children. In some cases, severe hypertriglyceridemia may be complicated by eruptive xanthomata and pancreatitis. This is a clinically useful way of distinguishing lipodystrophic syndromes from insulin receptoropathies as it essentially never occurs in the latter. Serum leptin and adiponectin concentrations are extremely low due to the lack of adipose tissue,[14] although not, in themselves, diagnostic. The presence of myopathy (typically associated with an elevated creatine kinase) should lead to consideration of the possibility of a recently described subtype of CGL caused by mutations in PTRF.[15] Although relatively few patients have been identified with this disorder, they also appear to exhibit less severe metabolic abnormalities than other forms of CGL.

Familial partial lipodystrophies (FPLD) have been classified into several subtypes of which the most prevalent include FPLD1 (Kobberling type; MIM #608600), FPLD2 (Dunnigan type; MIM #151660), and FPLD3 (MIM #603637). All three of these conditions are most readily detectable in postpubertal women where the loss of gluteal fat is particularly visually striking. They are very difficult to detect clinically in lean men. FPLD1 is characterized by loss of limb fat with preserved and frequently increased truncal fat.[16] While some of these patients do have affected family members, many do not, suggesting that not all cases are inherited, and clinical observation suggests that additional factors such as the menopause and hyperandrogenism may be contributory. FPLD2 is a face-sparing lipodystrophy, which usually becomes apparent during puberty, although careful study of children known to harbor the underlying genetic defect in the *LMNA* gene suggests that fat distribution may also be subtly abnormal during childhood. The lipodystrophy predominantly affects the limbs and gluteal fat depots with variable truncal involvement but with normal or excess fat on the face and neck and in the labia majora.[17] Metabolic abnormalities range from asymptomatic impaired glucose tolerance and mild dyslipidemia to severe insulin resistance with type 2 diabetes mellitus and severe dyslipidemia complicated by eruptive xanthomata and pancreatitis. As in the generalized forms of lipodystrophy, NAFLD/NASH is a common complication. Hypertension and accelerated atherosclerotic vascular disease have been reported in some kindreds.[18]

FPLD3 is another disorder characterized by a paucity of limb and gluteal fat. It differs from FPLD2 in that abdominal fat is generally preserved and facial fat is often normal. Insulin resistance and lipodystrophy may be apparent in young children with this disorder, although the lipodystrophy more commonly only becomes clinically discernible during puberty in girls. Affected individuals are typically severely insulin resistant and manifest all the features of the metabolic syndrome including hypertension. Indeed, the very high prevalence of early onset hypertension discriminates FPLD3 from FPLD2 to some extent. NAFLD/NASH is almost universal and some patients manifest severe hypertriglyceridemia.[19]

Recently, loss-of-function mutations have also been reported genes encoding lipid droplet coat proteins in even rarer patients with inherited partial lipodystrophies. The genes involved include *CIDEC* (only reported in one patient to date)[20] and *PLIN1*.[21] The phenotypes of these patients were most similar to that of FPLD3.

Complex Syndromes

In addition to these conditions where severe insulin resistance and/or lipodystrophy are the dominant clinical manifestations of the underlying genetic defect, there exist a group of more complex syndromes that feature severe insulin resistance, often with some degree of lipodystrophy only as part of a wider spectrum of clinical abnormality. These include (1) mandibuloacral dysplasia, a rare disorder featuring short stature, mandibular and clavicular hypoplasia, dental abnormalities, acro-osteolysis, stiff joints, skin atrophy, alopecia, and mottled pigmentation as well as partial lipodystrophy, and some forms of progeria[22] associated with biallelic *LMNA* or *ZMSPTE24* mutations;[23] (2) a somewhat similar syndrome characterized by subcutaneous lipodystrophy with sclerodermatous skin changes, sensorineural deafness, mandibular hypoplasia, and, in males, hypogonadism, due in all cases to *de novo* mutations in *POLD1*, encoding the dominant lagging strand DNA polymerase;[24] (3) Alström syndrome, a rare recessive disorder characterized by rod cone dystrophy, sensorineural hearing loss, heart failure, renal failure, and severely dyslipidemic insulin resistant diabetes, caused by mutations of a large centrosomal protein encoded by ALMS1;[25] and (4) several syndromes of primordial dwarfism and/or defective DNA damage repair, including Bloom[26] and Werner[27] syndromes, osteodysplastic primordial dwarfism of Majewski type 2,[28] and a recently described syndrome caused by mutation of NSMCE2.[29] Each of these commonly feature severely dyslipidemic insulin resistance. Other extremely rare causes of syndromic severe insulin resistance/lipodystrophy have also been characterized, but these are beyond the scope of this chapter.

GENETIC PATHOPHYSIOLOGY

Known *INSR* Mutations and Specific Phenotypes

The insulin receptor functions as a transmembrane dimer. The constituent monomers consist of disulfide-bonded alpha and beta subunits derived from the same allele, and the monomers in turn are also linked by disulfide bonds (Fig. 22.1). The alpha subunit contains the insulin binding domain, while the intracellular beta subunit contains the tyrosine kinase domain, which autophosphorylates as a consequence of binding of insulin to the alpha subunit, thereby triggering a complex network of intracellular signaling events. The first pathogenic mutations in the insulin receptor were described in 1988,[30,31] and since then over 150 different mutations have been discovered including missense, nonsense, and splice site mutations, insertions, and deletions.

Collectively, published studies suggest that the clinical syndrome resulting from an insulin receptor mutation depends on the overall loss of receptor function, but beyond this there appears to be little specific genotype–phenotype relationship. However, there is no single cell-based assay that gives an entirely reliable global index of insulin receptor function, and consequently it is not possible to project clinical phenotype with complete confidence from commonly undertaken cellular studies.

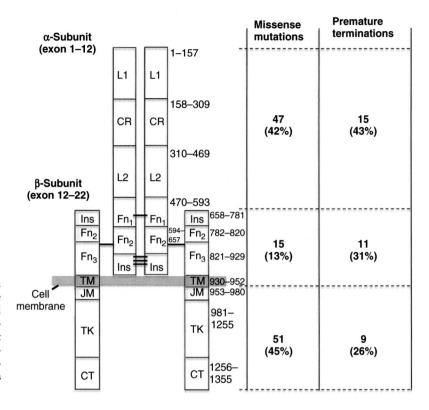

FIGURE 22.1   **Structure of the insulin receptor with distribution of known naturally-occurring missense and premature termination mutations (nonsense and frameshift).** L1,L2, leucine-rich domains; CR, cysteine-rich domain; Fn, fibronectin-like domain; Ins, insert domain; JM, juxtamembrane domain; TM, transmembrane domain; TK, tyrosine kinase domain; CT, C terminal domain. Numbering next to domains indicates amino acid boundaries.

TABLE 22.3   Examples of INSR Mutations with Associated Clinical Phenotypes

Syndrome	Mutations	Site
Donohue	*Homozygous* Entire INSR gene deletion: I29T, G31R, G84Q, R86P, R114W, K121X, Y134X, E124X, K148X, H209R, L233P, C259Y, S323L, del Val335, W412S, N431D, Q521X, R786X, L795P, V855X, R863X, T910M, R1092Q	1 Large deletion; 17 α subunit; 6 β subunit (5 extracellular, 1 intracellular)
	Heterozygous Q328X	1 α-Subunit
	+ 40 in compound heterozygous form	10 Missense/missense; 10 nonsense/nonsense; 16 missense/nonsense; 4 other
Rabson–Mendenhall	*Homozygous* R65W, R118C, I119M, P193L, C257Y, I321F, S323L, R735S, C807R, R1092Q	8 α Subunit (including 1 at cleavage site); 2 β subunit (1 extracellular, 1 intracellular)
	Heterozygous R1174W	1 β Subunit (intracellular)
	+ 22 in compound heterozygous form	14 Missense/missense; 4 missense/nonsense; 4 other
Type A IR IR-AN; HAIR-AN	*Homozygous* R118C, I119M, R252H, R252C, F382V, V1010L	5 α Subunit; 1 β subunit (intracellular)
	Heterozygous del exon 3, C225S, C253Y, R331X, S583W, S583N, Y675X, Y864X, del exon 14, del TK domain, del Leu999, G1008V, A1028V, A1028E, A1029G, A1048D, G1119R, A1121P, R1131W, R1131Q, A1134T, A1135E, A1137T, M1139K, G1149R, D1150H, M1153I, M1153K, R1155G, R1174Q, W1175R, P1178L, E1179D, R1182G, del T1187, W1193L, W1200S, P1209A, delT1214, Q1247X	9 α Subunit; 31 β subunit (intracellular)
	+ 28 in compound heterozygous form	14 Missense/missense; 12 missense/nonsense; 2 other

Furthermore, several reports have illustrated the potential of both genetic and environmental modifiers to influence the severity of the phenotype.[32] Table 22.3 shows examples of mutations that have been reported as associated with Donohue, Rabson–Mendenhall, and type A IR syndromes, while Fig. 22.1 illustrates the approximate distribution of known naturally occurring mutations. It should be noted that the insulin receptor has a 27 amino acid signal peptide, which is cleaved during processing, and published reports variably number residues either according to the sequence of the mature receptor (N) or according to the transcriptional start site (i.e., N + 27). In this chapter, all numbering refers to the mature receptor. Further numbering confusion may arise in the literature because of the existence of two splice variants of the insulin, one containing 12 extra amino acids (first residue at position 745 in the mature receptor) in the alpha subunit due to the inclusion of exon 11, which is omitted in the other isoform.

Pathophysiology of Genetic Insulin Receptoropathies

Normal function of the insulin receptor dimer involves its correct synthesis, assembly, and trafficking to the plasma membrane, its ability to bind insulin cooperatively and to transmit this to activation of its intracellular tyrosine kinase activity, its internalization after insulin binding, and either degradation or recycling to the plasma membrane from endosomes.[33] Disruption of any one of these steps may lead to clinical syndromes of severe IR, although more commonly more than one aspect of receptor function is affected by mutations. These observations led Taylor et al. to classify insulin receptor mutations according to the resulting functional deficit,[33]

as shown in Table 22.4, with examples of each type of mutation.

Both Donohue and Rabson–Mendenhall syndromes invariably show autosomal recessive inheritance, commonly featuring functionally null alleles in the former (e.g., alpha subunit nonsense mutations; Table 22.3). Although not systematically reported, limited published evidence and cumulative clinical experience suggest that heterozygotes for such functionally null alleles may have subtle evidence of IR and be at increased risk for diabetes;[34] however, this is heavily dependent on other factors such as obesity, and they are more commonly clinically silent. In contrast, heterozygosity for missense mutations in the tyrosine kinase domain of the receptor is the commonest cause of type A IR, which is usually autosomal dominant in inheritance. The penetrance of this type of mutation may be variable, depending on gender and BMI in particular. This *in vivo* difference between those carrying heterozygous tyrosine kinase domain missense mutations and heterozygous null alleles is thought to be evidence for a dominant negative function of tyrosine kinase mutations – that is, the mutant allele has the capacity to cross-inhibit the coexpressed wild-type allele. This phenomenon has been confirmed in heterologous transfection systems,[35,36] and most likely reflects loss of signaling activity of heterodimeric receptors carrying one mutant and one wild-type receptor monomer (Fig. 22.2). Assuming equal expression, this would be anticipated to lead to 75% loss of receptor function.

The metabolic components of the insulin receptoropathy phenotypes described previously are generally easy to account for in terms of loss of insulin action, though the protection from ketoacidosis in the first year or two of life remains perplexing. The growth perturbation is more difficult to account for precisely,

TABLE 22.4 Functional Classification of INSR Mutations with Examples

Class	Mechanism	Examples	Comment
I	Decreased levels of INSR mRNA	R372X, R897X, W133X, R1000X	Many examples of low mRNA levels due to nonsense-mediated decay or splicing defects. *Cis*-acting promoter mutations have been inferred from low mRNA levels with no abnormal INSR sequence, but no mutations proven
II	Impaired posttranslational processing/transport	G31R, L62P, L93Q, H209R, F382V, W412S, G359S, N431D, R735S, A1135E	Commonly combined with other defects in receptor function
III	Defective insulin binding	N15L, R86P, L233P, R252C, S323L, L460E, R735S	
IV	Impaired receptor tyrosine kinase activity	R993Q, G996V, G1008V, A1048D, L1068E, R1131Q, A1134T, A1135E, M1153I, R1164Q, R1174Q, P1178L, W1200S	Commonest class of mutation in type A IR
V	Mutations that accelerate receptor degradation	I119M, K460E, N462S, E1179D, W1193L	

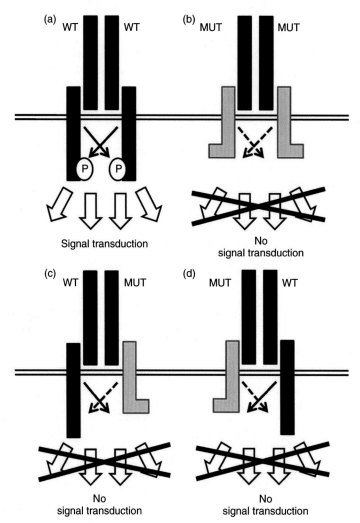

FIGURE 22.2 **Mechanism of dominant negative action of INSR beta subunit mutations.** Activation of the insulin receptor dimer on insulin binding requires autophosphorylation of the constituent monomers, which requires cooperation between two beta subunits of the dimer (a). An inactivating or truncating mutation in the beta subunit abolishes or severely impairs this activation both in homodimers (b) and heterodimers (c, d). If expression, trafficking, dimerization, and internalization of the mutant species were normal, this would reduce total receptor activity by 75%. In practice, mutant receptors are often inefficiently or abnormally processed, although this is often offset by prolonged residency at the cell surface due to impaired internalization.

however. It is most likely a result of extensive endocrine and paracrine disturbances of IGF1 and growth hormone (GH) signaling, accounted for by loss of insulin receptor mediated regulation of GH and IGF1 receptors in at least some tissues, derangement of levels of IGF binding proteins, and the ability of extremely elevated insulin concentrations to exert significant actions via the type 1 IGF1 receptor.

Known AGPAT2 and BSCL2 Mutations and Specific Phenotypes

BSCL is genetically heterogeneous, consisting of at least four different autosomal recessive conditions. BSCL1 is caused by homozygous or compound heterozygous mutations in 1-acylglycerol-3-phosphate-*O*-acyltransferase 2 (AGPAT2),[37,38] while BSCL2 is due to biallelic mutations in a gene encoding an endoplasmic reticulum protein named seipin.[39] Variants in both genes have been found in multiple ethnic groups. Nonsense and splice site variants, expected to completely abolish protein function, account for most of the identified pathogenic AGPAT2 and BSCL2 mutations (Fig. 22.3a and b).

It is currently not possible to distinguish BSCL1 and 2 confidently on clinical grounds; however, adipose tissue loss in mechanical fat pads such as the palms, soles, orbits, scalp, and periarticular regions has been reported as a specific feature of BSCL2.[40] Adiponectin

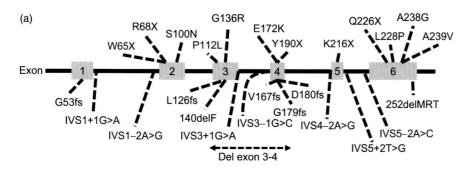

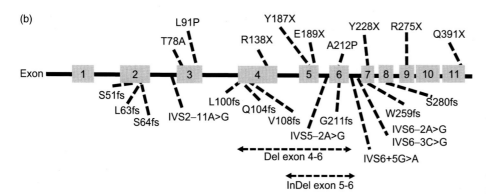

FIGURE 22.3 Distribution of naturally-occurring mutations in (a) AGPAT2 (transcript ENST00000371696) and (b) BSCL2 (transcript ENST00000403550) genes in CGL. Note that many new disease-causing mutations continue to be discovered.

levels are also reported to be lower in CGL patients with AGPAT2 mutations than in those with BSCL2 mutations.[41]

Pathophysiology of Congenital Generalized Lipodystrophy

AGPAT2 is an essential enzyme in glycerophospholipid and triacylglycerol synthesis. Following acylation of glycerol-3-phosphate at the first carbon (sn-1 position), AGPAT2 catalyzes acylation at the second carbon (sn-2) to form phosphatidic acid, an essential precursor to diacylglycerol and ultimately triacylglycerol. On the basis of this critical role in triacylglycerol biosynthesis, it is not surprising that genetic deficiency in AGPAT2 activity leads to a failure of development of lipidated adipose tissue. Early *in vitro* studies suggest that genetic variants associated with lipodystrophy significantly reduce AGAPT2 enzyme activity.[42] In contrast, the structure of seipin gives few clues to the pathogenesis of the severe global lipodystrophy associated with defects in its function. Seipin is highly expressed in white adipose tissue where it appears to be essential for adipogenesis.[43] Several studies have implicated seipin in lipid droplet biology, although precise mechanistic details remain unclear.[44–46] Seipin is also expressed in the brain, which may account for the reportedly higher prevalence of intellectual impairment in BSCL2 than in BSCL1. Heterozygous BSCL2 mutations, which appear to affect protein glycosylation and subsequent folding, have also been identified in two autosomal dominant motor neuron diseases (distal hereditary motor neuropathy type V (OMIM%182960) and Silver syndrome (OMIM#270685)). The aberrant folding is thought to precipitate endoplasmic reticulum (ER) stress-mediated cell death.[47]

Although mutations in AGPAT2 and BSCL2 are reported to account for the vast majority of cases of BSCL,[48] a single patient has been identified with BSCL due to homozygous loss-of-function mutations in the gene encoding caveolin 1 (CAV1), which is a critical component of plasma membrane caveoli, as a cause of BSCL,[49] and several papers have reported mutations in PTRF in patients with CGL and congenital myopathy. PTRF (polymerase I and transcript release factor) encodes cavin-1, another protein involved in the formation of caveolae, highlighting the importance of these membrane structures in adipocytes.[50]

Known LMNA Mutations and Specific Phenotypes

LMNA encodes lamin A/C, a structural component of the nuclear lamina, which is nearly ubiquitously

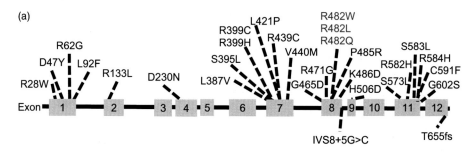

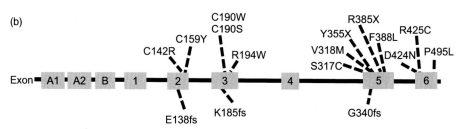

FIGURE 22.4 Distribution of naturally-occurring mutations in (a) LMNA (transcript ENST00000368300) and (b) PPARG (transcript ENST00000397012) genes in familial partial lipodystrophy or severe insulin resistance. Note that many new disease-causing mutations continue to be discovered. The R482 mutation hotspot in lipodystrophy is highlighted in red.

expressed. Remarkably mutations in this gene have been convincingly linked to several different disorders including muscular dystrophy, dilated cardiomyopathy, Charcot–Marie–Tooth neuropathy, premature aging syndromes (Hutchison–Gilford progeria syndrome (HGPS) and Werner's syndrome), restrictive dermopathy, and various overlapping syndromes. The vast majority of LMNA mutations associated with the "classical" FPLD2 phenotype are heterozygous missense variants clustered in exon 8, which encodes part of the globular carboxy terminal portion of lamin A and C (Fig. 22.4a). Although a large majority of FPLD2 mutations affect codon 482, these appear to be recurrent mutations as there is no evidence for common ancestry in reported families.[51] FPLD-associated LMNA mutations are shown in Fig. 22.4a.

There is significant heterogeneity in clinical phenotype even within FPLD2; for example, the S583L mutation is associated with a milder phenotype, and this has been attributed to its location in exon 10, which is specific to lamin A, whereas most of the other mutations affect both lamin A and C (lamin C lacks a C-terminal region found in lamin A).[52] A homozygous mutation (R527H) in lamin A/C is responsible for some cases of mandibuloacral dysplasia.[53] Heterozygous members of these kindreds were reportedly entirely normal. In addition to this genetic heterogeneity, there is considerable clinical phenotypic variability within and among kindreds with the same mutation, including reports of cardiac abnormalities and myopathic features in LMNA R482W kindreds.[54] In a single patient with severe insulin resistance the LMNA R482W mutation has been reported without clinically apparent lipodystrophy.[55]

Pathophysiology of LMNA-Associated Lipodystrophy

Detailed understanding of the mechanisms underlying the tissue-selective phenotypes associated with LMNA mutations is lacking, but proposed abnormalities include the following: (1) structural defects in the nuclear envelope; (2) altered binding of the nuclear lamina to chromatin and subsequent effects on the regulation of gene transcription; (3) altered binding of the nuclear lamina to transcription factors, for example, SREBP1. FPLD2-associated LMNA mutations are clustered in the globular carboxy terminal domain of lamin A/C, where they are not expected to alter three-dimensional protein structure, but are more likely to affect interactions with other proteins.

Known PPARG Mutations and Specific Phenotypes

PPARγ is a nuclear hormone receptor most highly expressed in adipose tissue where it is essential for adipocyte differentiation.[56] To date, all of the mutations (Fig. 22.4b) described were heterozygous, many displaying dominant negative activity *in vitro*. Mutations associated with partial lipodystrophy have been identified in the DNA and ligand-binding domains.[19,57] A single heterozygous variant in the PPARγ4 promotor has also been reported in a kindred with central obesity, limb lipodystrophy, and insulin resistance.[58]

Pathophysiology of PPARG-Associated Lipodystrophy

Of all the genes in which variants are associated with lipodystrophy, PPARG is arguably the most predictable candidate. It is a nuclear hormone receptor, most highly expressed in adipose tissue, where it is essential for adipocyte development and function. Overexpressing human dominant negative PPARγ mutants in preadipocyte cell lines inhibits differentiation, whereas adding chemical PPARγ agonists to cell media significantly enhances preadipocyte differentiation *in vitro* and increases fat mass *in vivo*. Thiazolidinediones, which are currently used as insulin sensitizers in the management of type 2 diabetes, are PPARγ agonists. Whether their considerable insulin-sensitizing properties and the severe insulin resistance noted in people with PPARγ mutations are solely attributable to the effects of PPARγ on adipocyte biology, or whether PPARγ expression in other insulin-sensitive tissues such as muscle and liver is also biologically relevant, remains unclear in humans.

Some of the genes more recently implicated in FPLD are known to be transcriptional targets of PPARγ, including CIDEC and perilipin 1. The fact that both the latter encode for proteins almost exclusively expressed in adipocytes provides compelling evidence for the notion that all the metabolic manifestations (i.e., other than obesity itself) of the metabolic syndrome can result from primary defects in adipose tissue metabolism.

Additional very rare forms of FPLD have been described but are beyond the scope of this chapter. Some of these have recently been reviewed.[23,59]

DIAGNOSIS, GENETIC TESTING, AND INTERPRETATION

Which Tests are Best to Order in Whom?

INSR gene sequencing should be requested in children and infants with clinical features of Donohue or Rabson–Mendenhall syndromes and biochemical proof of severe hyperinsulinemia (in these settings, this is usually at least one order of magnitude above the top of the resting range even in the fasting state, though it may be lower if late in the disease course and significant beta cell decompensation has occurred). The vast majority of patients with monogenic severe IR present later, however, usually in the first five years following puberty. Only a minority of these patients will have monogenic disease, so targeting of genetic testing only to the subsets most likely to harbor defects in the relevant gene is critical. For patients with partial lipodystrophy, careful assessment of the distribution of fat loss is critical, and to some extent guides genetic testing; however, patients with insulin receptoropathies present no such clinical clues. It has recently been shown,

however, that accurate triage may be achieved instead by biochemical testing: unlike other patients with severe IR, patients with insulin receptor defects have preserved or even elevated serum levels of adiponectin, a highly abundant and stable fat cell-derived hormone.[5] Similarly, many patients with receptoropathy have preserved or elevated SHBG and IGFBP1, in contrast to severely insulin resistant controls.[4] This observation has been developed using ROC analysis to generate suggested diagnostic cut-offs for these markers in severe IR.[4] For example, using one particular adiponectin immunoassay, an adiponectin level below 5 mg/L has a 97% negative predictive value for an insulin receptor defect, while a level above 7 mg/L has a 97% positive predictive value. It should be noted, however, that these precise diagnostic cut-offs are dependent on the assay used.[4]

AGPAT2 and BSCL2 sequencing should be requested in patients with CGL. As well as often having severe IR, such patients are usually severely dyslipidemic, and often have accelerated early childhood growth and a somewhat pseudoacromegalic appearance. The presence of short stature or myopathy in a patient with CGL should prompt consideration of CAV1 or PTRF sequencing.

The clinical distinction between FPLD2 and FPLD3 can be somewhat difficult to discern, particularly in lean patients and men. Pragmatically, bearing in mind this clinical overlap as well as the relatively much higher prevalence of LMNA than PPARG mutations, we suggest that any patient with face-sparing lipodystrophy should have exons 8 and 11 of LMNA sequenced in the first instance, proceeding to sequencing of PPARG and the remainder of LMNA where no pathogenic mutation is detected. However, where the adipose distribution is clearly more in keeping with FPLD3 than FPLD2, or where early-onset hypertension is noted, proceeding straight to PPARG sequencing is justified. It should be noted that both FPLD2 and FPLD3 are characterized by preservation of adipose tissue in the head and neck, and are thus quite distinct from acquired partial lipodystrophy, in which adipose tissue loss usually follows a craniocaudal sequence, usually arresting at or before the level of the umbilicus.[60] These observations are summarized in the proposed algorithm for the assessment of patients with severe IR shown in Fig. 22.5. Progeroid or other syndromic features in a patient with PLD should prompt sequencing of the relevant genes.

Labs Available for Testing

R, research service; C, clinically accredited genetic diagnostic service; CV, clinical validation of results from a research lab; PND, prenatal diagnosis.

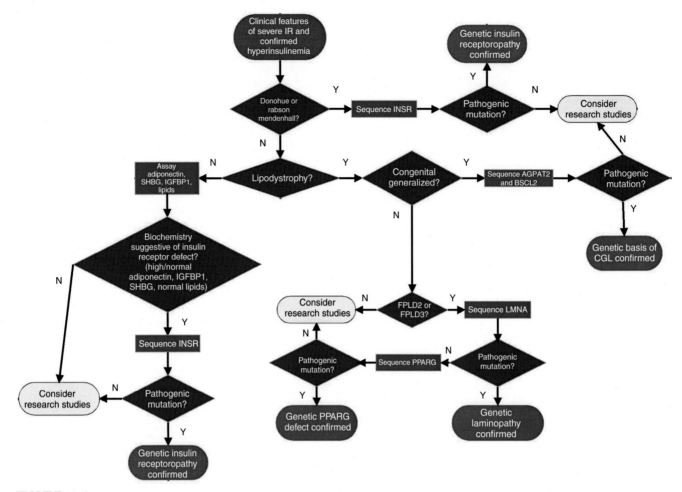

FIGURE 22.5   Suggested algorithm for the genetic investigation of syndromes of severe insulin resistance and/or lipodystrophy.

INSR

Dr. Robert Semple
Metabolic Research Laboratories
Institute of Metabolic Science
University of Cambridge
Addenbrooke's Hospital
Cambridge CB2 0QQ, UK
Tel: +44-1223-769-035
Email: rks16@cam.ac.uk
(R;C;CV;PND, research service for novel candidates)

Dr. Olivier Lascols
Unité de biologie et génétique moléculaires – Pôle de
Biologie Imagerie
CHU Hôpital Saint-Antoine
184 rue du Faubourg Saint-Antoine
75571 Paris Cedex 12, France
Tel: +33-(0)1-49-28-28-09
Email: olivier.lascols@sat.aphp.fr
(C;CV;PND)

Prof. Nicola Longo
University of Utah
2C412 SOM
50 N Medical Drive
Salt Lake City, UT 84132, USA
Tel: 801-585-2457
Email: Nicola.Longo@hsc.utah.edu
(C;CV)

Prof. Fabrizio Barbetti
Department of Laboratory Medicine
Tor Vergata University Hospital
Section D, Room 118
Viale Oxford, 81
00133 Rome, Italy
Tel: +39 06-2090-0672
Email: Fabrizio.Barbetti@uniroma2.it
(C;CV;PND)

Prof. Robert A. Hegele
Blackburn Cardiovascular Genetics Laboratory
Robarts Research Institute
406-100 Perth Drive

London, Ontario N6A 5K8, Canada
Tel: 519-663-3461
Email: hegele@robarts.ca
(R)

Lipodystrophy

Dr. David Savage
Metabolic Research Laboratories
Institute of Metabolic Science
University of Cambridge
Addenbrooke's Hospital
Cambridge CB2 0QQ, UK
Tel: +44-1223-769-023
Email: dbs23@medschl.cam.ac.uk
(C;CV for LMNA,PPARG,AGPAT2,BSCL2;R for novel candidates)

Prof. V.K.K. Chatterjee
Metabolic Research Laboratories
Institute of Metabolic Science
University of Cambridge
Addenbrooke's Hospital
Cambridge CB2 0QQ, UK
Tel: +44-1223-330191
Email: kkc1@mole.bio.cam.ac.uk
(R for PPARG)

Dr. Olivier Lascols
Unité de biologie et génétique moléculaires – Pôle de Biologie Imagerie
CHU Hôpital Saint-Antoine
184 rue du Faubourg Saint-Antoine
75571 Paris Cedex 12, France
Tel: +33-(0)1-49-28-28-09
Email: olivier.lascols@sat.aphp.fr
(C;CV;PND for LMNA,PPARG, AGPAT2, BSCL2)

Prof. Sian Ellard
Department of Molecular Genetics
Royal Devon & Exeter NHS Foundation Trust
Barrack Road, Exeter EX2 5DW, UK
Tel: +44-1392-402-910
Email: Sian.Ellard@rdeft.nhs.uk
(C;CV for LMNA)

Prof. Abhimanyu Garg
Contact: Sarah Gilmore
University of Texas Southwestern Medical Center Dallas
Lipodystrophy Laboratory
5323 Harry Hines Blvd, K5-214
Dallas, TX 75390-8537, USA
Email: sarah.gilmore@utsouthwestern.edu
Tel: (214) 648-0549
(R for AGPAT2, BSCL2, CAV1, LMNA, PPARG)

Prof. Robert A. Hegele, MD
Blackburn Cardiovascular Genetics Laboratory
Robarts Research Institute
406-100 Perth Drive
London, Ontario N6A 5K8, Canada
Tel: 519-663-3461
Email: hegele@robarts.ca
(R for AGPAT2, BSCL2, CAV1, LMNA, PPARG)

Additional clinically accredited laboratories may be identified via the GeneTests (http://www.genetests.org/) and Orphanet (http://www.orpha.net/) websites.

Lists of Consultants and Resources

Prof. Stephen O'Rahilly/Dr. Robert Semple/Dr. David Savage/Dr Anna Stears
Metabolic Research Laboratories
Institute of Metabolic Science
University of Cambridge
Addenbrooke's Hospital
Cambridge CB2 0QQ, UK
Tel: +44 1223 336855/769035/767923
Email: so104@medschl.cam.ac.uk
rks16@cam.ac.uk
dbs23@medschl.cam.ac.uk
insulinresistanceservice@addenbrookes.nhs.uk

(Specialized clinical and genetic consultation on insulin resistance and/or lipodystrophies)

Prof. Jacqueline Capeau
INSERM UMR_S893
Faculté de Médecine Pierre et Marie Curie
27 rue Chaligny
75571 Paris Cedex 12, France
Tel: +33-(0)1-40 01 13 32
Email: jacqueline.capeau@inserm.fr
(Genetic counseling on insulin resistance and/or lipodystrophy)

Dr. Corinne Vigouroux
Endocrinology Department
Saint-Antoine Hospital, Paris
33 1 49 28 24 09
corinne.vigouroux@inserm.fr
(Specialized consultation on insulin resistance and/or
lipodystrophies)

Prof. Abhimanyu Garg, MD
UT Southwestern Medical Center at Dallas
5323 Harry Hines Blvd, K5-214
Dallas, TX 75390-8537, USA
Tel: 214-648-2895
Email: Abhimanyu.Garg@UTSouthwestern.edu
(Specialized clinical and genetic consultation on
lipodystrophies)

Dr. Phillip Gorden, MD
Molecular and Cellular Physiology Section,
Investigator
Clinical Endocrinology Branch, NIDDK, NIH
Building 10, CRC 6-5940
Bethesda, MD 20892-1770, USA
Tel: 301-496-4658
Email: gordenp@extra.niddk.nih.gov
(Specialized consultation on insulin resistance and/or
lipodystrophies)

Prof. Robert A. Hegele, MD
Blackburn Cardiovascular Genetics Laboratory
Robarts Research Institute
406-100 Perth Drive
London, Ontario N6A 5K8, Canada
Tel: 519-663-3461
Email: hegele@robarts.ca
(Specialized genetic consultation on lipodystrophies
and/or insulin resistance)

Prof. Nicola Longo, MD, PhD
University of Utah
2C412 SOM
50 N Medical Drive
Salt Lake City, UT 84132, USA
Tel: 801-585-2457
Email: Nicola.Longo@hsc.utah.edu
(Specialized genetic consultation on insulin resistance)

Dr. Elif Arioglu Oral
University of Michigan
Department of Internal Medicine
Division of Endocrinology and Metabolism
3920 Taubman Center, Box 0354
1500 East Medical Center Drive
Ann Arbor, MI 48109-0354, USA
Tel: 734-615 7271
Email: eliforal@umich.edu
(Specialized consultation on lipodystrophies)

Predictive Value of Test

INSR

Even when new mutations are discovered, the availability of a large research literature on the structural determinants of insulin receptor function allied to crystal structures for both extracellular and intracellular domains means that assessment of likely pathogenicity of mutations is often relatively easy. Extrapolation to clinical consequences may be more complicated, due to the important environmental modifiers that influence phenotypes between individuals and kindreds with the same mutation, but some general rules apply:

1. Inheritance of two functionally null or nearly null alleles will lead to Donohue or Rabson–Mendenhall syndromes.
2. Inheritance of a single null or functionally null allele is unlikely to be clinically expressed beyond a mild predisposition to diabetes and insulin resistance in later life, though this has not been extensively reported.
3. Heterozygosity for an allele with dominant negative properties (usually a missense or truncating mutation in the beta subunit of the

receptor) is usually expressed as severe insulin resistance, although the clinical problems this poses will depend in turn upon environmental factors such as exercise and weight gain.

Lipodystrophy

A large majority of mutations causing BSCL are functionally null alleles, while no clear heterozygous phenotype has been reported for any of these BSCL-causing mutations. Penetrance of lipodystrophy in homozygotes or compound heterozygotes is effectively 100%, although the severity of metabolic complications varies markedly between individuals, as discussed.

Penetrance of heterozygous mutations in PPARG or LMNA is much more variable, and depends significantly on gender, lifestyle, and genetic modifiers. Furthermore, new mutations continue to be discovered and evaluated, and reporting of the natural history of the associated disease is incomplete. Nevertheless it is likely that most patients inheriting a loss-of-function mutation in LMNA or PPARG will clinically express FPLD to some extent, and at the very least inheritance of such an allele should be regarded as a potent genetic risk factor for metabolic disease when counseling patients.

Significance of Negative Test

The vast majority of cases of Donohue, Rabson–Mendenhall, and BSCL are accounted for by mutations in the *INSR*, *AGPAT2*, and *BSCL2* genes, while most cases of FPLD2 and 3 are accounted for by mutations in the *LMNA* or *PPARG* genes. In contrast, only a minority of cases of type A insulin resistance are caused by mutations in the *INSR* gene. Nevertheless, a large proportion of remaining cases of each of these syndromes in which no mutations of the relevant genes are discovered are likely to harbor defects in other genes. In these cases, research studies looking for alterations in novel candidate genes should be considered. Laboratories with an interest in such studies have been indicated earlier.

Should Family Members be Tested?

Donohue, Rabson–Mendenhall, and BSCL are usually caused by homozygous or compound heterozygous loss-of-function mutations. In most cases, there is little or no phenotype in heterozygotes, although for some insulin receptor mutations, there may be increased risk of diabetes and insulin resistance in later life. Because there are numerous different mutations in the *INSR*, *AGPAT2*, and *BSCL2* genes, and because the carrier frequency in the general population is very low, it is generally not necessary to screen family members in these settings. This may have to be reviewed in kindreds or populations in which there are high levels of consanguinity, and in which prenatal testing may be desirable.

Type A insulin resistance, FPLD2 and FPLD3, however, show autosomal dominant inheritance. Although penetrance is variable and greatly influenced by factors such as gender, age, food intake, and activity levels, it is likely that most carriers of mutations will express some clinical consequences of the mutation. A common scenario for each of these conditions is the identification of a pathogenic mutation in a peripubertal female proband, followed by diagnosis of older, often male, relatives with clinically silent severe insulin resistance or diabetes. As for all forms of type 2 diabetes, there is often a long period between the development of frank hyperglycemia and diagnosis of diabetes, during which time complications can accrue, and for this reason we believe that genetic testing of family members of patients with these conditions is desirable.

TREATMENT

Based on the Genetic Information, How Does That Affect Treatment, If At All?

The principles of managing severe insulin resistance include early and intensive use of insulin sensitizing agents and lifestyle modification to include as much aerobic exercise as reasonably possible. In lipodystrophies, close adherence to a low fat and, importantly, a low-calorie diet is also crucial in preventing or delaying dyslipidemia and diabetes. More recently, adjunctive use of subcutaneous leptin in patients who have secondary leptin deficiency due to lack of adipose tissue has proved highly effective (particularly in CGL patients where it should arguably be considered as first line medical therapy), and recombinant IGF1 or composite preparations have some utility in severe insulin resistance. Nevertheless, these therapies should be introduced based on clinical and biochemical criteria, and establishment of the genetic defect should not influence therapeutic decision making significantly. The FPLDs are minor exceptions to this: first, it is logical to suppose that use of potent thiazolidinedione PPARγ agonists in patients with PPARG mutations may be particularly efficacious; however, despite some limited evidence for this in the case of particular mutations and novel agonists,[61] this requires further study. Second, in the case of LMNA-associated FPLD2, it also now appears that there may be some overlap with other laminopathies, and that clinically silent myopathy and cardiomyopathy may be detected on careful screening, which should thus be considered in patients with LMNA mutations.[54]

Genetic Counseling and Prenatal Testing

As discussed, the clinical expression of almost all single gene defects in insulin action or adipose tissue function is strongly dependent on environmental and behavioral factors. Thus, while some patients with BSCL may suffer devastating clinical sequelae of unrestrained hyperglycemia and dyslipidemia, including microvascular complications of diabetes and recurrent pancreatitis, others – especially males – may have a fairly benign clinical course when they are adherent to a low fat diet and habitually undertake large amounts of aerobic exercise. Prospects are further enhanced with the advent of leptin therapy in those with severe secondary leptin deficiency.[62] Similar observations can be made for those with type A insulin resistance due to insulin receptor mutations. Severe insulin receptoropathies and those laminopathies where lipodystrophy occurs as part of a more severe multisystem syndrome, such as mandibuloacral dysplasia, present a different prospect, however, with death in infancy or childhood most common. However, as previously noted, even Donohue and Rabson–Mendenhall syndromes represent arbitrary points on a clinical spectrum of disease, with evidence of genetic and environmental modifiers of clinical severity even in these rare and extreme conditions. This clinical spectrum of disease, encompassing rare and severe recessive conditions, and less severe autosomal dominant conditions, and the close interplay of environmental factors with the single gene defects, must be borne in mind when counseling patients or parents.[32]

References

1. Warram JH, Martin BC, Krolewski AS, Soeldner JS, Kahn CR. Slow glucose removal rate and hyperinsulinemia precede the development of type II diabetes in the offspring of diabetic parents. *Ann Intern Med* 1990;**113**:909–15.

2. Kahn CR, Flier JS, Bar RS, Archer JA, Gorden P, Martin MM, Roth J. The syndromes of insulin resistance and acanthosis nigricans. Insulin-receptor disorders in man. *N Engl J Med* 1976;**294**:739–45.

3. Barbieri RL, Ryan KJ. Hyperandrogenism, insulin resistance, and acanthosis nigricans syndrome: a common endocrinopathy with distinct pathophysiologic features. *Am J Obstet Gynecol* 1983;**147**: 90–101.

4. Semple RK, Cochran EK, Soos MA, Burling KA, Savage DB, Gorden P, O'Rahilly S. Plasma adiponectin as a marker of insulin receptor dysfunction: clinical utility in severe insulin resistance. *Diabetes Care* 2008;**31**:977–9.

5. Semple RK, Soos MA, Luan J, Mitchell CS, Wilson JC, Gurnell M, Cochran EK, Gorden P, Chatterjee VK, Wareham NJ, O'Rahilly S. Elevated plasma adiponectin in humans with genetically defective insulin receptors. *J Clin Endocrinol Metab* 2006;**91**:3219–23.

6. Semple RK, Sleigh A, Murgatroyd PR, Adams CA, Bluck L, Jackson S, Vottero A, Kanabar D, Charlton-Menys V, Durrington P, Soos MA, Carpenter TA, Lomas DJ, Cochran EK, Gorden P, O'Rahilly S, Savage DB. Postreceptor insulin resistance contributes to human dyslipidemia and hepatic steatosis. *J Clin Invest* 2009;**119**:315–22.

7. George S, Rochford JJ, Wolfrum C, Gray SL, Schinner S, Wilson JC, Soos MA, Murgatroyd PR, Williams RM, Acerini CL, Dunger DB, Barford D, Umpleby AM, Wareham NJ, Davies HA, Schafer AJ, Stoffel M, O'Rahilly S, Barroso I. A family with severe insulin resistance and diabetes due to a mutation in AKT2. *Science* 2004;**304**: 1325–8.

8. Thauvin-Robinet C, Auclair M, Duplomb L, Caron-Debarle M, Avila M, St-Onge J, Le Merrer M, Le Luyer B, Heron D, Mathieu-Dramard M, Bitoun P, Petit JM, Odent S, Amiel J, Picot D, Carmignac V, Thevenon J, Callier P, Laville M, Reznik Y, Fagour C, Nunes ML, Capeau J, Lascols O, Huet F, Faivre L, Vigouroux C, Riviere JB. PIK3R1 mutations cause syndromic insulin resistance with lipoatrophy. *Am J Hum Genet* 2013;**93**:141–9.

9. Dyment DA, Smith AC, Alcantara D, Schwartzentruber JA, Basel-Vanagaite L, Curry CJ, Temple IK, Reardon W, Mansour S, Haq MR, Gilbert R, Lehmann OJ, Vanstone MR, Beaulieu CL, Majewski J, Bulman DE, O'Driscoll M, Boycott KM, Innes AM. Mutations in PIK3R1 cause SHORT syndrome. *Am J Hum Genet* 2013;**93**: 158–66.

10. Chudasama KK, Winnay J, Johansson S, Claudi T, Konig R, Haldorsen I, Johansson B, Woo JR, Aarskog D, Sagen JV, Kahn CR, Molven A, Njolstad PR. SHORT syndrome with partial lipodystrophy due to impaired phosphatidylinositol 3 kinase signaling. *Am J Hum Genet* 2013;**93**:150–7.

11. Garg A. Acquired and inherited lipodystrophies. *N Engl J Med* 2004;**350**:1220–34.

12. Seip M, Trygstad O. Generalized lipodystrophy, congenital and acquired (lipoatrophy). *Acta Paediatr Suppl* 1996;**413**:2–28.

13. Garg A, Fleckenstein JL, Peshock RM, Grundy SM. Peculiar distribution of adipose tissue in patients with congenital generalized lipodystrophy. *J Clin Endocrinol Metab* 1992;**75**:358–61.

14. Pardini VC, Victoria IM, Rocha SM, Andrade DG, Rocha AM, Pieroni FB, Milagres G, Purisch S, Velho G. Leptin levels, beta-cell function, and insulin sensitivity in families with congenital and

15. acquired generalized lipoatropic diabetes. *J Clin Endocrinol Metab* 1998;**83**:503–8.

15. Hayashi YK, Matsuda C, Ogawa M, Goto K, Tominaga K, Mitsuhashi S, Park YE, Nonaka I, Hino-Fukuyo N, Haginoya K, Sugano H, Nishino I. Human PTRF mutations cause secondary deficiency of caveolins resulting in muscular dystrophy with generalized lipodystrophy. *J Clin Invest* 2009;**119**:2623–33.

16. Herbst KL, Tannock LR, Deeb SS, Purnell JQ, Brunzell JD, Chait A. Kobberling type of familial partial lipodystrophy: an underrecognized syndrome. *Diabetes Care* 2003;**26**:1819–24.

17. Jackson SN, Howlett TA, McNally PG, O'Rahilly S, Trembath RC. Dunnigan–Kobberling syndrome: an autosomal dominant form of partial lipodystrophy. *QJM* 1997;**90**:27–36.

18. Hegele RA. Premature atherosclerosis associated with monogenic insulin resistance. *Circulation* 2001;**103**:2225–9.

19. Semple RK, Chatterjee VK, O'Rahilly S. PPAR gamma and human metabolic disease. *J Clin Invest* 2006;**116**:581–9.

20. Rubio-Cabezas O, Puri V, Murano I, Saudek V, Semple RK, Dash S, Hyden CS, Bottomley W, Vigouroux C, Magre J, Raymond-Barker P, Murgatroyd PR, Chawla A, Skepper JN, Chatterjee VK, Suliman S, Patch AM, Agarwal AK, Garg A, Barroso I, Cinti S, Czech MP, Argente J, O'Rahilly S, Savage DB. Partial lipodystrophy and insulin resistant diabetes in a patient with a homozygous nonsense mutation in CIDEC. *EMBO Mol Med* 2009;**1**:280–7.

21. Gandotra S, Le Dour C, Bottomley W, Cervera P, Giral P, Reznik Y, Charpentier G, Auclair M, Delepine M, Barroso I, Semple RK, Lathrop M, Lascols O, Capeau J, O'Rahilly S, Magre J, Savage DB, Vigouroux C. Perilipin deficiency and autosomal dominant partial lipodystrophy. *N Engl J Med* 2011;**364**:740–8.

22. Simha V, Garg A. Body fat distribution and metabolic derangements in patients with familial partial lipodystrophy associated with mandibuloacral dysplasia. *J Clin Endocrinol Metab* 2002;**87**: 776–85.

23. Garg A. Clinical review#: lipodystrophies: genetic and acquired body fat disorders. *J Clin Endocrinol Metab* 2011;**96**:3313–25.

24. Weedon MN, Ellard S, Prindle MJ, Caswell R, Lango Allen H, Oram R, Godbole K, Yajnik CS, Sbraccia P, Novelli G, Turnpenny P, McCann E, Goh KJ, Wang Y, Fulford J, McCulloch LJ, Savage DB, O'Rahilly S, Kos K, Loeb LA, Semple RK, Hattersley AT. An in-frame deletion at the polymerase active site of POLD1 causes a multisystem disorder with lipodystrophy. *Nat Genet* 2013;**45**:947–50.

25. Minton JA, Owen KR, Ricketts CJ, Crabtree N, Shaikh G, Ehtisham S, Porter JR, Carey C, Hodge D, Paisey R, Walker M, Barrett TG. Syndromic obesity and diabetes: changes in body composition with age and mutation analysis of ALMS1 in 12 United Kingdom kindreds with Alstrom syndrome. *J Clin Endocrinol Metab* 2006;**91**: 3110–6.

26. Diaz A, Vogiatzi MG, Sanz MM, German J. Evaluation of short stature, carbohydrate metabolism and other endocrinopathies in Bloom's syndrome. *Horm Res* 2006;**66**:111–7.

27. Epstein CJ, Martin GM, Schultz AL, Motulsky AG. Werner's syndrome a review of its symptomatology, natural history, pathologic features, genetics and relationship to the natural aging process. *Medicine (Baltimore)* 1966;**45**:177–221.

28. Huang-Doran I, Bicknell LS, Finucane FM, Rocha N, Porter KM, Tung YC, Szekeres F, Krook A, Nolan JJ, O'Driscoll M, Bober M, O'Rahilly S, Jackson AP, Semple RK. Genetic defects in human pericentrin are associated with severe insulin resistance and diabetes. *Diabetes* 2011;**60**:925–35.

29. Payne F, Colnaghi R, Rocha N, Seth A, Harris J, Carpenter G, Bottomley WE, Wheeler E, Wong S, Saudek V, Savage DB, O'Rahilly S, Carel J-C, Barroso I, O'Driscoll M, Semple RK. Hypomorphism in human NSMCE2 linked to primordial dwarfism and insulin resistance. *J Clin Invest* 2014;**124**(9):4028–38.

30. Kadowaki T, Bevins CL, Cama A, Ojamaa K, Marcus-Samuels B, Kadowaki H, Beitz L, McKeon C, Taylor SI. Two mutant alleles of

the insulin receptor gene in a patient with extreme insulin resistance. *Science* 1988;**240**:787–90.

31. Yoshimasa Y, Seino S, Whittaker J, Kakehi T, Kosaki A, Kuzuya H, Imura H, Bell GI, Steiner DF. Insulin-resistant diabetes due to a point mutation that prevents insulin proreceptor processing. *Science* 1988;**240**:784–7.

32. Taylor SI. Prenatal screening for mutations in the insulin receptor gene. How reliably does genotype predict phenotype? *J Clin Endocrinol Metab* 1995;**80**:1493–5.

33. Taylor SI, Cama A, Accili D, Barbetti F, Quon MJ, de la Luz Sierra M, Suzuki Y, Koller E, Levy-Toledano R, Wertheimer E, et al. Mutations in the insulin receptor gene. *Endocr Rev* 1992;**13**:566–95.

34. Kadowaki T, Kadowaki H, Rechler MM, Serrano-Rios M, Roth J, Gorden P, Taylor SI. Five mutant alleles of the insulin receptor gene in patients with genetic forms of insulin resistance. *J Clin Invest* 1990;**86**:254–64.

35. Odawara M, Kadowaki T, Yamamoto R, Shibasaki Y, Tobe K, Accili D, Bevins C, Mikami Y, Matsuura N, Akanuma Y, et al. Human diabetes associated with a mutation in the tyrosine kinase domain of the insulin receptor. *Science* 1989;**245**:66–8.

36. Treadway JL, Morrison BD, Soos MA, Siddle K, Olefsky J, Ullrich A, McClain DA, Pessin JE. Transdominant inhibition of tyrosine kinase activity in mutant insulin/insulin-like growth factor I hybrid receptors. *Proc Natl Acad Sci USA* 1991;**88**:214–8.

37. Garg A, Wilson R, Barnes R, Arioglu E, Zaidi Z, Gurakan F, Kocak N, O'Rahilly S, Taylor SI, Patel SB, Bowcock AM. A gene for congenital generalized lipodystrophy maps to human chromosome 9q34. *J Clin Endocrinol Metab* 1999;**84**:3390–4.

38. Agarwal AK, Arioglu E, De Almeida S, Akkoc N, Taylor SI, Bowcock AM, Barnes RI, Garg A. AGPAT2 is mutated in congenital generalized lipodystrophy linked to chromosome 9q34. *Nat Genet* 2002;**31**:21–3.

39. Magre J, Delepine M, Khallouf E, Gedde-Dahl Jr T, Van Maldergem L, Sobel E, Papp J, Meier M, Megarbane A, Bachy A, Verloes A, d'Abronzo FH, Seemanova E, Assan R, Baudic N, Bourut C, Czernichow P, Huet F, Grigorescu F, de Kerdanet M, Lacombe D, Labrune P, Lanza M, Loret H, Matsuda F, Navarro J, Nivelon-Chevalier A, Polak M, Robert JJ, Tric P, Tubiana-Rufi N, Vigouroux C, Weissenbach J, Savasta S, Maassen JA, Trygstad O, Bogalho P, Freitas P, Medina JL, Bonnicci F, Joffe BI, Loyson G, Panz VR, Raal FJ, O'Rahilly S, Stephenson T, Kahn CR, Lathrop M, Capeau J. Identification of the gene altered in Berardinelli–Seip congenital lipodystrophy on chromosome 11q13. *Nat Genet* 2001;**28**:365–70.

40. Simha V, Garg A. Phenotypic heterogeneity in body fat distribution in patients with congenital generalized lipodystrophy caused by mutations in the AGPAT2 or seipin genes. *J Clin Endocrinol Metab* 2003;**88**:5433–7.

41. Antuna-Puente B, Boutet E, Vigouroux C, Lascols O, Slama L, Caron-Debarle M, Khallouf E, Levy-Marchal C, Capeau J, Bastard JP, Magre J. Higher adiponectin levels in patients with Berardinelli–Seip congenital lipodystrophy due to seipin as compared with 1-acylglycerol-3-phosphate-o-acyltransferase-2 deficiency. *J Clin Endocrinol Metab* 2010;**95**:1463–8.

42. Haque W, Garg A, Agarwal AK. Enzymatic activity of naturally occurring 1-acylglycerol-3-phosphate-*O*-acyltransferase 2 mutants associated with congenital generalized lipodystrophy. *Biochem Biophys Res Commun* 2005;**327**:446–53.

43. Payne VA, Grimsey N, Tuthill A, Virtue S, Gray SL, Nora ED, Semple RK, O'Rahilly S, Rochford JJ. The human lipodystrophy gene BSCL2/Seipin may be essential for normal adipocyte differentiation. *Diabetes* 2008;**57**:2055–60.

44. Szymanski KM, Binns D, Bartz R, Grishin NV, Li WP, Agarwal AK, Garg A, Anderson RG, Goodman JM. The lipodystrophy protein seipin is found at endoplasmic reticulum lipid droplet junctions and is important for droplet morphology. *Proc Natl Acad Sci USA* 2007;**104**:20890–5.

45. Fei W, Shui G, Gaeta B, Du X, Kuerschner L, Li P, Brown AJ, Wenk MR, Parton RG, Yang H. Fld1p, a functional homologue of human seipin, regulates the size of lipid droplets in yeast. *J Cell Biol* 2008;**180**:473–82.

46. Boutet E, El Mourabit H, Prot M, Nemani M, Khallouf E, Colard O, Maurice M, Durand-Schneider AM, Chretien Y, Gres S, Wolf C, Saulnier-Blache JS, Capeau J, Magre J. Seipin deficiency alters fatty acid Delta9 desaturation and lipid droplet formation in Berardinelli–Seip congenital lipodystrophy. *Biochimie* 2009;**91**:796–803.

47. Ito D, Suzuki N. Molecular pathogenesis of seipin/BSCL2-related motor neuron diseases. *Ann Neurol* 2007;**61**:237–50.

48. Magre J, Delepine M, Van Maldergem L, Robert JJ, Maassen JA, Meier M, Panz VR, Kim CA, Tubiana-Rufi N, Czernichow P, Seemanova E, Buchanan CR, Lacombe D, Vigouroux C, Lascols O, Kahn CR, Capeau J, Lathrop M. Prevalence of mutations in AGPAT2 among human lipodystrophies. *Diabetes* 2003;**52**:1573–8.

49. Kim CA, Delepine M, Boutet E, El Mourabit H, Le Lay S, Meier M, Nemani M, Bridel E, Leite CC, Bertola DR, Semple RK, O'Rahilly S, Dugail I, Capeau J, Lathrop M, Magre J. Association of a homozygous nonsense caveolin-1 mutation with Berardinelli–Seip congenital lipodystrophy. *J Clin Endocrinol Metab* 2008;**93**:1129–34.

50. Liu L, Pilch PF. A critical role of cavin (polymerase I and transcript release factor) in caveolae formation and organization. *J Biol Chem* 2008;**283**:4314–22.

51. Shackleton S, Lloyd DJ, Jackson SN, Evans R, Niermeijer MF, Singh BM, Schmidt H, Brabant G, Kumar S, Durrington PN, Gregory S, O'Rahilly S, Trembath RC. LMNA, encoding lamin A/C, is mutated in partial lipodystrophy. *Nat Genet* 2000;**24**:153–6.

52. Garg A, Vinaitheerthan M, Weatherall PT, Bowcock AM. Phenotypic heterogeneity in patients with familial partial lipodystrophy (dunnigan variety) related to the site of missense mutations in lamin a/c gene. *J Clin Endocrinol Metab* 2001;**86**:59–65.

53. Novelli G, Muchir A, Sangiuolo F, Helbling-Leclerc A, D'Apice MR, Massart C, Capon F, Sbraccia P, Federici M, Lauro R, Tudisco C, Pallotta R, Scarano G, Dallapiccola B, Merlini L, Bonne G. Mandibuloacral dysplasia is caused by a mutation in LMNA-encoding lamin A/C. *Am J Hum Genet* 2002;**71**:426–31.

54. Vantyghem MC, Pigny P, Maurage CA, Rouaix-Emery N, Stojkovic T, Cuisset JM, Millaire A, Lascols O, Vermersch P, Wemeau JL, Capeau J, Vigouroux C. Patients with familial partial lipodystrophy of the Dunnigan type due to a LMNA R482W mutation show muscular and cardiac abnormalities. *J Clin Endocrinol Metab* 2004;**89**:5337–46.

55. Young J, Morbois-Trabut L, Couzinet B, Lascols O, Dion E, Bereziat V, Feve B, Richard I, Capeau J, Chanson P, Vigouroux C. Type A insulin resistance syndrome revealing a novel lamin A mutation. *Diabetes* 2005;**54**:1873–8.

56. Lehrke M, Lazar MA. The many faces of PPARgamma. *Cell* 2005;**123**:993–9.

57. Agostini M, Schoenmakers E, Mitchell C, Szatmari I, Savage D, Smith A, Rajanayagam O, Semple R, Luan J, Bath L, Zalin A, Labib M, Kumar S, Simpson H, Blom D, Marais D, Schwabe J, Barroso I, Trembath R, Wareham N, Nagy L, Gurnell M, O'Rahilly S, Chatterjee K. Non-DNA binding, dominant-negative, human PPARgamma mutations cause lipodystrophic insulin resistance. *Cell Metab* 2006;**4**:303–11.

58. Al-Shali K, Cao H, Knoers N, Hermus AR, Tack CJ, Hegele RA. A single-base mutation in the peroxisome proliferator-activated receptor gamma4 promoter associated with altered *in vitro* expression and partial lipodystrophy. *J Clin Endocrinol Metab* 2004;**89**:5655–60.

59. Vantyghem MC, Balavoine AS, Douillard C, Defrance F, Dieudonne L, Mouton F, Lemaire C, Bertrand-Escouflaire N, Bourdelle-Hego MF, Devemy F, Evrard A, Gheerbrand D, Girardot C, Gumuche S, Hober C, Topolinski H, Lamblin B, Mycinski B, Ryndak A, Karrouz W, Duvivier E, Merlen E, Cortet C, Weill J, Lacroix D, Wemeau JL.

How to diagnose a lipodystrophy syndrome. *Ann Endocrinol (Paris)* 2012;**73**:170–89.

60. Misra A, Peethambaram A, Garg A. Clinical features and metabolic and autoimmune derangements in acquired partial lipodystrophy: report of 35 cases and review of the literature. *Medicine (Baltimore)* 2004;**83**:18–34.

61. Agostini M, Gurnell M, Savage DB, Wood EM, Smith AG, Rajanayagam O, Garnes KT, Levinson SH, Xu HE, Schwabe JW, Willson TM, O'Rahilly S, Chatterjee VK. Tyrosine agonists reverse the molecular defects associated with dominant-negative mutations in human peroxisome proliferator-activated receptor gamma. *Endocrinology* 2004;**145**:1527–38.

62. Javor ED, Cochran EK, Musso C, Young JR, Depaoli AM, Gorden P. Long-term efficacy of leptin replacement in patients with generalized lipodystrophy. *Diabetes* 2005;**54**:1994–2002.

23

Lipodystrophies

Abhimanyu Garg

Division of Nutrition and Metabolic Diseases, Department of Internal Medicine,
Distinguished Chair in Human Nutrition Research, Center for Human Nutrition,
UT Southwestern Medical Center, Dallas, TX, USA

INTRODUCTION

During the last two decades, major advances have been made in elucidation of the molecular genetic basis of inherited lipodystrophy syndromes. The first step toward identification of the causal genes required careful phenotyping of the patients based on clinical features and in-depth characterization of the body fat distribution using conventional anthropometry and whole body magnetic resonance imaging, followed by proper classification of the various types and subtypes of lipodystrophies. Initially, discoveries of the causal genes were made using the classical linkage analysis approach followed by positional cloning. More recently, elucidation of the molecular genetic basis of some extremely rare syndromes has been made possible, due to availability and utilization of the next generation, whole exome sequencing. This knowledge of the underlying basis of genetic lipodystrophies has further improved categorization of the lipodystrophies and refinement of the clinical features associated with various subtypes.

The lipodystrophies are characterized by selective loss of body fat (Fig. 23.1) and a predisposition to developing insulin resistance and its complications, such as diabetes mellitus, hypertriglyceridemia, hepatic steatosis, polycystic ovarian syndrome, acanthosis nigricans, and hypertension. The loss of fat can be generalized (involving all the body fat depots), partial (affecting the limbs) or focal or localized (from discrete areas of the body). The severity of the metabolic complications is generally related to the extent of body fat loss. For example, patients with generalized lipodystrophies have more severe diabetes, hypertriglyceridemia, and hepatic steatosis than those with partial lipodystrophies.

The diagnosis of the various syndromes of lipodystrophies is mainly based on clinical findings, that is, history, physical examination, and anthropometry.

The clinical diagnosis can be confirmed by genetic screening for those subtypes for which the causal genes are known. However, for many syndromes, the molecular genetic basis remains unknown, and therefore at present, only clinical diagnostic criteria can be applied to achieve a diagnosis. Genetic lipodystrophies can be broadly classified based on the pattern of inheritance, that is, autosomal recessive or autosomal dominant (Table 23.1). The overall diagnostic approach to genetic lipodystrophies is presented in this chapter.

AUTOSOMAL RECESSIVE LIPODYSTROPHIES

Congenital Generalized Lipodystrophy (CGL)

This syndrome was first reported in 1954 in two young Brazilian boys who were two and six years old and presented with marked hepatosplenomegaly, acromegaloid gigantism, fatty liver, and hyperlipidemia.[4] Five years later, three additional patients from Norway were reported with generalized lack of body fat from birth and the disorder was termed congenital generalized lipodystrophy (CGL).[5] However, many subsequent investigators have also used the terminology Berardinelli–Seip congenital lipodystrophy in recognition of the physicians who initially reported this syndrome. Because of striking clinical appearance at birth, the diagnosis of CGL is usually made by neonatologists or pediatricians. Only a few patients may have a normal appearance at birth, and loss of body fat may become evident later during the first year of life.

The precise population prevalence of CGL is not known. In the literature to date, ~300 patients with CGL have been reported.[6–8] Some regions from Brazil and from Lebanon have reported a cluster of cases, likely due to

Genetic Diagnosis of Endocrine Disorders. http://dx.doi.org/10.1016/B978-0-12-800892-8.00023-3

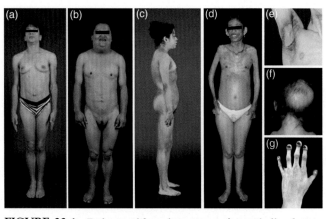

FIGURE 23.1 Patients with various types of genetic lipodystrophies. (a) Anterior view of a 33-year-old Hispanic female with congenital generalized lipodystrophy (also known as Berardinelli–Seip congenital lipodystrophy), type 1 due to homozygous c.IVS4-2A>G mutation in *AGPAT2*. The patient had generalized loss of subcutaneous fat with acanthosis nigricans in the axillae and neck. She has umbilical prominence and acromegaloid features (enlarged mandible, hands, and feet). (b) Anterior view of a 27-year-old Native American Hispanic female with familial partial lipodystrophy (FPL) of the Dunnigan variety due to heterozygous p.Arg482Trp mutation in *LMNA*. She had marked loss of subcutaneous fat from the limbs and anterior truncal region. The breasts were atrophic. She had increased subcutaneous fat deposits in the face, anterior neck, and vulvar regions. (c) Lateral view of a 20-year-old Hispanic woman with mandibuloacral dysplasia type A due to homozygous p.Arg527His mutation in the *LMNA* gene. She had loss of subcutaneous fat over the extremities leading to a muscular appearance with prominent veins and increased fat over the neck and face. Note the mottled skin hyperpigmentation over the trunk and the rounded finger tips from resorption of the terminal phalanges. (d) Anterior view of a 26-year-old female of Mexican origin with autoinflammatory lipodystrophy with a homozygous p.Thr75Met mutation in *PSMB8*. She had more marked loss of subcutaneous fat from the face, neck, chest, and upper extremities than from the abdomen, hips, and lower extremities, which were spared. The breasts were atrophic and the neck and chest showed many discrete, small, erythematous nodular skin lesions. She had mild contractures of the hand joints. (e) Acanthosis nigricans (brownish discoloration with thickening of the skin) in the axilla in a patient with lipodystrophy. (f) Loss of hair from the posterior scalp region in a 5-year-old girl with severe mandibuloacral dysplasia due to homozygous p.Arg527Cys mutation in *LMNA*. (g) Acro-osteolysis in the hand of patient shown in (c). The terminal digits appeared short and bulbous due to resorption of the terminal phalanges. The skin on the dorsum of the hand was atrophic, especially over the proximal interphalangeal and metacarpophalangeal joints. *The subparts (a), (b), (e–g) are reproduced with permission from Ref. [1]. Subpart (c) is reproduced with permission from Ref. [2]. Subpart (d) is reproduced with permission from Ref. [3].*

TABLE 23.1 Classification of Genetic Lipodystrophies

A. Autosomal recessive
  1. Congenital generalized lipodystrophy (CGL; Berardinelli–Seip syndrome)
      a. CGL1: *AGPAT2* (1-acylglycerol-3-phosphate *O*-acyltransferase 2) mutations
      b. CGL2: *BSCL2* (Berardinelli–Seip congenital lipodystrophy 2) mutations
      c. CGL3: *CAV1* (caveolin 1) mutation
      d. CGL4: *PTRF* (polymerase I and transcript release factor) mutations
      e. Other varieties: unknown genes
  2. Mandibuloacral dysplasia associated with *LMNA* mutations
      a. MADA: *LMNA* (lamin A/C) mutations
      b. MADB: *ZMPSTE24* (zinc metallopeptidase STE24) mutations
      c. Other varieties: unknown genes
  3. Familial partial lipodystrophy (FPLD)
      a. FPLD5: *CIDEC* (cell death-inducing DNA fragmentation factor a-like effector c) mutation
  4. Autoinflammatory lipodystrophy (JMP/CANDLE syndrome) *PSMB8* (proteasome subunit, beta-type, 8) mutations

B. Autosomal dominant
  1. Familial partial lipodystrophy (FPLD)
      a. FPLD1: Kobberling variety: unknown
      b. FPLD2: Dunnigan variety: *LMNA* (lamin A/C) mutations
      c. FPLD3: *PPARG* (peroxisome proliferator-activated receptor-gamma) mutations
      d. FPLD4: *PLIN1* (perilipin 1) mutations
      e. FPLD6: *AKT2* (v-AKT murine thymoma oncogene homolog 2) mutation
      f. Others: unknown
  2. Progeroid syndromes associated lipodystrophy
      a. Hutchinson–Gilford progeria syndrome (HGPS): *LMNA* mutations
      b. Atypical progeroid syndrome (APS): *LMNA* mutations
      c. Neonatal progeroid syndrome:
          – NPS1 (progeroid fibrillinopathy): *FBN1* (fibrillin 1) mutations
          – Others: unknown
      d. *Mandibular hypoplasia, deafness, progeroid features (MDP) syndrome-POLD1* (polymerase (DNA directed) delta 1 catalytic subunit) mutations
  3. SHORT syndrome associated lipodystrophy (autosomal dominant)
      a. SHORT1: *PIK3R1* (phosphatidylinositol 3-kinase, regulatory subunit 1 alpha) mutations
      b. Others: unknown

in-breeding.[9,10] If it is assumed that only a quarter of the actual number of cases may be reported in the literature, the estimated worldwide prevalence of CGL may be about one in 10 million.[8] Recent establishment of worldwide and regional patient registries may help in better estimates of prevalence of CGL and other genetic lipodystrophies in future. Further refinement of the diagnostic codes in the ICD-10 system may also be advantageous in estimating the prevalence of genetic lipodystrophies.

Patients with CGL present with muscular appearance at birth instead of the normal chubby appearance of the newborns. This is due to near complete absence of body fat (Fig. 23.1a). The face may look gaunt. It is likely that the parents and caregivers may not recognize this abnormal phenotype in male newborns, but will be able to recognize it in female infants. The affected children demonstrate an accelerated growth with an insatiable appetite.[11,12] They have acromegaloid features with enlarged mandible, hands, and feet. Prominence of umbilicus or umbilical hernias are noticed at birth likely due to underlying hepatomegaly and/or splenomegaly.[11,13]

Patients with CGL do not have acanthosis nigricans at birth; however, it appears during early childhood or after puberty, and can affect many body regions[11] including the typical neck, axillae, and groin regions, and other regions such as the trunk, hands, knees, elbows, and ankles (Fig. 23.1e).

Liver enlargement due to hepatic steatosis is usually noticed during infancy. A few patients develop steatohepatitis followed by fibrosis. Some do develop cirrhosis or end-stage liver disease and its complications later on in life requiring hepatic transplantation.[14,15] In females, mild hirsutism, clitoromegaly, and irregular menstrual periods are common and some present with primary or secondary amenorrhea and polycystic ovaries. In some women, clitoromegaly can occur without any signs of hirsutism. Breast development is normal but the overlying subcutaneous fat layer surrounding the mammary tissue is absent. Most affected women are unable to conceive; however, a few patients have had successful pregnancies. The pregnancy in these patients is high risk and may require management of diabetes as well as severe hypertriglyceridemia. Affected men usually have normal reproductive ability.

In some patients, radiographs reveal focal lytic lesions in the long bones such as the humerus, femur, radius, ulna, carpal, tarsal, or phalangeal bones after puberty, which may be confused with polyostotic fibrous dysplasia.[12,15–17] The lesions may develop due to lack of ability to make normal bone marrow fat to replace hematopoietic marrow during childhood and adolescence.[15] Instead, the marrow may be replaced by vascular tissue. The pathological fractures in the affected bones are likely and children with these lesions may be advised to avoid contact sports. Some patients have also been reported to have hypertrophic cardiomyopathy and mild mental retardation.[18,19]

Even during infancy and early childhood, patients with CGL develop extreme hyperinsulinemia,[11] and a few will also develop hyperglycemia. However, most of the patients develop diabetes during the pubertal years. An autopsy study revealed marked amyloidosis of pancreatic islets affecting more than 90% of the islets and β cell atrophy in a young adult female with CGL.[20] Thus, extreme insulin resistance from birth may induce premature and severe amyloidosis and β cell death, similar to that observed in patients with type 2 diabetes mellitus. Diabetes, however, is extremely challenging to manage in these patients and frequently requires extremely high doses of insulin for glycemic control.[21] Interestingly, diabetes is ketosis resistant likely due to endogenous hyperinsulinemia.

Most patients with CGL develop hypertriglyceridemia later during childhood and adolescence; however, some develop extreme hypertriglyceridemia during infancy. This type 5 hyperlipoproteinemia with chylomicronemia predisposes many patients to recurrent episodes of acute pancreatitis particularly those with uncontrolled diabetes.[13] The levels of high density lipoprotein (HDL) cholesterol also tend to be low. However, atherosclerotic vascular complications such as coronary heart disease, strokes, or peripheral vascular disease have not been reported in the literature. On the other hand, most of the reported subjects have been children. As we follow more patients with CGL into late adulthood, we will be able to ascertain cardiovascular implications of type 5 hyperlipoproteinemia in them.

Because of near total loss of fat in patients with CGL, the levels of serum leptin and adiponectin are markedly reduced.[22,23] However, extremely low levels of serum leptin should not be considered diagnostic. Serum leptin assays are not standardized and marked variability is observed in levels measured by various commercial kits. Severe hypoleptinemia may induce voracious appetite in patients with CGL, which may further contribute to metabolic complications.

To date, four distinct genetic subtypes of CGL have been reported. Type 1 CGL due to mutations in the 1-acylglycerol-3-phosphate-O-acyltransferase 2 (AGPAT2) gene and type 2 due to mutations in the Berardinelli–Seip congenital lipodystrophy 2 (BSCL2) gene are the most common subtypes.[10] Type 3 CGL due to a homozygous nonsense mutation in caveolin 1 (CAV1) was reported in a single patient[24] and type 4 CGL due to mutations in polymerase I and transcript release factor (PTRF) has been reported in about 30 patients.[25]

CGL Type 1: AGPAT2 Mutations

These patients have a characteristic body fat distribution on whole body magnetic resonance imaging. They have near total absence of adipose tissue from most subcutaneous areas, intra-abdominal and intrathoracic regions and bone marrow (metabolically active) but adipose tissue in the palms, soles, under the scalp, orbital, and periarticular regions is spared (mechanical adipose tissue).[13,26] They also show increased predisposition to develop lytic bone lesions; mostly affecting long bones of the extremities.

The AGPAT enzymes play an essential role in the biosynthesis of triglycerides and phospholipids.[27,28] Currently, 11 isoforms of AGPAT, each encoded by a different gene, are known.[29] These enzymes add an acyl (fatty acid moiety) to lysophosphatidic acid (1-acylglycerol-3-phosphate) to form phosphatidic acid (1,2-diacylglycerol-3-phosphate). Phosphatidic acid can be further acylated and processed to form triglycerides or other phospholipids. Each isoform has a unique tissue expression and the AGPAT2 isoform is highly expressed in the adipose tissue. Thus, deficiency of AGPAT2 may impair triglyceride and phospholipid biosynthesis in the adipocytes and result in lipodystrophy.

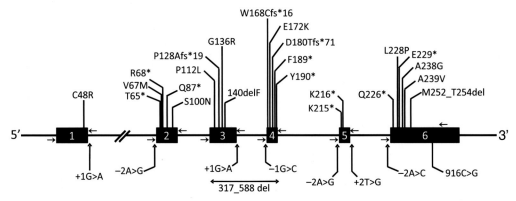

FIGURE 23.2 Structure of *AGPAT2* gene and mutations in patients with CGL type 1. Various mutations found in the affected individuals with CGL. Boxes represent exons and small arrows show location of primers used for amplification of exons.

Most of the patients harbor null mutations with no enzymatic activity demonstrable *in vitro*[30] (Fig. 23.2). However, some compound heterozygotes have a null and a missense mutation (with some residual enzymatic activity) and a few have homozygous missense mutations. However, so far the type of mutation does not seem to determine the phenotype or the loss of fat and all the patients lack nearly all body fat. Nearly all patients of African origin harbor the founder mutation, c.IVS4-2A>G (p.Gln196fsX228), on one or both alleles.[19]

CGL Type 2: BSCL2 Mutations

As compared to patients with type 1 CGL, patients with CGL type 2 have an increased prevalence of cardiomyopathy and mild mental retardation.[18,19] They have almost total lack of body fat with both metabolically active and mechanical adipose tissue being absent.[26] As compared to patients with type 1 CGL, their serum leptin levels are lower, but serum adiponectin levels are higher.[23] Recently, teratozoospermia was reported in a CGL type 2 patient with sperm defects including abnormal head morphology, bundled sperm with two or more sperm connected to each other with large ectopic lipid droplets.[31]

The *BSCL2* encodes a 398 amino acid transmembrane protein called seipin[10,32] with no significant homology to any other known protein. It has a CAAX motif at the carboxy-terminus and a glycosylation site, NVS, at position 88-90. Recent data suggest the role of seipin in lipid droplet formation and in adipocyte differentiation.[30,33] Studies with the seipin homolog in the yeast suggest that it may be playing a role in fusion of lipid droplets. Seipin has been shown to bind phosphatidic acid phosphatase (lipin 1) and thus may also play a role in phospholipids and triglyceride synthesis.[34] Nearly all *BSCL2* mutations reported in patients with CGL2 are null (Fig. 23.3). However, a few have been missense mutations. So far, no phenotypic differences have been reported between those with null and missense mutations.

CGL Type 3: Caveolin 1 Mutations

A 20-year-old Brazilian girl with generalized lipodystrophy, short stature, and presumed vitamin D resistance[24] has been reported to have a homozygous *CAV1* mutation. She had hepatic steatosis and splenomegaly

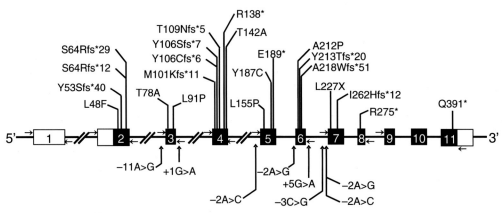

FIGURE 23.3 Structure of *BSCL2* gene and mutations in patients with CGL type 2. Various mutations found in the affected individuals with CGL. Boxes represent exons and small arrows show location of primers used for amplification of exons.

and developed diabetes at 13 years of age. Other clinical features included acanthosis nigricans, severe hypertriglyceridemia, primary amenorrhea, and functional megaesophagus. Mechanical adipose tissue and fat in the bone marrow was not affected by lipodystrophy.

Caveolin 1 is a major component of caveolae. Caveolae are specialized plasma membrane microdomains appearing as 50–100 nm vesicular invaginations.[35,36] *CAV1* is expressed ubiquitously but is highly expressed in the adipocytes, endothelial cells, and fibroblasts.[37] Caveolins play an important role in vesicular trafficking, homeostasis of cellular cholesterol, fatty acids and triglycerides, and signal transduction. Caveolae may contribute lipids to the lipid droplets and CAV1 mutation may cause lipodystrophy due to defective lipid droplet formation.

CGL Type 4: PTRF Mutations

About 30 patients have been reported to have CGL type 4 due to *PTRF*.[25,38,39] Besides having generalized lipodystrophy, they have congenital myopathy with high serum creatine kinase levels. Other clinical features include pyloric stenosis, atlantoaxial instability,[40,41] and predisposition to serious arrhythmias such as catecholaminergic polymorphic ventricular tachycardia, prolonged QT interval, and sudden death.[38,39] Mechanical and bone marrow fat is well preserved. These patients may not show severe lipodystrophy at birth and may lose body fat progressively during infancy. PTRF regulates the expression of caveolins 1 and 3 and thus plays an important role in biogenesis of caveolae.[25]

Molecular Diagnosis

The first step is to achieve a clinical diagnosis based on history and physical examination. Clinically the major differentiating clinical finding between the two major types of CGL, type 1 and 2, is the presence or absence of subcutaneous fat from the palms and soles, respectively. CGL type 4 can be easily distinguished from the other subtypes by the presence of myopathy and high serum creatine kinase levels. Molecular diagnosis at research laboratories is available for confirmation. Molecular diagnosis may be helpful for understanding the risk of having another child with CGL and can also be used for prenatal screening. Besides patients with known genotypes, there are a few patients with CGL who do not have mutations in any of the four known loci and thus there may be novel genes to be discovered for CGL.

Differential Diagnosis

CGL should be differentiated from acquired generalized lipodystrophy, leprechaunism, atypical progeroid syndrome, generalized lipodystrophy due to *LMNA* mutations, and neonatal progeroid syndrome. Patients with acquired generalized lipodystrophy develop fat loss after birth and are likely to have associated autoimmune diseases or panniculitis. Patients with generalized lipodystrophy or atypical progeroid syndrome due to *LMNA* mutations have normal body fat at birth and lose it later in childhood.

Mandibuloacral Dysplasia (MAD)-Associated Lipodystrophy

Mandibuloacral dysplasia is an extremely rare autosomal recessive disorder, which has been reported in about 40 patients so far. Clinical features include mandibular hypoplasia, resorption of lateral parts of the clavicles and acro-osteolysis (resorption of the terminal phalanges) (Fig. 23.1c,f,g).[2,42] Patients may also present with delayed closure of cranial sutures, joint contractures, mottled cutaneous pigmentation, short stature, and other skeletal deformities. Patients with MAD also develop "progeroid features" such as the bird-like facies, high-pitched voice, skin atrophy, pigmentation, alopecia, and nail dysplasia. They develop either partial lipodystrophy affecting the extremities (type A) or more generalized lipodystrophy, which affects the face, trunk, and extremities (type B). Metabolic complications such as hyperinsulinemia, insulin resistance, impaired glucose tolerance, diabetes mellitus, and hyperlipidemia are usually mild to moderate in severity.[2]

MAD Type A Due to LMNA Mutations

Novelli et al.[42] reported a homozygous missense p.Arg527His mutation in the lamin A/C (*LMNA*) gene in MAD patients with type A (partial) lipodystrophy of Italian origin, which appears to be a founder mutation.[42] *LMNA* has 12 exons and by alternative splicing in exon 10, it encodes for either a short protein, lamin C, or a full-length protein, prelamin A. While lamin C does not undergo posttranslational processing, prelamin A undergoes a series of modifications to form the mature protein called lamin A. This processing occurs at the extreme C-terminal of the prelamin A, which has a C-A-A-X motif. The first step in posttranslational processing of prelamin A involves farnesylation of cysteine residue in the C-A-A-X motif by a farnesyl transferase, which is followed by proteolysis of the tripeptide, A-A-X, by the enzyme zinc metalloprotease (ZMPSTE24). Then, the cysteine residue is carboxymethylated and finally ZMPSTE24 or another protease cleaves another 15 residues from the C-terminus forming mature lamin A. Lamins A and C, encoded by *LMNA*, and lamins B1 and B2, encoded by other genes, belong to the intermediate filament family of proteins and from heterodimeric or homodimeric coiled-coil structures, which constitute nuclear lamina, located between the inner nuclear membrane and chromatin.[43,44] Lamins provide structural integrity to the nuclear envelope and interact with chromatin and several other nuclear envelope proteins. So far, a total of 30 patients with MAD due to various *LMNA* mutations

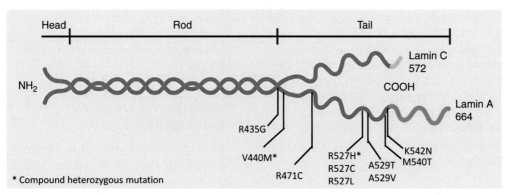

FIGURE 23.4   Structure of lamins A and C and mutations in patients with mandibuloacral dysplasia type A.

have been reported (Fig. 23.4).[2,42,45,46] Some patients have severe clinical manifestations including alopecia, loss of eyebrows, delayed sexual maturation, premature loss of teeth, as well as early death.[46,47] Most of the *LMNA* mutations causing MAD are located in the C-terminal region affecting exons 8–10 with p.R527H being the most common (Fig. 23.4). How these specific *LMNA* mutations cause resorption of bones such as the mandible, clavicles, and terminal phalanges remains unclear. Skin fibroblasts from some of the patients show nuclear morphological abnormalities such as budding, invaginations of the nuclear membrane, similar to those seen in patients with Hutchinson–Gilford progeria syndrome. Whether there is accumulation of prelamin A in skin fibroblasts from these patients is controversial.

MAD, Type B Associated with Zinc Metallopeptidase STE24 (ZMPSTE24) Mutations

Recognizing the critical role of ZMPSTE24 in prelamin A processing, our group considered *ZMPSTE24* as a candidate gene for MAD and reported compound heterozygous mutations in *ZMPSTE24* in a Belgian woman with MAD, progeroid features, and generalized lipodystrophy.[48] She died prematurely at age 24 years due to end-stage renal disease caused by focal segmental glomerulosclerosis.[48] It is suggested that accumulation of prelamin A and/or lack of mature lamin A in the cells may be the underlying mechanism of cellular toxicity in these patients. A total of eight patients with this subtype have been reported and most of them have been young children.[49,50] There are no reports of diabetes among them. As compared to MAD patients with *LMNA* mutations, those with *ZMPSTE24* mutations are premature at birth, have early-onset of skeletal defects including acro-osteolysis, have more severe progeroid appearance, and develop subcutaneous calcified nodules on the phalanges.[48,49] They are also prone to develop renal disease.

Molecular and Differential Diagnosis

Clinical diagnosis should be established first followed by molecular genetic diagnosis. Genotyping may help in predicting clinical course and complications in the patients. Differential diagnosis should include Hutchinson–Gilford progeria syndrome, and atypical progeroid syndrome (mostly due to heterozygous missense mutations in *LMNA*), and other disorders presenting with acro-osteolysis including Hajdu–Cheney (*NOTCH2* mutations), Haim–Munk, and Papillon–Lefevre syndromes (both due to Cathepsin C mutations).

Mandibular Hypoplasia, Deafness, Progeroid Features (MDP)-Associated Lipodystrophy Syndrome

This syndrome was reported by our group based on their unique clinical features including *m*andibular hypoplasia, *d*eafness, *p*rogeroid features (MDP)-associated lipodystrophy.[51] They showed distinct characteristics such as sensorineural hearing loss, and absence of clavicular hypoplasia and acro-osteolysis. All males with MDP had undescended testes and hypogonadism. One adult female showed lack of breast development. Two of the seven patients had diabetes mellitus. Recently, a recurrent *de novo* in-frame deletion (c.1812_1814delCTC, p.Ser605del) of a single codon in polymerase (DNA-directed) delta 1 catalytic subunit (*POLD1*) was reported in patients with MDP syndrome.[52] DNA polymerase δ cooperates with WRN, a DNA helicase, and serves an important role in maintaining genome stability.

Autoinflammatory Lipodystrophy Syndrome

Recently, we reported in depth a phenotype of an autosomal recessive, autoinflammatory, *j*oint contractures, *m*uscle atrophy, *m*icrocytic anemia, and *p*anniculitis-induced lipodystrophy (JMP) syndrome in two pedigrees.[3] Patients with JMP syndrome had onset of progressive panniculitis-induced lipodystrophy during childhood. Histopathology of the subcutaneous fat upon biopsy reveals infiltration of the adipose tissue with lymphocytes, neutrophils, and multinucleated giant cells called panniculitis. When these lesions with panniculitis heal, they result in localized loss of subcutaneous fat from a small area but with time it can affect the entire region such as the face, neck, upper trunk, and arms and legs (Fig. 23.1d). Previously, some patients with a similar but milder clinical phenotype were reported from Japan.[53,54] These patients also have overlapping features with recently reported patients with chronic atypical neutrophilic dermatosis with lipodystrophy and elevated temperature (CANDLE) syndrome.[53,54] All these patients with autoinflammatory lipodystrophy have variable clinical features including intermittent fever, hypergammaglobulinemia, elevated erythrocyte sedimentation rate, hepatosplenomegaly, and calcification of basal ganglia. A total of 20 patients (2–51 years old) have been reported so far.

We initially reported a homozygous, missense, loss of function, c.224C>T (p.Thr75Met) mutation in proteasome subunit, beta-type, 8 (PSMB8) gene in patients with JMP syndrome and subsequently the patients from Japan as well as those with CANDLE syndrome were reported also to have mutations in PSMB8.[55–58] PSMB8 encodes the β5i subunit of the immunoproteasome.[55] Immunoproteasomes play a critical role in proteolysis of antigens presented by major histocompatibility complex class I molecules and result in generation of immunogenic epitopes. The mutations in PSMB8 may trigger an autoinflammatory response resulting in panniculitis and other clinical manifestations.[56–59]

Short Stature, Hyperextensibility of Joints and/or Inguinal Hernia, Ocular Depression, Reiger Anomaly, and Teething Delay (SHORT) Syndrome

A total of 30 patients have been reported with SHORT syndrome. The pedigrees reveal both autosomal recessive[60,61] and dominant[62–64] modes of transmission.[60–62] Reiger anomaly consists of eye abnormalities such as iris hypoplasia, Schwalbe ring, iridocorneal synechiae, micro- or megalocornea, and dental anomalies such as hypodontia, microdontia, enamel hypoplasia, and atypical teeth. Other clinical features include intrauterine growth retardation, failure to thrive, delayed speech development, small head circumference, bilateral clinodactyly, and sensorineural hearing loss. Different patterns of fat loss have been reported. In many patients, lipodystrophy affects the face, upper extremities, and sometimes the trunk, with relative sparing of the lower extremities. Others have lipodystrophy affecting only the face, gluteal region, and elbows.[63,64] Diabetes occurs as early as the second and third decade of life. The pathogenesis of diabetes mellitus remains unclear. Recently, patients with SHORT syndrome were reported to harbor de novo or autosomal dominantly inherited heterozygous mutations in phosphatidylinositol 3-kinase regulatory subunit 1

(PIK3R1).[65–67] PIK3R1 encodes multiple regulatory subunits of phosphatidylinositol 3-kinase (PI3K), an intracellular enzyme with a central role in insulin signaling. The molecular genetic basis of the autosomal recessive variety of SHORT syndrome remains to be determined.

Neonatal Progeroid Syndrome (Wiedemann–Rautenstrauch Syndrome)

This syndrome has been described in about 25 cases.[68–71] The clinical features are evident at birth but are heterogeneous. Patients have a triangular, old-looking face with relatively large skull (progeroid appearance), prominent veins on the scalp, sparse scalp hair, large anterior fontanelle, and generalized lipodystrophy. However, subcutaneous fat in the sacral and gluteal areas is spared.[70–72] Approximately half of the patients reportedly died before the age of 6 years but some patients surviving up to the age of 16 years have been reported.[70,72–74] Recently, a few patients with overlapping features of progeroid and Marfan's syndromes were reported to harbor de novo heterozygous null mutations in the penultimate exon of fibrillin 1 (FBN1) gene. These patients have been suggested to be having progeroid fibrillinopathy.[75,76] Many others do not harbor FBN1 variants suggesting additional loci for this syndrome.

Familial Partial Lipodystrophy (FPLD) Due to CIDEC Mutation

A single patient has been reported to have autosomal recessive FPL FPLD, type 5 due to a homozygous missense mutation in cell death-inducing DNA fragmentation factor a-like effector c (CIDEC).[77] This 19-year-old girl from Ecuador developed diabetic ketoacidosis at age 14 and also had hypertriglyceridemia and hypertension. Interestingly, subcutaneous fat biopsy revealed multilocular, small lipid droplets in adipocytes, which had been previously reported in the Cidec knockout mice.[77,78]

AUTOSOMAL DOMINANT LIPODYSTROPHIES

Familial Partial Lipodystrophies (FPLD)

FPLD patients are characterized by the loss of body fat from both the upper and lower extremities. There may be variable loss of subcutaneous fat from the truncal region and the face and neck.[79,80] The diagnosis is easily made in affected women; however, men with FPLD may escape recognition because even some normal healthy men also appear very muscular and have markedly low subcutaneous fat on the limbs. Therefore, most of the FPLD families have been ascertained after diagnosis of a female proband. The index of suspicion should be high in patients with early onset of insulin resistance manifested by acanthosis nigricans, polycystic ovarian syndrome, diabetes mellitus, and severe hypertriglyceridemia. Generally these clinical features are seen in patients with obesity and if the patients appear to be nonobese, then physicians should examine carefully for fat loss from the extremities, particularly from the gluteal region. The diagnostic clinical feature of FPLD is the loss of subcutaneous fat from the extremities, but a Cushingoid appearance with increased subcutaneous fat deposition in the face, under the chin, posteriorly in the neck resulting in a dorsocervical hump, and in the intra-abdominal region may also prompt a suspicion of FPLD (Fig. 23.1b). Some women also gain fat in the perineal region, especially in the labia majora and pubic region. To date, four genetically distinct subtypes of autosomal dominant FPLD have been reported.

Type 1 Kobberling Variety (FPLD1-unknown)

Kobberling et al.[80] from Germany reported a phenotype of FPLD, which was distinct from that reported earlier by Dunnigan and colleagues.[76] The Kobberling variety has been reported in only two small pedigrees and four sporadic cases.[80–82] The age of onset and the mode of inheritance are not clear. The loss of subcutaneous fat is restricted to the extremities. Patients have been reported to have normal amounts of fat in the face area and may have normal, or even excess, subcutaneous fat in the truncal area. The molecular genetic basis of this subtype has not been identified.

Type 2 FPLD, Dunnigan Variety (FPLD2: LMNA Mutations)

Ozer et al.[83] reported a 52-year-old woman with "fat neck syndrome" but loss of subcutaneous fat from the limbs. Several members of her family also had similar features. All affected patients had hypertriglyceridemia, and some had impaired glucose tolerance and diabetes. Subsequently, Dunnigan et al.[79] from Ireland reported a detailed phenotypic description of two families with an autosomal dominant variety of FPLD. Initially, the investigators were only able to recognize the syndrome in affected females and considered an X-linked dominant inheritance pattern with lethality in hemizygous males.[81] Subsequent pedigrees, however, clearly showed an autosomal dominant inheritance pattern.[84] Since the original description, approximately 500 patients mostly of European origin have been reported to have FPLD, with a few Asian Indians and African-Americans.

The onset of lipodystrophy in FPLD2 patients usually occurs in late childhood or at puberty. The affected children have normal body fat distribution at birth and during the first decade. Concurrent with the onset of lipodystrophy, excess fat accumulation becomes noticeable at the face, neck, and in the intra-abdominal region. Patients also manifest acanthosis nigricans. Approximately one-third of women affected with FPLD2 have irregular periods, oligoamenorrhea, and hirsutism suggestive of polycystic ovarian syndrome.[85] While subcutaneous fat from the extremities is nearly totally lost but intermuscular fat between the muscle fascia is well preserved.[86] Affected women develop more severe metabolic complications than the affected men,[85,87] such as diabetes, hypertriglyceridemia, and coronary heart disease.

Diabetes usually develops during the third decade and multiparity and excess fat deposition in the nonlipodystrophic regions such as the chin may be predisposing factors.[88] Autopsy study in a patient revealed severe amyloidosis of pancreatic islets indicating the role of prolonged insulin resistance in the pathogenesis of diabetes in FPLD2.[89] Some patients with FPLD2 also develop cardiomyopathies, which manifest both as cardiac conduction system disturbances resulting in atrial fibrillation requiring pacemaker implantation and premature congestive heart failure requiring cardiac transplantation.[90,91]

We were the first to report the FPLD2 locus on chromosome 1q21-22.[84] Subsequently, Cao and Hegele[92] reported a single missense heterozygous mutation, p.Arg482Gln, in the lamin A/C (LMNA) gene in a Canadian pedigree. Since then, several missense mutations in LMNA have been reported in patients with FPLD2, most of them affecting the C-terminal amino acids.[93–95]

Some patients with FPLD2 harboring LMNA mutations show features of overlap syndrome and develop mild muscular dystrophy as well as cardiac conduction system disturbances.[90,91] However, how specific mutations in LMNA cause subcutaneous adipocyte loss mainly from the extremities remains unknown. It is hypothesized that disruption of interactions of lamins A and C with chromatin or other nuclear lamina proteins during cell division may lead to premature cell death or apoptosis of adipocytes. The accumulation of excess fat in nonlipodystropic regions may be a secondary phenomenon.

Most of the missense mutations in FPLD2 are in exon 8 of LMNA, which encodes the globular C-terminal (tail) portion of the protein (Fig. 23.5). Particularly, the arginine residue at position 482 seems to be a hot spot and about 75% of the patients with FPLD2 have a substitution of this residue either to tryptophan, glutamine, or leucine.[95] Some patients with mutations in exon 11, which can only affect lamin A and not lamin C, seem to have a milder, atypical FPLD2.[96] More importantly, patients with mutations in exon 1 or nearby exons encoding the amino terminal residues have associated cardiomyopathy.[90,91] Some of these patients also show evidence of mild muscular dystrophy with slightly increased serum creatine kinase levels. In some families, these mutations lead to

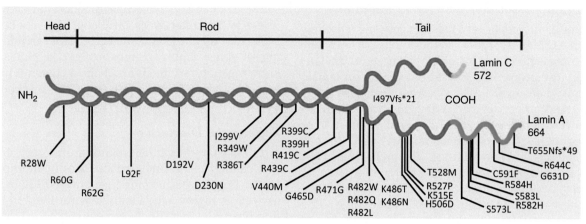

FIGURE 23.5 Structure of lamins A and C and mutations in patients with FPLD2, (Dunnigan variety).

severe congestive heart failure in the third decade resulting in a need for cardiac transplantation.[91]

By clinical examination it is difficult to diagnose affected men as well as prepubertal children with FPLD2. Therefore, molecular diagnosis may be helpful in characterizing these patients. Furthermore, genotyping for mutations, which are also associated with cardiomyopathies, may be particularly important for clinical management and also for predicting prognosis.

Type 3: FPLD Associated with PPARG Mutations

Using a candidate gene approach, our group reported a heterozygous missense mutation, p.Arg397Cys (p.Arg425Cys according to PPARγ2 protein), in the PPARG in a 64-year-old woman who presented with diabetes, hypertriglyceridemia, hypertension, and hirsutism.[97] She also had lipodystrophy of the face and extremities that was noticed much later in life. Since then, approximately 30 patients with FPLD3 due to PPARG mutations have been reported[98] (Fig. 23.6). The age of onset of this variety of lipodystrophy is not precisely known, but may range from the second decade or later. Also the pattern of progression of fat loss is not very clear. Patients have more fat loss from the distal extremities than from the proximal extremities and have variable loss of fat from the face. Some patients have normal or increased fat on the face. Given

the critical role of PPARγ as a transcription factor in adipogenesis and adipocyte differentiation, mutations in PPARG may result in lipodystrophy due to defective differentiation of adipocytes. Why fat loss in patients with PPARG mutations is restricted to the extremities remains unclear.

Type 4: FPLD Associated PLIN1 Mutations

Recently, Gandotra et al.[99] reported two heterozygous frameshift mutations in PLIN1 in five FPLD patients of European origin. All of them had hepatic steatosis, hypertriglyceridemia, and hyperinsulinemia and three had diabetes. Lipodystrophy was most striking in the lower limbs and femorogluteal depots. Acanthosis nigricans was present in all probands and two of them also had a Cushingoid appearance. Subcutaneous adipose tissue showed reduced size of adipocytes and increased macrophage infiltration and fibrosis.[99] Perilipin 1 is the most abundant protein coating lipid droplets in adipocytes.[100] Overexpression of mutant PLIN1 in 3T3-L1 preadipocytes resulted in small lipid droplets. PLIN1 is considered to be essential for formation and maturation of lipid droplets and storage of triglycerides as well as release of fatty acids from these lipid droplets. Mutant PLIN1 also failed to bind to AB-hydrolase containing 5, resulting in constitutive coactivation of adipose triglyceride lipase and increased basal lipolysis.[101]

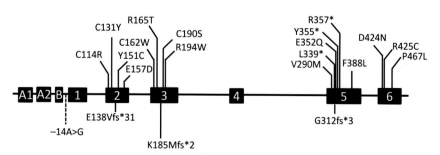

FIGURE 23.6 Structure of *PPARG* gene and mutations in patients with FPLD, type 3. Boxes represent exons.

Type 6 FPLD Associated with AKT2 Mutations

George et al.[102] reported a heterozygous missense mutation, p.Arg274His, in *AKT2* in four subjects from a family who presented with insulin resistance and diabetes mellitus. The proband, a 35-year-old Caucasian female, developed diabetes at age 30, whereas her affected mother and grandmother developed diabetes during their late thirties. A maternal uncle, a middle-aged person, had no diabetes but had hyperinsulinemia. Three of the four affected subjects had hypertension. The proband also had loss of subcutaneous fat from the extremities (Stephen O'Rahilly, personal communication). However, the precise pattern of subcutaneous fat loss is not clear. AKT2 is a phosphoinositide-dependent serine/threonine kinase and is also known as protein kinase B. AKT2 is predominantly expressed in insulin-sensitive tissues. Overexpression of the mutant form, p.Arg274His, in 3T3-L1 mouse preadipocytes resulted in markedly reduced lipid accumulation. AKT2 knockout mice also show features of lipodystrophy, insulin resistance, and diabetes with increasing age.[103] Thus, taken together, lipodystrophy in patients with *AKT2* mutations may be due to reduced adipocyte differentiation and dysfunctional postreceptor insulin signaling.

Other Types of FPLD

The four known FPLD genes are not able to explain the genetic basis of all the patients with FPLD and there is likelihood of additional loci.[97,104] In-depth characterization of the clinical phenotype related to the pattern of fat loss in FPLD patients with mutations in different genes may be helpful in identification of different phenotypes without resorting to molecular diagnosis.

Molecular Diagnosis

A clinical diagnosis should be made first based on history, physical examination, anthropometry, and whole body magnetic resonance imaging (MRI). At present, it is difficult to distinguish clinically between the various subtypes of FPLD. Molecular diagnosis at clinical and research laboratories is available for confirmation. Molecular diagnosis may be helpful for cascade screening of other family members. Besides patients with known genotypes, there are many patients with FPLD who do not have mutations in any of the four known loci and may have novel variants/genes.

Differential Diagnosis

FPLD should be differentiated from conditions such as Cushing's syndrome, truncal obesity, multiple symmetric lipomatosis, acquired generalized lipodystrophy as well as highly active antiretroviral therapy-induced lipodystrophy in HIV-infected patients.

Atypical Progeroid Syndrome Due to LMNA Mutations

About 30 patients have been reported to have partial or generalized lipodystrophy, insulin resistant diabetes, and progeroid features with heterozygous missense mutations in *LMNA*[105] (Fig. 23.7). Most of these patients have *de novo* mutations. Additional clinical features include mottling, pigmentation, and sclerosis of skin, hepatic steatosis, cardiomyopathy, short stature, cardiac valvular abnormalities, and deafness. Particularly striking is the lack of breast development in many females. Some women also develop premature ovarian failure. These patients are also predisposed to developing metabolic complications.

Hutchinson–Gilford Progeria Syndrome (HGPS)

Patients with HGPS appear normal at birth but develop features of early aging during the neonatal period or early childhood.[106–108] These features include severe alopecia, graying of hair, micrognathia, beaked nose, shrill voice, and extensive wrinkling of the skin due to loss of underlying adipose tissue, poor sexual development, joint contractures, and severe atherosclerosis.[106] Many of

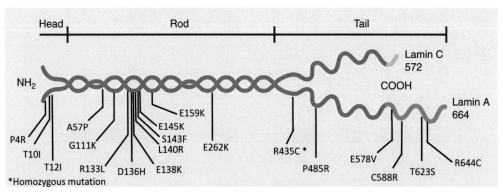

FIGURE 23.7   Structure of lamins A and C and mutations in patients with atypical progeroid syndrome.

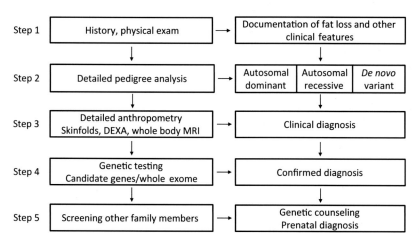

FIGURE 23.8 Diagnostic approach to genetic lipodystrophies.

them die between the ages of 6 and 20 years. Most of them have synonymous *de novo* heterozygous *LMNA* mutation p.Gly608Gly.[109] This mutation induces a cryptic splice site resulting in a mutant truncated form of prelamin A with 50 carboxy-terminal amino acids deleted. Occasional patients have other heterozygous missense *LMNA* mutations at the carboxy-terminal.[109–111] Patients develop severe lipodystrophy with increasing age.[108] Interestingly, diabetes among HGPS patients is rare.[107,108]

Other Extremely Rare Subtypes

A 30-year-old female with generalized and infantile onset lipodystrophy (CGL-like phenotype) has been reported to harbor compound heterozygous *PPARG* mutations[112]. Extremely rare cases with autosomal recessive FPLD-like phenotype have been reported with mutations in *LIPE*, *WRN* and *PCYT1A* genes[113–116]. *CAV1* mutations have been linked to a neonatal-onset lipodystrophy and (WRS-like) progeroid syndrome[117]. Finally, Keppen-Lubinsky syndrome associated lipodystrophy has been associated with *KCNJ6* mutation.[118]

Overall Diagnostic Approach

A stepwise diagnostic approach is presented in Fig. 23.8, which starts with history and physical examination with documentation of fat loss and other peculiar clinical features. A second step involves detailed pedigree analysis, which can suggest an autosomal dominant or autosomal recessive inheritance pattern or the possibility of *de novo* mutation. In the third step, detailed anthropometry using skinfold thickness, body fat distribution pattern using dual energy X-ray absorptiometry, and whole body magnetic resonance imaging can further substantiate clinical diagnosis. Finally, confirmation of diagnosis may be based on genetic testing. This confirmation can lead to screening of

other family members and genetic counseling. Another diagnostic algorithm is based on the age of onset of lipodystrophy (Fig. 23.9). CGL syndrome and Wiedemann-Rautenstrauch syndrome (WRS) present in children less than 1 year of age and MDP, SHORT, and autoinflammatory lipodystrophies (JMP) present in early childhood. Most laminopathies such as FPLD2, MAD, atypical progeroid syndrome, and HGPS present in late childhood or puberty as well as other types of FPLD and SHORT syndrome.

Availability of Genetic Testing

The best resource to find clinical and research laboratories available for genetic testing of various subtypes of lipodystrophies is at www.genetests.org. Many research laboratories are available in the United States, Canada, Europe, and the United Kingdom with interest in genetic lipodystrophies. In patients with complex phenotypes whole exome sequencing may be employed to find the molecular basis of the disease. Sometimes, however, interpretation of the data can be demanding and complex.

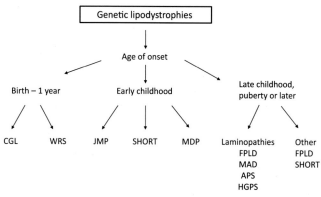

FIGURE 23.9 Diagnostic algorithm based on the age of presentation of genetic lipodystrophies.

Predictive Value of the Genetic Testing and Genetic Counseling

Penetrance of CGL in homozygotes or compound heterozygotes is 100%. The severity of metabolic complications, however, varies between males and females and with age. In our experience, disease-causing FPLD mutations in *LMNA* and *PPARG* are fully penetrant with respect to the phenotype of lipodystrophy among post-pubertal patients; however, clinical recognition is difficult in males. The severity of metabolic complications is much more variable, and depends significantly on gender and lifestyle and other modifiers. Since there is strong evidence for the presence of additional loci for genetic lipodystrophies, negative tests should not rule out a genetic condition. In fact, such patients should be referred to investigators interested in discovering novel loci for these syndromes. Many lipodystrophy syndromes such as CGL, MAD, and others have such striking phenotypes that there is no need to screen other family members. However, cascade screening should be conducted in FPLD pedigrees to identify those individuals who present with subtle clinical and metabolic complications. Genetic counseling must take into consideration that the understanding of the natural history of genetic lipodystrophies and the complications is incomplete. Prenatal screening should take into consideration the severity of lipodystrophy, parents' ethical beliefs, and interplay of environmental factors.

References

1. Garg A. Lipodystrophy. In: Goldsmith LA, Katz SI, Gilchrest BA, Paller AS, Leffell DJ, Wolff K, editors. *Fitzpatrick's Dermatology in General Medicine*. 8th ed. New York, NY: McGraw-Hill; 2012 p. 755–764.
2. Simha V, Garg A. Body fat distribution and metabolic derangements in patients with familial partial lipodystrophy associated with mandibuloacral dysplasia. *J Clin Endocrinol Metab* 2002;**87**:776–85.
3. Garg A, Hernandez MD, Sousa AB, Subramanyam L, Martinez de Villarreal L, dos Santos HG, Barboza O. An autosomal recessive syndrome of joint contractures, muscular atrophy, microcytic anemia, and panniculitis-associated lipodystrophy. *J Clin Endocrinol Metab* 2010;**95**:E58–63.
4. Berardinelli W. An undiagnosed endocrinometabolic syndrome: report of 2 cases. *J Clin Endocrinol Metab* 1954;**14**:193–204.
5. Seip M. Lipodystrophy and gigantism with associated endocrine manifestations: a new diencephalic syndrome? *Acta Paediatrica* 1959;**48**:555–74.
6. Garg A. Lipodystrophies. *Am J Med* 2000;**108**:143–52.
7. Garg A. Clinical review#: lipodystrophies: genetic and acquired body fat disorders. *J Clin Endocrinol Metab* 2011;**96**:3313–25.
8. Garg A. Acquired and inherited lipodystrophies. *N Engl J Med* 2004;**350**:1220–34.
9. Pardini VC, Victoria IM, Rocha SM, Andrade DG, Rocha AM, Pieroni FB, Milagres G, Purisch S, Velho G. Leptin levels, beta-cell function, and insulin sensitivity in families with congenital and acquired generalized lipoatrophic diabetes. *J Clin Endocrinol Metab* 1998;**83**:503–8.
10. Magre J, Delepine M, Khallouf E, Gedde-Dahl Jr T, Van Maldergem L, Sobel E, Papp J, Meier M, Megarbane A, Bachy A, Verloes A, d'Abronzo FH, Seemanova E, Assan R, Baudic N, Bourut C, Czernichow P, Huet F, Grigorescu F, de Kerdanet M, Lacombe D, Labrune P, Lanza M, Loret H, Matsuda F, Navarro J, Nivelon-Chevalier A, Polak M, Robert JJ, Tric P, Tubiana-Rufi N, Vigouroux C, Weissenbach J, Savasta S, Maassen JA, Trygstad O, Bogalho P, Freitas P, Medina JL, Bonnicci F, Joffe BI, Loyson G, Panz VR, Raal FJ, O'Rahilly S, Stephenson T, Kahn CR, Lathrop M, Capeau J. Identification of the gene altered in Berardinelli-Seip congenital lipodystrophy on chromosome 11q13. *Nat Genet* 2001;**28**:365–70.
11. Seip M, Trygstad O. Generalized lipodystrophy, congenital and acquired (lipoatrophy). *Acta Paediatrica Supplement* 1996;**413**:2–28.
12. Westvik J. Radiological features in generalized lipodystrophy. *Acta Paediatrica Supplement* 1996;**413**:44–51.
13. Garg A, Fleckenstein JL, Peshock RM, Grundy SM. Peculiar distribution of adipose tissue in patients with congenital generalized lipodystrophy. *J Clin Endocrinol Metab* 1992;**75**:358–61.
14. Anonymous: Case records of the Massachusetts General Hospital. Weekly clinicopathological exercises. Case 1-1975. *N Engl J Med* 1975;**292**:35–41.
15. Chandalia M, Garg A, Vuitch F, Nizzi F. Postmortem findings in congenital generalized lipodystrophy. *J Clin Endocrinol Metab* 1995;**80**: 3077–81.
16. Brunzell JD, Shankle SW, Bethune JE. Congenital generalized lipodystrophy accompanied by cystic angiomatosis. *Ann Intern Med* 1968;**69**:501–16.
17. Fleckenstein JL, Garg A, Bonte FJ, Vuitch MF, Peshock RM. The skeleton in congenital, generalized lipodystrophy: evaluation using whole-body radiographic surveys, magnetic resonance imaging and technetium-99m bone scintigraphy. *Skeletal Radiol* 1992;**21**: 381–6.
18. Van Maldergem L, Magre J, Khallouf TE, Gedde-Dahl Jr T, Delepine M, Trygstad O, Seemanova E, Stephenson T, Albott CS, Bonnici F, Panz VR, Medina JL, Bogalho P, Huet F, Savasta S, Verloes A, Robert JJ, Loret H, De Kerdanet M, Tubiana-Rufi N, Megarbane A, Maassen J, Polak M, Lacombe D, Kahn CR, Silveira EL, D'Abronzo FH, Grigorescu F, Lathrop M, Capeau J, O'Rahilly S. Genotype-phenotype relationships in Berardinelli-Seip congenital lipodystrophy. *J Med Genet* 2002;**39**:722–33.
19. Agarwal AK, Simha V, Oral EA, Moran SA, Gorden P, O'Rahilly S, Zaidi Z, Gurakan F, Arslanian SA, Klar A, Ricker A, White NH, Bindl L, Herbst K, Kennel K, Patel SB, Al-Gazali L, Garg A. Phenotypic and genetic heterogeneity in congenital generalized lipodystrophy. *J Clin Endocrinol Metab* 2003;**88**:4840–7.
20. Garg A, Chandalia M, Vuitch F. Severe islet amyloidosis in congenital generalized lipodystrophy. *Diabetes Care* 1996;**19**:28–31.
21. Oral EA, Simha V, Ruiz E, Andewelt A, Premkumar A, Snell P, Wagner AJ, DePaoli AM, Reitman ML, Taylor SI, Gorden P, Garg A. Leptin-replacement therapy for lipodystrophy. *N Engl J Med* 2002;**346**:570–8.
22. Haque WA, Shimomura I, Matsuzawa Y, Garg A. Serum adiponectin and leptin levels in patients with lipodystrophies. *J Clin Endocrinol Metab* 2002;**87**:2395–8.
23. Antuna-Puente B, Boutet E, Vigouroux C, Lascols O, Slama L, Caron-Debarle M, Khallouf E, Levy-Marchal C, Capeau J, Bastard JP, Magre J. Higher adiponectin levels in patients with Berardinelli-Seip congenital lipodystrophy due to seipin as compared with 1-acylglycerol-3-phosphate-o-acyltransferase-2 deficiency. *J Clin Endocrinol Metab* 2010;**95**:1463–8.
24. Kim CA, Delepine M, Boutet E, El Mourabit H, Le Lay S, Meier M, Nemani M, Bridel E, Leite CC, Bertola DR, Semple RK, O'Rahilly S, Dugail I, Capeau J, Lathrop M, Magre J. Association of a homozygous nonsense caveolin-1 mutation with Berardinelli-Seip congenital lipodystrophy. *J Clin Endocrinol Metab* 2008;**93**: 1129–34.
25. Hayashi YK, Matsuda C, Ogawa M, Goto K, Tominaga K, Mitsuhashi S, Park YE, Nonaka I, Hino-Fukuyo N, Haginoya K,

Sugano H, Nishino I. Human PTRF mutations cause secondary deficiency of caveolins resulting in muscular dystrophy with generalized lipodystrophy. *J Clin Invest* 2009;**119**:2623–33.

26. Simha V, Garg A. Phenotypic heterogeneity in body fat distribution in patients with congenital generalized lipodystrophy due to mutations in the *AGPAT2* or *Seipin* genes. *J Clin Endocrinol Metab* 2003;**88**:5433–7.

27. Agarwal AK, Garg A. Congenital generalized lipodystrophy: significance of triglyceride biosynthetic pathways. *Trends Endocrinol Metab* 2003;**14**:214–21.

28. Takeuchi K, Reue K. Biochemistry, physiology, and genetics of GPAT, AGPAT, and lipin enzymes in triglyceride synthesis. *Am J Physiol Endocrinol Metab* 2009;**296**:E1195–209.

29. Agarwal AK, Garg A. Enzymatic activity of the human 1-acylglycerol-3-phosphate-O-acyltransferase isoform 11: upregulated in breast and cervical cancers. *J Lipid Res* 2010;**51**:2143–52.

30. Szymanski KM, Binns D, Bartz R, Grishin NV, Li WP, Agarwal AK, Garg A, Anderson RG, Goodman JM. The lipodystrophy protein seipin is found at endoplasmic reticulum lipid droplet junctions and is important for droplet morphology. *Proc Natl Acad Sci USA* 2007;**104**:20890–5.

31. Jiang M, Gao M, Wu C, He H, Guo X, Zhou Z, Yang H, Xiao X, Liu G, Sha J. Lack of testicular seipin causes teratozoospermia syndrome in men. *Proc Natl Acad Sci USA* 2014;**111**:7054–9.

32. Agarwal AK, Garg A. Seipin: a mysterious protein. *Trends Mol Med* 2004;**10**:440–4.

33. Fei W, Shui G, Gaeta B, Du X, Kuerschner L, Li P, Brown AJ, Wenk MR, Parton RG, Yang H. Fld1p, a functional homologue of human seipin, regulates the size of lipid droplets in yeast. *J Cell Biol* 2008;**180**:473–82.

34. Sim MF, Dennis RJ, Aubry EM, Ramanathan N, Sembongi H, Saudek V, Ito D, O'Rahilly S, Siniossoglou S, Rochford JJ. The human lipodystrophy protein seipin is an ER membrane adaptor for the adipogenic PA phosphatase lipin 1. *Mol Metab* 2012;**2**:38–46.

35. Stan RV. Structure of caveolae. *Biochim Biophys Acta* 2005;**1746**:334–48.

36. Parton RG, Simons K. The multiple faces of caveolae. *Nat Rev Mol Cell Biol* 2007;**8**:185–94.

37. Razani B, Lisanti MP. Caveolin-deficient mice: insights into caveolar function human disease. *J Clin Invest* 2001;**108**:1553–61.

38. Shastry S, Delgado MR, Dirik E, Turkmen M, Agarwal AK, Garg A. Congenital generalized lipodystrophy, type 4 (CGL4) associated with myopathy due to novel PTRF mutations. *Am J Med Genet A* 2010;**152A**:2245–53.

39. Rajab A, Straub V, McCann LJ, Seelow D, Varon R, Barresi R, Schulze A, Lucke B, Lutzkendorf S, Karbasiyan M, Bachmann S, Spuler S, Schuelke M. Fatal cardiac arrhythmia and long-QT syndrome in a new form of congenital generalized lipodystrophy with muscle rippling (CGL4) due to PTRF-CAVIN mutations. *PLoS Genet* 2010;**6**:e1000874.

40. Simha V, Agarwal AK, Aronin PA, Iannaccone ST, Garg A. Novel subtype of congenital generalized lipodystrophy associated with muscular weakness and cervical spine instability. *Am J Med Genet A* 2008;**146A**:2318–26.

41. Rajab A, Heathcote K, Joshi S, Jeffery S, Patton M. Heterogeneity for congenital generalized lipodystrophy in seventeen patients from Oman. *Am J Hum Genet* 2002;**110**:219–25.

42. Novelli G, Muchir A, Sangiuolo F, Helbling-Leclerc A, D'Apice MR, Massart C, Capon F, Sbraccia P, Federici M, Lauro R, Tudisco C, Pallotta R, Scarano G, Dallapiccola B, Merlini L, Bonne G. Mandibuloacral dysplasia is caused by a mutation in LMNA-encoding lamin A/C. *Am J Hum Genet* 2002;**71**:426–31.

43. Fisher DZ, Chaudhary N, Blobel G. cDNA sequencing of nuclear lamins A and C reveals primary and secondary structural homology to intermediate filament proteins. *Proc Natl Acad Sci USA* 1986;**83**:6450–4.

44. Burke B, Stewart CL. Life at the edge: the nuclear envelope and human disease. *Nat Rev Mol Cell Biol* 2002;**3**:575–85.

45. Shen JJ, Brown CA, Lupski JR, Potocki L. Mandibuloacral dysplasia caused by homozygosity for the R527H mutation in lamin A/C. *J Med Genet* 2003;**40**:854–7.

46. Plasilova M, Chattopadhyay C, Pal P, Schaub NA, Buechner SA, Mueller H, Miny P, Ghosh A, Heinimann K. Homozygous missense mutation in the lamin A/C gene causes autosomal recessive Hutchinson-Gilford progeria syndrome. *J Med Genet* 2004;**41**:609–14.

47. Agarwal AK, Kazachkova I, Ten S, Garg A. Severe mandibuloacral dysplasia-associated lipodystrophy and progeria in a young girl with a novel homozygous Arg527Cys LMNA mutation. *J Clin Endocrinol Metab* 2008;**93**:4617–23.

48. Agarwal AK, Fryns JP, Auchus RJ, Garg A. Zinc metalloproteinase, ZMPSTE24, is mutated in mandibuloacral dysplasia. *Hum Mol Genet* 2003;**12**:1995–2001.

49. Agarwal AK, Zhou XJ, Hall RK, Nicholls K, Bankier A, Van Esch H, Fryns J-P, Garg A. Focal segmental glomerulosclerosis in patients with mandibuloacral dysplasia due to zinc metalloproteinase deficiency. *J Investig Med* 2006;**54**:208–13.

50. Ahmad Z, Zackai E, Medne L, Garg A. Early onset mandibuloacral dysplasia due to compound heterozygous mutations in ZMP-STE24. *Am J Med Genet A* 2010;**152A**:2703–10.

51. Shastry S, Simha V, Godbole K, Sbraccia P, Melancon S, Yajnik CS, Novelli G, Kroiss M, Garg A. A novel syndrome of mandibular hypoplasia, deafness, and progeroid features associated with lipodystrophy, undescended testes, and male hypogonadism. *J Clin Endocrinol Metab* 2010;**95**:E192–7.

52. Weedon MN, Ellard S, Prindle MJ, Caswell R, Allen HL, Oram R, Godbole K, Yajnik CS, Sbraccia P, Novelli G, Turnpenny P, McCann E, Goh KJ, Wang Y, Fulford J, McCulloch LJ, Savage DB, O'Rahilly S, Kos K, Loeb LA, Semple RK, Hattersley AT. An in-frame deletion at the polymerase active site of POLD1 causes a multisystem disorder with lipodystrophy. *Nat Genet* 2013;**45**:947–50.

53. Horikoshi A, Iwabuchi S, Iizuka Y, Hagiwara T, Amaki I. A case of partial lipodystrophy with erythema, dactylic deformities, calcification of the basal ganglia, immunological disorders and low IQ level (author's transl). *Clin Neurol* 1980;**20**:173–80.

54. Tanaka M, Miyatani N, Yamada S, Miyashita K, Toyoshima I, Sakuma K, Tanaka K, Yuasa T, Miyatake T, Tsubaki T. Hereditary lipo-muscular atrophy with joint contracture, skin eruptions and hyper-gamma-globulinemia: a new syndrome. *Intern Med* 1993;**32**:42–5.

55. Rivett AJ, Hearn AR. Proteasome function in antigen presentation: immunoproteasome complexes, peptide production, and interactions with viral proteins. *Curr Protein Pept Sci* 2004;**5**:153–61.

56. Arima K, Kinoshita A, Mishima H, Kanazawa N, Kaneko T, Mizushima T, Ichinose K, Nakamura H, Tsujino A, Kawakami A, Matsunaka M, Kasagi S, Kawano S, Kumagai S, Ohmura K, Mimori T, Hirano M, Ueno S, Tanaka K, Tanaka M, Toyoshima I, Sugino H, Yamakawa A, Niikawa N, Furukawa F, Murata S, Eguchi K, Ida H, Yoshiura K. Proteasome assembly defect due to a proteasome subunit beta type 8 (PSMB8) mutation causes the autoinflammatory disorder, Nakajo-Nishimura syndrome. *Proc Natl Acad Sci USA* 2011;**108**:14914–9.

57. Kitamura A, Maekawa Y, Uehara H, Izumi K, Kawachi I, Nishizawa M, Toyoshima Y, Takahashi H, Standley DM, Tanaka K, Hamazaki J, Murata S, Obara K, Toyoshima I, Yasutomo K. A mutation in the immunoproteasome subunit PSMB8 causes autoinflammation and lipodystrophy in humans. *J Clin Invest* 2011;**121**:4150–60.

58. Liu Y, Ramot Y, Torrelo A, Paller AS, Si N, Babay S, Kim PW, Sheikh A, Lee CC, Chen Y, Vera A, Zhang X, Goldbach-Mansky R, Zlotogorski A. Mutations in PSMB8 cause CANDLE syndrome with evidence of genetic and phenotypic heterogeneity. *Arthritis Rheum* 2011;**64**:895–907.

59. Agarwal AK, Xing C, DeMartino GN, Mizrachi D, Hernandez MD, Sousa AB, Martinez de Villarreal L, dos Santos HG, Garg A. PSMB8 encoding the beta5i proteasome subunit is mutated in joint contractures, muscle atrophy, microcytic anemia, and panniculitis-induced lipodystrophy syndrome. *Am J Hum Genet* 2010;**87**:866–72.

60. Sensenbrenner JA, Hussels IE, Levin LS. CC – a low birthweight syndrome, Rieger syndrome. *Birth Defects* 1975;**11**:423–6.

61. Gorlin RJ, Cervenka J, Moller K, Horrobin M, Witkop J. Rieger anomaly and growth retardation (the S-H-O-R-T syndrome). *Birth Defects* 1975;**11**:46–8.

62. Sorge G, Ruggieri M, Polizzi A, Scuderi A, Di Pietro M. SHORT syndrome: a new case with probable autosomal dominant inheritance. *Am J Med Genet* 1996;**61**:178–81.

63. Bankier A, Keith CG, Temple IK. Absent iris stroma, narrow body build and small facial bones: a new association or variant of SHORT syndrome? *Clin Dysmorphol* 1995;**4**:304–12.

64. Aarskog D, Ose L, Pande H, Eide N. Autosomal dominant partial lipodystrophy associated with Rieger anomaly, short stature, and insulinopenic diabetes. *Am J Med Gen* 1983;**15**:29–38.

65. Dyment DA, Smith AC, Alcantara D, Schwartzentruber JA, Basel-Vanagaite L, Curry CJ, Temple IK, Reardon W, Mansour S, Haq MR, Gilbert R, Lehmann OJ, Vanstone MR, Beaulieu CL, Consortium FC, Majewski J, Bulman DE, O'Driscoll M, Boycott KM, Innes AM. Mutations in PIK3R1 cause SHORT syndrome. *Am J Hum Genet* 2013;**93**:158–66.

66. Chudasama KK, Winnay J, Johansson S, Claudi T, Konig R, Haldorsen I, Johansson B, Woo JR, Aarskog D, Sagen JV, Kahn CR, Molven A, Njolstad PR. SHORT syndrome with partial lipodystrophy due to impaired phosphatidylinositol 3 kinase signaling. *Am J Hum Genet* 2013;**93**:150–7.

67. Thauvin-Robinet C, Auclair M, Duplomb L, Caron-Debarle M, Avila M, St-Onge J, Le Merrer M, Le Luyer B, Heron D, Mathieu-Dramard M, Bitoun P, Petit JM, Odent S, Amiel J, Picot D, Carmignac V, Thevenon J, Callier P, Laville M, Reznik Y, Fagour C, Nunes ML, Capeau J, Lascols O, Huet F, Faivre L, Vigouroux C, Riviere JB. PIK3R1 mutations cause syndromic insulin resistance with lipoatrophy. *Am J Hum Genet* 2013;**93**:141–9.

68. Rautenstrauch T, Snigula F, Krieg T, Gay S, Muller PK. Progeria: a cell culture study and clinical report of familial incidence. *Eur J Pediatr* 1971;**124**:101–11.

69. Wiedemann HR. An unidentified neonatal progeroid syndrome: follow-up report. *Eur J Pediatr* 1979;**130**:65–70.

70. Pivnick EK, Angle B, Kaufman RA, Hall BD, Pitukcheewanont P, Hersh JH, Fowlkes JL, Sanders LP, O'Brien JM, Carroll GS, Gunther WM, Morrow HG, Burghen GA, Ward JC. Neonatal progeroid (Wiedemann-Rautenstrauch) syndrome: report of five new cases and review. *Am J Med Genet* 2000;**90**:131–40.

71. O'Neill B, Simha V, Kotha V, Garg A. Body fat distribution and metabolic variables in patients with neonatal progeroid syndrome. *Am J Med Genet A* 2007;**143**:1421–30.

72. Korniszewski L, Nowak R, Okninska-Hoffmann E, Skorka A, Gieruszczak-Bialek D, Sawadro-Rochowska M. Wiedemann-Rautenstrauch (neonatal progeroid) syndrome: new case with normal telomere length in skin fibroblasts. *Am J Med Genet* 2001;**103**:144–8.

73. Hoppen T, Naumann A, Theile U, Rister M. Siblings with neonatal progeroid syndrome (Wiedemann-Rautenstrauch). *Klin Padiatr* 2004;**216**:70–1.

74. Thorey F, Jager M, Seller K, Krauspe R, Wild A. Kyphoscoliosis in Wiedemann-Rautenstrauch-syndrome (neonatal progeroid syndrome). *Z Orthop Ihre Grenzgeb* 2003;**141**:341–4.

75. Graul-Neumann LM, Kienitz T, Robinson PN, Baasanjav S, Karow B, Gillessen-Kaesbach G, Fahsold R, Schmidt H, Hoffmann K, Passarge E. Marfan syndrome with neonatal progeroid syndrome-like lipodystrophy associated with a novel frameshift mutation at the 3' terminus of the FBN1-gene. *Am J Med Genet A* 2010;**152A**:2749–55.

76. Goldblatt J, Hyatt J, Edwards C, Walpole I. Further evidence for a marfanoid syndrome with neonatal progeroid features and severe generalized lipodystrophy due to frameshift mutations near the 3' end of the FBN1 gene. *Am J Med Genet A* 2011;**155A**:717–20.

77. Rubio-Cabezas O, Puri V, Murano I, Saudek V, Semple RK, Dash S, Hyden CS, Bottomley W, Vigouroux C, Magre J, Raymond-Barker P, Murgatroyd PR, Chawla A, Skepper JN, Chatterjee VK, Suliman S, Patch AM, Agarwal AK, Garg A, Barroso I, Cinti S, Czech MP, Argente J, O'Rahilly S, Savage DB. Partial lipodystrophy and insulin resistant diabetes in a patient with a homozygous nonsense mutation in CIDEC. *EMBO Mol Med* 2009;**1**:280–7.

78. Nishino N, Tamori Y, Tateya S, Kawaguchi T, Shibakusa T, Mizunoya W, Inoue K, Kitazawa R, Kitazawa S, Matsuki Y, Hiramatsu R, Masubuchi S, Omachi A, Kimura K, Saito M, Amo T, Ohta S, Yamaguchi T, Osumi T, Cheng J, Fujimoto T, Nakao H, Nakao K, Aiba A, Okamura H, Fushiki T, Kasuga M. FSP27 contributes to efficient energy storage in murine white adipocytes by promoting the formation of unilocular lipid droplets. *J Clin Invest* 2008;**118**:2808–21.

79. Dunnigan MG, Cochrane MA, Kelly A, Scott JW. Familial lipoatrophic diabetes with dominant transmission. A new syndrome. *Q J Med* 1974;**43**:33–48.

80. Kobberling J, Willms B, Kattermann R, Creutzfeldt W. Lipodystrophy of the extremities. A dominantly inherited syndrome associated with lipatrophic diabetes. *Humangenetik* 1975;**29**:111–20.

81. Kobberling J, Dunnigan MG. Familial partial lipodystrophy: two types of an X linked dominant syndrome, lethal in the hemizygous state. *J Med Genet* 1986;**23**:120–7.

82. Kobberling J, Schwarck H, Cremer P, Fiechtl J, Seidel D, Creutzfeldt W. Partielle Lipodystrophie mit lipatrophischem Diabetes und Hyperlipoproteinamie. *Verhandlungen der deutschen Gesellschaft fur Innere Medizin* 1981;**87**:958–61.

83. Ozer FL, Lichtenstein JR, Kwiterovich PO, McKusick VA. A new genetic variety of lipodystrophy. *Clin Res* 1973;**21**:533 (abstract).

84. Peters JM, Barnes R, Bennett L, Gitomer WM, Bowcock AM, Garg A. Localization of the gene for familial partial lipodystrophy (Dunnigan variety) to chromosome 1q21-22. *Nat Genet* 1998;**18**:292–5.

85. Garg A. Gender differences in the prevalence of metabolic complications in familial partial lipodystrophy (Dunnigan variety). *J Clin Endocrinol Metab* 2000;**85**:1776–82.

86. Garg A, Peshock RM, Fleckenstein JL. Adipose tissue distribution in patients with familial partial lipodystrophy (Dunnigan variety). *J Clin Endocrinol Metab* 1999;**84**:170–4.

87. Hegele RA. Familial partial lipodystrophy: a monogenic form of the insulin resistance syndrome. *Mol Genet Metab* 2000;**71**:539–44.

88. Haque WA, Oral EA, Dietz K, Bowcock AM, Agarwal AK, Garg A. Risk factors for diabetes in familial partial lipodystrophy, Dunnigan variety. *Diab Care* 2003;**26**:1350–5.

89. Haque WA, Vuitch F, Garg A. Post-mortem findings in familial partial lipodystrophy, Dunnigan variety. *Diab Med* 2002;**19**:1022–5.

90. Garg A, Speckman RA, Bowcock AM. Multisystem dystrophy syndrome due to novel missense mutations in the amino-terminal head and alpha-helical rod domains of the lamin A/C gene. *Am J Med* 2002;**112**:549–55.

91. Subramanyam L, Simha V, Garg A. Overlapping syndrome with familial partial lipodystrophy, Dunnigan variety and cardiomyopathy due to amino-terminal heterozygous missense lamin A/C mutations. *Clin Genet* 2010;**78**:66–73.

92. Cao H, Hegele RA. Nuclear lamin A/C R482Q mutation in Canadian kindreds with Dunnigan-type familial partial lipodystrophy. *Hum Mol Genet* 2000;**9**:109–12.

93. Shackleton S, Lloyd DJ, Jackson SN, Evans R, Niermeijer MF, Singh BM, Schmidt H, Brabant G, Kumar S, Durrington PN, Gregory S, O'Rahilly S, Trembath RC. *LMNA*, encoding lamin A/C, is mutated in partial lipodystrophy. *Nat Genet* 2000;**24**:153–6.

94. Speckman RA, Garg A, Du F, Bennett L, Veile R, Arioglu E, Taylor SI, Lovett M, Bowcock AM. Mutational and haplotype analyses of families with familial partial lipodystrophy (Dunnigan variety) reveal recurrent missense mutations in the globular C-terminal domain of lamin A/C. *Am J Hum Genet* 2000;**66**:1192–8.

95. Jacob KN, Garg A. Laminopathies: multisystem dystrophy syndromes. *Mol Genet Metab* 2006;**87**:289–302.

96. Garg A, Vinaitheerthan M, Weatherall P, Bowcock A. Phenotypic heterogeneity in patients with familial partial lipodystrophy (Dunnigan variety) related to the site of mis-sense mutations in Lamin A/C (*LMNA*) gene. *J Clin Endocrinol Metab* 2001;**86**: 59–65.

97. Agarwal AK, Garg A. A novel heterozygous mutation in peroxisome proliferator-activated receptor-γ gene in a patient with familial partial lipodystrophy. *J Clin Endocrinol Metab* 2002;**87**: 408–11.

98. Semple RK, Chatterjee VK, O'Rahilly S. PPAR gamma and human metabolic disease. *J Clin Invest* 2006;**116**:581–9.

99. Gandotra S, Le Dour C, Bottomley W, Cervera P, Giral P, Reznik Y, Charpentier G, Auclair M, Delepine M, Barroso I, Semple RK, Lathrop M, Lascols O, Capeau J, O'Rahilly S, Magre J, Savage DB, Vigouroux C. Perilipin deficiency and autosomal dominant partial lipodystrophy. *N Engl J Med* 2011;**364**:740–8.

100. Olofsson SO, Bostrom P, Andersson L, Rutberg M, Levin M, Perman J, Boren J. Triglyceride containing lipid droplets and lipid droplet-associated proteins. *Curr Opin Lipidol* 2008;**19**: 441–7.

101. Gandotra S, Lim K, Girousse A, Saudek V, O'Rahilly S, Savage DB. Human frame shift mutations affecting the carboxyl terminus of perilipin increase lipolysis by failing to sequester the adipose triglyceride lipase (ATGL) coactivator AB-hydrolase-containing 5 (ABHD5). *J Biol Chem* 2011;**286**:34998–5006.

102. George S, Rochford JJ, Wolfrum C, Gray SL, Schinner S, Wilson JC, Soos MA, Murgatroyd PR, Williams RM, Acerini CL, Dunger DB, Barford D, Umpleby AM, Wareham NJ, Davies HA, Schafer AJ, Stoffel M, O'Rahilly S, Barroso I. A family with severe insulin resistance and diabetes due to a mutation in AKT2. *Science* 2004;**304**:1325–8.

103. Garofalo RS, Orena SJ, Rafidi K, Torchia AJ, Stock JL, Hildebrandt AL, Coskran T, Black SC, Brees DJ, Wicks JR, McNeish JD, Coleman KG. Severe diabetes, age-dependent loss of adipose tissue, and mild growth deficiency in mice lacking Akt2/PKB beta. *J Clin Invest* 2003;**112**:197–208.

104. Herbst KL, Tannock LR, Deeb SS, Purnell JQ, Brunzell JD, Chait A. Kobberling type of familial partial lipodystrophy: an underrecognized syndrome. *Diab Care* 2003;**26**:1819–24.

105. Garg A, Subramanyam L, Agarwal AK, Simha V, Levine B, D'Apice MR, Novelli G, Crow Y. Atypical progeroid syndrome due to heterozygous missense LMNA mutations. *J Clin Endocrinol Metab* 2009;**94**:4971–83.

106. Pollex RL, Hegele RA. Hutchinson-Gilford progeria syndrome. *Clin Genet* 2004;**66**:375–81.

107. Hennekam RC. Hutchinson-Gilford progeria syndrome: review of the phenotype. *Am J Med Genet A* 2006;**140**:2603–24.

108. Merideth MA, Gordon LB, Clauss S, Sachdev V, Smith AC, Perry MB, Brewer CC, Zalewski C, Kim HJ, Solomon B, Brooks BP, Gerber LH, Turner ML, Domingo DL, Hart TC, Graf J, Reynolds JC, Gropman A, Yanovski JA, Gerhard-Herman M, Collins FS, Nabel EG, Cannon 3rd RO, Gahl WA, Introne WJ. Phenotype and course of Hutchinson-Gilford progeria syndrome. *N Engl J Med* 2008;**358**:592–604.

109. Eriksson M, Brown WT, Gordon LB, Glynn MW, Singer J, Scott L, Erdos MR, Robbins CM, Moses TY, Berglund P, Dutra A, Pak E, Durkin S, Csoka AB, Boehnke M, Glover TW, Collins FS. Recurrent *de novo* point mutations in lamin A cause Hutchinson-Gilford progeria syndrome. *Nature* 2003;**423**:293–8.

110. D'Apice MR, Tenconi R, Mammi I, van den Ende J, Novelli G. Paternal origin of LMNA mutations in Hutchinson-Gilford progeria. *Clin Genet* 2004;**65**:52–4.

111. De Sandre-Giovannoli A, Bernard R, Cau P, Navarro C, Amiel J, Boccaccio I, Lyonnet S, Stewart CL, Munnich A, Le Merrer M, Levy N. Lamin a truncation in Hutchinson-Gilford progeria. *Science* 2003;**300**:2055.

112. Dyment DA, Gibson WT, Huang L, Bassyouni H, Hegele RA, Innes AM. Biallelic mutations at PPARG cause a congenital, generalized lipodystrophy similar to the Berardinelli-Seip syndrome. *Eur J Med Gen* 2014;**57**:524–6.

113. Farhan SM, Robinson JF, McIntyre AD, Marrosu MG, Ticca AF, Loddo S, Carboni N, Brancati F, Hegele RA. A novel LIPE nonsense mutation found using exome sequencing in siblings with late-onset familial partial lipodystrophy. *Can J Cardiol* 2014;**30**:1649–54.

114. Albert JS, Yerges-Armstrong LM, Horenstein RB, Pollin TI, Sreenivasan UT, Chai S, Blaner WS, Snitker S, O'Connell JR, Gong DW, Breyer 3rd RJ, Ryan AS, McLenithan JC, Shuldiner AR, Sztalryd C, Damcott CM. Null mutation in hormone-sensitive lipase gene and risk of type 2 diabetes. *N Engl J Med* 2014;**370**:2307–15.

115. Donadille B, D'Anella P, Auclair M, Uhrhammer N, Sorel M, Grigorescu R, Ouzounian S, Cambonie G, Boulot P, Laforet P, Carbonne B, Christin-Maitre S, Bignon YJ, Vigouroux C. Partial lipodystrophy with severe insulin resistance and adult progeria Werner syndrome. *Orphanet J Rare Dis* 2013;**8**:106.

116. Payne F, Lim K, Girousse A, Brown RJ, Kory N, Robbins A, Xue Y, Sleigh A, Cochran E, Adams C, Dev Borman A, Russel-Jones D, Gorden P, Semple RK, Saudek V, O'Rahilly S, Walther TC, Barroso I, Savage DB. Mutations disrupting the Kennedy phosphatidylcholine pathway in humans with congenital lipodystrophy and fatty liver disease. *Proc Natl Acad Sci USA* 2014;**111**:8901–6.

117. Garg A, Kircher M, Del Campo M, Amato RS, Agarwal AK. University of Washington Center for Mendelian Genomics. Whole exome sequencing identifies de novo heterozygous CAV1 mutations associated with a novel neonatal onset lipodystrophy syndrome. *Am J Med Genet A* 2015;**167**(8):1796–806.

118. Masotti A, Uva P, Davis-Keppen L, Basel-Vanagaite L, Cohen L, Pisaneschi E, Celluzzi A, Bencivenga P, Fang M, Tian M, Xu X, Cappa M, Dallapiccola B. Keppen-Lubinsky syndrome is caused by mutations in the inwardly rectifying K+ channel encoded by KCNJ6. *Am J Hum Genet* 2015;**96**:295–300.

PART IX

MULTISYSTEM DISORDERS

24

Multiple Endocrine Neoplasia Type 1 (MEN1)

Cornelis J. Lips*, Gerlof D. Valk*, Koen M. Dreijerink*,
Marc Timmers†, Rob B. van der Luijt**, Thera P. Links‡,
Bernadette P.M. van Nesselrooij**, Menno Vriens§,
Jo W. Höppener¶, Inne Borel Rinkes§,
Anouk N.A. van der Horst-Schrivers‡

*Department of Clinical Endocrinology, University Medical Center, Utrecht, the Netherlands
**Department of Medical Genetics, University Medical Center, Utrecht, the Netherlands
†Department of Molecular Cancer Research, University Medical Center, Utrecht, the Netherlands
‡Department of Clinical Endocrinology, University Medical Center, Groningen, the Netherlands
§Department of Surgery, University Medical Center, Utrecht, the Netherlands
¶Department of Molecular Cancer Research, Laboratory of Translational Immunology, University
Medical Center Utrecht, Utrecht, the Netherlands

INTRODUCTION

Background, Prevalence

Multiple endocrine neoplasia type 1 (MEN1) is an autosomal dominantly inherited syndrome. MEN1 is characterized by the occurrence of tumors of the parathyroid glands, the pancreatic islets, the anterior pituitary gland, and the adrenal glands, as well as neuroendocrine tumors (NETs) in the thymus, lungs, and stomach, often at a young age. Nonendocrine manifestations of MEN1 include angiofibromas, collagenomas, lipomas, leiomyomas, and meningiomas (Table 24.1).

The prevalence of MEN1 is 2–3 per 100,000, and is equal among males and females.

MEN1 and multiple endocrine neoplasia type 2 (MEN2) are two distinct syndromes. In MEN2, patients frequently develop medullary thyroid carcinoma and adrenal medullary tumors (pheochromocytoma).

Natural History

From family studies in the past, it appeared that if no treatment is given, life expectancy is considerably shortened.

In MEN1, the median life expectancy for patients with a malignant islet cell tumor is 46 years, for peptic ulcer disease 56 years, for malignant carcinoid 53 years, for hypercalcemia/uremia 42 years, and the overall median age at death for MEN1 is 47 years.[1–3]

MEN1 is caused by germline mutations of the *MEN1* gene. Since the discovery of the gene in 1997, DNA diagnosis has become available.[4,5] Carriers of a *MEN1* gene germline mutation can be monitored periodically by targeted clinical examination to identify MEN1-associated lesions at a presymptomatic stage.

In this review, the recent developments concerning the etiology of MEN1 as well as the current diagnostic and therapeutic options are presented. Furthermore, guidelines for *MEN1* gene mutation analysis and periodic clinical monitoring are provided.

Clinical Presentation, Diagnosis, and Treatment

According to the MEN consensus published in 2001 and updated in 2012, a practical definition of a MEN1 patient is a patient with at least two of the three main MEN1-related endocrine tumors (i.e., parathyroid adenomas, enteropancreatic endocrine tumors, and a

Genetic Diagnosis of Endocrine Disorders. http://dx.doi.org/10.1016/B978-0-12-800892-8.00024-5

TABLE 24.1 The Variable Expression of MEN1

A. Endocrine lesions	
Parathyroid adenomas	75–95%
Pancreatic islet cell tumors	55%
Gastrinomas	45%
Insulinomas	10%
No clear clinical picture (including PP-, SS-producing tumors)	10–80%
Other (VIP, GHRH, etc.)	2%
Pituitary adenomas	47%
Prolactinomas	30%
Nonfunctioning (i.e., not producing hormone)	10%
ACTH producing	1%
GH producing	3–6%
Adrenal cortical adenomas	20%
NETs	18%
Thymus	1%
Lungs	12%
Stomach	5%
B. Nonendocrine lesions	
Skin lesions	80%
Angiofibromas	75%
Collagenomas	5%
Lipomas	30%
Leiomyomas	5%
Meningiomas	25%

Percentages of MEN1-gene germline mutation carriers that develop a MEN1-associated tumor.
ACTH, adrenocorticotrophic hormone; GH, growth hormone; GHRH, growth hormone releasing horomone; NETs, neuroendocrine tumors; PP, pancreatic polypeptide; SS, somatostatin; VIP, vasoactive intestinal peptide.

pituitary adenoma).[6,7] Familial MEN1 is similarly defined as at least one MEN1 case plus at least one first degree relative with one of those main MEN1 tumors. Alternatively, because parathyroid and pituitary adenomas occur relatively frequently in the general population, a MEN1 patient can be defined more precisely as a patient with three or more of the five major MEN1-associated lesions (i.e., besides the three tumor types mentioned, adrenal gland tumors, and NETs in the thymus, lungs, and stomach). A MEN1-suspected patient is defined as having at least two major MEN1-associated lesions, multiple lesions within one MEN1-related organ, and/or a MEN1-associated lesion at a young age (<35 years).[8] Such a patient may be considered for DNA analysis (see under *MEN1* mutation analysis).

Below, for each tumor type the clinical presentation and the diagnostic and therapeutic options are listed. In Figure 24.1a–c, flow charts are shown for diagnosis and therapy of MEN1-associated tumors of the parathyroid glands, pancreatic islets, and pituitary adenoma.

It may explain the variable expression and the significance of general guidelines.[7,9]

Parathyroid Adenoma

Parathyroid adenomas are often the first manifestation of MEN1. About 75–95% of MEN1 patients develop parathyroid adenomas.[2] Usually, parathyroid adenomas in MEN1 are multiple and benign.

Phenocopies

Other forms of familial hyperparathyroidism (FHPT) or phenocopies of MEN1 are found in familial isolated hyperparathyroidism (FIHP). In some FIHP families, a CASR mutation may be identified.[10] The FHPT jaw tumor syndrome (HPT-JT, HRPT2) is caused by inactivating germline mutations in the parafibromin gene on chromosome 1 (1q31).[11] FHPT in MEN2A is caused by activating germline mutations in the *RET* proto-oncogene on chromosome 10 (10q11.2), and in MEN4, it is caused by inactivating germline mutations in the *CDKN1B* gene encoding the p27 protein.[12] *MEN1* genotyping appears worthwhile in FIHP families, as the finding of *MEN1* gene mutation(s) in this gene predicts possible involvement of other organs.[13]

The increased production of parathyroid hormone causes hypercalcemia. Symptoms and signs of hypercalcemia include fatigue, depression, constipation, nausea, symptoms caused by nephrolithiasis or nephrocalcinosis, bone pain, myalgia, and arthralgia as well as hypertension (see Fig. 24.1a).

Diagnosis

Laboratory investigation consists of measurement of (ionized) calcium, chloride, phosphate, and parathyroid hormone in blood. To distinguish primary hyperparathyroidism from hypocalciuric hypercalcemia, in addition to this, the 24-h calcium excretion in the urine is measured. Bone densitometry can be used to detect bone mass reduction.

Preoperative localization is useful, because adenomas may be situated in or behind the thyroid lobes.

Parathyroid adenomas can be effectively localized by a nuclear scan with Tc-99m sestamibi, which is retained selectively by parathyroid adenoma. To confirm the location, ultrasound (US) and/or computed tomography (CT) can be performed (see Fig. 24.1a).[14] Preoperative imaging in PHP in MEN1 is strictly not indicated as it will not alter the operative strategy (i.e., conventional neck exploration). One might consider Tc-99m sestamibi scintigraphy to exclude ectopic adenomas, but this is by no means standard practice.

Subtotal parathyroidectomy with bilateral transcervical thymectomy is the procedure of choice for MEN1-related hyperparathyroidism.[15] Total parathyroidectomy has lowest risk of persistent and recurrent pHPT. However, total parathyroidectomy had the highest risk of permanent hypoparathyroidism; a less than subtotal parathyroidectomy has an unacceptable high rate of recurrent and persistent pHPT.[16] Performing intraoperative confirmation by histologic examination and rapid parathyroid hormone assay with autologous graft of parathyroid tissue have the lowest recurrence rate.[17]

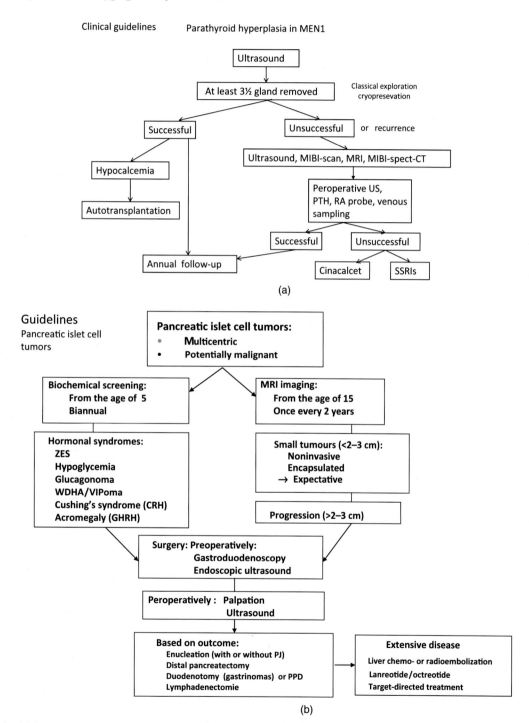

FIGURE 24.1 (a) Hyperparathyroidism results in increase of calcium levels; stimulation of G-cells overproduction of gastrin. In MEN1 patients, stimulation of gastrin production promotes cell division and development of gastrinoma and ECL-oma. Proton pump inhibitors increase gastrin levels. (b) Diagnosis and treatment of pancreatic islet cell tumors. PJ, pancreatic jejunostomy PPD, pylorus preserving pancreaticoduodenectomy. (c) Diagnosis and treatment of pituitary adenomas. NFPA, nonfunctioning pit adenoma.

(Continued)

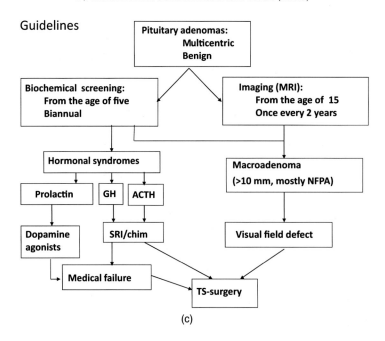

FIGURE 24.1 *(cont.)*

In most cases, this surgical approach was able to restore normal calcium/parathyroid hormone levels improved bone mineral density and ultimately lead to discontinuation of calcium and calcitriol supplementation.[18] Long-term follow-up is recommended.[19]

Tumors of the Endocrine Duodenum and Pancreatic Islets

Definition Neuroendocrine Tumors

In the past, the classical term "carcinoid" was well established in medical terminology; however, at present, it is not adequate to cover the entire spectrum of neuroendocrine neoplasms.

In 2000, the term NETs was suggested by the World Health Organization (WHO) in the new classification of tumors that differ in their morphological and functional features.[20]

Divergence in gene-expression patterns in the development of tumors from the gastroenteropancreatic (GEP) system (divided in NETs of jejunum and ileum and pancreaticoduodenal endocrine tumors (PETs)) was identified, when they were examined at a molecular level. On the basis of gene expression profiles, neuroendocrine lesions of jejunum and ileum and PETs need to be considered as two distinct entities within the group of NETs.[21]

PETs develop in about 55% of MEN1 patients.[22,23] Multicentric microadenomas are present in 90% of MEN1 patients.[24] In 44% of sporadically occurring PETs, a somatic inactivating mutation of the MEN1 gene was identified.

Symptomatology

Hormonal syndromes often occur late and indicate metastases in 50% of patients with this stage of functioning PETs.[25] Prospective screening with biochemical markers and endoscopic ultrasound (EUS) is therefore recommended. The use of biochemical tumor markers for diagnosing PETs has been recently debated.[26] Prospective endoscopic ultrasonic evaluation reveals that the frequency of nonfunctioning PETs is higher (55%) than previously thought (34%).[27]

Absence of Symptoms

It was demonstrated that high penetrance of NF-PETs occurs in 15–20-year-old MEN1 patients. The high percentage of the patients presenting consensus criteria for surgery for NF-PET alone or NF-PET/insulinoma suggests a potential benefit for the periodic surveillance of these tumors in this age group.[28]

Nonfunctioning pancreatic NETs may occur in asymptomatic children with MEN1 mutations, and screening for such enteropancreatic tumors in MEN1 children should be considered earlier than the age of 20 years, as is currently recommended by the international guidelines.[29]

Small asymptomatic neuroendocrine pancreatic tumors in MEN1 usually seem to grow slowly. Annual tumor incidence rate is low. However, faster growing tumors and patients with rapidly progressive disease can be observed. EUS is a sensitive method to detect these tumors.[30]

Diagnosis

Laboratory investigation includes determination of fasting plasma levels of glucose, insulin, C-peptide, glucagon, gastrin, pancreatic polypeptide, and chromogranin A.

Pancreatic islet cell tumors can be visualized by MRI, somatostatin receptor scintigraphy, Ga-68-DOTATATE scintigraphy CT, and F-DOPA positron emission tomography (PET) (see Fig. 24.1b).

EUS is useful for early detection of PETs and will allow early surgery before metastases have developed.[23,25]

EUS is a more sensitive technique than CT or transabdominal US for the detection and localization of potentially malignant lesions in patients with MEN1.[31]

Selective arterial secretagogue injection test (SASI test) may be useful for localizing functioning pNETs.[32] In case of a functioning pNET, the tumor should first be accurately located using the SASI test before an appropriate surgical method is selected.

Specific Syndromes

Zollinger–Ellison Syndrome (ZES)   In MEN1, gastrinomas are most often located submucosally in the antrum and duodenum.

The elevated levels of ectopic gastrin cause excessive gastric acid production. If untreated, this can lead to the ZES: ulcerations of the digestive tract, diarrhoea, and mucosal hypertrophy. Before treatment with proton pump inhibitors became available, the ZES was a frequent cause of death of MEN1 patients. Gastrinomas are still an important threat to MEN1 patients, because they are often multicentric and are able to metastasize to the lymph nodes and the liver.[2]

The diagnosis is delayed by the widespread use of proton pump inhibitors.

MEN1 patients frequently develop ZES. About 25% of all ZES patients have MEN1. Esophageal reflux symptoms are common resulting in strictures and Barrett's esophagus (BE). The frequency of severe peptic esophageal disease, including the premalignant condition BE, was higher in MEN1 patients with gastrinomas than in patients with sporadic gastrinomas. This higher incidence of severe esophageal disease in MEN1/ZES was due to delay of diagnosis, more frequent and severe esophageal symptoms, more frequent hiatal hernias, more common pyloric scarring, higher basal acid output, and underdiagnosed hyperparathyroidism.[33] There may be a relationship between hyperparathyroidism and the development of ZES and gastric NETs from enterochromaffin-like (ECL) cells (see Fig. 24.2).

Recurrence may occur by ectopic location. These findings suggest the importance of checking for the presence of ectopic gastrinomas in the biliary tree in MEN1 patients undergoing ZES surgery.[34]

Surgery

While small lesions may be treated conservatively, most larger lesions or those refractory to medical treatment should be considered for surgery. As submucosal duodenal lesions are often present, the standard approach involves a pylorus-preserving pancreaticoduodenectomy. In cases where less aggressive surgery is warranted, duodenal exploration should always be part of the procedure.

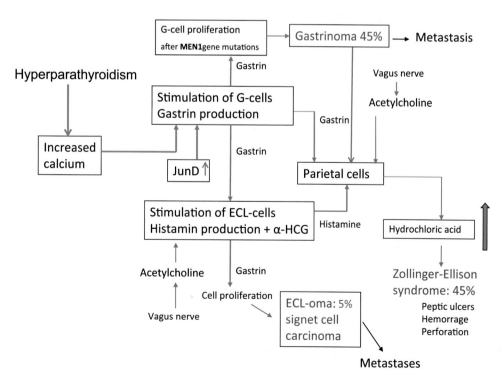

FIGURE 24.2   Effect of increased calcium levels on gastrin-producing cells in the antrum, duodenum, ectopic cells, and ECL cells of the stomach. Calcium receptor in Gastrin-producing G-cells; Gastrin receptor in Parietal and ECL cells.

A suggested procedure includes distal 80% subtotal pancreatic resection together with enucleation of tumors in the head of the pancreas, and in cases with ZES, excision of duodenal gastrinomas together with clearance of regional lymph node metastases. This strategy, with early and aggressive surgery before metastases have developed, is believed to reduce the risks for tumor recurrence and malignant progression.[35] Given the shortage of evidence from controlled clinical trials, the preferable surgical strategy for patients with gastrinoma is unclear. The treatment is controversial. The strategy should be adapted to the characteristics of the individual patient, for example, age, medical history, comorbidity, and preferably discussed within an experienced multidisciplinary team.[36] Improved long-term survival is obtained by curative surgery for patients with MEN1-associated GEPNET. The current surgical indications are expanding even in patients with hepatic metastases because of the improved surgical outcome.

Other Functioning Pancreatic Islet Cell Tumors

Insulinomas occur in about 10% of all MEN1 patients. They may present with symptoms of hypoglycemia, such as confusion or abnormal behavior, due to central nervous system dysfunction at times of exersise or fasting.[37,38]

Pasireotide (SOM230) demonstrates antisecretory, antiproliferative, and proapoptotic activity in a MEN1 knockout mouse model of insulinoma. Further studies of the effects of SOM230 in pNET patients with MEN1 mutations are needed to establish the role of SOM230 in patients.[39]

VIP, Somatstatin, Glucagonoma, and WDHA

Infrequently occurring syndromes are glucagonoma (1.6%), VIPoma (0.98%), and somatostatinoma (0.65%). Glucagonomas can cause skin lesions, whereas tumors producing vasoactive intestinal peptide (VIP), VIPomas, can cause the Verner–Morrison syndrome; also mentioned is the watery-diarrhea-hypokalemia-achlorhydria (WDHA) syndrome.[23] Surgical excision of insulinomas, glucagonomas, and VIPomas is usually curative. Tumors producing pancreatic polypeptide are common (>80% of pNETs), but only rarely cause symptoms and therefore do normally not require treatment.

Growth Hormone-Releasing Hormone (GHRH)-Producing Tumors

Acromegaly may be caused by growth hormone-releasing hormone (GHRH) produced by pancreatic islet cell tumors. GHRH may cause acromegaly indirectly through stimulation of growth hormone (GH) production by the pituitary gland. In more than 50% of MEN1 patients with acromegaly, a GHRH-producing pancreatic tumor is the cause of the disease.[40–42]

In MEN1 patients, acromegaly can also be caused by adenomas in the pituitary gland, primarily producing GH, and NETs in the thymus gland producing GHRH.

Pituitary Adenomas

Evidence of MEN1 is found in approximately 2.7% of all patients with pituitary adenomas. In addition, somatic mutations in the MEN1 gene do not play a prominent role in the pathogenesis of sporadic forms of pituitary adenoma (3.7%).

Pituitary tumors in MEN1 are larger in size and more aggressive than sporadic tumors. MEN1 must be considered in all children with tumors of the pituitary gland.[43,44]

The most frequently occurring pituitary tumors in MEN1 are prolactinomas. Prolactinomas occur in approximately 30% of patients with MEN1 and in this setting, they may be more aggressive than their sporadic counterparts.[45]

A MEN1 variant shows more frequent prolactinoma and less frequent gastrinoma than typical MEN1.[46] Gonadotrophic tumors are occurring infrequently but may cause ovarian hyperstimulation.[47]

Nonfunctioning tumors, GH, or adrenocorticotrophic hormone (ACTH) producing tumors[48] and mixed tumors are seen less frequently.

Pituitary tumors in patients with MEN1 syndrome tend to be larger, invasive, and more symptomatic, and they tend to occur in younger patients when they are the initial presentation of MEN1.[49]

Symptomatology

Elevated levels of prolactin may cause amenorrhoea, galactorrhoea, and lack of libido in females and hypogonadism in males.

Acromegaly, caused by a GH-producing tumor, is observed in 3–6% of MEN1 patients. Patients present with enlarged hands or feet, coarse facial features, or soft-tissue growth. Patients with acromegaly have an increased risk of developing cardiovascular disease, colon polyps, and cancer. Acromegaly may be associated with other tumors.

Nonfunctioning pituitary adenomas may grow large without symptoms. Due to compression of the surrounding tissues by the expanding tumor, complaints of visual field defects, headache, or an impairment of other pituitary functions may develop (Fig. 24.1c).

Diagnosis

The diagnosis is confirmed by determining plasma levels of prolactin (prolactinoma), one mg dexamethasone suppression test, and midnight salivary cortisol (Cushing's disease), or insulin-like growth factor I (IGF-I) and by an oral glucose tolerance test to demonstrate absence of suppression of GH production (acromegaly). Pituitary adenomas can be detected visually by MRI with gadolinium contrast.

Adrenal Tumors

About 20% of MEN1 patients develop adrenal tumors. These tumors are often detected early by imaging of the upper abdomen every two years. Like sporadic incidentalomas of the adrenals, these tumors usually do not produce hormones and are mostly benign.[48,49] Adrenal medulla tumors in MEN1 occur infrequently.[50]

Neuroendocrine Tumors of Thymus, Lungs, and Stomach

In MEN1, NETs arise from cells that are derived from the embryonic foregut. NETs in MEN1 can develop in the thymus (mostly in males), in the lungs, and in the stomach, duodenum, or the pancreas (PETs).

They do not cause symptoms until at an advanced stage. As these tumors are capable of infiltrating surrounding tissues and metastasizing, and treatment is very difficult, early detection of these tumors is of vital interest.[23]

NETs produce a vast spectrum of amines, peptides, and prostaglandins. NETs in MEN1 do not release serotonin (5HT), but do produce 5-hydroxytryptophan (5-HTP), the precursor of serotonin. The 5-HTP is partially converted to serotonin in the kidneys. Levels of platelet serotonin and chromogranin A are useful markers. The level of 5-hydroxyindoleacetic acid in the 24-h urine of MEN1 patients with NETs usually is not elevated.

Imaging

Tumors can be detected using MRI or CT scan, somatostatin receptor scintigraphy, or Ga-68-DOTATATE scintigraphy and endoscopy.

Thymus

Thymic NETs in MEN1 are associated with a very high lethality. Nearly all thymus carcinoid patients are male and smokers. Therefore, prophylactic thymectomy should be considered at neck surgery for primary hyperparathyroidism in male MEN1 patients, especially for smokers.

In 22 separate MEN1 families with thymic carcinoids, all but two (91%) have mutations coding for a truncated menin protein. There is clearly a high prevalence of truncating mutations in MEN1-related thymic carcinoids. Although when compared with the prevalence of truncating mutations among all reported MEN1 gene mutations, it is not significantly higher in MEN1 families with thymic NETs ($P = 0.39$).[51] In Japan, female patients are not rare exceptions.[52]

Screening every patient affected with a neuroendocrine thymic neoplasm for MEN1 syndrome is recommended. In some families, an association with smoking occurs in approximately 25%.[53] Khoo found a thymic NET in a father and son with MEN1.[54]

Cushing's syndrome due to ACTH-producing thymic carcinoid should also be considered as one phenotype of the MEN1 spectrum.[55]

In a 761-patient MEN1 cohort from the Groupe des Tumeurs Endocrines registry, the probability of occurrence was 2.6% (range 1.3–5.5%) at age 40 years; seven patients (33%) belonged to clustered MEN1 families. Several aims, objectives were suggested for future guidelines: (1) earlier detection of Th-NET in MEN1 patients is required; (2) screening of both sexes is necessary; (3) a prospective study comparing MRI versus CT scan in yearly screening for Th-NET is needed; (4) a reinforced screening program must be established for patients who belong to clustered families; and (5) thymectomies must be performed in specialized centers.[56]

Lung (NETs)

Bronchopulmonary NETs are relatively uncommon, occurring in approximately 3–8% of MEN1 patients. Multiplicity seems to be common in MEN-related lung NET. Very little is known of the natural course and prognosis of pulmonary carcinoids in MEN1. In one study, hypergastrinemia was significantly more common in patients with pulmonary nodules. No deaths or distant metastases occurred among these patients despite long-term follow-up. They did not appear to predict a poor prognosis in the majority of affected patients.[57–60]

Gastric (NETs)

In MEN1 patients, gastric NETs have their origin in enterochromaffin-like (ECL) cells. Longstanding tumors may become symptomatic and demonstrate aggressive growth. Patients may have hypergastrinemia and ZES. With increased long-term medical treatment and life expectancy, these tumors will become an important determinant of survival. They require surgical treatment before they metastasize to the liver.[61] Signet ring cell carcinomas may develop by gradual dedifferentiation from ECL cells[62,63] (see Fig. 24.2).

GENETIC PATHOPHYSIOLOGY OF MEN1

Known Mutations and Specific Phenotypes

MEN1 is caused by inactivating germline mutations of the MEN1 gene, which is located on chromosome 11 (11q13).[4,5] The MEN1 gene is a tumor suppressor gene. Biallelic inactivation of the MEN1 gene is required for the development of a tumor cell. Loss of the wild-type allele (loss of heterozygosity or LOH) is observed frequently in MEN1-associated tumors in MEN1 patients.

Since the discovery of the gene, more than 400 (459) different germline mutations have been identified in MEN1 families. These mutations are found scattered throughout the gene.[64]

Also in sporadic MEN1-associated tumors, mutations of the *MEN1* gene have been identified, which suggests that inactivation of the MEN1 gene contributes to the development of these tumors.

No clear genotype-phenotype correlation has been established. The expression of the disease is variable, even within families. However, some *MEN1* gene mutations seem to be causing FIHP or a variant MEN1 that is characterized by the frequent occurrence of prolactinoma.[46] Thus, additional genetic and/or epigenetic events may play a role in MEN1-associated tumorigenesis.

Pathophysiology of Mutations (How They Cause the Disease)

Normal Function of the Intact MEN1 Gene Product

The *MEN1* gene encodes the menin protein. Menin is expressed ubiquitously, and performs its tasks predominantly in the nucleus (see Fig. 24.3 and Table 24.2).

Recent observations indicate that normal menin functions in the regulation of gene transcription. This function is linked to mmodification of histones, cores of proteins wrapped in DNA wound in a double loop. In this capacity of transcription, intact menin has a dual function and may serve either as a corepressor or as a coactivator of gene expression. Corepressors and coactivators serve as adaptors between nuclear receptors and the general transcription machinery. Interaction of nuclear receptors with coactivators or corepressors takes place through LXXLL motifs (where L = leucine and X = any amino acid) present in the coactivator or LXXXIXXXL motifs (where I = isoleucine) present in the corepressor, respectively. In 2012 Huang et al. determined the crystal structure of human menin.[65]

Menin interacts with the AP1-family transcription factor JunD, changing it from an oncoprotein into a tumor-suppressor protein, putatively by recruitment of histone deacetylase complexes. Recently, the telomerase (*hTERT*) gene was identified as a menin target gene.[66] The ends of chromosomes in a cell, the telomeres, shorten after DNA replication. Eventually, after several cell divisions, the DNA loses its stability and the cell is subjected to apoptosis. Telomerase is an enzyme that maintains the length of the telomeres. Telomerase is not expressed in normal cells, but it is active in stem cells and tumor cells. Menin

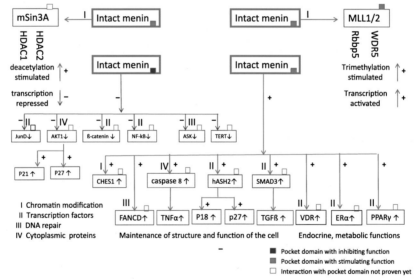

FIGURE 24.3   Normal function of menin. Menin is a dual regulator of gene expression by interconnecting modification of histone proteins with regulation of transcription. The normal function of menin is preservation of differentiation of the cell by modification of histone proteins and transcription of genes responsible for cell division. Arrows (+) indicate stimulation of apoptosis, cell differentiation, DNA repair, and endocrine metabolic functions. Arrows (−) indicate inhibition of cell division. In MEN1 tumors, inactivating mutations in the *MEN1* gene result in alteration of histone protein modifications: both deacetylation (left) and trimethylation (right) are repressed. Consequently, the normal function of menin acting as a corepressor (left) and coactivator (right) of gene transcription is disabled. AKT, serine/threonine protein kinase (also known as PKB); ASK, activator of S-phase kinase; ER, estrogen receptor; FANCD: Fanconi anemia group D2 protein; GSK3, glycogen synthase kinase; hAsh2, human absent, small or homeotic discs; HDAC, histone deacetylase; VDR, vitamin D receptor; hTERT, human telomerase gene.

In normal cells menin inhibits cell division and preserves the differentiated state by acting as:
* corepressor on transcription of other target genes, for example, Jun D and hTERT
* coactivator on transcription of target genes, for example, P18, P27 and pS2

In MEN-1 tumor cells inactivating mutations in the *MEN1* gene result in:
* corepressor function on a.o. JunD/c-Jun or hTERT is defective
* coactivator function on a.o. P18 and P27 is defective

TABLE 24.2 Intact Menin is Involved in Interaction with Distinct Intracellular Proteins (A, B, and C) and Cytoskeletal Proteins

A. Chromatin modification	
Histone deacetylation	Histone protein preservation
Histone trimethylation	Histone protein destabilization
B. Transcription regulation	
Embryonic development	Homeobox genes
Inhibition of cell proliferation	p18, p27
Apoptosis	Telomerase repression (hTERT)
Nuclear receptors for endocrine &	Steroid hormone receptors (ER, PR, GR, AR)
Metabolic functions	Vitamin D receptor (VDR)
	Fatty acids (PPARγ)
C. DNA-damage repair and stability	CHES1, FANCD2, RPA2, ASK
D. Cytoskeleton organization and cytoplasmatic processes	Vimentin Myosin

is a suppressor of the expression of the telomerase gene *hTERT*. Possibly, inactivation of menin could lead to cell immortalization by telomerase expression, which could allow a cell to develop into a tumor cell.

Besides this corepressor function, intact menin suppresses transforming growth factor-β (TGF-β)-mediated signal transduction involved in division of parathyroid hormone- and prolactin-producing cells.

Menin is an integral component of mixed lineage leukemia 1/2 (MLL1/MLL2) histone methyltransferase complexes. In this capacity menin is a coactivator of expression of p18[INK4C] and p27[Kip1] cyclin-dependent-kinase inhibitors. These proteins usually act as tumor suppressor by causing arrest of the cell cycle, that is, inhibition of cell division. Furthermore, menin serves as a coactivator of steroid receptor-mediated transcription, as illustrated by recruiting histone H3K4 methyltransferase activity to the estrogen-responsive *TFF1(pS2)* gene promoter (see Fig. 24.3).[67,68]

Menin links histone modification pathways to transcription-factor function. Evolutionary conservation indicates that menin plays a role in integrating multiple cellular stimuli at critical transcriptional loci during fundamental developmental processes.

Loss of Menin Function and Tumor Growth

Null Mutant Animals

Null mutant animals have indicated that menin is essential for viability. Inactivating mutations in both *MEN1* alleles disturb the tight link between chromatin modification and gene expression and are crucial for MEN1 tumorigenesis.

Familial Aberrant Expression (Burin)

Familial aberrant expression (Burin), a nonsense mutation in the *MEN1* gene, was found to be responsible for the disease in the affected members of the families.

More Malignant Forms or Aggresive Forms with Loss of Menin[69,70]

Genotype-phenotype correlation in 140 clinically affected MEN1 cases showed a tendency for truncating mutations, especially nonsense mutations, to be associated to GEP and carcinoids of the lungs and thymus. In view of the morbidity and frequency in familial cases an effective screening program should aim at an early diagnosis of GEP, particularly when truncating, especially nonsense, mutations are found.[71]

Sporadic MEN1-Related Tumors

Specific gene mutations from sporadic parathyroid adenomas, pituitary adenomas, and pancreatic NETs provide indications for a common origin in the NET pathway and risk for multistep carcinogenesis.

Sporadic Parathyroid Adenoma

A high rate of somatic *MEN1* gene mutations is observed in sporadic parathyroid adenoma. *In vitro* experiments have demonstrated that the presence of intact menin is required for Smad/TGF-β activation to effectively inhibit parathyroid cell proliferation.[72]

Somatic mutations in *CDKN1B* encoding p27 were found in 4.6% of a series of typical, sporadic parathyroid adenomas. Additionally, the identification of germline CDKN1B variants in patients with familial MEN4 provides evidence for CDKN1B as a susceptibility gene in the development of typical parathyroid adenomas. Most identified variants have reduced p27(kip1) protein levels or altered *in vitro* stability.[73] Recent next generation sequencing efforts showed that sporadic parathyroid adenomas harbor relatively few *MEN1* gene mutations. In these studies it was confirmed that *MEN1* gene mutations are the predominant genetic drivers in these tumors.[74,75]

Sporadic Pituitary Adenoma

MEN1 gene mutations are uncommon in sporadic pituitary tumors. Similarly, other tumor suppressor genes involved in familial pituitary adenomas, such the *PRKAR1A* gene responsible for Carney's syndrome, the *AIP* gene responsible for familial isolated pituitary adenomas, or the *CDKN1B* gene mutated in the MEN1-like syndrome MEN4, were found to be mutated infrequently. Epigenetic events, such as hypermethylation and/or microRNA-dependent impairment of protein translation, are likely to be responsible for the downregulation of gene and/or protein expression associated with pituitary tumorigenesis.

Somatic mutations identified in other neoplasias, such as the *BRAF* and *RAS* genes, are rare events in pituitary adenomas. However, the Raf/MEK/ERK and PI3K/AKT/TOR pathways are frequently overexpressed in pituitary tumors.[76]

Activating mutations of *Gsα* (the so-called *gsp* mutations) are the most important somatic mutation in pituitary adenomas, being present in up to 40% of GH-secreting adenomas. By exome sequencing of a discovery set of seven nonfunctioning pituitary adenomas, Newey suggested that there are no common somatic gene mutations, as only few mutations were found and the spectrum was heterogeneous.[77]

Recent studies have identified a crucial role for the high-mobility group A (HMGA) proteins in pituitary tumor development. A correlation was found between *Pit1* and *HMGA* gene overexpression in human pituitary adenomas, further supporting a role of HMGA-mediated *Pit1* overexpression in pituitary tumors.[78]

Activin is a member of the TGF-β superfamily. It represses transcription and expression of the pituitary transcription factor-1 (Pit-1) through the menin Smad pathway (activin inhibits prolactin and pituitary growth through SMAD, activin/TGF-β, and menin).[79,80] Pit-1 is essential for development of lactotrope cells and plays a key role in cell differentiation during organogenesis of the anterior pituitary. Pit-1 is a negative regulator of cell growth and prolactin production in pituitary lactotrope cells.

Vitamin D receptor (VDR) represses transcription of *Pit-1* and it has a repressive effect on *Pit-1* expression.[81]

Sporadic Pancreatic NETs

Somatic nonfunctioning pNETs have in 44% an inactivation of the *MEN1* gene; they have a relatively better prognosis than other somatic pNETs. Loss of menin expression is associated with overexpression of the Raf/MEK/ERK and PI3K/AKT/mTOR pathways in a mouse model.[82] Mutations in genes affecting the mTOR pathway are found in 14%.[83] In 2013, a study in a series of sporadic insulinomas demonstrated a recurrent T372R gain-of-function mutation of the Ying Yang 1 (YY1) transcripion factor gene in 30% of tumors. Only a single somatic *MEN1* gene mutation was found in a set of 39 insulinomas. YY1 is a target of mTOR signaling and can be a component of polycomb repressive histone modification complexes.[84]

Intact menin promotes the Wnt signaling pathway and β-catenin proteasome degradation in pancreatic endocrine cells.[85]

Loss of menin leads to Wnt/β-catenin signaling activation in the nucleus. Repressed p27 was found in pancreatic islet cell tumors.[86]

Common Pathways in Multistep Carcinogenesis

It appears that interaction of components of the PI3K/AKT pathway is involved in NET formation[87,88] (see Fig. 24.4). Gain of function occurs by oncogenic mutations of PI3K signaling. Loss of function occurs through deletion of the tumor suppressor *pTEN* gene or downstream loss of the tumor suppressors p18 and p27.

A combination of mutations in the *MEN1* disease gene with other mutations in the PI3K/AKT pathway may be associated with accelerated tumor growth. In addition, activating mutations of *RAS*, *MEK*, or *ERK* may accelerate tumor growth further.

Further insight into the mechanisms that underlie MEN1-associated tumorigenesis may provide opportunities for new therapeutic strategies.

MEN1 MUTATION ANALYSIS

Methods for MEN1 Mutation Analysis

Germline mutations identified in families with MEN1, which include all types of sequence alterations, appear scattered throughout the entire 1830-bp coding region and splice sites of the *MEN1* gene. The heterogeneous nature of the mutation spectrum in the *MEN1* gene reflects the inactivating character required for disease-associated mutations in tumor suppressor genes; virtually any position within the MEN1 sequence can be subject to a mutation, as long as the mutation leads to a loss of menin protein function. For this reason, mutation analysis of the entire coding region and the flanking intronic sequences must be performed to determine the family-specific *MEN1* gene mutation. Although several sensitive mutation scanning methods have been developed over the past decades, direct DNA sequence analysis remains the method of choice in most clinical diagnostic laboratories and is considered as the "gold standard." Large deletions encompassing one or more *MEN1* exons usually escape detection by DNA sequencing. Although several studies have indicated that gross deletion in the *MEN1* gene is a relatively uncommon cause of MEN1, clinical mutation testing should include an assay to detect such alterations. The multiplex ligation-dependent probe amplification (MLPA) method is a rapid, specific, and sensitive assay to detect large deletions in human disease genes.[89] MLPA has been adapted for its use in many hereditary disorders, including MEN1. Compared with the "classic" method for deletion detection in genomic DNA (Southern blot analysis) MLPA is simpler, requires much smaller amounts of DNA, allows nonradioactive signal detection, and is suitable for high-throughput analysis.

Interpretation of Results of MEN1 Gene Mutation Analysis

MEN1 gene mutations predicting a loss of function of menin due to the introduction of premature stop codons

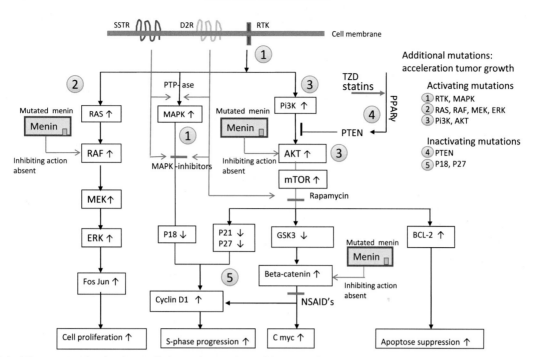

FIGURE 24.4 **The neuroendocrine intracellular pathways in multistep carcinogenesis. Additional mutations in MEN1 NETs enhance tumor growth.** Besides inactivation of the *MEN1* gene, additional mutations may be responsible for acceleration of tumor growth. Thus, these mutations involve a process of multistep tumorigenesis. Deregulation of normal neuroendocrine development or differentiation can occur through a range of processes, triggering a downstream cascade of events:

- upstream activation through receptor tyrosin kinase (RTK), mitogen-activated protein kinases (MAPK) signaling
- increased *RAS* expression or activating mutations of *RAF*, *MEK*, or *ERK* will accelerate cell proliferation
- activation of Pi3K or amplification of *AKT*
- loss of function of the tumor suppressor *PTEN* through gene deletion, mutation, miRNA expression, or epigenetic silencing
- downstream loss of the tumor suppressor genes *P18*, *P27*

Opportunities for target-directed treatment are phosphotyrosine phosphatases (PTP-ase), MAPK-inhibitors, rapamycin, statins, thiazolidinediones (TZD). D2R, dopamine-2 receptor agonists; SSTR, somatostatin receptor agonists.

in the protein-coding region (i.e., by protein truncation) as well as gross gene deletions are invariably regarded as disease causing. Protein-truncating mutations include nonsense mutations, frameshift mutations due to small deletions and insertions, and mutations at the conserved splice donor and acceptor sequences. However, if a missense mutation (i.e., amino acid substitution), an intronic change not affecting the most highly conserved parts of intron/exon junctions, an in-frame deletion or insertion or a synonymous codon change (i.e., a change in the DNA sequence that does not alter the corresponding amino acid) is found, it is crucial to discriminate such variants from neutral variants or normal polymorphisms, which also occur in the *MEN1* gene. Sequence changes in human disease genes with an unknown clinical relevance are usually referred to as "unclassified variants" (UVs). For many changes detected in the human genome, the clinical relevance is unknown and additional evidence is required to support their classification as being either disease causing or inconsequential. Functional analysis of UVs by studying their effect on protein/protein interactions and biochemical activity in model systems (cell culture, genetically modified organisms) is not applicable

in a clinical diagnostic setting. Supporting evidence in favor of causality (or neutrality) of a sequence change must therefore be obtained by genetic studies (cosegregation analysis in families, analysis of loss-of-heterozygosity in tumor-derived DNA, analysis of the *MEN1* gene transcript), by computational analysis of functional and structural aspects of such variants and by literature/database searches. An integrated statistical approach using genetic data in combination with evolutionary conservation analysis, amino acid properties, and location in functional domain(s) has successfully been applied to the assessment of missense UVs in the breast cancer susceptibility genes *BRCA1* and *BRCA2*.[90] A similar approach may be helpful in case of UVs detected in the *MEN1* gene. The assessment of UVs in disease genes has proven difficult, and for many genetic variants, definitive classification is not yet possible. However, predictive testing in unaffected family members of an index patient can only be offered after the disease-causing nature of the sequence change in this index patient has been unequivocally established. Such an extended analysis must be performed whenever an unknown, nonprotein truncating *MEN1* gene mutation is encountered in a patient.

Selection of Individuals Eligible for MEN1 Gene Mutation Testing

Who is considered to be a candidate to undergo genetic testing? What is the actual risk for close relatives and what is the optimal timing for genetic testing?

In a family, a sample from one subject already known to be affected should be tested (first) in order to determine the specific mutation for that family. If a pathogenetic mutation is identified within a family, family members at risk for carrying this mutation can opt for predictive DNA testing. Predictive DNA testing will be performed after genetic counseling, informing a person at risk about the advantages and disadvantages of testing, enabling an informed choice to be made. Since the screening for MEN1 symptoms starts at the age of 5, the parents of a child at risk for carrying the mutation can decide whether the child's DNA will be tested or not. Other first-degree family members, parents, brothers, and sisters of a patient share half of their genes with the proband and have an actual risk of 50% of carrying the MEN1 disease gene. Second and third degree relatives share one-quarter and one-eighth of their genes with the proband, respectively, and have a risk of 25 and 12.5%. Thus, also here, due to the involved risk, more extensive family investigation is indicated.

With regard to new case finding, unfortunately, only a small percentage in nonselected patients with apparently sporadic MEN1-associated tumors turn out to be carriers of a MEN1 gene germline mutation (at most 5%). To increase the sensitivity and specificity (cost-effectiveness) for mutation detection and to be able to identify all these MEN1 patients, without screening the entire group of patients with apparently sporadic tumors, the criteria for MEN1-suspected patients have been defined. In 60% of such MEN1-suspected patients, a germline mutation of the MEN1 gene was found.[8] The risk of MEN1 in patients with sporadic endocrine tumors can be predicted according to the age of onset, type of tumor, and family history.[91]

Criteria for MEN1 gene mutation analysis are given in Table 24.3. Guidelines for periodical clinical monitoring are given in Table 24.4.

The earliest manifestation of MEN1 reported is a pituitary adenoma in a 5-year-old boy. Therefore, in principle, mutation analysis should be performed before the age of 5.[44]

TABLE 24.3 Criteria for MEN1 Gene Mutation Analysis

MEN1-gene mutation analysis is offered to:

- Clinically demonstrated MEN1 patients to confirm diagnosis: patients with three of the five major MEN1-associated lesions: parathyroid adenomas, pancreatic islet cell tumors, pituitary adenomas, adrenal adenomas, NETs
- MEN1-suspected patients: patients with two of the five major lesions, two MEN1- associated tumors within one organ and/or a MEN1-associated lesion at a young age (<35 years)
- Relatives from MEN1 patients with a confirmed MEN1-gene germline mutation, after risk estimation and genetic counseling
- Relatives from a patient with clinically proven or highly suspected MEN1, however, without an identified MEN1-germline mutation or who declined mutation analysis, after risk estimation and genetic counseling

TABLE 24.4 Guidelines for Periodic Clinical Monitoring

Eligible for periodic clinical monitoring are:

- MEN1 patients
- MEN-gene germline mutation carriers and
- MEN1-suspected patients without a confirmed mutation

Periodical screening includes:

- From the age of 5, biannual clinical examination: Laboratory investigation including measurement in blood of ionized calcium, chloride, phosphate, parathyroid hormone, glucose, insulin, c-peptide, glucagon, gastrin, pancreatic polypeptide, prolactin, insulin-like growth factor 1, platelet serotonin, and chromogranin A

- From the age of 15: visualization once every 2 years (but no gamma radiation):
  - MRI of the upper abdomen
  - MRI of the pituitary with gadolinium contrast
  - MRI of the mediastinum in males

- EUS of the upper abdomen may permit earlier detection of pancreatic pathology. However, if there is cause of suspicion, that is, no complaints are present, does it have clinical consequences?

MRI, magnetic resonance imaging; EUS, endoscopic ultrasound.

Laboratories Available for Genetic Testing

A directory of US and international clinical laboratories offering in-house molecular genetic testing for MEN1 can be found at the National Institutes of Health-funded GeneTests website (http://www.genetests.org/). Information on laboratory contact details, test methodology, staff, and laboratory certification in GeneTests is provided by laboratory staff and is updated regularly. The following are two websites of centers offering *MEN1* genetic testing: http://www.ncbi.nlm.nih.gov/sites/GeneTests/ and http://www.orpha.net/consor/cgi-bin/index.php or http://www.eddnal.com.

Prenatal Testing

Prospective parents planning or carrying a pregnancy at risk for hereditary endocrine cancer syndromes face different options.

The couple may choose not to know until after the child is born and accept the 50% risk of having a child that carries the same *MEN1* gene mutation as the affected parent.

After genetic counseling, some couples choose prenatal diagnosis. DNA analysis can be performed utilizing a sample obtained by amniocentesis or chorionic villus sampling. Some couples wish to know the carrier status prior to birth in order to prepare, while others may elect to terminate a pregnancy if the fetus is affected. This last decision may have far-reaching ethical, emotional, and psychological consequences.

If prenatal genetic testing is not performed, then all at-risk children should be offered DNA testing in order to determine whether or not the hereditary endocrine cancer syndrome surveillance regimen is required (see Table 24.3).

Prospective parents should also be provided with information about reproductive technologies that greatly lower their risk of having a child with hereditary endocrine cancer syndromes, such as sperm or oocyte donation (depending on which parent is affected), and preimplantation genetic diagnosis. Preimplantation genetic diagnosis involves testing embryos fertilized *in vitro* for the familial mutation, usually on a single cell of a blastocyst, and selecting unaffected embryos for implantation. The various reproductive options available to prospective parents require thoughtful discussion and genetic counseling.

Predictive Value of Testing

MEN1 Genetic Screening in Apparently Sporadic Tumors

MEN1 gene germline mutations have been identified in patients with apparently sporadic MEN1-related tumors. However, at most 5% of patients with apparently sporadic primary hyperparathyroidism, pituitary adenomas, or insulinomas turn out to be MEN1 patients.[7] MEN1 gene germline mutations are found in 25% of patients with thymic carcinoid.[91] About 22% of patients with ZES had MEN1.[92] ZES was the initial clinical manifestation of MEN1 in 40%.[93] With the prediction model, the risk of MEN1 can be calculated in patients suspected for MEN1 with sporadically occurring endocrine tumors.[94]

Significance of a Negative Test

Although MEN1 is considered as a genetically homogeneous disorder, in some families fulfilling the clinical criteria no *MEN1* gene germline mutation can be found, even if testing for gross deletions is included. It is likely that some families may harbor mutations in regions of the gene not investigated on a routine basis, such as regulatory sequences. For this reason, in a MEN1-affected patient with a normal *MEN1* gene test result, the presence of a mutation cannot be completely excluded, and the test should be considered as noninformative. Only after a pathogenic mutation has been identified in an index case can the result of carriership analysis in unaffected family members be interpreted with virtual 100% reliability.

Several families with a MEN1-like disorder have been reported to carry germline mutations in genes other than the *MEN1* gene itself. These include *AIP* gene mutations in pituitary adenoma predisposition (PAP)[95] and mutations of the *CDKN1B* (*p27Kip1*) gene in patients with MEN4 (Ref. 96 and OMIM: 610755). The latter mutations appear to be extremely rare (at least in MEN1-like conditions), and the full phenotypic spectrum of CDKN1B mutations is probably not yet established. However, because some of these families were initially considered as possible MEN1 kindreds, mutational analysis of the *AIP* and *CDKN1B* genes should be considered in (a subset of the) *MEN1* gene mutation negative families.

TREATMENT

Based on Genetic Information: How Does it Affect Treatment?

MEN1 patients and their family members have to be monitored periodically. The clinical investigation is aimed at identification of MEN1-associated lesions and includes, besides the patient's history and physical examination, biochemical screening and imaging. The protocol for periodic clinical monitoring is shown in Table 24.4.

Genetic Counseling

As in all genetic tests, the advantages and disadvantages of DNA analysis should be discussed before actually performing the test. Also when the test is performed in an index patient (a symptomatic test) this patient should be informed preferably before the analysis, so they have the opportunity of accepting or declining the DNA test or knowledge of the result afterwards. Not only will personal arguments be discussed in the counseling, but also the consequences for family members. The medical intervention possibilities and their limitations have to be discussed, but also the psychosocial aspects. For example, the issues of obtaining (life) insurance after knowing test results, division of families into two camps, the carriers and the noncarriers, and the impact of these subjects for personal development and family life have to be discussed. After genetic counseling, a person can take a carefully considered decision about his testing. In the process of genetic counseling and the period after disclosure of the DNA result, psychosocial care has to be available.

The above mentioned is even more relevant when predictive DNA testing (performing DNA analysis when there are no clinical signs of MEN1) is considered. Since the screening for early symptoms starts preferably from the age of 5 years, parents should be given counseling in the process of considering the test for their child. In subsequent children, the counseling should be a moment of reflection: "What happened after the uncertainty about the previous child was elucidated? What is your perception or prediction in this child?" Especially when the predicted status differs from the actual DNA result, it will take time to level with the test result. Support by a psychosocial worker may be a valuable part of the counseling.

RECOMMENDATIONS FOR THE FUTURE

- In a motivated medical MEN1 center, a multidisciplinary team with knowledge and experience in MEN1 is responsible for the attendance of MEN1 patients and their families.
- A specialized nurse with expertise in the management of MEN1 may provide additional information to patients. Daily support has to be available and will be beneficial to patients.
- Guidelines for management and recommendations have to be available for patients, general practitioners, medical specialists, geneticists, psychologists, and social workers.
- Physicians and have to be aware of additional mutations that may occur and cause tumor acceleration. In this way target-directed treatment may become available.
- Pedigree studies have to be performed to detect second and third grade family members at risk.

- Finally, MEN patient interest groups or family alliances may be attended by patients and their family members to support them in urgent and emotional situations, and will champion common interests and be assisted by a medical advisory board.

CONCLUSIONS

MEN1 is an inherited disorder with a variable presentation, often already present at a young age. The initial symptoms of MEN1-associated lesions may be very general. By using stringent criteria, MEN1 patients can be identified efficiently. Mutation analysis enables *MEN1* disease gene carriers to be identified. Gastrinomas and other NETs have malignant potential. Periodic clinical monitoring makes presymptomatic detection and treatment of MEN1-associated tumors possible. This is beneficial for both life expectancy and quality of life of MEN1 patients.

Acknowledgment

Koen Dreijerink is supported by the Dutch Cancer Society.

References

1. Doherty GM, Olson JA, Frisella MM, Lairmore TC, Wells Jr SA, Norton JA. Lethality of multiple endocrine neoplasia type I. *World J Surg* 1998;**22**(6):581–6 discussion 586–587.
2. Geerdink EA, Van der Luijt RB, Lips CJ. Do patients with multiple endocrine neoplasia syndrome type 1 benefit from periodical screening? *Eur J Endocrinol* 2003;**149**:577–82.
3. Lairmore TC, Piersail LD, DeBenedetti MK, Dilley WG, Mutch MG, Whelan AJ, Zehnbauer B. Clinical genetic testing and early surgical intervention in patients with multiple endocrine neoplasia type 1 (MEN 1). *Ann Surg* 2004;**239**:637–45 discussion 645–647.
4. Chandrasekharappa SC, Guru SC, Manickam P, Olufemi SE, Collins FS, Emmert-Buck MR, Debelenko LV, Zhuang Z, Lubensky IA, Liotta LA, Crabtree JS, Wang Y, Roe BA, Weisemann J, Boguski MS, Agarwal SK, Kester MB, Kim YS, Heppner C, Dong Q, Spiegel AM, Burns AL, Marx SJ. Positional cloning of the gene for multiple endocrine neoplasia type 1. *Science* 1997;**276**:404–7.
5. Lemmens I, Van de Ven WJ, Kas K, Zhang CX, Giraud S, Wautot V, Buisson N, De Witte K, Salandre J, Lenoir G, Pugeat M, Calender A, Parente F, Quincey D, Gaudray P, De Wit MJ, Lips CJ, Höppener JW, Khodaei S, Grant AL, Weber G, Kytölä S, Teh BT, Farnebo F, Thakker RV, et al. Identification of the multiple endocrine neoplasia type 1 (MEN1) gene. The European Consortium on MEN1. *Hum Mol Genet* 1997;**6**(7):1177–83.
6. Brandi ML, Gagel RF, Angeli A, Bilezikian JP, Beck-Peccoz P, Bordi C, et al. Guidelines for diagnosis and therapy of MEN type 1 and type 2. *J Clin Endocrinol Metab* 2001;**86**:5658–71.
7. Thakker RV, Newey PJ, Walls GV, Bilezikian J, Dralle H, Ebeling PR, Melmed S, Sakurai A, Tonelli F, Brandi ML, Endocrine Society. Clinical practice guidelines for multiple endocrine neoplasia type 1 (MEN1). *J Clin Endocrinol Metab* 2012;**97**(9):2990–3011.
8. Roijers JF, de Wit MJ, van der Luijt RB, Ploos van Amstel HK, Höppener JW, Lips CJ. Criteria for mutation analysis in MEN 1-suspected patients: MEN 1 case-finding. *Eur J Clin Invest* 2000;**30**: 487–92.

9. Lips CJ, Dreijerink KM, Höppener JW. Variable clinical expression in patients with a germline MEN1 disease gene mutation: clues to a genotype-phenotype correlation. *Clinics* 2012;**67**(Suppl 1):49–56 Review.

10. Hendy GN, Cole DE. Genetic defects associated with familial and sporadic hyperparathyroidism. *Front Horm Res* 2013;**41**:149–65.

11. Haven CJ, Wong FK, van Dam EW, van der Juijt R, van Asperen C, Jansen J, Rosenberg C, de Wit M, Roijers J, Hoppener J, Lips CJ, Larsson C, Teh BT, Morreau H. A genotypic and histopathological study of a large Dutch kindred with hyperparathyroidism-jaw tumor syndrome. *J Clin Endocrinol Metab* 2000;**85**(4):1449–54.

12. Thakker RV. Multiple endocrine neoplasia type 1 (MEN1) and type 4 (MEN4). *Mol Cell Endocrinol* 2014;**386**(1–2):2–15.

13. Cetani F, Pardi E, Ambrogini E, Lemmi M, Borsari 5F S., Cianferotti L, Vignali E, Viacava P, Berti P, Mariotti S, Pinchera A, Marcocci C. Genetic analyses in familial isolated hyperparathyroidism: implication for clinical assessment and surgical management. *Clin Endocrinol* 2006;**64**:146–52.

14. Horiuchi K, Okamoto T, Iihara M, Tsukada T. Analysis of genotype-phenotype correlations and survival outcomes in patients with primary hyperparathyroidism caused by multiple endocrine neoplasia type 1: the experience at a single institution. *Surg Today* 2013;**43**(8):894–9.

15. Schreinemakers JM, Pieterman CR, Scholten A, Vriens MR, Valk GD, Rinkes IH. The optimal surgical treatment for primary hyperparathyroidism in MEN1 patients: a systematic review. *World J Surg* 2011;**35**(9):1993–2005.

16. Pieterman CR, van Hulsteijn LT, den Heijer M, van der Luijt RB, Bonenkamp JJ, Hermus AR, Borel Rinkes IH, Vriens MR, Valk GD, Dutch MEN1 Study Group. Primary hyperparathyroidism in MEN1 patients: a cohort study with longterm follow-up on preferred surgical procedure and the relation with genotype. *Ann Surg* 2012;**255**(6):1171–8.

17. Locchi F, Cavalli T, Giudici F, Brandi ML, Tonelli F. Intraoperative PTH monitoring: a new approach based on the identification of the "true" time origin of the decay curve. *Endocr J* 2014;**61**(3):239–47.

18. Coutinho FL, Lourenco Jr DM, Toledo RA, Montenegro FL, Toledo SP. Post-surgical follow-up of primary hyperparathyroidism associated with multiple endocrine neoplasia type 1. *Clinics* 2012;**67**(Suppl 1):169–72 Review.

19. Montenegro L, Lourenço Jr DM, Tavares MR, Arap SS, Nascimento Jr CP, Massoni Neto LM, D'Alessandro A, Toledo RA, Coutinho FL, Brandão LG, de Britto e Silva Filho G, Cordeiro AC, Toledo SP. Total parathyroidectomy in a large cohort of cases with hyperparathyroidism associated with multiple endocrine neoplasia type 1: experience from a single academic center. *Clinics* 2012;**67**(Suppl 1):131–9.

20. Klöppel G, Anlauf M. Epidemiology, tumour biology and histopathological classification of neuroendocrine tumours of the gastrointestinal tract. *Best Pract Res Clin Gastroenterol* 2005;**19**:507–17.

21. Zikusoka MN, Kidd M, Eick G, Latich I, Modlin TM. The molecular genetics of gastroenteropancreatic neuroendocrine tumors. *Cancer* 2005;**104**:2292–309.

22. Kouvaraki MA, Shapiro SE, Cote GJ, Lee JE, Yao JC, Waguespack SG, Gagel RF, Evans DB, Perrier ND. Management of pancreatic endocrine tumors in Multiple Endocrine Neoplasia Type 1. *World J Surg* 2006;**30**:643–53.

23. Levy-Bohbot N, Merle C, Goudet P, Delemer B, Calender A, Jolly D, Thiefin G, Cadiot G, Groupe des Tumeurs Endocrines. Prevalence, characteristics and prognosis of MEN 1-associated glucagonomas, VIPomas, and somatostatinomas: study from the GTE (Groupe des Tumeurs Endocrines) registry. *Gastroenterol Clin Biol* 2004;**28**:1075–81.

24. Anlauf M, Schienger R, Perren A, Bauersfeld J, Koch CA, Dralle H, Raffel A, Knoefel WT, Weihe E, Ruszniewski P, Couvelard A,

Komminoth P, Heitz PU, Klöppel G. Microadenomatosis of the endocrine pancreas in patients with and without the multiple endocrine neoplasia type 1 syndrome. *Am J Surg Pathol* 2006;**30**:560–74.

25. Akerström G, Hessman O, Heliman P, Skogseid B. Pancreatic tumours as part of the MEN-1 syndrome. *Best Pract Res Clin Gastroenterol* 2005;**19**:819–30.

26. de Laat JM, Pieterman CR, Weijmans M, Hermus AR, Dekkers OM, de Herder WW, van der Horst-Schrivers AN, Drent ML, Bisschop PH, Havekes B, Vriens MR, Valk GD. Low accuracy of tumor markers for diagnosing pancreatic neuroendocrine tumors in multiple endocrine neoplasia type 1 patients. *J Clin Endocrinol Metab* 2013;**98**(10):4143–51.

27. Thomas-Marques L, Murat A, Delemer B, Penfornis A, Cardot-Bauters C, Baudin E, Niccoli-Sire P, Levoir D, et al. Prospective endoscopic ultrasonographic (EUS) evaluation of the frequency of nonfunctioning pancreaticoduodenal endocrine tumors in patients with Multiple Endocrine Neoplasia Type 1. *Am J Gastroenterol* 2006;**101**:266–73.

28. Gonçalves TD, Toledo RA, Sekiya T, Matuguma SE, Maluf Filho F, Rocha MS, Siqueira SA, Glezer A, Bronstein MD, Pereira MA, Jureidini R, Bacchella T, Machado MC, Toledo SP, Lourenço Jr DM. Penetrance of functioning and nonfunctioning pancreatic neuroendocrine tumors in multiple endocrine neoplasia type 1 in the second decade of life. *J Clin Endocrinol Metab* 2014;**99**(1):E89–96.

29. Newey PJ, Jeyabalan J, Walls GV, Christie PT, Gleeson FV, Gould S, Johnson PR, Phillips RR, Ryan FJ, Shine B, Bowl MR, Thakker RV. Asymptomatic children with multiple endocrine neoplasia type 1 mutations may harbor nonfunctioning pancreatic neuroendocrine tumors. *J Clin Endocrinol Metab* 2009;**94**(10):3640–6.

30. Kann PH, Balakina E, Ivan D, Bartsch DK, Meyer S, Klose KJ, Behr T, Langer P. Natural course of small, asymptomatic neuroendocrine pancreatic tumours in multiple endocrine neoplasia type 1: an endoscopic ultrasound (EUS) imaging study. *Endocr Relat Cancer* 2006;**13**(4):1195–202.

31. Hellman P, Hennings J, Akerström G, Skogseid B. Endoscopic ultrasonography (EUS) for evaluation of p tumours in multiple endocrine neoplasia type 1. *Br J Surg* 2005;**92**:1508–12.

32. Niina Y, Fujimori N, Nakamura T, Igarashi H, Oono T, Nakamura K, Kato M, Jensen RT, Ito T, Takayanagi R. The current strategy for managing pancreatic neuroendocrine tumors in multiple endocrine neoplasia type 1. *Gut Liver* 2012;**6**(3):287–94.

33. Hoffmann KM, Gibril F, Entsuah LK, Serrano J, Jensen RT. Patients with multiple endocrine neoplasia type 1 with gastrinomas have an increased risk of severe esophageal disease including stricture and the premalignant condition, Barrett's esophagus. *J Clin Endocrinol Metab* 2006;**91**:04–12.

34. Tonelli F, Giudici F, Fratini G, Brandi ML. Pancreatic endocrine tumors in multiple endocrine neoplasia type 1 syndrome: review of literature. *Endocr Pract* 2011;**17**(Suppl 3):33–40.

35. Akerström G, Stålberg P, Hellman P. Surgical management of pancreatico-duodenal tumors in multiple endocrine neoplasia syndrome type 1. *Clinics* 2012;**67**(Suppl 1):173–8 Review.

36. Hanazaki K, Sakurai A, Munekage M, Ichikawa K, Namikawa T, Okabayashi T, Imamura M. Surgery for a gastroenteropancreatic neuroendocrine tumor (GEPNET) in multiple endocrine neoplasia type 1. *Surg Today* 2013;**43**(3):229–36.

37. Sakurai A, Yamazaki M, Suzuki S, Fukushima T, Imai T, Kikumori T, Okamoto T, Horiuchi K, Uchino S, Kosugi S, Yamada M, Komoto I, Hanazaki K, Itoh M, Kondo T, Mihara M, Imamura M. Clinical features of insulinoma in patients with multiple endocrine neoplasia type 1: analysis of the database of the MEN Consortium of Japan. *Endocr J* 2012;**59**(10):859–66.

38. Bartsch DK, Albers M, Knoop R, Kann PH, Fendrich V, Waldmann J. Enucleation and limited pancreatic resection provide long-term cure for insulinoma in multiple endocrine neoplasia type 1. *Neuroendocrinology* 2013;**98**(4):290–8.

39. Quinn TJ, Yuan Z, Adem A, Geha R, Vrikshajanani C, Koba W, Fine E, Hughes DT, Schmid HA, Libutti SK. Pasireotide (SOM230) is effective for the treatment of pancreatic neuroendocrine tumors (PNETs) in a multiple endocrine neoplasia type 1 (MEN1) conditional knockout mouse model. *Surgery* 2012;**152**(6):1068–77.

40. Biermasz NR, Smit JW, Pereira AM, Frölich M, Romijn JA, Roelfsema F. Acromegaly caused by growth hormone-releasing hormone-producing tumors: long-term observational studies in three patients. *Pituitary* 2007;**10**:237–49.

41. Sala E, Ferrante E, Verrua E, Malchiodi E, Mantovani G, Filopanti M, Ferrero S, Pietrabissa A, Vanoli A, La Rosa S, Zatelli MC, Beck-Peccoz P, Verga U. Growth hormone-releasing hormone-producing pancreatic neuroendocrine tumor in a multiple endocrine neoplasia type 1 family with an uncommon phenotype. *Eur J Gastroenterol Hepatol* 2013;**25**(7):858–62.

42. Borson-Chazot F, Garby L, Raverot G, Claustrat F, Raverot V, Sassolas G, GTE group. Acromegaly induced by ectopic secretion of GHRH: a review 30 years after GHRH discovery. *Ann Endocrinol* 2012;**73**(6):497–502.

43. Rix M, Hertel NT, Nielsen FC, Jacobsen BB, Hoejberg AS, Brixen K, Hangaard J, Kroustrup JP. Cushing's disease in childhood as the first manifestation of multiple endocrine neoplasia syndrome type 1. *Europ J Endocr* 2004;**15**(1):709–15.

44. Stratakis CA, Schussheim DH, Freedman SM, Keil MF, Pack SD, Agarwal SK, Skarulis MC, Weil RI, Lubensky IA, Zhuang Z, Edward H, Oldfield EH, Marx SJ. Pituitary macroadenoma in a 5-year-old: an early expression of multiple endocrine neoplasia type 1. *J Clin Endocrinol Metab* 2000;**85**:4776–80.

45. Ciccarelli A, Daly AF, Beckers A. The epidemiology of prolactinomas. *Pituitary* 2005;**8**:3–6.

46. Hao W, Skarulis MC, Simonds WF, Weinstein LS, Agarwal SK, Mateo C, James-Newton L, Hobbs GR, Gibril F, Jensen RT, Marx SJ. Multiple endocrine neoplasia type 1 variant with frequent prolactinoma and rare gastrinoma. *J Clin Endocrinol Metab* 2004;**89**:3776–84.

47. Benito M, Asa SL, Livolsi VA, West VA, Snyder PJ. Gonadotroph tumor associated with multiple endocrine neoplasia type 1. *J Clin Endocrinol Metab* 2005;**90**:570–4.

48. Simonds WF, Varghese S, Marx SJ, Nieman LK. Cushing's syndrome in multiple endocrine neoplasia type 1. *Clin Endocrinol* 2012;**76**(3):379–86.

49. Syro LV, Scheithauer BW, Kovacs K, Toledo RA, Londoño FJ, Ortiz LD, Rotondo F, Horvath E, Uribe H. Pituitary tumors in patients with MEN1 syndrome. *Clinics* 2012;**67**(Suppl 1):43–8 Review.

50. Jamilloux Y, Favier J, Pertuit M, Delage-Corre M, Lopez S, Teissier MP, Mathonnet M, Galinat S, Barlier A, Archambeaud F. A MEN1 syndrome with a paraganglioma. *Eur J Hum Genet* 2014;**22**(2):283–5.

51. Lim LC, Tan MH, Eng C, Teh BT, Rajasoorya RC. Thymic carcinoid in multiple endocrine neoplasia 1: genotype-phenotype correlation and prevention. *J Intern Med* 2006;**259**:428–32.

52. Sakurai A, Imai T, Kikumori T, Horiuchi K, Okamoto T, Uchino S, Kosugi S, Suzuki S, Suyama K, Yamazaki M, Sato A, MEN Consortium of Japan. Thymic neuroendocrine tumour in multiple endocrine neoplasia type 1: female patients are not rare exceptions. *Clin Endocrinol* 2013;**78**(2):248–54.

53. Ferolla P, Falchetti A, Filosso P, Tomassetti P, Tamburrano G, Avenia N, Daddi G, Puma F, Ribacchi R, Santeusanio F, Angeletti G, Brandi ML. Thymic neuroendocrine carcinoma (carcinoid) I in multiple endocrine neoplasia type 1 syndrome: the Italian series. *J Clin Endocrinol Metab* 2005;**90**:2603–9.

54. Khoo J, Bee YM, Giraud S, Chen RY, Rajasoorya C, Teh BT. Novel association of thymic carcinoid with a germline mutation in a kindred with multiple endocrine neoplasia 1 (MEN1). *Exp Clin Endocrinol Diabetes* 2012;**120**(5):257–60.

55. Takagi J, Otake K, Morishita M, Kato H, Nakoa N, Yoshikawa K, Ikeda H, Hirooka Y, Hattori Y, Larsson C, Nogimori T. Multiple endocrine neoplasia type I and Cushing's syndrome due to an aggressive ACTH producing thymic carcinoid. *Internal Med* 2006;**45**:81–6.

56. Goudet P, Murat A, Cardot-Bauters C, Emy P, Baudin E, du Boullay Choplin H, Chapuis Y, Kraimps JL, Sadoul JL, Tabarin A, Vergès B, Carnaille B, Niccoli-Sire P, Costa A, Calender A, GTE Network. Thymic neuroendocrine tumors in multiple endocrine neoplasia type 1: a comparative study on 21 cases among a series of 761 MEN1 from the GTE (Groupe des Tumeurs Endocrines). *World J Surg* 2009;**33**(6):1197–207.

57. Sachithanandan N, Harle RA, Burgess JR. Bronchopulmonary carcinoid in multiple endocrine neoplasia type 1. *Cancer* 2005;**103**:509–15.

58. D'Adda T, Bottarelli L, Azzoni C, Pizzi S, Bongiovanni M, Papotti M, Pelosi G, Maisonneuve P, Antonetti T, Rindi G, Bordi C. Malignancy-associated X chromosome allelic losses in foregut endocrine neoplasms: further evidence from lung tumors. *Mod Pathol* 2005;**18**(6):795–805.

59. Pieterman CR, Conemans EB, Dreijerink KM, de Laat JM, Timmers M, Vriens MR, Valk GD. Thoracic and duodenopancreatic neuroendocrine tumors in multiple endocrine neoplasia type 1. *Endocr Relat Cancer* 2014;**21**:R121–42.

60. Swarts DR, Scarpa A, Corbo V, Van Criekinge W, van Engeland M, Gatti G, Henfling ME, Papotti M, Perren A, Ramaekers FC, Speel EJ, Volante M. MEN1 gene mutation and reduced expression are associated with poor prognosis in pulmonary carcinoids. *J Clin Endocrinol Metab* 2014;**99**(2):E374–8.

61. Norton JA, Melcher ML, Gibril F, Jensen RT. Gastric carcinoid tumors in multiple endocrine neoplasia-1 patients with Zollinger-Ellison syndrome can be symptomatic, demonstrate aggressive growth, and require surgical treatment. *Surgery* 2004;**136**:1267–74.

62. Bakkelund K, Fossrnark R, Nordrum I, Waidum H. Signet ring cells in gastric carcinomas are derived from neuroendocrine cells. *J Histochem Cytochem* 2006;**54**(6):15–21.

63. Berna MJ, Annibale B, Marignani M, Luong TV, Corleto V, Pace A, Ito T, Liewehr D, Venzon DJ, Delle Fave G, Bordi C, Jensen RT. A prospective study of gastric carcinoids and enterochromaffin-like cell changes in multiple endocrine neoplasia type 1 and Zollinger-Ellison syndrome: identification of risk factors. *J Clin Endocrinol Metab* 2008;**93**(5):1582–91.

64. Lemos MC, Thakker RV. Multiple endocrine neoplasia type 1 (MEN1): analysis of 1336 mutations reported in the first decade following identification of the gene. *Hum Mutat* 2008;**29**:22–32.

65. Huang J, Gurung B, Wan B, Matkar S, Veniaminova NA, Wan K, Merchant JL, Hua X, Lei M. The same pocket in menin binds both MLL and JUND but has opposite effects on transcription. *Nature* 2012;**482**(7386):542–6.

66. Hashimoto M, Kyo S, Hua X, Tahara H, Nakajima M, Takakura M, Sakaguchi J, Maida Y, Nakamura M, Ikoma T, Mizumoto Y, Inoue M. Role of menin in the regulation of telomerase activity in normal and cancer cells. *Int J Oncol* 2008;**33**(2):333–40.

67. Dreijerink KM, Mulder KW, Winkler GS, Höppener JW, Lips CJ, Timmers HT. Menin links estrogen receptor activation to histone H3K4 trimethylation. *Cancer Res* 2006;**66**:4929–35.

68. Dreijerink KMA, Höppener JWM, Timmers HThM, Lips CJM. Mechanisms of disease: multiple endocrine neoplasia type 1-relation to chromatin modifications and transcription regulation. *Nat Clin Pract Endocrinol Metab* 2006;**2**:562–70.

69. Raef H, Zou M, Baitei EY, Al-Rijjal RA, Kaya N, Al-Hamed M, Monies D, Abu-Dheim NN, Al-Hindi H, Al-Ghamdi MH, Meyer BF, Shi Y. A novel deletion of the MEN1 gene in a large family of multiple endocrine neoplasia type 1 (MEN1) with aggressive phenotype. *Clin Endocrinol* 2011;**75**(6):791–800.

70. Hasani-Ranjbar S, Amoli MM, Ebrahim-Habibi A, Gozashti MH, Khalili N, Sayyahpour FA, Hafeziyeh J, Soltani A, Larijani B. A new frameshift MEN1 gene mutation associated with familial malignant insulinomas. *Fam Cancer* 2011;**10**(2):343–8.

71. Schaaf L, Pickel J, Zinner K, Hering U, Höfler M, Goretzki PE, Spelsberg F, Raue F, von zur Mühlen A, Gerl H, Hensen J, Bartsch DK, Rothmund M, Schneyer U, Dralle H, Engelbach M, Karges W, Stalla GK, Höppner W. Developing effective screening strategies in multiple endocrine neoplasia type 1 (MEN 1) on the basis of clinical and sequencing data of German patients with MEN 1. *Exp Clin Endocrinol Diabetes* 2007;**115**(8):509–17.

72. Davenport C, Agha A. The role of menin in parathyroid tumorigenesis. *Adv Exp Med Biol* 2009;**668**:79–86 Review.

73. Costa-Guda J, Marinoni I, Molatore S, Pellegata NS, Arnold A. Somatic mutation and germline sequence abnormalities in CDKN1B, encoding p27Kip1, in sporadic parathyroid adenomas. *J Clin Endocrinol Metab* 2011;**96**(4):E701–6.

74. Cromer MK, Starker LF, Choi M, Udelsman R, Nelson-Williams C, Lifton RP, Carling T. Identification of somatic mutations in parathyroid tumors using whole-exome sequencing. *J Clin Endocrinol Metab* 2012;**97**(9):E1774–81.

75. Newey PJ, Nesbit MA, Rimmer AJ, Attar M, Head RT, Christie PT, Gorvin CM, Stechman M, Gregory L, Mihai R, Sadler G, McVean G, Buck D, Thakker RV. Whole-exome sequencing studies of nonhereditary (sporadic) parathyroid adenomas. *J Clin Endocrinol Metab* 2012;**97**(10):E1995–2005.

76. Dworakowska D, Grossman AB. The molecular pathogenesis of pituitary tumors: implications for clinical management. *Minerva Endocrinol* 2012;**37**(2):157–72.

77. Newey PJ, Nesbit MA, Rimmer AJ, Head RA, Gorvin CM, Attar M, Gregory L, Wass JA, Buck D, Karavitaki N, Grossman AB, McVean G, Ansorge O, Thakker RV. Whole-exome sequencing studies of nonfunctioning pituitary adenomas. *J Clin Endocrinol Metab* 2013;**98**(4):E796–800.

78. Palmieri D, Valentino T, De Martino I, Esposito F, Cappabianca P, Wierinckx A, Vitiello M, Lombardi G, Colao A, Trouillas J, Pierantoni GM, Fusco A, Fedele M. PIT1 upregulation by HMGA proteins has a role in pituitary tumorigenesis. *Endocr Relat Cancer* 2012;**19**(2):123–35.

79. Lacerte A, Lee EH, Reynaud R, Canaff L, De Guise C, Devost D, Ali S, Hendy GN. Lebrun Activin inhibits pituitary prolactin expression and cell growth through Smads, Pit-1 and menin. *Mol Endocrinol* 2004;**18**(6):1558–69.

80. Lebrun JJ. Activin, TGF-beta and menin in pituitary tumorigenesis. *Adv Exp Med Biol* 2009;**668**:69–78 Review.

81. Seoane S, Perez-Fernandez R. The vitamin D receptor represses transcription of the pituitary transcription factor Pit-1 gene without involvement of the retinoid X receptor. *Mol Endocrinol* 2006;**20**(4):735–48.

82. Wang Y, Ozawa A, Zaman S, Prasad NB, Chandrasekharappa SC, Agarwal SK, Marx SJ. The tumor suppressor protein menin inhibits AKT activation by regulating its cellular localization. *Cancer Res* 2011;**71**(2):371–82.

83. Jiao Y, Shi C, Edil BH, de Wilde RF, Klimstra DS, Maitra A, Schulick RD, Tang LH, Wolfgang CL, Choti MA, Velculescu VE, Diaz Jr LA, Vogelstein B, Kinzler KW, Hruban RH, Papadopoulos N. DAXX/ATRX, MEN1, and mTOR pathway genes are frequently altered in pancreatic neuroendocrine tumors. *Science* 2011;**331**(6021):1199–203.

84. Cao Y, Gao Z, Li L, Jiang X, Shan A, Cai J, Peng Y, Li Y, Jiang X, Huang X, Wang J, Wei Q, Qin G, Zhao J, Jin X, Liu L, Li Y, Wang W, Wang J, Ning G. Whole exome sequencing of insulinoma reveals recurrent T372R mutations in YY1. *Nat Commun* 2013;**4**:2810.

85. Chen GAJ, Wang M, Farley S, Lee LY, Lee LC, Sawicki MP. Menin promotes the Wnt signaling pathway in pancreatic endocrine cells. *Mol Cancer Res* 2008;**6**(12):1894–907.

86. Ishida E, Yamada M, Horiguchi K, Taguchi R, Ozawa A, Shibusawa N, Hashimoto K, Satoh T, Yoshida S, Tanaka Y, Yokota M, Tosaka M, Hirato J, Yamada S, Yoshimoto Y, Mori M. Attenuated expression of menin and p27 (Kip1) in an aggressive case of multiple endocrine neoplasia type 1 (MEN1) associated with an atypical prolactinoma and a malignant pancreatic endocrine tumor. *Endocr J* 2011;**58**(4):287–96.

87. Cully M, You H, Levine AJ, Mak TW. Beyond PTEN mutations: the PI3K pathway as an integrator of multiple inputs during tumorigenesis. *Nat Rev Cancer* 2006;**6**(3):184–92.

88. Pitt SC, Chen H, Kunnimalaiyaan M. Inhibition of phosphatidylinositol 3-kinase/Akt signaling suppresses tumor cell proliferation and neuroendocrine marker expression in GI carcinoid tumors. *Ann Surg Oncol* 2009;**16**(10):2936–42.

89. Schouten JP, McElgunn CJ, Waaijer R, Zwijnenburg D, Diepvens F, Pals G. Relative quantification of 40 nucleic acid sequences by multiplex ligation-dependent probe amplification. *Nucleic Acids Res* 2002;**30**(12):e57.

90. Easton DF, Deffenbaugh AM, Pruss D, Frye C, Wenstrup RJ, Allen-Brady K, Tavtigian SV, Monteiro AN, Iversen ES, Couch FJ, Goldgar DE. A systematic genetic assessment of 1,433 sequence variants of unknown clinical significance in the BRCA1 and BRCA2 breast cancer-predisposition genes. *Am J Hum Genet* 2007;**81**(5):873–83.

91. Teh BT, Zedenius J, Kytola S, Skogseid B, Trotter J, Choplin H, Twigg S, Farnebo F, Giraud S, Cameron D, Robinson B, Calender A, Larsson C, Salmela P. Thymic carcinoids in multiple endocrine neoplasia type 1. *Ann Surg* 1998;**228**:99–105.

92. Roy PK, Venzon DJ, Shojamanesh H, Abou-Saif A, Peghini P, Doppman JL, Gibril F, Jensen RT. Zollinger-Ellison syndrome. Clinical presentation in 261 patients. *Medicine* 2000;**79**:379–411.

93. Gibril F, Schumann M, Pace A, Jensen RT. Multiple endocrine neoplasia type 1 and Zollinger-Ellison syndrome: a prospective study of 107 cases and comparison with 1009 cases from the literature. *Medicine* 2004;**83**:43–83 Review.

94. de Laat JM, Tham E, Pieterman CR, Vriens MR, Dorresteijn JA, Bots ML, Nordenskjöld M, van der Luijt RB, Valk GD. Predicting the risk of multiple endocrine neoplasia type 1 for patients with commonly occurring endocrine tumors. *Eur J Endocrinol* 2012;**167**(2):181–7.

95. Georgitsi M, Raitila A, Karhu A, Tuppurainen K, Mäkinen MJ, Vierimaa O, Paschke R, Saeger W, van der Luijt RB, Sane T, Robledo M, De Menis E, Weil RJ, Wasik A, Zielinski G, Lucewicz O, Lubinski J, Launonen V, Vahteristo P, Aaltonen LA. Molecular diagnosis of pituitary adenoma predisposition caused by aryl hydrocarbon receptor-interacting protein gene mutations. *Proc Natl Acad Sci USA* 2007;**104**:4101–5.

96. Georgitsi M, Raitila A, Karhu A, van der Luijt RB, Aalfs CM, Sane T, Vierimaa O, Mäkinen MJ, Tuppurainen K, Paschke R, Gimm O, Koch CA, Gündogdu S, Lucassen A, Tischkowitz M, Izatt L, Aylwin S, Bano G, Hodgson S, De Menis E, Launonen V, Vahteristo P, Aaltonen LA. Germline CDKN1B/p27Kip1 mutation in multiple endocrine neoplasia. *J Clin Endocrinol Metab* 2007;**92**:3321–5.

25

Genetics of Polyglandular Failure

George J. Kahaly

Department of Medicine I, Johannes Gutenberg University Medical Center, Mainz, Germany

DEFINITION, INCIDENCE, PREVALENCE

The autoimmune polyglandular failure syndromes (APS), also known as multiple endocrine abnormalities (MEA), define the autoimmune-induced failure of at least two glands, and comprise a wide spectrum of autoimmune disorders.[1,2] They encompass a rare juvenile type (APS1) and more frequent adult types (APS2 and APS3).[3] APS1 is also known as autoimmune polyendocrinopathy candidiasis ectodermal dystrophy (APECED) because it is formed by three main disorders, which are chronic mucocutaneous candidiasis, autoimmune hypoparathyroidism, and autoimmune Addison's disease, AD.[2,4,5] In contrast to APS1, APS2 and APS3 primarily manifest in adult age. APS2 is defined as the association of autoimmune AD and a further autoimmune endocrine disorder.[1,4,6,7] The APS3 variant is defined by the presence of autoimmune thyroid disease, AITD, and type 1 diabetes, T1D.[8] In contrast to APS1, chronic candidiasis is not present in APS2 and APS3. Further autoimmune endocrine and nonendocrine component disorders may occur in all APS types. Due to the tremendous overlap of phenotypes in APS2 and APS3, for daily use, it is clinically relevant to differentiate the more common adult type encompassing both APS2 and APS3 from the rare juvenile type APS1.[9,10]

APS1 manifests in infancy or early childhood. The female-to-male ratio approximates one. Prevalence of APS1 is generally rare: 2–3:1,000,000 in Great Britain,[11] but it occurs more frequently in three genetically isolated populations who are Finnish 1:25,000,[12] Iranian Jews 1:9000,[13] and Sardinians 1:14,000.[14] Prevalence in other populations is 1:43,000 in Slovenia,[15] 1:80,000 in Norway,[16] and 1:129,000 in Poland.[17] The adult type APS2/3 is more common, but still a rare syndrome. Its prevalence is 1:20,000.[18] It occurs more frequently in women. The male-to-female ratio is 1:3. The incidence of APS2/3 peaks at ages 20–60 years, mostly in the third or fourth decade.

CLINICAL SPECTRUM

APS1 is characterized by three major disease components.[1,2,4] These are chronic mucocutaneous candidiasis (chronic susceptibility to candida yeast infection), autoimmune hypoparathyroidism (parathyroid gland failure, which affects calcium metabolism including nails and tooth enamel), and AD (autoimmune adrenal failure). The first manifestation occurs in infancy or early childhood and is typically mucocutaneous candidiasis. The second manifestation is hypoparathyroidism around age seven years and the third manifestation is AD around age 13 years.[2,6] Clinical presentation of the diseases varies and comprises further endocrine and nonendocrine minor component disorders.[4,12,19] These include autoimmune endocrinopathies (foremost primary hypogonadism, T1D and AITD are less prevalent), gastrointestinal disorders with malabsorption, pernicious anemia, autoimmune hepatitis, autoimmune skin disorders (vitiligo, alopecia), ectodermal dysplasia (dental enamel hypoplasia, nail dystrophy), and keratoconjunctivitis. The minor components of the disorder do not manifest until the fifth decade of life.[4] Clinical diagnosis of APS1 requires the presence of two of the three major component diseases: AD and/or primary hypoparathyroidism, and/or chronic mucocutaneous candidiasis. Most patients have four disease manifestations and may display up to 10 component diseases. Some patients have the full constellation of diseases, while others do not.[20] However, in some patients, primary hypoparathyroidism was the only manifestation.[13]

APS2 is characterized by the presence of autoimmune adrenalitis (AD) and at least one other autoimmune endocrine disorder.[3,4] The second disease component may either be primary hypogonadism or AITD (Graves' disease, GD, or Hashimoto's thyroiditis, HT) or T1D, or both.[2,7] Further endocrine (primary hypogonadism, primary hypoparathyroidism) and nonendocrine component diseases (pernicious anemia, autoimmune hepatitis, alopecia, vitiligo, celiac disease with malabsorption, and myasthenia gravis) may be present.[6,8] The associated minor autoimmune diseases are less frequent than in APS1. APS2 mostly occurs in adulthood during the third and fourth decades. About half of patients with APS2 initially have T1D developing additional AD (APS2) within a few years.[21] In adults, the manifestation of one autoimmune endocrine disorder increases the risk of developing other autoimmune disorders while several years may separate the onset of different component diseases.[22] Diseases resulting in autoimmune-induced tissue destruction have a prolonged phase of cellular loss preceding overt autoimmune glandular failure. Silent autoantibodies are prevalent in families with APS2 and antibody screening is predictive for the early diagnosis of adrenal failure, especially in children.

The APS3 variant is defined by the presence of AITD and T1D without adrenal involvement.[8] Nonendocrine component diseases include type A autoimmune gastritis, pernicious anemia, vitiligo, alopecia, myasthenia gravis, Sjögren's syndrome, systemic lupus erythematosus, and rheumatoid arthritis. AITD peaks in the fourth decade for GD or fifth and sixth decade for HT. The simultaneous occurrence of autoimmune induced hypothyroidism (HT) and T1D is often accompanied by hypoglycemia due to decreased insulin requirement and increased insulin sensitivity. Glucose intolerance accompanies autoimmune hyperthyroidism in 50% of patients. In contrast to patients with APS1, patients with adult APS do not develop mucocutaneous candidiasis. Also, primary hypoparathyroidism, which is indicative of APS1, is rare in APS2/3. In adult APS, circulating organ-specific autoantibodies are present in each of the component diseases. Occasionally, antibodies will cross-react with more than one gland (e.g., steroid-producing cells). Antithyroid peroxidase and antiparietal cell antibodies are prevalent in healthy relatives of APS patients. Antibodies usually precede clinical disease; however, in contrast to anti-islet antibodies, antithyroid antibodies can be present for decades without progression to overt disease. Antibodies against steroidal enzymes, for example, 21-hydroxylase, are of high predictive and prognostic value. They will aid identifying patients at risk for developing AD.[23,24]

Patients with the adult APS could be exposed to many limitations of their illness in daily life. To objectify the degree of physical and emotional distress, the psychometric profile of patients with APS2/3 was prospectively evaluated and happened to be severely impaired.[25] Treatment modalities that would improve their well-being are warranted. In summary, current diagnosis of adult APS involves serological measurement of organ-specific autoantibodies and subsequent functional testing. Management of patients with APS, including their family relatives, is best performed in centers with special expertise in autoimmune endocrine disorders.

GENETIC PATHOPHYSIOLOGY

Known Mutations/Polymorphisms and Specific Phenotypes

APS1 is a monogenic disease due to a defect in a single gene. Haplotype analyses suggest that APS1 is caused in different populations by a number of different mutations in a single gene.[26] This gene is the *autoimmune regulatory (AIRE)* gene on chromosome 21, location 21q22.3.[27] The gene was identified in 1997 by positional cloning.[28,29] The *AIRE* gene is approximately 13 kb in length and the coding sequence comprises 14 exons. Currently, more than 50 different mutations in the *AIRE* gene causing APS1 have been detected.[14,15,28–43] These mutations are distributed over the whole coding region of the gene (Fig. 25.1). They comprise point mutations (nonsense and missense mutations), insertions, and deletions resulting in frameshifts and splice site mutations. The most frequent mutations are R257X, exon 6[28,29] and 967-979del13bp (also denominated in the literature as 1094-1106del13bp, exon 8).[30] The R257X mutation is characterized by a C→T substitution at position 769 of the *AIRE* gene, resulting in a TGA codon (stop codon) instead of CGA (coding for arginine), leading to a truncated regulator protein. It is present in 83% of Finnish APS1 patients[35] and predominates in Italian[31] and Polish APS1 patients.[17] Also, it is found in APS1 patients from Great Britain, Germany, France, Sweden, the Netherlands, Switzerland, Austria, Hungary, Croatia, Serbia, Slovenia, Russia, the USA (Caucasians), and New Zealand.[38,39] It was detected in patients on different chromosomal haplotypes suggesting various mutational origins. The frequent 13-bp deletion accounted for 70% of British and

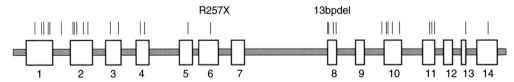

FIGURE 25.1 Mutations in the *AIRE* gene on chromosome 21 causing APS type 1. The *AIRE* gene comprises 14 exons, which are shown as rectangles. Lines indicate known mutations in the *AIRE* gene causing APS1. Most frequent mutations are R257X (exon 6) and 967-979del13bp (exon 8).

53% of North American (Caucasian) APS1 alleles but also occurred in Finland, Sweden, Norway, the Netherlands, Germany, Italy, Hungary, Canada, New Zealand, Russia, and other countries.[30,33,38,39] Thus, both R257X and 967-979del13bp have been noted in patients of different geo-ethnic origins, and both were associated with multiple different haplotypes using closely flanking polymorphic markers showing likely multiple mutation events.[31] The mutations causing APS1 are inherited in an autosomal recessive way.[28,44] One mutation (G228W) in the AIRE gene observed in an Italian family has a dominant inheritance.[37] In this family, only one heterozygous mutation was found in the entire coding sequence of the AIRE gene.

In contrast to APS1, APS2 and 3 are genetically complex and multifactorial syndromes. Several genetic loci possibly interact with environmental factors. Adult APS types are strongly associated with certain alleles of the human leucocyte antigen (HLA) genes within the major histocompatibility complex (MHC) (Table 25.1). The HLA-DRB1*03 allele was strongly increased in patients with adult APS (50.7%) versus both controls (21.8%, $P < 0.0001$; RR 2.32, 95% CI 1.62–3.33) and monoglandular autoimmune disease (11.4%, $P < 0.0001$). HLA-DRB1*03 was highly prevalent in APS2/3 patients with early versus late disease onset ($P < 0.05$, logistic regression analysis). HLA-DRB1*04 allele carriers were more present in adult APS versus controls (53.4% vs. 22.4%, $P < 0.0001$; RR 2.38, 95% CI 1.68–3.38). Further,

HLA-DQB1*02 was increased in adult APS versus controls ($P < 0.01$), whereas HLA-DQB1*06 was decreased ($P < 0.001$). Thus, HLA-DRB1*03 is a stronger genetic marker in adult APS than in monoglandular autoimmune diseases, foremost in those with early disease onset.[45]

APS2 frequently clusters in families. Several generations are often affected by one or more component diseases.[1] The inheritance pattern is autosomal dominant with incomplete penetrance in some patients.[6] Two genes have been shown to be associated with APS2. These are HLA genes on chromosome 6 and the cytotoxic T lymphocyte antigen (CTLA-4) gene on chromosome 2 (location 2q33). Of these, HLA appears to have the strongest gene effect.[4] Patients with APS2 had significantly more the HLA alleles DRB1*03 ($P < 0.0001$), DRB1*04 ($P < 0.000005$), DQA1*03 ($P < 0.0001$), DQB1*02 ($P < 0.05$) when compared to controls. Less frequent in APS2 were DRB1*15 ($P < 0.05$), DQA1*01 ($P < 0.0005$), DQB1*05 ($P < 0.005$). With regard to frequency and linkage of these alleles, the susceptible haplotypes DRB1*0301-DQA1*0501-DQB1*0201 and DRB1*0401/04-DQA1*0301-DQB1*0302 were deduced. Protective haplotypes in this study[46] were DRB1*1501-DQA1*0102-DQB1*0602 and DRB1*0101-DQA1*0101-DQB1*0501. Patients with T1D as a singular disease had the same susceptible and protective HLA alleles and haplotypes. The prevalence of DRB1*03 and DRB1*04 in APS2 patients was not due to the presence of diabetes, since the APS2 without T1D had the same allele distribution. These data suggest a common immunogenetic

TABLE 25.1 Putative Susceptibility Genes for Autoimmune Polyglandular Failure

Gene locus	Chromosomal location	Exons	Mutations/polymorphisms
IMMUNE-MODIFYING GENES			
MHC	6p21	–	MHC class II region
MICA	6p21.3	6 exons	Exon 5: GCT repeat (transmembrane region)
CTLA-4	2q33	4 exons	Exon 1: SNP +49 A/G STR (AT)$_n$ repeat within 3'UTR, allele 106 Intron 1: SNP +1822 C/T 3'UTR: SNPs CT60/G, JO31/G, JO30/G
CD40	20q11	9 exons	SNP C/T (position-1, designated as CD40-E1SNP)
CYTOKINE-RELATED GENES			
IL-1RA	2q14.2	6 exons	Intron 2: VNTR*
IL-4	5q31.1	4 exons	Promoter-590 C/T
TNF-α	6p21.3	4 exons	Promoter-863 C/A, 1031 T/C
Autoimmune regulator gene AIRE-1	21q22.3	14 exons	>45 mutations, most frequent mutations: Exon 6: R257X, Arg257stop Exon 8: 13bpdel, deletion of 13 bp (964del13)

CTLA-4, cytotoxic T-lymphocyte-associated protein 4; CD40, B cell-associated molecule CD40; IL-1, interleukin 1; IL-4, interleukin 4; IL-1RA, interleukin-1 receptor antagonist; MICA, *MHC class I chain-related gene A*; SNP, single nucleotide polymorphism; STR, short tandem repeat; TNF-α, tumor necrosis factor alpha; VNTR, variable number of an 86-tandem repeat.

* A penta-allelic 86-bp tandem repeat (VNTR) occurs in intron 2, of which allele 2 (IL1RN*2) is associated with autoimmune conditions.

pathomechanism for T1D and APS2, which might be different from the immunogenetic pathomechanism of other autoimmune endocrine diseases.

Many APS2 component disorders are associated with an increased frequency of the *HLA* haplotype A1, B8, DR3, DQA1*0501, DQB1*0201.[47] AD is strongly associated with DR3 and DR4[48]; the observed relative risks are 6.0, 4.6, and 26.5 for DR3, DR4, and DR3/DR4, respectively. AD is also correlated with DQ2/DQ8 with DRB1*0404, both as a single disease as well as within APS2.[49] T1D is positively associated with DR4-DQB1*0302, DRB1*04-DQA1*0301-DQB1*0302, or DRB1*03-DQA1*0501-DQB1*0201 (DR3-DQ2), and negatively associated with DRB1*15-DQA1*0102-DQB1*0602.[50,51] In APS2 patients without islet cell autoimmunity, only the haplotype DR3-DQB1*0201 occurred more frequently.[47] The DRB1*04-DQB1*0301 haplotype increases the risk for developing HT.[52]

Very recently, the HLA class II alleles, haplotypes, and genotypes were determined in a large cohort of patients with adult APS, AITD, T1D, and in healthy controls by the consistent application of high resolution typing at a four digit level. Inclusion of family members of APS patients enabled the assignment of segregation-derived HLA class II haplotypes.[53] Comparison of the allele and haplotype frequencies significantly discriminated patients with APS versus AITD and controls. The alleles DRB1*03:01, *04:01, DQA1*03:01, *05:01, and DQB1*02:01, *03:02 were more prevalent ($P < 0.001$) in adult APS than in AITD and controls. The DRB1*03:01-DQA1*05:01-DQB1*02:01 (DR3-DQ2) and DRB1*04:01-DQA1*03:01:DQB1*03:02 (DRB1*04:01-DQ8) haplotypes were overrepresented in adult APS ($P < 0.001$). The combination of both haplotypes to a genotype was highly prevalent in adult APS versus AITD and controls ($P < 0.001$). Dividing the APS collective into those with AD and T1D (APS2) and those without AD but including T1D and AITD (APS3) demonstrated DR3-DQ2/DRB1*04:01-DQ8 as a susceptibility genotype in APS3 ($P < 0.001$), whereas the DR3-DQ2/DRB1*04:04-DQ8 genotype correlated with APS2 ($P < 0.001$). The haplotypes DRB1*11:01-DQA1*05:05-DQB1*03:01 and DRB1*15:01-DQA1*01:02-DQB1*06:02 were protective in APS3 but not in type 2 ($P < 0.01$). Thus, the association of HLA class II haplotypes with APS2/3 depends on the contribution of either T1D or AD. Susceptible haplotypes favor the development of polyglandular autoimmunity in AITD patients.

The *CTLA-4* gene is assigned to chromosome 2 (location 2q.33). It comprises four exons and encodes a protein that acts as an important modifier of T-cell activation with downregulatory properties. An A49G substitution in exon 1 of the *CTLA-4* gene with more G alleles has been associated with GD in Caucasians and Asians[54–59] and with HT.[60] With respect to GD, family studies give evidence for an increased transmission of the G allele

from heterozygous parents to affected offspring compared to unaffected offspring.[61] A 3' microsatellite $(AT)_n$ repeat of the *CTLA-4* gene is relevant. The 106 bp allele of the AT repeat was more frequently observed in Caucasian patients with GD than in healthy controls.[58] *CTLA-4* alleles have been linked to AITD, and to a weaker extent to T1D. AD was also associated with *CTLA-4* alleles, particularly in a subgroup showing *HLA*-DQA1*0501. Data indicate interactions between *HLA* and *CTLA-4* genes, further unidentified genes and environmental factors.

APS3 is also characterized by a complex inheritance pattern. Family and population studies showed that the APS3 variant syndrome has a strong genetic background. Whole genome and candidate gene approaches identified several gene variations that are present in both AITD and T1D. Most important common disease susceptibility genes are *HLA*, *CTLA-4*, *the protein tyrosine phosphatase non-receptor type 22* (*PTPN22* on chromosome 1), the *forkhead box P3* (*FOXP3* on the X chromosome) and the *interleukin-2 receptor alpha (IL-2Ralpha)/CD25* gene region (chromosome 10), all of them contributing to the susceptibility for APS3.[62] With respect to the underlying pathogenetic mechanisms, these genes are altogether involved in the immune regulation, in particular in the immunological synapse and T-cell activation. In addition to these common genes, there are further candidate genes with joint risk for AITD and T1D, in particular the *v-erbb2 erythroblast leukemia viral oncogene homolog 3* (*ERBB3*) gene on chromosome 12 and the *C-type lectin domain family 16 member A* (*CLEC16A* on chromosome 16). The latter one might be involved in pathogen recognition.

HLA class II is a potential gene locus for combined susceptibility to T1D and AITD as has been shown in Caucasians and Asians.[47,63–68] Most family studies gave evidence that the haplotype *HLA*-DR3-DQB1*0201 is the primary haplotype conferring susceptibility to both T1D and AITD within families.[67] Here, DR3 seems to be the primary allele conferring risk to both T1D and AITD, whereas DQB1*0201 is less relevant. Many population studies indicate that both *HLA* haplotypes DR3-DQB1*0201 and DR4-DQB1*0302 contribute to the APS3 variant of combined T1D and AITD.[47,67,69]

The causative *CTLA-4* gene polymorphism for autoimmunity may be located in the 3'UTR (untranslated region) of the *CTLA-4* gene.[70] Here, an $(AT)_n$ microsatellite polymorphism occurs with longer and shorter repeats of AT (Fig. 25.2). The longer repeats are associated with decreased inhibitory function of *CTLA-4*.[71] Longer repeats were correlated with a shorter half-life of the *CTLA-4* mRNA than shorter repeats.[72] *CTLA-4* CT60, another *CTLA-4* gene polymorphism, was analyzed in patients with APS3, AITD, T1D, and healthy controls.[73] The CT60 G/G genotype was significantly more common in patients with APS3 than in healthy controls (48.6% vs. 32.0%, OR 2.01, 95% CI 1.07–3.77, $P = 0.038$). The CT60

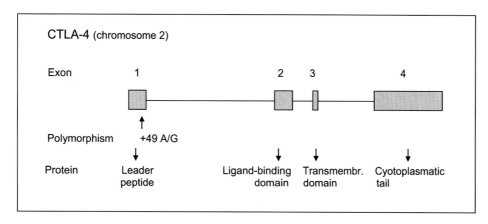

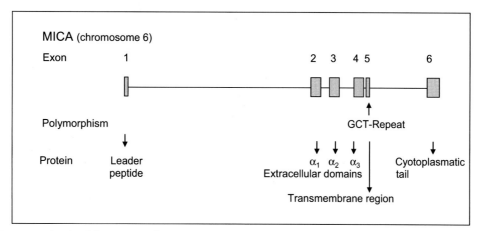

FIGURE 25.2   Gene organization (chromosome lines with exons as vertical boxes), important polymorphisms and mutations, as well as a schematic representation of protein are shown for selected putative susceptibility and immunoregulatory genes. Cytotoxic T-lymphocyte-associated protein 4 (CTLA-4) and *MHC class I chain-related gene A (MICA)*. The CTLA-4 exon 1 A49G dimorphism results in a threonine to alanine amino acid substitution at codon 17 in the leader peptide of the protein. For MICA, over 50 alleles have been described and most polymorphisms are SNPs encoding the extracellular domains. There is an additional triplet repeat microsatellite polymorphism (GCT) in exon 5 with five most common alleles (A4, A5, A5.1, A6, and A9) corresponding to 4, 5, 6, and 9 GCT repetitions. The A5.1 allele contains an additional G insertion (GGCT) causing a frameshift mutation resulting in a premature termination of the MICA protein.

allele frequencies differed as well between APS3 patients and controls, with the predisposing G allele being increased in APS3 (OR 1.63, 95% CI 1.03–2.55, $P = 0.042$). Patients with APS3 did not differ from those with AITD ($P = 0.602$) or T1D ($P = 0.362$).

The *PTPN22* gene maps on chromosome 1, location 1p13.[74] This gene encodes the lymphoid tyrosine phosphatase (LYP) protein. Alternative splicing of this gene results in two transcript variants encoding distinct isoforms of the protein. A single nucleotide polymorphism (SNP) in the *PTPN22* gene, an 1858C→T transition, results in an arg620-to-trp (R620W) substitution in the LYP protein.[75] The minor T allele was found to be associated with T1D,[75] AITD,[76] and other autoimmune diseases. This is involved in altered T lymphocyte activation. In Asian patients, a novel SNP in the promoter region of the *PTPN22* gene, G1123C, has been recently identified and associated with T1D and AITD.[77] Therefore, the promoter SNP is a further possible causative variant

for endocrine autoimmunity. Additional candidate polymorphisms may be also causative.[78]

In an association study, 310 white subjects with APS3, AITD, T1D, or healthy controls were genotyped for the C1858T polymorphism.[79] All subjects were also typed for HLA-DRB1. The PTPN22 1858 minor T-allele frequency was strongly increased in patients with APS3 (23.6%) compared with controls (8.0%, $P < 0.001$), with patients with AITD only (8.6%, $P < 0.006$), or with T1D only (10.7%, $P < 0.028$). T-allele carriers were also more frequently present in the group with APS3 vs. controls (41.4% vs. 14.0%, OR 4.35, 95% CI 2.08–9.09), AITD (17.1%, OR 3.42, 95% CI 1.56–7.48), and T1D (21.4%, OR 2.59, 95% CI 1.23–5.45). Especially in subjects with HT + T1D, T-allele carriers were mostly frequent (50% vs. 14%, OR 6.14, 95% CI 2.62–14.38, $P < 0.001$). Considering all included patients with AITD, T-allele carriers were 29.3% vs. 14.0% in controls ($P < 0.008$, OR 2.54, 95% CI 1.30–4.98). Patients carrying the PTPN22 1858 T allele had a twofold increased frequency of the HLA-DRB1*03 allele (64.7% vs. 37.3%, $P < 0.034$). In conclusion, the PTPN22 gene is a joint susceptibility locus for joint AITD (especially HT) and T1D (APS3).

Data regarding polymorphisms of immunoregulatory genes in polyglandular failure are lacking. Recently, the putative association between a polymorphism of the pro-inflammatory cytokine gene *tumor necrosis factor TNF-α -308* and mutations of the *AIRE* gene with APS in adults was analyzed.[80] The *TNF-α -308*A* allele occurred more frequently in patients (0.269) than in controls (0.163, $P = 0.008$). Also, *TNF-α -308*A* carriers were more frequent in patients than controls (47.8% vs. 31.1%, OR 1.89, 95% CI 1.19–3.00). The frequency of the AA genotype was increased in adult APS ($P = 0.014$). APS2/3 patients with AITD and the *TNF-α -308 AA* genotype showed the highest prevalence of thyroid autoantibodies. HLA-DRB1*03 and *TNF-α -308*A* alleles were strongly associated in patients with APS 2/3 (87.5%, $P < 0.00001$). In contrast, the *AIRE R257X* and *13bpdel* mutations were not observed in patients with adult APS.

Finally, a deficiency in the DNase enzyme, and thereby a failure to remove DNA from nuclear antigens, promotes disease susceptibility to autoimmune disorders. Recent studies[81,82] examined in patients with AITD and adult APS whether a reduced DNase activity is associated with sequence variations in the *DNASE1* gene. In patients with adult APS, a novel mutation (1218G>A, exon 5) and multiple polymorphisms were identified in the *DNASE1* gene. The allele frequency of the mutation was increased in patients vs. controls ($P < 0.001$). In contrast to controls, the novel mutation was present in all five members of a family with adult APS and AITD, showing decreased DNase activity. The mutation resulted in the replacement of highly conserved valine with methionine at amino acid position 89 of the DNase enzyme. It was related to lowered heat stability and lowered activity of the enzyme. Mean expression of the *DNASE1* mRNA in patients was 0.52 ± 0.22 and in controls 0.95 ± 0.22. The expression level of the *DNASE1* gene was strongly decreased in patients, amounting to only 54.7% of that in controls ($P < 0.001$). The identified new mutation and numerous polymorphisms, noted for the first time in patients with APS and AITD, may alter transcription and translation of the *DNASE1* gene, thereby decreasing the stability and activity of the corresponding enzyme.

Pathophysiology of Mutation

APS1

The *AIRE* gene is expressed in tissues that are involved in the maturation of the immune system such as thymus, lymph nodes, and fetal liver. It is expressed in epithelial antigen-presenting cells in the thymus where it is involved in the central induction of self-tolerance.

Thus, the *AIRE* gene is an important mediator of central tolerance. *AIRE* may regulate negative selection of organ-specific T cells.[83] The *AIRE* gene encodes a 545 amino acid protein of 57.5 kDa, which comprises several domains involved in nuclear transport, DNA binding, homomultimerization, and transcriptional activity.[84] *AIRE* shows several motives indicative of a transcription factor, and it includes two zinc fingers. *AIRE* upregulates transcription of certain organ-specific self-antigens in medullary thymic epithelial cells and plays a role in the negative selection of organ-specific thymocytes. At least three splice variant mRNAs products have been described,[85] including one resulting in a premature stop codon in the *AIRE* protein and a transcript predicted to be a candidate for nuclear-mediated decay (NMD). The mutated *AIRE* gene results in defect *AIRE* proteins, causing autoimmune destruction of target organs by disturbing the immunological tolerance of the patients.[84] Many *AIRE* mutations alter the nucleus-cytoplasm distribution of *AIRE*, thereby disturbing its association with nuclear dots and cytoplasmic filaments.[84] The R257X mutation results in a stop codon instead of CGA (coding for arginine), leading to a truncated regulator protein. Because APS1 patients homozygous for R257X display a considerable phenotypic disease variation, further genetic or environmental factors may determine the manifestation of the syndrome.

APS2/3

The gene products of the *HLA* class II genes are involved in immune reactions. The different *HLA* class II alleles are characterized by different affinities for peptides. As a consequence, some autoantigenic peptides may be recognized by T lymphocyte receptors, whereas others may not.[86] The *CTLA-4* gene encodes a negative regulator of T-cell activation, which is expressed on the surface of activated T lymphocytes (Table 25.2). It is involved in the interaction between T lymphocytes and antigen-presenting cells, APCs.[87] APCs present to the T lymphocyte receptor an antigenic peptide bound to an *HLA* class II protein on the cell surface, thus activating T lymphocytes. Further, costimulatory signals on the APC surface interact with receptors (e.g., *CTLA-4*) on the surface of CD4+ T lymphocytes during antigen presentation.[87] *CTLA-4* downregulates T lymphocyte activation.[88] *CTLA-4* polymorphisms are associated with several autoimmune disorders, particularly with AITD[89] but also with AD. In contrast, findings are inconsistent with respect to the association of *CTLA-4* and T1D suggesting a weak effect.[54,70,90–92] A 3'UTR (AT)$_n$ microsatellite polymorphism with longer and shorter repeats of AT may be related to autoimmunity while longer repeats are associated with decreased inhibitory function of *CTLA-4*.[71] Longer repeats were correlated with a shorter half-life of the *CTLA-4* mRNA than shorter repeats.[72] The *CTLA-4* AT repeat affects the inhibitory function

TABLE 25.2 Proposed Functions of the Putative Susceptibility Genes for Autoimmune Polyglandular Failure

Gene	Putative function
IMMUNE-MODIFYING GENES	
MHC	Modulation of binding and presentation of TSH-R peptides by MHC-DR sequences
	Aberrant expression of MHC class II molecules on thyrocytes
MICA	Binds NKG2D receptors on NK cells, gamma/delta T cells, and CD8$^+$ alpha/beta T cells
	Engagement of NKG2D receptors stimulates NK-cells and T-cell effector functions
	MIC-A5.1 allele → frameshift → premature termination codon (transmembrane region of MICA protein) → protein lacks for cytoplasmic domain → altered ligands for NK cells?
CTLA-4	Different effects of SNP +49 A/G alleles on inhibiting function of CTLA-4
	AT microsatellite repeat within 3′ UTR influences half-life of CTLA-4 m RNA
CD40	Interacts with CD40 ligand on T cells
Autoimmune regulator gene AIRE-1	Encodes protein containing zinc finger motives → transcriptional regulatory activity suggestive of transcription factor

AA, amino acid; Abs, autoantibodies; NK cells, natural killer cells; NKG2, group of genes that are expressed primarily in NK cells encoding a family of C-type lectins. The NKG2D gene is expressed as a major 1.8- and a minor 3.2-kb transcript in NK cell lines and in some T-cell lines.

of *CTLA-4* in that the long AT repeat allele is associated with a reduced control of T-cell proliferation in patients with GD.[71]

The *PTPN22* gene encodes the LYP, which is expressed primarily in lymphoid tissues. It is expressed in both immature and mature B and T lymphocytes. This enzyme associates with the molecular adapter protein CBL and may be involved in regulating CBL function in the T-cell receptor signaling pathway. LYP inhibits the T-lymphocyte antigen receptor signaling pathway.[93] It binds to the protein kinase (Csk), thereby limiting the response to antigens.[94] A mutation in *PTPN22* causing a tryptophan for arginine substitution in the LYP protein (R620W) has been reported to be associated with T1D, AITD, and vitiligo.[89,94–96] In contrast, AD is not associated with *PTPN22*.[96] With respect to AITD, the *PTPN22* variant is associated with GD as well as with HT.[57,76,97]

DIAGNOSIS, GENETIC TESTING, AND INTERPRETATION

Genetic testing may identify patients with APS1, but not those with APS2/3. For APS2/3, only susceptibility genes might be identified, which increase the risk for developing autoimmune diseases, but must not be causative for the disease. Therefore, the following narrative applies to genetic testing for APS1. Clinical diagnosis of APS1 is based on the presence of at least two of the three primary component diseases and is often delayed until severe complications associated with an endocrinopathy have developed. The diagnosis is confirmed by genetic

testing. Genetic testing for APS1-associated sequence variants in the *AIRE* gene provides the diagnosis of APS1 when only one of the characteristic component diseases has manifested. Also, in a child without overt disease or clinical manifestations, a predisposition for APS1 before any symptoms appear may be detected.

Which Tests are Best to Order in Whom?

Serological tests include autoantibodies against 17- and 21-hydroxylase (cytochrome 450) and autoantibodies against calcium sensing receptor for early testing of autoimmune-induced primary hypogonadism, autoimmune adrenal failure, and primary hypoparathyroidism, respectively. Definitive diagnosis of glandular failure within APS1 is obtained by functional testing and measurement of serum levels of gonadotropins, male (testosterone) or female (estradiol) sex hormones, baseline cortisol, parathormone, and the electrolytes.

Laboratories Available for Genetic Testing

Clinical laboratories available for genetic testing for identifying mutations in the *AIRE* gene related to APS1 are summarized in Table 25.3. The following different tests are provided by these laboratories:

• *Sequence analysis of the entire coding region* of the *AIRE* gene provides the order and nature of the nucleotide bases in the *AIRE* gene. Thus, pathogenic sequences reported in the literature as well as DNA sequence alteration predicted to be pathogenic but not reported in the literature may be identified.

TABLE 25.3   Clinical Laboratories Available for Genetic Testing of the *AIRE* Gene Causing APS1

Address of laboratory	Sequence analysis of entire *AIRE* coding region	Mutation scanning of entire *AIRE* coding region	Targeted mutation analysis	Prenatal diagnosis	DNA carrier testing
EUROPE					
Amplexa Genetics Odense, Denmark http://www.amplexa.com/		X			
HUSLAB Laboratory of Molecular Genetics Helsinki, Finland http://www.hus.fi/ Path = 1;28;824;2049;2265;2264;3663;4582			X		X
Oxford Medical Genetics Laboratories Oxford, United Kingdom Email: anneke.seller@orh.nhs.uk	X			X	X
ISRAEL					
Wolfson Medical Center Molecular Genetics Laboratory Holon, Israel Email: lev@wolfson.health.gov.il			X	X	X
USA					
Athena Diagnostics Inc. Reference Lab Worcester, MA http://athenadiagnostics.com	X				
Baylor College of Medicine Medical Genetics Laboratories Houston, TX http://www.bcm.edu/geneticlabs/	X			X	X
Correlagen Diagnostics, Inc. Waltham, MA http://www.correlagen.com/index.jsp	X				X
GeneDx, Inc. Gaithersburg, MD http://www.genedx.com/	X			X	X

- *Mutation scanning of the entire coding region* is applied when sequence analysis would be excessively time-consuming due to the size of a given gene. It identifies variant gene regions. This method is often applied when mutations are distributed throughout a gene and when families display different mutations.
- *Targeted mutation analysis* (allele-specific mutation analysis) means testing for the presence of common *AIRE* mutations found in most or a high proportion of individuals with APS1.
- *Carrier testing* is based on simple blood tests. It allows detecting recessive *AIRE* mutations. Often, abnormal recessive *AIRE* genes are passed from generation to generation without causing any clinical

overt disease and/or phenotype. Asymptomatic carriers of recessive genes could have a child with APS1 if both parents have the same negative *AIRE* recessive gene. For candidate parents this crucial information is highly relevant prior to children planning.

Most laboratories perform *sequence analysis* of the entire *AIRE* gene (Oxford, UK; Worcester, MA; Houston, TX; Waltham, MA; Gaithersburg, MD). The Baylor College of Medicine in Houston, TX, fully sequences the *AIRE* gene in both directions. All 14 exons of the *AIRE* gene are amplified by a PCR-based assay. The clinical sensitivity is approximately 80%, and the analytical sensitivity is over 98%. Also, an analysis of at-risk family members for identified mutations can be performed. The Correlagen Diagnostics, Inc. in Waltham (MA) analyses 59 pathogenic variants (known, possibly, and probably pathogenic variants) in the *AIRE* gene. Indications for testing are chronic mucocutaneous candidiasis, two or more autoimmune manifestations of APS1 and a family history of APS1 or chronic mucocutaneous candidiasis. The Amplexa Genetics laboratory in Odense, Denmark, performs *mutation scanning* of the *AIRE* gene. It scans the *AIRE* coding region for mutations. *Targeted mutation analysis* is performed in Helsinki (Finland) and Holon (Israel). *Carrier testing* is done in Helsinki (Finland), Oxford (UK), Holon (Israel), Houston (TX), Waltham (MA), and Gaithersburg (MD).

Prenatal Testing

Prenatal testing for mutations in the *AIRE* gene related to APS1 is done in laboratories in Oxford (UK), Holon (Israel), Houston (TX), and Gaithersburg (MD). These laboratories are listed in Table 25.3. The Medical Genetics Laboratories at Baylor College of Medicine (Houston, TX) perform prenatal diagnosis with respect to known mutations in the *AIRE* gene.

Lists of Resources

APS1 Resources

National Adrenal Diseases Foundation (NADF): 505 Northern Boulevard, Great Neck, NY 11021, USA; Phone: 516-487-4992, Email: nadfmail@aol.com, Web: http://www.nadf.us/
National Foundation for Ectodermal Dysplasias (NFED): 6 Executive Dr., Suite 2, Fairview Heights, IL 62208-1360; USA; Phone: 618-566-2020, Fax: 618-566-4718, Web: http://www.nfed.org/

Predictive Value of Test

Genetic testing for mutations in the *AIRE* gene associated with APS1 should be done when there is a highly likely presumption of APS1. Genetic testing may identify APS1 and permits fast and accurate diagnosis of APS1 after only one of the primary component diseases of APS1 has developed. It may inform patients and physicians to the risk of developing other manifestations of APS1 and will support genetic counseling. Also, mutational analysis of the *AIRE* gene may help identifying patients with atypical phenotypes resembling to APS1. Early diagnosis of APS1 by genetic testing may facilitate an early intervention because *de novo* autoimmune endocrine and nonendocrine disorders may develop. Also, it may help to prevent severe complications.

Significance of Negative Test

In some laboratories, patients can only be tested for known mutations in the *AIRE* gene. Therefore, a negative test may not exclude APS1. Also, an *AIRE* mutation may occur *de novo*[30] and pathogenic mutations were reported in the promoter and intron regions of the *AIRE* gene. However, some genetic tests are restricted to the coding region of the *AIRE* gene. In the *AIRE* promoter region, methylation patterns and regulatory elements have been detected indicating that expression of the *AIRE* gene may be modulated through modifications in chromatin methylation and acetylation.[98] This will not be detected by current genetic testing.

Should Family Members be Tested?

For family members of APS1 patients, genetic diagnosis allows avoiding unnecessary follow-up of family members, if *AIRE* mutations are not present. Genetic testing for mutations in the *AIRE* gene identifies APS1-associated mutations in relatives of APS1 patients before the occurrence of symptoms and signs, allowing early intervention and helping prevent potentially fatal complications from untreated endocrine insufficiencies. Children below the age of 10 years with monoglandular autoimmune disease, for example, AD, autoimmune hypoparathyroidism, and autoimmune primary hypogonadism, should be tested for *AIRE* mutations. In the presence of a positive *AIRE* gene mutation, the risk for developing polyglandular failure is extremely high. The relative risks should be discussed during genetic counseling of the families at risk.

MANAGEMENT

How Does the Genetic Information Affect Treatment, if at All?

The treatment of APS1 is based on treatment of the component diseases. It is directed at replacing the various hormones that are in short supply, treating the yeast infections. Immunosuppressive therapy may be indicated in severe cases of APS1. However, there is no current cure for APS1. Prognosis depends on whether infections can be successfully controlled, and whether the critical hormone deficiencies are remedied. Each of the component disorders of the adults APS types is characterized by several stages beginning with active organ-specific autoimmunity and followed by metabolic abnormalities with a clinically manifest or overt disease. Circulating organ-specific autoantibodies are observed in the various component diseases of APS2/3. The presence of such antibodies precedes clinical overt disease. Since these autoantibodies are predictive for the development of future autoimmune polyglandular diseases, kindreds and siblings of affected APS patients should be regularly screened. Several endocrine component disorders can be adequately treated with hormonal replacement therapy if the disease is recognized early. Regular follow-up of patients with monoglandular autoimmune disease, most especially those with AD and T1D, and to a lesser degree HT, is warranted, since a second autoimmune glandular disease may occur between 1 and 20 years after the manifestation of the first glandular failure.[99] Furthermore, serological screening of the first-degree relatives of patients with APS2/3 is recommended due to the high prevalence of various autoantibodies in these kindreds. Presence of organ autoantibodies in these relatives should be completed by functional diagnosis of an eventual glandular dysfunction.[100]

Genetic Counseling

Genetic counseling is recommended for families and kindreds of patients with APS1 since we are dealing with a monogenetic disease. Counseling should be done before performing genetic testing of the *AIRE* gene. For the adult APS type encompassing APS2 and APS3, genetic counseling is optional as several genes and environmental factors may be involved in the pathogenesis of the disease, contributing to the loss of immune self-tolerance. Based on a genetic predisposition, external factors such as pathogens, that is, viral or bacterial infections,[101] and psychosocial factors might induce autoimmune reactions. Counseling of the adult type should emphasize the rationale for a regular follow-up of the kindreds and first-degree relatives of these patients.

References

1. Neufeld M, MacLaren N, Blizzard R. Autoimmune polyglandular syndromes. *Pediatr Ann* 1980;9:154–62.
2. Betterle C, Greggio NA, Volpato M. Clinical review 93: autoimmune polyglandular syndrome type 1. *J Clin Endocrinol Metab* 1998;83:1049–55.
3. Eisenbarth GS, Gottlieb PA. Autoimmune polyendocrine syndromes. *N Engl J Med* 2004;350:2068–79.
4. Neufeld M, MacLaren NK, Blizzard RM. Two types of autoimmune Addison's disease associated with different polyglandular autoimmune (PGA) syndromes. *Medicine* 1981;60:355–62.
5. Obermayer-Straub P, Strassburg CP, Manns MP. Autoimmune polyglandular syndrome type 1. *Clin Rev Allergy Immunol* 2000;18:167–83.
6. Betterle C, Dal Pra C, Mantero F, Zanchetta R. Autoimmune adrenal insufficiency and autoimmune polyendocrine syndromes: autoantibodies, autoantigens, and their applicability in diagnosis and disease prediction. *Endocr Rev* 2002;23:327–64.
7. Betterle C, Lazzarotto F, Presotto F. Autoimmune polyglandular syndrome 2: the tip of an iceberg? *Clin Exp Immunol* 2004;137:225–33.
8. Betterle C, Volpato M, Greggio AN, Presotto F. Type 2 polyglandular autoimmune disease (Schmidt's syndrome). *J Pediatr Endocrinol Metab* 1996;9:113–23.
9. Kahaly GJ, Dittmar M. Polyglandular failure syndromes. In: Jamieson L, editor. *In: Harrison's online updates. Disorders affecting multiple endocrine systems*. New York, NY, USA: McGraw-Hill; 2004 Chapter 330.
10. Kahaly GJ, Dittmar M. Autoimmune polyglandular syndrome type 2. In: Weetman AP, editor. *In Autoimmune Diseasesin Endocrinology*. Totowa, NJ: Humana Press; 2007. p. 377–391.
11. Pearce SH, Cheetham TD. Autoimmune polyendocrinopathy syndrome type 1: treat with kid gloves. *Clin Endocrinol* 2001;54:433–5.
12. Ahonen P, Myllarniemi S, Sipila I, Perheentupa J. Clinical variation of autoimmune polyendocrinopathy-candidiasis-ectodermal dystrophy (APECED) in a series of 68 patients. *N Engl J Med* 1990;322:1829–36.
13. Zlotogora J, Shapiro MS. Polyglandular autoimmune syndrome type I among Iranian Jews. *J Med Genet* 1992;29:824–6.
14. Rosatelli MC, Meloni A, Meloni A, Devoto M, Cao A, Scott HS, Peterson P, Heino M, Krohn KJE, Nagamine K, Kudoh J, Shimizu N, Antonarakis SE. A common mutation in Sardinian autoimmune polyendocrinopathy-candidiasis-ectodermal dystrophy patients. *Hum Genet* 1998;103:428–34.
15. Podkrajsek KT, Bratanic N, Krzisnik C, Battelino T. Autoimmune regulator-1 messenger ribonucleic acid analysis in a novel intronic mutation and two additional novel AIRE gene mutations in a cohort of autoimmune polyendocrinopathy-candidiasis-ectodermal dystrophy patients. *J Clin Endocrinol Metab* 2005;90:4930–5.
16. Myhre AG, Halonen M, Eskelin P, Ekwall O, Hedstrand H, Rorsman F, Kampe O, Husebye ES. Autoimmune polyendocrine syndrome type 1 (APS I) in Norway. *Clin Endocrinol* 2001;54:211–7.
17. Stolarski B, Pronicka E, Korniszewski L, Pollak A, Kostrzewa G, Rowinska E, Wlodarski P, Skorka A, Gremida M, Krajewski P, Ploski R. Molecular background of polyendocrinopathy-candidiasis-ectodermal dystrophy syndrome in a Polish population: novel AIRE mutations and an estimate of disease prevalence. *Clin Genet* 2006;70:348–54.
18. Ten S, New M, MacLaren N. Clinical review 130: Addison's disease. *J Clin Endocrinol Metab* 2000;86:2909–22.
19. Perheentupa J. APS-I/APECED: the clinical disease and therapy. *Endocrinol Metab Clin North Am* 2002;31:295–320.
20. Dittmar M, Kahaly GJ. Genetics of autoimmune hypoparathyroidism. *Clin Cases Min Bone Metab* 2004;1:113–6.
21. Meyerson J, Lechuga-Gomez EE, Bigazzi PE, Walfish PG. Polyglandular autoimmune syndrome: current concepts. *Canad Med Assoc J* 1988;138:605–12.
22. Dittmar M, Kahaly GJ. Polyglandular autoimmune syndromes – immunogenetics and longterm follow-up. *J Clin Endocrinol Metab* 2003;88:2983–92.
23. Chen S, Sawicka J, Betterle C, Powell M, Prentice L, Volpato M, Smith BR, Furmaniak J. Autoantibodies to steroidogenic enzymes in autoimmune polyglandular syndrome, Addison's disease, and premature ovarian failure. *J Clin Endocrinol Metab* 1996;81:1871–6.
24. Hrdá P, Sterzl I, Matucha P, Korioth F, Kromminga A. HLA antigen expression in autoimmune endocrinopathies. *Physiol Res* 2004;53:191–7.
25. Storz SM, Wylenzek SAM, Matheis N, Weber MM, Kahaly GJ. Impaired psychometric testing in polyglandular autoimmunity. *Clin Endocrinol* 2011;74:394–403.
26. Björses P, Aaltonen J, Vikman A, Perheentupa J, Ben-Zion G, Chiumello G, Dahl N, Heideman P, Hoorweg-Nijman JJG, Mathivon L, Mullis PE, Pohl M, Ritzen M, Romeo G, Shapiro MS, Smith CS, Solyom J, Zlotogora J, Peltonen L. Genetic homogeneity of autoimmune polyglandular disease type I. *Am J Hum Genet* 1996;59:879–86.
27. Aaltonen J, Björses P, Sandkuijl L, Perheentupa J, Peltonen L. An autosomal locus causing autoimmune disease: autoimmune polyglandular disease type I assigned to chromosome 21. *Nat Genet* 1994;8:83–7.
28. Nagamine K, Peterson P, Scott HS, Kudoh J, Minoshima S, Heino M, Krohn KJ, Lalioti MD, Mullis PE, Antonarakis SE, Kawasaki K, Asakawa S, Ito F, Shimizu N. Positional cloning of the APECED gene. *Nat Genet* 1997;17:393–8.

29. Finnish-German APECED Consortium. An autoimmune disease, APECED, caused by mutations in a novel gene featuring two PHD-type zinc-finger domains. Autoimmune polyendocrinopathy–candidiasis–ectodermal dystrophy. *Nat Genet* 1997;**17**:399–403.

30. Pearce SHS, Cheetham T, Imrie H, Vaidya B, Barnes ND, Bilous RW, Carr D, Meeran K, Shaw NJ, Smith CS, Toft AD, Williams G, Kendall-Taylor P. A common and recurrent 13-bp deletion in the autoimmune regulator gene in British kindreds with autoimmune polyendocrinopathy type 1. *Am J Hum Genet* 1998;**63**:1675–84.

31. Scott HS, Heino M, Peterson P, Mittaz L, Lalioti MD, Betterle C, Cohen A, Seri M, Lerone M, Romeo G, Collin P, Salo M, Metcalfe R, Weetman A, Papasavvas MP, Rossier C, Nagamine K, Kudoh J, Shimizu N, Krohn KJ, Antonarakis SE. Common mutations in autoimmune polyendocrinopathy-candidiasis-ectodermal dystrophy patients of different origins. *Molec Endocr* 1998;**12**:1112–9.

32. Wang CY, Davoodi-Semiromi A, Huang W, Connor E, Shi JD, She JX. Characterization of mutations in patients with autoimmune polyglandular syndrome type 1 (APS1). *Hum Genet* 1998;**103**:681–5.

33. Heino JM, Scott HS, Chen Q, Peterson P, Mäebpää U, Papasavvas MP, Mittaz L, Barras C, Rossier C, Chrousos GP, Stratakis CA, Nagamine K, Kudoh J, Shimizu N, Maclaren N, Antonarakis SE, Krohn K. Mutation analyses of North American APS-1 patients. *Hum Mutat* 1999;**13**:69–74.

34. Ward L, Paquette J, Seidman E, Huot C, Alvarez F, Crock P, Delvin E, Kämpe O, Deal C. Severe autoimmune polyendocrinopathy–candidiasis–ectodermal dystrophy in an adolescent girl with a novel AIRE mutation: response to immunosuppressive therapy. *J Clin Endocrinol Metab* 1999;**84**:844–52.

35. Björses P, Halonen M, Palvimo JJ, Kolmer M, Aaltonen J, Ellonen P, Perheentupa J, Ulmanen I, Peltonen L. Mutations in the AIRE gene: effects on subcellular location and transactivation function of the autoimmune polyendocrinopathy–candidiasis–ectodermal dystrophy protein. *Am J Hum Genet* 2000;**66**:378–92.

36. Ishii T, Suzuki Y, Ando N, Matsuo N, Ogata T. Novel mutations of the autoimmune regulator gene in two siblings with autoimmune polyendocrinopathy–candidiasis–ectodermal dystrophy. *J Clin Endocrinol Metab* 2000;**85**:2922–6.

37. Cetani F, Barbesino G, Borsari S, Pardi E, Cianferotti L, Pinchera A, Marcocci C. A novel mutation of the autoimmune regulator gene in an Italian kindred with autoimmune polyendocrinopathy-candidiasis-ectodermal dystrophy acting in a dominant fashion and strongly cosegregating with hypothyroid autoimmune thyroiditis. *J Clin Endocr Metab* 2001;**86**:4747–52.

38. Cihakova D, Trebusak K, Heino M, Fadeyev V, Tiulpakov A, Battelino T, Tar A, Halász Z, Blümel P, Tawfik S, Krohn K, Lebl J, Peterson P. Novel AIRE mutations and P450 cytochrome autoantibodies in Central and Eastern European patients with APECED. *Hum Mutat* 2001;**18**:225–32.

39. Heino M, Peterson P, Kudoh J, Shimizu N, Antonarakis SE, Scott HS, Krohn K. APECED mutations in the autoimmune regulator (AIRE) gene. *Hum Mutat* 2001;**18**:205–11.

40. Saugier-Veber P, Drouot N, Wolf LM, Kuhn JM, Frébourg T, Lefebvre H. Identification of a novel mutation in the autoimmune regulator (AIRE-1) gene in a French family with autoimmune polyendocrinopathy–candidiasis–ectodermal dystrophy. *Eur J Endocrinol* 2001;**144**:347–51.

41. Lintas C, Cappa M, Comparcola D, Nobili V, Fierabracci A. An 8-year-old boy with autoimmune hepatitis and *Candida onychosis* as the first symptoms of autoimmune polyglandular syndrome (APS1): identification of a new homozygous mutation in the autoimmune regulator gene (aire). *Eur J Pediatr* 2008;**167**:949–53.

42. Wolff AS, Erichsen MM, Meager A, Magitta NF, Myhre AG, Bollerslev J, Fougner KJ, Lima K, Knappskog PM, Husebye ES. Autoimmune polyendocrine syndrome type 1 in Norway: phenotypic variation, autoantibodies, and novel mutations in the autoimmune regulator gene. *J Clin Endocrinol Metab* 2007;**92**:595–603.

43. Bhui RD, Lewis DB, Nadeau KC. A novel mutation associated with autoimmune polyendocrinopathy-candidiasis-ectodermal dystrophy. *Ann Allergy Asthma Immunol* 2008;**100**:169–73.

44. Ahonen P. Autoimmune polyendocrinopathy-candidosis-ectodermal dystrophy (APECED): autosomal recessive inheritance. *Clin Genet* 1985;**27**:535–42.

45. Dittmar M, Ide M, Wurm M, Kahaly GJ. Early onset of polyglandular failure is associated with HLA-DRB1*03. *Eur J Endocrinol* 2008;**159**:55–60.

46. Weinstock C, Matheis N, Barkia S, Haager MC, Janson A, Markovic A, Bux J, Kahaly GJ. Autoimmune polyglandular syndrome type 2 shows the same HLA class II pattern as type 1 diabetes. *Tissue Antigens* 2011;**77**:317–24.

47. Huang W, Connor E, Rosa TD, Muir A, Schatz D, Silverstein J, Crockett S, She JX, Maclaren NK. Although DR3-DQB1*0201 may be associated with multiple component diseases of the autoimmune polyglandular syndromes, the human leukocyte antigen DR4-DQB1*0302 haplotype is implicated only in beta cell autoimmunity. *J Clin Endocrinol Metab* 1996;**81**:2559–63.

48. Maclaren NK, Riley WJ. Inherited susceptibility to autoimmune Addison's disease is linked to human leukocyte antigens-DR3 and/or DR4, except when associated with type I autoimmune polyglandular syndrome. *J Clin Endocr Metab* 1986;**62**:455–9.

49. Robles DT, Fain PR, Gottlieb PA, Eisenbarth GS. The genetics of autoimmune polyendocrine syndrome type II. *Endocrinol Metab Clin North Am* 2002;**31**:353–68.

50. Tisch R, McDevitt H. Insulin-dependent diabetes mellitus. *Cell* 1996;**85**:291–7.

51. Sanjeevi CB, Lybrand TP, DeWeese C, Landin-Olsson M, Kockum I, Dahlquist G, Sundkvist G, Stenger D, Lernmark A. Polymorphic amino acid variations in HLA-DQ are associated with systematic physical property changes and occurrence of IDDM. Members of the Swedish Childhood Diabetes Study. *Diabetes* 1995;**44**:125–31.

52. Petrone A, Giorgi G, Mesturino CA, Capizzi M, Cascino I, Nistico L, Osborn J, Di Mario U, Buzzetti R. Association of DRB1*04-DQB1*0301 haplotype and lack of association of two polymorphic sites at CTLA-4 gene with Hashimoto's thyroiditis in an Italian population. *Thyroid* 2001;**11**:171–5.

53. Flesch BK, Matheis N, Alt T, Weinstock C, Bux J, Kahaly GJ. HLA Class II haplotypes differentiate between the adult autoimmune polyglandular syndrome types II and III. *J Clin Endocrinol Metab* 2014;**99**:E177–82.

54. Donner H, Rau H, Walfish PG, Braun J, Siegmund T, Finke R, Herwig J, Usadel KH, Badenhoop K. CTLA4 alanine-17 confers genetic susceptibility to Graves' disease and to type 1 diabetes mellitus. *J Clin Endocrinol Metab* 1997;**82**:143–6.

55. Yanagawa T, Hidaka Y, Guimaraes V, Soliman M, DeGroot LJ. CTLA-4 gene polymorphism associated with Graves' disease in a Caucasian population. *J Clin Endocrinol Metab* 1995;**80**:41–5.

56. Yanagawa T, Taniyama M, Enomoto S, Gomi K, Maruyama H, Ban Y, Saruta T. CTLA4 gene polymorphism confers susceptibility to Graves' disease in Japanese. *Thyroid* 1997;**7**:843–6.

57. Heward JM, Brand OJ, Barrett JC, Carr-Smith JD, Franklyn JA, Gough SC. Association of PTPN22 haplotypes with Graves' disease. *J Clin Endocrinol Metab* 2007;**92**:685–90.

58. Kouki T, Gardine CA, Yanagawa T, Degroot LJ. Relation of three polymorphisms of the CTLA-4 gene in patients with Graves' disease. *J Endocrinol Invest* 2002;**25**:208–13.

59. Bednarczuk T, Hiromatsu Y, Fukutani T, Jazdzewski K, Miskiewicz P, Osikowska M, Nauman J. Association of cytotoxic T-lymphocyte-associated antigen-4 (CTLA-4) gene polymorphism and non-genetic factors with Graves' ophthalmopathy in European and Japanese populations. *Eur J Endocrinol* 2003;**148**:13–8.

60. Donner H, Braun J, Seidl C, Rau H, Finke R, Ventz M, Walfish PG, Usadel KH, Badenhoop K. Codon 17 polymorphism of the cytotoxic T lymphocyte antigen 4 gene in Hashimoto's thyroiditis and Addison's disease. *J Clin Endocrinol Metab* 1997;**82**:4130–2.

61. Heward JM, Allahabadia A, Armitage M, Hattersley A, Dodson PM, Macleod K, Carr-Smith J, Daykin J, Daly A, Sheppard MC, Holder RL, Barnett AH, Franklyn JA, Gough SC. The development of Graves' disease and the CTLA-4 gene on chromosome 2q33. *J Clin Endocrinol Metab* 1999;**84**:2398–401.

62. Dittmar M, Kahaly GJ. Genetics of the autoimmune polyglandular syndrome type 3 variant. *Thyroid* 2010;**20**:737–43.

63. Chikuba N, Akazawa S, Yamaguchi Y, Kawasaki E, Takino H, Yoshimoto M, Ohe N, Yamashita K, Yano A, Nagataki S. Immunogenetic heterogeneity in type 1 (insulin-dependent) diabetes among Japanese-class II antigen and autoimmune thyroid disease. *Diabetes Res Clin Pract* 1995;**27**:31–7.

64. Chuang LM, Wu HP, Chang CC, Tsai WY, Chang HM, Tai TY, Lin BJ. HLA DRB1/DQA1/DQB1 haplotype determines thyroid autoimmunity in patients with insulin-dependent diabetes mellitus. *Clin Endocrinol* 1996;**45**:631–6.

65. Kim EY, Shin CH, Yang SW. Polymorphisms of HLA class II predispose children and adolescents with type 1 diabetes mellitus to autoimmune thyroid disease. *Autoimmunity* 2003;**36**:177–81.

66. Levin L, Ban Y, Concepcion E, Davies TF, Greenberg DA, Tomer Y. Analysis of HLA genes in families with autoimmune diabetes and thyroiditis. *Hum Immunol* 2004;**65**:640–7.

67. Golden B, Levin L, Ban Y, Concepcion E, Greenberg DA, Tomer Y. Genetic analysis of families with autoimmune diabetes and thyroiditis: evidence for common and unique genes. *J Clin Endocrinol Metab* 2005;**90**:4904–11.

68. Hashimoto K, Maruyama H, Nishiyama M, Asaba K, Ikeda Y, Takao T, Iwasaki Y, Kumon Y, Suehiro T, Tanimoto N, Mizobuchi M, Nakamura T. Susceptibility alleles and haplotypes of human leukocyte antigen DRB1, DQA1, and DQB1 in autoimmune polyglandular syndrome type III in Japanese population. *Horm Res* 2005;**64**:253–60.

69. Holl RW, Bohm B, Loos U, Grabert M, Heinze E, Homoki J. Thyroid autoimmunity in children and adolescents with type 1 diabetes mellitus. Effect of age, gender and HLA type. *Horm Res* 1999;**52**:113–8.

70. Ueda H, Howson JM, Esposito L, Heward J, Snook H, Chamberlain G, Rainbow DB, Hunter KM, Smith AN, Di Genova G, Herr MH, Dahlman I, Payne F, Smyth D, Lowe C, Twells RC, Howlett S, Healy B, Nutland S, Rance HE, Everett V, Smink LJ, Lam AC, Cordell HJ, Walker NM, Bordin C, Hulme J, Motzo C, Cucca F, Hess JF, Metzker ML, Rogers J, Gregory S, Allahabadia A, Nithiyananthan R, Tuomilehto-Wolf E, Tuomilehto J, Bingley P, Gillespie KM, Undlien DE, Ronningen KS, Guja C, Ionescu-Tirgoviste C, Savage DA, Maxwell AP, Carson DJ, Patterson CC, Franklyn JA, Clayton DG, Peterson LB, Wicker LS, Todd JA, Gough SC. Association of the T cell regulatory gene CTLA4 with susceptibility to autoimmune disease. *Nature* 2003;**423**:506–11.

71. Takara M, Kouki T, DeGroot LJ. CTLA-4 AT-repeat polymorphism reduces the inhibitory function of CTLA-4 in Graves' disease. *Thyroid* 2003;**13**:1083–9.

72. Wang XB, Kakoulidou M, Giscombe R, Qiu Q, Huang D, Pirskanen R, Lefvert AK. Abnormal expression of CTLA-4 by T cells from patients with myasthenia gravis: effect of an AT-rich gene sequence. *J Neuroimmunol* 2002;**130**:224–32.

73. Dultz G, Matheis N, Dittmar M, Bender K, Kahaly GJ. CTLA-4 CT60 polymorphism in thyroid and polyglandular autoimmunity. *Horm Metab Res* 2009;**41**:426–9.

74. Cohen S, Dadi H, Shaoul E, Sharfe N, Roifman CM. Cloning and characterization of a lymphoid-specific, inducible human protein tyrosine phosphatase. *Lyp Blood* 1999;**93**:2013–24.

75. Bottini N, Musumeci L, Alonso A, Rahmouni S, Nika K, Rostamkhani M, MacMurray J, Meloni GF, Lucarelli P, Pellecchia M, Eisenbarth GS, Comings D, Mustelin T. A functional variant of lymphoid tyrosine phosphatase is associated with type I diabetes. *Nat Genet* 2004;**36**:337–8.

76. Smyth D, Cooper JD, Collins JE, Heward JM, Franklyn JA, Howson JM, Vella A, Nutland S, Rance HE, Maier L, Barratt BJ, Guja C, Ionescu-Tirgoviste C, Savage DA, Dunger DB, Widmer B, Strachan DP, Ring SM, Walker N, Clayton DG, Twells RC, Gough SC, Todd JA. Replication of an association between the lymphoid tyrosine phosphatase locus (LYP/PTPN22) with type 1 diabetes, and evidence for its role as a general autoimmunity locus. *Diabetes* 2004;**53**:3020–3.

77. Kawasaki E, Awata T, Ikegami H, Kobayashi T, Maruyama T, Nakanishi K, Shimada A, Uga M, Kurihara S, Kawabata Y, Tanaka S, Kanazawa Y, Lee I, Eguchi K. Japanese Study Group on Type 1 Diabetes Genetics. Systematic search for single nucleotide polymorphisms in a lymphoid tyrosine phosphatase gene (PTPN22): association between a promoter polymorphism and type 1 diabetes in Asian populations. *Am J Med Genet* 2006;**140A**:586–93.

78. Dittmar M, Kahaly GJ. Immunoregulatory and susceptibility genes in thyroid and polyglandular autoimmunity. *Thyroid* 2005;**15**:239–50.

79. Dultz G, Matheis N, Dittmar M, Bender K, Röhrig B, Kahaly GJ. The protein tyrosine phosphatase non-receptor type 22 C1858T polymorphism is a joint susceptibility locus for immunothyroiditis and autoimmune diabetes. *Thyroid* 2009;**19**:143–8.

80. Dittmar M, Kaczmarczyk A, Bischofs C, Kahaly GJ. The proinflammatory cytokine TNF-alpha 308 AA genotype is associated with polyglandular autoimmunity. *Immunol Invest* 2009;**38**:255–67.

81. Dittmar M, Bischofs C, Matheis N, Poppe R, Kahaly GJ. A novel mutation in the DNASE1 gene is related with protein instability and decreased enzyme activity in thyroid autoimmunity. *J Autoimmunity* 2009;**32**:7–13.

82. Dittmar M, Woletz K, Kahaly GJ. Reduced DNASE1 gene expression in thyroid autoimmunity. *Horm Metab Res* 2013;**45**:257–60.

83. Liston A, Lesage S, Wilson J, Peltonen L, Goodnow CC. AIRE regulates negative selection of organ-specific T cells. *Nature Immunol* 2003;**4**:350–4.

84. Halonen M, Kangas H, Rüppell T, Ilmarinen T, Ollila J, Kolmer M, Vihinen M, Palvimo J, Saarela J, Ulmanen I, Eskelin P. APECED-causing mutations in AIRE reveal the functional domains of the protein. *Hum Mutat* 2004;**23**:245–57.

85. Kogawa K, Nagafuchi S, Katsuta H, Kudoh J, Tamiya S, Sakai Y, Shimizu N, Harada M. Expression of AIRE gene in peripheral monocyte/dendritic cell lineage. *Immunol Lett* 2002;**80**:195–8.

86. Faas S, Trucco M. The genes influencing the susceptibility to IDDM in humans. *J Endocrinol Invest* 1994;**17**:477–95.

87. Teft WA, Kirchhof MG, Madrenas J. A molecular perspective of CTLA-4 function. *Annu Rev Immunol* 2006;**24**:65–97.

88. Brunner MC, Chambers CA, Chan FK, Hanke J, Winoto A, Allison JP. CTLA-4-Mediated inhibition of early events of T cell proliferation. *J Immunol* 1999;**162**:5813–20.

89. Jacobson EM, Tomer Y. The CD40, CTLA-4, thyroglobulin, TSH receptor, and PTPN22 gene quintet and its contribution to thyroid autoimmunity: back to the future. *J Autoimmun* 2007;**28**:85–98.

90. Kavvoura FK, Akamizu T, Awata T, Ban Y, Chistiakov DA, Frydecka I, Ghaderi A, Gough SC, Hiromatsu Y, Ploski R, Wang PW, Ban Y, Bednarczuk T, Chistiakova EI, Chojm M, Heward JM, Hiratani H, Juo SH, Karabon L, Katayama S, Kurihara S, Liu RT, Miyake I, Omrani GH, Pawlak E, Taniyama M, Tozaki T, Ioannidis JP. Cytotoxic T-lymphocyte associated antigen 4 gene polymorphisms and autoimmune thyroid disease: a meta-analysis. *J Clin Endocrinol Metab* 2007;**92**:3162–70.

91. Ban Y, Taniyama M, Tozaki T, Yanagawa T, Yamada S, Maruyama T, Kasuga A, Tomita M, Ban Y. No association of type 1 diabetes with a microsatellite marker for CTLA-4 in a Japanese population. *Autoimmunity* 2001;**34**:39–43.

92. Cinek O, Drevinek P, Sumnik Z, Bendlova B, Kolouskova S, Snajderova M, Vavrinec J. The CTLA4 +49 A/G dimorphism is not associated with type 1 diabetes in Czech children. *Eur J Immunogenet* 2002;**9**:219–22.

93. Cloutier JF, Veillette A. Cooperative inhibition of T-cell antigen receptor signaling by a complex between a kinase and a phosphatase. *J Exp Med* 1999;**189**:111–21.

94. Vang T, Miletic AV, Bottini N, Mustelin T. Protein tyrosine phosphatase PTPN22 in human autoimmunity. *Autoimmunity* 2007;**40**:453–61.

95. Bottini N, Vang T, Cucca F, Mustelin T. Role of PTPN22 in type 1 diabetes and other autoimmune diseases. *Semin Immunol* 2006;**18**:207–13.

96. Kahles H, Ramos-Lopez E, Lange B, Zwermann O, Reincke M, Badenhoop K. Sex-specific association of PTPN22 1858T with type 1 diabetes but not with Hashimoto's thyroiditis or Addison's disease in the German population. *Eur J Endocrinol* 2005;**153**:895–9.

97. Velaga MR, Wilson V, Jennings CE, Owen CJ, Herington S, Donaldson PT, Ball SG, James RA, Quinton R, Perros P, Pearce SH. The codon 620 tryptophan allele of the lymphoid tyrosine phosphatase (LYP) gene is a major determinant of Graves' disease. *J Clin Endocrinol Metab* 2004;**89**:5862–5.

98. Murumagi A, Vahamurto P, Peterson P. Characterization of regulatory elements and methylation pattern of the autoimmune regulator (AIRE) promoter. *J Biol Chem* 2003;**278**:19784–90.

99. Kahaly GJ. Polyglandular autoimmune syndromes. *Eur J Endocrinol* 2009;**161**:11–20.

100. Kahaly GJ. Polyglandular autoimmune syndrome type II. *Presse Med* 2012;**41**:e663–70.

101. Gianani R, Sarvetnick N. Viruses, cytokines, antigens, and autoimmunity. *Proc Natl Acad Sci USA* 1996;**93**:2257–9.

PART X

GROWTH

26

Genetic Diagnosis of Growth Failure

Ron G. Rosenfeld*, Vivian Hwa**

Oregon Health & Science University, Portland, OR, USA
**Cincinnati Children's Hospital, Cincinnati, OH, USA

INTRODUCTION

Background, Incidence, Prevalence

For starters, short stature is *not* a disease; nor, for that matter, is growth failure. However one chooses to define short stature, it is, ultimately, a statistical definition, based (probably fallaciously) on the assumption that stature follows a perfect Gaussian distribution. Accordingly, 3% of children (or adults) fall below the third percentile of stature, and approximately 1.2% fall below −2.25 standard deviations, the FDA-approved definition of "idiopathic short stature."[1,2]

It is stated above that stature is not likely to follow a perfect bell-shaped curve because it is clear that there are many more pathological conditions that result in growth failure than in overgrowth. Many chronic diseases of childhood (e.g., chronic renal failure, inflammatory bowel disease, cystic fibrosis, rheumatoid arthritis, immunodeficiency states, chronic infection, etc.) are characterized by growth failure. The alert clinician recognizes that short stature may be an important symptom of underlying diseases outside of the endocrine system and, when appropriate, short patients should be evaluated for such conditions.

Additionally, many chromosomal disorders are characterized by short stature, as commonly observed in Turner syndrome (45,X or various abnormalities of the short arm of the X chromosome) and trisomies 21, 13, and 15. To such conditions, one may add a wide range of chromosomal deletions, inversions, or translocations. These pathologic states are characterized, generally, by characteristic dysmorphic features, which will help steer the clinician in the direction of a chromosomal anomaly, but short stature may, at times, be the presenting sign. A good general rule, for example, is that any female with unexplained short stature warrants a karyotype to rule out Turner syndrome, even in the absence of some of the typical signs or symptoms. Similarly, when short stature is associated with dysmorphic features and/or intrauterine growth failure, the possibility of a chromosomal defect should be excluded.

Any evidence of disproportionate short stature raises the question of an inborn skeletal dysplasia. In the 2010 revision of the International Nosology and Classification of Genetic Skeletal Disorders, 456 conditions, divided among 40 groups, were defined by molecular, biochemical and/or radiographic criteria[3] (http://isds.ch/uploads/pdf_files/Nosology2010.pdf). Some of these conditions result in drastic growth failure, and frequently, early mortality. Others can be relatively subtle and may not be evident without careful ascertainment of skeletal proportions, as well as radiologic evaluation of the long bones, spine, extremities, and skull. While specific mutations have now been identified for many of these conditions, and diagnostic tests are increasingly available, a discussion of skeletal dysplasias is beyond the scope of this chapter. An International Skeletal Dysplasia Registry has been created, and the reader is directed to an excellent review by Rimoin et al.[4] and to the website www.csmc.edu/skeletaldysplasia.

Disturbances in each of two hormonal systems may result in profound growth failure. Chronic hypothyroidism, either primary or secondary, may lead to severe stunting of growth, and all children with significant growth failure should have their thyroid status evaluated. Growth hormone (GH) deficiency, on either a hypothalamic or pituitary basis, can also result in dramatic postnatal growth failure, and is the subject of Chapter 7 (Congenital Defects of Thyroid Hormone Synthesis).

This chapter focuses on genetic defects of the GH-insulin-like growth factor (IGF) axis, distal to the production of bioactive GH. As such, it encompasses defects of the GH receptor (GHR), the GH signaling cascade, the genes for IGF-I and IGF-II, the IGF binding proteins, the IGF-I

Genetic Diagnosis of Endocrine Disorders. http://dx.doi.org/10.1016/B978-0-12-800892-8.00026-9

receptor (IGF1R), and the structurally-related insulin receptor (IR), and signaling defects of either (or both) the IGF1R and the IR.[5,6] The combination of targeted animal gene knockout studies and human mutational analysis has unequivocally demonstrated the critical and major roles that the IGF system plays in mammalian growth. In murine models, approximately two-thirds of total prenatal and postnatal growth is IGF dependent, and the preponderance of the growth-promoting actions of GH are attributable to its stimulation of IGF-I gene transcription. In light of the central role of IGF-I in mammalian postnatal growth, and by analogy with other pituitary-based endocrine systems, it is logical to divide IGF deficiency (IGFD) into secondary and primary forms: "secondary IGFD" encompasses IGF deficiency resulting from a failure of production or secretion of bioactive GH on either a hypothalamic or pituitary basis (see Chapter 7); "primary IGFD" encompasses IGF deficiency, existing in the presence of normal or elevated GH production or serum concentrations (see Table 26.1).

The prevalence of primary IGFD is difficult to assess; as is often the case, ascertainment bias, a significant factor in dealing with the phenotype of short stature, enters into the picture. To date, the total number of reported cases of the conditions listed in Table 26.1 is approximately 300−400, with the majority reflecting genetic abnormalities of *GHR*. It is apparent, however, that such cases are likely to represent the tip of the iceberg. It has been estimated, for example, that 25−40% of children with short stature (defined as heights less than −2.25 SD for age) have serum concentrations of IGF-I below −2 SD. The molecular basis for these observations is unclear at this time, although the subject of active investigation.

Homozygous disruption of *IGF1R* in mice resulted in severe intrauterine growth failure and early demise. To date, the majority of reported patients with *IGF1R* mutations have defects limited to one allele (heterozygous). Several patients have been described, however, with compound heterozygous or homozygous *IGF1R* defects. While such patients should not be described as IGF deficient, their growth failure further confirms the critical role of the IGF axis in prenatal and postnatal growth failure.

The IGFD Research Center, established in 2006 and located in Portland, Oregon, has served as a worldwide referral center for patients with unexplained growth failure and potential molecular defects resulting in primary IGFD or IGF resistance. The Center has received approximately 75−100 cases annually, from over 25 different countries, and has identified novel molecular defects at all levels of the GH-IGF axis described in the subsequent sections. The IGFD Research Center is now part of a coordinated international effort to construct, develop, and curate an international database and website for molecular defects of the GH-IGF axis.

Clinical Presentation

Short stature is the defining characteristic, naturally, but defining short stature can be challenging. When growth failure commences *in utero*, intrauterine growth retardation (IUGR) occurs, resulting in newborns who are small for gestational age. Intrauterine growth failure must be differentiated from postnatal growth failure, as the two conditions may or may not accompany one another.

As in the case of GH deficiency, GH insensitivity, either at the level of the GHR[7] or the postreceptor GH signaling cascade,[8,9] is characterized, typically, by relatively normal intrauterine growth, but severe postnatal growth failure, commencing within the first few months of life.[10] Mutations or deletions of the *IGF1* gene,[11] or mutations affecting *IGF1R*,[12] on the other hand, combine intrauterine growth retardation with poor postnatal growth. Thus, while GH does not appear to be essential for intrauterine growth, IGF-I itself is critically involved in both prenatal and postnatal growth. Presumably, IGF-I production *in utero* is largely GH independent, but switches to profound GH dependence sometime near birth. Defects of *IGF1* are also characterized by microcephaly, developmental delay and, in some cases, sensorineural deafness.

In addition to postnatal growth failure, patients with *GHR* defects have infantile facies, cephalofacial disproportion, truncal adiposity, bluish sclerae and delayed dentition, and skeletal maturation, much as is commonly observed in GH deficient children. A tendency to fasting hypoglycemia is observed in patients with either GHD or GHI, presumably reflecting the loss of counterregulatory actions of GH in glucose metabolism.

Table 26.2 displays characteristic biochemical findings of patients with defects of the GH-IGF axis. The combination of clinical phenotype and biochemical findings

TABLE 26.1 Molecular Defects Resulting in Primary IGF Deficiency

1. Mutations or deletions of *GHR* resulting in defective binding of GH
2. Mutations or deletions of *GHR* resulting in defective receptor dimerization
3. Mutations or deletions of *GHR* resulting in defective receptor anchoring in the cell membrane
4. Mutations or deletions of *GHR* resulting in defective signal transduction
5. Mutations in *Signal Transducer and Activator of Transcription (STAT)5B* resulting in defective GH signaling
6. Mutations of *Acid Labile Subunit (IGFALS)* resulting in rapid IGF clearance
7. Deletions of *IGF1*
8. Mutations of *IGF1* resulting in bioinactive IGF-I

TABLE 26.2 Biochemical Characteristics

	HT	SGA	GH	GHBP	IGF-I	BP-3	ALS	imm
				Features of ISS/primary IGFD				
GHRec	↓↓	—	↑↑	↓↓	↓↓	↓↓	↓↓	—
GHRtm	↓↓	—	↑↑	N-↑	↓↓	↓↓	↓↓	—
GHRic	↓↓	—	↑↑	N	↓↓	↓↓	↓↓	—
STAT5	↓↓	—	↑↑	N	↓↓	↓↓	↓↓	±
IGFdel	↓↓	↓↓	↑↑	N	↓↓↓↓↓	↑	↑	—
IGFbi	↓↓	↓↓	↑↑	N	↑↑	↑	↑	—
ALS	↓	—	↑	N	↓↓	↓↓	↓↓↓↓↓	—

Clinical and biochemical characteristics of various molecular causes of primary IGFD. GHRec, mutations/deletions affecting the extracelluar domain of the GHR; GHRtm, transmembrane domain of GHR; GHRic, intracellular domain of the GHR; IGFdel, IGF1 gene deletion; IGFbi, bioinactive IGF-I; HT, height; SGA, small for gestational age (birth size); imm, immunological defects.

generally can helps direct the clinician toward the most appropriate molecular diagnoses to consider. While many of the features of these conditions overlap, each molecular defect carries characteristic biochemical and clinical features, resulting in a continuum of findings.[10]

GENETIC PATHOPHYSIOLOGY

Known Mutations and Specific Mutations

Mutations/Deletions of GHR

Abnormalities of the GHR constitute both the earliest and the most prevalent identified molecular defects resulting in primary IGFD. Although the original genetic abnormalities reported were large deletions of *GHR*, involving exons 3, 5, and 6 (we now know that exon 3 may be spliced out in approximately one-third of the normal population), the majority of gene abnormalities have turned out to be point mutations. The reported deletions of exons 5 and 6 lead to a frameshift, with a consequent premature translational stop signal in exon 7 and a resultant aberrant protein lacking both the transmembrane and intracellular domains.

At least 70 mutations of *GHR* have been identified to date, including missense, nonsense, frameshift, and splice defects. These mutations are summarized in a recent review.[10] Figs. 26.1 and 26.2 show homozygous and compound heterozygous mutations of *GHR* identified in the IGFD Research Center. Approximately 90% of these involve the extracellular, GH-binding domain of the GHR and almost all are characterized by reduced serum

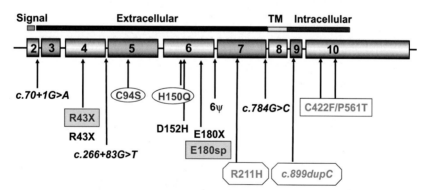

FIGURE 26.1 **Previously published homozygous or compound heterozygous mutations of *GHR* identified at the IGF Deficiency Research Center, Portland, Oregon.** The 10 exons of *GHR* are depicted; the GHR protein domains are indicated earlier the schematic of the *GHR* gene. TM, transmembrane domain. Mutations indicated in similar boxes, ovals, or hexagons indicate compound heterozygous mutations; all others are homozygous mutations. Splicing mutations are in italics. 6ψ, a single nucleotide change in intron 6 that results in an aberrant splicing event with subsequent in-frame insertion of 108 nucleotides (36 amino acid residues). Numbering is for the mature protein; for prepeptide numbering, add 18 residues for the signal peptide.

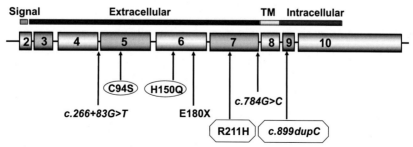

FIGURE 26.2 **Novel homozygous or compound heterozygous mutations of *GHR* identified at the IGF Deficiency Research Center, Portland, OR.** *See legend to Fig. 26.1.*

concentrations of GH binding protein (GHBP). Of interest is an intronic base change that results in activation of a pseudoexon sequence of 108 nucleotides and the insertion of 36 new amino acids within the extracellular domain of the GHR.[13] Phenotypes associated with this pseudoexon activation are highly variable and include patients whose stature and serum IGF-I concentrations fall within the lower portion of the normal range.

Perhaps of greater interest are the rarer cases of abnormalities involving the transmembrane or intracellular domains of the GHR, as these patients have normal or even elevated serum concentrations of GHBP and may not, consequently, be initially identified as having GHR abnormalities (Fig. 26.3). A heterozygous C→A transversion at position c.785-3 at the acceptor site of intron 7 results in the excision of exon 8 of the *GHR*, impaired anchoring of the receptor to the cell membrane, and markedly elevated serum GHBP.[14] Three splice-site mutations in exon 9 have been reported to behave in a dominant negative manner.[15] This has been attributed to the production of truncated GHR molecules, lacking most of the intracellular domain, but retaining extracellular and transmembrane domains, allowing them to dimerize with normal GHR molecules and, thereby, perturb GH signaling. Three deletions in exon 10 have been identified: 309delC results in GHR 1-330; a 22bp deletion results in GHR 1-449; and a 1776delG leads to a GHR molecule with the first 560 amino acids, followed by a nonsense sequence from 560 to 581. A number of polymorphisms and potential heterozygous missense mutations of the intracellular domain have been identified, but their functional significance remains uncertain at this time.

Indeed, the issue of clinical significance of heterozygous mutations of *GHR* remains a matter of debate. While it appears that some mutations are capable of behaving in a dominant negative manner, heterozygosity for most missense mutations does not appear to be associated with a clinical phenotype of growth failure. Reports describing heterozygous mutations of *GHR* in

short children often lack adequate functional studies of the mutations to convincingly prove an effect on GH action.[16] On the other hand, there are several reports where the combination of modest growth failure and *in vitro* studies consistent with GHR dysfunction has supported an effect of heterozygosity on growth.[17] At this point, it seems fair to say that such reports should be evaluated on a case-by-case basis, and that genotype-phenotype data on family members, as well as functional studies, are necessary to support any claim of heterozygous effect.

Mutations of STAT5B

Patients with homozygous mutations of *STAT5B* were identified through their clinical and biochemical resemblance to patients with GH insensitivity resulting from GHR abnormalities.[8,9,18]Through study of these cases, it has become apparent that, at least in humans, the STAT5b pathway represents the major (if not the sole) mechanism for GH stimulation of IGF-I gene transcription. Clinically, affected patients have normal or near-normal size at birth, but early in life manifest dramatic growth failure, with reported heights between −5.6 and −9.9 SD. When treated with GH, no growth response has been observed and serum IGF-I concentrations fail to rise. All cases identified to date have had evidence of immune compromise, with histories of chronic infections (especially pulmonary) in most, but also with unexplained arthritis; presumably, this reflects the use of the JAK-STAT signaling pathway by multiple cytokines. As in cases of *GHR* mutations, serum concentrations of IGF-I, IGFBP-3 and ALS are markedly low, despite normal-elevated serum GH concentrations. GHBP concentrations, however, are normal, as is sequencing of *GHR*.

Seven different mutations have been identified in the 10 reported patients; this includes two sets of siblings, one from Kuwait and one from Brazil (Fig. 26.4). Mutations have been identified in diverse areas of the STAT5b protein. Two mutations are in the CCD domain of STAT5b, two are in the DBD domain, one is in the L

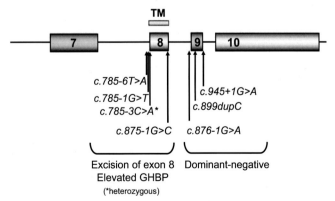

FIGURE 26.3 Published dominant-negative mutations of *GHR*. *See legend to Fig. 26.1.*

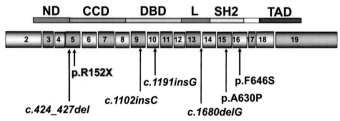

FIGURE 26.4 Published mutations of *STAT5B*. All mutations are homozygous, and include missense (A630P, F646S), nonsense (R152X), and frameshifts due to nucleotide deletion (*1680delG*) or insertion (*1191insG, 1103insC*). ND, N-terminal domain; CCD, coiled-coiled domain; DBD, DNA binding domain; L, linker; SH2, src-homology 2 domain; TAD, tranactivating domain.

domain, and two are in the SH2 domain. At this point, clinical and molecular experience is too limited to permit meaningful genotype:phenotype correlations. One recently reported mutation (F646S) is of particular interest, as it results in a STAT5b protein that was capable of being phosphorylated upon stimulation with either GH or interferon-δ, but was still incapable of driving IGF-I gene transcription.[19]

Mutations/Deletions of IGF1

To date, only three convincing homozygous mutations of *IGF1* have been reported (Fig. 26.5). All three cases are characterized by the combination of prenatal and postnatal growth failure, microcephaly, and developmental delay; two of the three cases also have sensorineural deafness.

Woods et al.[20] reported a 15-year-old male with a homozygous deletion of exons 4 and 5 of the *IGF1* gene, resulting in a mature IGF-I peptide truncated from 70 to 25 amino acids, followed by an additional nonsense sequence of eight residues. Walenkamp et al.[21] identified a 55-year-old male homozygous for a V44M mutation of *IGF1*. The resulting protein had a 90-fold lower affinity for the IGF-I receptor. Nine of 24 relatives studied were heterozygous for this mutation and were found to have lower birth weights, adult heights, and head circumferences than family members who were noncarriers, although only one heterozygote had an adult height below −2 SD. More recently, a child with IUGR, postnatal growth failure, and microcephaly was shown to be homozygous for an R36Q mutation, resulting in a two- to threefold lower affinity for the IGF-I receptor.[22]

Several recent publications have described cases of familial short stature associated with heterozygous mutations of *IGF1*. While it is unclear what contribution, if any, the disordered IGF-I allele made to the patients'

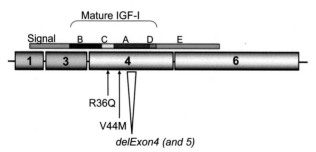

FIGURE 26.5 **Published mutations of *IGF1*.** All mutations are homozygous. The IGF-I gene is comprised of six exons and five introns, spanning 100 kb of chromosomal DNA. The presence of two functional promoters and alternative splicing events results in multiple mRNA variants encoding different signal peptides (exon 1 or 2), E domain variants that are posttranslationally removed (exon 5 and 6), and 3′ untranslated region (exon 5 and 6). The mature IGF-I protein, encompassing domains designated B, C, A, and D, is encoded by part of exon 3 and exon 4. Homozygous mutations identified to date are indicated.

growth failure, it is of note that mild short stature associated with heterozygous mutations of *IGFALS* have also been reported recently. Van Duyvenvoorde et al.[23] reported two children, their mother and their maternal grandfather, all of whom carried a heterozygous duplication of four nucleotides within *IGF1*, resulting in a frameshift and premature termination (Het *c.243-246dup-CAGC*). The two children had heights of −4.1 and −4.6 SD; affected adults had heights of −2.5 SD, compared to −1.6 SD for family noncarriers. Fuqua et al.[24] described an extended family in which five members had short stature associated with two novel heterozygous variants of *IGF1*: *c.207G>A* in exon 3 and *c.402+1G>C* in the donor splice site of intron 4. Although the *IGF1* gene was shown to be normal in 11 normal-statured family members, the contribution of these heterozygous mutations to the growth failure remains speculative.

Mutations of IGFALS

ALS (acid-labile subunit) is an 85 kDa protein that, together with IGFBP-3 (and to a lesser extent IGFBP-5), forms a ternary complex with a molecule of IGF-I or IGF-II for transport in serum. This complex acts to greatly extend the half-life of IGF molecules in the circulation and modulates IGF bioavailability by serving as a reservoir for IGF in serum. ALS belongs to the leucine-rich repeat (LRR) superfamily of proteins; the LRR domain in human ALS encompasses 75% of the protein and includes 20 LRRs, which form a donut-shaped, closed structure.

To date, over 20 cases, involving at least 15 families, have been reported of homozygous or compound heterozygous mutations of *IGFALS* (Fig. 26.6).[25] Growth impairment has been, in general, much milder than in other molecular causes of IGFD, with most heights in the −2 to −3 SD range, but with some affected individuals with heights within the normal range. Some short children with *IGFALS* mutations have exhibited delayed puberty, with attainment of normal adult stature, despite short stature in childhood. Insulin resistance, with hyperinsulinemia, has also been reported. The natural history of *IGFALS* mutations is difficult to determine, however, as ascertainment bias is clearly evident, since serum IGF-I (and ALS) concentrations are rarely measured in individuals of normal stature.

Patients have been identified, typically, when evaluation of growth failure has demonstrated low serum concentrations of IGF-I and IGFBP-3, despite normal GH levels. Upon administration of GH, serum levels of neither IGF-I nor IGFBP-3 rise significantly. Serum ALS levels, either by assay or by immunoblot, are markedly reduced. While it is presumed that serum concentrations of free IGF-I are relatively normal in a situation of reduced IGFBP-3 and ALS (hence, the near normal growth), assays for free IGF-I are not sufficiently reliable to confirm this hypothesis.

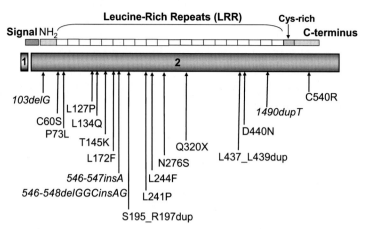

FIGURE 26.6 **Published mutations of *IGFALS*.** The *IGFALS* gene consists of only two exons, spanning 3.3 kb of chromosomal DNA, with the mature protein encoded solely by exon 2. The protein domain includes 20 leucine-rich repeats (LRR) and a cysteine-rich region (cys-rich). Both homozygous and compound heterozygous mutations have been identified.

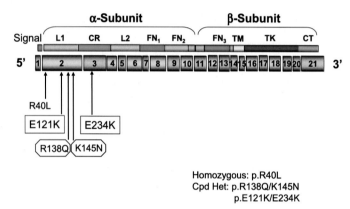

FIGURE 26.7 **Published homozygous (R40L) and compound heterozygous (E121K/E234K, R138Q/K145N) mutations of *IGF1R*.** Numbering is for the prepeptides; for the mature proteins, subtract 30 nucleotides for the signal peptide. L1, L2, leucine-rich domains; CR, cysteine-rich, furin-like domain; FN1,2,3, fibronectin type III; TM, transmembrane domain; TK, tyrosine kinase catalytic domain; CT, carboxy-terminal tail.

Most of the mutations reported to date involve the portion of *IGFALS* that encodes the LRR section of the protein. No genotype:phenotype correlations have been identified so far, which probably reflects the straightforward physiological implications of greatly reduced or nonfunctional ALS concentrations in serum.

The possibility that heterozygous mutations of *IGFALS* could result in growth impairment has received recent attention. In an analysis of 21 patients with homozygous or compound heterozygous mutations, the mean height of individuals with two affected alleles was −2.36 SD, while those with one affected allele was −0.83 SD.[26] When patients were compared with wild-type first degree relatives, individuals with two affected alleles were 2.13 SD shorter; those with one affected allele were 0.9 SD shorter.

Mutations of IGF1R

Knockout of both IGF1R alleles in the mouse resulted in dramatic intrauterine growth retardation and,

typically, early death from respiratory failure. On this basis, it was believed, at first, that humans carrying homozygous mutations of *IGF1R* would not be capable of surviving infancy. Abuzzahab et al.[12] described a child with compound heterozygous mutations of *IGF1R*, who presented with moderate IUGR and postnatal growth failure (initially reported as R108Q/K115N; using prepeptide numbering, this is R138Q/K145N) (Fig. 26.7). Most of the subsequent cases described proved to be heterozygotes (Fig. 26.8).

As described with mutations of *GHR* rigorous confirmation of the significance of heterozygous mutations of *IGF1R* requires functional *in vitro* studies and, where possible, genotype:phenotype correlations in family members. Of particular interest are the three reported mutations affecting the tyrosine kinase domain of the IGF1R: E1080K, G1155A, and *3348-3366dup*, which results in a frameshift and premature termination.[27]

The clinical presentation, in general, involves a combination of prenatal and postnatal growth failure, although

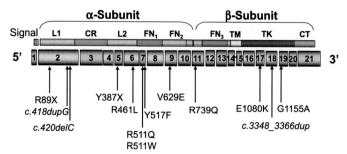

FIGURE 26.8 Published heterozygous mutations of IGF1R. *See legend to Fig. 26.7.*

almost certainly this reflects ascertainment bias issues. Serum concentrations of IGF-I are normal or high. Data are inadequate at this time to address issues concerning intellectual development, glucose intolerance, osteoporosis, and other features that might reflect resistance to IGF-I action.

In addition to the original compound heterozygote reported by Abuzzahab et al., siblings have been reported with compound heterozygosity (E121K/E234K),[28] and one patient homozygous for the mutation R40L has been described[29] (Fig. 26.7). It is assumed that these mutations, while impairing receptor function, permit sufficient IGF-I action as to allow for survival and some growth, although growth impairment appears to be more severe than that described in typical heterozygous cases.

DIAGNOSIS: GENETIC TESTING AND INTERPRETATION

Testing for molecular defects of the GH-IGF axis is not commercially available at the current time. There are a number of academic laboratories that perform DNA sequencing studies and molecular analysis of the GH gene, as well as pituitary transcription factors (see Chapter 9).

Similarly, molecular analysis of the *GHR*, *JAK-STAT*, *IGFI*, *IGFALS*, and *IGF1R* genes is not, generally, commercially available. These investigations are available, on a case-by-case basis, at the IGF Deficiency Research Center at Oregon Health & Science University in Portland, Oregon. Interested parties should contact the author by letter, telephone, or email.

A database and website for molecular defects of the GH-IGF axis has been developed and is accessible at www.growthgenetics.com.[30] The database displays clinical, biochemical, genetic, and molecular details on all published cases of defects of the following genes: *GHR*, *STAT5B*, *IGF1*, *IGF2*, *IGFALS*, and *IGF1R*.

Which Tests are Best to Order in Whom?

The defining characteristics of primary IGFD are growth failure, low serum IGF-I, and a normal or elevated serum GH concentration. In the case of defects at the level of the *IGF1R*, however, serum concentrations of IGF-I may be normal or even elevated, consistent with presumptive IGF resistance.

Each of the molecular causes of IGFD or IGF resistance has a characteristic clinical and biochemical phenotype (Table 26.2). Often the clinical presentation and family history, combined with serum studies, can dictate the priority and hierarchy of molecular tests to be performed. The absence of intrauterine growth failure in a child with postnatal growth retardation suggests a defect at the level of *GHR*, *STAT5B*, or, possibly, *IGFALS*. The combination of prenatal and postnatal growth failure, on the other hand, suggests that the defect might be at the level of *IGFI* or *IGF1R*.

Measurement of serum concentrations of GHBP is of value, as low levels are highly suggestive of a molecular defect impacting the extracellular, GH-binding domain of the GHR. The presence of normal (or even increased) serum concentrations does not, however, exclude the possibility of a mutation or a deletion of *GHR*.

Normal serum concentrations of IGFBP-3, despite very low serum IGF-I, suggests the possibility of a defect of *IGF1*. Extremely low serum concentrations of IGFBP-3, on the other hand, are consistent with a defect of *ALSIGF*, especially if the growth failure is modest.

An IGF generation test is often of value in cases of primary IGFD.[31,32] In such tests, GH is administered daily for four to seven days, and serum concentrations of IGF-I and IGFBP-3 measured before and after GH. Failure of serum IGF-I and IGFBP-3 to rise supports a diagnosis of GH resistance, and is consistent with molecular defects at the level of *GHR*, *STAT5B*, *IGFI*, and *IGFALS*.

Labs Available for Testing

For further information concerning molecular analysis of patients with IGF deficiency or resistance, please contact the IGF Deficiency Research Center at Oregon Health & Science University, Portland, Oregon:

Ron Rosenfeld, MD
Tel: 503-412-9671
Email: ron@stat5consulting.com

Prenatal Testing?

Prenatal testing has not been performed for the specific disorders described in this chapter, but should be feasible. The majority of the disorders described are transmitted in an autosomal recessive manner and should be amenable to prenatal testing, as well as family counseling.

Lists of Consultants and Resources

See above.

Predictive Value of a Test?

DNA sequencing of candidate genes should be relatively definitive. As discussed, however, results should be evaluated, ideally within the context of family history and *in vitro* functional studies.

Significance of a Negative Test?

The GH-IGF axis involves a complex cascade of molecular and biochemical events, which we are just beginning to understand. While negative tests provide important information, they can only be used to interpret those genes (and those portions of the genes) that are sequenced. If this is limited to coding exons, important genetic information may be missed and abnormalities overlooked.

It is important to note, additionally, that we are limited currently to those candidate genes that have been demonstrated to be relevant to GH-IGF action. There are many components of the GH and IGF signaling cascade that remain poorly understood and are legitimate candidates for harboring functionally significant mutations and/or deletions. Thus, the absence of identifiable mutations of the candidate genes described earlier cannot rule out the possibility of a molecular abnormality of the GH-IGF axis.

Should Family Members be Tested?

Ideally, family members should be tested. For starters, identified mutations in the patient of interest can best be understood in the context of genotype:phenotype correlations with other family members. We have begun to appreciate that even within a nuclear family, members harboring the same mutations may have a range of phenotypic expression. Additionally, for many of the autosomal recessive disorders described earlier, the issue of heterozygous expression remains unsettled and additional data should prove to be invaluable.

TREATMENT

Based on the Genetic Information, How Does that Affect Treatment, if at All?

Differentiation between secondary IGFD (i.e., secondary to GHD) and primary IGFD is critical in identifying optimal therapy.[33] For patients with secondary IGFD, the initial treatment of choice is GH, although patients who prove to be poorly responsive to GH warrant reevaluation and consideration of alternative therapeutic approaches. For patients with primary IGFD, the relative benefits and likelihood of success of GH versus IGF-I

therapy need to be considered. In situations of severe primary IGFD, especially with documented molecular abnormalities of *GHR*, *STAT5B*, or *IGFI*, IGF-I therapy would be the treatment of choice. In cases of mutations of *IGFALS*, we have inadequate data at this time to determine whether either GH or IGF-I therapy is likely to be of benefit.

Genetic Counseling

Genetic counseling should be made available to patients and families with identified molecular defects of the GH-IGF axis. As stated earlier, most of the conditions identified to date are transmitted in an autosomal recessive manner, but heterozygous expression and dominant negative mutations have also been observed for some of these genetic defects.

References

1. Backeljauw PF, Dattani M, Cohen P, Rosenfeld RG. Disorders of growth. In: Sperling MA, editor. *Pediatric endocrinology*. 5th ed. Philadelphia: Saunders; 2014.
2. Reiter EO, Rosenfeld RG. Normal and aberrant growth. In: Larsen PR, Kronenberg HM, Melmed S, Polonsky KS, editors. *Williams textbook of endocrinology*. 11th ed. Philadelphia: Elsevier; 2007. p. 849–968.
3. Superti-Furga A, Unger S. Nosology and classification of genetic skeletal disorders: 2006 revision. *Am J Med Genet* 2007;**143**:2082–3.
4. Rimoin DL, Cohn D, Krakow D, et al. The skeletal dysplasias: clinical-molecular correlations. *Ann NY Acad* 2007;**1117**:302–9.
5. Rosenfeld RG. The molecular basis of idiopathic short stature. *Growth Horm IGF Res* 2005;**15**(Suppl. A):S3–5.
6. Rosenfeld RG. Molecular mechanisms of IGF-I deficiency. *Horm Res* 2006;**65**(Suppl. 1):15–20.
7. Rosenfeld RG, Rosenbloom AL, Guevara-Aguirre J. Growth hormone (GH) insensitivity due to primary GH receptor deficiency. *Endocr Rev* 1994;**15**:369–90.
8. Hwa V, Nadeau K, Wit JM, Rosenfeld RG. STAT5b deficiency: lessons from *STAT5b* gene mutations. *Best Prac Res Clin Endocrinol Metab* 2011;**25**:61–75.
9. Nadeau K, Hwa V, Rosenfeld RG. STAT5b deficiency: an unsuspected cause of growth failure, immunodeficiency and severe pulmonary disease. *J Pediatr* 2011;**158**:701–10.
10. David A, Hwa V, Metherell LA, et al. Evidence for a continuum of genetic, phenotypic, and biochemical abnormalities in children with growth hormone insensitivity. *Endocr Rev* 2011;**32**:472–97.
11. Walenkamp MJ, Wit JM. Genetic disorders in the growth hormone-insulin-like growth factor-I axis. *Horm Res* 2006;**66**:221–30.
12. Abuzzahab MJ, Schneider A, Goddard A, et al. IGF-I receptor mutations resulting in intrauterine and postnatal growth failure. *N Engl J Med* 2003;**349**:2211–22.
13. Metherell LA, Akker SA, Munroe PB, et al. Pseudoexon activation as a novel mechanism for disease resulting in atypical growth hormone insensitivity. *Am J Hum Genet* 2001;**69**:641–6.
14. Aalbers AM, Chin D, Pratt KL, et al. Extreme elevation of serum growth hormone-binding protein concentrations resulting from a novel heterozygous splice site mutation of the growth hormone receptor gene. *Horm Res* 2009;**71**:276–84.
15. Aisenberg J, Auyeung V, Pedro HF, et al. Atypical GH insensitivity syndrome and severe insulin-like growth factor-I deficiency resulting from compound heterozygous mutations of the GH

receptor, including a novel frameshift mutation affecting the intracellular domain. *Horm Res Paediatr* 2010;**74**:406–11.

16. Goddard AD, Covello R, Luoh SM, et al. Mutations of the growth hormone receptor in children with idiopathic short stature. The Growth Hormone Insensitivity Study Group. *N Engl J Med* 1995;**333**:1093–8.

17. Fang P, Riedl S, Amselem S, et al. Primary growth hormone (GH) insensitivity and insulin-like growth factor deficiency caused by novel compound heterozygous mutations of the GH receptor gene: genetic and functional studies of simple and compound heterozygous states. *J Clin Endocrinol Metab* 2007;**92**:2223–31.

18. Kofoed EM, Hwa V, Little B, et al. Growth hormone insensitivity associated with a STAT5b mutation. *N Engl J Med* 2003;**349**:1139–47.

19. Scaglia PA, Martinez AS, Feigerlova E, et al. A novel mutation of the SH2 domain of the *STAT5B* gene results in a transcriptionally inactive STAT5b associated with severe IGF-I deficiency and immune dysfunction. *J Clin Endocrinol Metab* 2012;**97**:E830–9.

20. Woods KA, Camcho-Hubner C, Savage MO, et al. Intrauterine growth retardation and postnatal growth failure associated with deletion of the insulin-like growth factor I gene. *N Engl J Med* 1996;**335**:1363–7.

21. Walenkamp MJ, Karperien M, Pereira AM, et al. Homozygous and heterozygous expression of a novel insulin-like growth factor-I mutation. *J Clin Endocrinol Metab* 2005;**90**:2855–64.

22. Netchine I, Azzi S, Houang M, et al. Partial primary deficiency of insulin-like growth factor (IGF)-I activity associated with IGF-I mutation demonstrates its critical role in growth and brain development. *J Clin Endocrinol Metab* 2009;**94**:3913–21.

23. van Duyvenvoorde HA, van Setten PA, Walenkamp MJ, et al. Short stature associated with a novel heterozygous mutation of the insulin-like growth factor 1 gene. *J Clin Endocrinol Metab* 2010;**95**:E363–7.

24. Fuqua JS, Derr M, Rosenfeld RG, Hwa V. Identification of a novel heterozygous *IGF1* splicing mutation associated with familial short stature. *Horm Res Paediatr* 2012;**78**:59–66.

25. Domene HM, Hwa V, Argente J, et al. Human acid-labile subunit deficiency and growth failure: clinical, endocrine and metabolic consequences. *Horm Res* 2009;**72**:129–41.

26. Fofanova-Gambetti OV, Hwa V, Wit JM, et al. Impact of heterozygosity for acid-labile subunit (IGFALS) gene mutations on stature: results from the International Acid-Labile Subunit Consortium. *J Clin Endocrinol Metab* 2010;**95**:4184–91.

27. Fang P, Schwartz MAD, Johnson BD, et al. Familial short stature caused by haploinsufficiency of the insulin-like growth factor-I receptor due to nonsense-mediated mRNA decay. *J Clin Endocrinol Metab* 2009;**94**:1740–7.

28. Fang P, Choi YH, Derr MA, et al. Severe short stature caused by novel compound heterozygous mutations of the insulin-like growth factot I receptor (IGF1R). *J Clin Endocrinol Metab* 2012;**97**:E243–7.

29. Gannage-Yared MH, Klammt J, Chourey E, et al. Homozygous mutation of the IGF-I receptor gene in a patient with severe pre- and postnatal growth failure and congenital malformations. *Eur J Endocrinol* 2013;**168**:K1–7.

30. Rosenfeld RG, von Stein T. A database and website for molecular defects of the GH-IGF axis: www.growthgenetics.com. *Horm Res Paediatr* 2013;**80**:443–8.

31. Buckway CK, Guevara-Aguirre J, Pratt KL, et al. The IGF-I generation test revisited: a marker of GH sensitivity. *J Clin Endocrinol Metab* 2001;**86**:S176–83.

32. Selva KA, Buckway CK, Sexton G, et al. Reproducibility in patterns of IGF generation with special reference to idiopathic short stature. *Horm Res* 2003;**60**:237–46.

33. Savage MO, Camacho-Hubner C, David A, et al. Idiopathic short stature: will genetics influence the choice between GH and IGF-I therapy? *Eur J Endocrinol* 2007;**157**(Suppl. 1):S33–7.

PART XI

MISCELLANEOUS

27

Cost-Effectiveness of Genetic Testing for Monogenic Diabetes

Rochelle N. Naylor, *Siri Atma W. Greeley*, *Elbert S. Huang***

*Departments of Pediatrics and Medicine, Section of Adult and Pediatric Endocrinology, Diabetes and Metabolism, The University of Chicago, Chicago, IL, USA

**Department of Medicine, Section of General Internal Medicine, The University of Chicago, Chicago, IL, USA

COST OF DIABETES CARE

Diabetes mellitus is a group of heterogeneous disorders with diverse etiologies, affecting at least 285 million people worldwide. The economic costs are exorbitant, with $376 billion USD spent in 2010, reflecting 12% of health expenditures.[1] In the United States alone, 25.8 million people are affected at an annual cost of $245 billion USD.[2] The high costs of diabetes reflect both the rising costs of routine diabetes care as well as the increasing prevalence of diabetes. In the face of fiscal constraints, the substantial economic burden of diabetes has led to increased scrutiny of new interventions by health care payers. New interventions are systematically subjected to economic evaluations to ensure that the benefits of these interventions are produced at reasonable costs. In this chapter, we will discuss the cost-effectiveness analysis of genetic testing strategies for the diagnosis and classification of monogenic forms of diabetes.

HETEROGENEITY OF DIABETES MELLITUS

To appreciate the role of genetic testing in diabetes, one has to realize that diabetes is not one disease, but many, all characterized by sustained hyperglycemia due to varied underlying pathophysiology. Type 2 diabetes, resulting from a combination of insulin resistance and beta-cell failure and usually driven by obesity, accounts for over 90% of diabetes cases. Five to 10% of diabetes is type 1 diabetes, caused by immune-mediated or idiopathic beta-cell destruction. Other rare causes of diabetes together account for the remaining ~5% of cases, including monogenic forms of diabetes.

MONOGENIC DIABETES

Monogenic diabetes is due to single gene defects or chromosomal abnormalities that are sufficient to cause diabetes. There are nearly 30 different genes that can cause monogenic diabetes, together accounting for ~2% of all cases of diabetes.[3] Monogenic diabetes can be divided into two main types: neonatal diabetes and maturity-onset diabetes of the young (MODY), with MODY representing the majority of monogenic diabetes cases. Neonatal diabetes is defined as persistent hyperglycemia requiring treatment, with onset within the first 6 months of life.[4] MODY is classically defined as autosomal dominant, nonketotic, noninsulin dependent diabetes with onset typically before 25 years of age with two to three consecutively affected generations.[5] Patients typically lack significant obesity or metabolic features and do not have pancreatic autoantibodies. Studies show that less than 50% of individuals with a genetic diagnosis of MODY fit the classic description.[6]

PRECISION MEDICINE IN MONOGENIC DIABETES

The discovery of monogenic forms of diabetes creates an opportunity for precision medicine. Precision medicine includes the application of genomic information to personalize disease prevention, diagnosis, and treatment in order to improve patient outcomes. Genetic testing to diagnose monogenic diabetes is an example of precision medicine. While monogenic forms of diabetes are rare, the most common genetic causes have specific therapies with proven efficacy that differ from conventional therapy for type 1 and type 2 diabetes. Importantly, these therapies are substantially cheaper and less burdensome

Genetic Diagnosis of Endocrine Disorders. http://dx.doi.org/10.1016/B978-0-12-800892-8.00027-0

TABLE 27.1   Established First-Line Therapy for Selected Genetic Causes of Monogenic Diabetes

Genetic subtype	Established first-line therapy	Cost relative to type 1 diabetes	Cost relative to type 2 diabetes
NEONATAL DIABETES			
KCNJ11 mutations	High doses of sulfonylureas	↓↓↓	N/A
ABCC8 mutations	High doses of sulfonylureas	↓↓↓	N/A
MODY			
HNF1A mutations	Low doses of sulfonylureas	↓↓↓	↓/→
HNF4A mutations	Low doses of sulfonylureas	↓↓↓	↓/→
GCK mutations	No pharmacologic therapy	↓↓↓↓	↓↓

than insulin therapy (conventional therapy for type 1 diabetes) and have similar costs as metformin (conventional therapy for type 2 diabetes) with often improved glycemic control, which may translate into lower rates of diabetes complications (Table 27.1).[7] Thus, correct diagnosis of subtypes of monogenic diabetes may be an important source of cost saving *at the individual patient level* due to genetically driven treatment changes. Diagnosis of monogenic diabetes also aids in identifying at-risk family members for diagnostic or predictive genetic testing, particularly for MODY, which is dominantly inherited.[8,9]

Neonatal Diabetes

Neonatal diabetes occurs in 1:90,000–260,000 live births.[10,11] Approximately half of neonatal diabetes cases are transient and half are permanent.[12] An uncertain fraction of transient cases will have diabetes relapse in later life, typically during adolescence or pregnancy. Transient neonatal and relapsed transient neonatal diabetes may be amenable to treatment with sulfonylureas, but the best treatment approach is not clear.[13]

The most common causes of permanent neonatal diabetes are heterozygous activating mutations in the genes *KCNJ11* and *ABCC8*, which encode the Kir6.2 and SUR1 subunits of the ATP-sensitive potassium (K_{ATP}) channel. Such mutations abrogate the usual hyperglycemia-induced closure of this channel that allows insulin secretion from beta cells. Sulfonylureas can close this channel in an ATP-independent manner, representing a genetically targeted therapy for patients with diabetes due to K_{ATP} channel mutations. The majority of these patients are able to transition from insulin to oral sulfonylureas, which almost always results in decreased frequency of hypoglycemia and overall better glycemic control, including improved glycated hemoglobin.[14–17]

MODY

There are at least 13 described MODY genes but mutations in just three, *HNF1A*, *HNF4A*, and *GCK*, together account for >90% of all diagnosed MODY.[6] MODY due to *GCK* mutations results in mild, nonprogressive hyperglycemia where treatment is not needed, except possibly during pregnancy.[18] Studies have shown that discontinuation of pharmacologic therapy does not alter glycated hemoglobin.[19] *HNF1A*-MODY and *HNF4A*-MODY typically are sensitive to low doses of sulfonylureas, showing stable or improved glycemic control as compared to insulin therapy.[20–23]

CONSIDERATIONS IN GENETIC TESTING FOR MONOGENIC DIABETES

While there are published practice guidelines for utilizing genetic testing in order to diagnose monogenic forms of diabetes, it is estimated that at least 80% of diagnoses are missed.[6,7,24] The importance of genetic testing for monogenic diabetes in infants with persistent hyperglycemia seems to be increasingly accepted among neonatologists and pediatricians. However, screening for MODY is not a routine practice for diabetes classification, and genetic testing is generally pursued only in those with classic features of MODY, which may miss more than 50% of all cases.[6] Reasons for this include varying clinician knowledge of monogenic diabetes, frequent medical insurance denial of coverage for genetic testing, and significant clinical overlap between MODY and type 1 and type 2 diabetes. The lack of clinical distinction makes it difficult to identify individuals who are very likely to have monogenic diabetes and would benefit from genetic testing. The latter deserves special consideration in the discussion of cost-effectiveness of genetic testing for monogenic diabetes.

From an individual patient perspective, diagnosing subtypes of monogenic diabetes is cost-saving due to cheaper therapies as described earlier. Schnyder et al. illustrated how identifying MODY affects clinical care on an individual patient basis. A molecular genetic diagnosis of *GCK*-MODY in a 5-year-old female initially treated with insulin therapy allowed for discontinuation of treatment and decreased clinic visits from quarterly to annually. They evaluated the cost-effectiveness of genetic testing in their patient by comparing the cost of genetic analysis and treatment of *GCK*-MODY to the "costs for intensive diabetes control and treatment as recommended for type 1 diabetes." Their analysis showed that 1,140 euros would be saved in the first year and 1,640 euros annually thereafter. Additionally, the cost of genetic testing for relatives was ~25% of index patient testing costs, and testing in the patient's mother revealed that she was heterozygous for the same *GCK* mutation.[25]

However, from the societal perspective, the individual patient benefit must be balanced against the cost of genetic testing in those who do not have monogenic forms of diabetes. The cost of genetic testing in general is rapidly decreasing and is expected to continue to fall as next-generation sequencing approaches increase efficiency. However, commercial costs for testing a single monogenic diabetes gene still approach $1,000 USD. This cost currently prohibits universal diabetes genetic testing to be incorporated at diabetes diagnosis for correct classification. Thus, precision medicine for monogenic diabetes, and particularly for MODY, is hindered by difficulty in identifying appropriate patients for testing. Despite a number of clinical characteristics, family history and biomarkers that are highly suggestive of an underlying MODY diagnosis, there is no singular feature that can infallibly distinguish all affected individuals from type 1 or type 2 diabetes.[6,26–29] Testing criteria based on atypical features for type 1 or type 2 diabetes may identify MODY in 10–25% of the tested population.[23] While this percentage is substantial, it still means that genetic testing in up to 90% will be negative, and this cost must be considered against the benefit of targeted therapy for those with monogenic forms of diabetes. Thus, formal cost-effectiveness analysis is needed to understand how to integrate monogenic diabetes diagnosis into clinical practice.

THE ROLE OF COST-EFFECTIVENESS ANALYSIS IN HEALTHCARE

Health economic evaluations have become essential for policymakers and clinical leaders because we live in an era where there are limited healthcare resources and an almost limitless and expanding number of potential healthcare interventions. Because of the mismatch of available resources and potential interventions, we require a conceptual framework to determine how we should allocate our resources to improve health. Each proposed health intervention has the expectation to maximize "the years of healthy life gained for its population in return for a given level of investment."[30] To compare the value of one intervention to another, we need a way to predict and define the benefits produced and the costs incurred by each intervention. We also need to be able to interpret these benefits and costs in a tangible way that takes the aggregate health outcomes of a population into consideration, and not only the individual. Cost-effectiveness analysis is one of the most well-established methods for assessing the economic value of new diagnostic tests and treatments. Cost-effectiveness analysis "shows the tradeoffs involved in choosing among interventions... giv[ing] decision makers in diverse settings – physician's office, health maintenance organizations...or state or federal programs – important data for making informed judgments about interventions."[30] The practice of cost-effectiveness analysis can be summarized as follows: the first step is to calculate the mean costs incurred and the mean benefits gained by each intervention. States of health are assigned a value or utility ranging from 0 (death) to 1 (full health). Health benefits are determined using health utilities for given disease states, accounting for changes produced by the intervention to extend life and improve health. Health benefits are typically expressed as life years or most often as quality-adjusted life years (QALYs). QALYs do not only consider the ability of an intervention to extend life but to improve the quality of life. Next, an incremental cost-effectiveness ratio (ICER) is calculated by dividing the difference in the mean costs over the difference in the mean benefits between the new and the old interventions. The ICER represents the additional costs required by the new intervention to produce one extra unit of benefit over that produced by the old intervention. The ICER is then compared with a threshold value that represents the maximum a decision maker is willing to pay for an additional unit of benefit. If the ICER is lower than this threshold value then the new intervention is deemed to be cost-effective. Established guidelines, such as those developed by the Panel on Cost-Effectiveness in Health and Medicine, have helped to bring uniformity to the conduct of cost-effectiveness analysis.[31]

The cost-effectiveness threshold varies among different healthcare systems. In the Unites States, the threshold for cost-effectiveness is typically set at an ICER of $50,000 per QALY.[32,33] However, it has been noted that higher values ranging from $109,000 to $297,000 may be more reflective of what US society is willing to pay for healthcare interventions routinely.[33] But, when considering the specific context of diabetes, routine processes of care such as intensive glucose control and statin therapy for hyperlipidemia have ICERs of ~$41,000 and $52,000, respectively.[34] Thus, it is likely that the conventional benchmark of cost-effectiveness of $50,000 should be applied when considering genetic testing in diabetes.

Cost-effectiveness analysis requires a number of assumptions to be made and the validity of results will be determined by how appropriate assumptions are.

Varying values of model assumptions to reflect uncertainty through sensitivity analyses allows models to fully examine the quality of various data sources. The common theme of these techniques is that uncertainty regarding model parameters is made explicit. Cost-effectiveness analyses must also determine which perspective to represent, which impacts which benefits and costs must be considered. For example, the time spent by a patient or caregiver to deliver a health intervention is accounted for when cost-effectiveness analysis is performed from the societal perspective, but this "cost" is not considered in analyses from the health care system perspective.

COST-EFFECTIVENESS ANALYSIS OF MONOGENIC DIABETES

Our research group has conducted two studies to assess the cost-effectiveness of genetic testing strategies in monogenic diabetes. The first study of neonatal monogenic diabetes showed that genetic testing for monogenic diabetes in children with type 1 diabetes was cost saving, owing to significant improvements in quality of life and substantial savings in diabetes treatment costs. Our second study assessing cost-effectiveness of genetic testing for MODY in incident cases of type 2 diabetes demonstrated that such testing could be cost-effective if the prevalence of MODY was sufficiently high in the tested population. Both studies are described in greater detail in the subsequent sections.

Cost-Effectiveness Analysis in Neonatal Diabetes

The first study, published in 2011, assessed genetic testing for permanent neonatal diabetes among children diagnosed with type 1 diabetes.[35] The study was carried out from the societal perspective over 30 years of follow-up. The main outcomes were QALYs, based on assigned health utilities for type of diabetes treatment and diabetes-related complications as well as costs, with results expressed as the ICER ($/QALY). All costs and benefits were discounted at 3%, the standard practice in US cost-effectiveness analyses.

Policy Decision

A policy of genetic testing for mutations in *KCNJ11* and *ABCC8* with subsequent therapy change from insulin to sulfonylureas was compared to a policy of no genetic testing with continued insulin therapy for all individuals (Fig. 27.1). In the model, all children with permanent neonatal diabetes diagnosed before 6 months of age underwent one-time genetic testing for mutations in *KCNJ11* and *ABCC8* at 6 years of age at a combined cost of $2815 USD (expressed in 2008 US dollars). The genetic test was assumed to have 100% sensitivity and specificity. Hypothetical patients found to have a mutation underwent an attempt to switch from insulin to sulfonylurea therapy. A 4-day hospital stay was factored into costs for this transition. A 90% success rate was assumed, based on the literature. Successful conversion to sulfonylureas was modeled to lead to a lifetime hemoglobin A1c of 6.4% compared to a lifetime hemoglobin A1c of

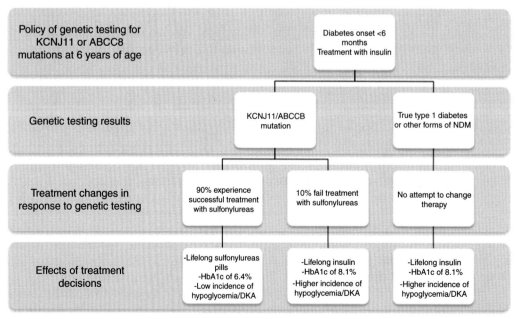

FIGURE 27.1   Policy decision for genetic testing for neonatal diabetes.

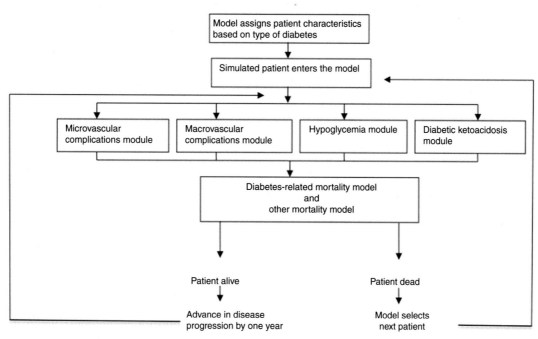

FIGURE 27.2 Basic simulation model structure for complications of diabetes.

8.1% in those with true type 1 diabetes and those failing conversion to sulfonylureas, who were treated with replacement doses of insulin. A health utility of 0.86 was assigned to life with diabetes treated with insulin, while a health utility of 0.96 was assigned to life with diabetes treated with sulfonylureas.

Model for Diabetes Complications

A simulated model of diabetes cost and complications was used to compare the policy of universal genetic testing to the policy of no genetic testing as the reference or base case (Fig. 27.2). The model began by assigning characteristics of hypothetical patients, including those with neonatal monogenic diabetes versus true type 1 diabetes. The model accounted for microvascular diabetes complications of retinopathy, nephropathy, and neuropathy, as well as macrovascular complications of ischemic heart disease, myocardial infarction, congestive heart failure, and stroke. It also accounted for complications of diabetic ketoacidosis and hypoglycemia, both of which were significantly more prevalent with insulin treatment as compared to those with mutations successfully treated with sulfonylurea therapy. The diabetes model was framed by simultaneous progression of disease through microvascular and macrovascular complications and mortality and their associated Markov states. The impact of pharmacologic interventions following genetic diagnosis was represented separately through the modification of transition probabilities over each one year cycle. Within the cycle length of one year, patients moved from one disease state to another or stayed in the current disease state.

Impact of Neonatal Diabetes Genetic Testing

Results showed that a routine genetic testing policy was actually cost saving with dramatic improvements in QALYs and reductions in costs. These benefits were seen as early as 10 years with a savings of $12,528/QALY and increased to $30,437/QALY at 30 years. Sensitivity analyses showed that the testing policy would remain cost saving if neonatal diabetes prevalence was >3%.[35] Few health interventions are capable of providing health benefits while actually saving money. Formal cost-effectiveness analysis was able to demonstrate this coveted outcome in neonatal diabetes diagnosis.

Cost-Effectiveness Analysis in MODY

The second study, published in 2014, assessed MODY genetic testing in new-onset apparent type 2 diabetes in adults 25–40 years of age, where overlap with MODY can occur. Analysis was conducted from a healthcare system perspective over the lifetime of the population. The main outcomes were QALYs and costs, expressed as the ICER. All costs and benefits were discounted at 3%.[36]

Policy Decision

We compared a policy of routine genetic testing for mutations in *HNF1A*, *HNF4A*, and *GCK* causing MODY among incident cases of type 2 diabetes with subsequent therapy change appropriate for each genetic subtype to a policy of no genetic testing as the reference case

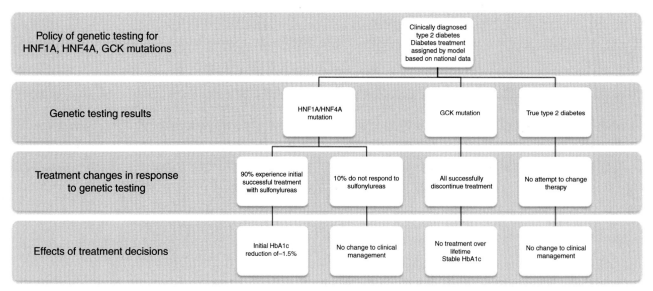

FIGURE 27.3　Policy decision for genetic testing for maturity-onset diabetes of the young (MODY).

(Fig. 27.3). The genetic test was assumed to have perfect performance (100% sensitivity and 100% specificity) to detect mutations in the three tested genes. Genetic testing cost for the three genes was $2,580 USD per individual, with costs expressed in 2011 US dollars. Based on the published literature, in the reference case it was assumed that 2% of the tested population had MODY (65% of those positive for mutations had MODY due to mutations in *HNF1A* or *HNF4A* and 35% had MODY due to *GCK* mutations) and 98% had true type 2 diabetes. Those with true type 2 diabetes were treated with either insulin, oral medications, insulin and oral medications, or no pharmacologic therapy based on data on distribution of diabetes treatment in national studies. We assumed that 90% of patients with *HNF1A*-MODY and *HNF4A*-MODY would be treated successfully with sulfonylureas and would experience a 1.5% decrease in hemoglobin A1c at the time of therapy initiation. The model accounted for increasing rates of sulfonylurea failure over time with subsequent insulin therapy, because the durability of sulfonylureas in *HNF1A*-MODY and *HNF4A*-MODY is not necessarily lifelong.[20] We modeled therapy discontinuation in 100% of those with *GCK*-MODY with no change in hemoglobin A1c, consistent with the natural course of MODY due to mutations in *GCK*. Annual costs of medication were accounted for each hypothetical patient. A health utility of 0.64 was assigned to life with diabetes treated with insulin, while a health utility of 0.77 was assigned to life with diabetes treated with oral medications.

Models for Diabetes Complications

After assigning initial patient characteristics, including the etiology of diabetes, each hypothetical patient

advanced through one of two distinct diabetes complications models modified to account for the natural history of MODY due to mutations in *HNF1A*, *HNF4A*, and *GCK*. In this study, those with *GCK*-MODY were not modeled to experience diabetes-related complications, consistent with the natural course of *GCK*-MODY.[37] Cardiovascular disease and background mortality was modeled using Framingham cardiovascular risk equations. Costs and health decrement of diabetes complications was accounted for in all other hypothetical patients. Type 2 diabetes and monogenic diabetes due to mutations in *HNF1A* and *HNF4A* experienced microvascular and macrovascular diabetes complications based on UK Prospective Diabetes Study (UKPDS) risk equations. Within each one year cycle, patients progressed through the complication modules as well as a mortality module.

Impact of MODY Genetic Testing

The base case analysis, where only 2% of the tested population had MODY, resulted in an ICER of $205,000/QALY, exceeding typical thresholds of cost-effectiveness. Sensitivity analyses showed that the prevalence of MODY and the cost of genetic testing were the main variables driving costs. A small increase in MODY prevalence from 2% to 6% or decreased genetic testing costs of $700 resulted in an ICER of ~$50,000/QALY. A prevalence >30% made the testing policy cost savings. This study has helped to provide important initial perspectives on what populations may benefit from MODY genetic testing, underscoring the importance of prevalence in the tested cohort to make testing economically appropriate.[36]

FUTURE STUDIES OF COST-EFFECTIVENESS ANALYSIS IN MONOGENIC DIABETES

The National Health Service in England and Wales published their planned model structure to assess the economic value of diagnosing monogenic diabetes caused by mutations in the GCK, HNF1A, or HNF4A genes. They identified five test–treatment strategies of: (1) no genetic testing; (2) genetic testing based on clinical features noted by clinicians (the current standard); (3) referral for genetic testing based on the results of the MODY probability calculator; (4) referral for genetic testing based on a combination of positive UCPCR and negative pancreatic autoantibody testing; and (5) referral for genetic testing for all patients with a diagnosis of diabetes under the age of 30 years.[38] The results of their study will enhance previous efforts to identify the best cost-effective health intervention strategy to identify MODY. Future cost-effectiveness analyses of monogenic diabetes also need to assess MODY diagnosis in the specific context of type 1 diabetes, where the minimum estimated prevalence in pediatric cases is 1.2%.[39] A specific cost-effectiveness analysis of genetic screening in type 1 diabetes is required as the MODY prevalence, age at onset, disease progression, and costs of therapy are distinct from that of type 2 diabetes. Studies also need to account for the impact of screening first-degree relatives (FDRs) on cost-effectiveness of monogenic diabetes genetic testing. Establishing an approach to assess cost-effectiveness of genetic testing in both probands and relatives is very important for accurate economic evaluation of health interventions involving genetic testing for diagnosis of inherited diseases. Testing of first-degree relatives (FDRs) following proband diagnosis has been shown to increase cost-effectiveness of genetic testing in certain autosomal dominant disorders.[40,41] In MODY, first-degree family members have a 50% chance of carrying the disease-causing gene mutation. Genetic testing cost for relatives of probands using the identified family mutation is nearly ten fold cheaper than the cost of propositus testing. These factors may lead to enhanced cost-effectiveness of a MODY genetic testing policy that includes relatives. The potential benefits of parental testing in neonatal diabetes or predictive testing of future offspring of parents with an affected child are questionable because 80% of cases are sporadic.[42,43] Again, additional cost-effectiveness analysis of neonatal diabetes will help to clarify the best strategies for diagnosis.

CONCLUSIONS

The expense of genetic testing represents an important challenge to integrating precision medicine for monogenic diabetes into clinical practice. The results of previous and future research formally studying the long-term clinical and economic consequences of monogenic diabetes genetic screening strategies can inform clinical recommendations and coverage decisions for this relatively new technology. This will allow the translation of diabetes genetics knowledge into improved health outcomes for affected patients in a cost-effective manner. This knowledge is crucial in this time of constrained healthcare resources to guide coverage policies for monogenic diabetes genetic testing and to allocate resources effectively.

References

1. Hu FB. Globalization of diabetes: the role of diet, lifestyle, and genes. *Diab Care* 2011;**34**(6):1249–57.
2. American Diabetes Association. Economic costs of diabetes in the U.S. in 2012. *Diab Care* 2013;**36**(4):1033–46.
3. Frayling TM, Evans JC, Bulman MP, Pearson E, Allen L, Owen K, et al. Beta-cell genes and diabetes: molecular and clinical characterization of mutations in transcription factors. *Diabetes* 2001;**50**(Suppl. 1): S94–S100.
4. Greeley SAW, Tucker SE, Naylor RN, Bell GI, Philipson LH. Neonatal diabetes mellitus: a model for personalized medicine. *Trends Endocrinol Metab* 2010;**21**(8):464–72.
5. Tattersall RB, Fajans SS. A difference between the inheritance of classical juvenile-onset and maturity-onset type diabetes of young people. *Diabetes* 1975;**24**(1):44–53.
6. Shields BM, Hicks S, Shepherd MH, Colclough K, Hattersley AT, Ellard S. Maturity-onset diabetes of the young (MODY): how many cases are we missing? *Diabetologia* 2010;**53**(12):2504–8.
7. Murphy R, Ellard S, Hattersley AT. Clinical implications of a molecular genetic classification of monogenic beta-cell diabetes. *Nat Clin Pract Endocrinol Metab* 2008;**4**(4):200–13.
8. Shepherd M, Hattersley AT, Sparkes AC. Predictive genetic testing in diabetes: a case study of multiple perspectives. *Qual Health Res* 2000;**10**(2):242–59.
9. Shepherd M, Ellis I, Ahmad AM, Todd PJ, Bowen-Jones D, Mannion G, et al. Predictive genetic testing in maturity-onset diabetes of the young (MODY). *Diabet Med* 2001;**18**(5):417–21.
10. Iafusco D, Massa O, Pasquino B, Colombo C, Iughetti L, Bizzarri C, et al. Minimal incidence of neonatal/infancy onset diabetes in Italy is 1:90,000 live births. *Acta Diabetol* 2012;**49**(5):405–8.
11. Wiedemann B, Schober E, Waldhoer T, Koehle J, Flanagan SE, Mackay DJ, et al. Incidence of neonatal diabetes in Austria-calculation based on the Austrian Diabetes Register. *Pediatr Diabetes* 2010;**11**(1):18–23.
12. Edghill EL, Hattersley AT. *Genetic disorders of the pancreatic beta cell and diabetes (permanent neonatal diabetes and maturity-onset diabetes of the young)*. Tokyo: Springer Japan; 2008. p. 399–430.
13. Loomba-Albrecht LA, Glaser NS, Styne DM, Bremer AA. An oral sulfonylurea in the treatment of transient neonatal diabetes mellitus. *Clin Ther* 2009;**31**(4):816–20.
14. Støy J, Greeley SAW, Paz VP, Ye H, Pastore AN, Skowron KB, et al. Diagnosis and treatment of neonatal diabetes: a United States experience. *Pediatr Diabetes* 2008;**9**(5):450–9.
15. Pearson ER, Flechtner I, Njølstad PR, Malecki MT, Flanagan SE, Larkin B, et al. Switching from insulin to oral sulfonylureas in patients with diabetes due to Kir6.2 mutations. *N Engl J Med* 2006;**355**(5): 467–77.
16. Rafiq M, Flanagan SE, Patch A-M, Shields BM, Ellard S, Hattersley AT, et al. Effective treatment with oral sulfonylureas in patients with diabetes due to sulfonylurea receptor 1 (SUR1) mutations. *Diab Care* 2008;**31**(2):204–9.

17. Begum-Hasan J, Polychronakos C, Brill H. Familial permanent neonatal diabetes with KCNJ11 mutation and the response to glyburide therapy – a three-year follow-up. *J Pediatr Endocrinol Metab* 2008;**21**(9):895–903.

18. Colom CC, Corcoy RR. Maturity onset diabetes of the young and pregnancy. *Best Pract Res Clin Endocrinol Metab* 2010;**24**(4):605–15.

19. Stride A, Shields B, Gill-Carey O, Chakera AJ, Colclough K, Ellard S, et al. Cross-sectional and longitudinal studies suggest pharmacological treatment used in patients with glucokinase mutations does not alter glycaemia. *Diabetologia* 2014;**57**(1):54–6.

20. Fajans SS, Brown MB. Administration of sulfonylureas can increase glucose-induced insulin secretion for decades in patients with maturity-onset diabetes of the young. *Diab Care* 1993;**16**(9):1254–61.

21. Pearson ER, Starkey BJ, Powell RJ, Gribble FM, Clark PM, Hattersley AT. Genetic cause of hyperglycaemia and response to treatment in diabetes. *Lancet* 2003;**362**(9392):1275–81.

22. Shepherd M, Pearson ER, Houghton J, Salt G, Ellard S, Hattersley AT. No deterioration in glycemic control in HNF-1alpha maturity-onset diabetes of the young following transfer from long-term insulin to sulphonylureas. *Diab Care* 2003;**26**(11):3191–2.

23. Thanabalasingham G, Pal A, Selwood MP, Dudley C, Fisher K, Bingley PJ, et al. Systematic assessment of etiology in adults with a clinical diagnosis of young-onset type 2 diabetes is a successful strategy for identifying maturity-onset diabetes of the young. *Diab Care* 2012;**35**(6):1206–12.

24. Ellard S, Bellanné-Chantelot C, Hattersley AT. European Molecular Genetics Quality Network (EMQN) MODY group. Best practice guidelines for the molecular genetic diagnosis of maturity-onset diabetes of the young. *Diabetologia* 2008;**51**(4):546–53.

25. Schnyder S, Mullis PE, Ellard S, Hattersley AT, Flück CE. Genetic testing for glucokinase mutations in clinically selected patients with MODY: a worthwhile investment. *Swiss Med Wkly* 2005;**135**(23–24):352–6.

26. Shields BM, McDonald TJ, Ellard S, Campbell MJ, Hyde C, Hattersley AT. The development and validation of a clinical prediction model to determine the probability of MODY in patients with young-onset diabetes. *Diabetologia* 2012;**55**(5):1265–72.

27. Bellanné-Chantelot C, Lévy DJ, Carette C, Saint-Martin C, Riveline J-P, Larger E, et al. Clinical characteristics and diagnostic criteria of maturity-onset diabetes of the young (MODY) due to molecular anomalies of the HNF1A gene. *J Clin Endocrinol Metab* 2011;**96**(8):E1346–51.

28. Thanabalasingham G, Shah N, Vaxillaire M, Hansen T, Tuomi T, Gašperíková D, et al. A large multi-centre European study validates high-sensitivity C-reactive protein (hsCRP) as a clinical biomarker for the diagnosis of diabetes subtypes. *Diabetologia* 2011;**54**(11):2801–10.

29. Besser REJ, Shepherd MH, McDonald TJ, Shields BM, Knight BA, Ellard S, et al. Urinary C-peptide creatinine ratio is a practical outpatient tool for identifying hepatocyte nuclear factor 1-{alpha}/hepatocyte nuclear factor 4-{alpha} maturity-onset diabetes of the young from long-duration type 1 diabetes. *Diab Care* 2011;**34**(2):286–91.

30. Gold MR, Siegel JE, Russell LB, Weinstein MC, editors. *Cost-Effectiveness in Health and Medicine*. New York: Oxford University Press; 1996.

31. Weinstein MC, Siegel JE, Gold MR, Kamlet MS, Russell LB. Recommendations of the Panel on Cost-effectiveness in Health and Medicine. *JAMA* 1996;**276**(15):1253–8.

32. Evans C, Tavakoli M, Crawford B. Use of quality adjusted life years and life years gained as benchmarks in economic evaluations: a critical appraisal. *Health Care Manag Sci* 2004;**7**(1):43–9.

33. Braithwaite RS, Meltzer DO, King JT, Leslie D, Roberts MS. What does the value of modern medicine say about the $50,000 per quality-adjusted life-year decision rule? *Med Care* 2008;**46**(4):349–56.

34. CDC Diabetes Cost-Effectiveness Group. Cost-effectiveness of intensive glycemic control, intensified hypertension control, and serum cholesterol level reduction for type 2 diabetes. *JAMA* 2002;**287**(19):2542–51.

35. Greeley SAW, John PM, Winn AN, Ornelas J, Lipton RB, Philipson LH, et al. The cost-effectiveness of personalized genetic medicine: the case of genetic testing in neonatal diabetes. *Diab Care* 2011;**34**(3):622–7.

36. Naylor RN, John PM, Winn AN, Carmody D, Greeley SAW, Philipson LH, et al. Cost-effectiveness of MODY genetic testing: translating genomic advances into practical health applications. *Diab Care* 2014;**37**(1):202–9.

37. Steele AM, Shields BM, Wensley KJ, Colclough K, Ellard S, Hattersley AT. Prevalence of vascular complications among patients with glucokinase mutations and prolonged, mild hyperglycemia. *JAMA* 2014;**311**(3):279–86.

38. Peters JL, Anderson R, Hyde C. Development of an economic evaluation of diagnostic strategies: the case of monogenic diabetes. *BMJ Open* 2013;**3**(5):e002905.

39. Pihoker C, Gilliam LK, Ellard S, Dabelea D, Davis C, Dolan LM, et al. Prevalence, characteristics and clinical diagnosis of maturity onset diabetes of the young due to mutations in HNF1A, HNF4A, and glucokinase: results from the SEARCH for Diabetes in Youth. *J Clin Endocrinol Metab* 2013;**98**(10):4055–62.

40. Ladabaum U, Wang G, Terdiman J, Blanco A, Kuppermann M, Boland CR, et al. Strategies to identify the Lynch syndrome among patients with colorectal cancer: a cost-effectiveness analysis. *Ann Intern Med* 2011;**155**(2):69–79.

41. Nherera L, Marks D, Minhas R, Thorogood M, Humphries SE. Probabilistic cost-effectiveness analysis of cascade screening for familial hypercholesterolaemia using alternative diagnostic and identification strategies. *Heart* 2011;**97**(14):1175–81.

42. Gloyn AL, Pearson ER, Antcliff JF, Proks P, Bruining GJ, Slingerland AS, et al. Activating mutations in the gene encoding the ATP-sensitive potassium-channel subunit Kir6.2 and permanent neonatal diabetes. *N Engl J Med* 2004;**350**(18):1838–49.

43. Babenko AP, Polak M, Cavé H, Busiah K, Czernichow P, Scharfmann R, et al. Activating mutations in the ABCC8 gene in neonatal diabetes mellitus. *N Engl J Med* 2006;**355**(5):456–66.

28

Genetic Counseling: The Role of Genetic Counselors on Healthcare Provider and Endocrinology Teams

Sarah M. Nielsen*, Shelly Cummings**

*Center for Clinical Cancer Genetics and Global Health, Department of Medicine, University of Chicago, Chicago, IL, USA
**Myriad Genetic Laboratories, Inc., Salt Lake City, UT, USA

INTRODUCTION

The completion of the Human Genome Project in 2003 ushered in a new era in genetics research and medicine and garnered much hope for "discovering the genetic basis for health and the pathology of human disease."[1] Indeed, the Human Genome Project laid the foundation for the next generation of investigations into Mendelian and complex disease through genome-wide association studies and, most recently, whole-genome and exome sequencing. However, the amount of scientific information acquired from sequencing (nearly) the entire human genome is immense, and we are now at a crossroads, where our technology and ability to amass genetic information has far surpassed our interpretation and application of the derived material. Nonetheless, genetics is becoming increasingly integrated into healthcare in the name of "personalized medicine," and as such, healthcare providers must understand the sensitive and unique nature of communicating genetic information to patients. Genetic counselors can assist with this communication process. They are healthcare professionals trained in medical genetics and counseling that work with individuals and families to help them understand and adapt to the implications of genetic contributions to disease, make decisions about genetic testing, and consider reproductive choices and other aspects that might affect their life and future.[2,3] Genetic counseling can make a difference for families, whether it is a one-time visit or multiple sessions over a long period.[4] The role of genetic counselors continues to expand and adapt to the new developments in genetic testing, and the growing role of genetic counseling in endocrinology will be highlighted here.

THE GENETIC COUNSELING PROFESSION

The practice of advising people about inherited traits began around 1906, shortly after Bateson named this new medical and biological study of hereditary "genetics."[5] However, it was not until Sheldon Reed coined the term "genetic counseling" in 1947 that the mechanism for translating the advances in genetics to clients had a name.[6] Scientists and the public were becoming fascinated by the idea that this new science might identify hereditary elements contributing not only to medical disorders, but also to social and behavioral conditions such as crime and mental illness. This enthusiasm took an awful turn when the eugenics (Greek term meaning "well-born") movement took hold and sought to institute "agencies under social control to improve or impair racial qualities of future generations, either physically or mentally."[7] These governmental agencies not only collected data on human traits but sometimes provided that information to affected families, usually with the intent of convincing them not to reproduce. In many cases, the data were scientifically unsound or tainted by political or social agendas. The granddaddy of all atrocities was the legalization of euthanasia for the "genetically defective" in Germany in 1939. This led to the deaths of over 70,000 people with hereditary disorders, in addition to the hundreds of thousands of Jews and others killed in the Holocaust.[8] It is through these early history lessons that the art of "nondirective" counseling grew and is the approach instilled in every practicing genetic counselor today. This approach ensures that clients are aware of

Genetic Diagnosis of Endocrine Disorders. http://dx.doi.org/10.1016/B978-0-12-800892-8.00028-2

concerns relevant to their situation and helps them make decisions to fit their lifestyle and belief system.[3]

Today, genetic counselors interact with clients and other healthcare professionals in a variety of clinical and nonclinical settings, including, but not limited to, university-based medical centers, private hospitals, private practice, and industry settings. Most genetic counselors have a master's degree from an accredited genetic counseling training program. The first class of genetic counselors graduated from Sarah Lawrence College in 1971. Currently, there are over 3,000 board-certified genetic counselors in the United States.[9] A limited number of genetic counselors also practice in Canada, Australia, and throughout Europe. Genetic counselors are board-certified by the American Board of Genetic Counseling. Board eligibility or certification is required for employment in many positions, and many states now license genetic counselors. The National Society of Genetic Counselors, Inc. (NSGC) was incorporated in 1979 and is the only professional society dedicated solely to the field of genetic counseling. Its mission is to "advance the various roles of genetic counselors in healthcare by fostering education, research, and public policy to ensure the availability of quality genetic services."[10]

There are currently over 2,500 clinical genetic tests available to clinicians, and providers are increasingly expected to be aware of the advances in genetics to better serve and counsel their clients. Because the majority of people seek their medical care in the community and may not have access to genetic counselors or genetic specialists, the burden of genetic interpretation and guidance is being shared by the wider clinical community.[11] Other healthcare professionals that are trained to provide genetic counseling services include medical geneticists and clinical nurse specialists trained in genetics. Although all healthcare professionals should be competent in genetics, data shows that many medical providers have difficulty interpreting even

the most basic pedigrees and genetic test results.[12,13] For this reason, it is preferable for individuals who warrant evaluation to be seen by a genetic counselor or genetics expert. Access to these services is now widely available through internet, telephone, and satellite-based telemedicine services, and several major health companies cover these services.[14] Table 28.1 lists organizations and their websites for locating professional genetic experts.

THE ROLE OF GENETIC COUNSELORS ON THE HEALTHCARE PROVIDER TEAM

The role of the genetic counselor has evolved greatly since 1971. Initially, genetic counselors worked almost exclusively in clinical settings under physician supervision, seeing clients who had been diagnosed as having a genetic disorder, were at risk for developing a genetic disorder, or were at risk for having a child with a genetic disorder. They would assess risk, provide information, and discuss available testing options and provide appropriate supportive counseling. The variety of clients and the information and testing options offered by genetic counselors was greatly restricted by the limited technology and genetic knowledge of the time. Today, because of the Human Genome Project and other advances, genetic counselors are now able to offer a wider array of services and options. They are able to specialize in a particular area of interest, such as cancer, prenatal, pediatric, assisted reproduction, and metabolic or neurogenetic disorders to name a few. Most genetic counselors still work in the clinical setting, either in a hospital or in private practice. However, advances in genetics have enabled genetic counselors to work in a variety of other settings including research, public health, education, private practice, and industry.[3]

TABLE 28.1   How to Find a Genetics Expert[3]

Organization	Website
American Board of Genetic Counseling	www.abgc.net
American College of Medical Genetics	www.acmg.net
Canadian Association of Genetic Counselors	www.cagc-accg.ca
GeneClinics	www.geneclinics.org
Genetic Centers in the British Isles	www.cafamily.org.uk/gencentr
Informed Medical Decisions	www.informedDNA.com
International Society of Nurses in Genetics	www.isong.org
March of Dimes	www.marchofdimes.com
National Cancer Institute	http://www.cancer.gov/about-cancer/causes-prevention/genetics/directory
National Society of Genetic Counselors	www.nsgc.org

The Role of Genetic Counselors on the Endocrinology Team

Given the complexity of endocrine disorders, a multidisciplinary approach with coordination of care with multiple healthcare specialists is the ideal medical management scenario. Such a team should include endocrinologists, surgeons, radiology/nuclear medicine, pathologists, and genetic counselors. The inclusion of genetic counselors on these teams is critical as our understanding of the hereditary nature of endocrine tumors increases, and options for genetic testing are expanding. Genetic counselors can assist the endocrinology team by eliciting a detailed pedigree, determining the appropriate genetic test to order, obtaining informed consent, interpreting complex genetic test results, providing psychosocial and family counseling, and assessing which family members are at risk.[15]

Genetic testing for the *RET* oncogene has long been considered the standard of care for managing individuals and family members suspected of having multiple endocrine neoplasia type 2 (MEN2).[16,17] Increasingly, the importance of genetic testing *and* counseling is being recognized for other endocrine tumor syndromes, including well-established syndromes such as multiple endocrine neoplasia type 1 (MEN1), and newer entities such as hereditary pheochromocytoma (PCC), paraganglioma (PGL), and familial nonmedullary thyroid cancer (FNMTC). Table 28.2 highlights features of endocrine disorders that should prompt referral for genetic counseling and testing, while Table 28.3 describes the associated hereditary syndromes and their related benign and malignant tumors of the endocrine system. Any individual with two or more features of these syndromes should also be referred for genetic counseling.

In MEN1, individuals diagnosed through genetic testing have been shown to fare better than those diagnosed

by clinical expression, with fewer clinical manifestations and lower frequencies of malignancies.[18] Individuals with MEN1 managed by a multidisciplinary team that included medical geneticists and/or genetic counselors also had improved long-term outcomes, including increased detection of early-stage disease, which led to more efficacious treatment and reduced morbidity and mortality from hormone excess and malignancy.[19,20] Affected individuals may also suffer less psychological distress when managed in these types of settings.[19] Recently published clinical practice guidelines for MEN1 strongly advise that all individuals offered *MEN1* mutation testing undergo genetic counseling before testing.[21]

Genetic testing and counseling for individuals diagnosed with PCC or PGL is particularly adventitious because up to one-third of these tumors are caused by germline mutations in one of 10 genes.[22] This high proportion of hereditary causes may warrant genetic testing in every individual presenting with a PCC or PGL, regardless of age or family history.[23] There have been numerous algorithms developed for genetic testing that incorporate certain clinical, biochemical, and histological features of the tumors, including our algorithm in Fig. 28.1. Immunohistochemical staining of tumor tissue for SDH proteins, particularly SDHB, is a relatively new technique that is not yet widely available but demonstrates high specificity.[22,24] The advent of next-generation sequencing has opened up new avenues of germline testing that can incorporate all the clinically available genes on a single targeted gene panel. Next-generation sequencing technology uses massively parallel sequencing to analyze thousands to millions of simultaneous sequences.[25] If there is more than one differential diagnosis being considered, this type of testing can be very cost-effective and time saving. Currently, Ambry Genetics offers the only clinically available gene panel that specifically targets PCC and PGL. The test is called PGLNext™ and analyzes 10 genes associated with an increased risk of developing PGL and/or PCC, including *MAX, NF1, RET, SDHA, SDHAF2, SDHB, SDHC, SDHD, TMEM127,* and *VHL*.[26] The current cost of the test is $3,900, which represents significant cost savings as testing for just six of these genes (*RET, VHL, NF1, SDHD, SDHB,* and *SDHC*) sequentially could cost up to $7,700.[15] Genetic counselors can help decide which type of testing is most appropriate, which could potentially decrease the costs of testing.

Genetic counselors that are part of an endocrinology team will frequently see patients with thyroid cancer, as it is the most common endocrine malignancy. As mentioned previously, genetic counseling for medullary thyroid cancer/MEN2B is well established, but genetic counseling for nonmedullary thyroid cancer is less frequently utilized. Approximately 5–10% of papillary and follicular thyroid cancer is thought to be hereditary, and is often referred to as FNMTC. With the exception of

TABLE 28.2 Indications for Genetic Counseling Referral of Endocrine Tumors

Medullary thyroid cancer (any age)

Familial nonmedullary thyroid cancer (2+ affected family members)

Multigland primary hyperthyroidism before age 40

Parathyroid carcinoma

Multifocal pancreatic neuroendocrine tumors

Pheochromocytoma or paraganglioma diagnosed before age 50

Malignant or multiple pheochromocytoma or paraganglioma (any age)

Adrenocortical carcinoma diagnosed before age 40

Adapted with permission from Thereasa Rich, certified genetic counselor.

TABLE 28.3 Hereditary Tumor Syndromes and Their Associated Risk for Endocrine Neoplasia

Syndrome (gene)	Thyroid	Pituitary	PHPT	Adrenal cortex	PCC/PGL	PNET	Foregut carcinoid	Other main features
Multiple endocrine neoplasia type 1 (*MEN1*)	Low (adenomas)	20–60%	>95%	20–50% (benign)		50–75%	5–10%	Facial angiofibromas, collagenomas, and lipomas
Multiple endocrine neoplasia type 2A/B, familial medullary thyroid cancer (*RET*)	>95% (medullary)		Up to 30%		Up to 50%			Mucosal neuromas, marfanoid habitus in MEN2B
Von Hippel-Lindau (*VHL*)					10–60%	10%		Hemangioblastomas, renal cancer/cysts
Familial PCC/PGL (*SDHB, SDHC, SDHD*, others)	Low (nonmedullary)				Up to 80%			Renal cancer GIST
Li Fraumeni (*TP53*)	Low (nonmedullary)			Low				Breast cancer, sarcoma brain tumor, and leukemia.
Cowden syndrome (*PTEN*)	35% (follicular)							Benign skin tumors breast/uterine/ renal/colon cancer
Carney complex (*PRKAR1A*)	Nodules common, low cancer risk	10%		25% (PPNAD)				Lentigines, myxomas Schwannoma
HPT-jaw tumor syndrome (*HRPT2* aka *CDC73*)			>90% (15% cancer)					Jaw, uterine tumors, and kidney cysts/ tumors
FAP (*APC*)	1–2% (cribiform morular)			Low				Colon polyposis and cancer
Neurofibromatosis type 1 (*NF1*)					<5%		Low	Café au lait spots and neurofibromas

PHPT, primary hyperparathyroidism; PCC, pheochromocytoma; PGL, paraganglioma; PNET, pancreatic neuroendocrine tumor; PPNAD, primary pigmented nodular adrenal dysplasia.

Adapted with permission from Thereasa Rich, certified genetic counselor.

a few rare familial syndromes, the underlying genetic predisposition to FNMTC is largely unknown. However, recent efforts have been able to identify multiple genomic regions by linkage analysis that may harbor predisposition genes.[27] Still, there is currently no clinical genetic testing for FNMTC, and therefore identification of at-risk families must rely on the personal and family histories of any individual presenting with thyroid cancer or disease. Aspects of a personal history that suggest FNMTC include early-onset bilateral/multifocal thyroid tumors (especially in males) with a more aggressive clinical course and association with benign thyroid pathologies. FNMTC is ultimately a diagnosis of exclusion in the sense that other familial cancer predisposition syndromes associated with nonmedullary thyroid cancer must first be ruled out, such as Cowden syndrome, familial adenomatous polyposis (FAP), or

Carney's complex, among others.[27] Genetic counselors can assist in the work-up of FNMTC families by obtaining detailed personal and family histories, researching the latest tumor associations, and identifying and consenting patients to appropriate research studies.[15]

The involvement of genetic counselors in the testing process for endocrine disorders may help decrease the cost of testing by ensuring that the appropriate test is being ordered through their astute analysis of the personal and family history.[15] In addition, genetic counselors have expertise in the interpretation of genetic testing results, which is particularly crucial in the realm of next-generation sequencing, as results are often not straightforward, and misinterpretation can have devastating effects on the patient and family.[28] While positive genetic test results seem relatively straightforward, there are many issues to consider once a diagnosis has been made. Genetic

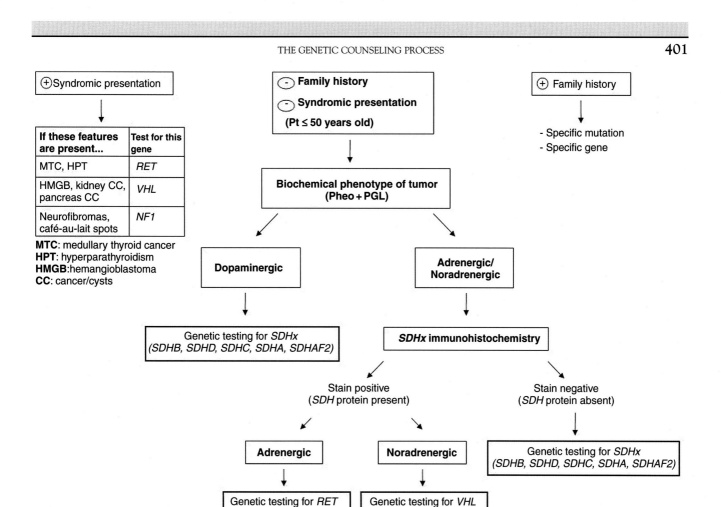

FIGURE 28.1 Targeted genetic screening approach for apparently sporadic PCC and PGL.

counselors can help individuals cope with a diagnosis and develop strategies to share this information with at-risk family members. In diseases like MEN2 where there are strong genotype–phenotype correlations, the specific *RET* mutation identified informs risks and guides management recommendations.[15] Conversely, in many endocrine disorders, management guidelines are not as well-established. When a genetic mutation is identified in a gene such as *SDHB*, medical management recommendations are currently based on expert opinion and for even newer genes, such as *TMEM127*, *MAX*, *SDHA*, and *SDHAF2*, there are no formal management recommendations at all. These are all nuances of a positive result that a genetic counselor can help work through with a patient. A negative genetic test result is considered uninformative unless a mutation proven to segregate with disease has been previously identified in the family, in which case the same result is considered a true negative. An uninformative genetic test result must be interpreted in the context of the individual's personal and family history; it *does not* rule out a clinical diagnosis (such as having two of three primary MEN1-related tumors) or a hereditary cause for the cancer/disease in the family. Lastly, variants of

uncertain significance (VUS) identified through genetic testing need to be carefully considered. A VUS means a genetic change is identified, but the significance of that change is unknown. These results are particularly common on next-generation sequencing panels where newer, less studied genes (such as *MAX* and *TMEM127* in hereditary PCC) are included. The patient needs to understand that these results do not immediately affect medical management recommendations, which are always based on personal/family history until that VUS is reclassified, a process that could take several months or several years. Genetic counselors can help the laboratories with the reclassification process by working with families to coordinate testing of affected or informative family members.

THE GENETIC COUNSELING PROCESS

The genetic counseling process encompasses risk assessment, informed consent, psychosocial support, and a discussion of the risks and benefits of testing, including the potential risks of genetic discrimination.[15] This communication process can take place anywhere along the continuum of life, from preconception counseling and prenatal diagnosis, to the diagnosis of a genetic condition in a newborn or toddler, and predisposition testing for inherited conditions that do not present themselves until adulthood like many heritable endocrine disorders.[3] The diagnosis of a condition, no matter what the cause, can be very difficult. The diagnosis of a disease can be even more devastating when it is inherited, since the entire family can be affected. The shock of the diagnosis can elicit a multitude of questions from spouses, partners, extended family members, and friends.[3] Genetic counselors can address these questions and use their training in the medical, psychosocial, ethical, and legal aspects of genetic medicine to effectively guide patients through the genetic counseling process, empowering them to act on the information they have been given.

From the medical perspective, the approach to genetic counseling involves assessing the family history, personal medical and reproductive history, environmental exposures (e.g., embryonic teratogens or occupational hazards), and lifestyle habits (e.g., smoking, excessive alcohol use, illicit drug use). This information will aid in determining the risk of disease, help in identifying a hereditary condition, assist in whether genetic testing is appropriate, and offer diagnosis and disease prevention and management. The most obvious difference with inherited diseases versus other conditions is that the diagnosis can have significant implications for the family and not just the individual with the diagnosis; the whole family, siblings, parents, children, grandparents, and more extended relatives become the patient. This expansion of the definition of who is the actual patient complicates the communication process and may present challenges for medical management (e.g., the ideal person for genetic testing in the family is the sister of the client in front of you who is not your patient, but whose results could influence your client's healthcare). We also have to take into account the privacy and confidentiality of other at-risk individuals while balancing the duty to inform. Frequently, we solicit the help of the client to begin the discussion of risk assessment and possibly genetic testing with their family members and encourage them to contact a genetic specialist. In many autosomal dominant conditions, testing an affected person first can be more informative before presymptomatic testing can be offered to unaffected relatives.[3]

Traditionally, genetic counselors have operated under the tenet of providing information in a nondirective way that leaves decision making up to the client after they have been fully informed of all the factual information regarding the condition; however, this model has been under some scrutiny.[29,30] While individuals or families typically come to genetic counseling hoping to learn information, their emotional state needs to be evaluated and considered as well. In most cases, individuals cannot

effectively process or act on what they learn until they have dealt with the strong reactions that this information can invoke. Therefore, an integral part of the genetic counseling process is the exploration of clients' past experiences, emotional responses, goals, cultural and religious beliefs, financial and social resources, family and interpersonal networks, and coping styles. This more interactive process not only educates individuals about risk, but also helps them with the complex task of exploring issues related to the disorder in question and making decisions about reproduction, genetic testing, and medical management that are consistent with their own needs and value system.[3]

Families with genetic disease often need support that is outside the realm of the healthcare provider. Genetic counselors can facilitate this process and offer individuals the option of support groups so that they will have the opportunity to speak to others who are grappling with the same situation as themselves or pair them up with other clients they know in their practice who are willing to speak one-on-one.[3] The Genetic Alliance is a coalition of more than 600 advocacy organizations serving 25 million people affected by thousands of conditions. The organization not only serves as a resource for families with genetic diseases, but also is very active in educating policymakers and working with genetic leaders in the community to serve as a collective voice for individuals living with genetic conditions.[31]

Genetic counselors serve as client advocates by remaining informed of ethical and legal issues regarding the use of genetic testing and incorporate pertinent information into the counseling session. For example, the decision to undergo genetic testing may involve controversial issues, such as prenatal diagnosis for an adult onset condition or sex selection. Depending on the type of test and the disorder present, testing will most likely have implications for other family members, insurance eligibility or coverage, employment, and quality of life.

One of the most frequently stated barriers to genetic testing is the fear of discrimination.[32] However, there is little evidence proving that the magnitude of this intense fear is warranted based on factual cases.[33,34] Clients and their families may worry about health and life insurance discrimination, being stigmatized based on their genetic status, or the fear of being labeled in their community. Not only can these fears make them not participate in potentially life-saving genetic testing but also they are wary about participating in clinical research studies to assist in furthering understanding of the disorder. In extreme situations, the fear may be so significant that clients withhold information from their healthcare providers. Most states have laws protecting discrimination and some state laws even define penalties for each infarction. Protection from employment and health discrimination exists under federal law in the form of the Genetic Nondiscrimination Act of 2007 (GINA), which was signed into law in May 2008. GINA prohibits the use of an individual's genetic information in setting eligibility or premiums or contribution amounts by group and individual health insurers. However, it does not prohibit medical underwriting based on current health status or mandate coverage for particular medical tests or treatments. It prohibits health insurers from requesting or requiring an individual to take a genetic test. It also prohibits the use of an individual's genetic information by employers in employment decisions such as hiring, firing, job assignments, and promotions (GINA). These protections are intended to encourage Americans to take advantage of genetic testing as part of their medical care. Providers should be aware of the laws in their state and federal legislation so they can adequately address client concerns and provide reassurance.[3]

The Patient Protection and Affordable Care Act, commonly called the Affordable Care Act (ACA), was signed into law in March 2010 and upheld by the Supreme Court in June 2012. One of the most significant changes with the ACA is the fact that individuals with pre-existing conditions can no longer be denied coverage, charged more, or denied treatment based on health status. This is particularly relevant for individuals with a diagnosis of cancer or genetic disease because it also prevents insurance companies from putting lifetime or annual dollar limits on screenings, preventions, and treatments for most life-threatening or chronic illnesses. In regards to genetic counseling and testing, the ACA specifies these provisions specifically for "women whose family history is associated with an increased risk for deleterious mutations in the *BRCA1* or *BRCA2* genes." Genetic testing is included along with genetic counseling as part of the US Preventive Services Task Force (USPSTF) B-rated preventive benefit with no cost–sharing for nongrandfathered insurance plans. For most insurance plans, this applies to unaffected women only, but Medicare still does not cover preventive *BRCA1/BRCA2* testing at all for individuals without a personal history of cancer.[35] There are currently no provisions in the ACA for genetic counseling and testing for any other hereditary cancer susceptibility syndromes or genetic diseases.

THE PEDIGREE: MEDICINE AND ART

The fascination with family links and origins has been of interest since biblical times. It is believed that the pedigree was developed in the fifteenth century as a tool for illustrating ancestry.[36] However the pedigree was not used to demonstrate the inheritance of disease and traits until the mid-nineteenth century, when Pliny Earle described the inheritance of colorblindness. In 1912, Pearson stated, "a complete pedigree is often work of great labor, and in its finished form is frequently a work of art."[37] The pedigree

has taken many forms, styles, and symbols, reflecting differences in individual preferences, training, and interests. There was concern among the genetic counseling community about the inconsistencies in the use of pedigree symbols and nomenclature. If pedigrees cannot be interpreted, then the value of the family history in establishing an accurate diagnosis and risk assessment is diminished. To address these inconsistencies, a universal set of standardized human pedigree nomenclature was developed and is now used internationally.[38]

Standard symbols are used to draw pedigrees. For example, males are represented by squares and females by circles. The individuals who brought the family to the attention of the medical professional is called a "proband" and is identified on the pedigree by an arrow pointing towards the symbol representing that individual. The proband usually has the disorder of interest. The informant is the individual who was interviewed to obtain the pedigree. The informant may or may not be the same person as the proband. Pedigrees are hand-drawn or created using special computer software. The standard pedigree typically includes at least three generations, with each generation arranged horizontally and connected to the other generations by lines. Family members who have the genetic disorder in question are colored in or shaded. Unique symbols represent carriers, miscarriages, people of unknown sex, twins, or other categories of individuals. Furthermore, different patterns within the pedigree symbol may represent variable expression (variation in symptoms of individuals with the same disorder).[3] For example, Fig. 28.2 is a pedigree using standardized nomenclature suggestive of familial papillary thyroid cancer, with a shaded symbol representing papillary thyroid cancer and a quadrant of the symbol representing benign thyroid disease, most commonly goiter in this family.

An important aspect of the genetic counseling session is the collection of the family medical history or pedigree. The pedigree is a diagram that records family history information, the tool for converting information provided by the client and/or obtained from the medical record into a standardized format. It demonstrates biological relationships in the family by using specific symbols, lines, and abbreviations. The completed pedigree serves as a quick visual depiction of the family structure and allows the genetic counselor to assess more accurately the client's risk of disease. Pedigree analysis aids in deciphering if there is a pattern of inheritance indicative of hereditary disease and may provide information in making a diagnosis. The pedigree is also pivotal in identifying family members who are at an increased risk of developing a disorder and for estimating risk of recurrence in other relatives, including future offspring. Sometimes the pedigree can reveal other conditions present in the family that might require an evaluation or counseling by a specialist.[3]

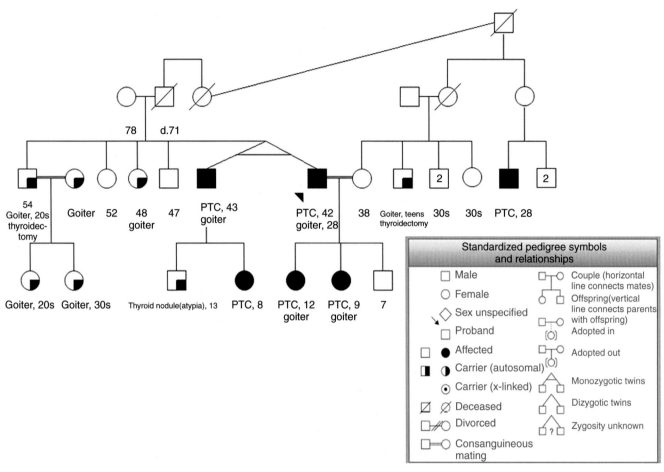

FIGURE 28.2   Example pedigree of a consanguineous family with papillary thyroid cancer.

In additional to the medical information obtained through pedigree construction, one can get a glimpse of the social dynamics and relationship of the client with other family members. Information about divorce, adoption, fertility, pregnancy loss, and death of family members is all recorded as part of the collection process for constructing a pedigree. Building a pedigree is an exercise of give-and-take with the client, since the information collected is contingent on them actively obtaining and sharing some of the most intimate family details with a total stranger. Before collecting the family history information, it is helpful to let the client know why this information is needed and how it will be used. If people know that this information is essential to determine whether there is an inherited syndrome in the family or if their condition is related to an altered gene, they may be more forthcoming when asked to answer a series of personal questions. Many individuals enjoy the process, as they feel dynamically involved in their health care since they are actively working with the healthcare provider, which is somewhat different from traditional medical care. One can uncover family myths and beliefs of how disease occurs, who gets it and who does not,

as well as traditions and cultural mores. The assembly of a pedigree is typically done either over the telephone before the client's appointment or at the beginning of the office visit. This interaction is the initial rapport builder between the client and the counselor, and as the pedigree emerges, so will the family relationships, which may highlight sensitive issues that might be of concern for the client or their family.[3]

PEDIGREE ANALYSIS AND RISK PERCEPTION

The purpose of drawing the pedigree is to identify inheritance patterns that may aid risk analysis and clinical diagnosis (e.g., distinguish a sporadic thyroid cancer from a case that is part of a hereditary syndrome). All types of Mendelian inheritance, as well as chromosomal aberrations and multifactorial predisposition, have been demonstrated in the heritable disorders of the endocrine glands.[39] These patterns of inheritance, associated endocrine disorders, and variables that can disguise recognition of these patterns are detailed in Table 28.4. It is

TABLE 28.4 Patterns of Inheritance and Variables that can Diagnose Those Patterns[3]

Inheritance patterns	Modes of transmission	Pedigree features	Confounding factors	Examples of disorders
Autosomal dominant (AD)	50% risk of disease to each offspring.	Presence of condition in successive generations. Females/males affected. Often see male-to-male transmission. Often variability in disease severity. Homozygous state may be lethal.	Reduced penetrance, missed diagnosis in mildly affected relatives, and new mutations may be mistaken for sporadic disease if limited family structure or size.	Tuberous sclerosis (60% due to new mutations), multiple endocrine neoplasia, type 2 (2A and 2B), VHL disease, and FAP.
Autosomal dominant with maternal imprinting	50% chance for each offspring to be carrier, but only offspring of male carriers are at risk for manifesting disease.	Paternal family history of disease. Females/males affected. Only males can transmit disease.	May be missed if few males in family.	Hereditary PCC and PGL (*SDHD, SDAF2,* and possibly *MAX* genes)
Autosomal recessive (AR)	25% risk of disease to each offspring. 50% risk of passing on carrier state to each offspring. Parents are carriers – not affected by disease and no symptoms.	Usually one generation. Females/males affected. Often seen in newborn, infancy, or childhood. May be more common in certain ethnic groups. Parental consanguinity.	May look like AD if carrier frequency is high. Can be mistaken as sporadic if family size is small.	Wilson disease, cystic fibrosis, congenital adrenal hyperplasia, and Schmidt's syndrome
X-linked dominant (XLD)	Heterozygous women are affected and their daughters have 50% chance of being affected/sons have 50% chance of being affected (lethal).	No male-to-male transmission. Often lethal in males so see few males in pedigree. May see multiple miscarriages (due to male fetal lethality). Females usually have milder symptoms of disease than males.	Small family size.	Hypophosphatemic or vitamin D resistant rickets.
X-linked recessive (XLR)	Sons of female carriers have a 50% chance of being affected. Daughters of female carriers have a 50% chance of being carriers.	No male-to-male transmission. Males affected. Females may be affected but often with milder and/or later onset than males.	May be missed if few females in family. Symptoms in female carriers thought to be caused by imbalance of X inactivation (lionization).	Nephrogenic diabetes insipidus, fabry disease, fragile X syndrome, and Kallman syndrome
Chromosomal	Increased risk for trisomy seen with advanced maternal age. Risk for affected fetus depends on specific chromosomal rearrangement (ranges from 1% to 15% or greater).	Suspect if individual has 2+ major birth defects or 3+ minor defects. Fetuses with structural abnormalities. Unexplained mental retardation, especially with dysmorphic features. Ambiguous genitalia. Couples with 3+ pregnancy losses. Unexplained infertility. Individuals with multiple congenital abnormalities and family history of mental retardation.		Trisomy 21 (Down's syndrome), trisomy 18, Klinefelter syndrome (45,X), Turner syndrome (47,XXY), and chromosomal translocations
Mitochondrial	1–100%	Only maternal transmission to offspring, no male transmission. Highly variable clinical expression. Often central nervous system disorders. Females/males affected.	Rare.	Mitochondrial encephalopathy with ragged-red fibers (MERRF), mitochondrial encephalopathy, lactic acidosis, and strokes (MELAS).
Multifactorial	Empiric risk estimates.	Females/males affected. Skips generations. Few affected family members.	May be a single gene disorder.	Diabetes, obesity, and epilepsy.

critical for the healthcare provider to evaluate carefully the pedigree and all associated findings (e.g., history of high blood pressure, benign tumors, and glandular dysfunction). Three important considerations in pedigree interpretation are reduced penetrance, variable expressivity, and the value of negative family history.

Penetrance

Penetrance refers to the proportion of people with a particular genetic change (such as a mutation in a specific gene) who exhibit signs and symptoms of a genetic disorder. If some people with the mutation do not develop features of the disorder, the condition is said to have reduced (or incomplete) penetrance. Reduced penetrance often occurs with familial cancer syndromes. For example, many people with a mutation of the MEN1 gene will develop cancer during their lifetime, but some people will not. Doctors cannot predict which people with these mutations will develop cancer or when tumors will develop. Reduced penetrance probably results from a combination of genetic, environmental, and lifestyle factors, many of which are unknown. This phenomenon can make pedigree analysis challenging, but it is important to recognize and understand this concept. It is also important to remember that the penetrance of some disorders is age dependent and ages of family members should be considered in the risk assessment process.[3]

Expressivity

Although some genetic disorders exhibit little variation, most have signs and symptoms that differ among affected individuals. Variable expressivity refers to the range of signs and symptoms that can occur in different people with the same genetic condition. For example, the features of MEN2A vary widely and some people only have mild symptoms (such as benign parathyroid tumors or high blood pressure), while others also experience potentially life-threatening pheochromocytomas and thyroid cancer. Although the features are highly variable, most people with this disorder have a mutation in the same gene (RET). In diseases with variable expressivity, affected individuals always express some of the symptoms of the disease and vary from very mildly affected to very severely affected. Since conditions may present diversely in families, the provider should inquire about the presence of physical features or conditions associated with the condition in other family members, in an effort to compile data that would support or refute the presence of the suspected condition. As with reduced penetrance, variable expressivity is probably caused by a combination of genetic, environmental, and lifestyle factors, most of which have not been identified.[3]

Negative Family History

An extended family history demonstrating no features or signs of the condition in question can be just as informative as a history of a genetic disorder in multiple relatives throughout several generations. Some conditions are the result of new mutations (de novo) in which case the pedigree does not demonstrate any other cases or suggestive findings of that condition in the family. Specifically, MEN2B, which is characterized by medullary thyroid carcinoma, pheochromocytomas, mucosal neuromas, ganglioneuromas, and skeletal and ophthalmic abnormalities, has been observed as both an inherited and sporadic disease, with an estimated 50% of cases arising de novo.[40] Such a history provides valuable information in establishing an estimate of risk for the proband and for other members of the family.[3]

Risk Perception

The concept of risk is a challenging component of the genetic counseling process. Clients have limited understanding of the meaning of risk figures and therefore these values may have little value to them.[41] This lack of understanding can affect their decision to undergo genetic testing, comprehension of the chance of recurrence, and even their screening practices. One person may consider a 10% risk to be very high, while another person may feel that is an insignificant risk. It is imperative that the healthcare providers do not display any personal bias as to whether they consider this risk high or low, as that will influence the client. Risk figures should be stated as fact and not couched with judgment statements like "The risk is only…." It is not the actual number that is important to explore with the client, but rather their perception of risk. This preconceived notion can be strongly influenced by their personal and family history, prior experiences with the condition, and their social, religious, and cultural views.[3,42] Communicating the concept of risk is one of the many roles where genetic counselors excel.

While risk communication to clients is challenging even for genetic counselors, they are uniquely trained to explain the meaning of risk in multiple ways that are the most effective means for assisting client understanding. Counselors may use diagrams, pictures, and bar graphs to demonstrate relative risk, absolute risk, five fold increase, and general population risk. Using multiple points of view allows counselors to frame the magnitude of risk in a way that is meaningful to the client. A client might understand that they have a one in two chance of passing the MEN1 gene associated with MEN1 to each of their offspring, rather than stating a 50% risk for this autosomal dominant condition. In this example, the counselor should also state that there is a 50% chance that the altered gene *will not* be passed down. While this is implied, it is important from a psychosocial aspect to emphasize the possibility of not passing the altered gene down, as well as the possibility of passing it down.[3]

SUMMARY

The Human Genome Project paved the way for many advances in genetic research and genetic testing that are increasingly becoming incorporated into patient care. This information is incredibly complex and can be difficult to interpret and needs to be explained to patients in a sensitive manner that they understand. Genetic counselors are critical to this communication process; they facilitate patient decision making to promote informed choices and help individuals understand and adapt to genetic conditions. Genetic counseling addresses the medical, psychosocial, and ethical/legal aspects of genetic predisposition to disease. It involves obtaining and accurately interpreting the family pedigree, educating how genetics plays a role in disease, explaining inheritance patterns and risk to other individuals, explaining the option of genetic testing and medical management, and discussing disease prevention, risk reduction, and reproductive options, This multifaceted medical profession not only includes the provision of accurate risk of disease occurrence or recurrence to individuals, but also a discussion of the impact genetic diagnosis, genetic testing, and results of genetic testing can have on a person's/family's life. A careful discussion both before and after genetic testing is an important component in aiding the patient's understanding of the condition and how genetic testing may influence medical management for them as well as other family members. The interpretation of the genetic test result must be discussed in the context of medical and family history and it should be emphasized that it is only one component of risk assessment. Ultimately, the client's decision should be supported in the context of that client's individual values, beliefs, and goals.[3]

Genetic counselors are experts at serving as resources for families, professional societies, and other healthcare providers, due to their wide breadth of knowledge and expertise. They are specifically trained to provide psychosocial support, guidance, and care for individuals and their families who are faced with many complex issues surrounding genetic diagnosis.[3]

References

1. The Human Genome Project Completion: Frequently Asked Questions. www.genome.gov. March 2014.
2. Resta R, Biesecker B, Bennett R, et al. A new definition of genetic counseling: National Society of Genetic Counselors' Task Force report. *J Genet Couns* 2006;**15**:77–83.
3. Cummings S. Genetic counseling. In: Weiss R, Refetoff S, editors. *Genetic Diagnosis of Endocrine Disorders*. 1st edition London: Elsevier; 2010. p. 293–302.
4. Bennett R, Hampel H, Mandell J, Marks J. Genetic counselors: translating genomic science into clinical practice. *J Clin Invest* 2003;**112**: 1274–9.
5. Bateson W. *Materials for the Study of Variation: Treated with Special Regard to Discontinuity in the Origin of Species.* London: Macmillan and Company; 1894.
6. Reed S. *Counseling in Medical Genetics.* Philadelphia: Saunders; 1955.
7. Carr-Sanders A. Eugenics. 14th edition *The Encyclopedia Britannica*, vol. 8. London: Encyclopedia Britannica; 1929.
8. Neel J. *Physician to the gene pool: genetic lessons and other stories.* New York: Wiley; 1994.
9. American Board of Genetic Counseling website. www.abgc.net. March 2014.
10. NSGC website. www.nsgc.org. March 2014.
11. Welch B, Kawamoto K. Clinical decision support for genetically guided personalized medicine: a systematic review. *J Am Med Inform Assoc* 2012;**20**:388–400.
12. Greendale K, Pyeritz R. Empowering primary care health professionals in medical genetics: how soon? How fast? How far? *Am J Med Genet* 2000;**106**:223–32.
13. Wideroff L, Vadaparampil S, Greene M, Taplin S, Olson L, Freedman A. Hereditary breast/ovarian and colorectal cancer genetics knowledge in a national sample of US physicians. *J Med Genet* 2005;**42**: 749–55.
14. Matloff E, Bonadies D. Cancer genetic counseling. In: Pine JW Jr, editor. *Cancer Principles & Practice in Oncology: Handbook of Clinical Cancer Genetics.* Philadelphia: Lippincott Williams & Wilkins; 2013.
15. Matloff E, Barnett R. The growing role for genetic counseling in endocrinology. *Curr Opin Oncol* 2010;**23**:28–33.
16. Raue F, Frank-Raue K. Update multiple endocrine neoplasia type 2. *Fam Cancer* 2010;**9**:449–57.
17. Pilarski R, Nagy R. Genetic testing by cancer site: endocrine system. *Cancer J* 2011;**18**:364–71.
18. Pieterman C, Schreinemakers J, Koppeschaar H, et al. Multiple endocrine neoplasia type 1 (MEN1): its manifestations and effect of genetic screening on clinical outcome. *Clin Endocrinol* 2009;**70**: 575–81.
19. Lips C, Höppener J, Van Nesselrooij B, Van der Luijt R. Counselling in multiple endocrine neoplasia syndromes: from individual experience to general guidelines. *J Int Med* 2004;**257**:69–77.
20. White H, Blair J, Pinkney J, et al. Improvement in the care of multiple endocrine neoplasia type 1 through a regional multidisciplinary clinic. *QJM* 2010;**103**:337–45.
21. Thakker R, Newey P, Walls G, et al. Clinical practice guidelines for multiple endocrine neoplasia type 1 (MEN1). *J Clin Endocrinol Metab* 2012;**97**:2990–3011.
22. Neumann H, Eng C. The approach to the patient with paraganglioma. *J Clin Endocrinol Metab* 2009;**94**:2677–83.
23. Petri B-J, van Eijck C, de Herder W, Wagner A, de Krijger R. Phaeochromocytomas and sympathetic paragangliomas. *Br J Surg* 2009;**96**:1381–92.
24. van Nederveen F, Gaal J, Favier J, et al. An immunohistochemical procedure to detect patients with paraganglioma and phaeochromocytoma with germline SDHB, SDHC, or SDHD gene mutations: a retrospective and prospective analysis. *Lancet Oncol* 2009;**10**:764–71.
25. Stadler Z, Schrader K, Vijai J, Robson M, Offit K. Cancer genomics and inherited risk. *J Clin Oncol* 2014;**32**:687–98.
26. Ambry Genetics website. www.ambrygenetics.com. March 2014.
27. Morrison P, Atkinson A. Genetic aspects of familial thyroid cancer. *Oncologist* 2009;**14**:571–7.
28. Brierley K, Campfield D, Ducaine W, et al. Errors in delivery of cancer genetics services: implications for practice. *Conn Med* 2010;**74**: 413–23.

29. Wolff G, Jung C. Nondirectiveness and genetic counseling. *J Genet Couns* 1995;**4**:3–25.

30. Raz A, Atar M. Nondirectiveness and its lay interpretations: the effect of counseling style, ethnicity and culture on attitudes towards genetic counseling among Jewish and Bedouin respondents in Israel. *J Genet Couns* 2003;**12**:313–32.

31. Genetic Alliance. www.geneticalliance.org. March 2014.

32. Lee S-C, Bernhardt B, Helzlsouer K. Utilization of BRCA1/2 genetic testing in the clinical setting: report from a single institution. *Cancer* 2002;**94**:1876–85.

33. Hall M, Rich S. Laws restricting health insurers' use of genetic information: impact on genetic discrimination. *Am J Hum Genet* 1999;**66**:293–307.

34. Pollitz K, Peshkin B, Bangit E, Lucia K. Genetic discrimination in health insurance: current legal protections and industry practices. *Inquiry* 2006;**44**:350–68.

35. Affordable Care Act Implementation FAQs. www.cms.gov. March 2014.

36. Resta R. The crane's foot: the rise of the pedigree in human genetics. *J Genet Couns* 1993;**2**:235–60.

37. Pearson K, editor. *The Treasure of Human Inheritance (Parts I and II)*. London: Dulau and Company; 1912.

38. Bennett R, Steinhaus K, Uhrich S, et al. Recommendations for standardized human pedigree nomenclature. *J Genet Couns* 1995;**4**: 267–79.

39. Lessick M. Genetic counseling of families with endocrine disorders. *Issues Compr Pediatr Nurs* 1980;**4**:27–40.

40. Carlson K, Bracamontes J, Jackson C, et al. Parent-of-origin effects in multiple endocrine neoplasia type 2B. *Am J Hum Genet* 1994;**55**: 1076–82.

41. Walter F, Emery J. "Coming down the line" – patients' understanding of their family history of common chronic disease. *Ann Fam Med* 2004;**3**:405–14.

42. d'Agincourt-Canning L. The effect of experiential knowledge on construction of risk perception in hereditary breast/ovarian cancer. *J Genet Couns* 2005;**14**:55–69.

29

Setting Up a Laboratory

Loren J. Joseph

Molecular Diagnostics Laboratory, The University of Chicago Medical Center, The University of Chicago, Chicago, IL, USA

INTRODUCTION

Calvin: Hmph. Where are the flying cars? …You call this the future?? HA!

Hobbes: I'm not sure people have the brains to manage the technology they've got.[1]

In the first edition, this chapter noted that Calvin would find neither flying cars nor the fabled $1,000 genome at that time. Even with this edition, neither flying cars nor a $1,000 genome is a reality, *but* today $1,000 *can* procure a "research grade" exome. More importantly, tens of thousands of whole genomes, exomes, and transcriptomes *have* been completed using next-generation sequencing (NGS).[2] The genetic causes for many rare inherited diseases have been discovered; for some there had been no clue to the causative gene. The ability to sequence large panels of genes or exomes is now within the capability of many clinical genetics laboratories.

It is also increasingly clear that many diseases with a genetic component will require more than "merely" sequencing the entire genome. Some "pseudogenes" are now known to be expressed; this requires RNA analysis.[3] A variation leading to increased expression of a noncoding element like lncRNA can have dramatic effect.[4] Synonymous ("silent") single nucleotide variants can alter RNA splice sites or miRNA binding sites; the latter can have numerous subtly ramifying effects.[5,6] Some epigenetic changes appear to be transgenerational.[7] NGS can identify variation between genotypically identical twins.[8] All this is without considering multiple-gene interactions or gene-environment interactions. Determining the clinical significance of thousands of newly identified genetic variants now makes attaining the sequence look straightforward.[9,10]

REGULATIONS FOR DIAGNOSTIC GENETIC LABORATORIES

The equipment and methods (Table 29.1) are identical for diagnostic and research genetic laboratories; however, the former must comply with a thicket of federal and state regulations. Physician-scientists, accustomed to federal regulation in healthcare, might still be surprised to learn that federal regulations specify the acceptable variation for measurement of serum sodium. Many laboratory requirements seem onerous but all share a goal: minimizing errors. For example, for a research laboratory it is good practice to regularly calibrate pipettes; for a clinical laboratory accredited by CAP, it is mandatory.

The Sources of Regulation

Clinical laboratory practice is covered by the Clinical Laboratory Improvement Amendments of 1988 (CLIA '88) (authorized under the Public Health Service Act: Section 353, Subpart 2), published in 2003 (Code of Federal Regulations Title 42, Section 493). CLIA regulations vary in detail. Most provide general goals such as requiring that each assay run includes positive and negative controls, but leave it to the laboratory to determine appropriate details.

Enforcement of Regulations

To oversee compliance with CLIA, the Department of Health and Human Services grants "deemed" status to several organizations and public health departments. The College of American Pathologists (CAP) is among the most active. CAP's Laboratory Accreditation

Genetic Diagnosis of Endocrine Disorders. http://dx.doi.org/10.1016/B978-0-12-800892-8.00029-4

TABLE 29.1   Capital Equipment

Item	Comments
GENERAL DNA LABORATORY AND SANGER SEQUENCING	
Capillary electrophoresis analyzer	DNA sequencing, amplicon size analysis
Real-time PCR	Genotyping, copy number, gene expression
Thermocyclers	General PCR, sequencing, NGS reactions
Thermistor array	Verify PCR temperature control
Gel electrophoresis rigs	Consider microfluidic capillary analyzers
UV camera	For imaging DNA gels. Consider "safelight" transilluminator
Desktop centrifuge	Spin down blood samples
Refrigerated high speed microfuges	General DNA protocols
Compact scanning spectrophotometer OR fluorometer	DNA quantitation
DNA analysis software	Sequencing, fragment size analysis
Tissue culture (optional)	For cell line control samples
NEXT-GENERATION SEQUENCING	
Second-generation sequencer	
Sonicator (optional)	Used for randomly shearing DNA
Computers – analysis, data storage	Consider cloud services, at least for backup
NGS analysis software suite	Commercial or shareware or cloud-based
Data and lab management software	Consider cloud services

Program (LAP) is based upon a series of standards rooted in CLIA that help to ensure accredited laboratories provide testing that meet the needs of patients, physicians, and other healthcare practitioners. CAP maintains accreditation "checklists," which are detailed lists of requirements the clinical laboratory should use to maintain good practices and compliance with CLIA requirements. The Molecular Genetics Checklist includes items for NGS. All checklists are available online but access requires participation with CAP (see Table 29.2).

Inspections

A clinical lab must undergo an unannounced inspection at least every 2 years with a documented self-inspection in intervening years. CAP inspector training emphasizes that inspection is an educational rather than an adversarial proceeding, but failure can have significant consequences. Cited deficiencies must be appealed or remedied.

Proficiency Testing

The laboratory must engage in proficiency testing, at least twice per year, from a Health and Human Services-approved proficiency testing provider for every analyte tested, when proficiency testing is available. CAP offers many proficiency tests but few for endocrine genetics. If there is no formal proficiency testing available, the laboratory must have a written policy detailing an alternative. A sample exchange with one or more clinical laboratories is a common approach; it must be carried out with a defined assessment plan.[11]

Laboratory-Developed Tests

The FDA recognizes several categories of diagnostic "kits" and reagents: FDA-approved, FDA-cleared, research use only, investigational use only, and analyte specific reagent. Endocrine genetic disorders are uncommon, so testing will probably involve what the FDA designates as a "laboratory developed test" (LDT) rather than an FDA-approved commercial kit. LDTs have been called "home-brews," a term which should be deprecated. An LDT *may* combine diverse reagent kits and instruments, for example, a kit for DNA preparation and a kit for real-time PCR DNA amplification from different manufacturers.[12] The results report for an LDT *must* carry the following specific disclaimer (the font size is not specified):

"This test was developed and its performance characteristics determined by (*the name of the laboratory*). It has not been cleared or approved by the U.S. Food and Drug Administration."

TABLE 29.2 Useful Websites

URL	Entity	Comments
GUIDELINES		
http://www.cap.org	College of American Pathologists	Definitive source of clinical checklists
http://www.clsi.org	Clinical Laboratory Standards Institute	Source of clinical laboratory guidelines
http:/www.acmg.net	American College of Medical Genetics	Source of clinical laboratory guidelines
http://www.amp.org	Association of Molecular Pathologists	Source of clinical laboratory guidelines, active listserv
GENOMIC DATABASE/PORTALS		
http://genome.ucsc.edu/cgi-bin/hgGateway	University of California Santa Cruz	Indispensable browser/portal for genome, transcriptome, chromatin, data across species
http://www.ensembl.org/index.html	European Molecular Biology Laboratory	Browser especially notable for visualization tools for exons and transcripts
http://www.ncbi.nlm.nih.gov/	National Institutes of Health (NIH)	
GENERAL GENETICS AND CLINICAL GENETIC INFORMATION		
http://www.genetests.org/by-genereview/		
http://www.genenames.org/	Human Gene Nomenclature Committee	
ONLINE SOFTWARE SOURCES – GENERAL DNA UTILITIES		
http://www.wageningenur.nl/en/Expertise-Services/Chair-groups/Plant-Sciences/Bioinformatics.htm	Wageningen University Laboratory of Bioinformatics	Primer design
http://www.ncbi.nlm.nih.gov/tools/primer-blast/	National Center Biotechnology Information	Complements Primer 3 program
http://biologylabs.utah.edu/jorgensen/wayned/ape/	Shareware Mac and PC	Superb general DNA utilities
http://www.mbio.ncsu.edu/bioedit/bioedit.html	Shareware for PC	General DNA utilities
http://genome.unipro.ru/	Shareware Mac and PC, "Ugene"	General DNA utilities, including some for NGS
http://www.ebi.ac.uk/services	European Bioinformatics Institute of EMBL	Numerous DNA software utilities
http://www.geospiza.com/Products/finchtv.shtml	Freeware	Sanger chromatogram viewer
WEB-BASED OR OPENSOURCE SOFTWARE FOR NGS ANALYSIS		
http://galaxyproject.org/	Freeware	Comprehensive NGS software
www.broadinstitute.org/gatk/	Freeware	Standard NGS variant calling suite
www.**bioconductor**.org/	Freeware/opensource	Comprehensive NGS software
www.broadinstitute.org/igv/home	Freeware	NGS data visualization
VARIANT ANALYSIS/INTERPRETATION SOFTWARE		
http://genetics.bwh.harvard.edu/pph2/	Polyphen-2	Variant pathogenicity prediction
http://sift.bii.a-star.edu.sg/index.html	Sorting Intolerant From Tolerant (SIFT)	Variant pathogenicity prediction
http://mendel.stanford.edu/SidowLab/downloads/gerp/	Genomic Evolutionary Rate Profiling (GERP)	Variant pathogenicity prediction
http://www.mutationtaster.org/		Variant pathogenicity prediction
http://cadd.gs.washington.edu/		Variant pathogenicity prediction

(Continued)

TABLE 29.2 Useful Websites *(cont.)*

URL	Entity	Comments
VARIANT ONLINE DATABASES		
http://www.ncbi.nlm.nih.gov/snp/	NIH	Comprehensive SNP database
http://www.ncbi.nlm.nih.gov/dbvar	NIH	Structural variant database
http://www.ncbi.nlm.nih.gov/clinvar/	NIH	Variants with clinical associations
http://www.ncbi.nlm.nih.gov/medgen	NIH	Database of genetic diseases
http://www.1000genomes.org/	1000 Genome Project	Variation database
http://huvariome.erasmusmc.nl/	Erasmus Medical Center	Variation database
http://www.hgmd.org/	Human Gene Mutation Database	Database for published associations of mutations with inherited disease
http://www.openbioinformatics.org/annovar/	Annovar	Variant database search tool

HIPAA

The Health Insurance Portability and Accountability Act includes provisions for ensuring privacy of all patient information, including test results. Clinical endocrinology laboratories are already required to comply with HIPAA regulations. Storage of NGS information raises logistic concerns – data files can run to hundreds of gigabytes per patient. In a specific uncommon circumstance, access to whole genome was used to identify research subjects.[13] Although "hacking" such data requires IT resources and bioinformatics skill much greater than illegally accessing the final report, hospital IT services are obligated to treat genetic sequence like any other protected health data.[14] Given the size of the data files this is an issue that must be addressed by any laboratory doing large-scale NGS.[15,16]

Intellectual Property

A clinical genetics laboratory must be attentive to intellectual property rights, such as the patent for real-time PCR (expires November 2016). Patents are often waived for research. Roughly 20% of genes have been patented. A typical patent claim encompassed any method (DNA, RNA, or protein based) that detected a variant, known or newly discovered, pathogenic or not. Whether or not a gene sequence or a mutation *should* be patentable has been controversial.[17] The 2013 Supreme Court decision in *Association of Molecular Pathology v. Myriad Genetics* held that "genes and the information they encode" are *not* patentable, but made a curious exception for "cDNA."[18]

THE PREANALYTIC PHASE

The Testing Process Overview

Based on CLIA (Section 493.1200), clinical laboratorians conceptualize the testing process into three distinct phases:

- Preanalytical – all steps prior to testing such as patient preparation, sample collection, sample preparation, etc.
- Analytical – the actual analytical process.
- Postanalytical – steps after completion of the test such as interpretation and results reporting.

The CLSI (Clinical Laboratory Standards Institute) publishes guidelines for many clinical laboratory activities, including nucleic acid extraction and DNA sequencing, including NGS (see Table 29.2).[19] Chen et al. provide a concise overview of general quality assurance measures; several recent documents specifically address NGS.[20–23]

Requisition

For clinical work, a properly labeled sample *must* be accompanied by a testing requisition completed by a licensed provider, which in most states means a physician. The sample must be labeled with patient name and unique identifier, while the requisition *must* include a minimum amount of information:

First and last name
Unique patient identifier (e.g., hospital medical record number)
Ordering physician name
Test requested
Test request date
Date of specimen collection
Type of specimen if other than blood

There is not a federal requirement for informed consent; however, hospitals and laboratories may require one. The process of consenting offers an opportunity to discuss potential results, including possible incidental findings, with the patient.

Sample Types

Peripheral blood is the standard sample type for genetic testing:

- *Blood*: Collection tubes that contain ethylenediaminetetraacetic acid (EDTA), which inhibit coagulation by chelating cations, are acceptable. Use of heparinized tubes can lead to interference; not all DNA preparative methods remove the heparin, an inhibitor of PCR. If a "serum" tube was used, which has no anticoagulant, some DNA can still be extracted from serum. DNA can be recovered from EDTA tubes after at least four weeks of refrigeration.[24,25]
- *Tissues*: Tissue is important for assessing tissue-limited mosaicism. Fresh tissue provides dramatically higher quality DNA and RNA than does the standard surgical pathology specimen, which is formalin fixed and paraffin embedded (FFPE). Cells from needle biopsies in alcohol-based fixatives are satisfactory. DNA can be routinely recovered from FFPE; in our experience, amplicons from 200 to 400 base pairs are routinely recoverable.
- *Chorionic villus sampling, amniotic fluid*: These are critical sources for evaluating a fetus for an inherited disorder. Contamination by maternal cells is always a concern and must be assessed by analysis of "identity" markers, such as short tandem repeats ("DNA fingerprinting") from the sample and from the mother.
- *Circulating fetal DNA*: The level of cell-free fetal DNA in the maternal circulation increases as pregnancy progresses. In some settings, routine genotyping techniques can detect significant variants, such as an Rh-positive allele from the father in an Rh-negative mother. NGS methods can detect fetal aneuploidy by testing maternal blood.[26]

Specimen Identification and Log-in

The laboratory must have a system for tracking primary samples including the time of receipt. The log number should be uniquely associated with the sample. A label printer, which can print small adhesive labels with the patient's name and log number date, is important for labeling derived sample tubes (such as DNA preparations).

DNA PREPARATION

Cell Handling

Nucleated cells can be isolated from whole blood by a variety of methods. Our laboratory osmotically lyses red blood cells in a whole blood sample, spins down the remaining cells, and washes the pellet with saline several

times before proceeding with nucleic acid preparation. Fresh tissue must be minced or homogenized (ultrasonic or mechanical) before being incubated in lysis buffer. FFPE sections must be dewaxed in serial xylene bathes, rehydrated with graded ethanol washes, and then incubated up to several days in a protease containing lysis buffer.

DNA/RNA Purification

"Home-brew" procedures are inexpensive but often include toxic organic chemicals (phenol, chloroform, guanidinium isothiocyanate). Most types of RNA other than miRNA are dramatically more susceptible to degradation than is DNA. This is attributed to the ubiquity and sturdiness of several RNases (some renature after boiling). RNA is *not* usually analyzed for genetic diagnosis; however, in cases where variants (including methylation) are predicted to affect splicing or transcription, confirmation by analysis of RNA could be sought.[6,27] Tissue specificity of expression will dictate the sample source. Silencing of one allele can be demonstrated *if* there is a "coding" single nucleotide polymorphism (SNP) distinguishing alleles.

Automated Extractors

Commercial kits typically forego toxic chemicals in favor of small spin columns or suspensions of charged magnetic particles, which reversibly bind nucleic acids. The reproducibility and labor saving provided by commercial kits can outweigh the additional cost relative to "lab-developed" methods.[28] Our laboratory uses an instrument that employs a spin column for purification. It is neither the smallest nor the fastest system, but an identical manual method is available that we could use in the event of an instrument problem. Reagent costs for automated extractors are higher per extraction than for the corresponding manual commercial methods. Overall yields are also typically lower because of the smaller input volume, but a single extraction from several hundred microliters should be sufficient for most purposes. Many automated procedures do *not* routinely include an RNase step for DNA purification or a DNase step for RNA purification; this can lead to an overestimate of the concentration.

DNA (RNA) Quantitation

- *UV spectrophotometry*: UV spectrophotometry is the traditional method for quantification of DNA (and RNA) and demonstration of purity. The absorption at 260 nm correlates with the concentration of nucleotides; it does *not* distinguish long double-stranded DNA from free nucleotides. Absorption at 280 nm gives a measure of residual protein. The A260/A280 ratio is a measure of purification; it should fall in the range 1.8–2.0. A 260/280 ratio greater than 2.0 does *not* indicate extra-high quality DNA; it most often reflects residual contaminants such as phenol. The A260 can be used to quantitate DNA or RNA. Our laboratory uses the Nanodrop™ spectrophotometer. Each reading requires one μL of sample applied directly to the analysis surface (no cuvette). Preparation for the next sample consists of applying a tissue to the reading surface to remove the prior sample. The instrument performs scanning spectrophotometry over a broad range, including A230, A260, and A280.

- *Dye-binding*: Picogreen™ and Ribogreen™ are representative fluorescent DNA-specific and RNA-specific dyes, respectively. The fluorescence is proportional to nucleic acid concentration. These assays are an order of magnitude more sensitive than UV spectrophotometric analysis but do *not* detect free nucleotides or short duplexes. The signal can be measured with UV ELISA plate readers, real-time PCR instruments, or readers specifically designed for DNA/RNA quantitation such as the Qubit™ or Fluorodrop™.

Nucleic Acid Integrity

High molecular weight (intact) DNA, intact RNA, degraded DNA and RNA, and nucleotides all absorb at 260 nm with similar efficiency. For routine preparation from fresh blood this check is *not* necessary except as a troubleshooting measure.

- *Agarose gel electrophoresis*: High quality genomic DNA should show only high molecular weight (at least 10–20 kB), which has barely migrated out of the sample well. DNA from FFPE routinely shows a diffuse smear in the sample lane. Badly degraded DNA might not show staining above a few hundred base pairs.
- *Microfluidic (chip or capillary) analyzers*: These perform the equivalent of gel electrophoresis, sizing the DNA or RNA products with high resolution, and measuring the concentration, using only microliter samples and run times of a few minutes. The chip is more expensive than agarose gel electrophoresis but saves time, sample, and labor. They can be used to assess quality of NGS library preps before an expensive sequencing run.

Sample Storage

- *Primary samples*: The remaining sample should be retained until analysis is complete. Consider spotting aliquots on filter paper or using commercial systems such as DNAStable Blood™ (BioMatrica) for storage at room temperature indefinitely.
- *DNA samples*: All analytes (DNA/RNA) should be stored in buffer, most commonly 10 mM or one mM Tris pH 8.0 supplemented with 0.1 mM or one mM EDTA. DNA can be stored at 4°C at least for months, indefinitely at –20°C. RNA should be stored at –80°C. DNA samples for clinical genetic testing must be stored for at least 20 years (consult state and local authorities as well).

ANALYTIC PHASE

Assay Validation

Validation of a new assay is a requirement for diagnostic labs. An assay should be shown (validated) to be able to detect all frequent mutations ("frequent" is at the discretion of the director). This requires testing *at least* one independently confirmed "positive" sample for every "common" mutation. The designation of a validation sample as "positive" (or negative) is based on prior analysis at a separate diagnostic laboratory or by an independent method. Cell lines carrying a mutation are acceptable. Another option which *might* be acceptable is synthetic DNA. DNA greater than 1,000 base pairs can be designed to match the region of interest and include one or more mutations at specified allelic frequencies. The synthetic DNA should include defined sequences at the 5′ and 3′ end for PCR amplification. It is advisable to include several mutations and several deoxyuridine-triphosphatase (dUTP) residues into such a control to limit the risk of contaminating patient samples.

Procedure Manual

For the diagnostic lab, a written procedure must be in place not only for each assay, but also for all phases of testing such as specimen login and sample storage. CLSI provides an excellent guideline. All staff trained to perform a given assay must read, sign, and date the protocol and be documented to show competence performing the assay.

Reagents

The diagnostic lab should track all reagents by lot including the date when opened and the expiration date. Assay worksheets should indicate when a new lot of a reagent has been introduced. Prior to introducing *any* new reagent lot, crossover validation should be performed and documented with at least one positive sample.

Controls

Ideally every run of every assay should include "positive" and "negative" controls; how to apply this rule is not always obvious. Consider a real-time PCR assay to detect a specific point mutation. The run should include a known normal (negative) sample, a sample known to carry the mutation (positive control), and a "no template control" (NTC) with water or buffer substituting for the DNA sample; this is another "negative" control. The NTC is intended to detect contamination with amplicons from previous reactions. If one is looking for a somatic mosaic, the "positive" control should be diluted by mixing with normal DNA to the level desired as the lower limit of detection. For a target gene that could have any one of numerous mutations a common approach is to have a set of samples with different mutations and rotate usage as controls.

Confirmatory Assays

Our laboratory processes 10–20 samples at a time for a clinically significant point mutation in the factor V gene. Typically one sample shows a mutation. A "positive" result often leads to long-term anticoagulation. All tested patients have histories of coagulopathy compatible with a mutation. For any patient who tests "positive" our policy is to process and test a second aliquot of blood from the original stock tube, confirming that the mutation is present in that subject; this is intended to catch labeling errors. For multiexon sequencing, it would be sufficient to retest only the exon(s) showing a mutation. This is *not* a required policy, but it is recommended in American College of Medical Genetics (ACMG) guidelines.

Data Retention and Storage

"Data" and reports must be kept for 10 years. This includes primary data (computer) files. There are several large data files for an NGS sample, which must be kept (for the diagnostic lab) remains to be clarified but certainly includes the variant file.

METHODS – GENERAL PCR

Thermocyclers

Thermocyclers vary in ramping speed, which can affect efficiency, with more expensive metal blocks giving better performance, but for most purposes most thermocyclers are adequate.

- *Thermal cycling profile validation*: For the clinical laboratory every well of every thermocycler must be shown to have the expected thermal cycle profile. Our laboratory uses a thermistor array that sends data by wire to a laptop. Validating

real-time thermocyclers is more challenging, because of the difficulty accommodating the array: wireless thermistor arrays exist. Demonstrating a reproducible C_t (threshold crossing point) for the amplification curve of a real-time PCR assay in every well can be used as evidence of reproducible performance even if the "true" temperature is unknown.
- *Automation*: Flexible programmable instruments are available, which can set up PCR and DNA sequencing reactions. Although slower and more expensive than manual set-up, for a small number of samples, reproducibility is excellent.

Minimizing PCR Amplicon Contamination

Contamination of the working environment by PCR amplicons from an earlier test is an ever present concern. Several measures can reduce the risk.

- *Spatial separation*: PCR involves three phases, which, ideally, are spatially separated: processing the specimen, setting up the PCR reaction mastermixes, and running the assay. The PCR mastermixes are brought into the specimen processing area, the samples added, and the completed reaction mixtures taken to the instrumentation room with the thermocycler. Post-PCR steps such as DNA sequencing should only be performed in the instrumentation room. The laboratory should have separate sets of pipettes, filtered pipette tips, PCR tubes, and gloves for each of the three work areas. Ideally, the instrumentation area should have a ventilation system separate from that for the sample prep and PCR mastermix prep areas, but in practice, this is expensive and uncommon.
- Although one does not want to undercut an argument to have at least three rooms, if necessary, setup of the master mixes and addition of samples can be performed in the same room but preferably on separate dedicated benches. The use of PCR hoods ("dead boxes" with UV lights to "sterilizing" rogue PCR amplicons) is a relatively inexpensive important additional precaution. A laminar flow hood is not necessary.
- *Unidirectional workflow*: PCR mastermixes may only go from the PCR set-up room to the sample setup room; reactions with samples added may only go to the PCR instrumentation room. Color-coded labcoats are helpful so that coats from the analysis room are never worn into the setup areas. Amplicons probably adhere well to labcoats, but this remains a common policy. The completed PCR reaction and instruments like pipetters never leave the PCR instrumentation room.

- *Use of dUTP and uracil-N-glycosylase (UNG):*
  Substitution of dUTP for a proportion of
  deoxythymidine triphosphate in the PCR
  deoxynucleotide triphosphate mix makes the
  resulting amplicons susceptible to cleavage by the
  enzyme UNG. Any prior amplicon contaminating a
  new reaction would be destroyed when the UNG is
  activated during the preincubation of the new PCR
  reaction. "Regular" UNG retains activity despite
  multiple PCR thermal cycles and can destroy new
  PCR amplicons. Use of the more expensive heat-
  labile UNG is strongly recommended.

METHODS – REAL-TIME AND DIGITAL PCR[29]

Real-time PCR is a fast, efficient, moderately scalable platform useful for several genetic applications: genotyping (better for single nucleotide changes than insertion-deletions), *methylation* analysis of specific nucleotide positions, measuring gene copy number, and quantifying gene expression.[30,31] Inherited variants found by second-generation sequencing are, at present, often confirmed by an "orthogonal" method. If the variant accounts for more than 30–40% of the allelic copies as determined by NGS, confirmation by Sanger sequencing offers the most straightforward confirmation. If the variant is present at a lower level, such as occurs in somatic mosaics, then a more sensitive method is needed. Common formats for real-time PCR can detect variants present at the 2–5% level. More sensitive "allele-specific" assays have been described, which are sensitive to the 0.1% level but these are harder to design and control. Although PCR, including real-time PCR, is a mature technology, useful improvements still come along such as "cooperative primers," which can sharply reduce background *and* incorporate a fluorescent probe.[32]

In real-time PCR a fluorescent signal is detected in "real time" by reading through UV transparent tubes, microtiter plates, or capillaries. The accumulation of amplicons is quantitatively monitored in each cycle without stopping the reaction. Because the tubes/plates are not opened, there is minimal risk of contamination of the lab space by PCR amplicons. There are three broad assay formats: double-stranded DNA-binding dyes, hydrolysis probes, and hybridization probes. All real-time PCR instruments should be compatible with the first two; not all will be compatible with the spectral requirements for hybridization probes.

Double-Stranded DNA Binding Dyes

This is the least expensive method.[33,34] They are most suitable for quantitative analysis of gene expression and gene copy number but can, with effort, be used for genotyping. Certain dyes, after binding to double-stranded DNA, fluoresce when excited at an appropriate wavelength. As more amplicons are generated in each cycle, more dye is bound by the end of each synthesis step. The higher the level of input target, the earlier the cycle in which fluorescence signal reaches a detectable level, permitting quantitation (with calibrators).

Hydrolysis Probes

These are often referred to by the more colorful but trademarked term "TaqMan" probes (derived from an early schematic in which the polymerase is depicted as a PacMan™ icon munching fluorescent molecules off probes). The probe is a DNA oligonucleotide complementary to the region of interest, labeled with a fluorescent reporter molecule at one end and a "quencher" moiety at the other end. When an intact probe molecule is excited by light of the appropriate wavelength, the fluor will transfer energy to the nearby quencher by a quantum mechanism (fluorescent energy resonance transfer) *without* emitting light. During the extension phase of PCR, as the polymerase copies the template strand and reaches the probe that is binding the complementary strand, it cleaves the fluor from the probe. The fluor, now free in solution, will fluoresce. With each cycle more free fluors accumulate.

Hydrolysis probes can be used for quantitation of gene expression, for gene copy number quantitation, and for genotyping. A typical genotyping assay format targeting a specific SNP would use two probes, their sequences differing by a single nucleotide at the position of interest, usually in the middle of the probe sequence. Each probe would carry a different fluor. If the sample is homozygous for nucleotide X, only one fluor is released by hydrolysis. If homozygous for nucleotide Y, only the other fluor is released. If the sample is heterozygous, both fluors are detected.

Hybridization Probes

These are often referred to by the trademarked name "LightCycler Probes." In addition to PCR primers, each assay has two probes: a long oligonucleotide, the "anchor probe" and a short oligonucleotide, the "sensor probe." Typically a fluor is at the 3′ end of the anchor probe and a reporter is at the 5′ end of the sensor. The anchor and sensor are designed to bind to the target with only a few nucleotides separating the fluor and the reporter. When they are bound to the target, and the fluor is excited, the energy is transferred to the reporter, which emits at a wavelength distinct from that expected for the fluor. The design is such that polymerase does *not* destroy the probes. As the temperature increases in each cycle, the probes come off the target. The fluor is no longer in proximity to the reporter, so the signal decreases. Hybridization probes give increasing signals like hydrolysis probes but can also generate melting curves. The sensor is short, so that at a low annealing temperature it can bind to either the wild-type sequence or to the variant sequence. The temperature at which the sensor probe comes off each allele (and signal decreases) depends on the sequence and can vary by as much as 10 degrees for a single nucleotide variant.

Real-Time PCR Instrumentation

Plate-based instruments typically excite the fluors and detect the resulting UV signal through the top of each tube or plate. Some models accommodate microplates with up to 1,536 wells. Heating and cooling samples is typically done by some combination of Peltier effect and forced-air cooling. Another design uses capillary tubes, resembling capillary tubes like those used for fingertip blood draws. The tubes are mounted in a carousel, which rotates rapidly within a cylindrical chamber heated/cooled by forced air. This format provides greater temperature uniformity than plate-based systems but has much lower throughput. Most real-time PCR instruments will perform most basic analyses satisfactorily. Performance characteristics to consider include the ease of sample data entry, analysis software features, and throughput.

Digital PCR

Digital *PCR* is a new PCR method using an old technique – limiting dilution analysis. Using various microfluidic approaches, the sample is aliquoted into thousands of individual isolated reactions, each of which is then scored separately. It is usually an end-point assay with a fluorescent readout: a well or bead is either positive or negative. This allows sensitive and precise measurement of low-level variants and of copy number for genes.[35–37]

METHODS – MICROARRAYS

Arrays are available with over 10^6 probes, usually a mixture of probes designed to detect common SNPs across the genome, and probes designed to detect specific regions of each chromosome. Large-scale (greater than 100,000 base pairs) deletions, gene amplifications, and isodisomy are efficiently detected by microarray.

- *Copy number variation (CNV):* It was not realized until 2004 that apparently healthy individuals can have hemizygous large deletions (100,000 bp or more), which are stably inherited. The medical significance for most CNV remains to be determined.[38]

- *Uniparental disomy (UPD):* In UPD, two copies of a gene or chromosomal region are inherited from one parent, none from the other, as a result of replication error. UPD has been described in several endocrine genetic disorders such as Russell–Silver syndrome, a short-stature syndrome that often presents with hypoglyemia.[39]

Microarrays are cost effective for detecting large CNV and for detection of UPD.[40] Microarray genotyping is now considered a first-line diagnostic tool in the work-up of pediatric genetic disorders; they are primarily used for subjects with developmental disabilities or congenital anomalies rather than isolated unexplained endocrine disorders.

METHODS – METHYLATION ANALYSIS

Methylation of specific deoxycytidine triphosphate (dCTP) bases in a promoter, creating 5-methylcytosines, can reduce or silence gene transcription from the corresponding allele ("in cis"). This effect is context dependent: methylation[30,41] of some promoters is associated with increased expression. In inherited disorders that show imprinting,[42] only one allele is expressed, in some cases, this has been associated with methylation of the promoter for the silenced allele.[43] For one disorder with imprinting, only the maternal allele is expressed; for another disorder, it might be only the paternal allele. Imprinting occurs in several endocrine disorders.[44–49] An unsuspected connection between endocrine genetic disorders and methylation occurs in paragangliomas associated with mutations in the succinate dehydrogenase genes. The mutations lead to accumulation of succinate, which in turn inhibits 2-oxoglutarate-dependent histone and DNA demethylases, leading to *hyper*methylation of some genes.[50]

Bisulfite treatment of DNA converts nonmethylated dCTP nucleotides into dUTP, leaving methylated dCTP unchanged. Traditional bisulfite treatment destroys as much as 90% of the template. Bisulfite treatment is within the scope of a typical genetics laboratory. Sequencing, single-base extension assays, and real-time PCR genotyping can determine which dCTP nucleotides were methylated and which changed to dUTP. An alternative method uses methylation-sensitive restriction enzymes to cleave genomic DNA, with and without bisulfite treatment, and from the resulting pattern infer the sites of methylation.[51]

Oxidation of 5-methylcytosine (5mC) by the TET proteins generates a variety of modified bases – 5-hydroxymethylcytosine (5hmC), 5-formylcytosine (5fC), and 5-carboxylcystosine (5caC). The latter two can be altered by thymine DNA glycosylase and base excision repair, restoring dCTP. Recently introduced protocols describe further chemical modification, which allows the modified bases to be detected and resolved at base pair resolution by NGS.[52–54] The frequency of these modified bases is much lower than that of 5mC, but the distribution and biological significance is just beginning to be explored.

Microarrays can be used to determine methylation status (after bisulfite treatment) at hundreds of thousands of sites. NGS can identify sites with base pair resolution, generating the "methylome." Immunoprecipitation with antibodies to various modified bases can enrich the sample for analysis of modified sites.[55]

A third-generation NGS system, Pac Bio, has been shown to recognize methylated dCTP directly, as well as other modifications.[56,57] The PacBio can read intact native DNA (PCR amplication not required), which preserves modified bases. Each of the 150,000 (or more) small wells (20 *zepto*liter volume) can accommodate a single DNA molecule and can read continuously 10–20,000 bases. The error rate for basic sequence determination was initially too high for diagnostic purposes, but accuracy is improving and it is the *only* system at present that can detect these modifications.

METHODS – SEQUENCING

First-Generation Sequencing

The capillary electrophoresis analyzer, better known as a "DNA sequencer," can be used for both traditional Sanger DNA sequencing *and* for detection of insertions/deletions by sizing of PCR products ("fragment size analysis"), offering single base resolution in the range from 20 nt to 1,000 nt.[58] The only active manufacturers are ABI and Beckman-Coulter, with ABI dominating the market. Sanger sequencing is a mature technology; improvements typically tweak performance or provide convenience, such as monitoring the number of runs for a given capillary array and preventing the user from accidentally squeezing an extra 200 runs out of an expensive capillary certified for 100 runs. Samples are assayed in parallel, eight to 96, depending on the number of capillaries in the instrument. The "read length" can reach 1,000 base pairs; the run time can range from 30 to 180 min, depending on the length of the capillary array and the choice of polymer.

Accuracy is typically cited as greater than 99.9%. Traditionally every amplicon is sequenced in both directions, so a random "sequencing" error in the same position in both directions should be very rare. A variant in only one direction would be flagged for further study. The software should give a "Phred" quality score, "*Q*," for every base. The Phred score is an estimate of the probability of an error in the base call where

$Q = -10\log_{10}P$. "*P*" is the probability of error (using an error model developed by the manufacturer). A *Q* score of 30 corresponds to a probability of an error in the base call of one per 1000 (>99.9%).

Sensitivity is low. Depending on the sequence context, a variant can be detected when it accounts for as little as 10% of the alleles, but a more typical limit is 20% of the alleles; this corresponds to the mutation being present in at least 40% of the cells (assuming diploidy). For analysis of inherited disorders this is usually not a problem, but for detection of somatic mosaicism and for mitochondrial disorders, this is a significant limitation.[59,60] Mosaicism has been reported in several endocrine genetic disorders.[49,61,62]

New models such as the ABI3500 have a smaller footprint, sleeker appearance, and greater ease of operation than prior models, *but* the price point remains noticeably higher than that for clinical-scale NGS systems. The advent of NGS does *not* mean the genetics laboratory can forego purchase of a "DNA sequencer." The capillary sequencer provides several key capabilities:

- *Confirmation of variants found by NGS*: For inherited disorders, it is currently standard practice to confirm detection of variants found by NGS and considered to be pathogenic by a second, "orthogonal" (different) method.[63] Designing amplicons for detection of point mutations and insertion-deletions by Sanger sequencing is usually straightforward. The main delay is in obtaining the primers.
- *Ease of setting up a new assay for a specific gene*: Again, designing primers for Sanger sequencing of a new gene is relatively straightforward and inexpensive. Modifying NGS panels often requires several rounds of optimization and for PCR resequencing panels might not be feasible.
- *Quickly checking a few exons in a single sample*: The cost and turnaround time for Sanger sequencing is significantly lower for this purpose than running even a small gene panel by NGS.

Second-Generation Sequencing (NGS)

Overview

In Sanger sequencing, a PCR amplicon is generated with a single pair of PCR primers in a single reaction *before* DNA sequencing. In second-generation sequencing, the target DNA molecules are spatially segregated and immobilized in wells or on beads *before* they are amplified and sequenced, hence the alternative name "massively parallel sequencing" (see Chapter 30 Whole Genome/Exome Sequencing). A target molecule with a variant sequence will have been amplified and sequenced with minimal background in its own "space."

Instrumentation Overview

At present, two companies dominate the NGS instrument space: Illumina and LifeTech. This platform war, like that between Apple and Microsoft, pits two capable systems against each other. Both offer "clinical-scale" instruments suitable for a diagnostic or research lab, rather than a core facility. A small number of comparative studies have appeared.[64–69] Because of rapid improvements, these comparisons, by the time of publication, do not necessarily reflect the instruments available.[70]

Startups and large biotech firms are entering the second-generation NGS space; a few have already exited. Meanwhile, third-generation systems, such as PacBio, are already in use, and at least one nanopore-based sequencing system is reported to (still) be about to enter the market.[57,71,72]

A distinguishing feature for third-generation systems is that they can analyze "native" DNA without PCR amplification, thereby retaining modifications, with speed several orders of magnitude greater than that for second-generation NGS. These are no longer avant garde: a fourth generation of systems is in development including one using electron microscopy for sequencing and one using quantum tunneling![73,74]

At present, the Ion Torrent PGM and Illumin MiSeq can each handle multiple samples per day with panels covering tens to hundreds of exons. The Ion Torrent Proton and the Illumina MiSeq can each sequence whole exomes. Whole genomes have been reported with the Proton. The next step up for Illumina, the HiSeq series, is by far the most widely used for whole genome sequencing. At present it is not clear if whole genome sequencing sufficiently outperforms whole exome sequencing to justify the increased cost except for selected cases at large centers.[75–80]

The most important consideration is how a given platform fits the needs of the laboratory: throughput, turnaround time (hours vs. days), hands-on time, cost per run, and the amount of input DNA required (a limitation for tissue samples). A laboratory interested in studying, for example, a panel of 10 genes will typically find both the PGM and the MiSeq satisfactory. If the laboratory is determined to find the cause for a patient in whom the usual genes are *not* mutated, then whole exome or genome sequencing might be an appropriate next step. For example, exome sequencing was recently used to identify new mutations associated with corticotropin-independent adrenal Cushing's syndrome.[81] Still, implementing whole genome or exome sequencing is nontrivial. For "one-off" cases, it would be cost-effective to have whole exome or genome sequencing performed at an established NGS service provider. Alternatively, at present a "research grade" exome can be obtained for just under $1,000; this includes the data files and the variant calls but neither orthogonal confirmation nor annotations of putative pathogenic variants. A molecular genetic endocrinology laboratory in receipt of these data could consider further analyzing the variants.

Library Preparation for NGS

Implementation varies markedly with the sequencing platform; however, certain broad features can be used to categorize the different approaches.

Whole Genome Sequencing

DNA is fragmented to a desired average size by one of several methods, including sonication, enzyme cleavage, or transposon activity, each of which is *assumed* to be random.[82,83] DNA in the desired size range is collected. The DNA fragments then have DNA adaptors added to each end. The adaptor includes a sequence that allows the amplicon to bind to a surface (bead or flow cell) with a complementary oligonucleotide. The adaptor can also include an additional unique sequence, barcoding the particular sample. After the amplicons are dispersed on the appropriate surface all the amplicons can be amplified and then sequenced with a single primer. After sequencing the amplicons in one direction one can choose to sequence the complementary strand ("paired-end" or "mate-pair" sequencing).

Pull-Down Including "Whole Exome" Sequencing

Genomic DNA is prepared and modified as described for whole genome sequencing, but instead of distributing *all* the fragments on the sequencing platform, the amplicons of interest are pulled out ("pull-down") by hybridization to a pool of "baits" and then distributed, thereby enriching for the regions of interest. A common "bait" is a long RNA sequence complementary to the region of interest. Each RNA bait includes a biotinylated moiety. The baits are pooled – pools can be large enough to cover the exome – and hybridized overnight with the denatured genomic library. Next the biotin-labeled baits, the hybridized regions of interest (and a nontrivial amount of off-target sequences), are "pulled-down," ready for distributing in wells or on flow cells for sequencing. Baits must avoid pseudogenes and undesired homologs so design is nontrivial. At present regions with high guanine-cytosine (GC) content are often missed. This is notable in the poor representation of the first exon for many genes; these exons tend to be GC rich.[84]

Amplicon Resequencing

In this approach a pool of gene-specific PCR primers is used to amplify the regions of interest. Our laboratory routinely uses a pool of 200 primer pairs. After several rounds of PCR with the gene-specific primer and the sample DNA, adaptor sequences (as described earlier) are added to the new amplicons. This allows the amplicons to bind to beads and then be further amplified and sequenced.

Notes on the Different Approaches

The resequencing format offers the fastest turnaround and generates the lowest proportion of off-target sequences. If an unsuspected single nucleotide variation (SNV) is present under a primer, it is possible that one allele will not amplify, possibly causing the assay to miss a significant variant and underestimate the gene copy number. With long pull-down baits, a single nucleotide variant should have much less effect.

Amplicon resequencing, by starting with several rounds of PCR, generates duplicates but unevenly among targets. With the whole genome and with the "pull-down" approach, as a result of random shearing, two amplicons reflecting the same target region but which originated from distinct input DNA molecules can be distinguished by the random ends of the amplicons. The ability to exclude duplicates makes copy number calculations more effective in whole genome and in pull-down assays.[85,86]

Library Automation

Manual preparation of NGS libraries for each of these methods is well-established.[87,88] Naturally automation has also been developed, already works pretty well, and is expensive, so the decision to implement will depend on the usual parameters, such as workload, number of samples/run, labor costs, and "buyer's remorse" – the concern that an even better solution will be released soon after your purchase.

BIOINFORMATICS FOR NGS

There are several basic sequential bioinformatics tasks:[89]

- Base calling (with quality control measures)
- Alignment (mapping)
- Variant/genotype calling (identifying variants)[90]
- Filtering variants/prioritizing/interpreting variants[91,92]
- Visualization of pathogenic variants (if available)

Each task can require invoking multiple programs. The entirety is often referred to as a "pipeline." Some steps such as variant calling and alignment are repeated iteratively ("recalibrated").

- *Base calling*: It is platform specific and includes an error model for the specific platform and chemistry calibrated against the human reference genome. This is performed by the instrument software. The raw

data files are large, especially for Illumina, which uses image files. After base calling is completed, the software determines which "sequence reads" show adequate quality, trimming away low-quality base calls. The resulting base calls and associated quality scores are stored in one of several formats. For the Ion Torrent we get FASTQ files, which go into the next step – alignment.

- *Alignment*: The file of acceptable "reads" is then aligned against the reference genome.
- *NGS and coverage*: Increased coverage is also important to make sure the entire region of interest is in fact "covered." In amplicon resequencing, every amplicon (in theory) covers the region of interest for that amplicon. For methods that randomly fragment genomic DNA, it is possible that with low fold coverage, a few nucleotides or more might not be covered by any of the amplicons that were pulled down. For inherited disorders, $6\times$–$40\times$ coverage is usually sufficient. For somatic mosaics, the greater the coverage, the greater the sensitivity: $1,000\times$ might be required to detect a mosaic at the 5% level.
- *NGS, error rates, and point mutations*: The error rate is greater than that of Sanger sequencing with respect to point mutations, but still is often cited as 98–99% accuracy. This might seem satisfactory, but 1% of 60,000,000 bases (an exome) gives a lot of false calls. By reading tens to thousands of distinct amplicons ("coverage") for each position, the error rate can be reduced. Generally NGS performs can detect point mutations at the 1–2% level (mosaicism, cancer).
- *NGS and copy number*: Bioinformatic algorithms are still evolving for this task. The larger the target region (especially exome or genome), the more reliable the statistics.[38] At present quantitative PCR or microarrays offer more straightforward ways to determine copy number for specific genes or regions.[38,93,94]
- *NGS and detection of translocations, insertion-deletions*: Also a work in progress. NGS clearly picks up many in each category but many algorithms also overcall PCR-chimera artifacts as variants or undercall translocations and insertion-deletions.[95]
- *Variant calling*: Variants are "called" and classified (point mutation, insertion, deletion, amplification, translocation). The software can tally up all the calls ("coverage") for a given nucleotide position in the reference genome, *and* the proportion accounted for by the expected nucleotide and by any variant.[96]

There is no one correct pipeline, but there are many programs, which are widely used in varying combinations in different pipelines. Numerous commercial software programs are available; the sequencing platform should come with some built-in software. A clever and low-cost feature of the (Ion) Torrent Suite screen report is that clicking on a variant call links out to an online program at the Broad Institute, the Integrated Genome Viewer (IGV), which gives a helpful graphic display of results. Graphical display is particularly helpful because it can highlight variants near the end of an amplicon where calls tend to be unreliable.

- *Performance*: To compare two instruments and/ or two pipelines on very large targets, one needs to know the "true" sequence of the target. The human genome "reference sequence" used by all databases is a composite from many individuals, so it cannot be used as a benchmark. The National Institute of Standards and the "Genome in a Bottle Consortium" are developing well-characterized whole genome reference materials. At present, the most advanced reference material is the pilot genome NA12878. To take into account the numerous disagreements among platforms and pipelines, Zook et al. integrated and arbitrated calls among five sequencing technologies, seven alignment programs, and three variant callers.[97] Soon a lab that wants to validate its performance, at least a little, will be able to sequence some or all of this reference genome material.

QUALITY ASSURANCE FOR NGS

As noted earlier, guidelines are now appearing from professional organizations regarding good practice and standards for NGS in clinical laboratories. Of course, the basic standards for all laboratory practice apply. Whenever software is upgraded on any instrument in the lab, performance must be certified; NGS just requires a lot more work and data analysis. The clinical laboratory must have a plausible written policy covering all aspects.[21–23,98]

GENERAL HARDWARE AND SOFTWARE CONSIDERATIONS

Computers

For analyzing real-time PCR and Sanger sequencing data, most modern desktop computers are adequate. Individual file sizes are on the order of 100 kb, so storage demands are small. At risk of stating the obvious, all files should be backed up regularly, frequently, and at some distance from the primary data storage area. NGS data storage is a challenge. Even small gene panels give multigigabyte files; these might fit on a DVD (which might not be readable a few years hence), but this will not be feasible for large panels. Cloud services are available for NGS data storage. As noted earlier, patient data privacy and security must be guaranteed. This can only be determined by consulting IT experts and legal authorities.

For NGS the more powerful the computer the less time spent drumming one's fingers. An eight-core 64 Gb RAM computer with one to two terabytes of memory should suffice for exome sequencing. Be sure to check that the configuration will work with the software selected. For exomes this will still be a lengthy analysis, so more than one computer could be required depending on the workload. There are now cloud services that offer complete analysis suites; services from the NGS platform manufacturer will offer easiest integration, but independent suites are available.[99]

Laboratory Management Software

There are a small number of lab management software suites that can accept variant call files from NGS. One open source solution is LabKeys (https://labkey.com/). The Galaxy Suite has an associated LIMS program but is not specifically clinically oriented.[100] Cloud-based suites, especially those from the platform manufacturers, can manage the NGS results; open-source cloud-based programs are becoming available.[101]

General Purpose DNA Analysis Software

There are many commercial suites. Our laboratory uses a shareware program, ApE (A Plasmid Editor), which includes an ABI (Sanger) chromatogram viewer. There are numerous freeware/shareware programs available; some with a specific focus such as PCR design (see Table 29.2).

Sanger Sequencing Software

Shareware programs such as FinchTV (see Table 29.2) are available for reading sequence, but a clinical laboratory should rely on clinically oriented, commercial, software, which can provide quality scores for each base "called" by the software; compare the sequence to a reference sequence, and provide an audit trail of who has accessed the file. Our laboratory uses two programs: SeqScape™ (ABI, commercial), excellent for aligning multiple patient samples, and Mutation Surveyor™ (SoftGenetics, commercial) Rapid performs well with insertion-deletions.

Several freeware/shareware programs are available for viewing sequence chromatograms and manipulating sequence (see Table 29.2). ABI itself has offered SeqScanner for sequence files and PeakScanner for fragment analysis. The current releases do *not* work with data from the ABI 3500 (http://resource.lifetechnologies.com/pages/WE28396/).

NGS Software

Data Formats

The output file for an NGS platform is typically a variation on FASTQ files. There are programs to convert files among formats. For a single read the FASTQ file might look like:

```
(Line 1)  @EMT56:00392:02184

(Line 2)  GAAGGGAGGTGGGGCTGGGAAGTGACCCTGGGTCAGT
          GGGGAGCAGGGAGCTGTCAGCC
(Line 3) +

(Line 4)  CB>DC9A>7::::+8>>>4>776:??94??C:@@>>
          9999+488<@A5@@>88:?7745
```

Line 1 is a flexible identifier, which might include the run name, the flowcell (Illumina), and the position of the reaction in the flow cell. Line 2 is the sequence. Line 3 is just a concatenation symbol. Line 4 shows the Q score for each base in the sequence using ASCII symbols to indicate increasingly high quality from the lowest level ("!") to the highest ("~"):

```
!"#$%&'()*+,-./0123456789:;<=>?@ABCDEFGHIJKLMNOPQRSTUV
WXYZ[\]^_`a

bcdefghijklmnopqrstuvwxyz~{|}
```

After alignment to the reference sequence, the reads can be exported for further analysis. Typical formats include SAM (sequence alignment/map format) and BAM, a binary form of SAM (http://en.wikibooks.org/wiki/Next_Generation_Sequencing_(NGS)).

Commercial Software

The NGS platform will come with a data analysis suite; we have used Torrent Suite with the Ion Torrent PGM. For diagnostic laboratories, the data analysis suite should ideally be "locked down" and provide an audit trail.

Platform-independent commercial NGS software suites are available. Our laboratory has used SoftGenetics' NextGene with satisfaction. Many software companies now use a leasing model, rather than a one-time purchase option. With each new panel, especially during the validation phase, consultation with a bioinformaticist is at present a worthwhile investment. With commercial software, it is feasible for a clinician or researcher to analyze small panels without extensive bioinformatics support, but it requires sitzfleisch. As the false positives or common but uninformative variants are catalogued, analysis and interpretation can become faster and more automated. As variant databases become more complete and detailed,

commercial software is likely to make the process of review and interpretation easier.

If you do not fear the Unix/Linux command line, you can take advantage of the Galaxy Suite freeware (Table 29.2). Many of the tools used by leading research labs are available as part of the Galaxy Suite. Galaxy can be installed on a laboratory computer, but if your institution has a bioinformatics core, it is likely that they offer the suite. Many online tutorials are available. Other widely used command-line programs are available online or for download, including the GATK suite (included in Galaxy), the IGV suite for visualizing data, and BioConductor, which uses the "R" statistics language (see Table 29.2).

Variant Annotation/Interpretation

Once your software is satisfied that a variant called by the alignment program is "real" rather than noise, one must determine if it is likely to be clinically significant.[90] With large panels, as well as with exomes and genomes, the number of analytically valid variants is so great that one must rely on software "filters" to analyze and exclude most variants as unlikely to be pathogenic. You could, for example, chose to ignore all synonymous (silent) SNVs except those at putative exon splice sites.[102]

For nonsynonymous variants, which alter the encoded amino acid, there are numerous programs that make predictions about pathogenicity. Examples include Polyphen 2 and SIFT2, which use physical chemical properties of the protein to predict altered function, or GERP, which uses evolutionary conservation of a base to predict pathogenicity (see Table 29.2).

Pipelines can be designed to query online databases to see if a variant has been reported in association with a disease or phenotype or as an SNP (Table 29.2).[103,104] "Annovar" is a freeware program that queries many of these databases as well as prediction algorithms (see Table 29.2). Some programs are specifically designed to search the research literature.[91]

Given the immense number of variants recently discovered, it is not surprising that most are classified as variants of uncertain significance (VUS). There are several schemes of similar-sounding categories (a.k.a. "bins" or "tiers") such as the following: "known deleterious," "presumed deleterious," VUS, presumed benign, known benign.[105] The ACMG has updated its recommendations for interpretation, but these are not yet in print.[106] Additional useful guidance, checklists, and databases are provided by Duzkale et al. and by Ramos et al.[92,107] Kircher et al. have applied a pathogenicity prediction program to all known SNVs and shown a high level of accuracy with known variants; however, performance remains to be confirmed by others.[108]

Performance

A consortium held a 2014 competition for analysis of three pediatric patient genomes. Thirty teams from around the world entered. The good news is that "entries reveal a general convergence of practices on most elements of the analysis and interpretation process;" the other news is that only two groups identified the expected variants in all disease cases, "demonstrating a need for consistent fine-tuning of the generally accepted methods."[109]

POSTANALYTIC PHASE

Assay Review

All controls must have performed satisfactorily. If not, the director (or designee) must annotate the worksheet and document troubleshooting measures and whether or not the failed control precludes analysis.

Sanger sequencing software includes quality indicators for every base, but visual inspection of sequence chromatograms, especially where mutations are identified, is essential to avoid overcalls.

For NGS, there is no chromatogram to inspect. There are quality indicators, such as gaps in the targeted region, coverage per base, the ratio of forward to reverse reads, and the Q score per base. The laboratory should have a policy in place specifying the required parameters and actions to take if requirements are not met. Although there is no chromatogram, graphic viewers such as IGV should be used to examine any putative pathogenic variants to make sure they do not represent artifacts typically found at ends of reads or at edges of indels.

Interpretation and Reporting

Interpretation

Mutations that lead to frameshifts and/or termination have a higher likelihood of being clinically significant than variants that change the amino acid (missense), but exome studies have already shown that many people carry loss-of-function (LOF) mutations without obvious consequence. The significance of LOF variants might depend on the presence of other variants and the clinical context (age, gender, medications, environment).

Assessment of variants of uncertain significance or which are presumed deleterious by software analysis might be helped by family studies, which could permit correlation of a variant with the presence of a clinical or laboratory finding. Since family members share a large proportion of alleles, the larger the family tree and the more numerous the affected members, the more members tested and the more discriminating the analysis.

For clinical results that are to be returned by a physician or a genetics counselor, important considerations are making sure that the results as reported are intelligible and that time is available.[105,110] For whole exome sequencing or whole genome sequencing results, even after filtering, adequate discussion with the patient could take several hours.

Incidentalomas

For exomes and whole genomes, it occasionally happens that a variant, known to be pathogenic, is identified but is not relevant to the clinical inquiry leading to the assay. These are "incidentalomas." There is no generally accepted policy on how to handle these.[111] The laboratory should have a policy in place before starting testing. The test consent form is an important entry point to address this, as is discussion with the ordering clinician.

Reporting

A standard template is helpful.[112,113] A CAP resource committee publication provides a detailed template with explanatory comments. The report must meet CLIA requirements (42 CFR §493.1291). Required features include:

- Unambiguous identification of the subject and the specimen
- Description of assay methodology
- A list of the genes/regions targeted
- A list of any genes/regions unexpectedly not covered
- The analytical result

The description of the genetic variation found should follow internationally accepted rules of nomenclature[114]:

- Meaningful interpretation integrating clinical and family genetic information where appropriate
- Limitations of the assay (such as sensitivity)
- A suggestion for genetic counseling, where appropriate

For NGS reports the gene names should be included in the report. Coordinates of the regions targeted should be available through the laboratory but not necessarily in the report. Any regions which were not adequately covered and are relevant to the diagnostic consideration, which led to testing, should be mentioned in the report in a manner intelligible to nonexperts (who are not going to look up gene coordinates). Dorschner et al. give a detailed format and checklist for a whole exome report.[115]

An interpretive report should help guide the clinician with respect to the significance of the finding. Penetrance and expressivity of a given mutation can be highly variable. Discussion with the ordering clinician, if possible, is almost always helpful to all discussants; what might be "obvious" to someone immersed in genetics might be opaque to an otherwise expert clinician. Pathogenic results found in a research study, even with a consent form that waived the right to see the result, can present or create a dilemma.[116]

SUMMARY

By the third edition, it could come to pass that a patient's entire genomic variant profile, generated from a drop of blood by a hand-held device minutes earlier, will be reviewed by the endocrinologist on her optical head-mounted display as her autonomously piloted hovercar takes her to the next housecall. At present a full-service molecular genetics endocrinology laboratory can be set up with a small number of instruments. A minimum configuration will include a real-time PCR thermocycler, a capillary electrophoresis system (DNA sequencer), a second-generation DNA sequencer, DNA analysis software, computers and data storage, and subscription to some databases. Agreement among second-generation sequencing systems (platforms, analysis software, and interpretation) is somewhat lower than desired but the method allows unprecedented insight into genetic disease. Whether or not second-generation sequencing becomes widely used for genomes or exomes, it represents a cost-effective approach to screening even small panels of disease relevant genes (and the annotated variant report *can* be accessed by smartphone in your parked car). Regardless of the methodological sophistication, the most important step of the process is regular communication among clinicians, molecular geneticists, bioinformaticists, and genetic counselors.

References

1. Watterson B. *The indispensable Calvin and Hobbes: a Calvin and Hobbes treasury*. Kansas City: Andrews and McMeel; 1992.
2. Metzker ML. Sequencing technologies – the next generation. *Nat Rev Genet* 2010;**11**(1):31–46.
3. Kalyana-Sundaram S, Kumar-Sinha C, Shankar S, et al. Expressed pseudogenes in the transcriptional landscape of human cancers. *Cell* 2012;**149**(7):1622–34.
4. Maass PG, Rump A, Schulz H, et al. A misplaced lncRNA causes brachydactyly in humans. *J Clin Invest* 2012;**122**(11):3990–4002.
5. Gartner JJ, Parker SC, Prickett TD, et al. Whole-genome sequencing identifies a recurrent functional synonymous mutation in melanoma. *Proc Natl Acad Sci* 2013;**110**(33):13481–6.
6. Sterne-Weiler T, Sanford JR. Exon identity crisis: disease-causing mutations that disrupt the splicing code. *Gen Biol* 2014;**15**(1):201.
7. Hackett JA, Surani MA. Beyond DNA: programming and inheritance of parental methylomes. *Cell* 2013;**153**(4):737–9.
8. Galetzka D, Hansmann T, El Hajj N, et al. Monozygotic twins discordant for constitutive BRCA1 promoter methylation, childhood cancer and secondary cancer. *Epigenetics* 2012;**7**(1):47–54.
9. Goldstein DB, Allen A, Keebler J, et al. Sequencing studies in human genetics: design and interpretation. *Nat Rev Genet* 2014;**14**(7):460–70.
10. Stanley CM, Sunyaev SR, Greenblatt MS, Oetting WS. Clinically relevant variants – identifying, collecting, interpreting, and disseminating: the 2013 Annual Scientific Meeting of the Human Genome Variation Society. *Hum Mutat* 2015;**35**(4):505–10.
11. Richards CS, Palomaki GE, Lacbawan FL, Lyon E, Feldman GL. Three-year experience of a CAP/ACMG methods-based external proficiency testing program for laboratories offering DNA sequencing for rare inherited disorders. *Genet Med* 2014;**16**(1):25–32.
12. Ferreira-Gonzalez A, Emmadi R, Day SP, et al. Revisiting oversight and regulation of molecular-based laboratory-developed tests: a position statement of the Association for Molecular Pathology. *J Mol Diagn* 2014;**16**(1):3–6.
13. Sweeney L, Abu A, Winn J. Identifying participants in the personal genome project by name (a re-identification experiment). *arXiv preprint arXiv:13047605*; 2013.
14. Taitsman JK, Grimm CM, Agrawal S. Protecting patient privacy and data security. *N Engl J Med* 2013;**368**(11):977–9.
15. Hazin R, Brothers KB, Malin BA, et al. Ethical, legal, and social implications of incorporating genomic information into electronic health records. *Genet Med* 2013;**15**(10):810–6.
16. Schweitzer EJ. Reconciliation of the cloud computing model with US federal electronic health record regulations. *J Am Med Inform Assoc* 2012;**19**(2):161–5.
17. Cook-Deegan R. Are human genes patentable? *Ann Intern Med* 2013;**159**(4):298–9.
18. Rai AK, Cook-Deegan R. Moving beyond "isolated" gene patents. *Science (New York, NY)* 2013;**341**(6142).
19. CLSI. *Nucleic acid sequencing methods in diagnostic laboratory medicine; approved guideline second edition.* 2nd ed. Wayne, PA: Clinical and Laboratory Standards Institute; 2014.
20. Chen B, Richards CS, Wilson JA, Lyon E. Quality assurance and quality improvement in US clinical molecular genetic laboratories. *Curr Prot Hum Genet* 2011;9.2.1–9.2.26.
21. Gargis AS, Kalman L, Berry MW, et al. Assuring the quality of next-generation sequencing in clinical laboratory practice. *Nat Biotechnol* 2012;**30**(11):1033–6.
22. Lubin IM, Kalman L, Gargis AS. Guidelines and approaches to compliance with regulatory and clinical standards: quality control procedures and quality assurance. Next generation sequencing. Springer; 2013 255-273.
23. Rehm HL, Bale SJ, Bayrak-Toydemir P, et al. ACMG clinical laboratory standards for next-generation sequencing. *Genet Med* 2013;**15**(9):733–47.
24. Haverstick DM, Groszbach AR. Specimen collection and processing. In: Bruns DE, Ashwood ER, Burtis CA, editors. *Fundamentals of molecular diagnostics*. St. Louis: Saunders Elsevier; 2007. p. 25–38.
25. Farkas DH, Kaul KL, Wiedbrauk DL, Kiechle FL. Specimen collection and storage for diagnostic molecular pathology investigation. *Arch Pathol Lab Med* 1996;**120**(6):591–6.
26. Twiss P, Hill M, Daley R, Chitty LS. Non-invasive prenatal testing for Down syndrome. *Semin FetalNeonat Med* 2014;**19**(1):9–14.
27. Zhang R, Li X, Ramaswami G, et al. Quantifying RNA allelic ratios by microfluidic multiplex PCR and sequencing. *Nat Meth* 2014;**11**(1):51–4.
28. Extraction and precipitation of DNA. *Current Protocols in Human Genetics*. John Wiley & Sons, Inc; 2001.
29. Kubista M, Andrade JM, Bengtsson M, et al. The real-time polymerase chain reaction. *Mol Aspects Med* 2006;**27**(2–3):95–125.
30. Bock C. Analysing and interpreting DNA methylation data. *Nat Rev Genet* 2012;**13**(10):705–19.
31. Ma L, Chung WK. Quantitative analysis of copy number variants based on real-time LightCycler PCR. *Curr Prot Hum Genet* 2014;7.21.1–7.21.8.
32. Satterfield BC. Cooperative primers: 2.5 million-fold improvement in the reduction of nonspecific amplification. *J Mol Diagn* 2014;**16**(2):163–73.
33. Monis PT, Giglio S, Saint CP. Comparison of SYTO9 and SYBR Green I for real-time polymerase chain reaction and investigation

of the effect of dye concentration on amplification and DNA melting curve analysis. *Anal Biochem* 2005;**340**(1):24–34.

34. Zhou L, Myers AN, Vandersteen JG, Wang L, Wittwer CT. Closed-tube genotyping with unlabeled oligonucleotide probes and a saturating DNA dye. *Clin Chem* 2004;**50**(8):1328–35.

35. Hindson BJ, Ness KD, Masquelier DA, et al. High-throughput droplet digital PCR system for absolute quantitation of DNA copy number. *Anal Chem* 2011;**83**(22):8604–10.

36. McDermott GP, Do D, Litterst CM, et al. Multiplexed target detection using DNA-binding dye chemistry in droplet digital PCR. *Anal Chem* 2013;**85**(23):11619–27.

37. Day E, Dear PH, McCaughan F. Digital PCR strategies in the development and analysis of molecular biomarkers for personalized medicine. *Methods* 2013;**59**(1):101–7.

38. de Ligt J, Boone PM, Pfundt R, et al. Detection of clinically relevant copy number variants with whole-exome sequencing. *Hum Mutat* 2013;**34**(10):1439–48.

39. Azzi S, Habib WA, Netchine I. Beckwith-Wiedemann and Russell-Silver syndromes: from new molecular insights to the comprehension of imprinting regulation. *Curr Opin Endocrinol, Diabetes Obesity* 2014;**21**(1):30–8.

40. Schaaf CP, Wiszniewska J, Beaudet AL. Copy number and SNP arrays in clinical diagnostics. *Annu Rev Genom Hum Genet* 2011;**12**:25–51.

41. Heyn H, Esteller M. DNA methylation profiling in the clinic: applications and challenges. *Nat Rev Genet* 2012;**13**(10):679–92.

42. Aw DK, Sinha RA, Tan HC, Loh LM, Salvatore D, Yen PM. Studies of molecular mechanisms associated with increased deiodinase 3 expression in a case of consumptive hypothyroidism. *J Clin Endocrinol Metab* 2014;**99**(11):3965–71.

43. Docherty LE, Rezwan FI, Poole RL, et al. Genome-wide DNA methylation analysis of patients with imprinting disorders identifies differentially methylated regions associated with novel candidate imprinted genes. *J Med Genet* 2014;**51**(4):229–38.

44. Elli FM, de Sanctis L, Peverelli E, et al. Autosomal dominant pseudohypoparathyroidism type Ib: a novel inherited deletion ablating STX16 causes loss of imprinting at the A/B DMR. *J Clin Endocrinol Metab* 2014;**99**(4):E724–8.

45. Kalish JM, Conlin LK, Mostoufi-Moab S, et al. Bilateral pheochromocytomas, hemihyperplasia, and subtle somatic mosaicism: the importance of detecting low-level uniparental disomy. *Am J Med Genet* 2013;**161**(5):993–1001.

46. Kalish JM, Conlin LK, Bhatti TR, et al. Clinical features of three girls with mosaic genome wide paternal uniparental isodisomy. *Am J Med Genet: Part A* 2013;**161**(8):1929–39.

47. Ball ST, Kelly ML, Robson JE, et al. Gene dosage effects at the imprinted GNAS cluster. *PLoS One* 2013;**8**(6):e65639.

48. Charalambous M, Hernandez A. Genomic imprinting of the type 3 thyroid hormone deiodinase gene: regulation and developmental implications. *Biochim Biophys Acta (BBA) – General Subjects* 2013;**1830**(7):3946–55.

49. Scholl UI, Goh G, Stolting G, et al. Somatic and germline CACNA1D calcium channel mutations in aldosterone-producing adenomas and primary aldosteronism. *Nat Genet* 2013;**45**(9):1050–4.

50. Letouze E, Martinelli C, Loriot C, et al. SDH mutations establish a hypermethylator phenotype in paraganglioma. *Cancer Cell* 2013;**23**(6):739–52.

51. Kristensen LS, Treppendahl MB, Grønbæk K. Analysis of epigenetic modifications of DNA in human cells. Current protocols in human genetics. John Wiley & Sons, Inc; 2013.

52. Song CX, Yi C, He C. Mapping recently identified nucleotide variants in the genome and transcriptome. *Nat Biotechnol* 2012;**30**(11):1107–16.

53. Yu M, Hon GC, Szulwach KE, et al. Tet-assisted bisulfite sequencing of 5-hydroxymethylcytosine. *Nat Prot* 2012;**7**(12):2159–70.

54. Booth MJ, Marsico G, Bachman M, Beraldi D, Balasubramanian S. Quantitative sequencing of 5-formylcytosine in DNA at single-base resolution. *Nat Chem* 2014;**6**(5):435–40.

55. Robertson AB, Dahl JA, Ougland R, Klungland A. Pull-down of 5-hydroxymethylcytosine DNA using JBP1-coated magnetic beads. *Nat Prot* 2012;**7**(2):340–50.

56. Flusberg BA, Webster DR, Lee JH, et al. Direct detection of DNA methylation during single-molecule, real-time sequencing. *Nat Meth* 2010;**7**(6):461–5.

57. Roberts RJ, Carneiro MO, Schatz MC. The advantages of SMRT sequencing. *Genome Biol* 2013;**14**:405.

58. Bosserhoff A, Hellerbrand C. Capillary Electrophoresis. Molecular diagnostics. Burlington: Elsevier Academic; 2005 p. 67–81.

59. O'Huallachain M, Karczewski KJ, Weissman SM, Urban AE, Snyder MP. Extensive genetic variation in somatic human tissues. *Proc Natl Acad Sci USA* 2012;**109**(44):18018–21823.

60. Biesecker LG, Spinner NB. A genomic view of mosaicism and human disease. *Nat Rev Genet* 2013;**14**(5):307–20.

61. Mamanasiri S, Yesil S, Dumitrescu AM, et al. Mosaicism of a thyroid hormone receptor-beta gene mutation in resistance to thyroid hormone. *J Clin Endocrinol Metab* 2006;**91**(9):3471–7.

62. Dutta RK, Welander J, Brauckhoff M, et al. Complementary somatic mutations of KCNJ5, ATP1A1, and ATP2B3 in sporadic aldosterone producing adrenal adenomas. *Endocr Related Cancer* 2013;**21**(1):L1–4.

63. Strom SP, Lee H, Das K, et al. Assessing the necessity of confirmatory testing for exome-sequencing results in a clinical molecular diagnostic laboratory. *Genet Med* 2014;**16**(7):510–5.

64. Li X, Buckton AJ, Wilkinson SL, et al. Towards clinical molecular diagnosis of inherited cardiac conditions: a comparison of benchtop genome DNA sequencers. *PLoS One* 2013;**8**(7):e67744.

65. Junemann S, Sedlazeck FJ, Prior K, et al. Updating benchtop sequencing performance comparison. *Nat Biotechnol* 2013;**31**(4):294–6.

66. Loman NJ, Misra RV, Dallman TJ, et al. Performance comparison of benchtop high-throughput sequencing platforms. *Nat Biotechnol* 2012;**30**(5):434–9.

67. Quail MA, Smith M, Coupland P, et al. A tale of three next generation sequencing platforms: comparison of Ion Torrent, Pacific Biosciences and Illumina MiSeq sequencers. *BMC Genomics* 2012;**13**:341.

68. Ratan A, Miller W, Guillory J, Stinson J, Seshagiri S, Schuster SC. Comparison of sequencing platforms for single nucleotide variant calls in a human sample. *PLoS One* 2013;**8**(2):e55089.

69. Boland JF, Chung CC, Roberson D, et al. The new sequencer on the block: comparison of Life Technology's Proton sequencer to an Illumina HiSeq for whole-exome sequencing. *Hum Genet* 2013;**132**(10):1153–63.

70. Ma Z, Lee RW, Li B, et al. Isothermal amplification method for next-generation sequencing. *Proc Natl Acad Sci* 2013;**110**(35):14320–3.

71. Eid J, Fehr A, Gray J, et al. Real-time DNA sequencing from single polymerase molecules. *Science* 2009;**323**(5910):133–8.

72. Haque F, Li J, Wu H-C, Liang X-J, Guo P. Solid-state and biological nanopore for real-time sensing of single chemical and sequencing of DNA. *Nano Today* 2013;**8**(1):56–74.

73. Mankos M, Shadman K, N'Diaye AT, Schmid AK, Persson HH, Davis RW. Progress toward an aberration-corrected low energy electron microscope for DNA sequencing and surface analysis. *J Vac Sci Technol B* 2012;**30**(6):6F402.

74. Boynton P, Balatsky A, Schuller I, Di Ventra M. Improving sequencing by tunneling with multiplexing and cross-correlations. *arXiv preprint arXiv:14017363*; 2014.

75. Blue Shield Association. Special report: exome sequencing for clinical diagnosis of patients with suspected genetic disorders. *Technology evaluation center assessment program executive summary*, vol. 28(3); 2013. p. 1.

76. Kaufman KM. The struggle to find reliable results in exome sequencing data: filtering out Mendelian errors. *Front Genet* 2014;**5**:16.

77. Koboldt DC, Larson DE, Sullivan LS, et al. Exome-based mapping and variant prioritization for inherited Mendelian disorders. *Am J Hum Genet* 2014;**94**(3):373–84.

78. Lohmueller KE, Sparsø T, Li Q, et al. Whole-exome sequencing of 2,000 Danish individuals and the role of rare coding variants in type 2 diabetes. *Am J Hum Genet* 2013;**93**(6):1072–86.

79. Patel ZH, Kottyan LC, Lazaro S, et al. The struggle to find reliable results in exome sequencing data: filtering out Mendelian errors. *Front Genet* 2014;**5**:16.

80. Sirmaci A, Edwards YJ, Akay H, Tekin M. Challenges in whole exome sequencing: an example from hereditary deafness. *PLoS One* 2012;**7**(2):e32000.

81. Beuschlein F, Fassnacht M, Assié G, et al. Constitutive activation of PKA catalytic subunit in adrenal Cushing's syndrome. *N Engl J Med* 2014;**370**(11):1019–28.

82. Caruccio N. Preparation of next-generation sequencing libraries using Nextera™ technology: simultaneous DNA fragmentation and adaptor tagging by *in vitro* transposition. High-throughput next generation sequencing. Springer; 2011 241-255.

83. Poptsova MS, Il'icheva IA, Nechipurenko DY, et al. Non-random DNA fragmentation in next-generation sequencing. *Sci Rep* 2014;**4**: 4532.

84. Dabney J, Meyer M. Length and GC-biases during sequencing library amplification: a comparison of various polymerase-buffer systems with ancient and modern DNA sequencing libraries. *Biotechniques* 2012;**52**(2):87–94.

85. Casbon JA, Osborne RJ, Brenner S, Lichtenstein CP. A method for counting PCR template molecules with application to next-generation sequencing. *Nucleic Acids Res* 2011;**39**(12):e81.

86. Zhou W, Chen T, Zhao H, et al. Bias from removing read duplication in ultra-deep sequencing experiments. *Bioinformatics* 2014.

87. Head SR, Komori HK, LaMere SA, et al. Library construction for next-generation sequencing: Overviews and challenges. *BioTechniques* 2013;**56**(2):61–77.

88. van Dijk EL, Jaszczyszyn Y, Thermes C. Library preparation methods for next-generation sequencing – tone down the bias. *Exp Cell Res* 2014;**322**(1):12–20.

89. Altmann A, Weber P, Bader D, Preuß M, Binder EB, Müller-Myhsok B. A beginners guide to SNP calling from high-throughput DNA-sequencing data. *Hum Genet* 2012;**131**(10):1541–54.

90. Pabinger S, Dander A, Fischer M, et al. A survey of tools for variant analysis of next-generation genome sequencing data. *Briefings Bioinform* 2014;**15**(2):256–78.

91. Neves M, Leser U. A survey on annotation tools for the biomedical literature. *Briefings Bioinform* 2012; bbs084.

92. Ramos EM, Din-Lovinescu C, Berg JS, et al. Characterizing genetic variants for clinical action. *Am J Med Genet* 2014;**166C**(1): 93–104.

93. Xi R, Lee S, Park PJ. A survey of copy-number variation detection tools based on high-throughput sequencing data. *Curr Prot Hum Genet* 2012;7.19.1–5.

94. Teo SM, Pawitan Y, Ku CS, Chia KS, Salim A. Statistical challenges associated with detecting copy number variations with next-generation sequencing. *Bioinformatics* 2012;**28**(21):2711–8.

95. Albers CA, Lunter G, MacArthur DG, McVean G, Ouwehand WH, Durbin R. Dindel: accurate indel calls from short-read data. *Genome Res* 2011;**21**(6):961–73.

96. Sims D, Sudbery I, Ilott NE, Heger A, Ponting CP. Sequencing depth and coverage: key considerations in genomic analyses. *Nat Rev Genet* 2014;**15**(2):121–32.

97. Zook JM, Chapman B, Wang J, et al. Integrating human sequence data sets provides a resource of benchmark SNP and indel genotype calls. *Nat Biotech* 2014;**32**(3):246–51.

98. Kalman LV, Lubin IM, Barker S, et al. Current landscape and new paradigms of proficiency testing and external quality assessment for molecular genetics. *Arch Pathol Lab Med* 2013;**137**(7):983–8.

99. Fischer M, Snajder R, Pabinger S, et al. SIMPLEX: cloud-enabled pipeline for the comprehensive analysis of exome sequencing data. *PLoS One* 2012;**7**(8):e41948.

100. Scholtalbers J, Rößler J, Sorn P, et al. Galaxy LIMS for next-generation sequencing. *Bioinformatics* 2013;**29**(9):1233–4.

101. Dander A, Pabinger S, Sperk M, Fischer M, Stocker G, Trajanoski Z. SeqBench: integrated solution for the management and analysis of exome sequencing data. *BMC Research Notes* 2014;**7**(1):43.

102. Worthey EA. Analysis and annotation of whole-genome or whole-exome sequencing-derived variants for clinical diagnosis. Current protocols in human genetics. John Wiley & Sons, Inc; 2013.

103. Landrum MJ, Lee JM, Riley GR, et al. ClinVar: public archive of relationships among sequence variation and human phenotype. *Nucleic Acids Res* 2014;**42**(D1):D980–5.

104. Stenson PD, Mort M, Ball EV, Shaw K, Phillips AD, Cooper DN. The Human Gene Mutation Database: building a comprehensive mutation repository for clinical and molecular genetics, diagnostic testing and personalized genomic medicine. *Hum Genet* 2014;**133**(1):1–9.

105. Berg JS, Khoury MJ, Evans JP. Deploying whole genome sequencing in clinical practice and public health: meeting the challenge one bin at a time. *Genet Med* 2011;**13**(6):499–504.

106. Richards CS, Bale S, Bellissimo DB, et al. ACMG recommendations for standards for interpretation and reporting of sequence variations: revisions 2007. *Genet Med* 2007;**10**(4):294–300.

107. Duzkale H, Shen J, McLaughlin H, et al. A systematic approach to assessing the clinical significance of genetic variants. *Clin Genet* 2013;**84**(5):453–63.

108. Kircher M, Witten DM, Jain P, O'Roak BJ, Cooper GM, Shendure J. A general framework for estimating the relative pathogenicity of human genetic variants. *Nat Genet* 2014;**46**(3):310–5.

109. Brownstein CA, Beggs AH, Homer N, et al. An international effort towards developing standards for best practices in analysis, interpretation and reporting of clinical genome sequencing results in the CLARITY Challenge. *Genome Biol* 2014;**15**(3):R53.

110. Scheuner MT, Edelen MO, Hilborne LH, Lubin IM. Effective communication of molecular genetic test results to primary care providers. *Genet Med* 2012;**15**(6):444–9.

111. Green RC, Berg JS, Grody WW, et al. ACMG recommendations for reporting of incidental findings in clinical exome and genome sequencing. *Genet Med* 2013;**15**(7):565–74.

112. Gulley ML, Braziel RM, Halling KC, et al. Clinical laboratory reports in molecular pathology. *Arch Pathol Lab Med* 2007;**131**(6): 852–63.

113. Powsner SM, Costa J, Homer RJ. Clinicians are from Mars and pathologists are from Venus. *Arch Pathol Lab Med* 2000;**124**(7):1040–6.

114. Ogino S, Gulley ML, den Dunnen JT, Wilson RB. Standard mutation nomenclature in molecular diagnostics: practical and educational challenges. *J Mol Diagn* 2007;**9**(1):1–6.

115. Dorschner MO, Amendola LM, Shirts BH, et al. Refining the structure and content of clinical genomic reports. *Am J Med Genet* 2014;**166C**(1):85–92.

116. Burke W, Evans BJ, Jarvik GP. Return of results: ethical and legal distinctions between research and clinical care. *Am J Med Genet* 2014;**166**(1):105–11.

30

Introduction to Applications of Genomic Sequencing

Stephan Zuchner

Department of Human Genetics and Neurology, Hussman Institute for Human Genomics,
University of Miami Miller School of Medicine, Miami, FL, USA

OVERVIEW

Since completion of the Human Genome Project in 2001,[1] dynamics in human genetics have greatly accelerated. Sequencing a human genome used to be a multimillion dollar project but is now reduced to less than $2,000 and increasingly becomes routine in research and clinical applications. Disease gene discovery has greatly accelerated and the importance of rare and "private" alleles is evident, which can only be identified by sequencing.[2] Clinical genetic testing is currently going through radical changes, leading to comprehensive, yet still more affordable tests. The optimism is such that the National Institutes of Health (NIH) and other organizations have targeted the identification of possibly all Mendelian disease genes in the coming decade.[3] New challenges have come into focus, including managing very large amounts of data, translation into clinical electronic record systems, genomic data sharing, data security, and others. Regardless, the enthusiasm resulting from decoding the entire human genome is beginning to play out in applications for research and patient care. This chapter will introduce a few key aspects of this, still new, age of genomics.

FROM HUMAN GENETICS TO GENOMICS

While the terms genetics and genomics are often used interchangeably, they have different meanings. Genetics refers to the science of inheritance, such as Mendelian laws of inheritance, and is generally more focused on chromosomal regions and specific genes. Genomics plays out on the scale of entire genomes and is equally important for classic inheritance traits and so-called complex disorders.[4,5] The relatively new technical applications of genomics include whole genome sequence, exome

sequencing, and genome-wide association studies. Before the availability of whole genome sequencing techniques, geneticists had to invent ingenious methods of mapping and linking smaller portions of the genome to a phenotype of interest.[6,7] For Mendelian phenotypes, the most successful approach was linkage studies. In brief, several hundred marker alleles are measured across a genome in single or multiple families. A genetic marker allele that cosegregates with the phenotype of interest should be in linkage disequilibrium with the actual disease-causing mutation. If several consecutive markers cosegregate, we speak of a haplotype. This can be statistically formalized and expressed with a LOD score.[8,9] Once established, a linked genomic region may be followed up in great detail by focused DNA sequencing studies. Until 2009, this was the only broadly applicable method of identifying novel disease genes and had enormous successes. However, the need for large families and homogeneous phenotypes contributed to the limits of this approach. In 2009, the new method of whole exome sequencing was broadly introduced (Fig. 30.1).[10]

Whole exome sequencing allows sequencing of most of the nucleotides of nearly all protein coding genes in a single individual.[11] With this single base pair resolution of measurement, the actual disease-causing variant, instead of a nearby marker allele, becomes the very target of the search.[12] A formal linkage analysis is not necessary anymore, but may still be performed as supporting evidence. A typical exome yields approximately 20,000 coding changes, with 10,000 being nonsynonymous changes altering protein sequence. Now the task becomes a filtering effort to zoom into the most likely candidate variant and gene. Popular filtering approaches include allele frequency (only very rare or truly novel alleles might be considered), conservation across species (changes of highly conserved nucleotides are more likely to have a

Genetic Diagnosis of Endocrine Disorders. http://dx.doi.org/10.1016/B978-0-12-800892-8.00030-0

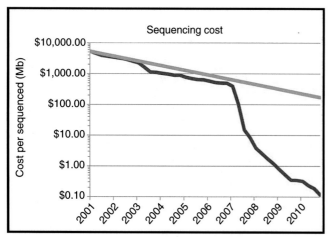

FIGURE 30.1 **The cost for DNA sequencing (lower curve) has dropped in an unprecedented fashion and will soon allow for the routine production of <$1000 whole genomes.** *Source: NHGRI.*

biological effect), or segregation studies according to the expected traits (a much smaller number of coding homozygous variants exist in a human genome).[13] Many more considerations may be used and typical annotation software packages provide dozens of annotation criteria for each single nucleotide variant.

An example of an innovative approach to the problem of sporadic patients with a suspected dominant trait is the search for *de novo* changes. *De novo* mutations arise in every meiosis at the order of less than a dozen variants per genome.[14,15] The likelihood that such a variant falls on a functionally highly relevant position is very small.[16,17] By exome sequencing both parents and an offspring with a phenotype (trio-sequencing), one can explore *de novo* variation systematically and at scale. While it can be daunting to sort out sequencing errors from the exceptionally rare *de novo* hits, recent studies in autism have been especially successful in applying this approach.[18]

To further improve these approaches, new *in silico* scores have been developed. Such scores predict consequences of genetic variation on protein function. Examples are SIFT,[19] MutationTaster,[20] and MutationAssessor.[21] In addition, statistical considerations may go into filtering approaches[22] and include new approaches such as the CADD[23] and VAAST score.[24]

As sequencing cost continues to decrease, whole genome sequencing will become the next big wave of exploring individual genomes in 2015 and thereafter. This will lead to the clarification as to what extent noncoding variation in known (and new) disease genes contributes to strong genetic effects. Today, noncoding variation is rarely tested in clinical settings because of limited understanding of the significance. In addition, whole genome sequencing will identify a very large number of small insertions/deletions. As of now, there is no comprehensive

knowledge on indels larger than ~50 bp and smaller than 1,000 bp in the population and in patients in particular. Yet, estimates suggest that this is the most abundant class of structural changes.[25,26]

To sift through the large number of variations that is detected by any larger-scale genome sequencing study, statistical approaches have been developed. The most successful ones still focus on genes and the protein coding regions. The general idea is to compare the "mutational load" in a given gene in a control sample to a set of cases. These "burden tests" also aim to reduce the background noise of neutral variation by prefiltering variation according to the same conservation and protein prediction algorithms mentioned above. While some studies were successful in identifying strong signals with these methods, it appears that very large sample sizes will be necessary to fully detect genetic signals.[27,28]

The limits of all these approaches lie in statistical power considerations (sample size), nongenetic contributors in the environment, and also epigenetic factors that are not measured by DNA sequencing. The combination of multiple different "genomic" datasets (whole genome sequencing, RNA sequencing, histone sequencing, proteomics, etc.) will therefore be a promising way forward.[29]

EXCESS OF RARE VARIATION IN THE HUMAN GENOME

Large-scale exome sequencing studies have shown that the human genome, on a population level, contains an enormous number of rare variants. "Rare variants" are not strictly defined, but refer to DNA changes that occur at levels from <1% allele frequency to changes identified only in a single person ("private changes"); such frequency assessments may be confounded by ancestry as some alleles are only rare or frequent in specific populations. While "rare variants" are infrequent individually, they actually provide by far the largest group of DNA variation. In a sample of 6500 exomes of European and African ancestry, the protein-coding regions contained a change every 21 nucleotides.[2] Approximately 86% of all single nucleotide variants in this study were measured in less than 0.5% of alleles and 82% were population specific (European/African ancestry).[2] Estimates have been published that predict that in 1 million exomes, every other coding nucleotide will be a variant compared to the NCBI reference in at least one exome of such a sample.[30] Given that the human species today encompasses more than 7 billion genomes (>14 billion copies of each chromosome), it is safe to assume that every nucleotide will be changed in at least one individual as long as its functional impact is not detrimental to life. This "excess" of rare variation in the human genome represents a

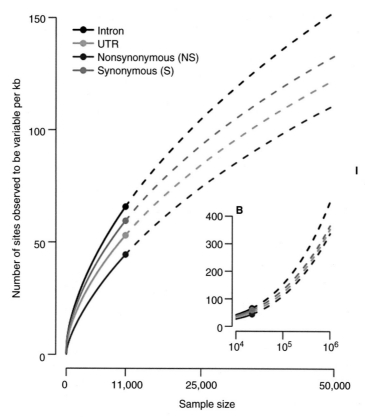

FIGURE 30.2 Whole exome and genome studies have precisely measured the amount of variation in the human genome in a population. To some surprise, rare variants are presenting in larger than expected numbers and represent by far the most common type of variation.[32]

rich opportunity to discover disease-associated alleles. On the other hand, it presents a high level of "noise" in large-scale statistical studies of rare variation. Pure statistical approaches will require very large numbers of participants and may still not be able to discern more rare causes of human disease.[29] It will therefore be important to develop innovative approaches to disease allele discovery. Combining different techniques to gather evidence from genetic, expression, *in silico* pathway modeling, and functional genomics studies is likely the most immediate way forward. More audacious concepts have proposed the large-scale functional mutational analysis in cell-based assays.[31] Enabled by more effective gene editing via CRISPR/Cas9, such assays may explore a large part of the theoretical "mutational space" of the human genome (Fig. 30.2).[31]

DATA SHARING BECOMES ESSENTIAL

Genome sequence data is increasingly recognized as a value beyond individual studies. Individual genomes may be used across different studies as controls or to search for broader phenotypic expressions of disease genes. Further, as rare variation is becoming the driving force for clinical applications, more genomes from

different ancestral backgrounds are available for comparison to specific alleles. The idea of data sharing on a global level is quickly catching up. NIH has begun to enforce DNA sequence data to be deposited on dbGAP.[33] The ClinVar[34] and HumanVariome[35] initiatives are collecting clinically relevant variation at a large scale and in a transparent fashion. The latter two initiatives place a strong focus on connecting phenotypes to alleles.

Researchers working on the identification of novel disease genes often do not have access to the necessary number of patients and extended families. With increasingly only small families available and new genes generally being very rare, the problem of finding "a second pedigree" with supporting evidence for a new and rare disease gene is increasingly recognized and has spurred several initiatives to address this issue. The underlying concept is that a rare disease gene still needs significant genetic evidence and should ideally be supported by more than one pedigree. Otherwise, the literature will soon be dominated by single patient/single-family gene reports of uncertain significance.

Over the past four years, the Zuchner lab at the University of Miami has developed a tool that is designed around data sharing.[36] This Genome Management Application (GENESIS/GEM.app) holds all variant data in a cloud-computing environment. This allows for worldwide access via any web browser, any internet-connected device, and response times comparable to a Google search. Individual investigators or large consortia are able to deposit genome-level data. The uploading investigator/organization stays in full control of their data; yet, easy sharing controls allow data access for other investigators with a few mouse clicks. The data never leave the database, making it possible also to revoke access rights and to create truly *ad hoc* collaborations. With now over 6,000 exomes and genomes from well-defined patients with over 100 different diseases, more than 500 researchers from 38 countries have already registered. This system has allowed for the discovery of over 60 novel disease genes in the past 30 months. The flexibility and ease of use offers a complementary approach to the often more rigid rules of formal consortia.

Since a small but growing number of tools and variant databases exist, the need for data searches across these resources grows. A new initiative supported by the Global Alliance for Genomics and Health is the Genomic Matchmaker Exchange (http://matchmakerexchange .org). This effort is creating a standard computer interface (API) that will allow for querying across databases in a secure fashion. Every database that contains phenotype or genotype information will be able to implement this API and become part of a global virtual genomics network (Fig. 30.3).

PNPLA6 GENE IDENTIFICATION – AN EXAMPLE FOR A NUMBER OF TRENDS IN GENOMICS

With sequencing becoming more affordable and also clinical sequencing now broadly available via gene panels a number of trends begin to emerge:

- The same gene may cause pure phenotypes, but also more complex syndromes.
- A broad phenotypic spectrum associated with a given gene may be the rule rather than the exception.
- Many disease phenotypes are impressively genetically heterogeneous.

A recent identification of the gene *PNPLA6* will serve as a useful example illustrating these observations. PNPLA6 related endocrine dysfunction includes the hypophysis and pituitary gland leading to complex phenotypic expressions. Originally, PNPLA6 was identified as underlying hereditary spastic paraplegia, a disease that leads to neurodegeneration of the first motoneuron.[37] More recently, it was reported that PNPLA6 does underlie a complex and very rare syndrome, Boucher–Neuhäuser syndrome (BNS).[38] BNS includes a classic triad of ataxia, hypogonadotropic hypogonadism, and chorioretinal dystrophy. This discovery was possible via

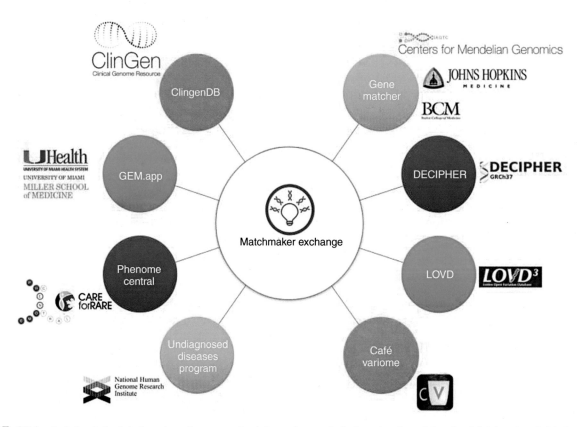

FIGURE 30.3 A federated global system for genomic data exchange is being developed by the Matchmaker initiative (**http:// matchmakerexchange.org**).

the efficient sharing of data on rare families and patients via GENESIS/GEM.app (see Conclusions). After the initial discovery, the data sharing abilities in this study connected investigators from Germany, Brazil, Venezuela, France, and the United States within days and without further sequencing required. It became then also clear that additional phenotypic elements may be included and thus some families were defined with alternative genetic syndrome names, such as Gordon Holmes syndrome (GHS), Oliver–McFarlane syndrome (OMS), and Laurence–Moon (LMS) syndrome.[39] It is now apparent that hereditary spastic paraplegia represents only a relatively special case along a multidimensional PNPLA6-associated spectrum of neurodegenerative/neuroendocrine disorders. This spectrum includes at least four clinical key features: ataxia, motor neuron disease (upper motor neuron disease with or without additional lower motor neuropathy), hypogonadism, and chorioretinal dystrophy (Fig. 30.4). More recently, it was shown that *PNPLA6* is also a major gene for a broader range of retinal defects.[40] Although these clinical features appear to be frequent in PNPLA6 disease, none of them is an obligate feature of the disease. In effect, multiple neuronal populations may be affected, and the extent of this in an individual patient will determine the exact phenotypic presentation. It will be very important to understand further what the determining factors are for this individual variability in affecting specific neuronal circuits. Candidates include mutation-specific effects, genetic modifiers, epigenetic modifiers, and environmental influences. Regardless, research into this topic will potentially identify novel strategies for medical diagnosis, patient management, and drug therapy.

PNPLA6 is only one example of a growing number of disease-related proteins that have been shown to cause a spectrum or even a multitude of seemingly unrelated phenotypic expression.[41,42] Driven by the more comprehensive screening capabilities of next-generation sequencing-based approaches, the near future will likely uncover a more complex structure of phenotype-genotype correlations. On a functional level, proteins like PNPLA6, on which different phenotypes converge, may represent functional platforms or hubs that connect a diversity of pathways with multiple biological functions. These "hub proteins" and intersecting nodes in phenotypic as well as cellular disease networks might offer promising opportunities for therapeutic intervention for a number of diseases.[43,44]

CONCLUSIONS

Taken together, the science of studying entire genomes at scale is very young and will certainly lead to very important discoveries at a rapid pace in the coming decade. Much to the surprise of many in the field, relatively simple "filtering" approaches can illuminate key features of individual exomes and their clinical relevance.[45] It will be most interesting to see how larger-scale whole genome sequencing will add to the complexities of interpreting the sheer number of variations present in every genome. This question will be answered in the very near future, as the $1000 genome effectively has arrived in 2015. These studies represent the necessary groundwork to eventually enable precision medicine.[46,47]

As the "low hanging fruit" of genome discoveries are quickly harvested, the need for collaboration is growing. Teams are increasingly combining their efforts to achieve larger sample sizes and to capture many ancestral backgrounds. Importantly, rare disease gene discoveries will need true global collaboration, as each gene may explain only a small number of families and patients in the discovery phase. Such collaborations do not necessarily have to be in the context of powerful, yet relatively inflexible consortia, but may play out in a new set of tools, such as GENESIS/GEM.app, which allow for ad hoc and instant data sharing.[36] It will be important to build infrastructures that allow for a dynamic interconnecting of the most creative investigators. It is far from clear which surprises are awaiting us by exploring our genomes in deeper detail, and investigator-driven research along with large-scale accessible resources produced by scientific consortia will allow for continued rapid progress.

Many recent discoveries were possible within months instead of years because of the existence of large exome databases.[48] It is newly appreciated how

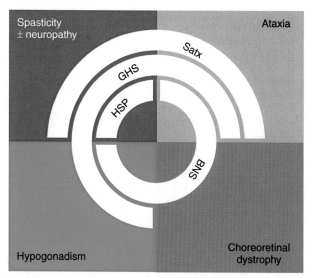

FIGURE 30.4 *PNPLA6* **disease spectrum is broad and caused by the variable degeneration of different neuronal populations, including hypophysis and pituitary gland.** This explains the broad and variable clinical picture.

importantly phenotypic details may influence our ability to make genetic discoveries. To make precision medicine a reality, we need to be able to predict not just the disease gene, but also the phenotypic consequences of each individual allele. This will require innovative biological assay development. Such assays should be able to screen larger portions of the genome in an automated fashion.

Because of broad genetic testing now increasingly available in clinics, the wider phenotypes associated with genetic mutations are becoming more apparent. This often causes situations of unknown pathogenic significance of genetic testing results. Thus, the drastically increased ability of genetic tests have been a big advance, but also created new challenges in data interpretation.[49]

The presented issues of genomic research and medicine hold true for all areas of medicine, from internal medicine and endocrinology to cancer and neurodegeneration. The work in this area in the coming decade will change our chance of success in finding truly novel therapies at a much accelerated pace.

References

1. Lander ES, et al. Initial sequencing and analysis of the human genome. *Nature* 2001;**409**:860–921.
2. Tennessen JA, et al. Evolution and functional impact of rare coding variation from deep sequencing of human exomes. *Science* 2012;**337**:64–9.
3. Bamshad MJ, et al. The Centers for Mendelian Genomics: a new large-scaleinitiative to identify the genes underlying rare Mendelian conditions. *Am J Med Genet* 2012;**158A**:1523–5.
4. Manolio TA, et al. Finding the missing heritability of complex diseases. *Nature* 2009;**461**:747–53.
5. Lupski JR, Belmont JW, Boerwinkle E, Gibbs RA. Clan genomics and the complex architecture of human disease. *Cell* 2011;**147**:32–43.
6. Lander ES, Botstein D. Strategies for studying heterogeneous genetic traits in humans by using a linkage map of restriction fragment length polymorphisms. *Proc Natl Acad Sci USA* 1986;**83**:7353–7.
7. Doniskeller H. A genetic linkage map of the human genome. *Cell* 1987;**51**:319–37.
8. Ott J. A computer program for linkage analysis of general human pedigrees. *Am J Hum Genet* 1976;**28**:528–9.
9. Ott J. Estimation of the recombination fraction in human pedigrees: efficient computation of the likelihood for human linkage studies. *Am J Hum Genet* 1974;**26**:588–97.
10. Ng SB, et al. Targeted capture and massively parallel sequencing of 12 human exomes. *Nature* 2009;**461**:272–6.
11. Hedges DJ, et al. Exome sequencing of a multigenerational human pedigree. *PLoS One* 2009;**4**:e8232.
12. Ng SB, et al. Exome sequencing identifies the cause of a Mendelian disorder. *Nat Genet* 2010;**42**:30–5.
13. Montenegro G, et al. Exome sequencing allows for rapid gene identification in a Charcot-Marie-Tooth family. *Ann Neurol* 2011;**69**:464–70.
14. Veltman JA, Brunner HG. De novo mutations in human genetic disease. *Nat Rev Genet* 2012;**13**:565–75.
15. Sanders SJ, Murtha MT, Gupta AR, Murdoch JD. De novo mutations revealed by whole-exome sequencing are strongly associated with autism. *Nature* 2012;**485**:237–41.
16. Kong A, et al. Rate of de novo mutations and the importance of father's age to disease risk. *Nature* 2012;**488**:471–5.
17. Vissers L, de Ligt J, Gilissen C, Janssen I. A de novo paradigm for mental retardation. *Nature* 2010;**42**:1109–12.
18. O'Roak BJ, et al. Exome sequencing in sporadic autism spectrum disorders identifies severe de novo mutations. *Nat Genet* 2011;**43**:585–9.
19. Ng PC, Henikoff S. Predicting deleterious amino acid substitutions. *Genome Res* 2001;**11**:863–74.
20. Schwarz JM, Rödelsperger C, Schuelke M, Seelow D. MutationTaster evaluates disease-causing potential of sequence alterations. *Nat Methods* 2010;**7**:575–6.
21. Reva B, Antipin Y, Sander C. Predicting the functional impact of protein mutations: application to cancer genomics. *Nucleic Acids Res* 2011;**39**:e118–e1118.
22. MacArthur DG, et al. Guidelines for investigating causality of sequence variants in human disease. *Nature* 2014;**508**:469–76.
23. Shihab HA, et al. An integrative approach to predicting the functional effects of non-coding and coding sequence variation. *Bioinformatics* 2015;**31**:1536–43.
24. Hu H, et al. VAAST 2.0: Improved variant classification and disease-gene identification using a conservation-controlledamino acid substitution matrix. *Genet Epidemiol* 2013;**37**:622–34.
25. Pang AW, MacDonald JR, Pinto D, Wei J. Towards a comprehensive structural variation map of an individual human genome. *Genome Biol* 2010;**11**(5):R52.
26. Wheeler DA, et al. The complete genome of an individual by massively parallel DNA sequencing. *Nature* 2008;**452**:872–6.
27. Cirulli ET, Goldstein DB. Uncovering the roles of rare variants in common disease through whole-genome sequencing. *Nat Rev Genet* 2010;**11**:415–25.
28. Bansal V, Libiger O, Torkamani A, Schork NJ. Statistical analysis strategies for association studies involving rare variants. *Nat Rev Genet* 2010;**11**:773–85.
29. Lander ES. Initial impact of the sequencing of the human genome. *Nature* 2011;**470**:187–97.
30. Fu W, et al. Analysis of 6,515 exomes reveals the recent origin of most human protein-coding variants. *Nature* 2013;**493**:216–20.
31. Findlay GM, Boyle EA, Hause RJ, Klein JC, Shendure J. Saturation editing of genomic regions by multiplex homology-directed repair. *Nature* 2014;**513**:120–3.
32. Nelson MR, et al. An abundance of rare functional variants in 202 drug target genes sequenced in 14,002 people. *Science* 2012;**337**:100–4.
33. Mailman MD, et al. The NCBI dbGaP database of genotypes and phenotypes. *Nat Genet* 2007;**39**:1181–6.
34. Landrum MJ, et al. ClinVar: public archive of relationships among sequence variation and human phenotype. *Nucleic Acids Res* 2014;**42**:D980–5.
35. Cotton RGH, et al. The Human Variome Project. *Science* 2008;**322**:861–2.
36. Gonzalez MA, et al. GEnomes Management Application (GEM.app): a new software tool for large-scale collaborative genome analysis. *Hum Mutat* 2013;**34**:842–6.
37. Rainier S, et al. Neuropathy target esterase gene mutations cause motor neuron disease. *Am J Hum Genet* 2008;**82**:780–5.
38. Synofzik M, et al. PNPLA6 mutations cause Boucher-Neuhauser and Gordon Holmes syndromes as part of a broad neurodegenerative spectrum. *Brain* 2014;**137**:69–77.
39. Hufnagel RB, et al. Neuropathy target esterase impairments cause Oliver-McFarlaneand Laurence-Moonsyndromes.*J Med Genet* 2015;**52**:85–94.

40. Kmoch S, et al. Mutations in PNPLA6 are linked to photoreceptor degeneration and various forms of childhood blindness. *Nat Commun* 2015;**6**:5614.

41. Nilius B, Voets T. The puzzle of TRPV4 channelopathies. *EMBO Rep* 2013;**14**:152–63.

42. Ho CY, Lammerding J. Lamins at a glance. *J Cell Sci* 2012;**125**: 2087–93.

43. Tong AHY, et al. Global mapping of the yeast genetic interaction network. *Science* 2004;**303**:808–13.

44. Barabási A-L, Oltvai ZN. Network biology: understanding the cell's functional organization. *Nat Rev Genet* 2004;**5**:101–13.

45. Biesecker LG. Exome sequencing makes medical genomics a reality. *Nat Genet* 2010;**42**:13–4.

46. Mirnezami R, Nicholson J, Darzi A. Preparing for precision medicine. *N Engl J Med* 2012;**366**:489–91.

47. Collins FS, Varmus H. A new initiative on precision medicine. *N Engl J Med* 2015;**372**:793–5.

48. NHLBI GO Exome Sequencing Project Exome Variant Server. http://evs.gs.washington.edu; 2011.

49. Berg JS, Khoury MJ, Evans JP. Deploying whole genome sequencing in clinical practice and public health: meeting the challenge one bin at a time. *Genet Med* 2011;**13**:499–504.

Index